Lecture Notes in Computer Science 15162

The series Lecture Notes in Computer Science (LNCS), including its subseries Lecture Notes in Artificial Intelligence (LNAI) and Lecture Notes in Bioinformatics (LNBI), has established itself as a medium for the publication of new developments in computer science and information technology research, teaching, and education.

LNCS enjoys close cooperation with the computer science R & D community, the series counts many renowned academics among its volume editors and paper authors, and collaborates with prestigious societies. Its mission is to serve this international community by providing an invaluable service, mainly focused on the publication of conference and workshop proceedings and postproceedings. LNCS commenced publication in 1973.

Yong Chen · Xiaofeng Gao · Xiaoming Sun ·
An Zhang
Editors

Computing and Combinatorics

30th International Conference, COCOON 2024 Shanghai, China, August 23–25, 2024 Proceedings, Part II

Springer

Editors
Yong Chen
Hangzhou Dianzi University
Hangzhou, China

Xiaofeng Gao
Shanghai Jiao Tong University
Shanghai, China

Xiaoming Sun
Chinese Academy of Sciences
Beijing, China

An Zhang
Hangzhou Dianzi University
Hangzhou, Zhejiang, China

ISSN 0302-9743 ISSN 1611-3349 (electronic)
Lecture Notes in Computer Science
ISBN 978-981-96-1092-1 ISBN 978-981-96-1093-8 (eBook)
https://doi.org/10.1007/978-981-96-1093-8

This Springer imprint is published by the registered company Springer Nature Singapore Pte Ltd.
The registered company address is: 152 Beach Road, #21-01/04 Gateway East, Singapore 189721, Singapore

Preface

The papers in these proceedings, which consist of three volumes, were presented at the 30th International Computing and Combinatorics Conference (COCOON 2024), on August 23–25, 2024, in Shanghai, China. COCOON 2024 provided an excellent venue for researchers working in the area of algorithms, theory of computation, computational complexity, and combinatorics related to computing. The technical program of the conference included 90 regular papers selected by the Program Committee from 277 full submissions received in response to the call for papers. Each submission was carefully reviewed by at least three Program Committee members and/or external reviewers in a double-blind process. Papers of high quality will be invited to submit their full versions to special issues of The Journal of Computer and System Sciences, Algorithmica, Theoretical Computer Science, and the Journal of Combinatorial Optimization.

The conference included four invited presentations, delivered by Gregory Gutin (Royal Holloway, University of London), Liu Yunhao (Tsinghua University), Uriel Feige (The Weizmann Institute), and My Thai (University of Florida). In addition, we also organized activities such as tutorials and workshops. We thank everyone who made this meeting possible: the authors for submitting papers, the Program Committee members, and the external reviewers for volunteering their time to review conference papers. We also appreciate the financial sponsorship from Springer for the Best Paper Award. We would also like to extend special thanks to the other chairs and the conference Organizing Committee for their work in making COCOON 2024 a successful event.

The Springer Nature Submission System EquinOCS was used to handle the submissions, to conduct the electronic Program Committee meetings, and to assist with the assembly of the proceedings.

August 2024

Yong Chen
Xiaofeng Gao
Xiaoming Sun
An Zhang

Organization

Honorary General Chair

Jiang, Changjun	Tongji University, China

General Chair

Chen, Guihai	Shanghai Jiao Tong University, China

TPC Co-chairs

Gao, Xiaofeng	Shanghai Jiao Tong University, China
Sun, Xiaoming	Institute of Computing Technology, Chinese Academy of Sciences, China

Vice TPC Chair

Shi, Yongtang	Nankai University, China

Publicity Co-chairs

Cai, Zhipeng	Georgia State University, USA
Huang, Zhiyi	University of Hong Kong, China
Zhang, Zhao	Zhejiang Normal University, China

Tutorial and Workshops Chair

Dai, Haipeng	Nanjing University, China

Web Chair

Gravin, Nick	Shanghai University of Finance and Economics, China

Publication Co-chairs

Chen, Yong	Hangzhou Dianzi University, China
Zhang, An	Hangzhou Dianzi University, China

Financial Chair

Guo, Jianxiong	Beijing Normal University, China

Local Chair

Zhang, Yuhao	Shanghai Jiao Tong University, China

Program Committee Chairs

Chen, Yong	Hangzhou Dianzi University, China
Gao, Xiaofeng	Shanghai Jiao Tong University, China
Shi, Yongtang	Nankai University, China
Sun, Xiaoming	Institute of Computing Technology, Chinese Academy of Sciences, China
Zhang, An	Hangzhou Dianzi University, China

Program Committee

Ammari, Habib	Texas A&M University-Kingsville, USA
An, Hyung-Chan	Yonsei University, Korea.
Bei, Xiaohui	Nanyang Technological University, Singapore
Bein, Wolfgang	University of Nevada, USA
Butenko, Sergiy	Texas A&M University, USA
Böckenhauer, Hans-Joachim	ETH Zurich, Switzerland
Calinescu, Gruia	Illinois Institute of Technology, USA
Cao, Yixin	Hong Kong Polytechnic University, China
Chang, Jou-Ming	National Taipei University of Business, Taiwan
Chen, Zhizhong	Tokyo Denki University, Japan
Chen, Xujin	Academy of Mathematics and Systems Science, Chinese Academy of Sciences, China
Chen, Ho-Lin	National Taiwan University, Taiwan
Cheng, Eddie	Oakland University, USA

Daescu, Ovidiu	The University of Texas at Dallas, USA
DasGupta, Bhaskar	University of Illinois Chicago, USA
De Bonis, Annalisa	Università degli Studi di Salerno, Italy
Du, Donglei	University of New Brunswick, Canada
Giancarlo, Raffaele	University of Palermo, Italy
Gu, Qianping	Simon Fraser University, Canada
Gutin, Gregory	University of London, UK
Han, Xin	Dalian University Of Technology, China
Hon, Wing-Kai	National Tsing Hua University, Taiwan
Hsieh, Sun-Yuan	National Cheng Kung University, Taiwan
Hung, Ling-Ju	National Taipei University of Business, Taiwan
Iwama, Kazuo	Kyoto University, Japan
Khachay, Michael	Krasovsky Institute of Mathematics and Mechanics, RAS, Russia
Klasing, Ralf	CNRS and University of Bordeaux, France
Köppl, Dominik	University of Yamanashi, Japan
Le, Van Bang	University of Rostock, Germany
Lee, Joong-Lyul	University of Rostock, Germany
Lei, Hui	Nankai University, China
Li, Shi	Nanjing University, China
Li, Tianyi	Aalborg University, Denmark
Li, Minming	City University of Hong Kong, China
Li, Meng	Nanjing University, China
Liao, Chung-Shou	National Tsing Hua University, Taiwan
Lin, Bingkai	Nanjing University, China
Lin, Guohui	University of Alberta, Canada
Liu, Hsiang-Hsuan (Alison)	Utrecht University, Netherlands
Mei, Lili Hangzhou	Dianzi University, China
Miyano, Eiji	Kyushu Institute of Technology, Japan
Nguyen, Viet Hung	Clermont Auvergne University, LIMOS - CNRS UMR 6158, France
Rafiey, Arash	Indiana State University, USA
Rahman, M. Sohel	Bangladesh University of Engineering and Technology, Bangladesh
Rangan, C. Pandu	Indian Institute of Technology Madras, India
Sadakane, Kunihiko	The University of Tokyo, Japan
Shi, Yangguang	Shangdong University, China
Skums, Pavel	Georgia State University, USA
Souza, Ueverton	Universidade Federal Fluminense, Brazil
Tian, Kun	California State University, USA
Uehara, Ryuhei	Japan Advanced Institute of Science and Technology, Japan

Wang, Haitao	University of Utah, USA
Weng, Paul	Duke Kunshan University, China
Wu, Xiaowei	University of Macau, China
Xiao, Mingyu	University of Electronic Science and Technology of China, China
Xu, Honghui	Kennesaw State University, USA
Yang, Kuan	Shanghai Jiao Tong University, China
Yang, Song	Beijing Institute of Technology, China
Yang, Wenguo	University of Chinese Academy of Sciences, China
Yang, Boting	University of Regina, Canada
Yap, Chee	NYU Courant Institute, USA
Ye, Deshi	Zhejiang University, China
Yin, Yitong	Nanjing University, China
Yuan, Jing	North Texas University, USA
Zaroliagis, Christos	University of Patras, Greece
Zhang, Xiaoyan	Nanjing Normal University, China
Zhang, Yong	Shenzhen Institute of Advanced Technology, China
Zhang, Huaming	University of Alabama in Huntsville, USA
Zhang, Louxin	National University of Singapore, Singapore
Zhang, Chihao	Shanghai Jiao Tong University, China
Zhang, Peng	Shandong University, China
Zhou, Rui	Swinburne University of Technology, Australia
Zhu, Yuqing	California State University, Los Angeles, USA
Zhu, Binhai	Montana State University, USA
Zissimopoulos, Vassilis	National and Kapodistrian University of Athens, Greece

Additional Reviewers

Baklanov, Artem
Bentert, Matthias
Cameron, Kathie
Chakraborty, Dibyayan
Chen, Xianrun
Chen, Xiuyang
Chu, Huairui
Dai, Guowei
Dai, Sijia
Dong, De
Eto, Hiroshi
Forghani, Majid
Francisco, Alexandre
Gan, Jiarui
Gong, Mingyang
Gu, Rong
Gu, Mei-Mei
Guo, Longkun
Guo, Heng
Harutyunyan, Hovhannes
He, Yuanyuan
Huang, Lingxiao

Huang, Chengyuan
Jiang, Shaofeng
Khodamoradi, Kamyar
Knop, Alexander
Kralovic, Richard
Lamprou, Ioannis
Lazaropoulos, Nikos
Lee, Chuan-Min
Lee, Chia-Wei
Li, Can
Lin, Joseph, Chuang-Chieh
Liptak, Laszlo
Liu, Shenghao
Liu, Shengxin
Ma, Ruikang
Ma, Meijie
Ma, Hengzhao
Mao, Yaping
Mazza, Lucian
Miao, Huahua
Miller, Lynn
Mömke, Tobias
Nguyen, Minh Hieu
Nishimura, Harumichi
Pai, Kung-jui
Pedrosa, Lehilton
Pu, Lianrong
Robinson, Peter
Runkang, Feng
Sigalas, Ioannis
Su, Yu-Heng
Sun, Xin
Tong, Weitian
Tong, Guangmo
Tsai, Meng-Tsung
Wang, Hancheng
Wang, Xiaoyu
Wang, Siqi
Wang, Hung-Lung
Wei, Tianze
Whittington, Philip
Wu, Chenchen
Xie, Jiahao
Xu, Xiaolong
Yamanaka, Katsuhisa
Yang, Hao-Tsung
Yang, Guangxu
Yang, Yang
Yao, Penghui
Yaping, Mao
Ye, Qingjie
Zhang, Yi
Zhang, Yaqian
Zhao, Yan
Zhong, Yihan
Zhou, Shuming
Zhu, Tongxin
Zhu, Xianbin
Zou, Yifei

Contents - Part II

Regular Papers Part II

VOABE: An Efficient Verifiable Outsourced Attribute-Based Encryption for Healthcare Systems

Junze Lu[1], Chunqiang Hu[1](✉), Tao Xiang[2], Wei Li[3], and Jiguo Yu[4,5]

[1] School of Big Data and Software Engineering, Chongqing University, Chongqing, China
{junze,chu}@cqu.edu.cn

[2] College of Computer Science, Chongqing University, Chongqing, China
txiang@cqu.edu.cn

[3] Department of Computer Science, Georgia State University, Atlanta, USA
wli28@gsu.edu

[4] School of Computer Science and Engineering, University of Electronic Science and Technology of China, Chengdu, China
jiguoyu@sina.com

[5] Big Data Institute, Qilu University of Technology, Jinan, China

Abstract. Healthcare edge devices (HEDs) can create effective Electronic Health Records (EHRs), thereby enhancing the patient care experience. Nonetheless, EHRs contain a vast amount of private information, necessitating strict access control. Attribute-based encryption (ABE) offers a promising solution for the secure storage and access control of EHRs. However, current implementations encounter several challenges. Firstly, extant schemes are hindered by inefficiency, requiring too many exponentiation calculations and bilinear pairing operations, imposing significant computational demands on HEDs. Secondly, due to the accuracy requirement of medical data and the semi-trustworthiness of the cloud, the verification of medical data decryption becomes imperative. In this paper, we propose an efficient outsourced EHR management scheme using verifiable attribute-based encryption, which ensures data verifiability and enables fine-grained access control. Our scheme significantly reduces the computational overhead for HEDs by outsourcing computation to the CSP, while also supporting a large universe of attributes. Our scheme is proven to be secure and efficient by rigorous security analysis and performance evaluation.

Keywords: Attribute-based · Outsourced computation · Cloud service · Security

1 Introduction

EHRs document patients' health data, serving as valuable assets for advancing clinical research and facilitating point-to-point healthcare applications [11].

Y. Chen et al. (Eds.): COCOON 2024, LNCS 15162, pp. 3–14, 2025.
https://doi.org/10.1007/978-981-96-1093-8_1

Through the integration of cloud computing technology, healthcare professionals and family members can efficiently access EHRs, thereby enhancing patient care [8]. However, EHR data comprises sensitive user information, necessitating compliance with legal regulations such as the European General Data Protection Regulation (GDPR) [1] and the U.S. Health Insurance Portability and Accountability Act (HIPAA) [2].

With the widespread use of various mobile devices, edge medical sensors are generating an increasing volume of medical data. The secure storage and sharing of this medical data is critical. However, assigning the task of encryption directly to these edge devices through local computation is challenging. Consequently, the majority of computational tasks need to be outsourced to the cloud. Moreover, while EHR offers advantages to both patients and medical professionals by delivering high-quality medical diagnoses and services, the EHR owner faces the risk of data leakage when outsourcing data to a cloud service provider (CSP), losing complete control over the data. Additionally, a majority of EHR systems depend on services provided by CSPs, making it paramount to ensure the integrity of data delivery [18]. These security and privacy concerns pose significant obstacles to the widespread adoption of healthcare services based on EHR.

In a multi-user environment, ensuring secure sharing of outsourced data presents a challenging task. Traditional encryption methods face challenges in providing adequate fine-grained access control. In contrast, ABE requires only a single encryption to generate a ciphertext that can be shared among specific groups. Although numerous schemes for outsourced ABE have been proposed [7,14], challenges remain in efficient encryption of medical data. Notably, edge medical devices, which often have limited computational resources, find it difficult to process large volumes of medical data using ABE only in local environment. Furthermore, the computational capabilities of the users themselves are typically inadequate. Consequently, outsourcing services to CSP becomes essential. Nevertheless, CSP is not entirely trustworthy and its potential curiosity about the data could lead to private information leakage.

The adoption of outsourced ABE with verifiable decryption presents an effective solution to alleviate the computational burden on healthcare edge devices and users, while simultaneously ensuring the security and verifiability of medical data. In recent years, ABE has attracted extensive attention in the field of cloud computing for its pivotal role in the secure sharing of EHR. The widespread use of ABE lies in its ability to provide fine-grained access control mechanisms while safeguarding patient data privacy. Through fine-grained access control, the data owner can customize access policies to achieve flexible data sharing, ensuring that unauthorized entities, including the cloud service provider, are unable to decrypt the data without requisite attributes. Further, with verifiable decryption, we can outsource the encrypted data to the CSP, thus achieving a reduced computational burden on HEDs and users.

1.1 Related Work

ABE is an encryption technology that enables fine-grained access control over encrypted data [17]. Unlike traditional identity-based encryption systems, ABE leverages attributes associated with users rather than fixed identities [17]. Its versatility makes ABE suitable for a wide range of applications, including the Internet of Things (IoT) [12,20], smart healthcare systems [12,20], etc. Rani et al. [15] employed an expert system to refine ABE access policies, albeit without offering a concrete ABE construction. Agrawal et al. [3] concentrated on enhancing efficiency and addressing storage issues. Ruan et al. [16] presented a method aimed at achieving partial policy hiding. Cui et al. [5] considered the computation burdens on both users and a single proxy server, and thus proposed a multi-proxy assisted ABE which supports revocation. Miao et al. [14] introduced an outsourced ABE that significantly reduces the computational demands on users, although it risks information exposure if CSP is curious about the data. Several researchers, including [9,14], have proposed their scheme to ensure the verifiability of ABE. The studies conducted by [9] utilized additional message encryption to achieve verifiable decryption, which, however, nearly doubles the encryption time and storage costs. Mahdavi et al. [13] proposed a new ABE method for IoT environments, introducing a strategy that uses pre-computation and outsourcing for categorized attributes, including commonly used attributes and rarely used attributes.

1.2 Our Contribution

In this paper, we present an efficient **V**erifiable **O**utsourced **A**ttribute-**B**ased **E**ncryption (VOABE) for Healthcare Systems that supports verifiable outsourced decryption. This scheme addresses the issue of potential data leakage found in some outsourced encryption schemes. Comparative analysis demonstrates the remarkable efficiency of our scheme when compared with the latest research. The contributions of our work can be summarized as follows:

1. The encryption and decryption processes require only a minimal number of bilinear pairing calculations, resulting in a significant enhancement in computational efficiency.
2. Outsourcing the encryption and decryption parts greatly reduces the computing costs for healthcare edge devices and users.
3. The computational and storage overheads of outsourced decryption in the verifiable component have been optimized.

2 Preliminaries

2.1 Bilinear Pairings

Consider two cyclic multiplicative groups $\mathbb{G}$ and $\mathbb{G}_{\mathbb{T}}$ of prime order p. The bilinear mapping $e : \mathbb{G} \times \mathbb{G} \rightarrow \mathbb{G}_{\mathbb{T}}$ satisfies three essential properties: bilinearity, ensuring that $e(a^x, b^y) = e(a, b)^{xy}$ for all $a, b \in \mathbb{G}$ and $x, y \in \mathbb{Z}_p$; non-degeneracy, which guarantees that there exist $e(a, b) \neq 1_{\mathbb{G}_\mathbb{T}}$ for $a, b \in \mathbb{G}$; and computability, indicating that there exists an efficient algorithm to compute $e(a, b)$ for any $a, b \in \mathbb{G}$.

Let $\mathcal{G}$ be a symmetric pairing group generator that on input λ, outputs the tuple $(e, \mathbb{G}, \mathbb{G}_\mathbb{T}, p)$ as described above.

2.2 Linear Secret Sharing Scheme

In a CP-ABE system, the complexity of encryption and decryption depends on the access structure defined by the data owner, always using a Linear Secret Sharing Scheme (LSSS) $\prod$ [19] as its foundation. This scheme is capable of constructing access formulas for any attribute set, embedding a secret value s within an access structure, denoted as $\Gamma = (\mathbb{M}_{n_1 \times n_2}, \rho(\cdot))$. Authorized data users can reconstruct s if their attributes match the conditions of Γ. The secret s is then reconstructed by calculating coefficients θ_k that satisfy the equation $\sum_{k \in K} \theta_k \mathbb{M}_k = (1, 0, 0, \cdots, 0)$. This process leads to the final reconstruction equation $\sum_{k \in K} \beta_k \theta_k = s$, where $\alpha = (s, y_2, y_3, \cdots, y_{n_1}) \in \mathbb{Z}_p^{*n_1}, \beta_k = \alpha \cdot \mathbb{M}_k$.

2.3 Complexity Assumptions

We state our complexity assumptions below.

Assumption 1 (Decisional Bilinear Diffie-Hellman (DBDH) Assumption). Let $(e, \mathbb{G}, \mathbb{G}_\mathbb{T}, p)$ be a tuple defining a prime order bilinear group system. Denote x, y, z, t as elements that are randomly selected from $\mathbb{Z}_p$. Considering g to be the generator of $\mathbb{G}$, under the decisional BDH assumption [4,17], it is stated that no probabilistic polynomial-time (PPT) algorithm $\mathcal{B}$ can distinguish, with more than a negligible advantage ϵ, between the tuples $(g, g^x, g^y, g^z, e(g, g)^{xyz})$ and $(g, g^x, g^y, g^z, e(g, g)^t)$.

Assumption 2 (Discrete Logarithm (DL) Assumption). Let $(e, \mathbb{G}, \mathbb{G}_\mathbb{T}, p)$ be a tuple defining a prime order bilinear group system. Select a random value $x \in \mathbb{Z}_p$. Given g, the generator of $\mathbb{G}$, under the DL assumption, it is stated that no PPT algorithm $\mathcal{B}$ can compute x from $(e, \mathbb{G}, \mathbb{G}_\mathbb{T}, p, g^x)$ with an advantage greater than a negligible value ϵ.

Assumption 3 (k - Subset Product Problem (k-SPP)). Let $\mathbb{P}$ be a set containing elements from $\mathbb{G}$ and $\mathbb{G}_\mathbb{T}$. Given an element x in $\mathbb{G}$ or $\mathbb{G}_\mathbb{T}$, under the SPP, it is stated that no PPT algorithm $\mathcal{B}$ can determine whether x is equal to the product of any subset S comprising k elements from $\mathbb{P}$ for appropriate values of $|\mathbb{P}|$ and k, with an advantage greater than a negligible value.

3 System Model

3.1 System Architecture

As depicted in Fig. 1, VOABE comprises data owner (DO), healthcare edge devices (HEDs), data users (DUs), trusted authority (TA) and the cloud service provider (CSP).

- **Trusted Authority**: TA generates and manages the security parameter and generates the private keys for users.
- **Cloud Service Provider**: The CSP offers medical data storage and computing services. Although they might exhibit curiosity about data, they ensure the accuracy of their operations as much as possible and function as a semi-trusted entity.
- **Data Owner**: Each HED owned by DO generates medical data. Furthermore, DO can establish encryption policy for HED and delegate computational tasks to CSP. Regularly generated encryption sequences are transmitted to HEDs to minimize their encryption overhead.
- **Data Users**: Authorized users, including hospital staff or family members, obtain partially decrypted ciphertext through CSP's outsourced decryption. Subsequently, they access the medical data using a small amount of computational operations.

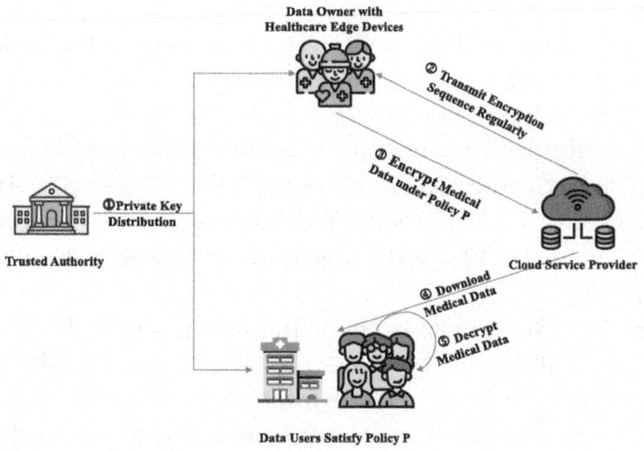

Fig. 1. System model of our scheme.

Figure 1 illustrates the comprehensive process of our scheme, which comprises five distinct steps. In Step 1, the TA generates private keys for the DO and DUs. In Step 2, the HED regularly acquires encryption sequences from the CSP. Subsequently, in Step 3, the HED encrypts the data and uploads it to the cloud. Then, in Step 4, users access the medical data. Lastly, in Step 5, users decrypt the medical data.

3.2 Security Definitions

In our scheme, we allow the adversary to execute private key generation queries. However, it's important to note that the outcome of these queries cannot be used to decrypt the specific ciphertext under challenge. To formally define the security, we introduce the following security game:

Game$_{\textbf{Semantic}}$. A VOABE scheme achieves semantic security for ciphertexts in the selective model if there is no PPT adversary $\mathcal{A}$ capable of winning the following game with a non-negligible advantage.

- **Setup**. During this step, input the security parameter λ. Subsequently, $\mathcal{C}$ executes the Setup to generate the public parameters PP and the master secret key msk. These public parameters PP are then sent to $\mathcal{A}$.
- **Query phase I.** Adaptively, $\mathcal{A}$ initiates multiple queries to the following oracles:
  1. $\mathcal{O}_{\text{KeyGen}}$: $\mathcal{A}$ submits the attribute set S to $\mathcal{C}$. Subsequently, $\mathcal{C}$ executes the KeyGen, generating the private key sk, which is then sent to $\mathcal{A}$. We denote the set of attributes queried by $\mathcal{A}$ as $\mathbb{S} = \mathbb{S} \cup \{S\}$.
  2. $\mathcal{O}_{\text{Blind Key}}$: $\mathcal{A}$ submits the attribute set S, and $\mathcal{C}$ calls KeyGen to generate the private key. $\mathcal{C}$ then runs KeyBlind to generate the blind key bk and blind factor ζ and sends bk to $\mathcal{A}$.
  3. $\mathcal{O}_{\text{Dec}}$: $\mathcal{A}$ is required to submit the ciphertext CT to be decrypted. Subsequently, $\mathcal{C}$ executes the Dec to obtain plaintext m, which is then transmitted to $\mathcal{A}$.

  During **Query phase I**, $\mathcal{A}$ is restricted from making the following queries:
  1. $\mathcal{O}_{\text{KeyGen}}$ if S satisfies the access structure Γ^*.
  2. $\mathcal{O}_{\text{Dec}}$ if CT is a derivative of CT^*
- **Challenge.** $\mathcal{A}$ submits two messages to be challenged, m_0 and m_1, which have the same length and access structure $\Gamma^* = (\mathbb{M}^*, \rho^*)$. It is noteworthy that access structure Γ^* is not satisfied for any $S \in \mathbb{S}$. $\mathcal{C}$ begins by randomly selecting a bit $\sigma \in \{0,1\}$, and subsequently invokes Enc to generate the ciphertext CT^* for $\mathcal{A}$.
- **Query phase II.** This phase is the same as the phase I.
- **Guess.** $\mathcal{A}$ outputs its guess bit $\sigma' \in \{0,1\}$, representing its prediction. The goal of the security game is to determine if σ' matches the actual value σ. If $\sigma' = \sigma$, we consider this a successful outcome for $\mathcal{A}$, indicating that the adversary has correctly guessed the target bit.

4 Our Scheme

4.1 Our VOABE Construction

- Setup(λ): TA runs $\mathcal{G}(\lambda)$ to obtain a bilinear pairing tuple $(e, \mathbb{G}, \mathbb{G}_{\mathbb{T}}, p)$ and chooses random values $a_1, a_2 \in \mathbb{Z}_p^*$, $g, L, Q \in \mathbb{G}$ and hash function $\mathcal{H}_1$, which maps to $\mathbb{G}$, and both $\mathcal{H}_2$ and $\mathcal{H}_3$ that map to $\mathbb{Z}_p$. It sets the master secret key $msk = g^{a_1}$ and public parameters $PP = (e, \mathbb{G}, \mathbb{G}_{\mathbb{T}}, p, g, L, Q, g^{a_2}, e(g,g)^{a_1}, \mathcal{H}_1, \mathcal{H}_2, \mathcal{H}_3)$.
- KeyGen(msk, S): Randomly choose an element $d \in \mathbb{Z}_p^*$. Then computes:

$$sk_1 = g^{a_1 + a_2 d}, sk_2 = g^d,$$

$$sk_x = \mathcal{H}_1(x)^d \; \forall x \in S.$$

  Output the private key as $sk = (sk_1, sk_2, \{sk_x\}_{x \in S})$.

- Activate($\mathbb{M}, \rho$): For every HED, activation is required first. On input an access structure $\Gamma = (\mathbb{M}_{n_1 \times n_2}, \rho)$, $\mathbb{M}_{n_1 \times n_2}$ is a matrix and ρ associates each row of $\mathbb{M}$ with an attribute. Randomly select a vector $\overrightarrow{\alpha}' = (r', y_2', \cdots, y_{n_1}') \in \mathbb{Z}_p^{*n_1}$. For each row $\mathbb{M}_k$ of $\mathbb{M}$, computes $\beta_k' = \overrightarrow{\alpha}' \cdot \mathbb{M}_k, k \in [1, n_1]$. Then compute:

$$C' = e(g,g)^{a_1 r'}, C_1' = g^{r'}, C_2 = L^{r'}, \tag{1}$$

$$C_{3,\rho(k)}' = g^{a_2 \beta_k'} \mathcal{H}_1(\rho(k))^{r'} \ \forall k \in [1, n_1]. \tag{2}$$

HED keeps $CT' = (C', C_1', C_2', \{C_{3,\rho(k)}'\}_{k \in [1,n_1]})$ for further encryption.
- Enc($m, (\mathbb{M}, \rho)$): Firstly, CSP regularly chooses a random vector $\overrightarrow{\alpha} = (r, y_2, \cdots, y_{n_1}) \in \mathbb{Z}_p^{*n_1}$. For each row $\mathbb{M}_k$ of $\mathbb{M}$, it computes $\beta_k = \overrightarrow{\alpha} \cdot \mathbb{M}_k, k \in [1, n_1]$. Subsequently, it computes:

$$\mathcal{P}_0 = e(g,g)^{a_1 r}, \mathcal{P}_1 = g^r, \mathcal{P}_2 = L^r,$$

$$\mathcal{P}_{3,\rho(k)} = g^{a_2 \beta_k} \mathcal{H}_1(\rho(k))^r \ \forall k \in [1, n_1].$$

After computing, CSP sends the encryption sequence $\mathcal{P} = (\mathcal{P}_0, \mathcal{P}_1, \mathcal{P}_2, \{\mathcal{P}_{3,\rho(k)}\}_{k \in [1,n_1]})$ to the HED. It should be noted that the frequency of encryption sequence generation is related to the encryption frequency of HED. We will further discuss it in Sect. 5.1. HED updates $\mathbb{P}$ as $\mathbb{P} \cup \{\mathcal{P}\}$ and randomly selects $\gamma \in {\mathbb{Z}_{|\mathbb{P}|}^{+}}^{l}$, where l is the number of elements selected in $\mathbb{P}$. Then HED computes:

$$\boldsymbol{C} = m \cdot C' \cdot \prod_{i=1}^{l} \mathbb{P}_{\gamma_i,0}, \boldsymbol{C}_1 = C_1' \cdot \prod_{i=1}^{l} \mathbb{P}_{\gamma_i,1}, \boldsymbol{C}_2 = C_2' \cdot \prod_{i=1}^{l} \mathbb{P}_{\gamma_i,2},$$

$$\boldsymbol{C}_{3,\rho(k)} = C_{3,\rho(k)}' \cdot \prod_{i=1}^{l} \mathbb{P}_{\gamma_i,3,\rho(k)} \ \forall k \in [1, n_1],$$

$$\mathcal{F} = L^{\mathcal{H}_2(m)} Q^{\mathcal{H}_3(m)},$$

where $\mathbb{P}_{\gamma_i,0}$ denotes the $\mathcal{P}_0$ in $\mathbb{P}_{\gamma_i}$. Then sends the ciphertext $CT = (\boldsymbol{C}, \boldsymbol{C}_1, \boldsymbol{C}_2, \{\boldsymbol{C}_{3,\rho(k)}\}_{k \in [1,n_1]}, \mathcal{F})$ to CSP.
- KeyBlind(sk): Randomly choose $\zeta, \tau \in \mathbb{Z}_p^*$ and compute:

$$bk_1 = sk_1^{\zeta} L^{\tau}, bk_2 = g^{\tau},$$

$$bk_3 = sk_2^{\zeta}, bk_{4,x} = sk_x^{\zeta} \ \forall x \in S.$$

Set $bk = (bk_1, bk_2, bk_3, \{bk_{4,x}\}_{x \in S})$. Output the key as $BK = (bk, \zeta)$. It should be noted that blind factor ζ is kept by DU, and blind key bk is sent to CSP for outsourced decryption.

– $\text{Dec}_{\text{out}}(bk, CT)$: Let $K = \{k : \rho(k) \in S\} \subset \{1, \cdots, n_1\}$, find elements $\theta_k \in \mathbb{Z}_p^*$, such that $\sum_{k \in K} \theta_k \mathbb{M}_k = (1, 0, \cdots, 0)$. Then compute:

$$bm = \frac{e(bk_1 \prod_{k \in K} bk_{4,\rho(k)}^{\theta_k}, \boldsymbol{C}_1)}{e(bk_2, \boldsymbol{C}_2) \cdot e(bk_3, \prod_{k \in K} \boldsymbol{C}_{3,\rho(k)}^{\theta_k})}.$$

Output the blind message as bm. Subsequently, send the blind message bm to DU.

– $\text{Dec}_{\text{u}}(\zeta, bm)$: On input the blind factor ζ and the blind message bm calculated by the CSP, the DU obtains the message m through the following calculation:

$$m = \frac{C}{bm^{\frac{1}{\zeta}}}.$$

Then DU verifies the correctness of m using the following calculation:

$$\mathcal{F}' = L^{\mathcal{H}_3(m)} Q^{\mathcal{H}_4(m)}.$$

If $\mathcal{F}' = \mathcal{F}$, the verification is successful; otherwise, it outputs $\perp$.

4.2 Security Proof

Theorem 1. *The proposed VOABE scheme is semantic secure under the DBDH assumption.*

Proof. If there exists a PPT adversary $\mathcal{A}$ capable of breaking the security of the proposed scheme in $\textbf{Game}_{\textbf{Semantic}}$, we can formulate a simulator $\mathcal{B}$ to break the DBDH assumption. $\mathcal{B}$ is provided with a tuple $v = (\mathbb{G}, \mathbb{G}_{\mathbb{T}}, g, g^x, g^y, g^z, T)$ as illustrated in Subsect. 2.3. Its goal is to determine whether T is equal to $e(g, g)^{xyz}$ or if T is randomly chosen from $\mathbb{G}_{\mathbb{T}}$. $\mathcal{B}$ tosses a fair binary coin μ. If $\mu = 0$, $\mathcal{B}$ sets $(X, Y, Z, T) = (g^x, g^y, g^z, g^{xyz})$; otherwise, it sets $(X, Y, Z, T) = (g^x, g^y, g^z, g^t)$, where x, y, z, t are random values.

– **Setup.** $\mathcal{B}$ chooses random values $\delta, a_2 \in \mathbb{Z}_p$ and $g, L, P \in \mathbb{G}$, and hash functions $\mathcal{H}_1$ mapping to $\mathbb{G}$, and both $\mathcal{H}_2$ and $\mathcal{H}_3$ mapping to $\mathbb{Z}_p$. $\mathcal{B}$ sets $e(X, Y) \cdot e(g, g^\delta) = e(g, g)^{xy+\delta}$, which is defined as $e(g, g)^{a_1}$. Finally, $\mathcal{B}$ outputs the public parameter PP.
– **Query phase I.** $\mathcal{A}$ queries:
  1. $\mathcal{O}_{\text{KeyGen}}$: $\mathcal{A}$ submits the attribute set S to $\mathcal{B}$. Subsequently, if S satisfies the access structure Γ^*, $\mathcal{B}$ outputs $\perp$. Otherwise, $\mathcal{B}$ randomly chooses $d \in \mathbb{Z}_p^*$ and computes:

$$sk_1 = g^{xy+\delta+a_2 d} = X^y g^{\delta a_2 d} \triangleq g^{a_1+a_2 d}, sk_2 = g^d, sk_x = \mathcal{H}_1(x)^d \; \forall x \in S.$$

  2. $\mathcal{O}_{\text{KeyBlind}}$: $\mathcal{A}$ submits the attribute set S to $\mathcal{B}$. $\mathcal{B}$ executes KeyGen to generate sk. Subsequently, $\mathcal{B}$ randomly chooses $\zeta, \tau \in \mathbb{Z}_p^*$ and computes:

$$bk_1 = sk_1^\zeta L^\tau = g^{(xy+\delta+a_2 d)\zeta} L^\tau, bk_2 = g^\tau, bk_3 = sk_2^\zeta, bk_{4,x} = sk_x^\zeta \; \forall x \in S.$$

  After computation, $\mathcal{B}$ sends bk to $\mathcal{A}$.

3. $\mathcal{O}_{\text{Dec}}$: $\mathcal{A}$ is required to submit the ciphertext CT to be decrypted. Subsequently, $\mathcal{B}$ runs KeyGen to obtain the private key sk satisfying the access policy, and then lets $K = \{k : \rho(k) \in S\} \subset \{1, \cdots, n_1\}$ and finds elements $\theta_k \in \mathbb{Z}_p^*$ such that $\sum_{k\in K} \theta_k \mathbb{M}_k = (1, 0, \cdots, 0)$. Then computes:

$$m = \frac{\boldsymbol{C} \cdot e(sk_2, \prod_{k\in K} \boldsymbol{C}_{3,\rho(k)}^{\theta_k})}{e(sk_1 \prod_{k\in K} sk_{3,\rho(k)}^{\theta_k}, \boldsymbol{C}_1)}. \tag{3}$$

After calculation, $\mathcal{B}$ sends m to $\mathcal{A}$.

- **Challenge.** Two plaintexts (m_0, m_1) of equal length are chosen by $\mathcal{A}$, after which $\mathcal{A}$ sends them to $\mathcal{B}$. Subsequently, $\mathcal{B}$ chooses a random bit $\sigma \in \{0, 1\}$ for following encryption.
If $\mu = 0$, a random vector is chosen such that $\overrightarrow{\boldsymbol{\alpha}}^* = (\boldsymbol{r}^*, \boldsymbol{y}_2^*, \cdots, \boldsymbol{y}_{n_1^*}^*) \in \mathbb{Z}_p^{n_1^*}$. For each row $\mathbb{M}_k^*$ of $\mathbb{M}^*$, computes $\boldsymbol{\beta}_k^* = \overrightarrow{\boldsymbol{\alpha}}^* \cdot \mathbb{M}_k^*, k \in [1, n_1^*]$. If we let $\boldsymbol{r}^* = z$. Then, $\mathcal{B}$ computes

$$\boldsymbol{C}^* = m_\sigma \cdot T \cdot e(g^z, g^\delta), \boldsymbol{C}_1^* = g^{\boldsymbol{r}^*} = g^z, \boldsymbol{C}_2^* = L^{\boldsymbol{r}^*} = L^z,$$

$$\boldsymbol{C}_{3,\rho(k)}^* = g^{a_2 \boldsymbol{\beta}_k^*} \mathcal{H}_1(\rho(k))^r = g^{a_2 \boldsymbol{\beta}_k^*} \mathcal{H}_1(\rho(k))^z \ \forall k \in [1, n_1^*],$$

where $T = e(g, g)^{xyz}$. Then we have:

$$\boldsymbol{C}^* = m_\sigma \cdot T \cdot e(g^z, g^\delta) = m_\sigma \cdot e(g, g)^{(xy+\delta)z} = m_\sigma \cdot e(g, g)^{a_1 z}.$$

Therefore, the ciphertext is a valid encryption of message m_σ.
Otherwise, if $\mu = 1$, then $T = e(g, g)^t$. We then have $\boldsymbol{C}^* = m_\sigma \cdot e(g, g)^t$. Since t is randomly chosen from $\mathbb{Z}_p$, $\boldsymbol{C}^*$ will be a random element of $\mathbb{G}_\mathbb{T}$ and $\boldsymbol{C}_1^*, \boldsymbol{C}_2^*, \boldsymbol{C}_{3,\rho(k)}^*$ will be a random element of $\mathbb{G}$ for $k \in [1, n_1]$.
- **Query phase II.** $\mathcal{A}$ can make queries the same as those in Query phase I.
- **Guess.** $\mathcal{A}$ outputs σ'. If $\sigma' = \sigma$, $\mathcal{B}$ outputs $\mu = 0$, indicating that T is equal to g^{xyz}. Otherwise, $\mathcal{B}$ outputs $\mu = 1$, indicating that T is randomly chosen from $\mathbb{G}_\mathbb{T}$.

Analysis. If $\mu = 1$, $\mathcal{A}$ cannot obtain σ-related information. Hence, we have $\Pr[\mu = 1|\sigma' = \sigma] = \frac{1}{2}$. If $\mu = 0$, then the adversary sees an encryption of m_σ. The adversary's advantage in this situation is ϵ by definition. Therefore we have $\Pr[\mu = 0|\sigma' = \sigma] = \frac{1}{2} + \epsilon$.

The adversary's overall advantage in breaking the game $\mathbf{Game_{Semantic}}$ is $|\frac{1}{2}\Pr[\mu = 1|\sigma' = \sigma] + \frac{1}{2}\Pr[\mu = 0|\sigma' = \sigma] - \frac{1}{2}| = \frac{\epsilon}{2}$.

Theorem 2. *Assuming that $\mathcal{H}_2$ and $\mathcal{H}_3$ are collision-resistant hash functions and that the DL assumption holds, then our scheme VOABE is verifiable in decryption.*

Proof. If adversary $\mathcal{A}$ can break the verifiability of our scheme with a non-negligible advantage, we build a simulator $\mathcal{B}$ that can obtain:

$$g^{x\mathcal{H}_2(m^*)+y\mathcal{H}_3(m^*)} = \mathcal{F}^* = \mathcal{F} = g^{x\mathcal{H}_2(m)+y\mathcal{H}_3(m)},$$

where we set $L = g^x$, $P = g^y$, and $m \neq m^*$. The probability of the adversary $\mathcal{A}$ breaking the verifiability of outsourced decryption is expressed as follows:

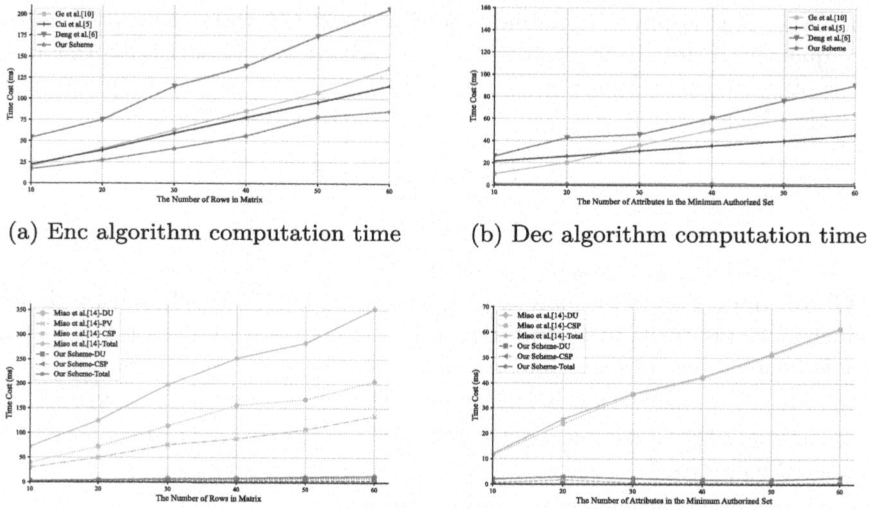

(a) Enc algorithm computation time (b) Dec algorithm computation time

(c) Enc-out algorithm computation time (d) Dec-out algorithm computation time

Fig. 2. Computation time of our proposed VOABE scheme

- *Case 1.* Given that $\mathcal{H}_2$ and $\mathcal{H}_3$ are collision-resistant hash functions, it is highly probable that $\mathcal{H}_2(m) \neq \mathcal{H}_2(m^*)$ and $\mathcal{H}_3(m) \neq \mathcal{H}_3(m^*)$. Then $\mathcal{B}$ outputs $x = \frac{y(\mathcal{H}_3(m)-\mathcal{H}_3(m^*))}{\mathcal{H}_2(m^*)-\mathcal{H}_2(m)}$ as the solution of the DL problem
- *Case 2.* If $\mathcal{H}_2(m) = \mathcal{H}_2(m^*)$ or $\mathcal{H}_3(m) = \mathcal{H}_3(m^*)$, then $\mathcal{B}$ finds a collision of hash function $\mathcal{H}_2$ or $\mathcal{H}_3$

Theorem 3. *Assuming that the k-SPP holds, then our scheme VOABE is semantic security in outsourced encryption.*

Proof. Straight-forward.

5 Performance Evaluation

5.1 Experiment Configuration

To demonstrate the actual performance of our scheme, we conducted experiments on Ubuntu 20.04 equipped with AMD Ryzen 7 5800X 8-Core 3.8 GHz using Go language and Pairing-Based Cryptography (PBC) library. We evaluate our scheme using the publicly available medical dataset known as the Medical Information Mart for Intensive Care (MIMIC).

During the testing phase, we divided the scenarios into those involving no outsourced computation and those with outsourced computation. In the performance evaluation without outsourced computation, the encryption part computes as follows:

$$C = m \cdot e(g,g)^{a_1 r}, C_1 = g^r, C_2 = L^r, C_{3,\rho(k)} = g^{a_2 \beta_k} \mathcal{H}_1(\rho(k))^r \ \forall k \in [1, n_1].$$

The above encryption process is similar to Eq. (1) and Eq. (2). For decryption, we use Eq. (3). As for performance evaluation with outsourcing computation, we set $l = 15$, which provides enough security under SPP. Additionally, the frequency of encryption sequence generation is set such that for every ten HED encryption operations, the CSP generates one encryption sequence.

5.2 Experimental Analysis

In Fig. 2(a), an efficiency comparison between our proposed scheme and schemes [5,6,10] for the encryption is depicted. The x-axis represents the number of rows included in the access matrix $\mathbb{M}$. The results indicate that our scheme exhibits similar encryption times to the latest results reported in [5,10], while surpassing the performance of [6]. In Fig. 2(b), a cost comparison between our scheme and schemes [5,6,10] for the decryption is presented. The x-axis represents the number of attributes involved in the decryption process. It is evident that our scheme outperforms the other three schemes significantly in terms of decryption time, achieving a speed that is more than ten times faster.

Figure 2(c) presents a cost comparison between our scheme and the scheme proposed by Miao et al. [14] for encryption operations in an outsourcing environment. It is evident that our scheme not only consumes less time overall but also exhibits lower user overhead and CSP overhead. In Fig. 2(d), a comparison of the decryption computation time costs between DU and CSP in an outsourcing environment is presented. Our scheme demonstrates certain advantages in the decryption process, resulting in lower total overhead and user overhead than [14].

6 Conclusion

This paper proposes an attribute-based secure and efficient verifiable outsourced EHR data management scheme. Specifically, compared with the latest schemes, our proposed scheme has greater advantages in efficiency. Moreover, the outsourced encryption has been optimized by generating the encryption sequence via CSP. This approach avoids submitting the secret to CSP, thereby enhancing security and resisting curious CSP. Furthermore, users are enabled to access data collected by HEDs at any time and at a low cost through verifiable outsourced decryption.

References

1. General data protection regulation. https://eur-lex.europa.eu/legal-content/EN/TXT/?uri=celex%3A32016R0679. 16 October 2023
2. Health insurance portability and accountability act. https://aspe.hhs.gov/reports/health-insurance-portability-accountability-act-1996. 16 October 2023
3. Agrawal, S., Chase, M.: Fame: fast attribute-based message encryption. In: Proceedings of the 2017 ACM SIGSAC Conference on Computer and Communications Security, pp. 665–682 (2017)

4. Boneh, D., Boyen, X.: Efficient selective-ID secure identity-based encryption without random oracles. In: Cachin, C., Camenisch, J.L. (eds.) EUROCRYPT 2004. LNCS, vol. 3027, pp. 223–238. Springer, Heidelberg (2004). https://doi.org/10.1007/978-3-540-24676-3_14
5. Cui, J., Li, B., Zhong, H., Xu, Y., Liu, L.: Achieving revocable attribute group-based encryption for mobile cloud data: a multi-proxy assisted approach. IEEE Trans. Dependable Secure Comput. **20**(4), 2988–3001 (2023)
6. Deng, H., Qin, Z., Wu, Q., Guan, Z., Zhou, Y.: Flexible attribute-based proxy re-encryption for efficient data sharing. Inf. Sci. **511**, 94–113 (2020)
7. Feng, C., Yu, K., Aloqaily, M., Alazab, M., Lv, Z., Mumtaz, S.: Attribute-based encryption with parallel outsourced decryption for edge intelligent IoV. IEEE Trans. Veh. Technol. **69**(11), 13784–13795 (2020)
8. Fennelly, O., Moroney, D., Doyle, M., Eustace-Cook, J., Hughes, M.: Key interoperability factors for patient portals and electronic health records: a scoping review. Int. J. Med. Inform. **183**, 105335 (2024)
9. Ge, C., Susilo, W., Baek, J., Liu, Z., Xia, J., Fang, L.: Revocable attribute-based encryption with data integrity in clouds. IEEE Trans. Dependable Secure Comput. **19**(5), 2864–2872 (2021)
10. Ge, C., Susilo, W., Liu, Z., Baek, J., Luo, X., Fang, L.: Attribute-based proxy re-encryption with direct revocation mechanism for data sharing in clouds. IEEE Trans. Dependable Secure Comput. **01**, 1–12 (2023)
11. Ghosheh, G.O., Li, J., Zhu, T.: A survey of generative adversarial networks for synthesizing structured electronic health records. ACM Comput. Surv. **56**(6), 1–34 (2024)
12. Girardi, F., De Gennaro, G., Colizzi, L., Convertini, N.: Improving the healthcare effectiveness: the possible role of EHR, IoMT and blockchain. Electronics **9**(6), 884 (2020)
13. Mahdavi, M., Tadayon, M.H., Haghighi, M.S., Ahmadian, Z.: IoT-friendly, pre-computed and outsourced attribute based encryption. Future Gener. Comput. Syst. **150**, 115–126 (2024)
14. Miao, Y., et al.: Verifiable outsourced attribute-based encryption scheme for cloud-assisted mobile e-health system. IEEE Trans, Dependable Secure Comput (2023)
15. Rani, S., Srivastava, G., et al.: Secure hierarchical fog computing-based architecture for industry 5.0 using an attribute-based encryption scheme. Expert Syst. Appl. **235**, 121180 (2024)
16. Ruan, C., Hu, C., Zhao, R., Liu, Z., Huang, H., Yu, J.: A policy-hiding attribute-based access control scheme in decentralized trust management. IEEE Internet Things J. **10**(20), 17656–17665 (2023)
17. Sahai, A., Waters, B.: Fuzzy identity-based encryption. In: Cramer, R. (ed.) EUROCRYPT 2005. LNCS, vol. 3494, pp. 457–473. Springer, Heidelberg (2005). https://doi.org/10.1007/11426639_27
18. Su, Y., Li, Y., Zhang, K., Yang, B.: A privacy-preserving public integrity check scheme for outsourced EHRs. Inf. Sci. **542**, 112–130 (2021)
19. Waters, B.: Ciphertext-policy attribute-based encryption: an expressive, efficient, and provably secure realization. In: Catalano, D., Fazio, N., Gennaro, R., Nicolosi, A. (eds.) PKC 2011. LNCS, vol. 6571, pp. 53–70. Springer, Heidelberg (2011). https://doi.org/10.1007/978-3-642-19379-8_4
20. Xiang, X., Zhao, X.: Blockchain-assisted searchable attribute-based encryption for e-health systems. J. Syst. Archit. **124**, 102417 (2022)

Topological Network-Control Games Played on Graphs

Zihui Liang(✉), Bakh Khoussainov, and Haidong Yang(✉)

University of Electronic Science and Technology of China, Chengdu, China
zihuiliang.tcs@gmail.com, bmk@uestc.edu.cn, haidongyang.tcs@outlook.com

Abstract. We study topological network-control games. These games belong to the class of scoring games in combinatorial game theory. In a topological network-control game, given a graph G, two players move alternately on G. During each move, a player selects an unclaimed vertex along with its unclaimed neighbors within the distance of t. The players should satisfy the topological condition that the set of all claimed vertices at each move stays connected. The objective is to decide which player can claim more vertices at the end of the game. We focus on deciding topological network-control games on unions of path graphs and unions of cycles. We fully solve our games by finding the winners and computing winning strategies on these graphs. These findings bring new insights into our understanding of prototypical examples of scoring games, – topological network-control games, and contribute to the development of combinatorial game theory.

Keywords: Combinatorial games · scoring games · combinatorics · algorithm design · deciding two-player games · extracting winning strategies

1 Introduction

We start with introducing topological network-control games played on graphs. Let $G = (V, E)$ be a graph. Our graphs are simple, undirected, and finite. For a vertex $x \in V$ and an integer $t \geq 0$, the t-neighbourhood of x, denoted by $\mathcal{N}_t(x)$, is the set of all vertices at distance at most t from x. So, $\mathcal{N}_0(x) = x$ and $\mathcal{N}_t(x) \subseteq \mathcal{N}_{t+1}(x)$. Vertices in $\mathcal{N}_t(x)$ are t-neighbours of x. For $X \subseteq V$, the t-neighbourhood of X is $\mathcal{N}_t(X) = \bigcup_{x \in X} \mathcal{N}_t(x)$.

The topological network-control game played on $G = (V, E)$ occurs between Player 1 and Player 2. At round 1, Player 1 selects an $x_1 \in V$ and claims all vertices in $S_1 = \mathcal{N}_t(x_1)$. Let $i \geq 1$ be the round at which a player selected x_i and claimed vertices in S_i. At round $i + 1$, the opponent faces with

1. the graph G_i on the vertex set $V_i = V \setminus \bigcup_{j=1}^{i} S_j$, and

This research work is partially supported by the NSFC grants 62350710215 and 62172077.

Y. Chen et al. (Eds.): COCOON 2024, LNCS 15162, pp. 15–26, 2025.
https://doi.org/10.1007/978-981-96-1093-8_2

2. the set $C_i = \mathcal{N}_{t+1}(\bigcup_{j=1}^{i} S_j) \setminus \bigcup_{j=1}^{i} S_j$, where the $(t+1)$-neighbourhood is considered in the graph G_i.

At this round the opponent selects an $x_{i+1} \in C_i$, and claims all vertices in $S_{i+1} = \mathcal{N}_t(x_{i+1})$ in G_i. This guarantees that the set $S_1 \cup \ldots \cup S_{i+1}$ of all claimed vertices is a connected subgraph of G. Call the pair (G_i, C_i) as defined above the **configuration** determined by the sequence (x_1, S_1), ..., (x_i, S_i). The initial configuration is G with $C = \emptyset$. In this case, any move x_1 by Player 1 is legal.

Thus, at round $2i$, where $i > 0$, Player 2 selects a vertex from C_{2i-1}, and then claims the selected vertex and its t-neighbour vertices in G_{2i-1}. At round $2i+1$, where $i > 0$, Player 1 selects a vertex in C_{2i}, and then claims the selected vertex and its t-neighbour vertices in G_{2i}. One convention is that, during a play, when a player makes a move (x, S), we might supress x and write S instead.

Let t be the round at which no vertices can be claimed. If $\bigcup_{j=1}^{t} S_j = V$, then the play stops. Otherwise, the player whose turn is at round t, selects a new vertex from a new component, and the play continues on just as above.

Let $Cl_{1,2i+1}$ and $Cl_{2,2i+2}$ be the set of vertices claimed by Player 1 and Player 2 at the end of rounds $2i+1$ and $2i+2$, respectively, where $i \geq 0$. In particular, if S_{2i+1} is the set of vertices claimed by Player 1 at round $2i+1$, then $Cl_{1,2i+1} = Cl_{1,2i-1} \cup S_{2i+1}$. Similarly, $Cl_{2,2i+2} = Cl_{1,2i} \cup S_{2i+2}$. Let X_1 and X_2 be all vertices claimed by Player 1 and Player 2, respectively, at the end of the play. The sets X_1 and X_2 partition G.

Definition 1. *We say that Player ϵ* **wins** *the play if $|X_\epsilon| > |X_{\bar{\epsilon}}|$, where $\bar{1} = 2$ and $\bar{2} = 1$. If $|X_\epsilon| = |X_{\bar{\epsilon}}|$, then we say that the play is a* **draw**.

If G is connected, then the set of claimed vertices is always connected. Connectedness is a topological property. Hence, we call our games *topological network-control game.* A player is **the winner** of game played on G if the player has a winning strategy. Our goal is twofold. First, we want to **decide** games by designing algorithms that given a game find the winner. Second, we want to **extract winning strategies** for the winners. When $t = 0$, Player 1 wins iff $|V|$ is odd. For this paper, we always assume that $t = 1$.

We recall a few classes of graphs. The **path graph** $Path_n$, where $n \geq 1$, is the graph with n vertices $\{v_1, \ldots, v_n\}$ and edges $\{v_1, v_2\}$, ..., $\{v_{n-1}, v_n\}$. The graph $Cycle_n$, $n \geq 3$, is obtained from $Path_n$ by adding the new edge $\{v_n, v_1\}$. A **linear forest** is a finite union of path graphs. The focus in this paper is on linear forests and unions of cycles. In combinatorial game theory, these type of simply structured graphs hold significance, as many bottleneck problems occur within such graphs, e.g., see [2,13,20,27,29,30]. For the reader here are two examples.

Example 1. Consider the topological network-control game played on $Path_n$. The first move, where the Player 1 selects the middle vertex, can guarantee a win for the player. Note that the first move made at any of the end points or their neighbours of $Path_n$ might not guarantee a win for Player 1.

Example 2. Consider the topological network-control game played on $Cycle_n$. Player 1 wins the topological network-control game on $Cycle_n$ iff $n \neq 0 \mod 6$.

Our games exhibit two complementary features: competition and collaboration. Just like Competetive Facility Location games [31], the players in our games compete. Namely, our games model scenarios where two parties, Player 1 and Player 2, compete to control the network. The more nodes the player claims the more control the player has over the network. At the same time, the players in our games collaborate. The collaboration side of our games is expressed in the fact that players aim to keep the set of all claimed vertices connected.

1.1 Brief Background

The topological network-control games belong to the area of combinatorial game theory [8,18,25,35]. The core of combinatorial game theory lies in the study of two classes of games: non-scoring games and scoring games. In non-scoring games, the winner is the player that puts the opponent into a deadlock (e.g., the opponent has no move left). In scoring games, players earn points, and the winner is decided by conditions on the points earned. So, the topological network-control games belong to the class of scoring games.

Examples of non-scoring games include GRIM [1], Nim on graphs [12,21], Kayles on graphs [10,11], the guard games on graphs [24], game 0.33 [4], Weighted Arc and generalized Kayles [17], and many more [35]. The class of non-scoring games has extensively been studied and there is a well-developed theory for this class of games. The example is the Sprague Grundy theorem showing that most of the non-scoring games are equivalent to the game of nim or its natural generalization [8,16,26,36]. There is also a deep algebraic theory connecting non-scoring games with groups and semigroups [9,16,35].

As opposed to non-scoring games, the theory of scoring games lacks a well-developed theory, particularly when viewed through an algorithmic lens. This is despite a substantial volume of work that includes the study of largest connected subgraph game [5,6], the competitive facility location (CFL) game [31], Smash and Grab game [19], Vertex-Capture Game [7], orthogonal colouring game [2,3], influence game [20], etc. We mention that Milnor [34] and Hanner [28] introduced scoring games. Ettinger followed up their work after almost 40 years [22,23]. Larsson et al. [32] investigated Guaranteed Scoring Games extending the work of Milnor and Ettinger. Stewart [37,38] developed an algebraic theory of scoring games and Sprague-Grundy theory on scoring octal games. This paper contributes new insights to the theory of scoring games through analysis of topological network-control games.

When solving combinatorial games on graphs, bottleneck problems arise in simple graph structures such as path graphs, cycles, trees, and their unions. [13,15,20,27,29,30]. In this respect, the configuration spaces of games contribute to the complexity of solving them more than the underlying graph structure. For instance, the paper [33] designs an algorithm for finding the winner in the topological network-control games on graphs $Path_n \cup Path_m$ by analysing over 40 cases. The proof in [33] cannot directly be extended to linear forests because with the number of paths increasing in the linear forests, the number of case analyses also increases. In general, a striking common feature of scoring games is that they are easy to describe, just like the topological network-control games, but hard to solve. This is attested by the fact that finding the winners in almost all scoring games, including topological network-control games [33], are PSPACE-complete problems [2,5,6,12,14]. Hence, any reasonable progress in developing the theory of scoring games requires intricate reasoning and novel techniques.

1.2 Optimal and Parity Driven Strategies

Intuitively, an optimal strategy is one for the given player that guarantees the best result under all possible behaviour of the opponent. Optimality is obviously determined by the number of vertices claimed by the player.

Definition 2. *A strategy for Player ϵ is w-**optimal**, where $w \geq 0$, if*

1. *The strategy can guarantee that Player ϵ wins any play with at least w more vertices independent on the opponent's strategies.*
2. *The opponent, Player $\bar{\epsilon}$, has a strategy that guarantees that no more than w vertices are lost independent on strategies of Player ϵ.*

Let (G, C) be a configuration. We define the value $Opt(G, C)$ such that player ϵ starting with configuration (G, C) can win with at least $Opt(G, C)$ vertices:

1. If $G = \emptyset$ then $Opt(G, C) = 0$.
2. Let $S_1, \ldots, S_m$ be all possible moves of player ϵ at (G, C). Let (G_k, C_k) be updated configuration after the move S_k and $n(S_k)$ be the cardinality of S_k, $k = 1, \ldots, m$. Set $Opt(G, C) = \max_{1 \leq k \leq m}\{n(S_k) - Opt(G_k, C_k)\}$.

Definition 3 (Optimal Strategy). *An* **optimal strategy** *is one that, given a configuration (G, C), selects a move (v_i, S_i) so that the value $n(S_i) - Opt(G_i, C_i)$ is maximized, that is, $n(S_i) - Opt(G_i, C_i) = \max_{1 \leq i \leq m}\{n(S_i) - Opt(G_i, C_i)\}$.*

Let (G, C) be a configuration. Let $Y_1, \ldots, Y_\ell$ be all components of G that contain a vertex in C. The players must partition the subgraph with vertex set $Y_1 \cup \ldots \cup Y_\ell$ to continue on playing the game in G. So, intuitively, the player who makes the last move in $Y_1 \cup \ldots \cup Y_\ell$ might have a disadvantage. Taking this observation into account, we now define two functions $F_{even}(G, S)$ and $S_{odd}(G, C)$.

The function $F_{even}(G, C)$ computes the maximal number of vertices won by the first player who starts the game at (G, C) under the assumption that the player can force the opponent to make the last move in the game. Similarly, $S_{odd}(G, C)$ computes the maximal number of vertices won by the second player in the game (G, C) under the assumption that the player can force the opponent to make the last move in the game. These functions are defined as follows:

- Set $F_{even}(\emptyset, \emptyset) = 0$ and $S_{odd}(\emptyset, \emptyset) = -\infty$.
- Let $S_1, \ldots, S_m$ be all possible moves of a player at (G, C). Let (G_i, C_i) be the configuration after move S_i, $i = 1, \ldots, m$. Set

$$F_{even}(G, C) = \max\{n(S_i) + S_{odd}(G_i, C_i) \mid i = 1, \ldots, m\}, \text{ and}$$
$$S_{odd}(G, C) = \min\{-n(S_i) + F_{even}(G_i, C_i) \mid i = 1, \ldots, m\}$$

Based on these recursive definitions, we define parity driven strategies:

Definition 4 (Even-parity Strategy). *An even-move strategy is one that, given* (G, C)*, selects a move* S_i *so that the value* $S_{odd}(G_i, C_i) + n(S_i)$ *is maximized, that is* $S_{odd}(G_i, C_i) + n(S_i) = \max_{1 \le i \le m}\{S_{odd}(G_i, C_i) + n(S_i)\}$.

Odd-parity strategies are defined similarly, where the objective consists of minimizing the value $-n(S_i) + F_{even}(G_i, S_i)$.

Finally we introduce the concept of dimension for configurations. Intuitively, the higher the dimensions, the harder it is to solve the games.

Definition 5. *The* **dimension** *of configuration* (G, C)*, where* $C \neq \emptyset$*, is the number of components of* G *that have non-empty intersection with* C.

The configuration spaces for games on finite union of cycles are one-dimensional. The configuration spaces for games on linear forests are typically two-dimensional.

1.3 Our Contribution

1. We solve topological network-control games on the unions of cycles. The proof involves a finite non-trivial modification of *mod* 6 relation on $\mathbb{Z}_{>0}$ of all positive integers. The proof employs a careful sequence of inductive and combinatorial reasoning, where many cases, even in the base case, should be thoroughly analyzed. Our proof has an algebraic implication in the sense that we fully characterise the unions of cycles where Player 2 guarantees a draw. See Theorem 2. This algebraic characterization has algorithmic consequences. Namely, we demonstrate that the winner can be decided in linear time based on the size of the representation. Moreover, we provide an efficient algorithm that computes the winning move for the victor in any given configuration of the game.
2. We fully solve network-control games on linear forests. Just as with the union of cycles, our analysis involves careful inductive and combinatorial reasoning. However, in comparison to the unions of cycles, our analysis is more intricate and complex because game configurations in linear forests can be two dimensional. Just like for the unions of cycles, we provide a full characterization of linear forests where Player 2 guarantees a draw. See Theorem 4.
3. One technically challenging aspect of our games is the following. Let $Y_1, \ldots, Y_\ell$ be all components of G that contain vertices in C in configuration (G, C). The players must partition the sub-graph with vertex set $Y = Y_1 \cup \ldots \cup Y_\ell$ in order to continue playing the game in the rest of G. So, intuitively, if a player makes the last move in $Y_1 \cup \ldots \cup Y_\ell$, then the opponent might have an advantage in the rest of the game. This implies that the players need to take into account the parity of the last move in Y. Controlling parity of the last move in the component of $Y_1 \cup \ldots \cup Y_\ell$ is a difficult problem even in such simple graphs as path graphs. For instance, the following is proved in [33]. Assume that the players play the topological network-control game in $Path_n$, where $n = 3$ modulo 5. Then through the analysis of even and odd parity strategies, see Definition 4,

Player 1 loses at least $n/5+1$ vertices in total if the player wants to ensure that the opponent makes the last move; moreover, if $n = 1$ modulo 5, then Player 2 can guarantee that Player 1 makes the last move. Our proofs of the main theorems overcome these parity-related challenges.

As we mentioned, solving scoring games, even in simple structured graphs such as linear forests, involves analysis of many cases. Simple methods, such as brute-force, do not usually solve such games because the number of cases to be analysed becomes prohibitively large. We have developed techniques to uniformly capture and reduce the number of cases to a manageable level. Consequently, we obtained our characterization results where Player 1 guarantees a win for linear forests and the union of cycles. To the best of our knowledge these types of full characterization results are rare for most combinatorial games.

Remark: Many lemmas and corollaries in our proofs compute optimal strategies for some graphs. Mostly, these are base cases of induction arguments. These computations were carried out by software available at Code Link.

2 Solving Games on Unions of Cycles

2.1 Main Lemmas

Let us fix the graph $G = Cycle_{x_1} \cup \ldots \cup Cycle_{x_k}$ that is a disjoint union of cycles. We represent the graph G as the multi-set set $Rep(G) = \{x_1, \ldots, x_k\}$. Our goal is to find the winner and compute a winning strategy for the winner.

When players play the game on G finite union of cycles, a typical configuration is of the form $(G \cup Path_n, C)$, where G is a union of cycles and C contains the two right end-points of $Path_n$ and the two left end-points of $Path_n$. So, configuration spaces of games on the union of cycles are at most **one-dimensional**. With no loss of generality, we can assume that $C = \{v_1, v_2, v_{n-1}, v_n\}$. In this section we prove several lemmas that will be used later. For the rest of this section, we let $m' = m \mod 6$ and $n' = n \mod 6$.

Lemma 1. *Let G' and $Path_n$, $n \geq 1$, be two disjoint graphs. If $1 \geq Opt(G', \emptyset) \geq 0$, then for $n' = 0$, $Opt(G' \cup Path_n, \{v_1, v_2, v_{n-1}, v_n\}) = Opt(G', \emptyset)$, for $n' = 1, 2, 3$, $Opt(G' \cup Path_n, \{v_1, v_2, v_{n-1}, v_n\}) = n' - Opt(G', \emptyset)$ and for $n' = 4, 5$, $Opt(G' \cup Path_n, \{v_1, v_2, v_{n-1}, v_n\}) = 6 - n' + Opt(G', \emptyset)$.*

Proof. Lemma 1 holds for all $n \in [1, 9]$. See Remark in Sect. 1.3. Assume that $n > 9$ and for all values $m \in [7, n-1]$, we have

$$Opt(G' \cup Path_m, \{v_1, v_2, v_{m-1}, v_m\}) = Opt(G' \cup Path_{m-6}, \{v_1, v_2, v_{m-7}, v_{m-6}\}).$$

We note the maximum of either $2 - Opt(G' \cup Path_{n-2}, \{v_1, v_2, v_{n-3}, v_{n-2}\})$ or $3 - Opt(G' \cup Path_{n-3}, \{v_1, v_2, v_{n-4}, v_{n-3}\})$ is $Opt(G' \cup Path_n, \{v_1, v_2, v_{n-1}, v_n\})$. Hence, $Opt(G' \cup Path_n, \{v_1, v_2, v_{n-1}, v_n\})$ equals to the maximum of the following two values: the value $2 - Opt(G' \cup Path_{n-8}, \{v_1, v_2, v_{n-9}, v_{n-8}\})$ or the value $3 - Opt(G' \cup Path_{n-9}, \{v_1, v_2, v_{n-10}, v_{n-9}\})$. Therefore, by hypothesis, we have

$$Opt(G' \cup Path_n, \{v_1, v_2, v_{n-1}, v_n\}) = Opt(G' \cup Path_{n-6}, \{v_1, v_2, v_{n-7}, v_{n-6}\}).$$

The lemma is proved.

Let G_1 and G_2 be two disjoint graphs and players play the game on graph $G_1 \cup G_2$. Assume that Player 1 starts on G_1 and G_1 is connected. With this assumption, we want to define the optimal strategy for Player 1. To do this, let $(v_1, S_1), \ldots, (v_m, S_m)$ be available moves on G_1 and (G_1, C_1), ..., (G_m, C_m) be the resulting configurations, respectively. Then the optimal strategy, denoted by $Opt_{G_1}(G_1 \cup G_2, \{\emptyset\})$, with the constraint that Player 1 starts on G_1, is this:

$$Opt_{G_1}(G_1 \cup G_2, \{\emptyset\}) = \max_{i=1}^{m}(|S_i| - Opt(G_i, C_i)).$$

Now note that, with the notation above, we have the following

$$Opt_{Cycle_n}(Cycle_n \cup G', \emptyset) = 3 - Opt(G' \cup Path_{n-3}, \{v_1, v_2, v_{n-4}, v_{n-3}\}).$$

Thus, we have the following corollary:

Corollary 1. *Let G' and $Cycle_n$ be disjoint graphs. If $1 \geq Opt(G', \emptyset) \geq 0$, then for $n' = 0$, $Opt_{Cycle_n}(Cycle_n \cup G', \emptyset) = Opt(G', \emptyset)$, for $n' = 1, 2, 3$, $Opt_{Cycle_n}(Cycle_n \cup G', \emptyset) = n' - Opt(G', \emptyset)$ and for $n' = 4, 5$, $Opt_{Cycle_n}(Cycle_n \cup G', \emptyset) = 6 - n' + Opt(G', \emptyset)$.* □

Clearly, if $n \in [1, 2, 3]$ then $Opt(Cycle_n, \emptyset) = n$. If we set $G' = \emptyset$ in Corollary 1, then we solve the topological network-control games on cycles:

Corollary 2. *On $Cycle_n$, we have $Opt(Cycle_n, \emptyset) = 2 \cdot \min(3, n') - n'$.* □

Lemma 1 assumes that $1 \geq Opt(G', \emptyset) \geq 0$. The next lemma assumes $n \geq 16$ and $3 \geq Opt(G', \emptyset) \geq 2$. The case when $n < 16$ is considered in Corollary 3.

Lemma 2. *Let G' and $Path_n$ be disjoint and $n \geq 16$. If $3 \geq Opt(G', \emptyset) \geq 2$, then for $n' = 0, 1, 2$, $Opt(G' \cup Path_n, \{v_1, v_2, v_{n-1}, v_n\}) = n' - 2 + Opt(G', \emptyset)$ and for $n' = 3, 4, 5$, $Opt(G' \cup Path_n, \{v_1, v_2, v_{n-1}, v_n\}) = 8 - n' - Opt(G', \emptyset)$.* □

Now note that, with the notation above, we have the following

$$Opt_{Cycle_n}(Cycle_n \cup G', \emptyset) = 3 - Opt(G' \cup Path_{n-3}, \{v_1, v_2, v_{n-4}, v_{n-3}\}).$$

Corollary 3. *Let G' and $Cycle_n$ be disjoint, where $n \geq 1$. Assume that $3 \geq Opt(G', \emptyset) \geq 2$. If $n \geq 19$ then for $n' = 0, 1, 2$, $Opt_{Cycle_n}(Cycle_n \cup G', \emptyset) = n' - 2 + Opt(G', \emptyset)$ and for $n' = 3, 4, 5$, $Opt_{Cycle_n}(Cycle_n \cup G', \emptyset) = 8 - n' - Opt(G', \emptyset)$.*

If $18 \geq n \geq 1$ then

$$Opt_{Cycle_n}(Cycle_n \cup G', \emptyset) = \begin{cases} 1 - Opt(G', \emptyset) & \text{if } n \in \{1, 7, 13\} \\ 2 - Opt(G', \emptyset) & \text{if } n \in \{2, 6, 8, 12, 18\} \\ 3 - Opt(G', \emptyset) & \text{if } n \in \{3, 11, 17\} \\ 4 - Opt(G', \emptyset) & \text{if } n = 16 \\ 1 + Opt(G', \emptyset) & \text{if } n \in \{5, 9, 15\} \\ 2 + Opt(G', \emptyset) & \text{if } n \in \{4, 10\} \\ Opt(G', \emptyset) & \text{if } n = 14 \end{cases}$$

□

Corollaries 1 and 3 are important in deciding the games on the unions of cycles in the next section. For part 2 of Corollary 3, see Remark in Sect. 1.3.

2.2 Solving the Games on Unions of Cycles

Let $G = Cycle_{x_1} \cup \ldots Cycle_{x_k}$ be a union of cycles and $Rep(G) = \{x_1, \ldots, x_k\}$ be its representation. In this section, we show how to find the winner and compute a winning strategy for the winner. For our analysis, we use the following notation:

For $S \subseteq \mathbb{Z}_{>0}$, set $cnt_S(G) = |\{x_i \mid x_i \in S\}|$, and $cnt_{odd}(G) = |\{x_i \mid x_i \text{ is odd}\}|$.

Our analysis will be based on the following partition of the set of positive integers $\mathbb{Z}_{>0}$. The partition is a finite modification of the *mod* 6 relation on $\mathbb{Z}_{>0}$.

Definition 6. *Define the partition $\mathcal{P} = \{P_0, \ldots, P_{12}\}$ of all positive integers:*

- $P_i = \{x \geq 19 \mid x \mod 6 = i\}$ *for* $i = 0, 1, 3$,
- $P_2 = \{x \geq 19 \mid x \mod 6 = 2\} \cup \{14\}$,
- $P_4 = \{x \geq 19 \mid x \mod 6 = 4\} \cup \{16\}$,
- $P_5 = \{x \geq 19 \mid x \mod 6 = 5\} \cup \{11, 17\}$,
- $P_6 = \{6, 12, 18\}$, $P_7 = \{1, 7, 13\}$, $P_8 = \{2, 8\}$, $P_9 = \{3\}$,
- $P_{10} = \{4, 10\}$, $P_{11} = \{5\}$, *and* $P_{12} = \{9, 15\}$.

Lemma 3. *If $n \in P_{10}$, then $F_{even}(Cycle_n, \emptyset) = 2$. If $n \in P_{11} \cup P_{12}$, then $F_{even}(Cycle_n, \emptyset) = 1$.* □

Thus, in the graph $G = \bigcup_{i=1}^{k} Cycle_{x_i}$ Player 1 follows the F_{even} strategy on $Cycle_{x_i}$, where $x_i \in P_{10} \cup P_{11} \cup P_{12}$. After winning each of these cycles, Player 1 faces with the configuration $G' = \bigcup_{j=1}^{k'} Cycle_{y_j}$, where $y_j \in \bigcup_{i=0}^{9} P_i$. Due to this, it suffices to focus on the graphs $G = \bigcup_{i=1}^{k} Cycle_{x_i}$, where $x_i \in \bigcup_{i=0}^{9} P_i$. For this graph we define the following set: $\mathcal{S}(G) = \{i \mid cnt_{P_i}(G) \mod 2 = 1\}$.

Theorem 1. *Let $G = Cycle_{x_1} \cup \ldots \cup Cycle_{x_k}$ be a disjoint union of cycles, where each $x_i \in \bigcup_{j=0}^{9} P_j$. Then $Opt(G, \emptyset)$ is computed as follows:*

- *Assume $cnt_{odd}(G) \mod 2 = 0$. If $\mathcal{S}(G) \cap \{1, 7, 8, 9\} \in \{\emptyset, \{1, 7\}\}$ and $Rep(G) \subset P_0 \cup P_1 \cup P_6 \cup P_7 \cup P_8 \cup P_9$, then $Opt(G, \emptyset) = 0$. Otherwise, $Opt(G, \emptyset) = 2$.*
- *Assume that $cnt_{odd}(G) \mod 2 = 1$. Then*
  1. *If $cnt_{P_3}(G) > 0$, then $Opt(G, \emptyset) = 3$.*
  2. *Assume that $\{1, 5, 7, 9\} \cap \mathcal{S}(G) \notin \{\{9\}, \{1, 7, 9\}\}$. If $4 \in \mathcal{S}(G)$, then $Opt(G, \emptyset) = 3$. Otherwise, $Opt(G, \emptyset) = 1$.*
  3. *otherwise:*
     - *(a) Assume that $cnt_{P_5}(G) > 0$. If $4 \in \mathcal{S}(G)$, then $Opt(G, \emptyset) = 3$. Otherwise, $Opt(G, \emptyset) = 1$.*
     - *(b) otherwise: If $\{4, 8\} \cap \mathcal{S}(G) \in \{\emptyset, \{4, 8\}\}$, then $Opt(G, \emptyset) = 3$. Otherwise, $Opt(G, \emptyset) = 1$.* □

Thus, we can decide our games on the union of cycles:

Theorem 2. *Player 2 has a draw in* $G = \bigcup_{i=1}^{k} Cycle_{x_i}$ *iff* $cnt_{odd}(G)$ *is even,* $\mathcal{S}(G) \cap \{1,7,8,9\} \in \{\emptyset, \{1,7\}\}$*, and* $Rep(G) \subset P_0 \cup P_1 \cup P_6 \cup P_7 \cup P_8 \cup P_9$. □

Corollary 4. *There exists an algorithm that, given* $G = \bigcup_{i=1}^{k} G_{x_i}$ *where for all* x_i *we have* $x_i \in \bigcup_{j=0}^{9} P_j$*, computes the* $Opt(G, \emptyset)$ *in time* $O(k \log(n(G)))$. □

Corollary 5. *There is an algorithm that, given the graph* $G = \bigcup_{i=1}^{k} Cycle_{x_i}$*, determines the winner of game played on* G *in* $O(k \log(n(G)))$. □

Corollary 6. *Given a configuration* (G, C) *in a play on* $\bigcup_{i=1}^{k} Cycle_{x_i}$ *where* $x_i \in \bigcup_{j=0}^{9} P_j$*, we can determine an optimal move in* $O(k \log(n(G)))$ *time.* □

Corollary 7. *There is an algorithm that, given that Player 1 wins* $\bigcup_{i=1}^{k} Cycle_{x_i}$*, determines the winning move in* $O(k \log(n(G)))$ *time at each round.* □

3 Linear Forests

In this section we solve the topological network-control games played on linear forests. Recall that a **linear forest** G is a finite union of paths $Path_{x_1} \cup \ldots \cup Path_{x_n}$. So, just like the union of cycles we can represent G as the multi-set $Rep(G) = \{x_1, \ldots, x_k\}$. Our solution idea follows the proof for disjoint union of cycles. However, the techniques for linear forests are more complex. When players play the game on a linear forest G, a typical configuration is of the form (G, C), where G is a linear forest and C has non-empty intersection with two components of G. Thus, typical configurations on linear forests are two dimensional while the configurations for the union of cycles are one dimensional. This is a reason why solving games on linear forests is technically more challenging than the one for union of cycles. The solution of the game for linear forests is the following.

Let $G = Path_{x_1} \cup \ldots Path_{x_k}$ be a union of Paths and $Rep(G) = \{x_1, \ldots, x_k\}$ be its representation. In this section, we find the winner and compute a winning strategy for the winner. For our analysis, we use the following notation:

For $S \subseteq \mathbb{Z}_{>0}$, set $cnt_S(G) = |\{x_i \mid x_i \in S\}|$, $cnt_{odd}(G) = |\{x_i \mid x_i \text{ is odd}\}|$.

Just like for union of cycles, our analysis will be based on the following partition of the set of positive integers $\mathbb{Z}_{>0}$. The partition is a finite modification of the *mod* 6 relation on the set of positive integers $\mathbb{Z}_{>0}$.

Definition 7. *Define the partition* $\mathcal{P} = \{P_0, \ldots, P_9\}$ *of all positive integers:*

- $P_i = \{x \geq 16 \mid x \mod 6 = i\}$ *for* $i = 0, 3, 4$,
- $P_1 = \{x \geq 16 \mid x \mod 6 = 2\} \cup \{13\}$,
- $P_2 = \{x \geq 16 \mid x \mod 6 = 2\} \cup \{8, 14\}$,
- $P_5 = \{x \geq 16 \mid x \mod 6 = 5\} \cup \{7, 11\}$,
- $P_6 = \{2, 6, 12\}$, $P_7 = \{1\}$, $P_8 = \{4, 10\}$, $P_9 = \{3, 5, 9, 15\}$,

The following is an easy lemma:

Lemma 4. *If $n \in P_8$, $F_{even}(Path_n, \emptyset) = 2$. If $n \in P_9$, $F_{even}(Path_n, \emptyset) = 1$.* □

Thus, in the graph $G = \bigcup_{i=1}^{k} Path_{x_i}$. Player 1 follows the F_{even} strategy on $Path_{x_i}$, where $x_i \in P_8 \cup P_9$. After winning each of these Paths, Player 1 faces with the configuration $G' = \bigcup_{j=1}^{k'} Path_{y_j}$, where $y_j \in \bigcup_{i=0}^{9} P_i$. Due to this, it suffices to focus on the graphs $G = \bigcup_{i=1}^{k} Path_{x_i}$, where $x_i \in \bigcup_{i=0}^{7} P_i$. For this graph we define the following set: $\mathcal{S}(G) = \{i \mid cnt_{P_i}(G) \mod 2 = 1\}$.

The theorem below tells us how to compute $Opt(G)$ on these graphs G.

Theorem 3. *Let $G = Path_{x_1} \cup \ldots \cup Path_{x_n}$ be a union of disjoint paths, where each $x_i \in \bigcup_{i=0}^{7}\{P_i\}$. Then $Opt(G, \emptyset)$ is computed as follows:*

- *Assume that $cnt_{odd}(G) \mod 2 = 0$. If for all x_i, we have $x_i \in P_0 \cup P_5 \cup P_6 \cup P_7$, and $\mathcal{S}(G) \cap \{0, 6, 5, 7\}) \in \{\emptyset, \{0, 6\}\}$, then $Opt(\bigcup_{i=1}^{k} Path_{x_i}, \emptyset) = 0$. Otherwise, $Opt(\bigcup_{i=1}^{k} Path_{x_i}, \emptyset) = 2$.*
- *Assume that $cnt_{odd}(G) \mod 2 = 1$. Then*
  1. *If $cnt_{P_3}(G) > 0$, then $Opt(G, \emptyset) = 3$.*
  2. *Assume that $cnt_{P_1}(G) > 1$. If $4 \in \mathcal{S}(G)$ then $Opt(G, \emptyset) = 3$. Otherwise, $Opt(G, \emptyset) = 1$.*
  3. *Assume that $cnt_{P_1}(G) <= 1$. Then*
     - *(a) Assume that $7 \in \mathcal{S}(G)$. If $4 \in \mathcal{S}(G)$ then $Opt(G, \emptyset) = 3$. Otherwise, $Opt(G, \emptyset) = 1$.*
     - *(b) If $\mathcal{S}(G) \cap \{0, 4, 6\} \in \{\emptyset, \{0, 4\}, \{0, 6\}, \{4, 6\}\}$, then $Opt(G, \emptyset) = 3$.*
     - *(c) otherwise, $Opt(G, \emptyset) = 1$.* □

Theorem 4. *Player 2 guarantees a draw on $G = \bigcup_{i=1}^{k} Path_{x_i}$ iff $cnt_{odd}(G)$ is even, $\mathcal{S}(G) \cap \{0, 6, 5, 7\} \in \{\emptyset, \{0, 6\}\}$, and $Rep(G) \subset P_0 \cup P_5 \cup P_6 \cup P_7$.* □

From Theorems 3 and 4 we get the following result:

Corollary 8. *There is an algorithm that, given $G = \bigcup_{i=1}^{k} Path_{x_i}$, where $x_i \in \bigcup_{j=0}^{7} P_j$ for all $k \geq i \geq 1$, computes the $Opt(G, \emptyset)$ in time $O(k \log(n(G)))$.* □

Corollary 9. *There is an algorithm that, given a linear forest $G = \bigcup_{i=1}^{k} Path_{x_i}$, determines the winner on game played on G in time $O(k \log(n(G)))$.* □

In conclusion, this paper fully solves topological network-control games on linear forests and unions of cycles. As we mentioned in the introduction, in combinatorial game theory these type of full solutions are rare as solving most scoring games falls within the class of PSPACE-complete problems. We conclude this paper with two open questions. The first question is if the methods of this paper can be extended to solving topological network-control games on trees or forests. Note that the dimensions of configurations of games on trees and forests are unbounded. Hence, this represents a challenge. For the second question, observe that there are configurations, where Player 1 starts and yet Player 2 wins. However, it is somewhat surprising (and annoying) that we still do not know if there are topological network-control games where Player 2 wins.

References

1. Adams, R., Dixon, J., Elder, J., Peabody, J., Vega, O., Willis, K.: Combinatorial analysis of a subtraction game on graphs. Int. J. Comb. **2016**, 1476359 (2016)
2. Andres, S.D., Dross, F., Huggan, M., Mc Inerney, F., Nowakowski, R.J.: The complexity of two colouring games. Research report, CISPA Helmholtz Center for Information Security, Saarbrücken, Germany (2021). https://hal.science/hal-02053265
3. Andres, S.D., Huggan, M., Mc Inerney, F., Nowakowski, R.J.: Corrigendum to "the orthogonal colouring game" Theor. Comput. Sci. 795, 312–325 (2019). Theoretical Computer Science **842**, 133–135 (2020)
4. Beaudou, L., et al.: Octal games on graphs: the game 0.33 on subdivided stars and bistars. Theoret. Comput. Sci. **746**, 19–35 (2018)
5. Bensmail, J., Fioravantes, F., Mc Inerney, F., Nisse, N.: The largest connected subgraph game. Algorithmica **84**(9), 2533–2555 (2022)
6. Bensmail, J., Fioravantes, F., Mc Inerney, F., Nisse, N., Oijid, N.: The maker-breaker largest connected subgraph game. Theoret. Comput. Sci. **943**, 102–120 (2023)
7. Bensmail, J., Mc Inerney, F.: On a vertex-capturing game. Theoret. Comput. Sci. **923**, 27–46 (2022)
8. Berkelamp, E., Conway, J., Guy, R.: Winning Ways for Your Mathematical Plays (4 Volumes). AK Peters, Wellesley, MA (2001)
9. Berlekamp, E.R., Conway, J.H., Guy, R.K.: Winning Ways for Your Mathematical Plays, vol. 3. CRC Press (2018)
10. Bodlaender, H.L., Kratsch, D.: Kayles and nimbers. J. Algorithms **43**(1), 106–119 (2002)
11. Brown, S., et al.: Nimber sequences of node-kayles games. J. Integer Sequences **23**, 20.3.5 (2020)
12. Burke, K., George, O.: A PSPACE-complete graph Nim. Games No Chance **5**, 259–269 (2011)
13. Calkin, N.J., James, K., Janoski, J.E., Leggett, S., Richards, B., Sitaraman, N., Thomas, S.M.: Computing strategies for graphical Nim. In: Proceedings of the Forty-First Southeastern International Conference on Combinatorics, Graph Theory and Computing, vol. 202, pp. 171–185. Citeseer (2010)
14. Cibulka, J., Kynčl, J., Mészáros, V., Stolař, R., Valtr, P.: Graph sharing games: complexity and connectivity. Theoret. Comput. Sci. **494**, 49–62 (2013)
15. Cohen, N., Martins, N.A., Mc Inerney, F., Nisse, N., Pérennes, S., Sampaio, R.: Spy-game on graphs: complexity and simple topologies. Theoret. Comput. Sci. **725**, 1–15 (2018)
16. Conway, J.H.: On Numbers and Games. CRC Press (2000)
17. Dailly, A., Gledel, V., Heinrich, M.: A generalization of arc-kayles. Internat. J. Game Theory **48**(2), 491–511 (2019)
18. Demaine, E.D.: Playing games with algorithms: algorithmic combinatorial game theory. In: Sgall, J., Pultr, A., Kolman, P. (eds.) MFCS 2001. LNCS, vol. 2136, pp. 18–33. Springer, Heidelberg (2001). https://doi.org/10.1007/3-540-44683-4_3
19. Duchêne, É., Gledel, V., Gravier, S., Mc Inerney, F., Mhalla, M., Parreau, A.: Smash and grab: the 0.6 scoring game on graphs (2021). https://hal.science/hal-03371099, working paper or preprint
20. Duchene, E., Gonzalez, S., Parreau, A., Rémila, E., Solal, P.: Influence: a partizan scoring game on graphs. Theoret. Comput. Sci. **878**, 26–46 (2021)

21. Duchêne, É., Renault, G.: Vertex Nim played on graphs. Theoret. Comput. Sci. **516**, 20–27 (2014)
22. Ettinger, J.M.: A metric for positional games. Theoret. Comput. Sci. **230**(1–2), 207–219 (2000)
23. Ettinger, J.M.: Topics in Combinatorial Games. The University of Wisconsin-Madison (1996)
24. Fomin, F.V., Golovach, P.A., Lokshtanov, D.: Guard games on graphs: keep the intruder out! Theoret. Comput. Sci. **412**(46), 6484–6497 (2011)
25. Fraenkel, A.S.: Combinatorial games: selected bibliography with a succinct gourmet introduction. More Games Chance **42**, 475–535 (2002)
26. Grundy, P.M.: Mathematics and games. Eureka **2**, 6–8 (1939)
27. Guignard, A., Sopena, É.: Compound node-kayles on paths. Theoret. Comput. Sci. **410**(21–23), 2033–2044 (2009)
28. Hanner, O.: Mean play of sums of positional games. Pac. J. Math. **9**(1), 81–99 (1959)
29. Huggan, M.A., Stevens, B.: Polynomial time graph families for arc kayles. Integers **16**, A86 (2016)
30. Huggan, M.A.: Impartial intersection restriction games. Ph.D. thesis, Carleton University (2015)
31. Kleinberg, J., Tardos, E.: Algorithm Design. Pearson Education India (2006)
32. Larsson, U., Nowakowski, R., Santos, C.P.: Scoring games: the state of play, games of no chance 5 (2017)
33. Liang, Z., Khoussainov, B., Yang, H.: Topological network-control games. In: Wu, W., Tong, G. (eds.) International Computing and Combinatorics Conference, vol. 14423, pp. 144–156. Springer, Cham (2023). https://doi.org/10.1007/978-3-031-49193-1_11
34. Milnor, J.: Sums of positional games. Contrib. Theory Games **II**(28), 291–301 (1953)
35. Siegel, A.N.: Combinatorial Game Theory, vol. 146. American Mathematical Soc. (2013)
36. Sprague, R.: Über mathematische kampfspiele. Tohoku Math. J. First Series **41**, 438–444 (1935)
37. Stewart, F.: Impartial scoring play games. arXiv preprint arXiv:1202.4655 (2012)
38. Stewart, F.: Scoring play combinatorial games. Games No Chance **5**, 447–467 (2012)

FAIR: Accurate Data Acquisition for Mobile Crowdsensing

Fuyuan Xia[1,2], Chenhao Ying[1,2], Wei Chen[1], Xikun Jiang[3](✉), Liwei Lin[4], and Yuan Luo[1,2](✉)

[1] Shanghai Jiao Tong University, Shanghai 200240, China
{fuyuanxia,yingchenhao,chenwei8,yuanluo}@sjtu.edu.cn
[2] Shanghai Jiao Tong University (Wuxi) Blockchain Advanced Research Center, Wuxi 214101, China
[3] University of Copenhagen, Copenhagen 2100, Denmark
xikun@di.ku.dk
[4] Fujian University of Technology, Fuzhou 350118, China

Abstract. Mobile crowdsensing, while a promising method for handling vast data, encounters two key challenges: accurately estimating the truth of collected data and setting enticing rewards to increase user participation. Most existing studies tackle these issues separately, but our proposed geometry-based data acquisition FrAmework for mobIle cRowdsensing, named FAIR, addresses both simultaneously. FAIR comprises two interlinked parts: FAIR-EST for truth estimation and FAIR-REW for reward determination. Unlike typical methods, FAIR-EST aims to minimize worst-case estimation error, effectively finding the Chebyshev center of the data, which leads to a non-convex quadratic optimization problem solved by semi-definite relaxation. Based on the truth estimated by FAIR-EST, FAIR-REW assigns rewards to maximize platform profit, using an innovative asymmetric all-pay auction model to account for user diversity. Moreover, we offer a rigorous theoretical analysis demonstrating that FAIR can achieve numerous desirable properties under more commonplace conditions.

Keywords: Mobile crowdsensing · Incentive mechanism · Tight semi-definite relaxation · Asymmetric all pay auction

1 Introduction

The rapid increase in smart devices and advances in wireless technology have made mobile crowdsensing a key method for gathering sensing data [1]. This

This work was supported by Shanghai Science and Technology Innovation Action Plan (No. 23511100400), Educational Scientific Research Project of Fujian Provincial Department of Education (No. JAT210291), and the Science Foundation of Fujian University of Technology (No. GY-Z220206).

F. Xia and C. Ying—Equal contribution.

Y. Chen et al. (Eds.): COCOON 2024, LNCS 15162, pp. 27–39, 2025.
https://doi.org/10.1007/978-981-96-1093-8_3

approach leverages numerous smart devices to collect data from the physical world using their built-in sensors. Recognizing the potential of these devices, mobile crowdsensing has been applied in various fields such as localization [2], environment monitoring [3] and navigation [4]. However, this approach encounters two main issues: First, various factors like noise can lead to differing data for the same task, making accurate truth estimation crucial. Second, participation incurs costs for users, such as battery use, computing resources, or privacy risks, leading to reluctance without adequate rewards. Thus, creating effective incentives is essential to encourage broader participation in crowdsensing tasks.

Numerous studies are dedicated to the estimation of truth values in crowdsensing scenarios. Cheng *et al.* [5] proposed a privacy-preserving and reputation-based truth discovery framework. Li *et al.* [6] proposed a general truth discovery framework. Zhang *et al.* [7] proposed an efficient and strong privacy-preserving truth discovery scheme. Wang *et al.* [8] proposed an adaptive Hypergraph-based EM truth discovery method. Yang *et al.* [9] formulated truth discovery as a semi-supervised learning model. However, all of these schemes might not obtain a good estimation until collecting enough data.

Moreover, there are also a lot of works about the incentive mechanism. Ji *et al.* [10] proposed incentive mechanisms for socially-aware and socially-unaware mobile crowdsensing. Yang *et al.* [11] investigated a participant-quantity-aware online task allocation problem. Jin *et al.* [12] considered the quality based prize mechanisms for crowdsensing. However, all of these works did not consider the fact that a user requires some costs once participating in the sensing task. Furthermore, all of these works designed the truth estimation schemes and prize mechanisms separately, which might lead to performance loss.

Unlike previous efforts, our work seeks to create a framework that simultaneously addresses truth estimation and reward determination. This involves designing both an accurate truth estimation scheme and a reward mechanism to increase user participation. Motivated by this question, we have to address several notable challenges.

First, designing a scheme to closely match estimates with actual truth values is difficult because of the absence of known truth values. Second, the lack of historical data complicates the creation of effective estimation schemes. Third, creating a reward mechanism that ensures users are compensated fairly for their costs (*individual rationality*) and discourages data manipulation for better rewards (*truthfulness*) [10] is challenging. Finally, traditional reward systems that offer fixed payouts do not consider the varying contributions of users [13]. Although tying rewards to contributions appears fair, accurately assessing and compensating for these contributions proves difficult.

In response to these challenges, we propose FAIR, a geometry-based data acquisition <u>F</u>r<u>A</u>mework for mob<u>I</u>le c<u>R</u>owdsensing, comprising two key components: FAIR-EST for truth estimation and FAIR-REW for reward allocation. FAIR-EST, leveraging the bounded nature of data noise, aims to minimize the worst-case estimation error, effectively identifying the Chebyshev center of the data geometrically. This approach elegantly sidesteps the need for historical data, rendering it suitable for real-time sensing tasks. To address the issue of equitable reward distribution, FAIR-REW evaluates data quality by measuring the proximity of collected data to the estimated truth, using this quality as a proxy for each user's contribution. This mechanism not only maximizes platform profitability but also ensures fairness and truthfulness through a novel asymmetric all-pay auction model that accounts for the varied data collection capabilities of users. The main contributions of this paper are summarized as follows.

- *Truth Estimation:* FAIR-EST estimates the truth values by minimizing the estimation error of the worst case, and a closed-form expression of estimated truth is obtained by applying a tight semi-definite relaxation. To the best of our knowledge, FAIR-EST is the first attempt to use the geometry method to estimate the truth values.
- *Reward Determination:* Utilizing the truth values estimated by FAIR-EST, FAIR-REW introduces an innovative reward mechanism based on an asymmetric all-pay auction that accounts for user heterogeneity, aiming to maximize platform profit. Furthermore, we provide theoretical proofs demonstrating that FAIR satisfies the properties of *individual rationality* and *truthfulness.*

The rest of this paper is organized as follows. Technical preliminaries are presented in Sect. 2. FAIR-EST and FAIR-REW are introduced in Sect. 3 and Sect. 4, respectively. Finally, Sect. 5 concludes this paper.

2 Preliminaries

2.1 System Overview and Basic Notations

A typical mobile crowdsensing system consists of a mobile data platform and some mobile users, where the platform announces a set of points of interests (PoIs) and collects the corresponding data from users. Let $\mathbb{N} \triangleq \{1, 2, \ldots, N\}$ and $\mathbb{U}_i \triangleq \{i_1, i_2, \ldots, i_{M_i}\}$ denote the sets of N PoIs and M_i users at the i-th PoI, respectively, where user i_ℓ means that she is the ℓ-th user at the i-th PoI. For ease of illustration, we assume that each user prefers exactly one PoI, and the numbers of users at all PoIs are equal, *i.e.*, $\mathbb{U}_i \bigcap \mathbb{U}_j = \emptyset$ and $M_i = M_j = M$ for $i \neq j$. Furthermore, we assume that $M < N$ and these N PoIs are independent due to the large number of PoIs.

In this paper, we apply a geometry based scheme to obtain an accurate estimated truth from collected data. For ease of illustration, we denote x_i and $\omega_{i,\ell}$ as the truth value of data at the i-th PoI and the noise mixed in user i_ℓ's collected data, respectively. Therefore, the data collected by user i_ℓ can be denoted as $y_{i,\ell} = x_i + \omega_{i,\ell}$. The truth values at all PoIs consist of a truth vector expressed as $\mathbf{x} = (x_1, \ldots, x_N)^T$, where $(\cdot)^T$ means the transpose.

2.2 Solution Concepts

In this paper, we use an all pay auction model to characterise this property in our prize mechanism. Informally, in this auction model, the user with the highest bid wins the reward, but all users require to tender their respective bids. Furthermore, the following definitions are critical for an all pay auction model.

Definition 1 (Bayesian Nash Equilibrium). *A (pure strategy) Bayesian Nash equilibrium of the all pay auction model is a strategy profile (*i.e.*, the user's data quality in this paper)* $\boldsymbol{q}_i^* = (q_{i,1}^*, q_{i,2}^*, \ldots, q_{i,M}^*)$ *that satisfies*

$$u_{i,\ell}(q_{i,\ell}^*, \boldsymbol{q}_{i,-\ell}^*) \geq u_{i,\ell}(q_{i,\ell}, \boldsymbol{q}_{i,-\ell}^*),\ \forall q_{i,\ell},$$

where $u_{i,\ell}(\cdot)$ *is the utility of user* i_ℓ *and* $\boldsymbol{q}_{i,-\ell} = (q_{i,1}, \ldots, q_{i,\ell-1}, q_{i,\ell+1}, \ldots, q_{i,M})$.

Definition 2 (Individual Rationality). *A mechanism is individual rational if in equilibrium, all participating users expect higher surplus than from not participating. It means that for each user i in equilibrium, her utility satisfies* $u_{i,\ell} \geq 0$.

Definition 3 (Truthfulness). *A mechanism is truthful if no user can improve her utility by submitting her type different from its truth valuation.*

3 Truth Estimation

3.1 Detailed Estimation Procedure

For the i-th PoI, we denote the data collected by user i_ℓ as $y_{i,\ell}$, where $i_\ell \in \mathbb{U}_i$ and $i \in \mathbb{L}$. Due to the noise, $y_{i,\ell}$ has the form as $y_{i,\ell} = x_i + \omega_{i,\ell}$, where x_i and $\omega_{i,\ell}$ are the truth value of the i-th PoI and the noise of user i_ℓ, respectively. Therefore, the data of the ℓ-th user at each PoI can be further written as a vector

$$\mathbf{y}_\ell = \mathbf{x} + \mathbf{w}_\ell, \tag{1}$$

where $\mathbf{y}_\ell = (y_{1,\ell}, y_{2,\ell}, \ldots, y_{N,\ell})$, $\mathbf{x} = (x_1, x_2, \ldots, x_N)$ and $\mathbf{w}_\ell = (\omega_{1,\ell}, \omega_{2,\ell}, \ldots, \omega_{N,\ell})$ for $\ell = 1, \ldots, M$. Hence, it is rational to assume that $||\mathbf{w}_\ell||^2 \leq \rho^2$ for $\ell = 1, \ldots, M$, which can also be written as

$$||\mathbf{y}_\ell - \mathbf{x}||^2 \leq \rho^2. \tag{2}$$

This means that the truth vector $\mathbf{x}$ belongs to an N-dimensional ball $\mathcal{B}(\mathbf{y}_\ell, \rho)$, whose center and radius are $\mathbf{y}_\ell$ and ρ, respectively. Since there are M N-dimensional balls, the truth vector $\mathbf{x}$ lies in the intersection $\mathcal{I}$ of these M balls, which can be denoted as

$$\mathcal{I} \triangleq \bigcap_{\ell=1}^{M} \mathcal{B}(\mathbf{y}_\ell, \rho). \tag{3}$$

An appropriate estimation scheme $\hat{\mathbf{x}} = (\hat{x}_1, \ldots, \hat{x}_N)$ on the truth values $\mathbf{x}$ of collected data $\mathbf{y}$ aims to obtain an estimation with small errors. To reach this goal, FAIR-EST provides an optimal estimation $\hat{\mathbf{x}}^* = (\hat{x}_1^*, \ldots, \hat{x}_N^*)$ by minimizing the worst error over all feasible vectors, which is given by

$$\hat{\mathbf{x}}^* = \arg\min_{\hat{\mathbf{x}}} \max_{\mathbf{x} \in \mathcal{I}} ||\hat{\mathbf{x}} - \mathbf{x}||^2. \tag{4}$$

In order to solve this problem, FAIR-EST investigates by developing geometry interpretation of $\hat{\mathbf{x}}^*$. In fact, Eq. (4) can be written equivalently as

$$\min_{\hat{\mathbf{x}}}\{r : ||\hat{\mathbf{x}} - \mathbf{x}||^2 \leq r \text{ for all } \mathbf{x} \in \mathcal{I}\}. \tag{5}$$

For a given r, the set of all values of $\mathbf{x}$ satisfying $||\hat{\mathbf{x}} - \mathbf{x}||^2 \leq r$ defines a ball with radius $\sqrt{r}$ and center $\hat{\mathbf{x}}$. Therefore, the constraint in Eq. (5) is equivalent to the ball defined by r and $\hat{\mathbf{x}}$ encloses the set $\mathcal{I}$. It means that Eq. (5) can be further written as

$$\begin{aligned} &\min r \\ &s.t.\ \mathcal{I} \subseteq \mathcal{B}(\hat{\mathbf{x}}, \sqrt{r}), \\ &\quad\ \hat{\mathbf{x}} \in \mathbb{R}^N,\ r \in \mathbb{R}_+. \end{aligned} \tag{6}$$

As shown in Eq. (6), the goal of FAIR-EST is to obtain the center of the minimum radius ball enclosing the intersection $\mathcal{I}$. Since the minimization is over the squared-radius, this center is also referred as the Chebyshev center of $\mathcal{I}$, while the corresponding squared-radius is the optimal minimax value of Eq. (4). This is illustrated in Fig. 1 with the filled area being the intersection of three data balls. The solid circle is the minimum enclosing circle of the intersection of the balls. Meanwhile, the corresponding center is the estimated truth which is also the Chebyshev center of the intersection.

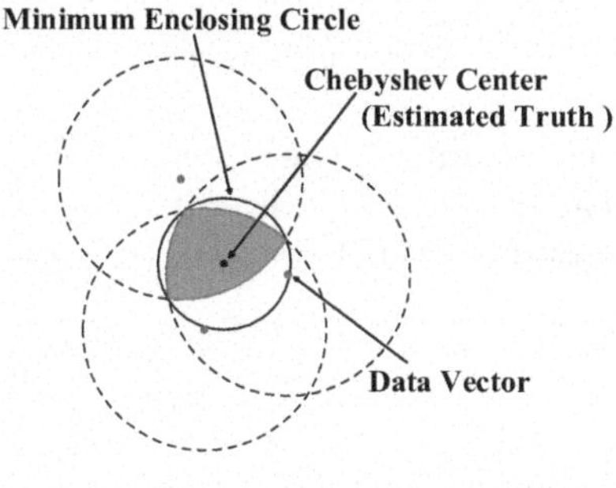

Fig. 1. Chebyshev Center of the Intersection of Three Balls

Computing the Chebyshev center of Eq. (6) is a hard optimization problem since it is a non-convex quadratic optimization problem. Therefore, FAIR-EST uses a tight semi-definite relaxation to obtain the center of the ball, *i.e.*, the estimated vector $\hat{\mathbf{x}}^*$. To achieve this goal, the following lemma is critical, which is similar to [14].

Lemma 1. *For data vectors $\boldsymbol{y}_\ell$ and corresponding N-dimensional balls $\mathcal{B}(\boldsymbol{y}_\ell, \rho)$ defined in Eq. (1) and Eq. (2), the following two statements are equivalent:*

i. $\mathcal{I} \subseteq \mathcal{B}(\hat{\boldsymbol{x}}, \sqrt{r})$, where $\mathcal{I}$ is given in Eq. (3).

ii. There exist $\lambda_1, \lambda_2, \ldots, \lambda_M \in \mathbb{R}_+$ such that the following linear matrix inequality is satisfied:

$$\boldsymbol{N}_0 - \sum_{\ell=1}^{M} \lambda_\ell \boldsymbol{N}_\ell \preceq \boldsymbol{0},$$

where

$$\boldsymbol{N}_0 = \begin{pmatrix} \boldsymbol{I} & -\hat{\boldsymbol{x}} \\ -\hat{\boldsymbol{x}}^T & ||\hat{\boldsymbol{x}}||^2 - r \end{pmatrix}, \quad \boldsymbol{N}_\ell = \begin{pmatrix} \boldsymbol{I} & -\boldsymbol{y}_\ell \\ -\boldsymbol{y}_\ell^T & ||\boldsymbol{y}_\ell||^2 - \rho^2 \end{pmatrix},$$

for $\ell = 1, \ldots, M$.

Due to space limitation, the details of proof is omitted. According to Lemma 1, it is easy to obtain the following theorem.

Theorem 1. *For the data vectors* $\boldsymbol{y}_\ell$, *noise radius* ρ *and their corresponding data balls* $\mathcal{B}(\boldsymbol{y}_\ell, \rho)$ *defined in Eq. (1) and Eq. (2), where* $\ell = 1, \ldots, M$, *the estimated truth vector* $\hat{\boldsymbol{x}}^*$ *is given by*

$$\hat{\boldsymbol{x}}^* = \sum_{\ell=1}^{M} \lambda_\ell \boldsymbol{y}_\ell, \tag{7}$$

where $\lambda = (\lambda_1, \ldots, \lambda_M)$ *is an optimal solution of the convex quadratic minimization problem given by*

$$\begin{aligned} &\min ||\sum_{\ell=1}^{M} \lambda_\ell \boldsymbol{y}_\ell||^2 - \sum_{\ell=1}^{M} \lambda_\ell(||\boldsymbol{y}_\ell||^2 - \rho^2), \\ &s.t.\ \lambda \in \Delta_M, \end{aligned} \tag{8}$$

and $\Delta_M \triangleq \{\lambda = (\lambda_1, \ldots, \lambda_M) \in \mathbb{R}^M : \lambda_\ell \geq 0, \sum_{\ell=1}^{M} \lambda_\ell = 1\}$.

Proof. (*sketch*) As shown in Lemma 1, the statement $\mathcal{I} \subseteq \mathcal{B}(\hat{\mathbf{x}}, \sqrt{r})$ in Eq. (6) can be reformulated as a linear matrix inequality. It means Eq. (6) can be rewritten as a semi-definite programming (SDP) problem [15], *i.e.*,

$$\begin{aligned} &\min\ r \\ &s.t. \begin{pmatrix} (\sum_{\ell=1}^{M} \lambda_\ell - 1)\mathbf{I} & \hat{\mathbf{x}} - \sum_{\ell=1}^{M} \lambda_\ell \mathbf{y}_\ell \\ (\hat{\mathbf{x}} - \sum_{\ell=1}^{M} \lambda_\ell \mathbf{y}_\ell)^T & r - ||\hat{\mathbf{x}}||^2 + \sum_{\ell=1}^{M} \lambda_\ell(||\mathbf{y}_\ell||^2 - \rho^2) \end{pmatrix} \succeq \mathbf{0}, \\ &\hat{\mathbf{x}} \in \mathbb{R}^N,\ \lambda \in \mathbb{R}_+^M. \end{aligned} \tag{9}$$

Therefore, it is easy to obtain Eq. (7) from Eq. (9). Furthermore, according to Eq. (9), we can use contradiction to obtain the condition $\lambda \in \Delta_M$ in Eq. (8) by assuming $\sum_{\ell=1}^{M} \lambda_\ell > 1$.

3.2 Data Quality Evaluation

After obtaining the estimated truth vector, the platform can easily derive the noise of each user's data by substituting $\hat{\mathbf{x}}$ for $\mathbf{x}$ in Eq (1). For user i_ℓ at the i-th PoI, the noise of her data derived from Eq. (1) can be denoted as $\hat{\omega}_{i,\ell}$. Generally, her data quality $q_{i,\ell}$ evaluated by the platform is a strictly decreasing function

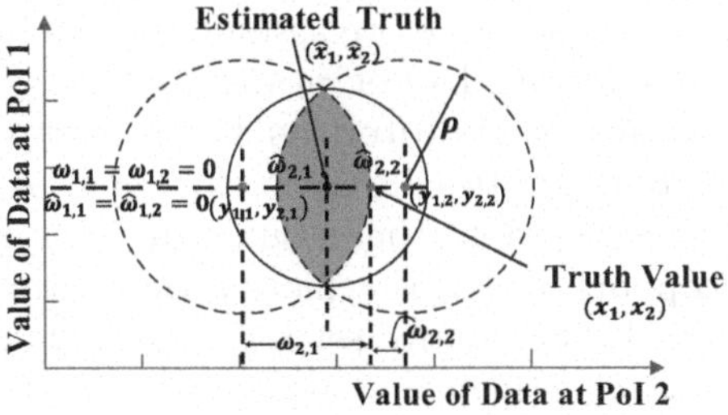

Fig. 2. Data Qualities Estimated by the Platform (Color figure online)

of her data noise, which is denoted as $\beta(\cdot) : \mathbb{R}^+ \rightarrow \mathbb{R}^+$, *i.e.*, $q_{i,\ell} = \beta(|\hat{\omega}_{i,\ell}|)$. Then, according to the data quality, the platform orders users' data at each PoI respectively to determine the reward of each user. Note that this paper does not set a specific form of function $\beta(\cdot)$.

Example 1. Let's see an example shown in Fig. 2, where there are two PoIs and each PoI has two users, *i.e.*, $\mathbb{N} = \{1, 2\}$, $\mathbb{U}_1 = \{1_1, 1_2\}$ and $\mathbb{U}_2 = \{2_1, 2_2\}$. At PoI 1, users 1_1 and 1_2 sample the data denoted as $y_{1,1}$ and $y_{1,2}$ with real noises $\omega_{1,1}$ and $\omega_{1,2}$, respectively. Additionally, at PoI 2, users 2_1 and 2_2 sample the data of PoI 2 denoted as $y_{2,1}$ and $y_{2,2}$ with real noises $\omega_{2,1}$ and $\omega_{2,2}$, respectively. Then, data $y_{1,1}$ and $y_{2,1}$ constitute a data vector $\mathbf{y}_1 = (y_{1,1}, y_{2,1})$, while data $y_{1,2}$ and $y_{2,2}$ constitute another vector $\mathbf{y}_2 = (y_{1,2}, y_{2,2})$ (as the red nodes in Fig. 2). We assume that the truth values of PoIs 1 and 2 are x_1 and x_2, respectively, which also constitute a truth vector $\mathbf{x} = (x_1, x_2)$ (as the yellow node in Fig. 2). Similarly, the corresponding noises also constitute two noise vectors $\mathbf{w}_1 = (\omega_{1,1}, \omega_{2,1})$ and $\mathbf{w}_2 = (\omega_{1,2}, \omega_{2,2})$. We assume that the noises satisfy $\omega_{1,1} = 0$, $\omega_{1,2} = 0$ and $|\omega_{2,1}| > |\omega_{2,2}|$, respectively.

FAIR-EST outputs an estimated truth vector $\hat{\mathbf{x}}^* = (\hat{x}_1^*, \hat{x}_2^*)$ of PoIs 1 and 2. By using Theorem 1, it is easy to calculate that $\hat{x}_1^* = \frac{1}{2}y_{1,1} + \frac{1}{2}y_{1,2}$ and $\hat{x}_2^* = \frac{1}{2}y_{2,1} + \frac{1}{2}y_{2,2}$ (as the black node in Fig. 2). The estimated noises of users 1_1, 1_2, 2_1 and 2_2 can be obtained by substituting $\hat{\mathbf{x}}^* = (\hat{x}_1^*, \hat{x}_2^*)$ for $\mathbf{x} = (x_1, x_2)$, which are denoted as $\hat{\omega}_{1,1}$, $\hat{\omega}_{1,2}$, $\hat{\omega}_{2,1}$ and $\hat{\omega}_{2,2}$, respectively, where $\hat{\omega}_{1,1} = \frac{1}{2}y_{1,1} - \frac{1}{2}y_{1,2}$, $\hat{\omega}_{1,2} = -\frac{1}{2}y_{1,1} + \frac{1}{2}y_{1,2}$, $\hat{\omega}_{2,1} = \frac{1}{2}y_{2,1} - \frac{1}{2}y_{2,2}$ and $\hat{\omega}_{2,2} = -\frac{1}{2}y_{2,1} + \frac{1}{2}y_{2,2}$. Finally, the platform evaluates their corresponding qualities $q_{1,1}$, $q_{1,2}$, $q_{2,1}$ and $q_{2,2}$ by applying function $\beta(\cdot)$, respectively. It is easy to confirm that $q_{1,1} = q_{1,2}$ and $q_{2,1} = q_{2,2}$ since $|\hat{\omega}_{1,1}| = |\hat{\omega}_{1,2}|$ and $|\hat{\omega}_{2,1}| = |\hat{\omega}_{2,2}|$.

4 Reward Determination

4.1 Structure of FAIR-REW

In FAIR-REW, the quality of user i_ℓ can be denoted as a function of ability value $v_{i,\ell}$, *i.e.*, $q_{i,\ell} = q_{i,\ell}(v_{i,\ell})$. Generally, the ability of each user is a *private information*, which means that each user only knows her own ability but does not know that of any others. Furthermore, we assume that **the ability value** $v_{i,\ell}$ of user i_ℓ **is uniformly distributed over a support** $[0, \bar{v}_{i,\ell}]$**, whose cumulative distribution function is denoted as** $F_{i,\ell}(v)$**, respectively**. Generally, different users have different abilities in real world scenarios. To characterise this heterogeneous nature of users, we further assume that $\bar{v}_{i,\ell} \neq \bar{v}_{j,\ell}$ for $i \neq j$. If the lower bound 0 in the support $[0, \bar{v}_{i,\ell}]$ is achieved, the corresponding user i_ℓ does not participate in the task. The above setting corresponds to an **incomplete information** setting and constitutes an asymmetric Bayesian game where the common priors of user abilities at the same PoI consist of M different distributions.

As mentioned above, for the i-th PoI, to attract users to collect data with high quality, user i_ℓ will win a corresponding reward $R(q_{i,\ell})$ only if her data has the highest quality. In real world scenarios, the same reward may have different values to different users. It means that the real value of the reward is more worth to him. Therefore, a value function $V(v_{i,\ell}, R(q_{i,\ell}))$ is used to characterize the real value of reward $R(q_{i,\ell})$ to user i_ℓ. An example of the value function is $V(v, R) = vR$, where R is a fixed reward and v is user's ability. In fact, $V(v, R)$ is a generalization of a standard all pay auction when the fixed reward is $R = 1$, *i.e.*, $V(v, R) = v$. Without loss of generality, we assume $V(v_{i,\ell}, R(q_{i,\ell}))$ is differentiable with respect to $v_{i,\ell}$.

A user i_ℓ collecting the data with evaluated quality $q_{i,\ell}$ incurs a corresponding cost, which is denoted as $c(v_{i,\ell}, q_{i,\ell})$. Generally, the utility function of user i_ℓ is formulated as

$$\tilde{u}_{i,\ell} = \begin{cases} V(v_{i,\ell}, R(q_{i,\ell})) - c(v_{i,\ell}, q_{i,\ell}) & \text{if } q_{i,\ell} > q_{i,\iota}, \forall \iota \neq \ell \\ -c(v_{i,\ell}, q_{i,\ell}) & \text{otherwise} \end{cases}. \tag{10}$$

Note that this paper does not set specific forms for $c(\cdot)$ and $V(\cdot)$. According to Eq. (10), the expected utility of user i_ℓ is given by

$$u_{i,\ell} = p_{i,\ell} V(v_{i,\ell}, R(q_{i,\ell})) - c(v_{i,\ell}, q_{i,\ell}), \tag{11}$$

where $p_{i,\ell}$ is the probability of user i_ℓ being a winner.

As to the platform, the corresponding profit can be written as

$$\phi_i = \sum_{\ell=1}^{M} q_{i,\ell} - U(\theta, R(q_{i,\kappa})) \tag{12}$$

where κ is a random variable for the index of the winner at the i-th PoI. Similar to users, the corresponding payment (*i.e.*, the user reward) also has a real value. For example, a fixed payment may have different values in the platform for different sensing tasks. To formulate this situation, a value function denoted as $U(\theta, R(q_{i,\kappa}))$ is applied, where θ is a parameter that characterises the real value of payment in different scenarios and $R(q_{i,\kappa})$ is the payment to the winner at the i-th PoI.

For the i-th PoI, FAIR-REW aims to maximize the expected profit Φ_i of platform, which is defined to be

$$\Phi_i = \mathbb{E}\left[\sum_{\ell=1}^{M} q_{i,\ell} - U(\theta, R(q_{i,\kappa}))\right], \tag{13}$$

where the expectation depends on the index random variable κ and some other parameters. Therefore, the total profit Φ of platform can be expressed as $\Phi = \sum_{i=1}^{N} \Phi_i$.

4.2 Analysis

Similar to [13], it can be proved that FAIR-REW has a Bayesian Nash equilibrium since the user abilities are drawn from different uniform distributions.

Lemma 2 (Existence, Uniqueness, Continuity, Monotonicity). *Based on the all pay auction mechanism with incomplete information, FAIR-REW has a unique continuous Bayesian Nash equilibrium. Furthermore, data quality $q_{i,\ell}$ of user i_ℓ at the equilibrium is strictly increasing with $v_{i,\ell}$.*

For convenience, the following parts use the notations without the superscript * to express the variables at the equilibrium since FAIR-REW has a unique equilibrium. Lemma 2 tells that the equilibrium data quality $q_{i,\ell}(v_{i,\ell})$ of user i_ℓ is a strictly increasing function of her ability $v_{i,\ell}$. Therefore, the ability $v_{i,\ell}$ of user i_ℓ is the inverse function of her data quality $q_{i,\ell}$ denoted as $v_{i,\ell} = v_{i,\ell}(q_{i,\ell})$, which is also monotone increasing with respect to $q_{i,\ell}$. This result is consistent with our common sense, *i.e.*, the user with the higher ability can collect higher quality data. Furthermore, the probability that the quality of user i_ℓ is larger than that of user i_ι can be denoted as

$$\Pr(q_{i,\ell} > q_{i,\iota}) = \Pr(v_{i,\iota}(q_{i,\ell}) > v_{i,\iota}) = F_{i,\iota}(v_{i,\iota}(q_{i,\ell})), \tag{14}$$

where

$$F_{i,\iota}(v_{i,\iota}(q_{i,\ell})) = \begin{cases} 1 \text{ if } q_{i,\ell} > q_{i,\iota}(\bar{v}_{i,\iota}), \ \forall \iota \neq \ell \\ \epsilon \qquad\qquad \text{otherwise} \end{cases} \tag{15}$$

for $\epsilon \in (0, 1)$.

Therefore, since the probability of user i_ℓ being the winner is

$$p_{i,\ell} = \prod_{\iota=1,\iota\neq\ell}^{M} \Pr(q_{i,\ell} > q_{i,\iota}),$$

Eq. (11) can be further written as

$$u_{i,\ell} = V(v_{i,\ell}, R(q_{i,\ell})) \prod_{\iota=1,\iota\neq\ell}^{M} F_{i,\iota}(v_{i,\iota}(q_{i,\ell})) - c(v_{i,\ell}, q_{i,\ell}). \tag{16}$$

Unfortunately, it is difficult to obtain a closed-form expression for data quality of each user at the equilibrium in this mechanism. Therefore, FAIR-REW will give the expression of user reward and the profit of platform without obtaining the closed form expression of equilibrium data quality. To determine the winner reward, FAIR-REW requires the explicit forms of the value function $V(\cdot)$ and $U(\cdot)$. In this section, let $V(v_{i,\ell}, R(q_{i,\ell})) = h(v_{i,\ell})R(q_{i,\ell})$ and $U(\theta, R(q_{i,\ell})) = g(\theta)R(q_{i,\ell})$ be the value functions for user i_ℓ and platform, respectively, where $h(\cdot)$ satisfies $h(0) = 0$. For convenience, let $\hat{c}(v_{i,\ell}, q_{i,\ell}) = c(v_{i,\ell}, q_{i,\ell})/h(v_{i,\ell})$.

Theorem 2. *In the FAIR-REW, whose structure is illustrated in Subsect. 4.1, the maximum expected profit of platform at the i-th PoI is given by*

$$\begin{aligned}\Phi_i = \sum_{\ell=1}^{M} \frac{1}{\bar{v}_{i,\ell}} \int_0^{\bar{v}_{i,\ell}} & \Big[q_{i,\ell}(v_{i,\ell}) - g(\theta)\hat{c}(v_{i,\ell}, q_{i,\ell}(v_{i,\ell})) \\ & + g(\theta)\hat{c}'_{v_{i,\ell}}(v_{i,\ell}, q_{i,\ell}(v_{i,\ell}))(\bar{v}_{i,\ell} - v_{i,\ell})\Big]\, dv_{i,\ell}.\end{aligned} \tag{17}$$

Furthermore the optimal reward that maximizes the profit of platform is given by

$$R(q_{i,\ell}) = \frac{\hat{c}(v_{i,\ell}(q_{i,\ell}), q_{i,\ell}) - \int_0^{q_{i,\ell}} \hat{c}'_{v_{i,\ell}}(\tilde{q}_{i,\ell}, v_{i,\ell}(\tilde{q}_{i,\ell}))\, dv_{i,\ell}(\tilde{q}_{i,\ell})}{\prod\limits_{\iota=1,\iota\neq\ell}^{M} \frac{v_{i,\iota}(q_{i,\ell})}{\bar{v}_{i,\iota}}} \tag{18}$$

where $v_{i,\ell}(q_{i,\ell})$ is the inverse function of $q_{i,\ell}(v_{i,\ell})$ and the optimal data quality $q_{i,\ell}(v_{i,\ell})$ is implicitly given by

$$\hat{c}'_{q_{i,\ell}}(v_{i,\ell}, q_{i,\ell}) = \frac{1}{g(\theta)} + \hat{c}''_{v_{i,\ell}, q_{i,\ell}}(v_{i,\ell}, q_{i,\ell})(\bar{v}_{i,\ell} - v_{i,\ell}). \tag{19}$$

Proof. (*Sketch*) Due to the assumption that N PoIs are independent, FAIR-REW can calculate the maximum expected profit of platform at each PoI separately. It means FAIR-REW only requires to maximize Eq. (13) denoted as

$$\Phi_i = \left[\sum_{\ell=1}^{M} \int_0^{\bar{v}_{i,\ell}} q_{i,\ell}(v_{i,\ell}) \mathrm{d}F_{i,\ell}(v_{i,\ell})\right] - \mathbb{E}[U(\theta, R(q_{i,w}))] \tag{20}$$

where

$$\mathbb{E}[U(\theta, R(q_{i,w}))] = \sum_{\ell=1}^{M} \left[\int_0^{\bar{v}_{i,\ell}} g(\theta)R(q_{i,\ell}(v_{i,\ell})) \cdot \prod_{\iota=1,\iota\neq\ell} F_{i,\iota}(v_{i,\iota}(q_{i,\ell}))\mathrm{d}F_{i,\ell}(v_{i.\ell})\right].$$

Furthermore, the winner reward at the i-th PoI can be expressed as

$$\arg\max_{R(q_{i,\kappa})} \Phi_i \Longleftrightarrow \arg\max_{R(q_{i,\kappa})} \mathbb{E}\left[\sum_{\ell=1}^{M} q_{i,\ell} - U(\theta, R(q_{i,\kappa}))\right]. \tag{21}$$

According to the envelope theorem [16], the maximum profit of platform and the winner reward can be obtained by setting the first derivative of Eq. (20) to be 0.

Example 2. Following the above example, let's see the case where the abilities $v_{1,1}$ and $v_{1,2}$ of two users 1_1 and 1_2 are independently drawn from $F_{1,1}(v) = \frac{v}{\bar{v}_{1,1}}$ and $F_{1,2}(v) = \frac{v}{\bar{v}_{1,2}}$, respectively. For convenience, assume $\bar{v}_{1,1} = 1$ and $\bar{v}_{1,2} = 2$. Furthermore, assume that $g(\theta) = \theta$, $c(v,q) = q^2$ and $h(v) = v$, *i.e.*, $\hat{c}(v,q) = q^2/v$. Using Eq. (19) in Theorem 2, it can be obtained that

$$\frac{2q_{1,1}}{v} = \frac{1}{\theta} - \frac{2q_{1,1}}{v^2}(1-v) \text{ and } \frac{2q_{1,2}}{v} = \frac{1}{\theta} - \frac{2q_{1,1}}{v^2}(2-v),$$

which mean the optimal equilibrium qualities of users 1_1 and 1_2 are

$$v_{1,1}(q_{1,1}) = \sqrt{2\theta q_{1,1}} \text{ and } v_{1,2}(q_{1,2}) = \sqrt{4\theta q_{1,2}}.$$

Accordingly, by using Eq. (18) in Theorem 2, the optimal reward of user 1_1 is

$$R(q_{1,1}) = \left.\frac{\frac{v_{1,1}^3}{4\theta^2} + \int_0^{v_{1,1}} \frac{\tilde{v}_{1,1}^2}{4\theta^2} \mathrm{d}\tilde{v}_{1,1}}{\frac{v_{1,1}}{2}}\right|_{v_{1,1}=\sqrt{2\theta q_{1,1}}} = \frac{4q_{1,1}}{3\theta},$$

while the reward of user 1_2 is

$$R(q_{1,2}) = \left.\frac{\frac{v_{1,2}^3}{16\theta^2} + \int_0^{v_{1,2}} \frac{\tilde{v}_{1,2}^2}{16\theta^2} \mathrm{d}\tilde{v}_{1,2}}{v_{1,2}}\right|_{v_{1,2}=\sqrt{4\theta q_{1,2}}} = \frac{q_{1,2}}{3\theta}.$$

The maximized profit of platform for the PoI 1 is $\Phi_1 = \xi_1 + \xi_2 = \frac{1}{12\theta}$ where

$$\xi_1 = \int_0^1 \left[\frac{v^2}{2\theta} - \frac{v^3}{4\theta} - \frac{v^2}{4\theta}(1-v)\right] \mathrm{d}v = \frac{1}{12\theta},$$

$$\xi_2 = \frac{1}{2}\int_0^2 \left[\frac{v^2}{2\theta} - \frac{v^3}{4\theta} - \frac{v^2}{4\theta}(2-v)\right] \mathrm{d}v = 0.$$

The rewards of users 1_1 and 1_2 are different even though their qualities $q_{1,1}$ and $q_{1,2}$ obtained from Example 1 are equal. This situation is called the **implicit punishment** in this paper since compared with user 1_1, user 1_2 has higher ability but contributes the same data quality.

Proposition 1. *FAIR-REW satisfies the individual rationality property. Furthermore, a user obtains positive utility once her data quality is positive.*

Proof. (*Sketch*) Combining Theorem 2 and Eq. (16), it is easy to yield

$$\begin{aligned}\frac{u_{i,\ell}}{h(v_{i,\ell})} &= -\int_0^{v_{i,\ell}} \hat{c}(\tilde{v}_{i,\ell}, q_{i,\ell}(\tilde{v}_{i,\ell}))\mathrm{d}\tilde{v}_{i,\ell} \\ &= -h(v_{i\ell})\int_0^{v_{i,\ell}} \frac{c'_{v_{i,\ell}}(\tilde{v}_{i,\ell}, q_{i,\ell})h(\tilde{v}_{i,\ell}) - c(\tilde{v}_{i,\ell}, q_{i,\ell})h'(\tilde{v}_{i,\ell})}{h^2(\tilde{v}_{i,\ell})}\mathrm{d}\tilde{v}_{i,\ell}.\end{aligned}$$

Based on the monotonicity of equilibrium (Lemma 2), it can conclude that $u_{i,\ell} \geq 0$ where the equality holds iff $v_{i,\ell} = 0$. However, $v_{i,\ell} = 0$ means that the user does not participate in the task. Therefore, the conclusion holds.

Truthfulness: Another critical property for a practical mechanism is called **truthfulness**, which ensures that users will report their ability truthfully. This property is technically irrelevant to FAIR-REW since the reward of each user is determined by her **observable** data quality evaluated by the platform rather than the **unobservable** ability of each user.

5 Conclusion

In this paper, by jointly considering the truth estimation and reward determination, we propose a geometry based data acquisition framework in mobile crowdsensing, namely FAIR. Specifically, FAIR-EST estimates the truth values by minimizing the estimated error of the worst case. Since this problem can be shown as a non-convex quadratic optimization problem, we further obtain a closed form expression of estimated truth in Theorem 1 by applying a tight semi-definite relaxation. Furthermore, according to the estimated truths obtained from FAIR-EST, FAIR-REW determines the user reward and profit of the platform in Theorem 2, simultaneously incorporating user diversity into the design of a reward mechanism.

References

1. Suhag, D., Jha, V.: A comprehensive survey on mobile crowdsensing systems. J. Syst. Archit. **142**, 102952 (2023)
2. Chen, L.W., Liu, J.X.: Mobile crowdsourced guiding and finding with precise target positioning based on internet-of-things localization. IEEE Trans. Syst. Man Cybern. Syst. **52**(8), 4849–4861 (2021)
3. Fascista, A.: Toward integrated large-scale environmental monitoring using WSN/UAV/crowdsensing: a review of applications, signal processing, and future perspectives. Sensors **22**(5), 1824 (2022)
4. Li, T., Han, D., Chen, Y., Zhang, R., Zhang, Y., Hedgpeth, T.: IndoorWaze: a crowdsourcing-based context-aware indoor navigation system. IEEE Trans. Wireless Commun. **19**(8), 5461–5472 (2020)
5. Cheng, Y., et al.: A privacy-preserving and reputation-based truth discovery framework in mobile crowdsensing. IEEE Trans. Dependable and Secure Comput. **20**, 5293–5311 (2023)
6. Li, Q., Li, Y., Gao, J., Zhao, B., Fan, W., Han, J.: Resolving conflicts in heterogeneous data by truth discovery and source reliability estimation. In: Proceedings of the 2014 ACM SIGMOD international conference on Management of data, pp. 1187–1198 (2014)
7. Zhang, C., Zhao, M., Zhu, L., Wu, T., Liu, X.: Enabling efficient and strong privacy-preserving truth discovery in mobile crowdsensing. IEEE Trans. Inf. Forensics Secur. **17**, 3569–3581 (2022)
8. Wang, P., Jiao, D., Yang, L., Wang, B., Yu, R.: Hypergraph-based truth discovery for sparse data in mobile crowdsensing. ACM Trans. Sens. Netw. **20**, 1–23 (2024)
9. Yang, S., Han, K., Zheng, Z., Tang, S., Wu, F.: Towards personalized task matching in mobile crowdsensing via fine-grained user profiling. In: IEEE INFOCOM 2018-IEEE Conference on Computer Communications, pp. 2411–2419. IEEE (2018)

10. Ji, G., Zhang, B., Zhang, G., Li, C.: Online incentive mechanisms for socially-aware and socially-unaware mobile crowdsensing. IEEE Trans. Mob. Comput. **23**, 6227–6242 (2023)
11. Yang, G., Guo, D., Wang, B., He, X., Wang, J., Wang, G.: Participant-quantity-aware online task allocation in mobile crowd sensing. IEEE Internet of Things J. **10**, 22650–22663 (2023)
12. Jin, H., Su, L., Chen, D., Nahrstedt, K., Xu, J.: Quality of information aware incentive mechanisms for mobile crowd sensing systems. In: Proceedings of the 16th ACM International Symposium on Mobile Ad Hoc Networking and Computing, pp. 167–176 (2015)
13. Parreiras, S.O., Rubinchik, A.: Contests with three or more heterogeneous agents. Games Econom. Behav. **68**(2), 703–715 (2010)
14. Beck, A.: On the convexity of a class of quadratic mappings and its application to the problem of finding the smallest ball enclosing a given intersection of balls. J. Global Optim. **39**(1), 113–126 (2007)
15. Vandenberghe, L., Boyd, S.: Semidefinite programming. SIAM Rev. **38**(1), 49–95 (1996)
16. Milgrom, P., Segal, I.: Envelope theorems for arbitrary choice sets. Econometrica **70**(2), 583–601 (2002)

The Hybrid Diagnosability of Hypercube Under the HMM* (Hybrid MM*) Model

Aoshuai Tan[1,2], Chen Guo[2,3(✉)], Shengbo Chen[1], Yaoyao Luo[1,2], Xin He[1], Ming Tu[2], and Zhonghao Yao[4]

[1] School of Software, Henan University, Kaifeng, China
[2] Key Laboratory of Jiangxi Electronic Data Control and Forensics, Nanchang, China
[3] Department of Computer Science, Jinggangshan University, Ji'an, China
guochen@jgsu.edu.cn
[4] School of Computer Science and Technology, Beijing Institute of Technology, Beijing, China

Abstract. The Preparata, Metze, and Chien (PMC) model and Maeng and Malek (MM*) model are the two most widely adopted models in system−level fault diagnosis. Both of them assume that processors occur permanent failure and communication link never fail. Communication links may experience failures for various reasons. Therefore, the assumption that only processors occur failure is purely theoretical. Zhu et al. proposed a hybrid fault diagnosis model based on the PMC model, known as the HPMC (Hybrid Preparata, Metze, and Chien) model. In this paper, we propose a novel hybrid fault comparison diagnostic model, which is named the HMM* (Hybrid Maeng and Malek) model. Under the HMM* model, a test is composed of one node, its two adjacent nodes, and the associated edges. Additionally, we provide the sufficient and necessary conditions for the distinguishability of two hybrid consistent fault sets. Finally, as applications, the hybrid diagnosability of the n−dimensional hypercube under the HMM* model is determined.

Keywords: Multiprocessor systems · Fault diagnosis · HMM* model · Hybrid Diagnosability · Hypercube

1 Introduction

With the rapid advancement of high-performance computing and semiconductor technologies, the number of processors in multiprocessor computer systems, exemplified by supercomputers, continues to expand, accompanied by increasingly intricate structures [1]. This trend has led to a dramatic increase in the probability of processor failures [2]. System−level fault diagnosis is a diagnostic technique that utilizes graph-theoretic methods to provide fault diagnosis services. Most of the traditional system−level fault diagnosis theories have been developed within the context of processor failures. A series of fault diagnosis strategies have been proposed, including one−step t−diagnosis,

Y. Chen et al. (Eds.): COCOON 2024, LNCS 15162, pp. 40–52, 2025.
https://doi.org/10.1007/978-981-96-1093-8_4

$g-$good$-$neighbor conditional diagnosis, etc. [3–6], which can only be used to diagnose faulty processors.

In addition, the diagnostic process needs to follow a certain fault diagnostic model. F.P. Preparatory et al. first proposed the classical PMC (Preparata, Metze, and Chien) model [4]. This model was widely studied [7,8]. To better evaluate the test results, Maeng et al. proposed the MM* (Maeng and Malek) model known as the comparison model [9]. However, both the PMC model and the MM* model also consider only process failures and ignore the impact of communication link failures on testing. Processor failures and communication link failures in multiprocessor systems often intertwine and occur simultaneously [10,11]. Therefore, the study of hybrid fault diagnosis strategies for the simultaneous existence of processor faults and communication link faults has become a current research hotspot [12,13].

In 2020, Zhu et al. amended the traditional PMC model and proposed the HPMC (Hybrid Preparata, Metze, and Chien) model for adapting to the hybrid fault scenarios [14]. Under the HPMC model, two hybrid diagnostic strategies, $h-$restricted node diagnosability and $r-$restricted edge diagnosability, are proposed. The hybrid diagnosability of networks is determined [15,16]. However, similar to the PMC model, the HPMC model also suffers from the problem of difficult test evaluation.

Therefore, in this paper, we propose a hybrid diagnostic model based on the MM* model, called HMM* (Hybrid Maeng and Malek) model, that can be applied to the hybrid fault diagnosis. Subsequently, we demonstrate the distinguishability of any two distinct hybrid fault sets under the HMM* model. Finally, we prove that the hybrid diagnosability of $n-$dimensional hypercubes is $n-1$ under this model.

The remainder of the paper is organized as follows. In Sect. 2, the relevant symbolic terminology as well as the hypercube network are introduced. We propose the HMM* model and provide the sufficient and necessary conditions for any two distinguishable hybrid consistent fault sets in Sect. 3. As an application, we determine the hybrid diagnosability of hypercubes under the HMM* model in Sect. 4. Finally, Sect. 5 concludes the paper.

2 Preliminaries

A graph typically serves as the model for a multiprocessor system, where nodes represent processors and edges symbolize communication links between processors. In a graph $G(V, E)$, $V(G)$ and $E(G)$ refer to the set of nodes and the set of edges, respectively. For any node $a \in V(G)$, $N(a)$ and $N_E(a)$ denote the set of neighbors and neighboring edges of node a, respectively. $d(a)$ represents the degree of node a. Let $\delta(G)$ denote the minimum degree of G. Clearly, $d(a) \geq \delta(G)$ for any $a \in V(G)$.

Given an arbitrary hybrid set $S \subset V(G) \cup E(G)$, S is called a hybrid cut of G if $G-S$ is disconnected or trivial. Let nodes $a, b \in V(G)$, $cn(a, b)$ denotes the common neighbor set of a and b. The maximum number of common neighbor nodes between any two nodes in graph G is denoted by $cn(G) = \max\limits_{a,b \in V(G)} |cn(a,b)|$. Furthermore, given two arbitrary distinct sets A and B, where the existence incident edges between set A to set B are denoted by $A — B$, and $A \overset{C}{—} B$ denotes that these existence incident edges belong to set C.

Hypercubes are extensively utilized in multiprocessor systems for their reliability, symmetry, and scalability [17]. An $n-$dimensional hypercube is a $n-$regular graph with 2^n nodes and $n \cdot 2^{n-1}$ edges, denoted by Q_n. Each node in Q_n can be represented by a binary string of length n. This implies that, $V(Q_n) = \{a_1 a_2 \cdots a_n\}$, where $a_i \in \{0, 1\}$ and $i \in [1, n]$. Let $a, b \in V(Q_n)$, a and b have at most two common neighbors [18].

In the MM* model, each node conducts a test by sending identical tasks to its two neighbors, determining the state of the tested node through result comparison. Define a test as $(b, c; a)$, with node a as the tester node and the other two nodes as the tested nodes. The test result is represented by $t(b, c; a)$. If node a is fault-free, $t(b, c; a) = 0$ implies both nodes b and c are fault-free, while $t(b, c; a) = 1$ suggests at least one of nodes b and c is faulty. Otherwise, the test is unreliable [9].

3 The HMM* Model and the Hybrid Diagnosability

This section proposes the HMM* model and demonstrates the distinguishability of any two hybrid fault sets under the HMM* model.

Like the MM* model, a test under the HMM* model involves three nodes. However, a test under the HMM* model needs to consider the edges' states between the tester node and its two neighbors. Consider nodes $a, b, c \in V(G)$. Nodes b and c represent distinct neighbors of node a. In the test $t(b, c; a)$, node a is the testing node, while nodes b and c are the tested nodes, with edges (a, b) and (a, c) designated as testing edges. The HMM* model must adhere to the following two assumptions [14].
1) All edges connected to faulty nodes are fault-free,
2) Test results are reliable when the test node is fault-free.

Under these assumptions, we give the test validation for the HMM* model. (See Table 1)

Table 1. Test validation for the HMM* Model.

Tester node a	Tester node (a, b)	Tested node b	Tester node (a, c)	Tested node c	t(b, c;a)
fault-free	*fault-free*	*fault-free*	*fault-free*	*fault-free*	0
fault-free	*fault-free*	*fault*	*fault-free*	*Any Status*	1
fault-free	*fault-free*	*fault-free*	*fault-free*	*fault*	1
fault-free	*fault*	*fault-free*	*fault-free*	*Any Status*	1
			fault	*fault-free*	
fault-free	*fault-free*	*Any Status*	*fault*	*fault-free*	1
fault	*fault-free*	*Any Status*	*fault-free*	*Any Status*	0 or 1

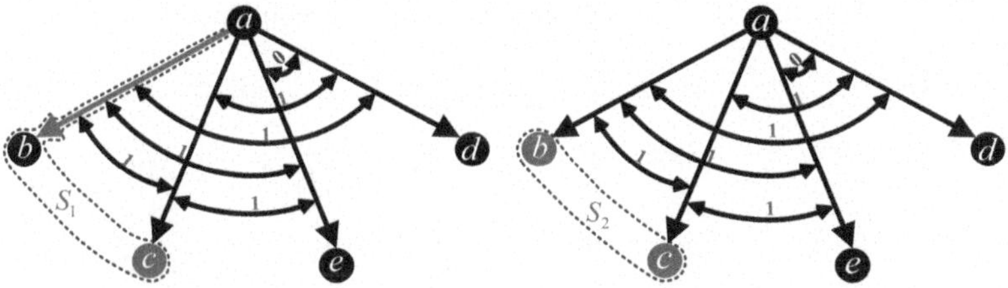

Fig. 1. An Illustration of two indistinguishable hybrid consistent fault sets.

According to Table 1, if $t(b, c; a) = 1$, it suggests that either node a, b, or c and test edges (a, b) or (a, c) are faulty. All nodes and the edges in the test are fault-free or test node a is faulty if $t(b, c; a) = 0$. A hybrid fault set $S \subset V(G) \cup E(G)$ is called a hybrid consistent fault set if none of the nodes in S is incident to any edge in S. The set of all test results is called a syndrome, denoted as σ. Since tests conducted by faulty nodes are unreliable, a hybrid fault set S may generate multiple syndromes. We use $\sigma(S)$ to denote the set of all syndromes that can be generated by S. For any two different hybrid consistent fault sets $S_1, S_2 \subset V(G) \cup E(G)$, S_1 and S_2 are distinguishable if and only if $\sigma(S_1) \cap \sigma(S_2) = \varnothing$. For example, given a graph consisting of 5 nodes and 4 edges (see Fig. 1). Let $S_1 = \{(a, b), c\}$, $S_2 = \{b, c\}$. Clearly, both S_1 and S_2 can produce a same syndrome $\sigma(b, \text{c}; a) = \sigma(b, e; a) = \sigma(b, d; a) = \sigma(c, \text{e}; a) = \sigma(c, d; a) = 1$, $\sigma(d, e; a) = 0$. Therefore, $\sigma(S_1) \cap \sigma(S_2) \neq \varnothing$, which means that S_1 and S_2 are indistinguishable. Since none of the faulty nodes are incident to faulty edges in S_1, S_2, (S_1, S_2) is a hybrid consistent fault pair.

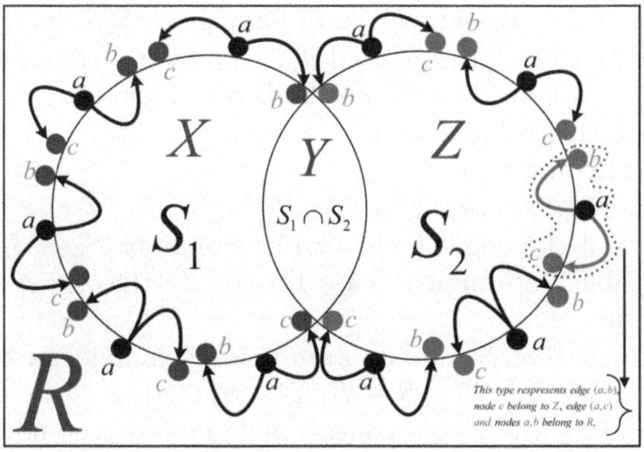

Fig. 2. Illustration of Theorem 1.

Definition 1. *Let $S_1, S_2 \subset V(G) \cup E(G)$ be any two distinct hybrid consistent fault sets in system G with $|S_1|, |S_2| \leq t_h$. If $S_1, S_2 \subset V(G) \cup E(G)$ are distinguishable, G is a hybrid t_h−diagnosable system.*

Table 2. The test table satisfying two conditions.

$NO.$	a	(a,b)	b	(a,c)	c	$NO.$	a	(a,b)	b	(a,c)	c
1	R	R	R	R	S_1-S_2	9	R	R	R	R	S_2-S_1
2	R	R	R	S_1-S_2	R	10	R	R	R	S_2-S_1	R
3	R	R	S_1-S_2	R	R	11	R	R	S_2-S_1	R	R
4	R	S_1-S_2	R	R	R	12	R	S_2-S_1	R	R	R
5	R	R	S_1-S_2	R	S_1-S_2	13	R	R	S_2-S_1	R	S_2-S_1
6	R	S_1-S_2	R	R	S_1-S_2	14	R	S_2-S_1	R	R	S_2-S_1
7	R	R	S_1-S_2	S_1-S_2	R	15	R	R	S_2-S_1	S_2-S_1	R
8	R	S_1-S_2	R	S_1-S_2	R	16	R	S_2-S_1	R	S_2-S_1	R

Definition 2. *The hybrid diagnosability of system G under the HMM^* model is maximum value of t_h such that G is t_h-diagnosable, denoted as $t_h^{HMM^*}(G)$.*

Theorem 1. *Let $S_1, S_2 \subset V(G) \cup E(G)$ be any two different hybrid consistent sets of G and $R = V(G) \cup E(G) - S_1 - S_2$, where $|S_1|, |S_2| \leq t_h$. S_1 and S_2 are distinguishable under the HMM^* model if and only if at least one of the following two conditions holds.(see Fig. 2):*
1. There exists a test $t(b, c; a)$ with $a \in R$, $b, c, (a, b), (a, c) \notin S_2$, and at least one of the nodes b, c and edges $(a, b), (a, c)$ out of the four belongs to $S_1 - S_2$.
2. There exists a test $t(b, c; a)$ with $a \in R$, $b, c, (a, b), (a, c) \notin S_1$, and at least one of the nodes b, c and the edges $(a, b), (a, c)$ out of the four belongs to $S_2 - S_1$.

Proof. Under the model assumptions, two conditions can be mapped to 16 test cases, as illustrated in Table 2.

Sufficiency: We now prove that if at least one of the above 2 conditions holds, S_1 and S_2 are distinguishable.

Case 1. Condition (1) holds.

There exists a test $t(b, c; a)$ with $a \in R$, $b, c, (a, b), (a, c) \notin S_2$, and at least one of the nodes b, c and the edges $(a, b), (a, c)$ belongs to $S_1 - S_2$ (corresponding to cases 1–8 in Table 2). Thus, $\sigma(b, c; a) = 1$ if S_1 is a hybrid consistent fault set, and $\sigma(b, c; a) = 0$ if S_2 is a hybrid consistent fault set. Therefore, $\sigma(S_1) \cap \sigma(S_2) = \varnothing$ and (S_1, S_2) is a distinguishable pair.

Case 2. Condition (2) holds.

There exists a test $t(b, c; a)$ with $a \in R$, $b, c, (a, b), (a, c) \notin S_1$, and at least one of nodes b, c and the edges $(a, b), (a, c)$ belongs to $S_2 - S_1$ (corresponding to cases 9-16 in Table 2). Similar to case 1, $\sigma(S_1) \cap \sigma(S_2) = \varnothing$ and (S_1, S_2) is a distinguishable pair.

Necessity: Assume that S_1 and S_2 are distinguishable if neither of the above two conditions are met. Let $X = S_1 - S_2$,$Y = S_1 \cap S_2$, $Z = S_2 - S_1$, as shown in Fig. 2. Since S_1 and S_2 are hybrid consistent fault sets, the nodes in $X \cup Y$ are not incident to any edge in $X \cup Y$, and the nodes in $Y \cup Z$ are not incident to any edge in $Y \cup Z$. Therefore, all edges connected to nodes in $X \cup Y$ belong to $Z \cup R$, and all edges connected to nodes in $Y \cup Z$ belong to $X \cup R$. The endpoints of edges in $X \cup Y$ belong to $Z \cup R$, and the endpoints of edges in $Y \cup Z$ belong to $X \cup R$. Then, edges incident to nodes in Y must exclusively be part of set R. Subsequently, we consolidate the 153 test categories, detailed in Table 3.

Table 3. A syndrome σ^* of G.

NO.	a	b	c	(a, b)	(a, c)	σ(b, c; a)	NO.	a	b	c	(a, b)	(a, c)	σ(b, c; a)	NO.	a	b	c	(a, b)	(a, c)	σ(b, c; a)
1	X	X	X	Z	Z	1	52	Y	R	R	R	R	1	103	R	X	R	R	Z	1
2	X	X	X	Z	R	1	53	Z	X	X	R	R	1	104	R	Y	X	R	Z	1
3	X	X	X	R	Z	1	54	Z	X	Y	R	R	1	105	R	Y	X	R	R	1
4	X	X	X	R	R	0	55	Z	X	Z	R	X	1	106	R	Y	Y	R	R	1
5	X	X	Y	Z	R	1	56	Z	X	Z	R	R	1	107	R	Y	Z	R	X	1
6	X	X	Y	R	R	0	57	Z	X	R	R	X	1	108	R	Y	Z	R	R	1
7	X	X	Z	Z	R	1	58	Z	X	R	R	R	1	109	R	Y	R	R	X	1
8	X	X	Z	R	R	1	59	Z	Y	X	R	R	1	110	R	Y	R	R	Y	1
9	X	X	R	Z	Z	1	60	Z	Y	Y	R	R	0	111	R	Y	R	R	Z	1
10	X	X	R	Z	R	1	61	Z	Y	Z	R	X	0	112	R	Y	R	R	R	1
11	X	X	R	R	Z	1	62	Z	Y	Z	R	R	0	113	R	Z	X	X	Z	1
12	X	X	R	R	R	0	63	Z	Y	R	R	X	1	114	R	Z	X	X	R	1
13	X	Y	X	R	Z	0	64	Z	Y	R	R	R	0	115	R	Z	X	R	Z	1
14	X	Y	X	R	R	0	65	Z	Z	X	X	R	1	116	R	Z	X	R	R	1
15	X	Y	Y	R	R	0	66	Z	Z	X	R	R	1	117	R	Z	Y	X	R	1
16	X	Y	Z	R	R	1	67	Z	Z	Y	X	R	1	118	R	Z	Y	R	R	1
17	X	Y	R	R	Z	1	68	Z	Z	Y	R	R	0	119	R	Z	Z	X	X	1
18	X	Y	R	R	R	0	69	Z	Z	Z	X	X	1	120	R	Z	Z	X	R	1
19	X	Z	X	R	Z	1	70	Z	Z	Z	X	R	1	121	R	Z	Z	R	X	1
20	X	Z	X	R	R	1	71	Z	Z	Z	R	X	1	122	R	Z	R	X	X	1
21	X	Z	Y	R	R	1	72	Z	Z	Z	R	R	0	123	R	Z	R	X	Y	1
22	X	Z	Z	R	R	1	73	Z	Z	R	X	X	1	124	R	Z	R	X	Z	1
23	X	Z	R	R	Z	1	74	Z	Z	R	X	R	1	125	R	Z	R	X	R	1
24	X	Z	R	R	R	1	75	Z	Z	R	R	X	1	126	R	Z	R	R	X	1
25	X	R	X	Z	Z	1	76	Z	Z	R	R	R	0	127	R	Z	R	R	Y	1
26	X	R	X	Z	R	1	77	Z	R	X	X	R	1	128	R	R	X	X	Z	1
27	X	R	X	R	Z	1	78	Z	R	X	R	R	1	129	R	R	X	Y	Z	1
28	X	R	X	R	R	0	79	Z	R	Y	X	R	1	130	R	R	X	Y	R	1
29	X	R	R	R	Z	1	80	Z	R	Y	R	R	0	131	R	R	X	Z	Z	1
30	X	R	Y	Z	R	1	81	Z	R	Z	X	X	1	132	R	R	X	Z	R	1
31	X	R	Y	R	R	0	82	Z	R	Z	X	R	1	133	R	R	X	R	Z	1
32	X	R	Z	Z	R	1	83	Z	R	Z	R	X	1	134	R	R	Y	X	R	1
33	X	R	Z	R	R	1	84	Z	R	Z	R	R	0	135	R	R	Y	Y	R	1
34	X	R	R	Z	Z	1	85	Z	R	R	X	X	1	136	R	R	Y	Z	R	1
35	X	R	R	Z	R	1	86	Z	R	R	X	R	1	137	R	R	Y	R	R	1
36	X	R	R	R	R	0	87	Z	R	R	R	X	1	138	R	R	Z	X	X	1
37	Y	X	X	R	R	1	88	Z	R	R	R	R	0	139	R	R	Z	X	R	1
38	Y	X	Y	R	R	1	89	R	X	X	Z	Z	1	140	R	R	Z	Y	X	1
39	Y	X	Z	R	R	1	90	R	X	X	Z	R	1	141	R	R	Z	Y	R	1
40	Y	X	R	R	R	1	91	R	X	X	R	Z	1	142	R	R	Z	Z	X	1
41	Y	Y	X	R	R	1	92	R	X	Y	Z	R	1	143	R	R	Z	R	X	1
42	Y	Y	Y	R	R	1	93	R	X	Y	R	R	1	144	R	R	R	X	Y	1
43	Y	Y	Z	R	R	1	94	R	X	Z	Z	X	1	145	R	R	R	X	Z	1
44	Y	Y	R	R	R	1	95	R	X	Z	Z	R	1	146	R	R	R	Y	X	1
45	Y	Z	X	R	R	1	96	R	X	Z	R	X	1	147	R	R	R	Y	Y	1
46	Y	Z	Y	R	R	1	97	R	X	Z	R	R	1	148	R	R	R	Y	Z	1
47	Y	Z	Z	R	R	1	98	R	X	R	Z	X	1	149	R	R	R	Y	R	1
48	Y	Z	R	R	R	1	99	R	X	R	Z	Y	1	150	R	R	R	Z	X	1
49	Y	R	X	R	R	1	100	R	X	R	Z	Z	1	151	R	R	R	Z	Y	1
50	Y	R	Y	R	R	1	101	R	X	R	Z	R	1	152	R	R	R	R	Y	1
51	Y	R	Z	R	R	1	102	R	X	R	R	Y	1	153	R	R	R	R	R	0

By the validations for HMM* model, we construct a syndrome σ^*, as shown in Table 3. Each row denotes the test result of one type of test in Table 3. For example, the second row in the table, the seventh column, indicates $\sigma(b, c; a) = 1$, and the test node and the tested node $a, b, c \in X$ and testing edges $(a, b), (a, c) \in Z$. By Table 3, we have $\sigma^* \in \sigma(S_1) \cap \sigma(S_2)$. Therefore, S_1 and S_2 are indistinguishable, which contradicts the assumption that G is a hybrid t_h−diagnosable system.

4 The Hybrid Diagnosability of Hypeycube

In this section, we determine the hybrid diagnosability of Q_n under the HMM* model.

Theorem 2. *$t_h^{\text{HMM}^*}(Q_n) \leq n-1$, for $n \geq 3$.*

Proof. Without loss of generality, let $a \in V(Q_n)$ and $N(a) = \{a_1, a_2, \cdots\cdots, a_n\}$. Then, we set $S_1 = \{a\}$ and $S_2 = \{(a, a_1), \cdots\cdots, (a, a_n)\}$. Clearly, $|S_1| = 1$ and $|S_2| = n$. Hence, (S_1, S_2) is a hybrid consistent fault pair. By Theorem 1, S_1 and S_2 are indistinguishable. Therefore, $t_h^{\text{HMM}^*}(Q_n) \leq n-1$, for $n \geq 3$.

Theorem 3. *$t_h^{\text{HMM}^*}(Q_n) \geq n-1$, for $n \geq 3$.*

Proof. Suppose, by contradiction. There exist two indistinguishable hybrid consistent fault sets $S_1, S_2 \subset V(Q_n) \cup E(Q_n)$ under the HMM* model, where $|S_1|, |S_2| \leq n-1$. Let $R = V(Q_n) \cup E(Q_n) - S_1 - S_2$, $X = S_1 - S_2$,$Y = S_1 \cap S_2$, $Z = S_2 - S_1$, $\forall = X \cup Y \cup Z \cup R$. Since $|V(S_1)| \leq |S_1| \leq n-1$, $|V(S_2)| \leq |S_2| \leq n-1$,

$$
\begin{aligned}
|V(R)| &= |V(Q_n)| - |V(S_1)| - |V(S_2)| \\
&\geq 2^n - 2(n-1) \\
&\geq 2^n - 2n + 2 \\
&\geq 2
\end{aligned}
$$

,for $n \geq 3$.

That is, $V(R) \neq \varnothing$. In addition, $S_1 \neq S_2$. Then, $X \Delta Z \neq \varnothing$. We consider the following two cases according to whether there exist nodes in $X \cup Z$.

Case 1. $V(X \cup Z) \neq \varnothing$.

Since $V(R) \neq \varnothing$, we discuss the following two scenarios based on whether there are neighbors of $V(X \cup Z)$ in R.

Case 1.1. $N(V(X \cup Z)) \cap R = \varnothing$.

Next, We discuss the following two subcases.

Case 1.1.1. $|V(X)| \geq 1, |V(Z)| = 0$ or $|V(Z)| \geq 1, |V(X)| = 0$.

Without loss of generality, let $|V(X)| \geq 1, |V(Z)| = 0$.

If $|V(X)| = 1$, let node $a \in V(X)$. Since $N(a) \cap R = \varnothing$ and $V(Z) = \varnothing$, $N(a) \in Y$. Then $|S_2| = |N(a)| \geq n$, which is a contradiction with $|S_2| \leq n-1$. If $|V(X)| > 1$, let nodes $a, b \in V(X)$. Since $N(V(X \cup Z)) \cap R = \varnothing$ and $V(Z) = \varnothing$, $N[a, b] \in S_1$, which is a contradiction with $|S_1| \leq n-1$.

Case 1.1.2. $|V(X)| \geq 1|$ and $|V(Z)| \geq 1$.

Let nodes $a \in V(X)$,$b \in V(Z)$. Since $N(V(X \cup Z)) \cap R = \varnothing$, $N(a) \cap R = \varnothing$ and $N(b) \cap R = \varnothing$. Hence $N(a) \in X \cup Y \cup Z$ and $N(b) \in X \cup Y \cup Z$. In addition, $|cn(a, b)| = 0$ if $(a, b) \in E(Q_n)$ or $|cn(a, b)| = 2$ [18]. Therefore, $|V(S_1) \cup V(S_2)| \geq |N(a) \cup N(b)| \geq 2n$, which is a contradiction with $|S_1| + |S_2| \leq 2n-2$.

Case 1.2. $N(V(X \cup Z)) \cap R \neq \varnothing$.

Since S_1, S_2 are hybrid consistent fault sets, the existence incident edges between X and R can only belong to $R \cup Z$, denoted by $X \overset{R \cup Z}{—} R$. Similarly, the existence edges between Z and R can only belong to $R \cup X$, denoted by $Z \overset{R\cup X}{-} R$. We discuss the following two subcases.

Case 1.2.1. There exists at least one edge of X—R belonging to R. or There exists at least one edge of Z—R belonging to R.

Without loss of generality, let there exist at least one edge of X—R belonging to R. Since $N(V(X \cup Z)) \cap R \neq \varnothing$, let node $a \in V(X)$, $b \in N(a) \cap R$, and assume that edge $(a, b) \in R$ (See Fig 3.(a)).

Since S_1, S_2 are hybrid consistent fault sets, we have $b \overset{R\cup X}{-} Z$, $b \overset{R}{-} Y$. Similarly, $b \overset{Y\cup Z}{-} R$. We know that S_1, S_2 are indistinguishable. If node b has additional neighbors(besides a) belonging to X, the edges between b and neighbors belong to Z, denoted by $b \overset{Z}{-} (N(b) - a)$).

Moreover, if there is an existence edge between b and a node in R that belongs to Z, there is no edge between b and nodes in Z that belong to R. Otherwise, it contradicts condition (2) in Theorem 1. Vice versa. That is, (1) and (2) in the blue dashed line in Fig. 3.(a) occur at most once. Otherwise, S_1, S_2 are distinguishable. Thus,

$$\begin{aligned}
|N(b)| - 1 &\geq |N_{V(X \cup Y \cup Z)}(b) - a| + |N_{E(X \cup Y \cup Z)}(b)| \\
&= |N_{V(X)}(b) - a| + |N_{V(Y \cup Z)}(b)| + |N_{E(X \cup Y \cup Z)}(b)| \\
&\geq n - 1
\end{aligned}$$

Next, we discuss the following two sub-cases based on whether node a has another neighbor (other than b) belonging to R.

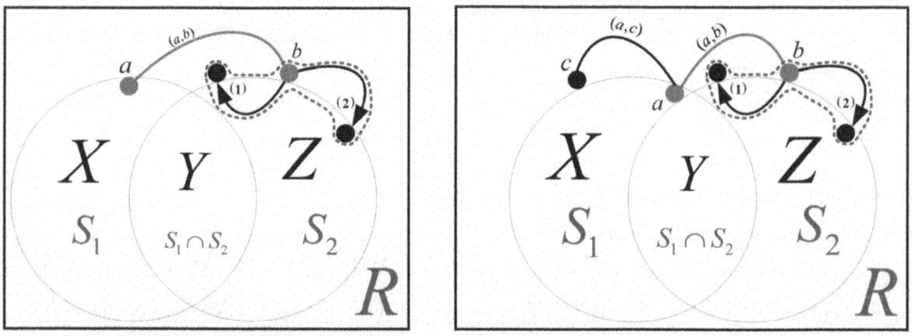

(a) Illustration of Subcases 1). (b) Illustration of Subcases 2).

Fig. 3. An illustration of **Case 1.2.1** (Color figure online)

Subcases 1). $(N(a) - b) \cap R \neq \varnothing$. Without loss of generality, let node $c \in (N(a) - b) \cap R$. Since there exists at least one edge of X—R belonging to R. Also since S_1, S_2 are hybrid consistent fault sets, the edge $(a, c) \in R$ or $(a, c) \in Z$.

Suppose, that $(a, c) \in R$. Since nodes b, c are in R and have a common neighbor a (See Fig. 3.(b)), nodes b and c have similar distributions of neighbors and neighboring edges if they have neighbors and neighboring edges in $X \cup Y \cup Z$.

That is, $|N_{V(X\cup Y\cup Z)}(c) - a| + |N_{E(X\cup Y\cup Z)}(c)| \geq n-1$. In addition, $|cn(a,b)| = 2$ [18]. Thus,

$$\begin{aligned}|S_1| + |S_2| &= |X| + |Y| + |Z| \\ &\geq |a| + |N_{V(X\cup Y\cup Z)}(b) - a \cup N_{E(X\cup Y\cup Z)}(b)| \\ &+ |N_{V(X\cup Y\cup Z)}(c) - a \cup N_{E(X\cup Y\cup Z)}(c)| \\ &\geq 1 + n - 1 + n - 1 \\ &\geq 2n - 1\end{aligned}$$

, which is a contradiction with $|S_1| + |S_2| \leq 2n-2$ for $n \geq 3$.

Therefore, $(a,c) \notin R$. Then, $(a,c) \in Z$. When $(N(a) - b) \cap R \neq \varnothing$, we have $a \overset{Z}{—} (R - b)$. That is, there exists only one edge (a,b) from X to R at node a. Then, $|N_{V(X\cup Y\cup Z)}(a)| + |N_{E(X\cup Y\cup Z)}(a)| \geq n-1$. Thus,

$$\begin{aligned}|S_1| + |S_2| &= |X| + |Y| + |Z| \\ &\geq |a| + |N_{V(X\cup Y\cup Z)}(b) - a \cup N_{E(X\cup Y\cup Z)}(b)| \\ &+ |N_{V(X\cup Y\cup Z)}(c) - a \cup N_{E(X\cup Y\cup Z)}(c)| \\ &\geq 1 + n - 1 + n - 1 \\ &\geq 2n - 1\end{aligned}$$

This is a contradiction. Next, we consider the case of $N_{V(R)}(x) - y = \varnothing$.

Since $N_{V(R)}(x) - y = \varnothing$, $N_V(x) - y \in F_1 \cup F_2$ and $|N_{V(F_1\cup F_2)}(x) - y| = n-1$. Therefore,

$$\begin{aligned}|F_1| + |F_2| &= |F_1 - F_2| + |F_1 \cap F_2| + |F_2 - F_1| \\ &\geq 1 + |N_{V(F_1\cup F_2)}(y) - x| + |N_{E(F_1\cup F_2)}(y)| + |N_V(x) - y| \\ &\geq 1 + n - 1 + n - 1 \\ &\geq 2n - 1\end{aligned}$$

, which is a contradiction with $|S_1| + |S_2| \leq 2n-2$ for $n \geq 3$.

Subcases 2). $(N(a) - b) \cap R = \varnothing$.

Since $(N(a) - b) \cap R = \varnothing$, $N(a) - b \in X \cup Y \cup Z$. We know $|N(a)| = n$, $|N_{V(X\cup Y\cup Z)}(a) - b| = n-1$ and $|N_{V(X\cup Y\cup Z)}(b) - a| + |N_{E(X\cup Y\cup Z)}(b)| \geq n-1$.

$$\begin{aligned}|S_1| + |S_2| &= |X| + |Y| + |Z| \\ &\geq 1 + |N_{V(X\cup Y\cup Z)}(b) - a| + |N_{E(X\cup Y\cup Z)}(b)| + |N_{V(X\cup Y\cup Z)}(a) - b| \\ &\geq 1 + n - 1 + n - 1 \\ &\geq 2n - 1\end{aligned}$$

, which is a contradiction with $|S_1| + |S_2| \leq 2n-2$.

Case 1.2.2. There is no edges of $X — R$ belonging to R. or There is no edges of $Z — R$ belonging to R.

Without loss of generality, There is no edges of $X — R$ belonging to R. Let node $a \in X$. Since S_1, S_2 are hybrid consistent fault sets, we have $a \overset{Z}{—} R$(See case

(1) in Fig. 4). Similarly, $a \overset{R}{—} (Y \cup Z)$(See case (2) in Fig. 4). Then, $a \overset{R \cup Z}{—} X$ (See cases (3) and (4) in Fig. 4)). Also since $N(a) = N_E(a) = n$, the relationship of number between the neighbors of node a in $X \cup Y \cup Z$ and the incident edges is $|N_{V(X \cup Y \cup Z)}(a)| + |N_{E(X \cup Y \cup Z)}(a)| \geq n$. Next, we discuss based on the existence

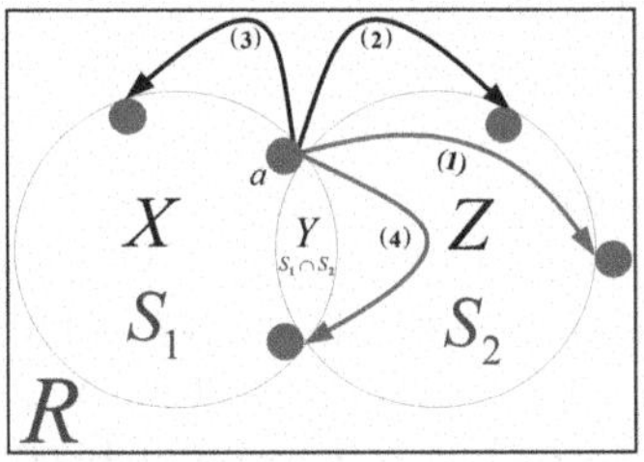

Fig. 4. An illustration of **Case 1.2.2**

location of the neighbors of node a.

Suppose, there exists node $b \in N(a) \cap X$ such that the incident edge $(a, b) \in R$. similar to case (3) in Fig. 4. There is no edges of $X — R$ belonging to R and S_1, S_2 are hybrid consistent fault sets, we have $b \overset{Z}{—} R$. And $b \overset{R}{—} (Y \cup Z)$, $b \overset{R \cup Z}{—} X$. Similarly, $|N_{V(X \cup Y \cup Z)}(b) - a| + |N_{E(X \cup Y \cup Z)}(b)| \geq n - 1$. Thus,

$$\begin{aligned} |S_1| + |S_2| &\geq |X| + |Y| + |Z| \\ &\geq |a| + |N_{V(X \cup Y \cup Z)}(a)| + |N_{E(X \cup Y \cup Z)}(a)| \\ &+ |N_{V(X \cup Y \cup Z)}(b) - a| + |N_{E(X \cup Y \cup Z)}(b)| \\ &\geq 1 + n + n - 1 \\ &\geq 2n \end{aligned}$$

, which is a contradiction with $|S_1| + |S_2| \leq 2n - 2$ for $n \geq 3$, and case (3) in Fig. 4 does not exist.

Then, if the node a has neighbors in X, $a \overset{Z}{—} X$. (See case (4) in Fig. 4) That is, $|S_2| \geq |N_{V(Y \cup Z)}(x)| + |N_{E(Z)}(x)| \geq n$, which is a contradiction. Thus, case (4) does not exist.

If case (3) and (4) in Fig. 4 do not exist, $(N(a) \cap X) = \varnothing$, $N(a) \subset (Y \cup Z \cup R)$. In this case, clearly, $a \overset{Z}{—} R$, $a \overset{R}{—} (Y \cup Z)$. Then, $|S_2| \geq |N_{V(Y \cup Z)}(a) \cup N_{E(Y \cup Z)}(a)| = n$, which is a contradiction. The cases (1) and (2) do not exist.

Case 2. $V(X \cup Z) = \varnothing$.

Since $X \cup Z \neq \varnothing$ and $V(X \cup Z) = \varnothing$, $E(X \cup Z) \neq \varnothing$. We consider the following two cases.

Case 2.1. $|E(X)| \geq 1, |E(Z)| = 0$ or $|E(X)| = 0, |E(Z)| \geq 1$.

Without loss of generality, let $|E(X)| \geq 1, |E(Z)| = 0$.

If $|E(X)| = 1, |E(Y)| = 0$, let $(u, v) \in E(X)$. Since S_1 is a hybrid consistent fault set, $u, v \notin V(X \cup Y)$. In addition, $V(X \cup Z) = \varnothing$. Then $u, v \in R$ and

$|N_{V(X\cup Z)}(u)|, |N_{V(X\cup Z)}(v)| = 0$. Since S_1, S_2 are indistinguishable and $|E(Z)| = 0$, as an example for the node u , we have $u \overset{R}{—} Y$, $u \overset{Y}{—} R$. Then,

$$\begin{aligned}|S_1| &= |X| + |Y| \\ &\geq |(u, v)| + |N_{V(Y)}(u)| + |N_{E(Y)}(u)| \\ &\geq 1 + n - 1 \\ &\geq n\end{aligned}$$

, which is a contradiction with $|S_1| \leq n - 1$.

If $|E(X)| > 1, |E(Z)| = 0$, let $(a, b), (u, v) \in E(X)$ and they have no common endpoints. Otherwise, it contradicts condition (1) in Theorem 1. Since $V(X \cup Z) = \varnothing$, $a, b, u, v \in R$ and $|N_{V(X\cup Z)}(a)|, |N_{V(X\cup Z)}(b)|, |N_{V(X\cup Z)}(v)|, |N_{V(X\cup Z)}(u)| = 0$, it follows that $|E(X)| = 1, |E(Y)| = 0$ is similarly contradicting, for any nodes a, b, u, v.

Taking any node a, b, u, v as an example, a similar contradiction can be obtained with $|E(X)| > 1, |E(Z)| = 0$.

Case 2.2. $|E(X)| \geq 1, |E(Z)| \geq 1$.

Without loss of generality, there exists an edge in $E(X)$ and $E(Z)$ respectively, and the two edges have a common endpoint. Otherwise, it is similar to Case 2.1. Since $V(X \cup Z) = \varnothing$ and $E(X \cup Z) \neq \varnothing$, let edges $(u, v) \in E(X)$, $(u, w) \in E(Z)$ and nodes $u, v, w \in R$. Since nodes v, w have common neighbor node u, $|cn(a, b)| = 2$ [18]. Also, since S_1, S_2 are hybrid consistent fault sets, we have $u \overset{\forall}{—} R$, $v \overset{\forall}{—} R$, $w \overset{\forall}{—} R$. And $u \overset{R}{—} Y$, $v \overset{R}{—} Y$, $w \overset{R}{—} Y$. Furthermore, since S_1, S_2 are indistinguishable and $(u, v) \in E(X)$, $(u, w) \in E(Z)$, $u \overset{Y}{—} R$. It's easy see that $v \overset{Y\cup Z}{—} R$ with $|N_{E(Z)}(v)| \leq 1$. And, $w \overset{X\cup Z}{—} R$ with $|N_{E(X)}(w)| \leq 1$. Otherwise, S_1, S_2 are distinguishable.

Thus, $|N_{V(Y)}(u) \cup N_{E(Y)}(u)| = n - 2$, $|N_{V(Y)}(v) \cup N_{E(Y)}(v)| \geq n - 2$, $|N_{V(Y)}(w) \cup N_{E(Y)}(w)| \geq n - 2$ for $n \geq 3$. Then,

$$\begin{aligned}|S_1| + |S_2| &= |X| + |Y| + |Z| \\ &\geq |(u, v)| + |N_{V(Y)}(u) \cup N_{E(Y)}(u)| \\ &+ |N_{V(Y)}(v) \cup N_{E(Y)}(v)| \\ &+ |N_{V(Y)}(w) \cup N_{E(Y)}(w)| \\ &+ |(u, w)| \\ &\geq 1 + n - 2 + n - 2 + n - 2 + 1 \\ &\geq 3n - 4\end{aligned}$$

, which is a contradiction with $|S_1| + |S_2| \leq 2n - 2$ for $n \geq 3$.

Thus $t_h^{\mathrm{HMM}^*}(Q_n) \geq n - 1$ for $n \geq 3$, the Theorem holds. Combining Theorems 2 and 3, we have the following theorem.

Theorem 4. $t_h^{\mathrm{HMM}^*}(Q_n) = n - 1$, *for* $n \geq 3$.

5 Conclusion

In this paper, we introduce a novel hybrid diagnostic model, termed the HMM* model, built upon the MM* model. It is used for the diagnosis capability analysis of multiprocessor systems with both processors and communication links failure. This model is designed for analyzing the diagnostic capabilities of multiprocessor systems in the event of failures in both processors and communication links. Under this model, A test initiated by a testing node consists of two adjacent nodes and their corresponding edges. Subsequently, we present a table outlining the validation of tests within this model. Additionally, we establish sufficient and necessary conditions for determining the distinguishability of a pair of hybrid faulty sets under the HMM* model. Finally, we demonstrate that $t_h^{\mathrm{HMM}^*}(Q_n) = n-1$, for $n \geq 3$. In the future, the hybrid diagnosability of other interconnection networks remains a subject for further investigation within this model.

Acknowledgements. This work was supported by the National Natural Science Foundation of China (No. 62262032, 61862035, 61562046, 62102133), the Henan University Graduate Student Excellence Program (No. SYLYC2023135), the Science and Technology Program of Jiangxi Provincial Department of Education (No. GJJ2201604), The Project of High-level and Urgently Needed Overseas Talents of Jiangxi Province (Approval No.: 20232BCJ25026), and Ji'an Financial Science Fund Project (Approval No.: 20211-085454, 20222-151746, 20222-151704). Additionally, funding was provided by the Kaifeng City Major Science and Technology Special Project (Approval No. 21ZD011) and the National Natural Science Foundation of Jiangxi Province (No. 20202BABBBL202042).

References

1. Zhang, J., Zhou, J.-X., Li, Y.-T., Kwon, Y.S.: The automorphisms of 2− token graphs. Appl. Math. Comput. **446**, 127872 (2023)
2. Preparata, F.P., Metze, G., Chien, R.T.: On the connection assignment problem of diagnosable systems. IEEE Trans. Electron. Comput. **EC-16**(6), 848–854 (1967)
3. Zhang, H., Zhou, S., Cheng, E.: The cyclic diagnosability of hypercubes under the *PMC* model and the MM^* model. Comput. J. **67**(2), 709–718 (2023)
4. Peng, S.-L., Lin, C.-K., Tan, J.J.M., Hsu, L.-H.: The g−good−neighbor conditional diagnosability of hypercube under *pmc* model. Appl. Math. Comput. **218**(21), 10406–10412 (2012)
5. Mei-Mei, G., Hao, R.-X., Feng, Y.-Q., Wei, E.: Conditional diagnosability of multiprocessor systems based on cayley graphs generated by transpositions. Discret. Appl. Math. **304**, 137–152 (2021)
6. Guo, C., Xiao, Z., Liu, Z., Peng, S.: Rg conditional diagnosability: a novel generalized measure of system-level diagnosis. Theoret. Comput. Sci. **814**, 19–27 (2020)
7. Tian, Y., Zhu, Q.: r−component diagnosability of hypercubes under the *pmc* model. Theoret. Comput. Sci. **933**, 114–124 (2022)
8. Guo, C., Chengzhong, W., Xiao, Z., Jianbo, L., Liu, Z.: The intermittent diagnosability for two families of interconnection networks under the *pmc* model and mm^* model. Discret. Appl. Math. **339**, 89–106 (2023)

9. Malek, M.: A comparison connection assignment for diagnosis of multiprocessor systems. ISCA 1980, pp. 31–36, New York, NY, USA (1980). Association for Computing Machinery
10. Yang, C.-L., Masson, G.M.: Hybrid fault diagnosability with unreliable communication links. IEEE Trans. Comput. **37**(2), 175–181 (1988)
11. Liang, J., Guo, Y., Xie, Y., Liang, X.: An evolutionary fault diagnosis algorithm for interconnection networks under the *pmc* model. J. Supercomput. **79**(9), 9964–9984 (2023)
12. Zhu, Q., Li, L., Liu, S., Zhang, X.: Hybrid fault diagnosis capability analysis of hypercubes under the *pmc* model and mm^* model. Theoret. Comput. Sci. **758**, 1–8 (2019)
13. Lian, G., Zhou, S., Cheng, E., Liu, J., Chen, G.: Persistence of hybrid diagnosability of regular networks under testing diagnostic model. Comput. J. **64**(9), 1401–1411 (2020)
14. Qiang, M.X., Zhu, K.T., Radhakrishnan, S.: Hybrid *pmc* (*hpmc*) fault model and diagnosability of interconnection networks. AKCE Int. J. Graphs Comb. **17**(3), 755–760 (2020)
15. Zhang, N., Zhu, Q.: Two important parameters of *hpmc* model. Theoret. Comput. Sci. **905**, 10–17 (2022)
16. Yin, S., Liqiong, X.: Hybrid diagnosis of regular networks under the *hpmc* fault model. J. Supercomput. **78**(15), 17247–17260 (2022)
17. Balasubramanian, K.: Topological indices, graph spectra, entropies, laplacians, and matching polynomials of n-dimensional hypercubes. Symmetry **15**(2), 557 (2023)
18. Qiang, Z., Xu, J.: On restricted edge connectivity and extra edge connectivity of hypercubes and folded hypercubes. J. Univ. Sci. Technol. China **36**, 246–253 (2006)

Sponsored Search Auction Design Beyond Single Utility Maximization

Changfeng Xu, Chao Peng(✉), Chenyang Xu(✉), and Zhengfeng Yang

Shanghai Key Laboratory of Trustworthy Computing, Software Engineering Institute, East China Normal University, Shanghai, China
51265902120@stu.ecnu.edu.cn, {cpeng,cyxu,zfyang}@sei.ecnu.edu.cn

Abstract. Auction design for the modern advertising market has gained significant prominence in the field of game theory. With the recent rise of auto-bidding tools, an increasing number of advertisers in the market are utilizing these tools for auctions. The diverse array of auto-bidding tools has made auction design more challenging. Various types of bidders, such as quasi-linear utility maximizers and constrained value maximizers, coexist within this dynamic gaming environment. We study sponsored search auction design in such a mixed-bidder world and aim to design truthful mechanisms that maximize the total social welfare. To simultaneously capture the classical utility and the value-max utility, we introduce an allowance utility model. In this model, each bidder is endowed with an additional allowance parameter, signifying the threshold up to which the bidder can maintain a value-max strategy. The paper distinguishes two settings based on the accessibility of the allowance information. In the case where each bidder's allowance is public, we demonstrate the existence of a truthful mechanism achieving an approximation ratio of $(1+\epsilon)$ for any $\epsilon > 0$. In the more challenging private allowance setting, we establish that a truthful mechanism can achieve a constant approximation. Further, we consider uniform-price auction design in large markets and give a truthful mechanism that sets a uniform price in a random manner and admits bounded approximation in expectation.

Keywords: Auction Design · Mixed Bidders · Value Maximizers

1 Introduction

Sponsored search advertising is one of the most classical and fundamental problems in game theory. In this context, advertisers bid for the placement of their ads on search engine result pages, and the auction mechanism determines the ad ranking and the cost per click for each advertiser. The design of sponsored search auctions holds considerable significance in online advertising and search engine monetization [5]. The efficiency and fairness of the auction mechanisms directly impact the advertising ecosystem.

This paper studies sponsored search auction design in the modern advertising market. Compared to the traditional auction environment, where bidders

Y. Chen et al. (Eds.): COCOON 2024, LNCS 15162, pp. 53–65, 2025.
https://doi.org/10.1007/978-981-96-1093-8_5

universally acknowledge quasi-linear utilities, the utility functions of bidders in the modern market are more complex and diverse. As highlighted in recent economical literature [2], an increasing number of advertisers are making use of auto-bidding tools [3,16] to optimize their profits, posing challenges in devising a model that fits reality well.

Auction design for value maximizers has emerged as a pivotal model to navigate this dynamic landscape. Recent literature [7,15] has extensively explored this domain. Bidders in the value-max model strive to maximize their total obtained values under various constraints, such as budget constraints [4,6,9] and return-on-investment (ROI) constraints [1,10,14]. Experimental studies [8,15] further highlight the effectiveness of the value maximizer model in accurately capturing the behaviors of a significant number of auto-bidding advertisers.

However, as argued in [11], the real-world market is a heterogeneous environment where various utility functions are at play. The classical quasi-linear utility model and the emerging value-max model each capture only a subset of advertisers. Recognizing this diversity, the authors initiate the study of a mixed-bidder environment and propose truthful mechanisms designed to accommodate both quasi-linear utility maximizers and value maximizers. We continue along this line of research and make a further generalization of their setting.

1.1 Our Contributions

This paper explores sponsored search auction design for mixed bidders. To gracefully model both quasi-linear bidders and (constrained) value maximizers, a novel *allowance utility function* is introduced. In this new model, each bidder is endowed with an additional allowance. If a bidder's payment is less than the allowance, she operates as a value maximizer – meaning her utility equals the obtained value. Conversely, if a bidder's payment exceeds the allowance, her utility becomes the obtained value minus the portion of payment exceeding the allowance. See a formal definition in Eq. (1).

We study sponsored search auction design for allowance utility maximizers and design truthful mechanisms that maximize the total social welfare. The utility function captures and generalizes the two prominent utility types in the literature. A quasi-linear bidder corresponds to the scenario where the allowance is set to 0, while a value-max bidder can be considered when the allowance is set to ∞. Besides, our model allows for more diverse choices of allowance values, accommodating a broader range of advertiser behaviors in the market.

Refer to the scenario where the allowance information is publicly disclosed as the *public allowance* setting and the setting where it is not disclosed as the *private allowance* setting. Both public allowance and private settings are considered in the paper. The main results of the paper are summarized as follows:

- For the public allowance setting, we first establish the two crucial properties essential for truthful mechanisms – *allocation monotonicity* and *unit-price strict-monotonicity* (Lemma 1). Leveraging them can prove that no deterministic truthful mechanism can guarantee $(1+\epsilon)$ approximation for arbitrarily

tunable ϵ. We then give a deterministic truthful mechanism parameterized by ϵ that admits an approximation ratio of $(1+\epsilon)$ (Theorem 1).
- For the private allowance setting, the multi-parameter nature poses significant challenges to the design of truthful mechanisms. Thus, we explore a class of allowance-independent mechanisms and show that even if not using the reported allowance profile, there exists a randomized mechanism that is truthful and admits a constant approximation (Theorem 2).
- We further consider uniform price auction design in large markets, where a single bidder's contribution (power) to the total market is very small. We show that there exists a randomized truthful mechanism (for the private allowance setting) capable of setting a uniform price to obtain an expected social welfare $\frac{3}{8}\left(1-\sqrt{1-\frac{\rho}{3k}}\right)^2 \cdot \mathrm{OPT}$, where k is the number of slots, OPT is the optimal social welfare and ρ is the ratio between OPT and the maximum contribution of a single bidder. When $\rho = O(k^{2/3})$, the mechanism is $O(k^{2/3})$-approximation.

1.2 Paper Organization

Section 2 provides a formal definition of our model and introduces the terminology necessary to understand the paper. Then we present our results for the public allowance setting in Sect. 3 and the private allowance setting in Sect. 4. The discussion on uniform-price auction design is deferred to the full version of the paper[1]. Section 5 finally concludes the paper.

2 Preliminaries

We consider the standard sponsored search environment, where the goods for sale are the k "slots" for sponsored links on a search results page. The bidder set in this setting consists of n advertisers, each of whom has a standing bid on the keyword relevant to the search. Each slot $j \in [k]$ is associated with click-through-rate (CTR) $\alpha_j \geq 0$, representing the probability that the end user clicks on this slot. Without loss of generality, assume that $\alpha_1 \geq \alpha_2 \geq ... \geq \alpha_k$ and $n \geq k$.

Allowance Utility Maximizer. Each bidder i desires to purchase at most one slot and has a private value $v_i > 0$ for one click, implying that the maximum amount she is willing to spend for slot j is $v_i\alpha_j$. Additionally, our model introduces an allowance γ_i for each bidder i, and defines the allowance utility function as follows: if the auctioneer assigns slot j to bidder i and charges her p_i, the utility

$$u_i := \begin{cases} v_i\alpha_j - (p_i - \gamma_i)^+ & \text{if } p_i \leq v_i\alpha_j, \\ -\infty & \text{otherwise,} \end{cases} \tag{1}$$

where $(p_i - \gamma_i)^+ := \max\{0, p_i - \gamma_i\}$.

[1] Full Version: https://arxiv.org/abs/2406.05988.

During the auction, each bidder i confidentially communicates her private information to the auctioneer (with the possibility of misreporting). Subsequently, the auctioneer determines the allocation of slots to bidders $\Pi = \{\pi_1, ..., \pi_n\}$ and the corresponding payment requirements $\mathcal{P} = \{p_1, ..., p_n\}$, where π_i, p_i represent the assigned slot and the charged money for bidder i, respectively. To facilitate the representation of bidders i who are not allocated any slot, we add a dummy slot $k+1$ with CTR 0 and let $\pi_i = k+1$. Note that each slot except the dummy slot can be assigned to at most one bidder. The goal of the auctioneer is to design a truthful mechanism that maximizes the total social welfare $\sum_{i\in[n]} v_i \alpha_{\pi_i}$.

Definition 1. *([13]) A mechanism is* truthful *(or DSIC) if for any bidder i, reporting the private information truthfully always maximizes the utility regardless of other bidders' profiles, and the utility of any truthtelling bidder is non-negative.*

3 Public Allowance

This section considers the public allowance setting and shows the following:

Theorem 1. *In the public allowance setting, there exists a truthful and* $(1+\epsilon)$*-approximation deterministic mechanism for any parameter* $\epsilon > 0$*. Further, no deterministic truthful mechanism can guarantee* $(1+\epsilon)$ *approximation for arbitrarily tunable* ϵ*.*

To derive the algorithmic intuition and layout the techniques employed, we begin by stating the negative result.

3.1 Lower Bound

In this subsection, we first characterize truthful mechanisms for (public) allowance utility maximizers and then prove a lower bound. Use $x_i(v_i, \mathbf{v}_{-i})$ and $p_i(v_i, \mathbf{v}_{-i})$ to denote the obtained CTR and payment of bidder i when she bids v_i and other agents bid $\mathbf{v}_{-i}$.

Lemma 1. *A mechanism is truthful for (public) allowance utility maximizers only if the following two properties hold for each bidder* $i \in [n]$ *and any* $\mathbf{v}_{-i}$*:*

- *(Allocation Monotonicity) As* v_i *increases,* $x_i(v_i, \mathbf{v}_{-i})$ *is non-decreasing.*
- *(Unit-Price Strict-Monotonicity) If bidder i increases* v_i *to* v_i' *such that* $x_i(v_i', \mathbf{v}_{-i}) > \frac{\gamma_i}{v_i} > x_i(v_i, \mathbf{v}_{-i}) > 0$*, the paid unit-price has to strictly increase:* $\frac{p_i(v_i', \mathbf{v}_{-i})}{x_i(v_i', \mathbf{v}_{-i})} > v_i$*.*

The lemma establishes two properties of truthful mechanisms for allowance utility maximizers. We observe that classical mechanisms such as VCG auction and Generalized Second Price (GSP) auction satisfy allocation monotonicity but fail in unit-price strict-monotonicity due to their treatment of ties. The basic idea is similar to the proof of Myerson's lemma [12]. Due to space limitations, we omit its proof. We build upon Lemma 1 to prove a lower bound.

Lemma 2. *For any deterministic truthful mechanism, there always exists a value* $\epsilon > 0$ *such that its approximation ratio is at least* $(1 + \epsilon)$.

Proof. We assume for contradiction that there exists a deterministic truthful mechanism $\mathcal{M}$ such that for any $\epsilon > 0$, its approximation ratio is less than $(1 + \epsilon)$. Consider a scenario where two bidders are competing for a single slot. The allowance of each bidder is set to be 1/2. Without loss of generality, we can assume that when both of them bid 1, $\mathcal{M}$ assigns the slot to the first bidder.

Supposing that the second bidder increases her bid to $(1 + \epsilon)$ for any ϵ, $\mathcal{M}$ has to assign the slot to her immediately; otherwise, its approximation ratio is at least $(1 + \epsilon)$ when the two bidder's private value profile is $< 1, 1 + \epsilon >$. Due to Lemma 1, in this case, the second bidder's payment should be strictly larger than 1. Denote by $(1 + \delta)$ the second bidder's payment. Clearly, $\delta \leq \epsilon$.

When the second bidder has a private value $(1 + \epsilon)$, her current utility is $1/2 + \epsilon - \delta$. Then the second bidder has the incentive to misreport a lower bid $(1 + \epsilon')$ for $\epsilon' < \delta$. According to the assumption, $\mathcal{M}$ still has to assign the slot to her but due to individual rationality, $\mathcal{M}$ can only charge her at most $(1 + \epsilon')$, implying that the bidder obtains a higher utility at least $(1/2 + \epsilon - \epsilon')$. Thus, $\mathcal{M}$ is untruthful, which contradicts the assumption and completes the proof. □

3.2 Optimal Deterministic Mechanism

This subsection gives a truthful deterministic mechanism. As mentioned earlier, a significant factor leading to the lack of truthfulness in classical mechanisms under allowance utilities lies in the handling of tie-breaking situations. To solve this issue, we first scale each bidder's bid to the nearest integer power of $(1 + \epsilon)$, and then design the allocation and payment rules based on the scaled bids.

We state the details in Algorithm 1. The mechanism allocates the slots to the bidders with top k rounded bids in a fixed tie-breaking manner. According to this allocation rule, we compute function $f_i(z)$ and z_m for each bidder i. Intuitively, the payment rule is running "first pricing" up to the allowance and then switching to the VCG-like payment.

Proof of Theorem 1. Lemma 2 has already proven the negative side of the theorem. We now show that Algorithm 1 is truthful and $(1 + \epsilon)$ approximation. According to our allocation rule, for each slot, the assigned bidder's value is guaranteed to be at least $\frac{1}{1+\epsilon}$ of the value of the bidder who obtains the slot in the optimal solution, establishing the approximation ratio.

The following analyzes the truthfulness. Consider a bidder i with value v_i and $t_i = \lfloor \log_{1+\epsilon} v_i \rfloor$. We distinguish two cases: (1) $t_i \leq z_m$, (2) $t_i > z_m$.

In the first case, submitting a truthful bid results in a utility of $(1+\epsilon)^{t_i} \cdot f_i(t_i)$. If the bidder intends to alter the utility, she must misreport a bid b_i' such that t_i' is either $< t_i$ or $> t_i$. When $t_i' < t_i$, the utility strictly decreases due to the allocation monotonicity; when $t_i' > t_i$, regardless of whether t_i' is larger than z_m or not, the mechanism charges her a unit price of at least $\min(z_m, t_i')$, violating the payment constraint and leading to a utility of $-\infty$.

Algorithm 1. Public Allowance Auction

Input: bids $\mathcal{B} = \{b_i\}_{i\in[n]}$, allowances $\gamma = \{\gamma_i\}_{i\in[n]}$, CTRs $\boldsymbol{\alpha} = \{\alpha_j\}_{j\in[k]}$ and parameter $\epsilon > 0$.
Output: allocation Π and payment $\mathcal{P}$.
Initialize $\pi_i \leftarrow k+1$ (the dummy slot) and $p_i \leftarrow 0$, $\forall i \in [n]$.
for each bidder $i \in [n]$ **do**
  /* Round b_i down slightly such that it is an exponential multiple of $(1+\epsilon)$. */
  Set $t_i \leftarrow \lfloor \log_{1+\epsilon} b_i \rfloor$ and $\bar{b}_i = (1+\epsilon)^{t_i}$.
end for
Reindex the bidders such that $\bar{b}_1 \geq \bar{b}_2 \geq ... \geq \bar{b}_n$ and break the ties in a fixed manner.
/* The Allocation Rule */
For each bidder $i \in [k]$, $\pi_i \leftarrow i$.
/* The Payment Rule */
for each bidder $i \in [k]$ **do**
  Based on the allocation rule, compute a function $f_i(z)$ representing the CTR obtained by bidder i when her rounded bid is $(1+\epsilon)^z$ and others' are $\bar{\mathbf{b}}_{-i}$.
  Find the maximum value z_m such that $(1+\epsilon)^{z_m} \cdot f_i(z_m) \leq \gamma_i$.
  if $t_i \leq z_m$ **then**
    $p_i \leftarrow \bar{b}_i \cdot f_i(t_i)$.
  else
    $p_i \leftarrow (1+\epsilon)^{z_m} \cdot f_i(z_m) + \sum_{z=z_m+1}^{t_i} (1+\epsilon)^z \cdot (f_i(z) - f_i(z-1))$.
  end if
end for
return $\Pi = \{\pi_i\}_{i\in[n]}$ and $\mathcal{P} = \{p_i\}_{i\in[n]}$.

In the second case, the truthtelling utility is

$$u_i = (1+\epsilon)^{t_i} f_i(t_i) + \gamma_i - (1+\epsilon)^{z_m} f_i(z_m) - \sum_{z=z_m+1}^{t_i} (1+\epsilon)^z \left(f_i(z) - f_i(z-1)\right)$$

If the bidder misreports a higher bid b_i' such that $t_i' > t_i$, the new utility is

$$\begin{aligned} u_i' &= (1+\epsilon)^{t_i} f_i(t_i') + \gamma_i - (1+\epsilon)^{z_m} f_i(z_m) - \sum_{z=z_m+1}^{t_i'} (1+\epsilon)^z \left(f_i(z) - f_i(z-1)\right) \\ &= u_i + (1+\epsilon)^{t_i} \left(f_i(t_i') - f_i(t_i)\right) - \sum_{z=t_i+1}^{t_i'} (1+\epsilon)^z \left(f_i(z) - f_i(z-1)\right) \leq u_i. \end{aligned}$$

On the other hand, if the bidder misreports a lower bid b'_i such that $t'_i < t_i$, we further distinguish two subcases: (i) $t_i \geq z_m$; (ii) $t_i < z_m$. For subcase (i), the new utility has

$$\begin{aligned} u'_i &= (1+\epsilon)^{t_i} f_i(t'_i) + \gamma_i - (1+\epsilon)^{z_m} f_i(z_m) - \sum_{z=z_m+1}^{t'_i} (1+\epsilon)^z \left(f_i(z) - f_i(z-1)\right) \\ &= u_i - (1+\epsilon)^{t_i}(f_i(t_i) - f_i(t'_i)) + \sum_{z=t'_i+1}^{t_i} (1+\epsilon)^z \left(f_i(z) - f_i(z-1)\right) \leq u_i. \end{aligned}$$

For subcase (ii), we have $u'_i = (1+\epsilon)^{t_i} f_i(t'_i)$. Thus,

$$\begin{aligned} u_i - u'_i = &(1+\epsilon)^{t_i}(f_i(t_i) - f_i(t'_i)) + \gamma_i - (1+\epsilon)^{z_m} f_i(z_m) \\ &- \sum_{z=z_m+1}^{t_i} (1+\epsilon)^z \left(f_i(z) - f_i(z-1)\right) \geq 0. \end{aligned}$$

The analysis above collectively proves the truthfulness. □

4 Private Allowance

This section considers the setting where the allowance information γ is private. Compared to the public allowance setting, this scenario extends beyond the single-parameter environment, introducing increased complexity and posing greater challenges. In Algorithm 1, the payment of bidder i is significantly dependent on the allowance γ_i. Such a strong correlation between the payment rule and γ indicates that when γ_i becomes private, a bidder could benefit from misreporting a lower allowance. To handle this issue, we explore *γ-independent* mechanisms, where both the allocation and payment rules are independent of the allowance profile of bidders, and show the following.

Theorem 2. *In the private allowance setting, there exists a γ-independent mechanism which is truthful and obtains a constant approximation ratio.*

To gain algorithmic intuition and streamline the analysis, we begin by considering the simplest case where $k = 1$. Even in the presence of only one slot, developing a truthful and γ-independent mechanism is not a trivial task.

4.1 Single-Slot Auction

In this subsection, we consider the single-slot environment. Without loss of generality, assume that the slot's CTR is 1. Given our aim for a truthful mechanism, the auction must be truthful for any quasi-linear utility maximizer, i.e., the bidder with $\gamma_i = 0$. For this type of bidders, Myerson's Lemma [12] dictates that the allocation rule must be monotone, and the payment rule is unique for a

Algorithm 2. γ-Independent Single-Slot Auction

Input: bids $\mathcal{B} = \{b_i\}_{i\in[n]}$ and parameter $\epsilon > 0$.
Output: allocation Π and payment $\mathcal{P}$.
Initialize $\pi_i \leftarrow 2$ (the dummy slot) and $p_i \leftarrow 0$, $\forall i \in [n]$.
/* Round b_i down slightly such that it is an exponential multiple of $(1+\epsilon)$. */
For each bidder $i \in [n]$, set $\bar{b}_i = (1+\epsilon)^{\lfloor \log_{1+\epsilon} b_i \rfloor}$.
Use $\mathcal{S}$ to denote the set of bidders with the highest $\bar{b}$.
if $|\mathcal{S}| > 1$ **then**
    Let k be the smallest index in $\mathcal{S}$. /* Break ties in a fixed manner. */
    Set $\pi_k \leftarrow 1$ and $p_k \leftarrow \bar{b}_k$.
else
    Let $\bar{b}_k$ be the highest (rounded) bid and $\bar{b}_\ell$ be the second-highest.
    Set $\pi_k \leftarrow 1$ and $p_k \leftarrow (1+\epsilon)\cdot \bar{b}_\ell$.
end if
return $\Pi = \{\pi_i\}_{i\in[n]}$ and $\mathcal{P} = \{p_i\}_{i\in[n]}$.

monotone allocation. Specifically, a bounded approximation mechanism should allocate the slot to one of the highest bidders and charge her the threshold bid.

Since we target a γ-independent mechanism, we have to treat bidders as if they do not disclose their allowance profiles. Consequently, we are unable to distinguish whether a bidder employs quasi-linear utility or not. In light of this, to maintain truthfulness and a bounded approximation ratio, the classical second-price mechanism seems the most viable choice, where the highest bidder wins (with ties broken arbitrarily) and is charged the second-highest bid.

We examine the truthfulness of the second-price auction for bidders with $\gamma_i > 0$. We observe that these bidders are truthful in most cases, except in situations involving multiple highest bidders. When there is more than one highest bidder, quasi-linear utility maximizers have no incentive to lie because their utilities, whether they obtain the slot or not, consistently amount to 0. However, the scenario differs for other types of bidders. Upon securing the slot, they gain a non-zero utility, leading to a potential misreport of a higher bid to obtain the slot while paying an amount that adheres to the constraint. To tackle this challenge, we employ a scaling technique. Due to space, we only give the mechanism (Algorithm 2) and the theorem here.

Theorem 3. *For any $\epsilon > 0$, Algorithm 2 is truthful and obtains an approximation ratio of* $(1+\epsilon)$.

Theorem 3 proves a constant approximation for single-slot auction. This result can further be extended to the multiple-slot setting. Specifically, if there exists a bidder making a significant contribution to the optimal social welfare OPT, we can choose to auction only the first slot and then apply Algorithm 2 to achieve a bounded approximation ratio:

Corollary 1. *For the multiple-slot setting, if there exists a bidder i with $v_i \cdot \alpha_1 \geq \frac{\mathrm{OPT}}{\rho}$ for some $\rho \geq 1$, Algorithm 2 is truthful and obtains $\rho\cdot(1+\epsilon)$ approximation.*

Algorithm 3. γ-Independent Large Market Auction

1: **Input:** bids $\mathcal{B} = \{b_i\}_{i\in[n]}$, CTRs $\boldsymbol{\alpha} = \{\alpha_j\}_{j\in[k]}$.
2: **Output:** allocation Π and payment $\mathcal{P}$.
3: Initialize $\pi_i \leftarrow k+1$ (the dummy slot) and $p_i \leftarrow 0$, $\forall i \in [n]$.
4: Independently divide all the bidders with equal probability into set $S^{(1)}$ and $S^{(2)}$; subsequently, also with equal probability, designate one of $\{S^{(1)}, S^{(2)}\}$ as the pricing benchmark set L and the other as the target bidder set R.
5: For each slot $j \in [k]$, let z_j be the j-th highest bid in L (if $|L| < j$, let z_j be 0).
6: Arrange the bidders in set R in an arbitrarily fixed order. As each bidder i arrives, offer all remaining slots to her at a unit price profile of $\left\{\frac{1}{2}z_j\right\}_{j\in[k]}$ and allow her to select the most profitable slot j. Set $\pi_i \leftarrow j$ and $p_i \leftarrow \frac{1}{2}z_j \cdot \alpha_j$.
7: **return** $\Pi = \{\pi_i\}_{i\in[n]}$ and $\mathcal{P} = \{p_i\}_{i\in[n]}$.

4.2 Large Market Auction

Corollary 1 gives a bounded approximation for the scenario where $\max_{i\in[n]} v_i \cdot \alpha_1 \geq \frac{\text{OPT}}{\rho}$. This subsection addresses the converse scenario – the large market case, where $\max_{i\in[n]} v_i \cdot \alpha_1 < \frac{\text{OPT}}{\rho}$. In a large market, no bidder or slot dominates. In addition, we also assume, w.l.o.g., that ρ is sufficiently large, e.g. $\rho \geq 4$. Consequently, we can leverage the random sampling technique to develop a truthful mechanism with a bounded approximation ratio.

The mechanism is described in Algorithm 3. Initially, it partitions all bidders into two subsets $S^{(1)}$ and $S^{(2)}$, padding them with zeros until their sizes are at least k. Subsequently, it sets the prices of slots based on one subset and allows the bidders in the other subset to purchase them. The algorithmic intuition is derived from the following concentration lemma:

Lemma 3. *Let $w_1 \geq w_2 \geq ... \geq w_\ell$ be positive real numbers, such that the sum $w = w_1 + ... + w_\ell$ satisfies $w_1 < w/4$. We independently partition each number with equal probability into two subsets A and B. With probability at least $1/2$, there exists a matching $\mathcal{M}$ between A and B such that $\sum_{(w_i,w_j)\in\mathcal{M}} \min\{w_i, w_j\} > \frac{w}{4}$.*

Due to space constraints, we omit the proof. Intuitively, this lemma asserts that with high probability, there exists a matching between k slots and R such that if for each slot $j \in [k]$, we enforce the matched bidder in R purchases it at a unit price of z_j once her bid is no less than z_j, the total obtained payment is at least $\frac{1}{4}$ of the optimal social welfare. However, the mechanism cannot enforce such actions as it would lead to untruthfulness. Therefore, we employ a technique of discounted sales to ensure both truthfulness and a constant approximation ratio simultaneously. We show the following theorem.

Theorem 4. *Under the large market assumption that $\max_{i\in[n]} v_i \cdot \alpha_1 < \frac{\text{OPT}}{4}$, Algorithm 3 is truthful and obtains an approximation ratio at most 64.*

Proof. The proof of truthfulness is straightforward. For bidders in L, they are assigned anything, and therefore, misreporting their information cannot improve the utilities; while for bidders in R, they are also truthtelling because their reported information determines neither the arrival order nor the market prices.

Now we analyze the approximation ratio. Let $\Pi^* = \{\pi_i^*\}_{i\in[n]}$ be an optimal assignment. Viewing $b_i\alpha_{\pi_i^*}$ as w_i in Lemma 3, we have with probability at least 1/2, there exists a matching $\mathcal{M}$ between $S^{(1)}$ and $S^{(2)}$ such that $\sum_{(i,h)\in\mathcal{M}} \min\{b_i\alpha_{\pi_i^*}, b_h\alpha_{\pi_h^*}\} > \mathrm{OPT}/4$. Refer to such a partition $\{S^{(1)}, S^{(2)}\}$ as a good partition. The following proof is conditioned on the occurrence of a good partition and analyzes the conditional expected performance of our algorithm. For brevity, we omit the notation for conditional expectations and use $\mathbb{E}[\cdot]$ to denote the conditional expectation.

Consider an assignment $\mathcal{A}$ of the slots to bidders in R as follows. For each slot j, we assign it to the bidder i who is matched with the j-th highest bidder among L in $\mathcal{M}$. Since L and R are designated randomly, if enforcing that each bidder in R either purchases the assigned slot at a unit price of z_j or opts out, the total obtained payment $\mathcal{P}(\mathcal{A})$ is at least $\sum_{(i,h)\in\mathcal{M}} \min\{b_i, b_h\} \cdot \alpha_j/2$ in expectation. Due to the optimality of Π^*, we have that $b_i\alpha_{\pi_i^*} \geq b_h\alpha_{\pi_h^*}$ iff $b_i \geq b_h$, and for the j-th highest bidder h in L, $\alpha_j \geq \alpha_{\pi_h^*}$. Thus,

$$\mathbb{E}[\mathcal{P}(\mathcal{A})] \geq \frac{1}{2} \cdot \sum_{(i,h)\in\mathcal{M}} \min\{b_i\alpha_{\pi_i^*}, b_h\alpha_{\pi_h^*}\} > \frac{\mathrm{OPT}}{8}. \tag{2}$$

Let ALG be the social welfare obtained by Algorithm 3. The last piece is to prove that $\mathbb{E}[\mathrm{ALG}]$ is at least a constant factor of $\mathbb{E}[\mathcal{P}(\mathcal{A})]$. Fix an arbitrary (good) pair of (L, R). Let $\mathcal{J}$ be the set of slots j that the assigned bidder $a(j) \in R$ has a bid at least z_j. We abuse the notion slightly and also let $a(j)$ represent the corresponding bid. Consider a slot $j \in \mathcal{J}$. We distinguish three cases according to its state in our mechanism:

(1) it is purchased by some bidder.
(2) no bidder purchases it and its corresponding bidder $a(j)$ picks a slot j' with a higher CTR, i.e., $j' < j$.
(3) no bidder purchases it and its corresponding bidder $a(j)$ picks a slot j' with a lower CTR, i.e., $j' > j$.

Using $\mathcal{J}_i$ to denote the set of slots in Case (i), we have

$$\mathcal{P}(\mathcal{A}) = \sum_{j\in\mathcal{J}_1} z_j \cdot \alpha_j + \sum_{j\in\mathcal{J}_2} z_j \cdot \alpha_j + \sum_{j\in\mathcal{J}_3} z_j \cdot \alpha_j.$$

Since all the slots in $\mathcal{J}_1$ are sold out, Algorithm 3 obtains a social welfare at least $\sum_{j\in\mathcal{J}_1} \frac{1}{2} z_j \cdot \alpha_j$. For the remaining two cases, we show that $z_j \cdot \alpha_j$ can be bounded by the bidder $a(j)$'s contribution in ALG, denoted as $a(j) \cdot \alpha_{j'}$. This claim is obvious for Case (2) since $a(j) \geq z_j$ and $\alpha_{j'} \geq \alpha_j$, while the proof for Case (3) requires leveraging the property of allowance utility.

Algorithm 4. γ-Independent Auction

1: **Input:** bids $\mathcal{B} = \{b_i\}_{i\in[n]}$, CTRs $\boldsymbol{\alpha} = \{\alpha_j\}_{j\in[k]}$ and parameter $\epsilon \in (0,1)$.
2: **Output:** allocation Π and payment $\mathcal{P}$.
3: With probability of 1/17 **begin**
4: Run Algorithm 2 parameterized with ϵ to sell the first slot.
5: **end**
6: With probability of 16/17 **begin**
7: Run Algorithm 3.
8: **end**
9: **return** $\Pi = \{\pi_i\}_{i\in[n]}$ and $\mathcal{P} = \{p_i\}_{i\in[n]}$.

Consider a slot $j \in \mathcal{J}_3$. When bidder $a(j)$ arrives, this slot is available but the bidder picks another slot j' with a lower CTR. According to the definition of allowance utility, we have

$$a(j)\cdot\alpha_j - \left(\frac{1}{2}z_j\cdot\alpha_j - \gamma_{a(j)}\right)^+ \le a(j)\cdot\alpha_{j'} - \left(\frac{1}{2}z_{j'}\cdot\alpha_{j'} - \gamma_{a(j)}\right)^+$$
$$\frac{1}{2}z_j\cdot\alpha_j \le a(j)\cdot\alpha_{j'},$$

where the second inequality is due to $\alpha_j \ge \alpha_{j'}$ and the last inequality uses the fact that $a(j) \ge z_j$. By summarizing the analysis of the three cases, we have

$$\mathcal{P}(\mathcal{A}) \le 4\cdot\text{ALG}. \tag{3}$$

By combining Eqs. (2), (3), and the probability of a good partition occurring, we can demonstrate that the expected approximation ratio of Algorithm 3 is at most 64 under the large market assumption. □

Final Mechanism. We builds on Algorithms 2 and 3 to give the final mechanism Algorithm 4. It is a simple random combination of the two mechanisms. Due to space, we omit the details and the proof.

5 Conclusion

The paper delves into sponsored search auctions in modern advertising markets, introducing a novel bidder utility model and exploring two distinct settings within the model. Specifically, for the public allowance setting, we demonstrate the achievability of an optimal mechanism, while in the private allowance setting, we propose a truthful mechanism with a constant approximation ratio and a uniform-price auction with bounded approximation in large markets.

There leave quite a lot of possibilities for future work. One direction is investigating the lower bounds of the allowance utility model. It is noteworthy that [11] establishes a lower bound of 5/4 in their mixed-bidder setting, which directly extends to our private allowance setting as a special case. Consequently, a constant gap exists between the upper and lower bounds in the private allowance setting, posing an open question regarding the potential closure of this gap.

Acknowledgments. This work is supported by the National Key Research Project of China under Grant No. 2023YFA1009402, the National Natural Science Foundation of China (No. 62302166), the Scientific and Technological Innovation 2030 Major Projects under Grant 2018AAA0100902, the Shanghai Science and Technology Commission under Grant No. 20511100200, the Dean's Fund of Shanghai Key Laboratory of Trustworthy Computing, ECNU, and the Key Laboratory of Interdisciplinary Research of Computation and Economics (SUFE), Ministry of Education.

References

1. Auerbach, J., Galenson, J., Sundararajan, M.: An empirical analysis of return on investment maximization in sponsored search auctions. In: AdKDD@KDD, pp. 1–9. ACM (2008)
2. Baisa, B.: Auction design without quasilinear preferences. Theor. Econ. **12**(1), 53–78 (2017)
3. Balseiro, S.R., Deng, Y., Mao, J., Mirrokni, V.S., Zuo, S.: Robust auction design in the auto-bidding world. In: NeurIPS, pp. 17777–17788 (2021)
4. Balseiro, S.R., Deng, Y., Mao, J., Mirrokni, V.S., Zuo, S.: Optimal mechanisms for value maximizers with budget constraints via target clipping. In: EC, p. 475. ACM (2022)
5. Birmpas, G., Celli, A., Colini-Baldeschi, R., Leonardi, S.: Fair equilibria in sponsored search auctions: the advertisers' perspective. In: IJCAI, pp. 95–101. ijcai.org (2022)
6. Caragiannis, I., Voudouris, A.A.: The efficiency of resource allocation mechanisms for budget-constrained users. Math. Oper. Res. **46**(2), 503–523 (2021)
7. Chen, Y., et al.: Coordinated dynamic bidding in repeated second-price auctions with budgets. In: ICML. Proceedings of Machine Learning Research, vol. 202, pp. 5052–5086. PMLR (2023)
8. Deng, Y., Zhang, H.: Prior-independent dynamic auctions for a value-maximizing buyer. In: NeurIPS, pp. 13847–13858 (2021)
9. Devanur, N.R., Weinberg, S.M.: The optimal mechanism for selling to a budget constrained buyer: the general case. In: EC, pp. 39–40. ACM (2017)
10. Lu, P., Xu, C., Zhang, R.: Auction design for value maximizers with budget and return-on-spend constraints. In: WINE. Lecture Notes in Computer Science, vol. 14413, pp. 474–491. Springer (2023)
11. Lv, H., et al.: Utility maximizer or value maximizer: mechanism design for mixed bidders in online advertising. In: AAAI, pp. 5789–5796. AAAI Press (2023)
12. Myerson, R.B.: Optimal auction design. Math. Oper. Res. **6**(1), 58–73 (1981)

13. Roughgarden, T., Iwama, K.: Twenty lectures on algorithmic game theory. Bull. EATCS **122** (2017)
14. Szymanski, B.K., Lee, J.S.: Impact of ROI on bidding and revenue in sponsored search advertisement auctions. In: Second Workshop on Sponsored Search Auctions (2006)
15. Wilkens, C.A., Cavallo, R., Niazadeh, R.: Mechanism design for value maximizers. CoRR arXiv:1607.04362 (2016)
16. Wilkens, C.A., Cavallo, R., Niazadeh, R.: GSP: the cinderella of mechanism design. In: WWW, pp. 25–32. ACM (2017)

An On-Orbit Data Balancing Online Algorithm For LEO Satellite Cluster: A Repeated Stochastic Game Approach

Ying Qiao, Juan Luo(✉), and Shuyang Teng

Hunan University, Changsha 410082, China
{qy2020,juanluo,S2110Z0004}@hnu.edu.cn

Abstract. Processing uneven data requests is crucial for Low Earth Orbit (LEO) satellites to effectively balance resources. Various multi-hop peer offloading methods have been proposed to address this challenge. Multi-hop offloading can offload data to satellite with sufficient resources by multi-hop transmission, which results in additional transmission energy consumption. Moreover, the dynamic network topology further exacerbates significant energy consumption. Hence, this increased energy consumption may not be sustainable for satellites with limited energy resources. In response, we propose an on-orbit data balancing online algorithm for a LEO satellite cluster, where overloaded satellites directly offload data to idle ones. Designing such a algorithm entails addressing three key questions: determining which LEO satellite to offload, how much data to offload, and how to manage the satellite's energy storage system. To tackle these questions, we formulate the offloading process as a mixed integer non-linear optimization problem aimed at minimizing overall energy consumption. Subsequently, we devise a repeated stochastic non-cooperative game and leverage an online data balancing algorithm based on Lyapunov to achieve coarse correlated equilibrium for the game. Finally, simulated experiments demonstrate that our proposed algorithm reduces energy consumption by 61.49%, 56.22%, and 55.29% compared to the baselines across three different received data sizes.

Keywords: LEO Satellite Cluster · Uneven Data Distribution · Repeated Stochastic Game · Satellite Edge Computing · Lyapunov

1 Introduction

Low Earth Orbit satellite-assisted Mobile Edge Computing (LEO-MEC) offers low-latency services globally, depending on enhanced satellite computation capabilities and the location advantage of LEO orbits. With on-orbit computing, LEO satellites can directly process data, alleviating the time burden for data transmission to ground stations. Moreover, the proximity of LEO satellites to Earth reduces transmission latency. However, the limited energy resources of

Y. Chen et al. (Eds.): COCOON 2024, LNCS 15162, pp. 66–78, 2025.
https://doi.org/10.1007/978-981-96-1093-8_6

LEO satellites results in significant energy consumption when handling large user requests. Furthermore, due to non-uniform distribution of users across the Earth's surface, some LEO satellites may receive high request volumes while others experience lower demand [1]. Consequently, certain satellites may face resource overuse, thereby compromising the stability and longevity of the satellite system.

In response to the above problems, existing works have done towards satellites peer offloading, routing strategy, adjusting beam, and caching strategy. The satellites peer offloading works [2,3] alleviated unevenly distributed requests by transferring data from overloaded satellites to idle ones through multiple hops. Notably, these methods have been simulated on the Walker constellation. Routing strategies [4] established optimal paths from users to satellites under the assumption that satellite networks are static for certain periods. Beam adjustment strategies [5,6] balance data flow among satellites by adjusting beam sizes, allocating smaller sub-beams for lower data flows and larger sub-beams for higher data flows. Caching strategies [7] pre-cache data on satellites to mitigate problems associated with dynamic topology and uneven data distribution.

In summary, existing approaches can be broadly categorized into two methods: collaborative satellite operations and single-satellite performance optimization. Collaborative operations address overloaded data on a single satellite by multi-hop offloading, while single-satellite optimization focuses solely on allocating communication resources to receive an appropriate data volume. While single-satellite optimization addresses the issue of uneven data distribution, it may struggle to enhance processed data volume. Conversely, as shown in Fig. 1(a), collaborative approaches can handle large data volumes but the dynamic nature of satellite networks leads to increased energy consumption for each satellite. As a result, these methods may not be applicable in scenarios involving satellite networks are dynamic.

In this paper, we present an on-orbit data balancing online algorithm for a LEO satellite cluster, where data overloaded satellites directly offload their data to idle satellites in the cluster. Designing such a algorithm involves the two key challenges.

1. Satellites moving in opposite directions form a dynamic network topology, leading to heightened communication and computation energy consumption. Satellites moving in opposite directions disconnect with each other, prompting data retransmission and reprocessing, thereby increasing energy and time consumption.
2. To ensure regular satellite operation and prolong satellite lifespan, efficient energy management is essential. Satellites frequently transition between light and shadow regions of their orbit. In light zones, solar photovoltaic panels harvest solar energy to power computing loads. Conversely, in shaded regions, satellite loads rely entirely on stored energy from onboard storage systems. Hence, satellites should manage energy to sustain operations in shaded zones.

Our algorithm resolves three questions for each satellite. Firstly, determining which idle satellites to offload. Secondly, determining the amount of tasks to offload. Thirdly, how to manage energy storage system. Our main contributions are outlined as follows:

1. We construct an on-orbit dynamic communication model for the LEO satellite cluster, defining constraints on dynamic connections among LEO satellites. We then formulate an objective problem aimed at minimizing energy consumption under such constraints.
2. We develop a repeated stochastic non-cooperative game to characterize the non-cooperation among satellites within the LEO satellite cluster. To optimize the performance of all satellites, we propose an online data balancing algorithm based on Lyapunov theory, obtaining coarse correlated equilibrium (CCE) equilibrium by transforming the repeated stochastic non-cooperative game into virtual static game.
3. We model energy storage system as energy queue for each satellite. The Lyapunov solution process optimize objective problem and energy queue jointly, achieving the minimization of the objective problem while stabilizing the energy queue. Therefore, we ensure that satellites maintain a stable energy queue when performing data offloading, thereby extending the service life of the satellites.

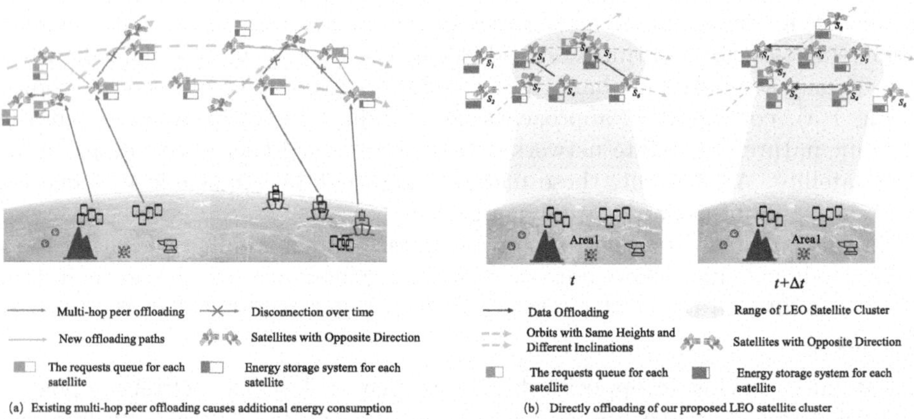

Fig. 1. The schematic of processing uneven data for existing methods and ours.

2 System Model

In this section, we model the offloading and computation process of LEO satellite cluster and then formulate an optimization problem to maximize energy utility.

2.1 System Architecture Model

We illustrate the deployment of LEO satellite cluster in Fig. 1(b), which consists of one head satellite and some ordinary satellites. The satellites within this cluster orbit at the same altitude but with varying inclinations. They receive tasks from the ground and execute them in real-time. Given the uneven distribution of data, overloaded satellites directly offload data to idle satellite.

We use $\mathcal{S}$ to denote the set of satellites in LEO satellite cluster, and $\mathcal{S} = \{1, \cdots, s, \cdots, S\}$. Let $\mathcal{T}$ denote the set of time of orbiting the Earth, $\mathcal{T} = \{1, 2, \cdots, t, \cdots, T\}$.

The LEO satellite cluster orbits the Earth and may cover several areas. The coverage time that LEO satellite cluster covers each area is denoted by $c_{\Delta t}$. The set of areas is represented by $\mathcal{C} = \{1, 2, \cdots, c, \cdots, C\}$. We use $D_{s,c}(t)$ to represent the arrival tasks that satellite s receive in area c. We assume that the network topology of LEO satellite cluster don't change when it covers an area, and changes across different areas. Therefore, we ignore the communication process between cluster and ground users and mainly focus on the computation process of LEO satellite cluster.

2.2 On-Orbit Dynamic Communication Model

The communication links among satellites in the LEO satellite cluster are primarily influenced by their visibility to each other. We model the on-orbit offloading and communication process, which encompasses offloading decisions, transmission rates, communication delays, and communication energy consumption.

Offloading Decision : When a satellite has a large number of data, it may decide to offload a portion of them to another satellite for processing. The offloading process is successful when the two satellites involved are visible to each other. We use $v_{s,j}(t)$ to denote the indicator whether two satellites are visible. $v_{s,j}(t) = 1$ represents that the two satellites are visible to each other, otherwise, $v_{s,j}(t) = 0$.

Each satellite in the cluster selects only one other satellite for offloading data, while the receiving satellite can accept data from multiple sending satellites. We use $\alpha_{s,j}$ to denote the selection strategy of satellite s, which is an indicator whether selects satellite j. If $\alpha_{s,j}(t) = 1$, the satellite s selects the satellite j to offloading when $v_{s,j}(t) = 1$, otherwise, $\alpha_{s,j}(t) = 0$. Additionally, the selection strategy of all satellites must satisfy certain constraints.

$$\alpha_{s,j}(t) \leq v_{s,j}(t), \forall s, j \in \mathcal{S}, \tag{1}$$

$$\sum_{j \in \mathcal{S}/\{s\}} \alpha_{s,j}(t) \leq 1, \forall s \in \mathcal{S}, \tag{2}$$

Transmission Rate : The satellite uses laser communication technology to communicate with other satellites. The transmission rate between them can be given by

$$R_{s,j}(t) = B_s(t)\log_2\left(1 + \frac{P_s^t G_s^2}{k_B \epsilon B_s(t) L_{s,j}(t)}\right), \forall s \in \mathcal{S}, \tag{3}$$

where $B_s(t)$ is the channel bandwidth of satellite s. P_s^t is the transmission power of satellite s. G_s is the gain between the transmitter and receiver antennas of satellites. k_B is the Boltzmann constant, ϵ is the thermal noise in Kelvin. $L_{s,j}(t)$ is the distance between satellite s and satellite j at t time slot.

The communication qualities between satellites are changing. The communication qualities are easily affected by atmospheric conditions, background light radiation and satellite vibration and other factors when using laser communication. We denote the communication quality using antenna gains, which is denoted as G_s within a range $\Omega_s = [G_s^{min}, G_s^{max}]$.

Communication Delay : The satellite offloads a portion of the received data to another satellite, resulting in communication delay. This delay can be defined as:

$$T_{s,c}^{tr}(t) = \frac{\alpha_{s,j}(t)\beta_{s,j}(t)D_{s,c}(t)}{R_{s,j}(t)}, \forall c \in \mathcal{C}, \forall s, j \in \mathcal{S}, s \neq j, \tag{4}$$

where $\beta_{s,j}(t) = [0, 1]$ is the ratio of data that the satellite s offloads to the satellite j at t time slot.

Communication Energy Consumption : The satellite consumes energy to offload data, thus the energy consumption can be defined as

$$E_{s,c}^{tr}(t) = P_s^t T_{s,c}^{tr}(t), \forall s \in \mathcal{S}. \tag{5}$$

2.3 Computation Model

Each satellite receives data from the ground and processes it, incurring computation delay and energy consumption. The computation delay is defined as:

$$T_{s,c}^{cm}(t) = \frac{(D_{s,c}(t) - \alpha_{s,j}(t)\beta_{s,j}(t)D_{s,c}(t))c_a + \sum_{i\in\mathcal{S}/\{s,j\}} \alpha_{i,s}(t)\beta i, s(t)D_{i,c}(t)c_a}{f_s(t)}, \quad \forall s \in \mathcal{S}, \tag{6}$$

where c_a is the number of CPU cycles required to process a unit bit of data, $f_s(t)$ is the CPU-cycle frequency of satellite s, which cannot exceed the maximum value f_s^{max}. For each satellite, it can offload data to other satellites and receive offloaded data from other satellite. Therefore, the total amount of data stored in satellite s cannot exceed its memory size,

$$\sum_{i\in\mathcal{S}/\{s\}} \alpha_{i,s}(t)\beta i, s(t)D_{i,c}(t) + D_{s,c}(t) - \alpha_{s,j}(t)\beta_{s,j}(t)D_{s,c}(t) \leq m_s, \quad \forall j \in \mathcal{S}/\{s\}, j \neq i, c \in \mathcal{C}. \tag{7}$$

The computation energy consumption is defined as

$$E_{s,c}^{cm}(t) = \omega_s (f_s(t))^3 T_{s,c}^{cm}(t), \forall s \in \mathcal{S}, \tag{8}$$

where ω_s represents the processing unit's effective switched capacitance of satellite s.

2.4 Energy Queue

We model the energy storage systems as an energy queue and use $\gamma_s(t) = \{0, 1\}$ to denote the status of whether the satellite is in the shadow or light zone. When $\gamma_s(t) = 0$, the satellite is in shadow zone, otherwise $\gamma_s(t) = 1$.

$$Q_s(t+1) = \begin{cases} Q_s(t) + e_s(t) - E_s^{total}(t), \text{if } \gamma_s(t) = 1, \\ Q_s(t) - E_s^{total}(t), \text{if } \gamma_s(t) = 0. \end{cases} \tag{9}$$

$e_s(t)$ is harvested energy of satellite s at t time slot, which cannot exceed the maximum value e_s^{max}. $E_s^{total}(t) = E_{s,c}^{tr}(t) + E_{s,c}^{cm}(t)$ is the total energy consumption of satellite s. Owing to the limited capacity of the satellite's battery, the sum of the energy acquired and the energy available to the satellite cannot exceed its battery capacity, which is denoted as

$$Q_s(t) + e_s(t) \leq \Lambda_s, \forall s \in \mathcal{S}. \tag{10}$$

2.5 Problem Formulation

To ensure the satellite's operation in shadow and light regions, the energy consumption for each satellite cannot exceed a threshold E_s. The LEO satellite cluster aims to maximize the energy utility. Thus, we formulate the above offloading and computing process as an optimization, which the object is to maximize the energy utility. The optimization is defined as follows:

$$(\mathcal{P}1) \quad \max_{\beta_{s,j}(t), \alpha_{s,j}(t), e_s(t)} \sum_{s \in \mathcal{S}} \sum_{c \in \mathcal{C}} E_s - (E_{s,c}^{tr}(t) + E_{s,c}^{cm}(t)) \tag{11a}$$

$$\text{s.t.} \quad \beta_{s,j}(t) \in [0, 1], \forall s, j \in \mathcal{S}, \tag{11b}$$

$$\alpha_{s,j}(t) \in \{0, 1\}, v_{s,j}(t) \in \{0, 1\}, \forall s, j \in \mathcal{S}, \tag{11c}$$

$$\alpha_{s,j}(t) \leq v_{s,j}(t), \forall s, j \in \mathcal{S}, \tag{11d}$$

$$\sum_{j \in \mathcal{S}/\{s\}} \alpha_{s,j}(t) \leq 1, \forall s \in \mathcal{S}, \tag{11e}$$

$$\sum_{s \in \mathcal{S}} T_s^{tr}(t) + T_s^{cm}(t) \leq c_{\Delta t} \tag{11f}$$

$$0 \leq e_s(t) \leq e_s^{max}, s \in \mathcal{S} \tag{11g}$$

$$Q_s(t) + e_s(t) \leq \Lambda_s, \forall s \in \mathcal{S} \tag{11h}$$

3 Repeated Stochastic Non-cooperative Game

In this section, we formulate a repeated stochastic non-cooperative game for the LEO satellite cluster and introduce the coarse correlated equilibrium of the game.

We use game theory to address the optimization problem. The reason is that the head satellite aims to maximize the energy utility for the LEO satellite cluster, determining to offload tasks from overloaded satellite to idle satellite. However, the idle satellite wants to maximize its own energy utility and may reject the offloading strategy. Therefore, the satellites in LEO satellite cluster form a game relationship.

The head satellite and the ordinary satellite form a non-cooperative game, while satellites have dynamic communication links and the game they formed is also a stochastic non-cooperative game. Meanwhile the LEO satellite cluster covers multiple areas and it repeats operations in each area. Therefore, the game for the satellites is a repeated stochastic non-cooperative game.

The repeated stochastic non-cooperative game is defined as $\mathcal{G}_c = \{\mathcal{S}, \mathcal{A}, \boldsymbol{\Omega}, U_s(t)\}$. The meaning of each variable is respectively described below.

1. Players $\mathcal{S}$: The satellites in LEO satellite cluster.
2. Action space $\mathcal{A}$: The three variables that the satellite determines.
3. Random Event $\boldsymbol{\Omega}$: The random event of communication quality in dynamic satellite network environment.
4. Utility $U(t)$: The energy utility for each satellites, i.e., $U_s(t) = E_s - (E_{s,c}^{tr}(t) + E_{s,c}^{cm}(t))$. Thus, we have $0 \leq U_s(t) \leq E_s$.

In the repeated stochastic non-cooperative game, the head satellite acts as the game manager, giving strategies to each ordinary satellites after observing the random event of dynamic network environment [8]. The random events are dynamic communication quality, which is $\Omega_s = [G_{min}, G_{max}]$, thus the set of random events is represented as $\boldsymbol{\Omega} = \Omega_1 \times \Omega_2 \times \cdots \times \Omega_s \times \cdots \Omega_S$. Let $\boldsymbol{G}(t) = \{G_1(t), \cdots, G_S(t)\}$ denote a vector of the communication conditions of all satellites, where $\boldsymbol{G}(t) \in \boldsymbol{\Omega}$.

Let $\boldsymbol{\beta}(t)$, $\boldsymbol{\alpha}(t)$, and $\boldsymbol{e}(t)$ be vectors of the decision variables, where $\boldsymbol{\beta}(t) = \{\beta_{s,1}(t), \cdots, \beta_{s,S}(t)\}$, $\boldsymbol{\alpha}(t) = \{\alpha_{s,1}(t), \cdots, \alpha_{s,S}(t)\}$, and $\boldsymbol{e}(t) = \{e_1(t), \cdots, e_S(t)\}$. Therefore, the utility can be rewrite as

$$U_s(t) = \hat{U}_s((\boldsymbol{\beta}(t), \boldsymbol{\alpha}(t), \boldsymbol{e}(t)), \boldsymbol{G}(t)), \forall s \in \mathcal{S}. \tag{12}$$

Specifically, let $Pr[(\boldsymbol{\beta}, \boldsymbol{\alpha}, \boldsymbol{e})|\boldsymbol{G}]$ be a conditional probability mass function over $\boldsymbol{G} \in \boldsymbol{\Omega}$, $\boldsymbol{\beta}, \boldsymbol{\alpha}, \boldsymbol{e} \in \mathcal{A}$. It is assumed that

$$Pr[(\boldsymbol{\beta}, \boldsymbol{\alpha}, \boldsymbol{e})|\boldsymbol{G}] \geq 0, \forall \boldsymbol{G} \in \boldsymbol{\Omega}, \forall \boldsymbol{\beta}, \boldsymbol{\alpha}, \boldsymbol{e} \in \mathcal{A} \tag{13}$$

$$\sum_{\boldsymbol{\beta}, \boldsymbol{\alpha}, \boldsymbol{e} \in \mathcal{A}} Pr[(\boldsymbol{\beta}, \boldsymbol{\alpha}, \boldsymbol{e})|\boldsymbol{G}] = 1, \forall \boldsymbol{G} \in \boldsymbol{\Omega} \tag{14}$$

Considering the compositional characteristics of the cluster satellites, we define the ordinary satellite correlate their strategies through the strategies from head satellites, indirectly reaching a relatively stable state, i.e., a coarse correlated equilibrium (CCE). The CCE is achieved when the probability mass function satisfies [9]:

$$\begin{aligned}&\sum_{\boldsymbol{G}\in\boldsymbol{\Omega}}\sum_{\boldsymbol{\beta},\boldsymbol{\alpha},\boldsymbol{e}\in\mathcal{A}} Pr[(\boldsymbol{\beta},\boldsymbol{\alpha},\boldsymbol{e})|\boldsymbol{G}]\hat{U}_s((\boldsymbol{\beta},\boldsymbol{\alpha},\boldsymbol{e}),\boldsymbol{G})\\ &\geq \sum_{\boldsymbol{G}\in\boldsymbol{\Omega}}\sum_{\boldsymbol{\beta},\boldsymbol{\alpha},\boldsymbol{e}\in\mathcal{A}} Pr[(\boldsymbol{\beta},\boldsymbol{\alpha},\boldsymbol{e})|\boldsymbol{G}]\hat{U}_s(p^{(y)}(G_s),\boldsymbol{G})\\ &\forall s\in\mathcal{S},\forall y\in\mathcal{Y}_s,\end{aligned} \tag{15}$$

where $p^{(y)}(G_s)$ is pure strategy function of ordinary satellite s that maps Ω_s to $\mathcal{A}_s$. The set of pure strategies of satellite s is $\mathcal{Y}_s \triangleq \{1,2,\cdots,|\mathcal{A}_s|^{|\Omega_s|}\}$, and $\mathcal{Y} = \mathcal{Y}_1 \times \cdots \times \mathcal{Y}_S$ is set of pure strategies for all satellites. The reason using pure strategy is that the probability mass function is a mixed strategy, which hardly achieve without prior knowledge of communication quality.

4 An Online Data Balancing Algorithm Based On Lyapunov

The use of general game solving methods, such as Reinforcement learning, Lagrange, metaheuristic algorithm, etc., does not guarantee the stabilization of the satellite energy queue in the context of obtaining a equilibrium. In this section, we convert the repeated stochastic non-cooperative game into virtual static game by using the pure strategy. Subsequently, using Lyapunov solves the virtual static game while guarantees stable energy queue of each satellite.

The decision variables $\boldsymbol{\beta},\boldsymbol{\alpha},\boldsymbol{e}$ are determined in each time according to the same $Pr[(\boldsymbol{\beta},\boldsymbol{\alpha},\boldsymbol{e})|\boldsymbol{G}]$ and random event $\boldsymbol{G}(t)$, the time average utility is denoted as

$$\bar{U}_s = \sum_{\boldsymbol{G}\in\boldsymbol{\Omega}}\sum_{\boldsymbol{\beta},\boldsymbol{\alpha},\boldsymbol{e}\in\mathcal{A}} \pi[\boldsymbol{G}]Pr[(\boldsymbol{\beta},\boldsymbol{\alpha},\boldsymbol{e})|\boldsymbol{G}]\hat{U}_s((\boldsymbol{\beta},\boldsymbol{\alpha},\boldsymbol{e}),\boldsymbol{G}), \tag{16}$$

where $\pi[\boldsymbol{G}] = Pr[\boldsymbol{G}(t) = \boldsymbol{G}]$, $\forall\boldsymbol{G}\in\boldsymbol{\Omega}$ is probability mass function of $\boldsymbol{G}(t)$.

The head satellite provides the strategies to achieve a fair allocation among ordinary satellites and minimize the utility. We define the fairness function as an decreasing and concave function, which is

$$\psi(\bar{U}_1,\bar{U}_2,\cdots,\bar{U}_S) = \sum_{s=1}^{S}\log(\bar{U}_s), \tag{17}$$

In the repeated stochastic game, the optimization problem can be represented as

$$(\mathcal{P}2) \quad \max_{Pr[(\boldsymbol{\beta},\boldsymbol{\alpha},\boldsymbol{e})|\boldsymbol{G}]} \lim_{t\to\infty} \inf \psi(\bar{U}_1, \bar{U}_2, \cdots, \bar{U}_S) \tag{18a}$$

$$\text{s.t.} \quad Pr[(\boldsymbol{\beta},\boldsymbol{\alpha},\boldsymbol{e})|\boldsymbol{G}] \geq 0, \forall \boldsymbol{G} \in \boldsymbol{\Omega}, \forall \boldsymbol{\beta},\boldsymbol{\alpha},\boldsymbol{e} \in \mathcal{A} \tag{18b}$$

$$\sum_{\boldsymbol{\beta},\boldsymbol{\alpha},\boldsymbol{e}\in\mathcal{A}} Pr[(\boldsymbol{\beta},\boldsymbol{\alpha},\boldsymbol{e})|\boldsymbol{G}] = 1, \forall \boldsymbol{G} \in \boldsymbol{\Omega} \tag{18c}$$

$$\bar{U}_s = \sum_{\boldsymbol{G}\in\boldsymbol{\Omega}} \sum_{\boldsymbol{\beta},\boldsymbol{\alpha},\boldsymbol{e}\in\mathcal{A}} \pi[\boldsymbol{G}] Pr[(\boldsymbol{\beta},\boldsymbol{\alpha},\boldsymbol{e})|\boldsymbol{G}] \hat{U}_s((\boldsymbol{\beta},\boldsymbol{\alpha},\boldsymbol{e}),\boldsymbol{G}) \tag{18d}$$

$$\text{CCE constraints (15) are satisfied} \tag{18e}$$

$$\text{Eq. (11b) to Eq. (11h)} \tag{18f}$$

The optimization problem for the repeated stochastic game is a convex program in the unknown $Pr[(\boldsymbol{\beta},\boldsymbol{\alpha},\boldsymbol{f})|\boldsymbol{G}]$. This is because that the satellites don't store information related to communication conditions considering the limited satellite's resources. Therefore, it's difficult to achieve the probability mass function $Pr[\boldsymbol{G}]$.

The utility $\bar{U}(t)$ can be seen as a real-valued stochastic process, which can be defined as

$$\bar{U}(t) \triangleq \frac{1}{t}\sum_{\tau=0}^{t-1} \mathbb{E}[(U(\tau))]. \tag{19}$$

Specifically, $U_s(t) \triangleq \hat{U}_s((\boldsymbol{\beta}(t),\boldsymbol{\alpha}(t),\boldsymbol{e}(t)),\boldsymbol{G}(t))$. $U_s^{(y)}(t) \triangleq \hat{U}_s^{(y)}((\boldsymbol{\beta}(t),\boldsymbol{\alpha}(t),\boldsymbol{e}(t)),\boldsymbol{G}(t))$, which is the utility of satellite s obtained by the pure strategy $p^{(y)}(G_s)$. Therefore, we have $\hat{U}_s^{(y)}((\boldsymbol{\beta}(t),\boldsymbol{\alpha}(t),\boldsymbol{e}(t)),\boldsymbol{G}(t)) \triangleq \hat{U}_s^{(y)}((p^{(y)}(G_s),(\boldsymbol{\beta}(t),\boldsymbol{\alpha}(t),\boldsymbol{e}(t))_{s^-}),\boldsymbol{G}(t))$. We can transform the $(\mathcal{P}2)$ into $(\mathcal{P}3)$ as follows:

$$(\mathcal{P}3) \quad \max_{Pr[(\boldsymbol{\beta},\boldsymbol{\alpha},\boldsymbol{e})|\boldsymbol{G}]} \lim_{t\to\infty} \inf \psi(\bar{U}_1(t), \cdots, \bar{U}_S(t)) \tag{20a}$$

$$\text{s.t.} \lim_{t\to\infty} \inf[\bar{U}_s(t) - \hat{U}_s^{(y)}(t)] \geq 0, \tag{20b}$$

$$\forall s \in \mathcal{S}, y \in \mathcal{Y}_s, (\boldsymbol{\beta},\boldsymbol{\alpha},\boldsymbol{e}) \in \mathcal{A}_s, t \in \mathcal{T} \tag{20c}$$

$$\text{Eq. (11b) to Eq. (11h)} \tag{20d}$$

It is proved in work [9]. If the probability mass function $Pr[(\boldsymbol{\beta},\boldsymbol{\alpha},\boldsymbol{e})|\boldsymbol{G}]$ can solve $\mathcal{P}2$, it must be able to solve $\mathcal{P}3$.

We then turn the nonlinear function of time average into the time average of nonlinear function by using auxiliary variable technique. Let $\xi(t) = (\xi_1(t), \cdots, \xi_S(t))$ be an auxiliary vector that the head satellite chooses at time slot t. And we define $J(t) \triangleq \psi(\xi_1(t), \cdots, \xi_S(t))$. The Jensen's inequality is

$$\bar{J}(t) \leq \psi(\bar{\xi}_1(t), \cdots, \bar{\xi}_S(t)) \tag{21}$$

Based on these vectors, problem $\mathcal{P}3$ can be transformed into $\mathcal{P}4$, which is denoted by

$$(\mathcal{P}4) \quad \max_{\xi_s(t),\boldsymbol{\beta},\boldsymbol{\alpha},\boldsymbol{e}} \ \liminf_{t\to\infty} \bar{J}(t) \tag{22a}$$

$$\text{s.t.} \lim_{t\to\infty} |\bar{\xi}_s(t) - \bar{U}_s(t)| = 0, \forall s \in \mathcal{S} \tag{22b}$$

$$\liminf_{t\to\infty} [\bar{U}_s(t) - \hat{U}_s^{(y)}(t)] \geq 0, \tag{22c}$$

$$\forall s \in \mathcal{S}, y \in \mathcal{Y}_s, (\boldsymbol{\beta}, \boldsymbol{\alpha}, \boldsymbol{e}) \in \mathcal{A}_s, t \in \mathcal{T} \tag{22d}$$

$$0 \leq \xi_s(t) \leq E_s, \forall t \in \mathcal{T}, s \in \mathcal{S} \tag{22e}$$

$$\text{Eq. (11b) to Eq. (11h)} \tag{22f}$$

The Jensen's inequality (21) ensures the optimal value of $\mathcal{P}4$ is less than or equal to that of $\mathcal{P}3$. Furthermore, $\mathcal{P}4$ can be achieved by choosing $\xi_s(t) = \bar{U}_s^*$ for all t, where $\bar{U}_1^*, \cdots, \bar{U_S}^*$ are optimal time average utilities for $\mathcal{P}3$. Therefore, The problem $\mathcal{P}3$ and $\mathcal{P}4$ are equivalent.

To enforce the constraints (22b), we define virtual queues $X_s^{(y)}(t)$ and $Z_s(t)$ for all satellites with update equation:

$$X_s^{(y)}(t+1) = \max[X_s^{(y)}(t) + U_s^{(y)}(t) - U_s(t), 0]. \tag{23}$$

$$Z_s(t+1) = Z_s(t) + \xi_s(t) - U_s(t). \tag{24}$$

Furthermore, we need to make the queues mean rate stable, which denoted as $\lim_{t\to\infty} \mathbb{E}[X_s^{(y)}(t)]/t = 0$ and $\lim_{t\to\infty} \mathbb{E}[Z_s(t)]/t = 0$, respectively.

Using Lyapunov optimization methods [10], we divide the problem $\mathcal{P}4$ into several sub-problems. The Lyapunov function is defined as

$$L(t) = \frac{1}{2}\sum_{s=1}^{S}\sum_{y\in\mathcal{Y}_s}[X_s^{(y)}(t)]^2 + \frac{1}{2}\sum_{s=1}^{S}[Z_s(t)]^2 + \frac{1}{2}\sum_{s=1}^{S}[\tilde{Q}_s(t)]^2, \tag{25}$$

where $\tilde{Q}_s(t) = \Lambda_s - Q_s(t)$ is the remaining energy in the satellite storage system of satellite s.

The algorithm is achieved by solving the following subproblems.

Auxiliary Variable :

$$\begin{aligned} &\max_{\xi_s(t)} \ V\psi(\xi_1(t), \cdots, \xi_S(t)) - \sum_{s=1}^{S} Z_s(t)\xi_s(t) \\ &\text{s.t.} \ \ 0 \leq \xi_s(t) \leq E_s. \end{aligned} \tag{26}$$

Offloading Decision Variables :

$$
\begin{aligned}
\min_{\alpha_{s,j}(t),\beta_{s,j}(t)} \quad & \sum_{s=1}^{S} X_s^{(y)}(t)E_s - U_s(t)\big[X_s^{(y)}(t) + Z_s(t)\big] \\
& - \big[X_s^{(y)}(t) + \tilde{Q}_s(t)\big]\Bigg[P_s^t \frac{\alpha_{s,j}(t)\beta_{s,j}(t)D_{s,c}(t)}{R_{s,j}(t)} \\
& + w_s(f_s(t))^2\big[(D_{s,c}(t) - \alpha_{s,j}(t)\beta_{s,j}(t)D_{s,c}(t))c_a \\
& + \sum_{i\in\mathcal{S}/\{s,j\}} \alpha_{i,s}(t)\beta_{i,s}(t)D_{i,c}(t)c_a\big]\Bigg] \\
\text{s.t.} \quad & \text{Eq. (11b) to Eq. (11f)}
\end{aligned}
\tag{27}
$$

Harvested Energy Variable :

$$
\begin{aligned}
\min_{e_s(t)} \quad & \sum_{s=1}^{S} \tilde{Q}_s(\tau)e_s(t) \\
\text{s.t.} \quad & \text{Eq. (11g) to Eq. (11h)}
\end{aligned}
\tag{28}
$$

5 Simulated Experiments

In this section, we evaluate the efficiency of the proposed algorithm using STK and Python.

5.1 Experimental Setup

Setting Parameters : The satellites in the satellite constellation are deployed at an altitude of 553 km and an inclination of 35 degrees. The constellation comprises 6 orbits, with each orbit hosting 15 satellites. And other computation parameters include $f_s = [100, 512]$ $cycle/s$, $P_s^t = [3, 10]$ W, $G_s = [0.5, 2]$, $D_{s,c}(t) = [50, 500]$ $kbit$, $e_s^{max} = [252, 300]$ W.

Comparison Algorithms : We compare the proposed algorithms with baselines. Multi-Hop Algorithm(MHA) [3]: each satellite offloads data to another satellite through multi-hop communication. Sing-Hop without Lyapunov(SH): each satellite offloads data to other satellite by sing-hop without using Lyapunov methods. No-offloading Algorithm(NA): each satellite processes data by itself. Random Selection Algorithm(RSA): each satellite randomly selects another satellite for offloading.

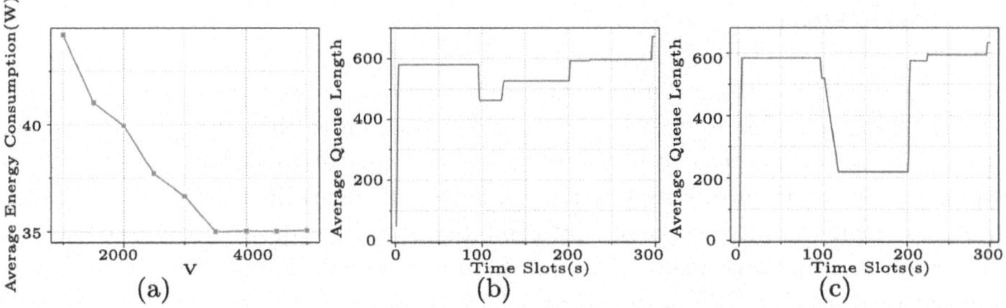

Fig. 2. Average energy consumption and average energy queue Length of LEO satellite cluster.

5.2 Performance Evaluation

Figure 2 illustrates the average energy consumption and average energy queue length of LEO satellite cluster. Figure 2(a) demonstrates the average energy consumption decreases with increasing value of V. In Fig. 2(b), the average energy queue length of the LEO satellite cluster in the light zone is depicted. The fluctuations in the line are attributed to the departure and joining of satellites. Figure 2(c) presents the average energy queue length of the LEO satellite cluster across both the light and shadow zones. When satellites in shadow zone, they maintenance a minimum energy queue to support their orbit. In conclusion, our proposed algorithm minimizes energy consumption while ensuring the stability of energy queue and adapting to the dynamic network environment.

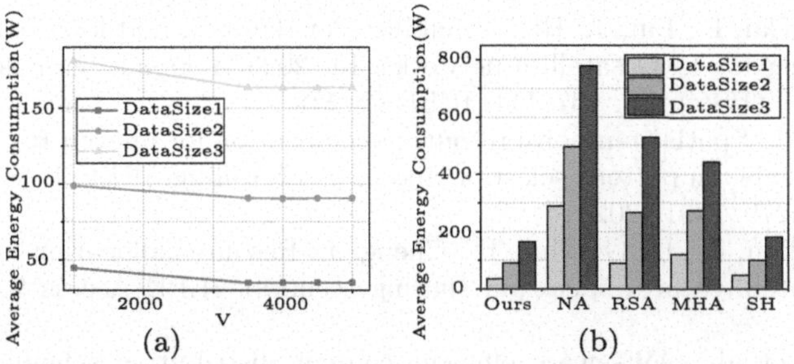

Fig. 3. Average energy consumption under different received data size.

Figure 3 shows the average energy consumption under different received data size, where DataSize1 < DataSize2 < DataSize3. Figure 3(a) illustrates the average energy consumption of our proposed algorithm under different received data size. The energy consumption increases with the increasing amount of data size. Figure 3(b) demonstrates the average energy consumption among different comparison algorithms. Our proposed algorithm reduces energy consumption by 61.49%, 56.22%, 55.29% compared to other algorithms on three DataSizes.

6 Conclusion

The LEO satellite cluster balance its data load when serving an ground area and meanwhile maximize energy utility of each satellite. Firstly, we constructed a system model to describe offloading process in cluster. Secondly, we developed a repeated stochastic non-cooperative game for LEO satellites within the cluster. Finally, we proposed the on-orbit data balancing online algorithm based on Lyapunov theory to achieve the CCE equilibrium, maintaining stable energy queue while performing data offloading for each satellite.

Acknowlegments. This work was supported in part by the National Natural Science Foundation of China under Grant 62372163, Natural Science Foundation of Chongqin under Grant ZE20210011, Key scientific and technological research and development plan of Hunan Province under Grant 2022NK2046, Yuelushan Industrial Innovation Center Cultivation Project under Grant 2013YC110130.

References

1. Hassan, S.S., Park, Y.M., Tun, Y.K., Saad, W., Han, Z., Hong, C.S.: Satellite-based its data offloading & computation in 6G networks: a cooperative multi-agent proximal policy optimization DRL with attention approach. IEEE Trans. Mob. Comput. **23**, 4956–4974 (2023)
2. Cui, G., Duan, P., Lexi, X., Wang, W.: Latency optimization for hybrid geo-LEO satellite-assisted IoT networks. IEEE Internet Things J. **10**(7), 6286–6297 (2022)
3. Zhang, X., et al.: Energy-efficient computation peer offloading in satellite edge computing networks. IEEE Trans. Mob. Comput. **23**, 3077–3091 (2023)
4. Wang, Y., Cai, L., Liu, J.: High-reliability, low-latency, and load-balancing multi-path routing for LEO satellite networks. In: 2023 Biennial Symposium on Communications (BSC), pp. 107–111. IEEE (2023)
5. Lei, L., et al.: Spatial-temporal resource optimization for uneven-traffic LEO satellite systems: beam pattern selection and user scheduling. IEEE J. Sel. Areas Commun. **42**, 1279–1291 (2023)
6. Tang, J., Bian, D., Li, G., Jing, H., Cheng, J.: Resource allocation for LEO beam-hopping satellites in a spectrum sharing scenario. IEEE Access **9**, 56468–56478 (2021)
7. Jiang, D., et al.: QoE-aware efficient content distribution scheme for satellite-terrestrial networks. IEEE Trans. Mob. Comput. **22**(1), 443–458 (2021)
8. Li, X., Huang, Q., Wu, D.: A repeated stochastic game approach for strategic network selection in heterogeneous networks. In: IEEE INFOCOM 2018-IEEE Conference on Computer Communications Workshops (INFOCOM WKSHPS), pp. 88–93. IEEE (2018)
9. Neely, M.J.: A lyapunov optimization approach to repeated stochastic games. In: 2013 51st Annual Allerton Conference on Communication, Control, and Computing (Allerton), pp. 1082–1089. IEEE (2013)
10. Tang, Q., et al.: Stochastic computation offloading for LEO satellite edge computing networks: a learning-based approach. IEEE Internet Things J. **11**, 5638–5652 (2023)

Resource-Limited Network Security Games with General Contagious Attacks

Rufan Bai[1], Chao Xu[2], Chenyang Xu[3], and Ruilong Zhang[4(✉)]

[1] School of Computer Science and Engineering, Southeast University, Nanjing, China
rfbai@seu.edu.cn

[2] School of Computer Science and Engineering, University of Electronic Science and Technology of China, Chengdu, China
the.chao.xu@gmail.com

[3] Shanghai Key Laboratory of Trustworthy Computing, Software Engineering Institute, East China Normal University, Shanghai, China
cyxu@sei.ecnu.edu.cn

[4] Department of Computer Science, City University of Hong Kong, Hong Kong, China
ruilzhang4-c@my.cityu.edu.hk

Abstract. In classic security games on the graph with contagious attacks, there is a defender and an attacker. The defender first distributes defending resources to the nodes of the graph, and each unit of defending resource incurs a cost. The attacker picks a node to attack to maximize the damage after the transmission of the attack. Generally, the attacker has two attack types—uniform attack and adaptive attack, where the uniform attacker attacks each node with uniform and the adaptive attacker can select one node to attack according to the defender's strategy. The usual objective in the literature is to minimize the total loss of the defender, including the damage caused by the attacker and the paid cost. However, we notice that in many real-world applications, the defending resources are often limited, and the existing objective fails to capture these scenarios.

The paper handles this issue by considering the resource-limited setting of the security game, where there is a given upper bound on the paid resource cost and the defender aims to minimize the caused damage without violating the resource cost constraint. We first discuss the hardness, proving that for both attack types, the problem is NP-hard on general graphs. Then, we focus on bounded treewidth graphs, a well-known graph class that includes many common graphs. We show that given any graph with bounded treewidth and any attack type, a dynamic programming algorithm always exists that solves the problem in polynomial time. Further, the algorithm can be easily extended to more general objective settings. The proposed algorithm is built on an interesting connection to a graph partition problem, which could be intrigued independently.

All authors (ordered alphabetically) have equal contributions and are corresponding authors.

Y. Chen et al. (Eds.): COCOON 2024, LNCS 15162, pp. 79–91, 2025.
https://doi.org/10.1007/978-981-96-1093-8_7

Keywords: Network Security Games · Bounded Treewidth · Dynamic Programming · Graph Partition · NP-hardness

1 Introduction

Recently, network security games have attracted much research attention within the artificial intelligence community and have been widely adopted for the computation of optimal allocation of security resources in many areas [5,7,9,17,23,30]. In a network security game, there is an underlying graph, where each node of the graph represents a target with a defending requirement and a value to protect. The game is played between a defender who allocates defensive resources to the nodes of the graph and an attacker who picks a node to attack, depending on how the nodes are defended.

Many existing works consider the network security games with the contagious attack [4,6,7,18]. In these games, the attack on a node u can spread to its neighbors. We need to count the damage caused by this attack on multiple nodes, rather than only considering the damage on a single node. In previous works, there exists mainly three deficiencies, which is summarized as follows.

Resource Supply Quantity. Aspnes et al. [4] consider a network security game model in which each node is attacked with equal probability. The node in the graph will suffer a loss if it is not covered by a defending resource. The spread level is infinity in their model, which means the propagation process will persist as long as there are undefended nodes expanding outward from the initially attacked node. Kumar et al. [18] study more generalized setting in which the defending requirement and value of each node can be different. And they also explore the results at different spread level κ, which is ranging from 1 to infinity. The aforementioned studies take into account contagious attacks but overlook the constraints on available resources. In some real-world applications, such as defending against COVID-19 virus, which is highly contagious virus, the resources like doctors and masks are limited and we need to design an allocation strategy under the resource constraint. Therefore, the existing results designed for scenarios without resource constraints are not applicable.

Spread Mode. Bai et al. [7] study the network security game with contagious attacks and consider the resource constraint. But in their model, the attack will spread to neighbors no matter how the node under attack is well-defended. To put it differently, they do not take into account the fact that a well-defended node can impede the spread of the attack. In network security games or the application of inoculation, it is more reasonable that attacks may propagate from other nodes only if they are successfully attacked.

Attack Types. Note that the attack will only happen once in these works. But there are two attack types, which are uniform attack and adaptive attack. The uniform attacker will attack each node in the graph with equal probability, which is applied in the works of Aspnes et al. [4] and Kumar et al. [18]. Differently, the adaptive attacker leverages the ability of observing the defending strategies and then chooses one node to attack. This category of games is referred to as Stackelberg security games (**SSG**), which is studied in the research by Bai et al. [7] and Bachrach et al. [6] et al. We note that the above studies primarily focus on either the uniform attack or the adaptive attack. They tend to neglect the alternate attack type, and it is important to highlight that the outcomes in these two scenarios diverge significantly. Therefore, the results and algorithm in one scenario may not be directly applicable to the other.

In this paper, we propose Resource-limited Network Security Games (**RNSG**), which comprehensively consider all three aspects mentioned above: the limitation of resources, a more reasonable spread mode and both attack types. In a network graph denoted as $G(V, E)$, the objective is to allocate limited defending resources strategically among the nodes in the graph to defend against attackers with contagious attacks in both attack types. Note that the well-defended node will obstruct the propagation of the attack in our model. We give the following example to illustrate the differences from previous works.

Example 1. Given a star graph with n nodes and the defending requirement and value of each node is 1. Given 1 unit of resource, the best allocation strategy for our problem is to put the resource to the center node to split the graph to $n-1$ disconnected components. In this example, for our model, we only consider the adaptive attack so that the loss is always 1 no matter which node is under attack.

But for the problem model in [7], they did not consider that the well-defended node can block the propagation of the attack. In other words, they assume that the attack will spread to the neighbor no matter how the node under attack is whether defended. Hence, the loss will be at least $n - 1$.

Regarding the model from Aspnes et al. [4] and Kumar et al. [18], they considered the total cost of all disconnected components caused by the attacker and the attacker is assumed to attack each node with equal probability $1/n$. So the loss is $1 + 1/n \cdot 1 \cdot (n - 1) = 2 - 1/n$, where the first 1 is the resource cost (they did not consider the limited resource).

1.1 Our Contributions

We study the problem of resource-limited network security games by considering both the uniform and adaptive attackers. We first propose a novel model for network security games with contagious attacks. Our model considers the setting of limited defense resources and assumes that if a node is well-defended, it can block the spread process of the attack when the attack spreads to this node. We show that the problem is NP-hard on general graphs with both attack types, while for two natural graph classes, the optimal solution can be obtained in polynomial time. In the following, we summarize the main results.

Main Result 1. Given an arbitrary graph, it is NP-hard to compute the optimal solution for the problem of resource-limited network security games under both uniform and adaptive attacks.

Our first result is on the hardness of the studied problem. We show that our problem is NP-hard on general graphs with both attack types by giving a reduction from the classical vertex cover problem. The reductions for uniform and adaptive attacks follow the same instance construction method, but the correctness relies on the different properties of the reduction. Furthermore, our hardness proof works for a more general spread level, i.e., a virus that successfully infects a particular machine will only spread to its κ-level-neighbor machines. Specifically, the virus can only infect the machines that are at most κ away from the current machine. Due to space limits, this part is omitted in this version.

Main Result 2. Given any graph with bounded treewidth, there is a polynomial time algorithm that computes the optimal solution to resource-limited network security games under both uniform and adaptive attacks.

Given the hardness result on general graphs, we turn to consider a graph class called *graphs with bounded treewidth (*tw*)*. This graph class includes many other well-known graphs, such as trees ($\mathsf{tw} = 1$), series-parallel graphs ($\mathsf{tw} = 2$), outerplanar graphs ($\mathsf{tw} = 3$), and so on. As a warm-up case, we first present our dynamic programming on trees (Sect. 3) and then extend it to graphs with bounded treewidth. We remark that our algorithm has much better running time on trees compared to graphs with bounded treewidth. To build the above result, we observe that RNSG with uniform and adaptive attacks have a strong connection with a fair graph partition problem (see Sect. 2.2), which aims to remove at most R nodes such that the size of each remaining connected component is balanced. We then present dynamic programming for this general fair graph partition problem. The correctness of our dynamic programming relies on the nice properties tree decomposition of the given graph.

Main Result 3. Given any graph with bounded treewidth, there is a polynomial time algorithm that computes the optimal solution to the network security games considered by Aspnes et al. [4]. Moreover, the fair graph partition problem with knapsack constraint can be solved in pseudo-polynomial time.

The last part of our result is an extension of our model and algorithm (omitted in this version). Let the network security games problem considered by Aspnes et al. [4], denote by ANSG. The main difference between ANSG and RNSG are the following two points: (i) ANSG includes the resources constraint in its objective, i.e., ANSG has unlimited resources; (ii) ANSG considers the uniform attacker. Based on the second main result, we can apply the standard searching technique to solve ANSG. This implies that the problem studied in [4] is polynomially solvable on graphs with bounded treewidth. To capture more applications, we further extend our model to include the knapsack constraints in which each node v shall be associated with a value (g_v) and a weight (w_v). The goal is to select a set of resource-feasible vertices such that the value of each remaining connected component is balanced, i.e., the total weight of the selected vertices is not allowed to exceed the given resource R. For this general problem, we show that our dynamic programming can also be applied, i.e., the balanced graph partition problem with ℓ_p-norm objective and knapsack constraint can be solved in pseudo-polynomial time on bounded treewidth graphs.

1.2 Other Related Works

Network Security Games. Network security games have also received a considerable attention in recent years. There are some works study the contagious attacks in network security games [1,7,10,12,25,28]. Acemoglu et al. [1] study the security investments in a network of agents and compute the equilibrium in both attack types, but they mainly focus on the centrality of the network structure. Bai et al. [7] consider the network security games with the resource constraint when solving for the optimal defense strategy. They, however, did not take into account that well-defended nodes can block the propagation of the attack. Gan et al. [16] and Li et al. [20] model the network security games as a graph with resource sharing, where the resource can be reallocated between neighboring nodes according to the attack. Yin et al. [28] study a network security games model that the resource can be transferred but transferring resources takes time. Moreover, there are some other works consider the mixed strategy for network security games [8,14,26,27].

Graph Partition. There are most existing works study the graph partition problem [2,3,11,13,19]. Andreev et al. [2] and Feldmann et al. [15] study the (k, v)-balanced graph partitioning problem that dividing the nodes of a graph into k almost equal size components (each of size less than $v \cdot \frac{n}{k}$). Andreev et al. focus on the approximation algorithm for any constant $v > 1$ and Feldmann et al. give some results on trees and related application. But they do not consider the node-cut graph partition, which is more similar to our problem. Moreover, the objective of balanced graph partition aims to minimize the differences in size between different components as much as possible, which is slightly different from that of our model. Leighton et al. [19] study the node-cut graph partition problem and establish max-flow min-cut theorems for several important classes of multicommodity flow problems. However, they overlook the impact of quantity constraints on graph partitioning. There are also some works that utilize the technique of deep learning to solve the graph partition problem [21,22,24,29].

2 Preliminaries

2.1 Problem Definition

Consider a graph $G = (V, E)$ with n nodes and m edges. Our resource-limited network security game (**RNSG**) consists of two sequential phases: First, the defender decides her defense strategy under a total resource of R—that is, she selects at most R nodes in the graph and marks them as *well-defended* or *defended* for short. Then, the attacker picks a starting node v to attack, and causes a loss $\mathsf{L}(v)$ to the defender. The loss incurred from an attack is influenced by the defense strategy (see Definition 1), and our goal is to find a defense strategy that minimizes the loss caused by the attacker.

Definition 1 (Loss). *When the attacker attacks a node v, the attack can spread across the graph starting from v. Once the attack reaches a node $u \in V$, it spreads* *to all the nodes in its neighbors $N(u)$ if u is not marked as defended. Given a defense strategy, the loss $\mathsf{L}(u)$ incurred from the starting node u is the total number of undefended nodes that are attacked as the attack spreads.*

As mentioned above, there are two types of attackers: *uniform attacker* and *adaptive attacker* in the literature. A uniform attacker attacks each node in the graph with equal probability $1/n$ and causes an expected loss of $\frac{1}{n}\sum_{u\in V}\mathsf{L}(u)$, while an adaptive attacker can adapt based on the defense strategy and pick the node u with the maximum $\mathsf{L}(u)$, causing a loss of $\max_{u\in V}\mathsf{L}(u)$. We refer to our security game model under these two attack types as **RNSG-U** and **RNSG-A**, respectively.

The paper considers both of the two models. Instead of treating them independently, we introduce a general fair graph partition problem that can capture RNSG-U and RNSG-A simultaneously.

2.2 General Fair Graph Partition

This subsection first gives the formal definition of the graph partition problem and then discusses the connections.

Definition 2 (Fair Graph Partition). *Given an undirected graph $G = (V, E)$ with n nodes and an integer $R \leq n$, let $\mathbb{C}(S)$ be the set of connected components C after removing a node subset $S \subseteq V$ from G. The goal is to select a subset S of at most R nodes such that the ℓ_p norm term $\left(\sum_{C\in\mathbb{C}(S)}(|C|)^p\right)^{\frac{1}{p}}$ is minimized.*

Observation 1 *Given a defense strategy for the RNSG problem, the loss $\mathsf{L}(v)$ of each defended node v is 0 while the loss $\mathsf{L}(u)$ of each undefended node u is the size of the connected component that contains u after removing all the defended nodes from the graph.*

The above observation establishes the strong connection between RNSG and the fair graph partition problem.

(ℓ_∞-norm vs Adaptive Attack) We claim that RNSG-A is equivalent to the fair graph partition problem with the objective of minimizing the ℓ_∞-norm $\max_{C\in\mathbb{C}(S)}|C|$. To see the connection, assume that we are given any feasible solution $O \subseteq V$ to the fair graph partition problem. Then, we shall assign a unit of resource to each node in O. According to Observation 1, the adaptive attacker will attack a node that belongs to the largest connected component after removing O. Thus, the total loss of the assignment is exactly equal to the objective value of solution O to the fair graph partition problem.

(ℓ_2-norm vs Uniform Attack) We claim that RNSG-U shares the same optimal solution as the fair graph partition problem with the objective of minimizing the squared ℓ_2-norm $\left(\sum_{C\in\mathbb{C}(S)}(|C|^2)\right)^{1/2}$. Recall that the goal in the RNSG-U model is to minimize the expected loss $\frac{1}{n}\sum_{v\in V}\mathsf{L}(u)$. Consider any defense strategy for RNSG-U and denote by Q the set of defended nodes. Assume that there are k connected components, denoted by $C_1, \ldots, C_k$, after removing the node set Q from the graph. According to Observation 1, the expected loss is $\frac{1}{n} \cdot \sum_{i \in [k]} |C_i|^2$. Clearly, these two problems share the same optimal solution.

Example 2. There is a graph $G = (V, E)$ with 6 nodes, as shown in the Fig. 1(a). Given 2 units of resources, we study how to evaluate the objective under different defending strategies.

When we allocate the resources on nodes a and b, we can remove nodes a and b, as well as the edges incident to them. Now the original graph is split into two components, which are $S_1 = \{d\}$, $S_2 = \{c, e, f\}$. The adaptive attacker will choose one node from S_2, and the attack will spread to all nodes in the component S_2. So the objective will be 3.

When we protect nodes a and d, the objective will be 4 since the adversarial attacker will attack one node in the component $\{b, c, d, e\}$.

If nodes b and e are well-defended, the original graph is split into two components, which are $S_1 = \{a, d\}$ and $S_2 = \{c, f\}$ (As shown in the Fig. 1(b)). The loss will be at most 2 no matter which node the attacker attacks. In this example, the optimal objective $OPT = 2$.

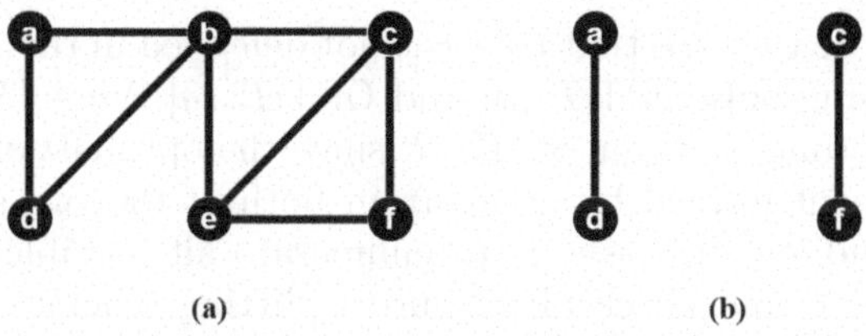

Fig. 1. An example for showing the loss due to an attack. (b) shows the attack graph, which is obtained from splitting the original graph after removing nodes b and e (nodes b, e are covered by a resource, respectively).

From this example, we observe that a better-defense strategy is to protect nodes in order to partition the graph into several components, each of which has nearly equal numbers of nodes.

Recall that the objective of RNSG is to distribute limited resources R to minimize the defense result, which can be transformed into the graph partition problem, i.e., removing R nodes to minimize the size of the largest component. In the latter section, the proposed algorithms and techniques are built on a connection to the graph partition problem, which could be intrigued independently.

3 Warm-Up: RNSG-A on Binary Trees

As a warm-up, this section considers RNSG-A on binary trees and shows that in this special case, the problem can be solved in polynomial time. Given a binary tree $T = (V, E)$ rooted at r and the total resource R, we can always assume that $R < n$ where n is the number of nodes because when $R \geq n$, all nodes can be defended and the problem is trivial. When the tree has a constant height, the optimal solution can be computed easily through brute force enumeration; therefore, if we are able to divide the problem on T into several subproblems on its subtrees recursively, the problem can be solved.

Use T_v and $|T_v|$ to denote the subtree rooted at v for each node $v \in V$ and the number of nodes in it respectively. Let $[T_v, q]$ be the subproblem on the subtree T_v with total resource $q \in [R]$ and $\mathsf{OPT}[T_v, q]$ be the optimal objective of subproblem $[T_v, q]$. The intuition of the dynamic programming algorithm is the following. For ease of description, we w.l.o.g. assume that the tree is full binary, where each non-leaf node has exactly two children.

Algorithmic Intuition. Let node u and w be the two children of root r. If root r is defended in the optimal solution of $[T_r, R]$, the problem is easy to divide. We can directly remove r from the graph, getting two subtrees T_u, T_w, and decrease the total resource to $R-1$. By enumerating all resource-splitting schemes among the two subtrees, we can construct an optimal solution of $[T_r, R]$ easily:

$$\mathsf{OPT}[T_r, R] = \min_{0 \leq q \leq R-1} \max \{\mathsf{OPT}[T_u, q], \mathsf{OPT}[T_w, R-1-q]\}.$$

However, it might be the case that root r is not defended in the optimal solution. In this case, knowing only $\mathsf{OPT}[T_u, q]$ and $\mathsf{OPT}[T_w, q]$ ($\forall q \in [R]$) is insufficient to calculate the optimal solution of $[T_r, R]$ since the optimal solution contains a connected component formed by nodes from both of the two subtrees. A trivial adaptation handling this issue is to enumerate all possible connected components containing r and check all resource-splitting schemes among the subtrees obtained by removing v's connected component. But this will lead to an exponential-time algorithm. To improve the running time, we claim that only the possible sizes of the connected component containing r need to be enumerated, not the specific nodes contained in the component. This is due to the connectivity property of a tree.

Let us formally state the algorithm. We use the following notation: (i) $[T_v, q, 0]$ represents the subproblem where we are required to defend node v and use $q-1$ remaining resources to defend other nodes; (ii) $[T_v, q, \ell]$ ($q \in [R], 1 \leq \ell \leq |T_v|$) represents the subproblem where we must first select and remove a connected component $C \subseteq T_v$ of size ℓ that contains v, and then use a total resource of q to defend the remaining graph. Clearly, the total number of subproblems is at most n^3. Based on the notion above, we define substate $\mathcal{S}[T_v, q, \ell]$ in the dynamic programming as follows. If $q > 0$, its value is the optimal objective of the subproblem $[T_v, q, \ell]$, i.e., the minimax connected component size after removing C and the defended nodes. If $q = 0$, we treat it as a special case since no resources are available. We enforce ℓ to take on either $|T_v|$ or 0 by letting the value be $|T_v| - \ell$ if $\ell \in \{|T_v|, 0\}$ and ∞ otherwise.

According to the definition, one can observe that $\mathcal{S}[\cdot]$ does not take the size of the connected component containing v into account when $\ell > 0$. Thus, we

---

Algorithm 1 State Transition for RNSG-A on Binary Trees

Input: A subtree T_v, a total resource of q and parameter ℓ.
Output: $\mathcal{S}[T_v, q, \ell]$.

if $\ell > |T_v|$ **then return** ∞
else if $q = 0$ **then**
  return $|T_v| - \ell$ if $\ell \in \{|T_v|, 0\}$ and ∞ otherwise.
else if v is a leaf node **then**
  return 0.
else
  Let node u, w be the two children of v.
  if $\ell = 0$ **then**
    /* When v is defended, enumerate all possible resource-splitting schemes. */
    return $\min_{0 \le q' \le q-1} \max \left\{ \begin{array}{c} \min_{0 \le j \le |T_u|} \max\{j, \mathcal{S}[T_u, q', j]\}, \\ \min_{0 \le j \le |T_w|} \max\{j, \mathcal{S}[T_w, q-1-q', j]\} \end{array} \right\}.$
  else
    /* When v is not defended, enumerate all possible resource-splitting schemes and component-size-splitting schemes. */
    return $\min_{0 \le q' \le q,\ 0 \le j \le \ell-1} \max \left\{ \begin{array}{c} \mathcal{S}[T_u, q', j], \\ \mathcal{S}[T_w, q-q', \ell-1-j] \end{array} \right\}$
  end if
end if

---

need to take the maximum of ℓ and $\mathcal{S}$ when backing to the original problem. More precisely, using OPT to denote the original problem's optimal objective, we have $\mathsf{OPT} = \min_{\ell \in [n]} \max\{\ell, \mathcal{S}[T_v, q, \ell]\}$.

Algorithm 1 describes the state transition algorithm of the dynamic programming. We claim the following.

Theorem 1. *Algorithm 1 computes $\mathcal{S}[T_v, q, \ell]$ correctly and the state transition time is $O(n^2)$.*

The proof is omitted in this version. Since the total number of subproblems is $O(n^3)$, Theorem 1 proves a polynomial-time algorithm for the binary tree case.

4 RNSG-A on Bounded Treewidth Graphs

This section considers the RNSG-A problem on graphs with bounded treewidth and generalizes the algorithmic idea in the last section. Given a graph $G = (V, E)$, use d to represent its treewidth and $\mathbf{T} = (\mathbf{V}, \mathbf{E})$ to represent its nice tree decomposition, where each vertex $\mathbf{v} \in \mathbf{V}$ corresponds to a node subset $X(\mathbf{v}) \subseteq V$ in the original graph. For each vertex $\mathbf{v} \in \mathbf{V}$, let $\mathbf{T_v}$ be the subtree rooted at $\mathbf{v}$. Denote by $X(\mathbf{T_v}) := \bigcup_{\mathbf{u} \in \mathbf{T_v}} X(\mathbf{u})$ the union of the corresponding node subsets in subtree $\mathbf{T_v}$, and $G[\mathbf{T_v}]$ the subgraph of the original graph G induced by $X(\mathbf{T_v})$.

Algorithmic Intuition. Recall that in the binary tree case, we enumerate the possible size of the connected component containing the root for each subtree. We want to apply a similar idea to the nice tree decomposition of the given graph. A nice tree decomposition also admits a binary tree structure; but the main challenge is that the root $\mathbf{v}$ of each subtree $\mathbf{T_v}$ is essentially a node set $X(\mathbf{v})$, rather than a single node. Thus, in order to capture the optimal substructure, we have to first enumerate all possible partitions of $X(\mathbf{v})$ and then the possible connected component sizes for each group in each partition. Since the graph has a bounded treewidth, this enumeration takes polynomial time.

Now we state the algorithm formally. For ease of description, we first introduce a tuple for each vertex $\mathbf{v} \in \mathbf{T}$ and then use them to construct dynamic programming substates. Define $\Gamma := [\mathbf{v}, q, A_0, (A_1, n_1), (A_2, n_2), ..., (A_{d+1}, n_{d+1})$ for each $\mathbf{v} \in \mathbf{T}$, $q \in [R]$, $\{A_i \subseteq X(\mathbf{v})\}_{i \in [d+1]}$ and $\{n_i \leq n\}_{i \in [d+1] \setminus \{0\}}$. Intuitively, we are trying to enumerate all possible partitions[1] by $\{A_i\}$ and all possible connected component size configurations by $\{n_i\}$. Say a tuple is feasible if it satisfies the following constraints:

1. **(Partition Feasibility)** $\{A_i\}_{i \in [d+1]}$ forms a partition of $X(\mathbf{v})$, i.e., these sets[2] do not intersect with each other, and their union is exactly $X(\mathbf{v})$;
2. **(Component Size Feasibility)** For each $i = 1, 2, ..., d+1$, if $A_i = \emptyset$, $n_i = 0$; otherwise, n_i is an integer at least $|A_i|$; $\sum_{i=1}^{d+1} n_i \leq \max\{0, |X(\mathbf{T_v})| - q\}$;
3. **(Resource Feasibility)** $|A_0| \leq q$.

Note that we can always assume that R is at most the number of nodes n; otherwise, we can directly defend all nodes. Thus, we have Lemma 1.

Lemma 1. *The number of feasible tuples is* $O(d \cdot (d+2)^{(d+1)} \cdot n^{d+3})$, *and the feasibility check of a tuple takes* $O(d)$ *time.*

The proof is omitted in this version. We use feasible tuples to enumerate all the partitions and connected component size configurations that could arise from feasible defending strategies. However, although a tuple is feasible, there may still not exist a defense strategy resulting in it due to the given graph's structure. Thus, we further define the validity of tuples. Say a feasible tuple is valid if there exists a valid node subset B in the induced graph $G[\mathbf{T_v}]$ that satisfies the following:
(Resource Validity) $A_0 = B \cap X(\mathbf{v})$, $|B| \leq q$;
(Defense Validity) After removing (defending) the nodes in B from $G[\mathbf{T_v}]$ (i.e., in graph $G[\mathbf{T_v}] \setminus B$), each A_i $(i \geq 1)$ belongs to a connected component with size n_i, and any two subsets A_i, A_j is disconnected.

Now we define a state $\mathcal{S}$ for each tuple Γ For completeness, if the tuple is infeasible or invalid, let $\mathcal{S}(\Gamma) := \infty$. When a tuple is valid, denote by $\mathcal{B}$

[1] Note that A_0 represents the defended nodes and the reason for using a maximum index of $d+1$ is that we have at most $d+1$ nodes in $X_\mathbf{v}$ due to treewidth's definition.
[2] Note that A_i is allowed to be empty.

the set of all its valid B's, and for each $B \in \mathcal{B}$, define $f_b(\Gamma)$to be the largest size among all the connected components in $G[\mathbf{T_v}] \setminus B$ that are disjoint with $X(\mathbf{v}) = \bigcup_{i \in [d+1]} A_i$. Then the state $\mathcal{S}$ is defined by the best $B \in \mathcal{B}$: $\mathcal{S}(\Gamma) = \min_{B \in \mathcal{B}} f_B(\Gamma)$.

One may be concerned that the validity check of a tuple may need exponential time when the induced subgraph $G[\mathbf{T_v}]$ is large. We remark that the final dynamic programming does not need to perform any exponential validity check on large subgraphs. The State Transition Algorithm of our dynamic programming is omitted in this version. For the sake of clarity, we use $\mathbb{C} \preceq \mathbb{C}'$ for two set collections $\mathbb{C}, \mathbb{C}'$ to represent that for any set $C \in \mathbb{C}$, there always exists a set $C' \in \mathbb{C}'$ such that $C' \supseteq C$; and use $\mathbb{C} \npreceq \mathbb{C}'$ otherwise.

Theorem 2. *There is an algorithm that computes $\mathcal{S}(\Gamma)$ correctly and the state transition time is $O((d+1)^{d+2} \cdot n^{d+1})$.*

Combining Lemma 1 and Theorem 2, we obtain a polynomial-time dynamic programming algorithm for the RNSG-A problem on bounded treewidth graphs. The proof of Theorem 2 is omitted in this version.

Acknowledgments. Chenyang Xu is supported by the National Key Research Project of China under Grant No. 2023YFA1009402, the National Natural Science Foundation of China (No. 62302166), the Scientific and Technological Innovation 2030 Major Projects under Grant 2018AAA0100902, the Shanghai Science and Technology Commission under Grant No. 20511100200, the Dean's Fund of Shanghai Key Laboratory of Trustworthy Computing, ECNU, and the Key Laboratory of Interdisciplinary Research of Computation and Economics (SUFE), Ministry of Education. Rufan Bai is supported by the National Natural Science Foundation of China (No. 62203123).

References

1. Acemoglu, D., Malekian, A., Ozdaglar, A.E.: Network security and contagion. J. Econ. Theory **166**, 536–585 (2016)
2. Andreev, K., Räcke, H.: Balanced graph partitioning. In: SPAA, pp. 120–124. ACM (2004)
3. Arora, S., Rao, S., Vazirani, U.: Expander flows, geometric embeddings and graph partitioning. J. ACM (JACM) **56**(2), 1–37 (2009)
4. Aspnes, J., Chang, K.L., Yampolskiy, A.: Inoculation strategies for victims of viruses and the sum-of-squares partition problem. J. Comput. Syst. Sci. **72**(6), 1077–1093 (2006)
5. Assimakopoulos, N.: A network interdiction model for hospital infection control. Comput. Biol. Med. **17**(6), 413–422 (1987)
6. Bachrach, Y., Draief, M., Goyal, S.: Contagion and observability in security domains. In: Allerton, pp. 1364–1371. IEEE (2013)

7. Bai, R., Lin, H., Yang, X., Wu, X., Li, M., Jia, W.: Defending against contagious attacks on a network with resource reallocation. In: AAAI, pp. 5135–5142. AAAI Press (2021)
8. Bai, R., Lin, H., Yang, X., Wu, X., Li, M., Jia, W.: Mixed strategies for security games with general defending requirements. In: IJCAI, pp. 46–52 (2022). https://www.ijcai.org/
9. Bai, R., Lin, H., Yang, X., Wu, X., Li, M., Jia, W.: Stackelberg security games with contagious attacks on a network: reallocation to the rescue. J. Artif. Intell. Res. **77**, 487–515 (2023)
10. Baldwin, A., Gheyas, I., Ioannidis, C., Pym, D.J., Williams, J.: Contagion in cyber security attacks. J. Oper. Res. Soc. **68**(7), 780–791 (2017)
11. Caragiannis, I., Micha, E., Shah, N.: A little charity guarantees fair connected graph partitioning. In: AAAI, pp. 4908–4916. AAAI Press (2022)
12. Cerdeiro, D., Dziubinski, M., Goyal, S.: Individual security, contagion, and network design. J. Econ. Theory **170**, 182–226 (2017)
13. Chen, G., Chen, Y., Chen, Z., Lin, G., Liu, T., Zhang, A.: Approximation algorithms for the maximally balanced connected graph tripartition problem. J. Comb. Optim. **44**(3), 1753–1773 (2022)
14. Durkota, K., Lisý, V., Bosanský, B., Kiekintveld, C.: Optimal network security hardening using attack graph games. In: IJCAI, pp. 526–532. AAAI Press (2015)
15. Feldmann, A.E., Foschini, L.: Balanced partitions of trees and applications. Algorithmica **71**(2), 354–376 (2015)
16. Gan, J., An, B., Vorobeychik, Y.: Security games with protection externalities. In: AAAI, pp. 914–920. AAAI Press (2015)
17. Gan, J., An, B., Vorobeychik, Y., Gauch, B.: Security games on a plane. In: AAAI, pp. 530–536. AAAI Press (2017)
18. Kumar, V.S.A., Rajaraman, R., Sun, Z., Sundaram, R.: Existence theorems and approximation algorithms for generalized network security games. In: ICDCS, pp. 348–357. IEEE Computer Society (2010)
19. Leighton, T., Rao, S.: Multicommodity max-flow min-cut theorems and their use in designing approximation algorithms. J. ACM (JACM) **46**(6), 787–832 (1999)
20. Li, M., Tran-Thanh, L., Wu, X.: Defending with shared resources on a network. In: AAAI, pp. 2111–2118. AAAI Press (2020)
21. Liu, Q., Xia, T., Cheng, L., van Eijk, M., Ozcelebi, T., Mao, Y.: Deep reinforcement learning for load-balancing aware network control in IoT edge systems. IEEE Trans. Parallel Distrib. Syst. **33**(6), 1491–1502 (2022)
22. Mofrad, M.H., Melhem, R., Hammoud, M.: Revolver: vertex-centric graph partitioning using reinforcement learning. In: 2018 IEEE 11th International Conference on Cloud Computing (CLOUD), pp. 818–821. IEEE (2018)
23. Schlenker, A., et al.: Deceiving cyber adversaries: a game theoretic approach. In: AAMAS, pp. 892–900 (2018)
24. Tanaka, M., Taura, K., Hanawa, T., Torisawa, K.: Automatic graph partitioning for very large-scale deep learning. In: 2021 IEEE International Parallel and Distributed Processing Symposium (IPDPS), pp. 1004–1013. IEEE (2021)
25. Tsai, J., Nguyen, T.H., Tambe, M.: Security games for controlling contagion. In: AAAI. AAAI Press (2012)
26. Wang, Y., et al.: Deep reinforcement learning for green security games with real-time information. In: AAAI, pp. 1401–1408. AAAI Press (2019)
27. Xue, W., An, B., Yeo, C.K.: NSGZero: efficiently learning non-exploitable policy in large-scale network security games with neural monte Carlo tree search. In: AAAI, pp. 4646–4653. AAAI Press (2022)

28. Yin, Y., Xu, H., Gan, J., An, B., Jiang, A.X.: Computing optimal mixed strategies for security games with dynamic payoffs. In: IJCAI, pp. 681–688. AAAI Press (2015)
29. Yuan, B., et al.: Decentralized training of foundation models in heterogeneous environments. In: NeurIPS (2022)
30. Zhang, Y., An, B., Tran-Thanh, L., Wang, Z., Gan, J., Jennings, N.R.: Optimal escape interdiction on transportation networks. In: IJCAI, pp. 3936–3944 (2017). https://www.ijcai.org/

HR-Tree: A Hybrid PMem-DRAM and Write-Optimized R-Tree for Spatial Data Storage

Rui Zhang, Yukai Huang, Lulu Chen, Shangyi Sun, Ming Yan, and Jie Wu(✉)

School of Computer Science, Fudan University, Shanghai, China
{zhangrui21,shangyisun21}@m.fudan.edu.cn,
{ykhuang20,llchen18,myan,jwu}@fudan.edu.cn

Abstract. R-tree is a popular dynamic spatial index structure in spatial databases. The traditional R-tree is easily constrained by the limited DRAM capacity when facing massive-scale spatial data today, making it difficult to achieve low latency. Emerging persistent memory (PM) offers DRAM-scale latency, byte addressability, large capacity, and low cost, bringing new opportunities for R-tree. However, integrating R-tree into PM has the following challenges:(1)PM has different access characteristics from DRAM, it has higher latency and different access granularity. (2) The overhead of persistent instructions is high, especially during split. (3) The overhead of concurrency control cannot be ignored, especially in PM-based systems.

In this paper, we propose HR-tree, a hybrid PM-DRAM R-tree to address the challenges above. HR-tree uses a collaborative PM-DRAM approach to minimize the read and write operation costs on PM while ensuring recoverability. It employs a write coalescing mechanism during split and selective persistence for leaf nodes to minimize persistence overhead. Finally, HR-tree proposes a lightweight concurrent control strategy to reduce locking overhead. Experimental results with the real-world datasets show that the HR-tree outperforms the state-of-the-art PM-based R-tree(FBR-tree) and hybrid FBR-tree by up to 3.2× and 1.4×, respectively.

Keywords: persistent memory · spatial index · concurrency control

1 Introduction

Spatial databases are widely used in diverse fields for fast spatial object indexing. With the proliferation of location-based services, such as Google Maps [4] and Yelp [14], there are hundreds of millions of spatial objects to manage nowadays. R-tree is one of the most popular dynamic spatial index structures. R-tree can quickly exclude objects or sets that are not relevant to access geographic location, which dramatically improves the efficiency of spatial operations. R-tree has aroused vast research [12] and has been applied in commercial database systems like Oracle Spatial, and PostGIS [10]. Nevertheless, the traditional R-tree is easily constrained by the limited DRAM capacity when facing massive-scale spatial data, making it difficult to meet low-latency requirements.

Y. Chen et al. (Eds.): COCOON 2024, LNCS 15162, pp. 92–103, 2025.
https://doi.org/10.1007/978-981-96-1093-8_8

Persistent memory (PM), is assumed to be with the characteristics of high storage density, byte-addressability, large capacity, DRAM-scale latency, and low power consumption. It bridges the capacity and performance gap between DRAM and SSD/Disk. Accompanying the advent of non-volatile memory technologies numerous studies have been concentrated on how to design fast persistent data structures [5,8] and PM-based systems [2,6].

To resolve R-tree limitations, a promising solution is to use PM-enabled R-tree to store large-scale spatial data on PM and execute real-time spatial operations. We observe the following three non-trivial challenges in this way. First, PM has different access characteristics from DRAM. PM is slower than DRAM. Existing PM-based R-tree [3] places the whole data structures in PM, which may expose the long write latency to R-tree operations and result in a significant degradation of performance. Second, ensuring persistence incurs significant overhead. Previous research [1] indicates that the overhead of flush and fence instructions is considerable. A higher number of persistent instructions can impact program performance. Third, the overhead of concurrency control in a PM-based R-tree cannot be ignored. Using a mutex lock will generate more small writes to PM. Lock-free based concurrent controls are too complex and need to be specialized for each data structure. We need to design a suitable concurrency control algorithm for R-tree.

This paper presents HR-tree, a hybrid R-tree based on PM-DRAM architecture, for large-scale spatial data processing. First, HR-tree uses a collaborative PM-DRAM approach to minimize read and write operations on PM. Internal nodes are stored in DRAM, while leaf nodes are kept in PM, aiming to ensure recoverability while maximizing performance. Second, HR-tree employs a combination of write merging and partial persistence to reduce the usage of persistent instructions. During the splitting process, data is placed in the same cache line whenever possible to minimize the number of refreshes. Metadata that can be recovered from other data for leaf nodes is not persisted, reducing the usage of persistent instructions. Third, a lightweight concurrency control algorithm is designed to minimize the impact on PM reads, writes, and cache synchronization. We use an optimistic lock-like approach to reduce unnecessary concurrent modifications, which greatly reduces the number of writes to PM.

The contributions of this paper are summarized as follows:

- We analyze the index performance of DRAM and PM respectively, and then propose HR-tree with a hybrid PM-DRAM architecture.
- We employ a selective meta-data persistence strategy and write coalescing mechanism, which aims to mitigate write amplification and reduce persistence overhead to achieve more desirable spatial indexing performance.
- We leverage a lightweight concurrency strategy, optimistic universal lock (OUL) to efficiently deal with multi-thread concurrent contention and reduce cache misses.
- We evaluate HR-tree with both synthetic and real-world datasets on Optane DCPM. The results show that the HR-tree significantly outperforms the state-of-the-art FBR-tree.

2 Background and Motivation

2.1 Persistent Memory

Persistent memory (PM) is a supplement to the memory/storage hierarchy, which combines the persistence of storage with DRAM-scale low latency and high bandwidth. The only available PM product is Intel's Optane DIMM, so our paper is based on this. It can be directly plugged into the standard DDR4 interface and supports two operational modes: Memory Mode and App Direct Mode. We adopt App Direct mode due to its persistent capabilities.

2.2 Spatial Data Structure R-Tree

R-tree is a fundamental data structure that is applied to store and query multidimensional data. It is a balanced tree in which all leaf nodes are at the same depth. All nodes of the R-tree except the root node comprise from m to M nodes, $m = M/2$. Where M is the max capacity of a node and m is the minimum number of objects per node to avoid the R-tree structure being too loose. The Minimum Bounding Rectangle (MBR) of the original data, serving as the key for irregular geometry, is stored in the R-tree index. It is simpler and easier to construct a spatial index for querying and storing compared to convex polygons. The MBR of the spatial object is included in the leaf nodes of the R-tree, and overlapping between sibling nodes is allowed. The R-tree query filters candidate spatial data by MBRs.

2.3 Challenges

It is becoming insufficient for DRAM to meet the demand for large memory due to density limitations. As a complement or substitute for DRAM, high-capacity PM, boasting superior performance, energy efficiency, and byte-addressability, emerges as the ideal choice for constructing large-capacity memory systems. Therefore, developing a R-tree optimized for PM is imperative and promising. However, due to the unique characteristics of PM, directly incorporating PM into an R-tree can lead to performance problems, which can be summarized into three main issues.

First, *unique access characteristics of PM.* (i) The latency of PM is higher than DRAM. (ii) The internal data access granularity of PM is larger, which is 256B. (iii) The bandwidth of PM is lower. The Optane DCPM features a read bandwidth of 39 GB/s (4KB sequential access with 6 Optane modules), significantly higher than the write bandwidth (13.2 GB/s). As Fig. 1 shows, PM write latency is 2.5X higher than that of DRAM. Storing all the data in PM like FBR-tree [3] may lead to high latency for R-tree operations, and significant performance degradation.

Second, *expensive costs associated with persistence.* The overhead of persistent instructions, such as flush and fence, is significant [1]. Existing R-tree-based

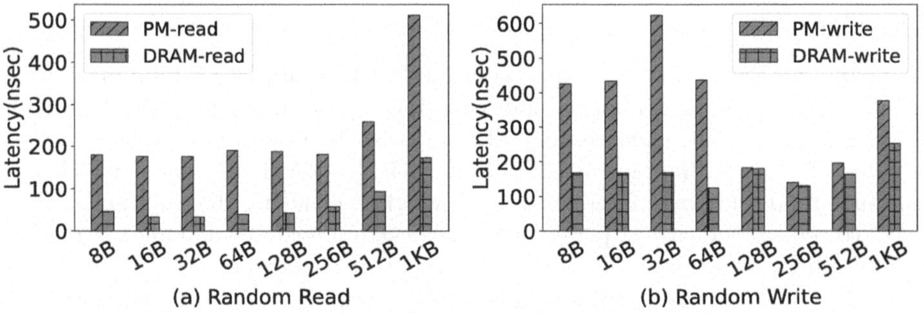

Fig. 1. Random read and write performance.

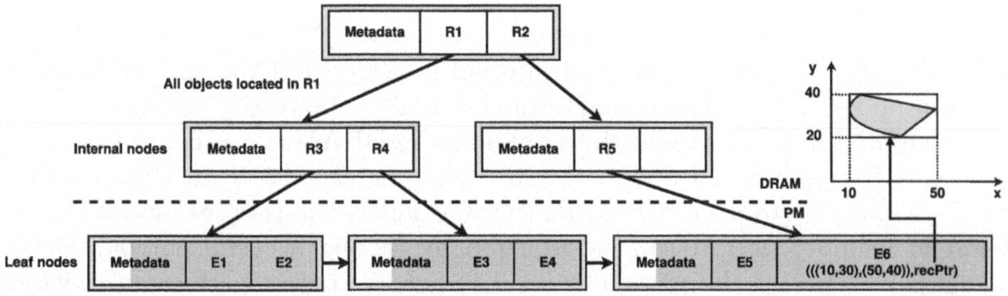

Fig. 2. Architecture of HR-tree. R for Rectangle, E for Entry. Blue for persistent regions, blank for volatile regions. (Color figure online)

implementations suffer from duplicate flushes [7]. We need to minimize the use of persistence instructions while ensuring data persistence and recoverability.

Third, *consistency overhead in concurrency control.* Existing persistent R-tree [3] uses mutex locks for concurrency control. However, as the number of threads increases, research [7] indicates that the overhead of locks in insertion and deletion operations accounts for 54% and 33.4%, respectively. Furthermore, the expense of frequent locking and unlocking in PM is considerably higher than in DRAM. Therefore, we need to design efficient concurrent control algorithms.

3 HR-Tree Design

To address the three issues outlined in Sect. 2.3, we propose the HR-tree. In this section, we present three design points based on the previous discussion, then elaborate on the overview and node structure of the HR-tree.

3.1 Overview

Through a hardware and software co-design, HR-tree employs Minimum Bounding Rectangles (MBR) for spatial indexing, leveraging the strengths of PM while minimizing the additional overhead. To address the issues about access characteristics of PM, we employ a collaborative PM-DRAM approach, which can minimize read and write operations on PM while ensuring the recoverability of data. To reduce the cost of persistence, we design a write coalescing mechanism during split. To mitigate consistency overload in concurrent control, we introduce a lightweight concurrent control strategy, aiming to minimize the overhead associated with PM operations and cache synchronization.

3.2 Collaborative PM-DRAM Approach

As shown in Fig. 2, the HR-tree uses a collaborative PM-DRAM approach, consisting of two parts: the upper-level internal nodes on DRAM and the bottom-level leaf nodes on PM. The actual data location can be traced through pointers on the leaf nodes. While placing internal nodes in DRAM can enhance the access speed of HR-tree internal nodes, this will also increase the time during recovery. To balance recovery overhead and performance, the ratio of internal nodes stored in memory or PM can be adjusted by setting a coefficient parameter. In this paper, we set the parameter to 0 to achieve better performance, which means only the leaf node is placed on PM. As leaf nodes are persisted, the internal nodes can always be recovered when the system restarts.

Inspired by the performance results in Fig. 1 and existing work [13], to maximize the throughput, it is essential to choose appropriate sizes for the nodes in the HR tree. Different from B+-tree, the key size of the spatial data is usually $\geq$16B if it is two-dimensional data or larger. When the node size is 256B, each node can only accommodate 15 entries and cause a high height for the tree. Higher trees result in more random accesses during traversal, which leads to more severe performance decrease in PM. Additionally, when inserting/deleting an entry, there is a higher probability of initiating a split or merge operation, resulting in SMOs and consequently raising the system overhead. Therefore, it is recommended that the node size start at least 512 bytes.

3.3 Selective Persistence and Write Coalescing

HR-tree uses two approaches to reduce persistence overhead, including selective persistence and write coalescing.

Selective Persistence. Firstly, some of the nodes are placed in DRAM to reduce the persistence overhead, as mentioned in Sect. 3.2. Secondly, as shown in Fig. 3, the leaf nodes in HR-tree are selective persistence. The blue fields in Fig. 3 are on PM and are persisted. Since the white fields can be reconstructible, no extra persistence operations are required. Note that the structure of the internal node is almost identical to the leaf node except that it is without *Nextpos* and is

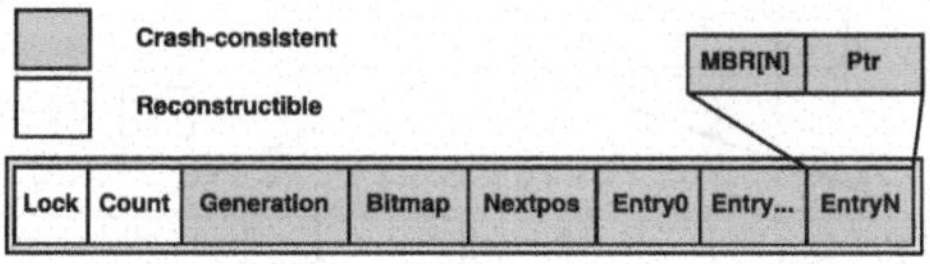

Fig. 3. Leaf Node structure of HR-tree.

stored on DRAM. The *Lock* which occupies 1 bit and *Generation* are designed for concurrency control. *Count* records the number of entries on the node. The *Bitmap* validates the entries in the corresponding position. The total size of *Bitmap* and *Generation* does not exceed 8B, so it can be persistent atomically. *Nextpos* is a pointer and points to the next adjacent leaf node of the current node. Each entry is a key-value pair, which is an MBR with a pointer.

Write Coalescing. Since the size of the persistence instructions granulates is the cache line of the current CPU, the persistence overhead is not affected by the number of elements in the same cache line. Inspired by this feature, we design a write coalescing mechanism to reduce the use of persistence instructions during SMO. Keeping entries in the same cache lines as much as possible during split can reduce the use of persistence instructions. We will talk about the implementation of write coalescing in detail in the Sect. 3.6.

3.4 Lightweight Concurrent Control Strategy

To enhance the performance of the R-tree, it is essential to support multi-threaded operations. There are two main solutions for concurrency control. The *Lock-based method* may lead to frequent lock contention with high concurrency, causing write amplification in the context of PM. The *Lock-free method* such as atomic operations, requiring complex design, and Hardware Transactional Memory (HTM) is limited by L1 cache size for Intel HTM.

Lightweight Concurrent Control Strategy. We adopt the ***optimistic universal lock (OUL)*** in HR-tree. OUL provides simple primitives like locks for concurrent reads and writes, in which the *Generation* is encapsulated along with the *Lock* of the node. We simply call OULRead before reading the data and call OULRead again before retrieving the data just to get a read lock. If the data is changed while reading the data, it will automatically perform a read retry. Thus, it is not necessary to make careful consideration and design to avoid inconsistencies as in the lock-free method. In brief, it is similar to encapsulating the original lock-free logic into a function. For write operations, it first performs a traversal of the tree. The node is not locked until the node requiring modification is located, then get the *Lock* by OULGetLock. Since the leaf nodes are stored on the PM, when the crash restarts, the system needs to scan the leaf nodes and invalidate all the locks. According to the OUL, the reader does not block the writer, and it checks whether the *Generation* is changed only before reading the next node. And the writer updates the *Generation* after obtaining the

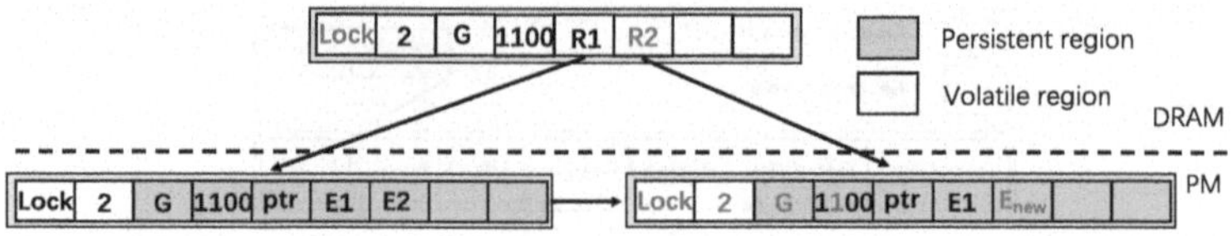

Fig. 4. Entry insertion in HR-tree.Red means modified. (Color figure online)

Lock of the corresponding node. In a nutshell, the OUL method avoids frequent persistent memory write operations, achieving better concurrency.

3.5 Basic Operations Supported by HR-Tree

Insert and update: As illustrated in Fig. 4, the incoming request will insert a new entry E_{new} into the HR-tree. ❶ Start traversing from the root node and return the leaf node where the entry should be recorded and its parent node. ❷ Check the *Lock* and acquire the lock of the parent node by OUL. Then judge whether the leaf node is full or not. If the *Count* is not equal to the M, update the covering rectangle R and release the *Lock*. Otherwise, perform the split. ❸ Acquire the *Lock* of the corresponding leaf node. ❹ Read the *Bitmap* and insert the new entry E_{new} and perform a persistent operation. ❺ Update metadata and release the *Lock*. Flip the corresponding *Bitmap* bit and increment the *Generation*. Then persist it.

Delete: ❶ First, the tree is traversed to retrieve the leaf node where the entry is to be deleted. ❷ Acquire the *Lock* of the leaf node. Otherwise, retry for the lock if failed. ❸ Update the metadata. Clear the corresponding bit of *Bitmap* and increment the *Generation*. Then perform a persistence operation. The *Count* decreases by one and releases the *Lock*. Due to each bit of the *bitmap* being used to indicate whether the entry at the corresponding position exists, when deleting an entry, only bitmap modification is needed. ❹ Then acquire the *Lock* of the parent node and update the covering rectangle of the parent node upwards. The range will be updated because it may be reduced after deletion, thus ensuring that all entries in the node are not redundant to improve search efficiency.

Search: As shown in step ❶ of insert, we start searching from the root node, find the entries stored in each node according to the bitmap, and find the contained child nodes to the next level after comparison. Note that after reading, we need to determine whether the *Generation* is the same, which is used to determine whether the data read is the latest.

3.6 Structural Modification Operations (SMOs)

Split: As shown in Fig. 5, the split needs to go through the following steps: ❶ Acquire the lock of N_p and perform split. ❷ Acquire the *Lock* of node N_{old}. ❸ Allocate and create a new persistent node N_{new} on PM and initialize metadata.

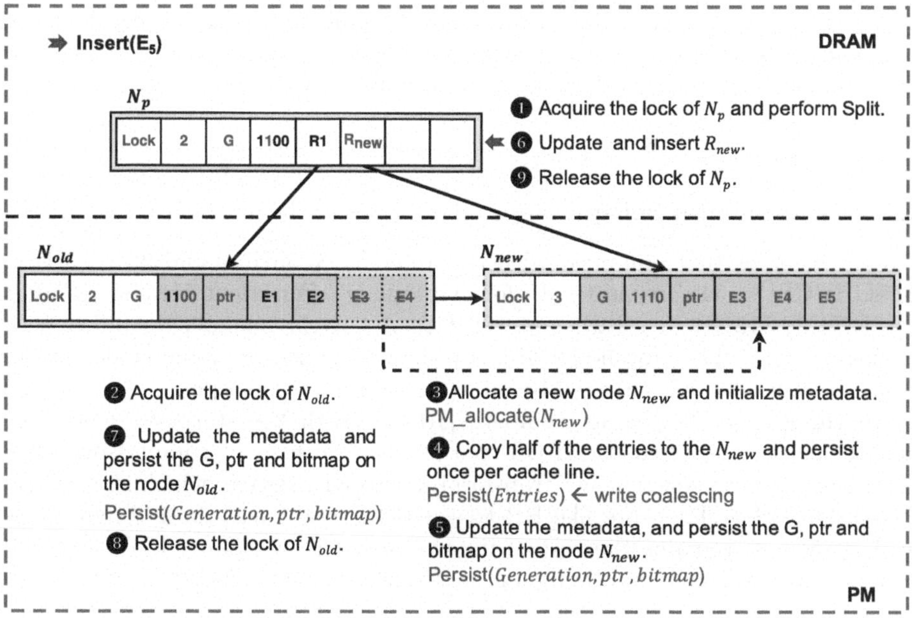

Fig. 5. Entry splitting in HR-tree.

❹ Copy half of the entries to the new node N_{new} and persist. Due to the granularity of persistent operation is one cache line, which is usually 64 bytes. So we utilize **write coalescing**, by placing all the data in one cache line, the flush times can be reduced. ❺ Update the metadata. Flip the *Bitmap* corresponding to these entries on the node N_{new}. Set the *ptr* of the current node N_{new} to point to the node pointed by the N_{old} and increment the *Generation* then follow a persistence operation; the *Count* is updated without persistence, meanwhile. ❻ Update the original covering rectangle and insert the covering rectangle of N_{new} on N_p; ❼ Clear the bit of the removed entry in the bitmap, update the *ptr* point to N_{new} and increment the generation following a persistence operation. The *Count* should also be modified. ❽ Release the *Lock* of node N_{old}. ❾ Release the *Lock* of N_p. Finally, the split operation ends.

3.7 Fast Recovery

During crash recovery, HR-tree can rebuild the volatile inner nodes from leaf nodes in PM. It consumes less time to restore an HR-Tree than to rebuild a transient tree from scratch. There is a persistent *status* bit to indicate whether the last shutdown was normal or not. If it is normal, the *status* will be set after the whole tree is persisted. Otherwise, it will not be set when suffering a system failure. When restarted, it scans leaf nodes, rebuilding HR-tree by starting from the leftmost leaf node that is stored on a static address. Recount the nodes, set the lock to 0, then find the next leaf node according to the *Nextpos* field. It takes only 1 min to recover 1.2 billion spatial data key-value pairs. The reconstruction time can be further reduced by placing more non-leaf nodes in PM, which is a trade-off between index performance and rebuild time.

4 Evaluation

4.1 Experimental Setup

We use a server that equipped with two Intel(R) Xeon(R) Gold 6242 CPU @ 2.80GHz CPUs (64 cores in total), four Optane DC Persistent Memory (128GB per DIMM, 512GB in total), and 128 GB of DRAM memory, installed with Ubuntu 18.04.6. The Optane DIMMs are the first generation, and configured in App Direct mode and managed by an ext4-DAX file system [9].

In this paper, we mainly compare HR-tree with the state-of-the-art persistent spatial data structure FBR-tree [3]. For fairness of comparison, since FBR-tree is fully persistent, we transform it into HFBR-tree, which is of hybrid PMem-DRAM architecture like HR-tree. HFBR-tree places the internal nodes on DRAM and remains the rest of its own structure unchanged. We conduct experiments using both synthetic and real-world datasets like [11].

4.2 Operation Performance

Insertion Performance. We first measured the average latency and throughput of insertion with different index sizes with a single thread, from 200 to 3200 million on uniform data, as shown in Fig. 6(a). The average insertion latency of the HR-tree achieves about 1.2∼1.8μs and is significantly lower than the FBR-tree as expected. HFBR-tree exhibits 1.9× higher insert latency since it is not optimized with respect to the characteristics of PM. As the number of records increases, the average insertion latency of the three methods gradually increases but is most prominent in FBR-tree. Due to the fact that the height of the tree expands in response to the increase in the number of records. PM features higher access latency than DRAM. Therefore, it suffers less impact than FBR-tree because of its hybrid placement architecture of HR-tree and HFBR-tree.

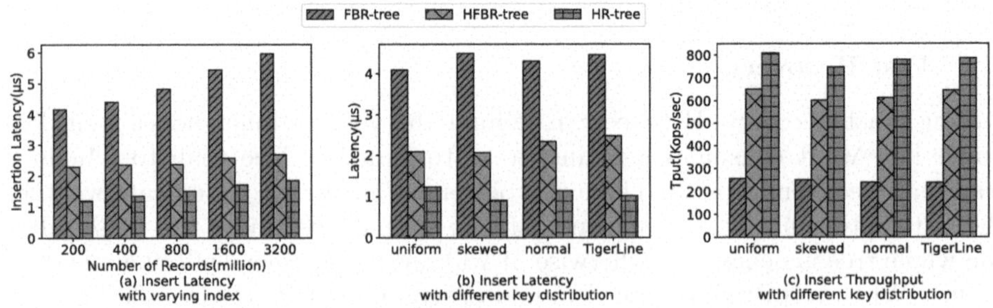

Fig. 6. Insertion performance.

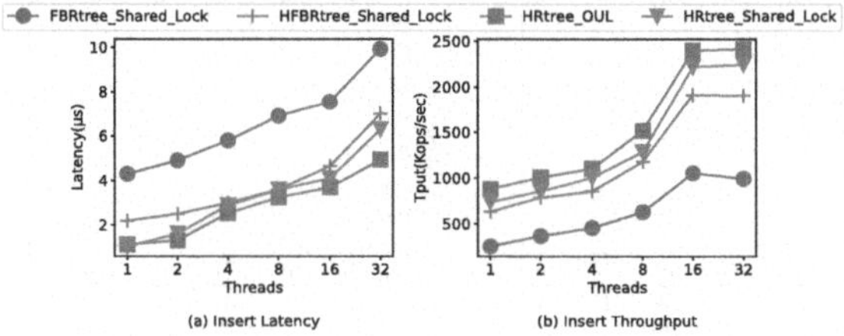

Fig. 7. Insert performance with multi-threads.

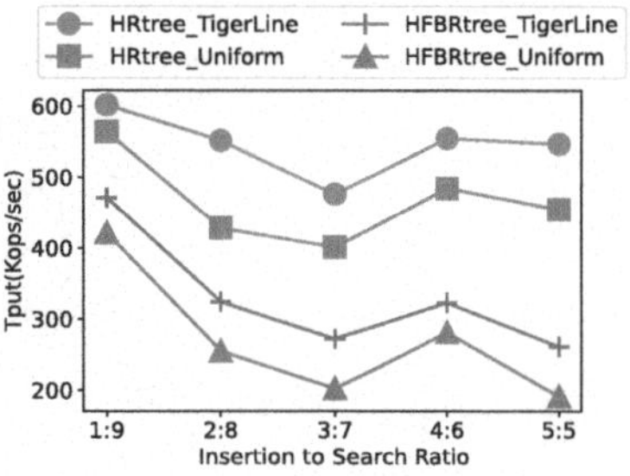

Fig. 8. Search performance with varying skewness and thread numbers.

Figure 6(b) and (c) illustrate the latency and throughput (Tput) for inserting both real and synthetic data to measure its sensitivity to skewness. Figure 6(b) shows that HR-tree can reduce latency than FBR-tree by up to 75% and around 41% to HFBR-tree. Here are some reasons to account for the improvement: 1) hybrid PM-DRAM architecture. 2) The node size in HR-tree is set to fit the internal granularity of Optane DCPM and employs the write coalescing mechanism to minimize persistence overhead.

In addition, we evaluate HR-tree by increasing the number of threads from 1 to 32 in Fig. 7. We compare HR-tree based on OUL (`HRtree_OUL`) with in-place FBR-tree (`FBRtree_Lock`), `HFBRtree_Lock` and `HRtree_Lock`. Figure 7(a) shows that the insertion latency of all four methods. The latency increases as the thread count increases, due to contention when accessing to the same resource. And the pure PM-based FBR-tree is the most affected because FBR-tree insertion adopts a lock-based method. It modifies shared memory on each concurrent read and write due to the CPU cache coherence mechanism, resulting in a large number of cache misses. Figure 7(b) shows `HRtree_OUL` scales well until the number of threads is 16 and then stabilizes, but for `FBRtree_Lock`, it suffers very high contention and the throughput even dropped than 16 threads. As a result, The HRtree_OUL achieves ~1.5× higher throughput than HFBRtree_Lock and ~1.2× higher than `HRtree_Lock`. Furthermore, it is up to 3.5× than `FBRtree_Lock`.

Search Performance. We first analyze how HR-tree behaves in terms of latency and throughput when the skewness varies. The results are shown in fig. 9(a), skewness varies from 2 to 8. We observe that the phenomenon is more noticeable in the case of accessing data with higher skewness. Its throughput peaks, when the skewness is 6. Then keeping stable basically with higher skewness. The latency is reduced by 14%, and the throughput is increased by 20%. To further analyze the search performance and scalability, we implement the following four methods with multiple threads: HR-tree based on OUL (`HRtree_OUL`), lock-free in-place FBR-tree (`FBRtree_Lockfree`), lock-free in-place HFBR-tree (`HFBRtree_Lockfree`), and HR-tree with the similar lock-free strategy on FBR-

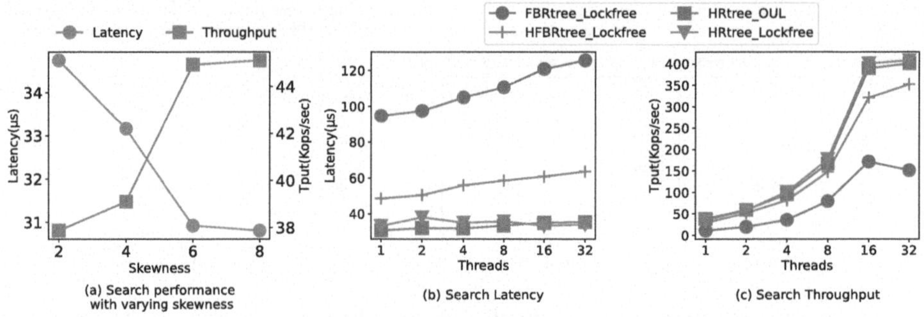

Fig. 9. Search performance with varying skewness and thread numbers.

tree (`HRtree_Lockfree`). Figure 9(b) and (c) show that the hybrid-based methods scale well across more threads due to the performance benefits of DRAM. The FBR-tree is more significantly affected by PM performance as it is fully persistent. Figure 9(b) shows its latency gradually increases. The hybrid architecture of the `HFBRtree_Lockfree` performs well with multiple threads. As for `HRtree_OUL` and `HFBRtree_Lockfree`, the performance is comparable as the number of threads increases. `HRtree_Lockfree` is even a little better. The lock-free method is fine-grained based on the characteristics of the tree structure, while the OUL method is not optimized for versatility.

In real-world scenarios, reads and writes are usually performed at different ratios simultaneously. We conduct mixed read and write performance experiments with varying ratios on both synthetic uniform data and the real-word Tiger-Line data under 32 threads, respectively. In Fig. 8, the experiments of HR-tree and HFBR-tree between the two datasets show a similar trend. As the insertion ratio grows, the throughput first tends to decrease and then followed by an increase. Initially, the rise in insertion ratio causes part of the search thread to be blocked, resulting in lower throughput. As the proportion of insertion becomes larger, an increase in throughput is observed because the insertion request is faster than the range query search process. Meanwhile, the throughput of the Tiger-Line data is always above the uniform data, which is due to the Tiger-Line dataset with a better locality having a higher cache hit rate and therefore the rate of decline is a bit slower compared to the uniform dataset.

5 Conclusion

In this paper, we present HR-tree, a hybrid PM-DRAM and write-optimized R-tree. It provides a solution for solving the problems of spatial data processing efficiency and storage capacity. HR-tree uses a collaborative PM-DRAM approach to minimize the read and write operations on PM while ensuring recoverability. It employs a write coalescing mechanism during split and selective persistence for leaf nodes to minimize persistence overhead. Finally, HR-tree proposes a lightweight concurrent control strategy to reduce locking overhead. Our experimental results show that HR-tree scales with up to ~3.2× and ~1.4× better performance than prior state-of-the-art FBR-tree and its variants HFBR-tree on real Intel Optane DCPM.

Acknowledgements. This research is supported in part by the National Key Research and Development Program of China (2021YFC3300600).

References

1. Chen, Y., Lu, Y., Yang, F., Shu, J.: Flatstore: an efficient log-structured key-value storage engine for persistent memory. In: Proceedings of the 25th International Conference on Architectural Support for Programming Languages and Operating Systems, pp. 1077–1091 (2020)
2. Chen, Y., Shu, J., Ou, J., Lu, Y.: Hinfs: A persistent memory file system with both buffering and direct-access. ACM Trans. Storage (ToS) **14**(1), 1–30 (2018)
3. Cho, S., Kim, W., Oh, S., Kim, C., Koh, K., Nam, B.: Failure-atomic byte-addressable R-tree for persistent memory. IEEE Trans. Parallel Distrib. Syst. **32**(3), 601–614 (2020)
4. Google: Google maps (2005). https://maps.google.com
5. Hwang, D., Kim, W.H., Won, Y., Nam, B.: Endurable transient inconsistency in byte-addressable persistent B+-Tree. In: Proceedings of the 16th USENIX Conference on File and Storage Technologies, pp. 187–200. FAST'18, USENIX Association, USA (2018)
6. Kadekodi, R., Lee, S.K., Kashyap, S., Kim, T., Kolli, A., Chidambaram, V.: Splitfs: reducing software overhead in file systems for persistent memory. In: Proceedings of the 27th ACM Symposium on Operating Systems Principles, pp. 494–508 (2019)
7. Lavinsky, B., Zhang, X.: Pm-rtree: A highly-efficient crash-consistent r-tree for persistent memory. In: Proceedings of the 34th International Conference on Scientific and Statistical Database Management. SSDBM '22, Association for Computing Machinery, New York, NY, USA (2022). https://doi.org/10.1145/3538712.3538713
8. Li, Y., Chan, H.H.W., Lee, P.P.C., Xu, Y.: Enabling Efficient Updates in KV Storage via Hashing: Design and Performance Evaluation. ACM Trans. Storage **15**(3) (aug 2019)
9. Peterson, J.L., Norman, T.A.: Buddy Systems. Commun. ACM **20**(6), 421–431 (jun 1977)
10. PostGIS: Spatial database extender for postgresql (2022). https://postgis.net/
11. Qi, J., Tao, Y., Chang, Y., Zhang, R.: Theoretically optimal and empirically efficient r-trees with strong parallelizability. Proc. VLDB Endow. **11**(5), 621–634 (jan 2018)
12. Qi, J., Zhang, R., Kulik, L., Lin, D., Xue, A.: The min-dist location selection query. Proceedings - International Conference on Data Engineering (04 2012)
13. Yang, J., Kim, J., Hoseinzadeh, M., Izraelevitz, J., Swanson, S.: An Empirical Guide to the Behavior and Use of Scalable Persistent Memory. In: Proceedings of the 18th USENIX Conference on File and Storage Technologies, pp. 169–182. FAST'20, USENIX Association, USA (2020)
14. Yelp: Yelp inc (2004). https://www.yelp.com

A Distributed Computation Offloading Scheme Based on Stackelberg Game in MEC

Weifeng Sun(✉), Yixing Qin, and Bowei Zhang

School of Software, Dalian University of Technology, Dalian, China
wfsun@dlut.edu.cn

Abstract. The computation offloading problem in Multi-access Edge Computing (MEC) has become one of the current important research topics. Most existing computation offloading solutions mainly focus on the offloading within the edge and terminal structures. This paper proposes a new computation offloading scheme that considers the resource utilization of cloud servers in edge computing, treating cloud servers as supplements to edge server resources. This approach can alleviate insufficient computing resources of edge server. Additionally, we establish a Stackelberg game to solve the computation offloading problem, and then analysis the strategies of cloud servers, edge servers, as well as mobile devices. Then, we determine the optimal strategies for all participants and prove the existence and uniqueness of Nash equilibrium solutions in the game. Simulation results and analysis also validate the feasibility of the computation offloading scheme based on Stackelberg games. The proposed method is more efficient and suitable for scalable edge computing environments compared with the two existing offloading schemes.

Keywords: Multi-Access edge computing · Computation offloading · Stackelberg game

1 Introduction

Multi-access Edge Computing (MEC) with emerging technique such as 6G [1] and NOMA [2], performs processing and storage tasks closer to users, which can significantly reduce the amount of data transmitted over the network, thereby improving efficiency for users [3], and reducing the load on the core network [4]. Its main structure includes three parts: the central cloud, edge servers, and end devices. Since the resources of MEC servers are limited, it is even more necessary to allocate resources reasonably. The computation offloading model can mainly be divided into binary computation offloading and partial computation offloading models. Binary computation offloading assumes that tasks are indivisible and can only be executed locally or remotely, while the partial computation offloading model assumes that tasks can be divided into many small tasks, and offload a portion of the tasks to remote servers. When more tasks are received by limited edge, tasks are queued for processing, causing significant latency and energy consumption. Therefore, offloading the tasks queued in edge is necessary.

Y. Chen et al. (Eds.): COCOON 2024, LNCS 15162, pp. 104–115, 2025.
https://doi.org/10.1007/978-981-96-1093-8_9

This paper proposes a Computation Offloading Algorithm for MEC based on Stackelberg Game (COASG) to address the aforementioned issues, which adopts a partial offloading approach and considers the utilization of resources across three layers. This solution addresses the issue of insufficient resources in edge servers in MEC. Then, we solve the optimal strategies for each participant in the Stackelberg game and prove the existence and uniqueness of Nash equilibrium. We also compare the cost functions of participants under different methods to validate the feasibility and efficiency of the proposed solution.

2 Related Work

Researchers have proposed numerous computation offloading methods. For example, Fu et al. [5] study the minimization of energy costs for mission-critical Internet of Things (IoT) tasks in MEC systems, and designed an iterative algorithm to optimize two problems: offloading strategy and resource allocation. Huang et al. [6] propose a Distributed Deep Learning-based Offloading (DDLO) algorithm suitable for MEC networks, where multiple parallel DNNs are used to generate offloading decisions. Shared replay memory is employed to store newly generated offloading decisions for further training and improvement of all DNNs.

Game theory [7] has become a popular approach in current research due to its distributed and equilibrium-guaranteed characteristics. As a new branch of modern mathematics, game theory [8] has been widely applied in the field of economics, enabling the representation of competitive relationships in practical problems as mathematical models and analyzing them to obtain optimization results. Therefore, it has extensive applications in many fields, especially in the field of computer science. Game theory can naturally address the problem of computing offloading. For example, Xu et al. [9] investigate partial offloading of multiple mobile devices in a cloud collaborative scenario. Based on game theory, they first obtained the optimal offloading decision from edge to devices, and then obtained the optimal offloading decision from edge to cloud for all terminal devices. This ultimately achieves the goal of optimal offloading to the cloud, and proposes a Two-Stage Computation Offloading Algorithm (TSGTA). Base on the visibility and accessibility of the cloud to UE, Ding et al. [10] propose End-Edge-Cloud Computing (EECC) that is divided into two types of computing architecture. A potential game is constructed for the EECC environment, where each UE selfishly minimizes its payoff.

Current research mostly focuses on offloading queued tasks from edge servers to cloud servers, without considering queued tasks on terminal devices and overall resource coordination between edge servers and cloud servers. Cloud servers desire to handle more tasks, edge servers in MEC also want more tasks, while users expect to complete tasks with the lowest energy consumption and latency. There is competition between cloud servers and edge servers, as well as among users. Establishing a reasonable mathematical model and finding a computation offloading solution satisfactory to all parties involved is a significant issue.

3 Problem Description

3.1 System Model

We consider scenarios involving numerous tasks at the user end, where nearby edge servers may struggle to handle the high workload. So we propose a joint offloading approach for the resources at cloud, edge, and terminal, treating the cloud server and edge servers as remote servers. Figure 1 introduces our three-tier offloading model with a cloud server, multiple edge servers and multiple users. Each edge server is directly connected to a base station, and through this connection, mobile devices select whether to offload computations to the nearby edge servers or the cloud server acting as a remote server.

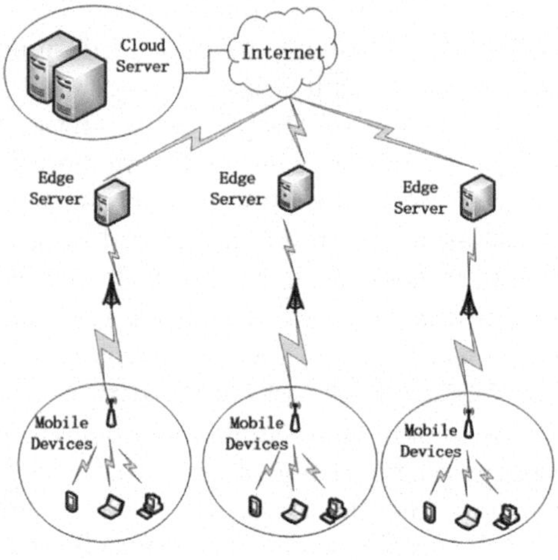

Fig. 1. The scenario of the proposed COASG strategy.

As illustrated in Fig. 2, in the first-tier Stackelberg game, both the cloud server and the edge server strive to handle more tasks. The cloud server, acting as a follower, sends its available resource quantity and computing capacity to the edge server. The edge server, as a leader, receives this information and aims to minimize latency and energy consumption. It then calculates the optimal offloading scheme for tasks between the cloud server and the edge server. The leader in the first-tier Stackelberg game, namely the edge server, stores the optimal schemes for tasks of different sizes between the cloud server and the edge server in a table. This table is then transmitted to the mobile devices participating in the second-tier Stackelberg game. After the game, the mobile devices send their allocation strategies to the remote server. In the second-tier Stackelberg game, when mobile devices send offloading requests, the base station calculates channel interference based on the number of devices it serves and the distance to each device. It then sends this channel interference information along with information about the remote server to each mobile device. Each mobile device can calculate data transmission over the channel to determine energy consumption and task completion time under different decision strategies. The first terminal device acts as the leader, while the rest act as followers. Each mobile device sequentially selects a strategy with minimal latency and energy consumption. The dynamic updating of the entire decision process is divided into multiple time slots, with only one user able to update its offloading strategy in each time slot.

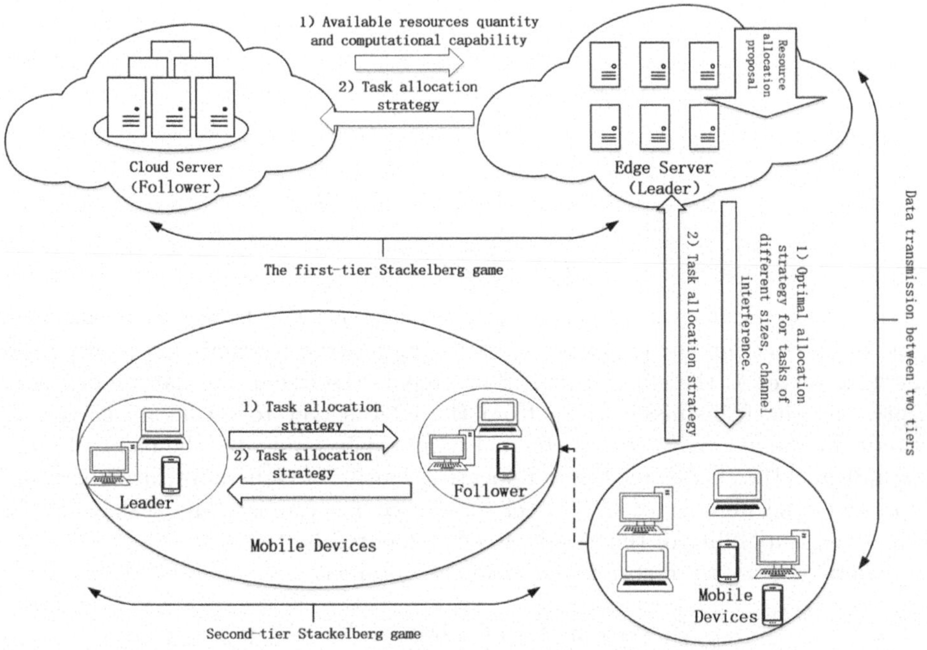

Fig. 2. COASG strategy process.

3.2 Cloud Server Strategy Model

In this model, assuming the task flow set is $s = \{1, 2, ..., s\}$, the set of tasks offloaded to the edge server is $n = \{1, 2, ..., n\}$, the resource capacity of the cloud server is R^C, and the cloud server provides resources to the edge server with a quantity denoted by $h^j (j \in n)$, where h^j is less than the total resources of the cloud server k^j, i.e., $h^j \leq k^j$. The first layer of game strategy is represented by $A^j = \{a^1, a^2, ..., a^j\}$, indicating all the offloading strategies of tasks to remote servers in the task.

$$A^j = \begin{cases} 0 & \text{Offload to edge server} \\ 1 & \text{Offload to cloud server} \end{cases} \tag{1}$$

The cost function of the cloud server comprises both the latency $T_n^C(a^j)$ and the energy consumption $E_n^C(a^j)$. The latency consists of the execution time of tasks on the cloud server $t_{n_i}^c = \frac{d \cdot y_{n_i}}{b^c}$ and the transmission time of tasks between the cloud server and the edge server $t_{n_i}^{cc} = \frac{d}{V_{cloud}}$. The energy consumption includes the energy consumption of task execution on the cloud server and the transmission energy consumption. Consequently, the total cost of the cloud server is obtained as $U^C(a^j)$.

$$U^C(a^j) = E_n^C(a^j) + T_n^C(a^j) \tag{2}$$

$$E_n^C = P^C \cdot t_{n_i}^{cc} + \frac{P^{CC} \cdot d}{V_{cloud}} \tag{3}$$

$$V_{cloud} = B \log_2 \left(1 + \frac{P^{cc} \cdot g^2}{N_0 \cdot B}\right) \tag{4}$$

$$T_n^C(a^j) = t_{n_i}^{cc} + t_{n_i}^c \tag{5}$$

where P^c represents the power of the cloud server, $t_{n_i}^c$ denotes the execution time of the task on the cloud server, P^{cc} signifies the transmission power from the edge server to the cloud server, d represents the size of the data, and V_{cloud} represents the transmission rate from the edge server to the cloud server. B stands for the average wireless channel bandwidth, g denotes the channel gain, N_0 indicates the single-sided power spectral density of Gaussian white noise. y_{n_i} represents the number of CPU cycles required to complete the task, and the computing capability of the cloud server is denoted by b^c.

Therefore, the total cost $U_E^C(a^j)$ can be obtained.

$$U_E^C(A^j) = a^c \sum_{j \in n} E_n^C(A^j) + b^c \sum_{j \in n} T_n^C(A^j) \tag{6}$$

where a^c and b^c respectively represent the weighting factors for latency and energy consumption, with $0 \leq a^c, b^c \leq 1$ and $0 \leq a^c + b^c \leq 1$. These factors vary in different scenarios.

The optimization model for the cloud server can be derived $\min_{h^j} U_E^C(A^j)$.

3.3 Edge Server Strategy Model

In the cloud-edge game, it is necessary to compare the total costs of executing tasks at the edge server and offloading tasks to the cloud server. Therefore, the cost of the edge server is composed of the latency $T_n^E(a^j) = t_{n_i}^e$ and energy consumption $E_n^E(a^j)$. The latency is represented by the execution time of tasks on the edge server $t_{n_i}^e = \frac{d^* \cdot y_{n_i}}{b^e}$, and the energy consumption is the energy expended by tasks executed on the edge server. Hence, the total cost $U^E(a^j)$ of the edge server is given by:

$$U^E(a^j) = E_n^E(a^j) + T_n^E(a^j) \tag{7}$$

$$V_{\text{edge}} = B \log_2 \left(1 + \frac{P^{EE} \cdot g^2}{N_0 \cdot B}\right) \tag{8}$$

where $E_n^E = P^E \cdot t_{n_i}^{ee}$ and P^E is the edge server's power. P^{EE} is the transmission power of the task sent to the edge server. The computing capability of the edge server is denoted by b^e. $t_{n_i}^{ee} = \frac{d}{V_{\text{edge}}}$ represents the transmission latency of the task from the mobile device to the edge server. $\frac{P^E \cdot d}{V_{\text{edge}}}$ represents the transmission energy consumption of the task from the user to the edge server.

The total cost $U_E^E(a^j)$ has been obtained, and the optimization model for the edge server is $\min_{R^e} U_E^E(A^j)$.

$$U_E^E(a^j) = a^e \sum_{j \in n} E_n^E(a^j) + b^e \sum_{j \in n} T_n^E(a^j) \tag{9}$$

The model of the first-layer Stackelberg game obtained by:

$$\min_{h^j, R^e} \{U_E^C(A^j) + U_E^E(A^j)\} \tag{10}$$

3.4 Users Strategy Model

Assuming there are N mobile devices within the service range of MEC servers, where $N = \{1, 2, \ldots, N\}$ represents the set of users. All users are to execute the same type of task, implying that tasks are homogeneous. Each user offloads tasks to the nearest edge server. Every task comprises S subtasks, where $\Omega = \{1, 2, \ldots, m\}$ represents the set of all subtask nodes.

When tasks are executed locally, the energy consumption of a mobile device equals the energy consumption of task processing. The local computation time $t_{n_i}^l$ represents the task execution time, where $t_{n_i}^l = \frac{y_{n_i}}{b_n^l}$. Here, b_n^l denotes the computing capability of the mobile device, and q_n^l denotes the user's ability to process tasks locally. Thus, the energy consumption $E_{n_i}^l$ of subtask S_{n_i} when processed locally is given by $E_{n_i}^l = q_n^l \cdot t_{n_i}^l$.

Let q_n^b represent the capability of each mobile device when idle. When tasks are offloaded to remote servers, the idle energy consumption of the mobile device during task execution is $E_{n_i}^e = q_n^b \cdot t_{n_i}^e$, and $E_{n_i}^c = q_n^b \cdot t_{n_i}^c$, respectively.

According to the task model, the initial start time of the task is 0. The execution time of the task is the sum of the execution time of subtasks and the transmission time.

$$t_{n_i}^z = \begin{cases} t_{n_i}^l & \text{if } d_{n_i} = 0 \\ t_{n_i}^e + t_{n_i}^{ee} & \text{if } d_{n_i} = 1 \\ t_{n_i}^c + t_{n_i}^{cc} + t_{n_i}^{ee} & \text{if } d_{n_i} = -1 \end{cases} \tag{11}$$

where the mobile device uses the vector $D_n = \{d_{n_1}, d_{n_2}, \ldots, d_{n_S}\}$ to represent the offloading strategy for all subtasks in the task. In other words, D_{n_i} represents the offloading decision for Ω_{n_i}.

$$D_n = \begin{cases} 0 & \text{Local execution} \\ 1 & \text{Offload to edge server} \\ -1 & \text{Offload to cloud server} \end{cases} \tag{12}$$

The total latency of the task is the sum of transmission latency and execution latency and When task Ω_{n_i} is offloaded, the total energy consumption $E_{n^i}^{zz}$ of the mobile device is the sum of the transmission energy consumption during uploading and the energy consumption when the mobile device is idle during task execution. The energy consumed during task execution is denoted by $E_{n_i}^{z}$.

$$t_n(D_n) = \frac{1-d_{n_i}}{2} \cdot (t_{n_i}^{cc} + t_{n_i}^{c} + t_{n_i}^{ee}) + \frac{1+d_{n_i}}{2} \cdot (t_{n_i}^{ee} + t_{n_i}^{e}) \tag{13}$$

$$E_{n_i}^{zz} = \frac{1-d_{n_i}}{2} \cdot (E_{n_i}^{c} + E_{n_i}^{cc} + E_{n_i}^{ee}) + \frac{1+d_{n_i}}{2} \cdot (E_{n_i}^{e} + E_{n_i}^{ee}) \tag{14}$$

$$E_{n_i}^{z} = |d_{n_i}| \cdot E_{n_i}^{zz} + (1 - |d_{n_i}|) \cdot E_{n_i}^{l} \tag{15}$$

$|d_{n_i}| \cdot E_{n_i}^{zz}$ denotes offloading energy for task Ω_{n_i}, while $(1 - |d_{n_i}|) \cdot E_{n_i}^{l}$ is local execution energy. The mobile device's energy consumption under strategy D_n is:

$$E_n(D_n) = \sum_{i,j \in S} E_{(n_i,n_j)}^{z} + \sum_{i \in S} E_{(n_i,n_j)}^{zz} \tag{16}$$

Let d_{n_S} be the decision space. $d_{n_S} = 0$ represents task execution locally; $d_{n_m} = 1$ represents task offloading to the edge server; $d_{n_S} = -1$ represents task offloading to the cloud server. $D_n = \{d_{n_1}, d_{n_2}, \ldots, d_{n_S}\}$ is the decision vector of the mobile device, and $\Xi = \{D_1, D_2, \ldots, D_N\}$ is the decision set. $D_{(-n)} = \{D_{(-1)}, D_{(-2)}, \ldots, D_{(n-1)}, D_{(n+1)}, \ldots, D_N\}$ is the offloading decision set of other participants. When other participants have already decided on strategy $D_{(-n)}$, the optimal offloading strategy can be determined by calculating the upload rate in the current environment.

$$\min_{d_{n_S} \in \{0,1\}} \varphi_n(D_n, D_{(-n)}), \quad \forall n \in N, \forall s \in S \tag{17}$$

where $\varphi_n(D_n, D_{(-n)})$ represents the cost function of the mobile device.

$$\varphi_n(D_n, D_{(-n)}) = \begin{cases} E_n(D_n) + t_n(D_n) & \text{if } d_{n_S} = 0 \\ c_1 E_n(D_n) + c_2 t_n(D_n) & \text{if } \exists d_{n_m} = 1 \end{cases} \tag{18}$$

where $c_1 + c_2 \in [0, 1]$ is the balancing factor. In the game, if the mobile device is in the offloading state, the energy consumption and delay are balanced by adjusting the balancing factor to a certain ratio, which tends to favor offloading.

This encourages more mobile devices to offload subtasks to remote servers. The cost of tasks offloaded to remote servers is reduced to near-optimal values.

We construct the second-layer Stackelberg game model $G = \{N, \{\Xi\}, \varphi_n, n \in N\}$, where Ξ represents the strategy set of the participants. The Nash equilibrium strategy $\Xi^* = \{D_1^*, D_2^*, \ldots, D_N^*\}$ is the Nash equilibrium point of the task offloading game, where each participant's cost is $\varphi_n(D_n^*, D_{-n}^*)$, and $D_n^* = \text{argmin}\varphi_n(D_n, D_{-n})$ is the best strategy for the participant.

4 COASG Strategy and Equilibrium Analysis

To minimize the latency and energy consumption of offloading while ensuring fairness, we propose an offloading algorithm considering the three-tier resources at the cloud-edge and prove the existence of Nash equilibrium among them.

Algorithm 1. Remote server optimal strategy

Input: R^e,h^j,b^{ee} and b^{cc};
Output: A_n^*;
```
for d in n do
    calculate edge server cost U_E^E(a^j);
    calculate cloud server cost U_E^C(a^j);
    a^max ← a^j;
    if U_E^E(a^j) ≤ U_E^C(a^j) then
        a^j = 0;
    else
        a^j = 1;
    end if
end for
```

The Optimal Strategy for the Cloud Server. The cost of the cloud server includes the total latency and energy consumption. As a follower in the cloud-edge game, the cloud server receives information from the edge servers and the resource demand of each task. Then, it calculates its cost and determines the best strategy through cost comparison.

Theorem 1. *The strategy set of the cloud server exists and is non-empty. By taking the first-order and second-order partial derivatives with respect to* h_j *of the function* $U_E^C(h^j) = \sum_{j\in n} E_n^C(h^j) + \sum_{j\in n} T_n^C(h^j)$, *it can be observed that* $U_E^C(h^j)$ *is a concave function with respect to* h_j. *Therefore, by setting* $\frac{\partial U_E^C}{\partial h_j} = 0$, *we can obtain the optimal solution* h_j^* *for* h_j.

Theorem 2. *Given the cost of the edge server, the optimal strategy of the cloud server exists and is unique.*

The Optimal Strategy for the Edge Server. The primary objective of the edge server is to minimize the system cost associated with task execution, i.e., to minimize $U_E^E(a^j)$. Since the strategy set is finite, there must exist an optimal solution $a^{(j^*)}$. Algorithm 1 finds the optimal strategy for users, ensuring that all tasks of users are allocated.

Theorem 3. *Given the strategy of the cloud server, there exists a unique optimal strategy* $a^{(j^*)}$ *for the edge server.*

From the above theorem, it can be concluded that there exists a unique Nash equilibrium solution in the cloud-edge game between the cloud and edge server.

4.1 The Optimal Strategy for Users

Algorithm 2. Users optimal strategy

Input: y_{n_i},g_{n_i},b_n^l,q_n^l,b^{ee},$dist_n$,r,K,b^{cc} and V_c;
Output: D_n^*;
initialization: $d_{n_s} = 0$;
while users did not receive the END message **do**
  get channel interference information from the edge server;
  calculate V_{edge};
  while there is currently a user device strategy **do**
    caculate $\phi_n(D_n, D_{-n})$ and D_n^*;
    if $\phi_n(D_n, D_{-n}) < \phi_n(D_n^*, D_{-n})$ **then**
      $D_n^* \leftarrow D_n$;
      $\phi_n(D_n^*, D_{-n}) = \phi_n(D_n, D_{-n})$;
    end if
  end while
  if $D_n^* \neq D_n$ **then**
    send new request;
  end if
  if receive update information from control node **then**
    start strategy update for the next time slot $D_n(slot_{t+1}) = D_n^*$;
  else
    strategy for next time slot $D_n(slot_{t+1}) = D_n(slot_t)$;
  end if
  start strategy update for the next time slot
end while

In Algorithm 2, the cost incurred by each mobile device depends not only on its own unloading decision but also on the decisions made by other mobile devices. Therefore, the entire game consists of a non-cooperative game between followers and leaders. If it is proven that the non-cooperative game between the followers and leaders in the second layer game can achieve a Nash equilibrium, then there must be at least one Stackelberg equilibrium in the entire game.

If D_N^* satisfies $D_N^* = \text{argmin}\,\phi_n(D_n, D_{-n})$, then D_N^* is the optimal decision for the current mobile device, and the Nash equilibrium strategy $\xi^* = \{D_1^*, D_2^*, \ldots, D_N^*\}$ is the Nash equilibrium point for all mobile devices in the game $G = \{N, \{\xi\}, \phi_n, n \in N\}$. When mobile devices' computation offloading strategies reach Nash equilibrium, no user can lower its cost by changing its strategy, i.e., $\phi_n(D_n^*, D_{-n}^*) \le \phi_n(D_n^*, D_{-n})$. It's a stable state where no participant can benefit from deviating from the equilibrium.

Theorem 4. *If there exists a function that for any $n \in N$ and decisions $D_n, D_n' \in D_n$ satisfying $D_N^* = \arg\min \phi_n(D_n, D_{-n})$, then in this game, keeping the decisions D_{-n} of other mobile devices unchanged, the change in the cost for any device when adopting different decisions equals the change in the function.*

$$\phi_n(D_n, D_{-n}) - \phi_n(D_n', D_{-n}) = \Phi(D_n, D_{-n}) - \Phi(D_n', D_{-n}) \tag{19}$$

Theorem 5. *By constructing a function to transform the problem of minimizing user costs into a game, there exists at least one Nash equilibrium point in the iterations of the game.*

5 Experimentation and Evaluation

5.1 Experimental Setup

The experiment ran on a computer with a 12th Gen Intel(R) Core(TM) i7-1255U processor at 1.70 GHz and 16 GB of memory. MATLAB 2020b was used for simulation. Specific parameters are provided in Table 1.

Table 1. Experiment parameters.

Parameters	Values	Parameters	Values
N	[1,50]	f_n^b	0.5 Ghz
Ω	[1,60]	B	20 Mhz
b^{cc}	50 Ghz	d	[60,90] Mbit
b^{ee}	20 Ghz	N_0B	$2 * 10^{-27}$ W
b_n^l	[2,4] Ghz	y_{n_i}	[2000, 4000] Megacycles

5.2 Experimental Results and Analysis

Firstly, analyzing Algorithm 2, as illustrated in Fig. 3(a), it can be observed that after 200 iterations, the total user cost gradually stabilizes. In other words, Algorithm 2 converges to a stable cost value after a finite number of iterations, which is the Nash equilibrium solution. And as illustrated in Fig. 3(b), it can

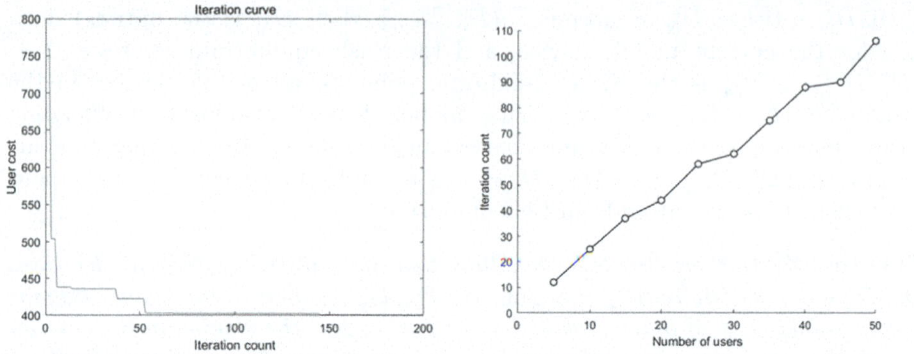

(a) The relationship between iteration count and user cost

(b) The relationship between iteration count and number of users

Fig. 3. Convergence of cost.

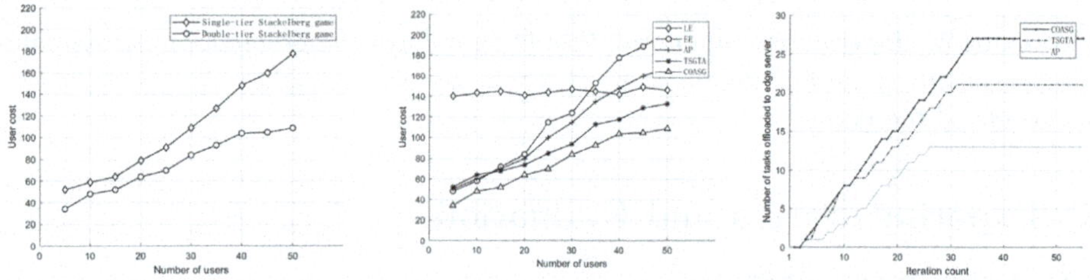

(a) User cost for different numbers of users

(b) User cost under different environment

(c) Comparative experiment on user cost

Fig. 4. Comparative experiment on user cost.

be seen that the number of iterations in the Stackelberg game is essentially linearly related to the number of users, approximately at a ratio of 2.08. This demonstrates the rationality of the algorithm. In the cloud-edge game, the cost functions of the cloud server and edge server are compared iteratively with a time complexity of $O(1)$. Let L_1 be the iterations for the first-layer game. Then, the time complexity of the optimal strategy algorithm for the cloud-edge game is $O(L_1)$. In the second-layer optimal strategy algorithm, each terminal device task is divided into D subtasks, resulting in a time complexity of $2D$. Assuming the number of iterations for the game is L_2, the second-layer's optimal strategy has $O(L_2 * 2D)$ complexity. Overall, the total time complexity based on the Stackelberg game algorithm is $O(L_1 + L_2 * 2D)$.

Then, comparing the cost curve of the Stackelberg game with that without the cloud-layer game shows our solution's feasibility and efficiency. As seen in Fig. 4(a), with more tasks, the cost of this solution remains lower than the single-layer Stackelberg game, proving its efficiency. Then comparative experiments are conducted using the following four main methods, Local Execution (LE) represents all tasks are executed locally without offloading; Edge Execution (EE) denotes all tasks are offloaded to the edge server for execution; One Access Point (AP) means one-tier Stackelberg game strategy is applied with one server.

Figure 4(b) shows that as the number of users increases, the average cost of the algorithm rises and efficiency decreases due to resource competition among users. However, our algorithm has a slower cost growth. Figure 4(c) illustrates that as iterations increase, more tasks are offloaded to edge server until stabilization due to limited resources of remote server. Our proposed algorithm offloads more tasks to the edge server, adapting better to changes in the number of users, while reducing mobile device resource utilization and maintaining low cost.

6 Conclusion

For the computation offloading problem in MEC, we propose a Stackelberg-based game solution that considers resource allocation across cloud, edge, and terminal. We use a Stackelberg game model, utilizing partial offloading, to address the computation offloading problem in a distributed manner. We analyze the strategies and costs of cloud servers, edge servers, and users, and ultimately determine the optimal strategies for all participants, proving the existence of Nash equilibrium solutions for this game. Finally, we conduct simulation experiments to demonstrate the rationality and efficiency of the COASG strategy.

References

1. Wei, Z., et al.: Integrated sensing and communication signals toward 5g-a and 6g: a survey. IEEE Internet Things J. **10**(13), 11068–11092 (2023)
2. Ding, Z., Xu, D., Schober, R., Poor, H.V.: Hybrid Noma offloading in multi-user MEC networks. IEEE Trans. Wirel. Commun. **21** (7) (2022)
3. Hao, Y., Jiang, Y., Hossain, M.S., et al.: Learning for smart edge: cognitive learning-based computation offloading. Mobile Networks Appl. **25**, 1016–1022 (2020)
4. Ndikumana, A., Tran, N.H., Ho, T.M., et al.: Joint communication, computation, caching, and control in big data multi-access edge computing. IEEE Trans. Mob. Comput. **19**(6), 1359–1374 (2019)
5. Fu, Y., et al.: Energy-efficient offloading and resource allocation for mobile edge computing enabled mission-critical internet-of-things systems. EURASIP J. Wirel. Commun. Netw. **2021**(1), 1–16 (2021). https://doi.org/10.1186/s13638-021-01905-7
6. Huang, L., Feng, X., Feng, A.: Distributed deep learning-based offloading for mobile edge computing networks. Mob. Networks Appl. **27**(3), 1123–1130 (2022)
7. Verma, T., Kumar, A.: Fuzzy Solution Concepts for Non-cooperative Games. Springer, New York (2020)
8. Fudenberg, D., Tirole, J.: Game Theory. MIT Press (1991)
9. Xu, F., Xie, Y., Sun, Y., et al.: Two-stage computing offloading algorithm in cloud-edge collaborative scenarios based on game theory. Comput. Electr. Eng. **97**, 107624 (2022)
10. Ding, Y., Li, K., Liu, C., et al.: A potential game theoretic approach to computation offloading strategy optimization in end-edge-cloud computing. IEEE Trans. Parallel Distrib. Syst. **33**(6), 1503–1519 (2021)

Target Influence Maximization Against Overexposure Under Threshold-Dependent Model in Online Social Networks

Xiaoping Zhu[1], Jianming Zhu[1(✉)], Guoqing Wang[2], and Peikun Ni[1]

[1] School of Emergency Management Science and Engineering, University of Chinese Academy of Sciences, Beijing 100049, China
{zhuxiaoping22,nipeikun18}@mails.ucas.ac.cn, jmzhu@ucas.ac.cn
[2] School of Engineering Science, University of Chinese Academy of Sciences, Beijing 100049, China
wgq94@ucas.ac.cn

Abstract. Recently, viral marketing campaigns leverage the power of word-of-mouth, a strategy whose efficacy can be negatively affected by overexposure. Existing studies exploited the extended independent cascade (IC) model, while the essence of overexposure highly relies on the cumulative impact from its neighbors, which is a feature of the linear threshold (LT) model. Therefore, this paper proposes a novel threshold-dependent(TD) model that captures overexposure in target influence maximization(TIM) problem, aiming to select at most k nodes as advocates such that the expected positive influence on target users is maximized eventually. In this model, each node randomly generates a threshold window including activation and overexposure threshold. An inactive node is positively activated if the total weight of its active in-neighbors is within its threshold window.

Since the resulting influence function in TD model is non-monotone and non-submodular, we analyze the approximation ratio of incremental greedy algorithm with suboptimal performance. Then, we develop a variant of sandwich approximation algorithm with theoretical analysis. This novel sandwich framework exploits an upper bound function, holding the form of the difference of two submodular functions, and a lower submodular bound function.

Keywords: Overexposure · Threshold-Dependent · Target Influence Maximization

1 Introduction

Recently, many social platform giants represented by Weibo, Xiaohongshu, TikTok and WeChat in China as well as Facebook and Twitter in other countries have experienced rapid development, with their influence steadily expanding.

Y. Chen et al. (Eds.): COCOON 2024, LNCS 15162, pp. 116–127, 2025.
https://doi.org/10.1007/978-981-96-1093-8_10

During viral marketing campaigns, some influential users are identified to advertise the promoted product to its neighbors so that a large number of adoptions in an expected sense are achieved [14,15]. Furthermore, in real-world marketing scenarios, it is typical to tailor promotions for specific user groups with high potential value. For instance, consider an event organizer who shares an event advertisement on a social platform, aiming to capture the interest of a large local audience.

Scholars have observed saturation effect when individuals receive information from their neighbors. When approaching a saturation point, the effect of multiple recommendations quickly saturates and any further attempts to activate become ineffective. In other words, advertising campaigns on various social platforms give people frequent ad impression and exposure, and academic research indicates that overall impact of media surpassing a certain level can have a negative effect on product evaluations [11], which can be defined as overexposure. Overexposure also arises from some other realistic cases, such as insights into being personalized and over-expected effect. In these instances, users might be less inclined to adopt the promoted product if inundated with an "excessive" number of recommendations from their friends, as they may perceive the product to have lost its uniqueness and trendiness and not being special and fashionable. For example, at friend invitation in mobile game platform, about four times invitations can trigger the invitee to sign up for the game, while excessive invitations will become an annoyance to make the invitee turn a blind eye [8]. On the other hand, overambitious seeding can result in people adopting in suboptimal contexts, having bad experiences due to over-expectation [7].

Therefore, inspired by realistic scenario and academic studies, research on target influence maximization (TIM) based on threshold-dependent (TD) model is an urgent and critical problem. TD model implies that each node has a threshold interval named threshold window containing both upper and lower bounds on the threshold. Activation attempts fail when the cumulative effect falls below the lower bound or exceeds the upper bound. Additionally, TD model also can be applied to binary choice system. When presented with two options, X or Y, individuals make decisions based on the cumulative level of evidence. When the accumulated evidence reaches the upper threshold, they choose option X, and if it falls below the lower threshold, they opt for choice Y [16].

Summary of Contributions. In order to formulate overexposure and mitigate it, we introduce a novel threshold-dependent model in Sect. 3 and analyze its computational complexity in Sect. 4. Due to its non-monotone and non-submodular property, incremental greedy algorithm can only obtain γ/k approximation ratio, where γ is the submodular ratio. Then, in Sect. 5, we propose a novel sandwich algorithm, in which the upper bound is presented as the difference of two submodular functions. This algorithm produces a $\sigma^{\mathcal{D}}(S_{sand}) \geq max\{\frac{\mu(S^*)}{\sigma^{\mathcal{D}}(S^*)} \cdot (1 - 1/e) \cdot \sigma^{\mathcal{D}}(S^*), \frac{\sigma^{\mathcal{D}}(S_\lambda)}{\lambda(S_\lambda)}((1-1/e) \cdot \sigma^{\mathcal{D}}(S^*) + (1/e - c_\tau - 1)\sigma_\tau(S_\lambda))\}$ approximation ratio in expectation.

2 Related Works

In the influence maximization problem on social media, IC model and LT model are two commonly used diffusion models, earliest proposed by Kempe [9] in 2003. They both adopt a discrete time stochastic process to quantify the information diffusion process. However, in recent years, there exists extensive literature on the extension of IC model, while only a few studies on the in-depth research of LT model. LT model has the following two distinguishing features: i) Random diffusion process: threshold θ of each node is randomly selected from $[0,1]$, reflecting our lack of understanding of user attributes. ii) Cumulative effect: LT model places emphasis on the cumulative impact exerted by seed set S on v.

Some studies, starting from the influence maximization problem, proposed extended LT model.(i)The general threshold model enjoys broader practical applications compared to LT model. In 2012, Bhagat et al. [2] defined a novel activation function $f_v(A)$ for general threshold diffusion model named LT-C in order to maximize product adoption. The activation criterion for node v is defined as $f_v(A) \geq \theta$. In 2024, Wang et al. [17] also studied product promotion problem and defined a activation function $f_v(U_t(v), S_t)$ with the form $\left(c_v - d_v e^{-S_t/\tilde{c}}\right) \sum_{u \in U_t(v)} b_{u,v}$, representing each preceding review has the potential to unveil more information than subsequent ones. (ii)The extended LT model is also designed for competitive influence diffusion. Lu et al. [12] introduced K-LT model, in which a node v would be positively activated at time t with a probability given by $\sum_{u \in A^+_{t-1} \backslash A^+_{t-2}} b_{uv} / \sum_{u \in A_{t-1} \backslash A_{t-2}} b_{uv}$, reflecting recency effect in the final decision. In 2017, Bozorgi et al. [3] proposed a decidable competitive model(DCM). In DCM, users can turn to thinking state for at most d timesteps and then decide to be activated by the neighbor with the majority of adoption. In 2020, Gao et al. [5] analyzed event-based social network and considered fair-aware competitive influence maximization under LT model, aiming to maximizing the number of participants of all events in a competitive situation. It can be referred that existing studied on extended LT model have not taken into account overexposure effect.

Recently, scholars are showing an escalating interest in the overexposure effect after overambitious seeding on the influence maximization problem and formalized the overexposure problem [1,7]. Loukides et al. [11] focused on the extended IC model by overexposure score. Abebe et al. [1] and Irfan et al. [6] assumed that each product has a appeal parameter ϕ_i and each agent has a criticality threshold θ_i. User i accepts the product and spreads to its neighbors iff $\phi \geq \theta_i$. This formulation for overexposure focuses on the product attractiveness and user expectations in order to avoid catching unintended recipients who do not view the product favorably. However, they neglect portrayal of user saturation patterns. Meanwhile, the risk of overexposure highly relies on the cumulative impact from its neighbors, which is a distinguishing feature of the LT model. Therefore, this paper constructs target influence maximization problem based on threshold-dependent model. Table 1 highlights notations frequently used in the paper.

Table 1. Frequently Used Notations

Notations	Description
$G(V, E)$	A social network
$\mathcal{D}$	The set of target users who hold high potential value
S	The seed users who spread influence and remain positive state
$[\theta_v^{\kappa}, \theta_v^{\tau}]$	The threshold window of node v where θ_v^{κ} implies activation threshold and θ_v^{τ} represents overexposure threshold.
ω	$\omega : V \times V \rightarrow [0, 1]$. A weight function such that $\omega_{uv} = 0$ iff $(u, v) \notin E$ *and* $\sum_{u \in V} \omega_{uv} \leq 1$
$\delta(v, t)$	Exposure score of node v until time t
$\sigma^{\mathcal{D}}(S)$	The number of activated target users when the seed set is S
$\lambda(S)$	The upper bound function of $\sigma^{\mathcal{D}}$, where $\lambda(\cdot) = \sigma_{\kappa}(\cdot) - \sigma_{\tau}(\cdot)$
$\sigma_{\kappa}(\cdot)$	The number of activated target nodes under threshold θ_{κ}
$\sigma_{\tau}(\cdot)$	The number of activated target nodes under threshold θ_{τ}
$\mu(S)$	The lower bound function of $\sigma^{\mathcal{D}}$
c_f, γ	The curvature and submodularity ratio of function f, resp.
$\mathcal{P}(u, S)$	The probability of node u being positively activated

3 Threshold-Dependent Model

In the LT, each edge $e = (u, v)$ has a weight ω_{uv} and each node v chooses an activated threshold θ_v uniformly at random from the interval $[0, 1]$. Currently, this paper introduces the extended LT model, which can be described as threshold-dependent (TD) model. This model adopts an overexposure threshold to depict this negative interference effect. Hence, this article assumes that each node v has a threshold window $[\theta_v^{\kappa}, \theta_v^{\tau}]$. θ_v^{τ} and θ_v^{κ} are random selected from the interval $[0, 1]$ with $\theta_v^{\tau} \geq \theta_v^{\kappa}$. θ_v^{κ} refers to activation threshold, which models the inherent tendency of a consumer to adopt the new product or information. θ_v^{τ} is an overexposure threshold, which depicts the negative externalises caused by overexposure phenomenon. Therefore, the corresponding relations of attitude state and total weight of its active in-neighbors as follows.

Suppose that each node has an exposure score $\delta(v, t)$ representing the proportion of recommendations received by node v from its neighbors until timestamp t, where $\delta(v, t) = \sum_{u \in N_+^{in}(v,t)} \omega_{uv}$ and $N_+^{in}(v, t)$ represents the positively activated and ever positively activated in-neighbors of node v by timestamp t. In the process of information diffusion, the users' status may be one of the following: (i)Inactive state (In-s): the user is not activated by its active neighbors, that is, $\delta(v, t) < \theta_v^{\kappa}$. (ii)Positive state (Ps-s): the user accepts the information corresponding to the target product and is appropriately activated by its active neighbors, that is, $\theta_v^{\kappa} \leq \delta(v, t) \leq \theta_v^{\tau}$. (iii)Negative state (Ng-s): the user receives excessive information corresponding to the target product, is influenced by overexposure effect and finally holds the negative attitude about the product. This kind of user rejects it and does not advertise it to neighbors. The procedure of TD model is as follows:

1) At timestamp $T = 0$, the seed set S is activated, spreads positive information and remains positive state in subsequent time.
2) At timestamp $T = t$, an inactive node v is positively activated if the total weight of its active in-neighbors is within its threshold window, that is, $\theta_v^{\kappa} \leq \delta(v,t) \leq \theta_v^{\tau}$. At the same time, node v in inactive or positive active state is negatively activated if the total weight of its active in-neighbors is great than its overexposure threshold, namely, $\delta(v,t) > \theta_v^{\tau}$.
3) If a node becomes negatively activated, its status remains in the subsequent time steps. The above spreading activation process repeat until there are no nodes change the state.

Definition 1 (Target Influence Maximization under Threshold-dependent Model). *Given a social network $G(V,E)$, target users set $\mathcal{D}$, threshold windows for nodes in V, this problem aims to choose seed set S with at most k users from $V\backslash\mathcal{D}$ as advocates who publish advertising information corresponding to the product in order to maximize the total expected number of positively activated target users, that is,*

$$S^* = arg \max_{S \subseteq V\backslash\mathcal{D}, |S| \leq k} \mathbb{E}\left(\sigma^{\mathcal{D}}(S)\right) \tag{1}$$

$\sigma^{\mathcal{D}}$ is the resulting influence under TD model, where threshold window for each node is random selected from the interval $[0,1]^2$. Hence, the final activated target user is probability event and the objective function can be expressed as $\mathbb{E}\left(\sigma^{\mathcal{D}}(S)\right) = \sum_{u\in\mathcal{D}} \mathcal{P}(S,u)$, where $\mathcal{P}(S,u)$ is the probability of node u being activated with the seed set S.

4 Problem Property

This section explores the computational complexity of TIM under TD model.

Lemma 1. *Denote $\delta(v,t) = \sum_{u\in N_+^{in}(v,t)} w_{uv}$ as the total weight of the positively activated in-neighbors by time step t. In the TD model, the probability of v being positively activated at time step t can be proved to $2 \cdot \delta(v,t) \cdot (1 - \delta(v,t))$.*

Proof. Drawing on the activated probability in the LT model [9,13], this lemma can be explained by graphical method generalized to probability theory. In TD model, if $\theta_v^{\kappa} \leq \delta(v,t) \leq \theta_v^{\tau}$, then v is activated at time step t. Because $\theta_v^{\tau}, \theta_v^{\kappa}$ are random selected from the interval $[0,1]$ with $\theta_v^{\tau} \geq \theta_v^{\kappa}$, the probability that $\delta(v,t)$ falls in the interval $[\theta_v^{\tau}, \theta_v^{\kappa}]$ is $2 \cdot \delta(v,t) \cdot (1 - \delta(v,t))$.

Theorem 1. *Finding an optimal solution for the TIM problem under TD model is NP-hard.*

Proof. Essentially, this problem is a variation of traditional influence maximization problem under LT model, which can be reduced to a set covering problem.

Theorem 2. *For an arbitrary set $S \subseteq V$, the objective function $\sigma^{\mathcal{D}}(S)$ under TD model is a non-monotone function.*

Proof. The positive activation of a target node is contingent upon the balance of the total weight of activated incoming neighbors, implying that insufficient or excessive activation among its incoming neighbors would inhibit its own positive activation. $\mathcal{P}_t(u, S) = 2 \cdot \delta(v, t) \cdot (1 - \delta(v, t))$ is the probability node u being activated when the seed set is S at time step t. Due to the monotone non-decreasing property of $\delta(v, t)$, $\mathcal{P}_t(u, S)$ is non-monotone and shows the property that target node u cannot be positively activated when it has fewer incoming neighbors activated or has more incoming neighbors activated.

Theorem 3. *Given a social network $G = (V, E)$ and target users set $\mathcal{D}$, for a fixed choice of threshold window for node in social network, the number of final activated target nodes is a neither submodular nor supermodular function.*

Proof. We leverage a counterexample of social network shown in Fig. 1(a) to prove this Theorem, in which the target set $\mathcal{D} = \{d, e, h\}$. The positive activated set can be obtained, that is, $A(a, b, c) = \{e\}$, $A(a, b) = \{d, h\}$, $A(a, c) = \{d, h\}$, $A(b, c) = \emptyset$, $A(c) = \{d, e, h\}$, $A(b) = \{d, e, h\}$, $A(a) = \{e\}$.

$\sigma^{\mathcal{D}}(\{a, b, c\}) - \sigma^{\mathcal{D}}(\{a, b\}) \geq \sigma^{\mathcal{D}}(\{b, c\}) - \sigma^{\mathcal{D}}(\{b\})$ and $\sigma^{\mathcal{D}}(\{a, c\}) - \sigma^{\mathcal{D}}(\{a\}) \leq \sigma^{\mathcal{D}}(\{c\}) - \sigma^{\mathcal{D}}(\emptyset)$. Therefore, for fixed threshold windows of the nodes, the objective function is neither submodular nor supermodular.

Theorem 4. *Given a social network $G = (V, E)$ and target users set $\mathcal{D}$, for an arbitrary instance of the TD model, the number of final positive activated target nodes is a neither submodular nor supermodular function.*

Proof. A simple counterexample shown in Fig. 1(b) is constructed. Lemma 1 provides the method for probability calculation. Denote target set $\mathcal{D} = \{\tau\}$, $S = \{a\}$ and $T = \{a, c\}$. $\Delta_u \mathcal{P}(T, \tau) = \mathcal{P}(T \cup \{u\}, \tau) - \mathcal{P}(T, \tau) = 0$. $\Delta_u \mathcal{P}(S, \tau) = 2\omega(1 - \omega) \cdot 2(x + y)(1 - (x + y)) + (1 - 2\omega(1 - \omega)) \cdot 2x(1 - x) - 2x(1 - x) = 4\omega(1 - \omega)(y - y^2 - 2xy)$. When $2x + y < 1$, $\Delta_u \mathcal{P}(S, \tau) > 0$. When $2x + y > 1$, $\Delta_u \mathcal{P}(S, \tau) < 0$. Because $x + y \leq 1$, The two cases described above are both possible. Hence, when $x + 2y < 1$, this problem is submodular. And conversely, this problem is supermodular. Therefore, TIM problem under TD model is neither submodular nor supermodular.

5 Approximation Algorithm Analysis

5.1 Incremental Greedy Algorithm(IGA)

Denote S_i as the seeds after i_{th} iteration in IGA. At each iteration of IGA, it selects v_i such that $max_{v_i \in V \backslash S_i} \left(\sigma^{\mathcal{D}}(S_i \cup v_i) - \sigma^{\mathcal{D}}(S_i)\right)$ and $\sigma^{\mathcal{D}}(S_i \cup v_i) - \sigma^{\mathcal{D}}(S_i) \geq 0$. Considering that this problem does not follow monotonicity and submodularity, the IGA for influence maximization problem under TD model is intuitively prone to highly suboptimal performance. Essentially, we regularize this intuition through the following result.

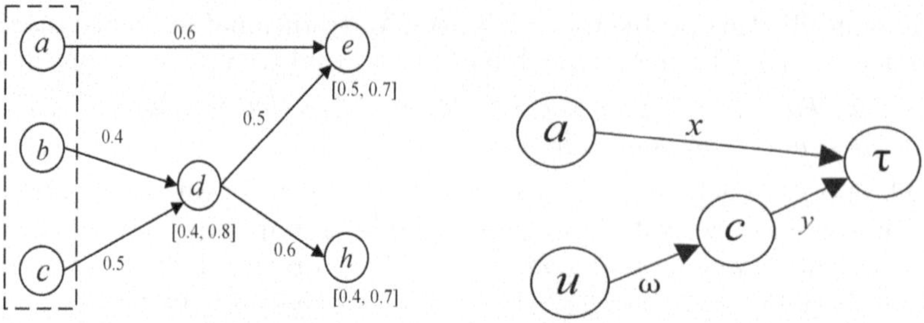

(a) Fix Threshold Graph (b) Random Threshold Graph

Fig. 1. Counterexample of Modularity Analysis

Theorem 5. *For the TIM problem under TD model, the approximation ratio for the incremental greedy algorithm(IGA) is* γ/k.

Proof. Denote S^{Ω} as the solution constructed by the incremental greedy algorithm. Recall that IGA selects the item that increases the objective value by the most. Let node v be the first node added by IGA. So, for each node $u \in S$, satisfying $\sigma^{\mathcal{D}}(\{v\}) \geq \sigma^{\mathcal{D}}(\{u\})$. Then, we can obtain the following result.

$$\sigma^{\mathcal{D}}\left(S^{\Omega}\right) \geq \sigma^{\mathcal{D}}\left(\{v\}\right) \geq \frac{1}{|S|} \cdot \sum\nolimits_{u \in S} \sigma^{\mathcal{D}}\left(\{u\}\right) \geq \frac{\gamma}{|S|} \cdot \sigma^{\mathcal{D}}\left(S\right) \tag{2}$$

where S is any subset of V. The last inequality comes from submodularity ratio γ.

5.2 Algorithm Design Based on Sandwich Framework

In order to find an optimal solution S with size at most k such that $\sigma^{\mathcal{D}}(S)$ is maximized, this section illustrates our approximation algorithm based on sandwich strategy, such that $\mu(\cdot) \leq \sigma^{\mathcal{D}}(\cdot) \leq \lambda(\cdot)$.

Upper Bound. We formulate an upper bound of $\sigma^{\mathcal{D}}(\cdot)$ by using the difference of the two functions, that is, $\sigma_{\mathcal{K}}(\cdot) - \sigma_{\mathcal{T}}(\cdot)$. In the TD model, each node v has a threshold window $[\theta_v^{\mathcal{K}}, \theta_v^{\mathcal{T}}]$. We suppose that $\sigma_{\mathcal{K}}(\cdot)$ $(\sigma_{\mathcal{T}}(\cdot))$ is the resulting influence function under LT model, in which any node v is activated for which the total weight of its active neighbors is at least $\theta_v^{\mathcal{K}}$ $(\theta_v^{\mathcal{T}})$:$\sum_{u \in N_+^{in}(v)} \omega_{uv} > \theta_v^{\mathcal{K}}$ $(\theta_v^{\mathcal{T}})$. Denote $\lambda(\cdot) = \sigma_{\mathcal{K}}(\cdot) - \sigma_{\mathcal{T}}(\cdot)$.

Theorem 6. *For any* $S \in V$, *we have* $\lambda(S) \geq \sigma^{\mathcal{D}}(S)$.

Proof. We introduce a small case to prove this theorem, shown in Fig. 2. In Fig. 2(a), each node has a threshold window $[\theta_v^{\mathcal{K}}, \theta_v^{\mathcal{T}}]$. When $S = \{u, v\}$ and target set $\mathcal{D} = \{a, b, c, e\}$, the positively activated node set refers to $A = \{a, b\}$ under TD model. Correspondingly, we construct this problem into the difference of $\sigma_\kappa(\cdot)$ and $\sigma_\tau(\cdot)$. The lower layer of Fig. 2(b) illustrates the influence spreading using LT model under the threshold θ_v^κ in which the activated set is $A_\kappa = \{a, c, b, e\}$. The upper layer of Fig. 2(b) shows the influence spreading using LT model under the threshold θ_v^κ in which the activated set is $A_\tau = \emptyset$. Here, $|A| \leq |A_\kappa| - |A_\tau|$.

Because node a is not activated in the upper layer social network, the influence cannot be transmitted to node c. Then, the node c, which surpasses the threshold, will also remain inactive. The same goes for node e. It stands to reason that influence spreading is cascading process, such that, $\sigma_\kappa(\cdot)$ is unable to cover all nodes in the resulting influence $\sigma_\tau(\cdot)$ that exceed the upper bound of window θ_v^κ. The theorem holds.

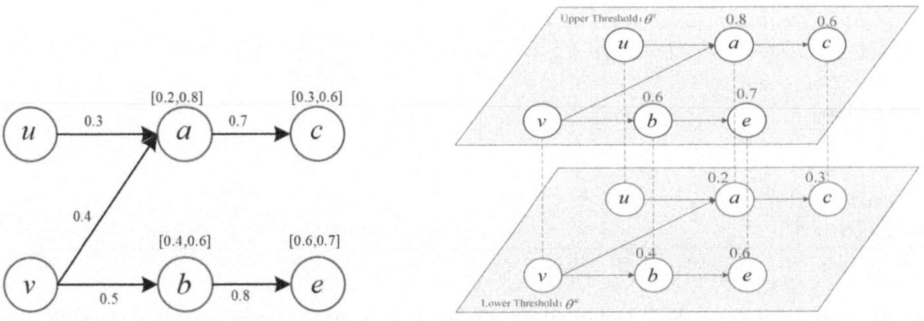

(a) Threshold-dependent Graph (b) Decomposition Map

Fig. 2. Illustration of Upper Bound Analysis

Lemma 2. *For an arbitrary threshold window* $[\theta^\kappa, \theta^\tau]$*, the resulting influence function* $\sigma_\kappa(\cdot)$ *and* $\sigma_\tau(\cdot)$ *are both monotone non-decreasing and submodular.*

Proof. According to the LT model, it is obviously that $\sigma_\kappa(\cdot)$ and $\sigma_\tau(\cdot)$ are monotone non-decreasing functions. Then, we will prove the submodular property. On the one hand, we fix the upper bound θ^τ of threshold window. Each node v selects an upper bound θ_v^τ uniformly at random from interval $[0, 1]$. Then, we can obtain the resulting influence $\sigma_\tau(\cdot)$. On the other hand, each node v selects an lower bound θ_v^κ is uniformly at random from the interval $[0, \theta_v^\tau]$. Then, we can calculate the value of resulting influence $\sigma_\kappa(\cdot)$.

The calculation of $\sigma_\tau(\cdot)$ indeed falls within the general framework of LT model, thereby belonging to the submodular function. On the other hand, due to threshold θ^κ with a restricted range, $\sigma_\kappa(\cdot)$ is a special case of general threshold model [13]. In general threshold model, each node is provided with a monotone non-decreasing function f_v. Initially, every node v randomly chooses a threshold θ from $[0, 1]$ with uniform distribution. If $f_v \geq \theta$, then its state turns to active. Else, it remains inactive state. Mossel and Roch [13] had proven that, given that f_v is monotone non-decreasing and submodular for every node v, it follows that the resulting influence, $\sigma(S)$, is likewise monotone non-decreasing and submodular with respect to the seed set S. For the resulting influence $\sigma_\kappa(\cdot)$, $f_v^\kappa(S)$ can be formulated as $f_v^\kappa(S) = \begin{cases} \sum_{u \in S} w_{uv}/\theta_v^\tau, \sum_{u \in S} w_{uv} \leq \theta_v^\tau \\ 1, \sum_{u \in S} w_{uv} > \theta_v^\tau \end{cases}$, which is inspired by $f_v = \sum_{u \in S} w_{uv}$ in traditional LT model. Clearly, f_v^κ is submodular. $\sigma_\kappa(\cdot)$ holds submodularity property.

Algorithm 1. DS Greedy Framework

Input: Directed Graph $G = (V, E)$ and Cadinality Constraint k
Output: Seed Set S^*
1: Initialize $S_0 \rightarrow \emptyset$
2: **while** $|S_i| \leq k$ **do**
3: $\quad v_i \leftarrow argmax_{v \in V \setminus S_i} \sigma_\kappa(v_i|S_i) - \sigma_\tau(v_i|V \setminus v_i)$
4: $\quad$ **if** $\sigma_\kappa(v_i|S_i) - \sigma_\tau(v_i|V \setminus v_i) \geq 0$ **then**
5: $\quad\quad S_{i+1} \leftarrow S_i \cup \{v_i\}$
6: $\quad$ **else**
7: $\quad\quad S_{i+1} \leftarrow S_i$
8: $\quad$ **end if**
9: **end while**
10: Return S

The structure of $\lambda(\cdot)$ is the difference of two monotone non-decreasing and submodular functions,which is studied as the DS decomposition [4]. Then, we present the DS greedy algorithm in Algorithm 1 and its approximation ratio in Theorem 7.

Theorem 7. *[4] Assume that S_k is the optimal solution of Algorithm 1, then we can obtain the following approximation ratio.*
$\sigma_\kappa(S_k) - (e^{-1} - c_\tau)\, \sigma_\tau(S_k) \geq (1 - e^{-1})[\sigma_\kappa(S^*) - \sigma_\tau(S^*)]$, where S^* is the optimal solution and c_τ is is the curvature of $\sigma_\tau(\cdot)$.

Lower Bound. In scalable IM problem, activation probability in LT model satisfies that $\mathcal{P}(u, S) = \sum_{u_l \in N^{in}(u)} \omega_{u_l,u} \mathcal{P}(u_l, S)$ for DAG. Inspired by this equation and combining Lemma 1, we present the following estimated recursion rule. Our estimation is inspired by utilizing the expected number of activated nodes as an estimate of the true number of activated nodes [10,11] in tree structure. However, obtaining an exact and submodular lower bound is still an open problem.

Theorem 8. *Denote $\mathcal{P}_t(u, S)$ as the activation probability of u given the seed set S at time step t. Under TD model, for any $u \in V$, we can obtain,*

$$\mathcal{P}_t(u, S) \simeq \begin{cases} 1, u \in S \\ 2\left[\sum_{u_l \in N^{in}(u)} \omega_{u_l,u} \mathcal{P}_{t-1}(u_l, S) - \left(\sum_{u_l \in N^{in}(u)} \omega_{u_l,u} \mathcal{P}_{t-1}(u_l, S)\right)^2\right] \\ , u \in V \setminus S \end{cases} \quad (3)$$

The above theorem holds on the basis of the tree network structure. Given that there might be multiple paths originating from S that pass through more than one in-neighbor of u, calculating the probability of u being activated becomes complex. The tree network structure can make sure that the events "in-neighbors of u are activated" are mutually independent. Eliminating some redundant paths to get a tree network have dual effects. Hence, generating a tree network guaranteed to be a lower bound is an urgent issue to address. Here, we propose a trivial method for generating a lower bound with tree structure. First, we adopt minimum spanning tree approach to obtain a network. Second, adjusting the weight of each edge.

Theorem 9. *Let* $[x]^+ = \max(x, 0)$. *Denote* $\mu(S) = \sum_{u \in V} \mathcal{P}^-(u, S)$, *where* $\mathcal{P}^-(u, S) = \sum_{u_l \in N^{in}(u)} 2 \cdot \left[\omega_{u_l,u} - \sum_{v \in N^{in}(u)} \omega_{v,u}^2\right]^+ \cdot \mathcal{P}^-(u_l, S)$. *For any* $S \in V$, *it holds* $\sigma^{\mathcal{D}}(S) \geq \mu(S)$.

Proof. According to Cauchy inequality, it holds that

$$\left(\sum_{u_l \in N^{in}(u)} \omega_{u_l,u} \mathcal{P}(u_l, S)\right)^2 \leq \left(\sum_{u_l \in N^{in}(u)} \omega_{u_l,u}^2\right) \left(\sum_{u_l \in N^{in}(u)} (\mathcal{P}(u_l, S))^2\right) \leq \left(\sum_{u_l \in N^-(u)} \omega_{u_l}^2\right) \left(\sum_{u_l \in N^-(u)} \mathcal{P}(u_l, S)\right).$$

Then, we can obtain a lower bound of $\sigma^{\mathcal{D}}$, which can be reformulated as $\mu(S)$. The structure of the lower bound is the same as the LT model after adjusting the weight of each edge. Because the activation probability in LT model satisfies that $\mathcal{P}(u, S) = \sum_{u_l \in N^{in}(u)} \omega_{u_l,u} \mathcal{P}(u_l, S)$ for DAG. Hence, it can be proved by extrapolate backwards method that the $\mu(\cdot)$ is submodular.

Approximation Ratio Based on Sandwich Framework. From above section, we can respectively obtain an upper bound $\lambda(S)$ and a submodular lower bound $\mu(\cdot)$. The following theorem discusses the effectiveness of proposed algorithm in terms of finding a solution with good spread $\sigma^{\mathcal{D}}$.

Theorem 10. *The solution* S_{sand} *obtained by Sandwich Approximation Framework using the result in Theorem 7:*

$$\sigma^D(S_{sand}) \geq max \left\{ \begin{array}{l} \frac{\mu(S^*)}{\sigma^{\mathcal{D}}(S^*)} \cdot (1 - 1/e) \cdot \sigma^D(S^*), \\ \frac{\sigma^D(S_\lambda)}{\lambda(S_\lambda)} ((1 - 1/e) \cdot \sigma^D(S^*) + (1/e - c_\tau - 1) \cdot \sigma_\tau(S_\lambda)) \end{array} \right\} \quad (4)$$

Proof. Let S_λ^* and S_μ^* be the optimal solution to maximize $\lambda(S)$ and $\mu(S)$ subject to $|S| \leq k$, respectively.

For submodular lower bound $\mu(\cdot)$:

$$\begin{aligned} \sigma^{\mathcal{D}}(S_\mu) &\geq \mu(S_\mu) \geq (1 - 1/e) \cdot \mu(S_\mu^*) \\ &\geq (1 - 1/e) \cdot \mu(S^*) \\ &\geq \frac{\mu(S^*)}{\sigma^{\mathcal{D}}(S^*)} \cdot (1 - 1/e) \cdot \sigma^{\mathcal{D}}(S^*) \end{aligned} \quad (5)$$

For upper bound $\lambda(\cdot)$:

$$\begin{aligned}\sigma^{\mathcal{D}}(S_\lambda) &= \frac{\sigma^{\mathcal{D}}(S_\lambda)}{\lambda(S_\lambda)} \cdot \lambda(S_\lambda) \geq \frac{\sigma^{\mathcal{D}}(S_\lambda)}{\lambda(S_\lambda)} \cdot \{(1-1/e)\lambda(S_\lambda^*) + (1/e - 1 - c_\tau)\sigma_\tau(S_\lambda)\} \\ &\geq \frac{\sigma^{\mathcal{D}}(S_\lambda)}{\lambda(S_\lambda)} \cdot \{(1-1/e)\lambda(S^*) + (1/e - 1 - c_\tau)\sigma_\tau(S_\lambda)\} \\ &\geq \frac{\sigma^{\mathcal{D}}(S_\lambda)}{\lambda(S_\lambda)} \cdot \{(1-1/e)\sigma^{\mathcal{D}}(S^*) + (1/e - 1 - c_\tau)\sigma_\tau(S_\lambda)\} \end{aligned} \quad (6)$$

6 Conclusion

Viral marketing campaigns have received increasing attention in recent years. This paper provides a new perspective on the formulation of overexpsure effect by introducing another overexposure threshold into the linear threshold model. In this paper, we propose target influence maximization problem under threshold-dependent model, which aims to trigger the largest influence spread or product adoption among target users. In threshold-dependent model, activation attempts fail when the cumulative effect falls below the lower bound or exceeds the upper bound. This problem is non-monotone and non-submodular, presenting significant challenges for algorithm design. Hence, we analyze the approximation ratio of incremental greedy algorithm with highly suboptimal performance. Then, we propose a novel sandwich framework and present its approximation ratio. Furthermore, threshold-dependent model also can be extended to model other social systems, such as accumulated evidence with the available options.

Acknowledgements. This work was supported in part by the National Natural Science Foundation of China under Grant No. 72074203 and the Fundamental Research Funds for the Central Universities.

References

1. Abebe, R., Adamic, L.A., Kleinberg, J.: Mitigating overexposure in viral marketing. In: Proceedings of the Thirty-Second AAAI Conference on Artificial Intelligence and Thirtieth Innovative Applications of Artificial Intelligence Conference and Eighth AAAI Symposium on Educational Advances in Artificial Intelligence. AAAI'18/IAAI'18/EAAI'18, AAAI Press (2018)
2. Bhagat, S., Goyal, A., Lakshmanan, L.V.: Maximizing product adoption in social networks. In: Proceedings of the Fifth ACM International Conference on Web Search and Data Mining, Seattle Washington USA, pp. 603–612. ACM (2012)
3. Bozorgi, A., Samet, S., Kwisthout, J., Wareham, T.: Community-based influence maximization in social networks under a competitive linear threshold model. Knowl.-Based Syst. **134**, 149–158 (2017)
4. Chen, S., Yang, W., Gao, S., Jin, R.: Novel algorithms for maximum DS decomposition. Theoret. Comput. Sci. **857**, 87–96 (2021)

5. Gao, S., Zhang, Z., Su, S., Wen, J., Sun, L.: Fair-aware competitive event influence maximization in social networks. IEEE Trans. Network Sci. Eng. **7**(4), 2528–2540 (2020)
6. Irfan, M.T., Hancock, K., Friel, L.M.: Cascades and overexposure in social networks: the budgeted case. In: Proceedings of the 21st International Conference on Autonomous Agents and Multiagent Systems (2022)
7. Iyer, S., Adamic, L.A.: When can overambitious seeding cost you? Appl. Network Sci. **4**(1), 38 (2019)
8. Jeong, J., Moon, S.: Invite your friends and get rewards: dynamics of incentivized friend invitation in Kakaotalk mobile games. In: Proceedings of the Second ACM Conference on Online Social Networks, Dublin Ireland, pp. 151–160. ACM (2014)
9. Kempe, D., Kleinberg, J., Tardos, E.: Maximizing the spread of influence through a social network. In: Proceedings of the Ninth ACM SIGKDD International Conference on Knowledge Discovery and Data Mining, Washington, D.C., pp. 137–146. ACM (2003)
10. Liu, B., Cong, G., Xu, D., Zeng, Y.: Time constrained influence maximization in social networks. In: 2012 IEEE 12th International Conference on Data Mining, Brussels, Belgium, pp. 439–448. IEEE (2012)
11. Loukides, G., Gwadera, R., Chang, S.W.: Overexposure-aware influence maximization. ACM Trans. Internet Technol. **20**(4), 1–31 (2020). number: 4
12. Lu, W., Bonchi, F., Goyal, A., Lakshmanan, L.V.: The bang for the buck: fair competitive viral marketing from the host perspective. In: Proceedings of the 19th ACM SIGKDD International Conference on Knowledge Discovery and Data Mining, Chicago Illinois USA, pp. 928–936. ACM (2013)
13. Mossel, E., Roch, S.: On the submodularity of influence in social networks. In: Proceedings of the Thirty-Ninth Annual ACM Symposium on Theory of Computing, San Diego California USA, pp. 128–134. ACM (2007)
14. Ni, P., Guidi, B., Michienzi, A., Zhu, J.: Equilibrium of individual concern-critical influence maximization in virtual and real blending network. Inf. Sci. **648**, 119646 (2023)
15. Ni, P., Zhu, J., Wang, G.: Activity-oriented production promotion utility maximization in metaverse social networks. IEEE/ACM Transactions on Networking, pp. 1–15 (2023)
16. Tump, A.N., Pleskac, T.J., Kurvers, R.H.J.M.: Wise or mad crowds? The cognitive mechanisms underlying information cascades. Sci. Adv. **6**(29), eabb0266 (2020)
17. Wang, Z., Song, J., Zhao, J.: Enhancing product promotion in social networks: leveraging adaptive discount-offering and the crowd effect. IEEE Trans. Eng. Manage. **71**, 1613–1628 (2024)

A Partition-and-Merge Algorithm for Solving the Steiner Tree Problem in Large Graphs

Ming Sun[1], Xinyu Wu[1], Yi Zhou[1(✉)], Jin-Kao Hao[2], and Zhang-Hua Fu[3,4]

[1] University of Electronic Science and Technology of China, Chengdu, China
{sunm,zhou.yi}@uestc.edu.cn, wu.xinyu@outlook.com

[2] LERIA, Université d'Angers, 2 Boulevard Lavoisier, 49045 Angers, France
jin-kao.hao@univ-angers.fr

[3] Shenzhen Institute of Artificial Intelligence and Robotics for Society, Shenzhen, China

[4] The Chinese University of Hong Kong, Shenzhen, China
fuzhanghua@cuhk.edu.cn

Abstract. The Steiner tree problem aims to determine a minimum edge-weighted tree that spans a given set of terminal vertices from a given graph. In the past decade, a considerable number of algorithms have been developed to solve this computationally challenging problem. However, existing algorithms typically encounter difficulties for solving large instances, i.e., graphs with a high number of vertices and terminals. In this paper, we present a novel partition-and-merge algorithm to effectively solve this problem in large graphs. The algorithm breaks the input network into small subgraphs and then merges the subgraphs in a bottom-up manner. In the merging procedure, partial Steiner trees in the subgraphs are also created and optimized by efficient local optimization. When the merging procedure ends, the algorithm terminates and reports the final solution for the input graph. We evaluated the algorithm on a wide range of benchmark instances, showing that the algorithm outperforms the best-known algorithms on large instances and competes favorably with them on small or medium-sized instances.

Keywords: Steiner tree problem · Network Design · Local search · Partition-and-merge · Large graphs

1 Introduction

The Steiner tree problem (STP) is a fundamental network design problem to determine the topology of optical networks [14], telephone networks [7], or even multimedia networks [17]. The input of the classic STP consists of an undirected edge-weighted graph (or network) G with a vertex set $V(G)$, an edge set $E(G)$, a non-negative weight function on its edges $c_G : E(G) \to \mathbb{R}^+$, and a set of *terminals* $A \subseteq V(G)$. The problem is to determine a minimum Steiner tree, that is, a

Y. Chen et al. (Eds.): COCOON 2024, LNCS 15162, pp. 128–139, 2025.
https://doi.org/10.1007/978-981-96-1093-8_11

tree T spanning all terminals in A (and possibly other vertices) and minimizes $\sum_{e \in E(T)} c_G(e)$. In fact, this optimization problem has also been studied in many communities, including operations research and theoretical computer science.

The decision version of STP is one of Karp's 21 NP-Complete problems. In recent decades, a rich number of theoretical studies have been devoted to this problem. For instance, it is known that STP can be solved in time $O(2^t n^2 + nm)$ [3] where n, m and t denote the number of vertices, edges, and terminals, respectively. The current best approximation ratio is 1.39 [5], and there is no 1.01 approximation algorithm (unless P = NP) [9]. In practice, except in the area of network design, STP is also modeled in many other fields like computational biology [8], VLSI design [13], and social network analysis [18]. Notably, two international competitions, the 11th DIMACS Implementation Challenge in 2014 [15] and the 3rd Parameterized Algorithms and Computational Experiments (PACE) Challenge in 2018 [4] have been dedicated to solving STP practically, which demonstrates the interest of the research community for this challenging problem and its variants such as the directed Steiner tree problem, the prize-collecting Steiner tree problem, etc.

Due to the great importance of STP, there exists a considerable number of both exact and heuristic solution algorithms for the problem. Exact algorithms typically rely on two search frameworks, dynamic programming and integer linear programming. To our knowledge, exact algorithms can solve problem instances with up to thousands of vertices or hundreds of terminals in a reasonable time, mostly thanks to some new ILP techniques like vertex formulation and local branching [6,10]. However, to deal with larger instances whose optimality cannot be determined by exact algorithms, heuristic algorithms become indispensable [10,11,20–22]. For example, for some VLSI graphs with up to ten thousand vertices, no exact algorithms can produce a feasible solution. According to the final report of the 11th DIMACS Challenge, PUW [20] and Staynerd [10] are considered the best global heuristics compared to other competing algorithms (*AB* [1], mozartballs [10], polito [2], and scipjack [12]) in the heuristic track. Interestingly, the exact algorithm mozartballs, which also uses some search techniques of Staynerd, is the winner in the exact track during the 11th DIMACS Challenge. As for the more recent PACE Competition, an evolutionary algorithm developed by the CIMAT Team ranked first in the heuristic track [4]. Considering that PUW and Staynerd were absent from PACE, we consider that PUW, Staynerd and CIMAT are among the most competitive heuristics for STP so far. Notably, all these heuristic algorithms use local search as their key optimization component.

Today, large STP instances with at least several thousand vertices appear ubiquitously in many applications. For example, the well-known GEO instances have more than 100,000 vertices or up to 5,000 terminals. We observed that existing heuristic algorithms, including the aforementioned DIMACS and PACE winners, attain their limit when they are applied to such large instances, leading to large gaps to the best lower bounds. In this work, our aim is to propose a novel heuristic algorithm that is able to better solve large-scale STP instances. The contributions can be summarized as follows.

- We design a partition-and-merge (PM) algorithm to deal with the large-scale nature of the given instance. PM first breaks the input graph into a number of smaller subgraphs. Then it finds partial solutions for these subgraphs and pairwise merges the subgraphs and solutions at the same time. The algorithm provides a high-quality complete solution when all the subgraphs are merged into the input graph. Meanwhile, we investigate different merging heuristics.
- We conduct extensive experiments, with a focus on very large instances, to evaluate the PM algorithm and different techniques. We show that our algorithm is very competitive in solving large and hard instances. Remarkably, for the largest publicly available STP instances, the EFST family, PM outperforms the known heuristic solvers including the winners of the 11th DIMACS and 3rd PACE competitions.

2 Basic Notations

A STP instance is composed of an edge-weighted undirected graph G and a subset of terminal vertex A. We use $V(G)$ to denote the vertex set, $E(G)$ to denote the edge set, and $c_G : E(G) \to \mathbb{R}^+$ to denote the edge-weight function of G. So $A \subseteq V(G)$. When the context is clear, we use n, m and t to denote the number of vertices, edges, and terminals, respectively. For two arbitrary vertices $u, v \in V(G)$, the *distance* $d_G(u, v)$ between them is the smallest total weight among the paths between u and v in G. For a vertex subset $S \subseteq V(G)$, $G[S]$ denotes the subgraph induced by S.

A *partition* of the graph G, $\mathcal{P} = \{H_1, ..., H_{|\mathcal{P}|}\}$, is a set of vertex-induced subgraphs of G such that $V(H_i) \cap V(H_j) = \emptyset$ for any $1 \leq i < j \leq |\mathcal{P}|$ and $\bigcup_{1 \leq i \leq |\mathcal{P}|} V(H_i) = V(G)$. $(u, v) \in E(G)$ is a *cutting edge* of $\mathcal{P}$ if u and v are in different subgraphs of $\mathcal{P}$.

Give an input instance with graph G and terminal set $A \subseteq V(G)$, a *Voronoi diagram* $\mathcal{D} = \{G_1, ..., G_t\}$ is a special partition of G such that each subgraph G_i contains exactly one terminal, denoted by a_i, and for all other vertex $u \in V(G_i) \setminus \{a_i\}$, a_i is the nearest terminal of u in G, that is, $d_G(u, a_i) = min_{a \in A} d_G(u, a)$. Clearly, the number of subgraphs in a Voronoi diagram is exactly t. For an undirected graph $G = (V, E)$, the Voronoi diagram can be computed by Dijkstra's shortest path algorithm in $O(m + n \log n)$ time (with ties breaking randomly) [22].

3 The Partition-and-Merge Algorithm

Our Partition-and-Merge (PM) algorithm is described in Alg. 1. The input of PM consists of an input graph $G = (V, E)$, a terminal set $A \subseteq V$, two parameters $\rho \leq |A|$ and $\gamma > 1$, and two maximum iteration numbers I_{short} and I_{long}.

PM maintains a partition of G, that is, $\mathcal{P} = \{H_1, ..., H_{|\mathcal{P}|}\}$. As PM runs, the subgraphs in $\mathcal{P}$ are merged in pairwise. The *merge operation* concatenates H_i and H_j into a larger graph H' by adding the cutting edges between H_i and H_j. In H', the edge weights are the same as their weights in $E(H_i)$ and $E(H_j)$ from H_i and H_j, respectively, but the weights of the cutting edges between H_i and H_j are the same as in the original input graph G. It is important to notice that the weight of the edge e in $E(H_i)$ is possibly different from its weight in $E(G)$. The reason is that PM modifies the weights of the edges of the subgraphs in $\mathcal{P}$ as the search runs.

Algorithm 1: The Partition-and-Merge Algorithm

Input: Graph G, terminal set $A \subseteq V(G)$, parameters $\rho \in [0, |A|]$, $\gamma \in [1, \infty)$, the maximum iteration numbers I_{short} and I_{long} for local search.
Output: Steiner tree T

1 Compute a Voronoi diagram $\mathcal{P} = \{H_1, ..., H_{|\mathcal{P}|}\}$ with respect to G and A.
2 Initially, $level \leftarrow 1$
3 **while** $|\mathcal{P}| > \rho$ **do**
4     $\mathcal{P} \longleftarrow$ OneRoundMerge($G, \mathcal{P}, \lceil \frac{n\gamma^{level}}{t} \rceil$)
5     $level \leftarrow level + 1$
6 **while** $|\mathcal{P}| > 1$ **do**
7   **for** *each* $H_i \in \mathcal{P}$ **do**
8         $T_i \longleftarrow$ LocalOptimize($G[V(H_i)]$, H_i, $A \cap V(H_i)$, I_{short})
9         $H_i' \leftarrow$ SolutionStoring(H_i,T_i)
10         $\mathcal{P} \leftarrow \mathcal{P} \setminus \{H_i\} \cup \{H_i'\}$
11     $\mathcal{P} \longleftarrow$ OneRoundMerge($G, \mathcal{P}, \lceil \frac{n\gamma^{level}}{t} \rceil$).
12     $level \leftarrow level + 1$
13 There is only one graph H_1 in $\mathcal{P}$.
14 $T \longleftarrow$ LocalOptimize(G, H_1, A, I_{long})
15 **return** T

3.1 Two-Phase Merging Procedure

Initially, as shown in line 1 of Alg. 1, PM builds $\mathcal{P}$ by computing the Voronoi diagram, so each subgraph in $\mathcal{P}$ contains one terminal vertex. The algorithm then performs a two-phase merging procedure, which gradually merges the subgraphs of $\mathcal{P}$ into a whole graph. The parameter ρ is used to set the time to start the second phase and γ controls the size gap between different subgraphs in $\mathcal{P}$.

In the first phase (lines 3–5), *OneRoundMerge* is used to pairwise merge the graphs in $\mathcal{P}$. In general, *OneRoundMerge* selects two subgraphs by a specific heuristic rule and joins them to form a larger graph. We introduce different merging rules in Sect. 3.2. *OneRoundMerge* also makes sure that the subgraph after a round of merge has no more than $\lceil \frac{n\gamma^{level}}{t} \rceil$ vertices (so that the subgraphs in $\mathcal{P}$ have roughly the same scale). Clearly, $\lceil \frac{n\gamma^{level}}{t} \rceil$ increases geometrically with a factor of γ as *level* increases. For example, if $\gamma = 2$, this value doubles at every level.

The second phase, as in lines 6–12, begins when there are no more than ρ subgraphs in $\mathcal{P}$. In this phase, for each subgraph H_i, we additionally carry out a local search algorithm, LocalOptimize, to find a (high-quality) Steiner tree in $G[V(H_i)]$. It is worth noting that LocalOptimize finds a starting solution from H_i, but improves the solution in $G[V(H_i)]$. Assume T_i is a Steiner tree obtained by LocalOptimize($G[V(H_i)]$, H_i, $A \cap V(H_i)$, I_{short}). T_i spans terminals $A \cap V(H_i)$ in both H_i and $G[V(H_i)]$. Our intuition is to keep the information that T_i is a high-quality solution of graph H_i, so that the further search will benefit from this information. This is implemented by SolutionStoring method, which essentially changes the edge weights of H_i. We will introduce two different ways for SolutionStoring in Sect. 3.3. After local optimization and solution storing for each subgraph, we continue using *OneRoundMerge* with the same heuristic in the first phase to merge the subgraphs.

Example. We consider the example in Fig. 1, which describes the partition-and-merge procedure. We set $\gamma = 2$ and $\rho = 6$. The size of each subgraph approximately doubles after each level of merging. The Steiner tree solution is computed when there are no more than 6 subgraphs.

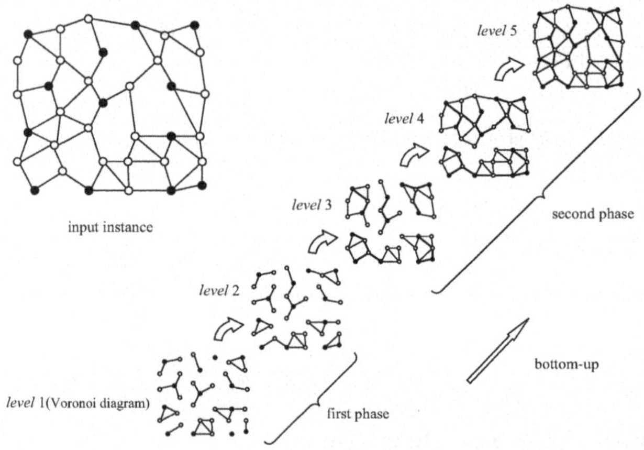

Fig. 1. The bottom-up partition-and-merge procedure. Terminal vertices are marked by black dots. The sub-solutions computed for subgraphs are marked by bold lines.

3.2 Merging Heuristics

As mentioned, the *OneRoundMerge* procedure repeatedly selects a pair of subgraphs (H_i, H_j) from $\mathcal{P}$ and merges them into one subgraph. In Alg. 2, we show the general procedure of *OneRoundMerge* using the so-called *random rule*. We will also discuss the *random-edge* and *min-distance* rules afterward.

Algorithm 2: One round of merging by random heuristic

Input: Graph G, a partition $\mathcal{P}$, subgraph size limit ω
Output: A new partition of G

1 **while** *true* **do**

> **Random rule**:
> 2 From all pairs $(H_k, H_l) \in \mathcal{P} \times \mathcal{P}$ such that $|V(H_k)| + |V(H_l)| \leq \omega$ and $cut_{\mathcal{P}}(H_k, H_l) \neq \emptyset$, randomly pick one (H_i, H_j).

3 Merge H_i and H_j to H'.
4 $\mathcal{P} \leftarrow \mathcal{P} \setminus \{H_i, H_j\} \cup \{H'\}$
5 **if** *none of the pairs of graphs in* $\mathcal{P}$ *meet the random heuristic* **then**
6 break
7 **return** $\mathcal{P}$

Random Rule. By this rule, we randomly select two subgraphs that are *adjacent*, that is, there is at least one cutting edge between them, and the total number of vertices in the two subgraphs does not exceed ω. Then we merge the two subgraphs and repeat the procedure until no such pair of graphs exits. The intuition behind this heuristic is to keep each subgraph in $\mathcal{P}$ connected and balanced in terms of vertex number after merging.

Random-Edge Rule. The random-edge rule, as shown in below, resembles the random rule except that it picks a cutting edge randomly rather than a pair of adjacent subgraphs randomly. Clearly, this rule concerns the number of cutting edges between every pair of adjacent subgraphs. Using this rule to replace line 3 in Alg. 2, we can obtain another version of *OneRoundMerge* with the random-edge heuristic.

> **Random-edge rule**:
> From all cutting edges (x, y) such that $x \in V(H_k)$, $y \in V(H_l)$ and $|V(H_k)| + |V(H_l)| \leq \omega$, randomly pick one edge (u, v).
> Suppose $u \in V(H_i)$ and $v \in V(H_j)$ without loss of generality.

We note that this rule is inspired by Karger's minimum-cut randomized algorithm [16]. As we know, Karger's algorithm finds the minimum cut of a graph with high probability by repeatedly merging a randomly selected edge. Thus, it is believed that this heuristic gives preference to the pair of adjacent subgraphs that are *closely connected*, that is, there are more cutting edges between them.

Min-Distance Rule. Let $A_i = A \cap V(H_i)$ and $A_j = A \cap V(H_j)$ where A is the input terminal set, H_i and H_j are two subgraphs in $\mathcal{P}$. For each pair of H_i and H_j, we compute $d_G(A_i, A_j)$, the minimum value over all the shortest lengths of path between any two terminals in $A_i \times A_j$, i.e., $d_G(A_i, A_j) = \min_{(u,v) \in A_i \times A_j} d_G(u, v)$. Then, the min-distance is shown as follows. Using this rule to replace line 3 in Alg. 2, we obtain *OneRoundMerge* with the min-distance heuristic.

Min-distance rule:
Find all pairs $(H_k, H_l) \in \mathcal{P} \times \mathcal{P}$ such that $|V(H_k)| + |V(H_l)| \leq \omega$ and $cut_{\mathcal{P}}(H_k, H_l) \neq \emptyset$, pick a pair (H_i, H_j) with smallest $d_G(A_i, A_j)$, ties breaking randomly

The running time of the min-distance rule bounded by $O(|E|)$ due to the early computation of the Voronoi diagram. For each non-terminal vertex $u \in H_i \subseteq A_i$, we already know its nearest terminal $a_u = argmin_{a \in A_i} d_G(a, u)$ after computing the Voronoi diagram. Thus, for each cutting edge $(u, v) \in E_{\mathcal{P}}$, we first obtain $d(u, v) = d_G(u, a_u) + c_G(u, v) + d_G(v, a_v)$ where a_u and a_v are the nearest terminals of u and v, respectively. Then, we randomly pick up an (u, v) of the smallest values $d(u, v)$ and merge its corresponding subgraphs H_i and H_j. Because we only iterate over the edges in $E_{\mathcal{P}}$, the above running time follows.

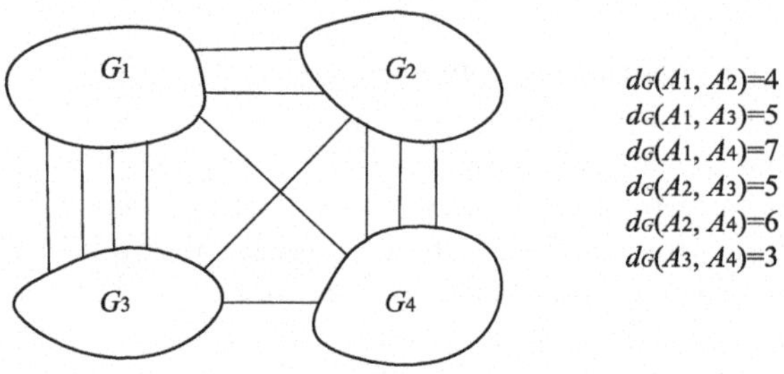

Fig. 2. Example for three merging rules. $G_1 - G_4$ are four subgraphs. Lines connecting the subgraphs represent the cutting edges between subgraphs. The minimum distance between the terminals of each two subgraphs is shown in the right-side of the figure.

Example. In Fig. 2, there are currently four subgraphs $G_1, ..., G_4$ and any two subgraphs have fewer than ω vertices. Now, we need to select a pair of adjacent subgraphs to merge. If the random rule is used, the probability that each pair of adjacent subgraphs is selected is the same. If the random-edge rule is used, the probability of selecting subgraphs (G_1, G_3) is the highest, followed by (G_2, G_4), (G_1, G_2) and (G_3, G_4). If the min-distance rule is used, (G_3, G_4) is definitely selected for merging.

3.3 Solution Storing

The motivation of solution storing is to guide future search by the fact that T_i is a high-quality solution in H_i. To achieve this, we simply change the edge weight of H_i so that we can apply the same local search algorithm. Here, we introduce a random way to reweight the edges in H_i.

For any edge $e \in E(H_i)$, we score the edge weight $c_{H_i}(e)$ as follows, where α is a random real number from $[0, 0.75]$.

$$c_{H_i}(e) = \begin{cases} (1-\alpha)c_G(e) & e \in E(T_i) \\ (1+\alpha)c_G(e) & \text{otherwise} \end{cases}$$

Clearly, the weights of edges in $E(T_i)$ are reduced, while the weights of edges not in $E(T_i)$ are increased, implying that the edges of $E(T_i)$ are preferred.

3.4 Local Optimization

The local optimization procedure, LocalOptimize(G, H, A, I_{max}) is a population-based search algorithm modified from the state-of-the-art PUW solver [20]. The original graph G, the edge reweighted graph H, terminal set A, maximum number of iterations I_{max}.

In summary, LocalOptimize follows the scheme of genetic programming. It maintains a pool of elite solutions $\mathcal{Q}$. In each generation, it produces new solutions to update the elite pool – It first builds a new Steiner tree T from H (Note that H is a graph with biased edge weights by solution storing). Then, T is improved by local search, but the local search runs on the initial graph G without biased edge weights. Afterward, the improved solution T exchanges information with another solution from the pool by combination, resulting in a new solution T'. Finally, T and T' are used to update the pool.

For detailed description of the local optimization algorithm, please refer to the full version.

4 Experiments

4.1 Settings

Our algorithm is programmed in C++ and compiled in g++ with the optimization option '-O3'. [1] All experiments were conducted on a Linux server with Intel Xeon Gold 6130 (2.10 GHz) processor.

Parameters Setting. In the experiments, ρ is set to 64 and γ is set to 2. For all test instances, we set the total time limit $t_{total} = 3600$ seconds (1h). The two parameters for controlling the number of iterations in LocalOptimize, I_{short} and I_{long}, are set to 256 and 2048, respectively. However, with these configurations, it is possible that we may not obtain a complete solution after 3600 s for some graphs. This happens when there are still more than one subgraph in $\mathcal{P}$ after this time frame. Thus, we additionally require that each LocalOptimize runs at most $\frac{0.7(\gamma-1)}{\gamma(\rho-1)} \times t_{total}$ seconds for each subgraph. This implies that at least $0.3 \times t_{total}$ seconds are allocated for running LocalOptimize with the final graph, that is, line 14 in Alg. 1.

4.2 Performance Assessment

Results on the Set of 113 Large Instances We first test the algorithm in 113 Large Instances[2]. They are grouped in 8 datasets shown in Table 1, which indicates

[1] Source codes are publicly available at https://github.com/xyu03/PM.git.

[2] All instances are downloaded from site http://dimacs11.zib.de/downloads.html.

the number of instances, average vertex number, average edge number, average terminal number, and source of the dataset. GEO, I and EFST contain larger instances (with more vertices, edges, and terminals) compared to ES, VLSI and TSPFST.

Table 1. Information about the 8 large datasets

Dataset	Num	ave.n	ave.m	ave.t	Source
ES	16	4245	6235	1563	GeoSteiner generator
TSPFST	10	4687	7043	1871	TSPLIB
VLSI	10	17926	30973	452	VLSI applications
GEO	22	105906	150767	1845	Telecommunication networks
I	10	65120	106450	3211	Telecommunication networks
EFST(R25K)	15	39483	47893	25000	GeoSteiner generator
EFST(R50K)	15	79128	96293	50000	GeoSteiner generator
EFST(R100K)	15	158147	192190	100000	GeoSteiner generator

For this set of instances, we adopt the following best STP heuristic algorithms as our reference algorithms.

- PUW [20], a multi-start local search algorithm which ranks first at Formula 1 in the single-thread heuristic track of the 11th DIMACS Competition.
- Staynerd [10], a winner of most category of instances in the single-thread heuristic track of the 11th DIMACS Competition.
- CIMAT [4], the champion of the heuristic track of the PACE Challenge (2018).
- MPCH [19], a partition-based algorithm focusing on solving large graphs.

For each instance, we perform each of the above algorithms (except MPCH[3]) 10 times independently. For each algorithm and each instance, we take the best solution obtained in the 10 trials. We summarize these results for each dataset in Table 2. Detailed results of these 113 instances are provided in the full version of the paper. For each dataset and each algorithm, we indicate the average solution gaps of the algorithm[4] and the number of instances for which the algorithm finds the best solution. In terms of solution gap, our PM algorithm performs better than other algorithms on the largest instances of the GEO, I, and EFST family datasets. It is also remarkable that PM is the best on all I and EFST instances and 16/22 on the GEO instances. Meanwhile, for the relatively smaller instances in this group, the algorithms compete with each other.

[3] Because the code of MPCH is unavailable, we extract its results from [19].

[4] For an instance, the gap of a solution T is $\frac{\sum_{e\in E(T)} c_G(e) - lb}{lb} \times 100\%$, lb being the best-known lower bound for the instance.

Table 2. The best solutions found by each algorithms on the set of 113 large instances. The smallest average gaps are marked in bold.

Dataset	ave.gap(%)					#best				
	MPCH	PUW	Staynerd	CIMAT	PM	MPCH	PUW	Staynerd	CIMAT	PM
ES	0.84	0.0082	**0.0069**	0.0987	0.0078	–	0	3	0	1
TSPFST	0.96	**0.0100**	0.0203	0.1068	0.0380	–	2	2	0	0
VLSI	1.64	**0.0292**	0.3963	0.0665	0.0730	–	2	0	1	1
GEO	-	0.8136	1.0044	1.0753	**0.7744**	–	4	0	0	16
I	0.02	0.0130	0.0154	0.0394	**0.0107**	–	0	0	0	10
EFST(R25K)	–	0.0798	0.0347	0.2761	**0.0101**	–	0	0	0	15
EFST(R50K)	–	0.1131	–	0.3387	**0.0188**	–	0	0	0	15
EFST(R100K)	–	0.1344	–	0.3644	**0.0353**	–	0	0	0	15

Results on the Set of 30 DIMACS Instances. We first test 30 instances that are used in the final ranking phase for the single thread heuristic challenge of the 11th DIMACS Implementation Challenge[5]. They are representative instances from the well-known SteinLib, Vienna, Copenhagen, PUCN, and GAPS libraries.

For this set of instances, we compare our results with those of six best performing reference algorithms: AB [1], mozartballs [10], polito [2], PUW [20], scipjack [12] and Staynerd [10], which participated in the final heuristic track of the 11th DIMACS Challenge. Following the time constraints in the competition results of the DIMACS challenge heuristic track (1 thread), all algorithms are run under a cutoff time of 2 h.

According to Table 3, in terms of finding the best solutions, polito and PM each have 6 solutions that outperform the remaining algorithms. Mozartballs performs best on 'Average' and 'ave. gap(%)'. In particular, a detailed data analysis reveals that PM has the best solution for fnl4461fst-p, alue7080-p, es10000fst01-p, cc3-12p-p, G106ac-p, I064ac-p, which are the largest instances in the group.

Table 3. The best solutions found by each algorithms on DIMACS competition instances

Instance	AB	mozartballs	polito	PUW	scipjack	staynerd	PM
ave.gap(%)	776.0807	**0.3640**	3.8920	0.4174	1.5015	0.3737	0.6219
#best	0	0	**6**	4	0	1	**6**

The Effectiveness of Different Merging Rules. We show the average solution gaps obtained by using each rule in Table 4. We use the Fredman test to check the statistical significance of the differences in terms of solution quality, which

[5] https://dimacs11.zib.de/contest/results/SPGp.1.7200.heur.html.

reveals a p-value of 0.00. The p-value (< 0.05) clearly indicates the differences between the variants with the three different merging rules. The random-edge and min-distance strategies perform better than the random strategy, while min-distance shows a slight advantage over random-edge.

Table 4. The average gaps (%) of PM variants using different merging rules.

Dataset	random	random-edge	min-distance
ES	0.02069	0.00922	**0.00783**
TSPFST	0.01832	**0.01831**	0.03799
VLSI	0.09954	0.08684	**0.07299**
GEO	0.80160	0.78055	**0.77435**
I	0.01144	0.01090	**0.01074**
EFST(R25K)	0.01423	**0.00960**	0.01008
EFST(R50K)	0.02740	**0.01868**	0.01877
EFST(R100K)	0.04900	0.03646	**0.03533**

5 Conclusion

In this paper, we introduced a new partition-and-merge algorithm to effectively solve large-scale STP instances. To ensure the effectiveness of the algorithm, we investigated different merging heuristics, local search algorithms, and new solution storing techniques.

This is the first STP algorithm based on a multi-level merging and search strategy. The algorithm has some attractive features for parallel computing. For example, we can run local optimizing procedures on different threads in parallel as they are independent.

Acknowledgements. We thank the authors of Staynerd, PUW, CIMAT and other solvers in the paper for providing their codes. We also thank the organizers of the 11th DIMACS challenge and the 3rd PACE competition for hosting the data. This work is partially funded by the National Natural Science Foundation of China under grant No. 62372093 and the Natural Science Foundation of Sichuan Province of China under grant No. 2023NSFSC1415.

References

1. Althaus, E., Blumenstock, M.: Algorithms for the maximum weight connected subgraph and prize-collecting Steiner tree problems (2014). http://dimacs11.cs.princeton.edu/workshop.html
2. Biazzo, I., Braunstein, A., Zecchina, R.: Performance of a cavity-method-based algorithm for the prize-collecting Steiner tree problem on graphs. Phys. Rev. E **86**(2), 026706 (2012)

3. Björklund, A., Husfeldt, T., Kaski, P., Koivisto, M.: Fourier meets möbius: fast subset convolution. In: Proceedings of the 39th ACM Symposium on Theory of Computing, pp. 67–74 (2007)
4. Bonnet, É., Sikora, F.: The pace 2018 parameterized algorithms and computational experiments challenge: the third iteration. In: IPEC 2018 (2018)
5. Byrka, J., Grandoni, F., Rothvoß, T., Sanità, L.: Steiner tree approximation via iterative randomized rounding. J. ACM **60**(1), 1–33 (2013)
6. Carrabs, F., Cerulli, R., Pentangelo, R., Raiconi, A.: Minimum spanning tree with conflicting edge pairs: a branch-and-cut approach. Ann. Oper. Res. **298**(1), 65–78 (2021)
7. Cheng, X., Du, D.Z.: Steiner Trees in Industry, vol. 11. Springe (2013)
8. Cheng, X., Li, Y., Du, D.Z., Ngo, H.Q.: Steiner trees in industry. In: Handbook of Combinatorial Optimization, pp. 193–216. Springer (2004)
9. Chlebík, M., Chlebíková, J.: The Steiner tree problem on graphs: Inapproximability results. Theoret. Comput. Sci. **406**(3), 207–214 (2008)
10. Fischetti, M., et al.: Thinning out Steiner trees: a node-based model for uniform edge costs. Math. Program. Comput. **9**(2), 203–229 (2017)
11. Fu, Z.H., Hao, J.K.: Swap-vertex based neighborhood for Steiner tree problems. Math. Program. Comput. **9**(2), 297–320 (2017)
12. Gamrath, G., Koch, T., Maher, S.J., Rehfeldt, D., Shinano, Y.: SCIP-jack-a solver for STP and variants with parallelization extensions. Math. Program. Comput. **9**(2), 231–296 (2017)
13. Held, S., Korte, B., Rautenbach, D., Vygen, J.: Combinatorial optimization in VLSI design. Combin. Optim.-Meth. Appl. **31**, 33–96 (2011)
14. Hsieh, C.Y., Liao, W.: All-optical multicast routing in sparse splitting WDM networks. IEEE J. Sel. Areas Commun. **25**(6), 51–62 (2007)
15. Johnson, D.S., Koch, T., Werneck, R.F., Zachariasen, M.: 11th Dimacs implementation challenge in collaboration with Icerm: steiner tree problems (2014)
16. Karger, D.R.: Global min-cuts in RNC, and other ramifications of a simple min-cut algorithm. In: SODA, vol. 93, pp. 21–30. Citeseer (1993)
17. Kun, Z., Yong, Q., Hong, Z.: Dynamic multicast routing algorithm for delay and delay variation-bounded Steiner tree problem. Knowl.-Based Syst. **19**(7), 554–564 (2006)
18. Lappas, T., Liu, K., Terzi, E.: Finding a team of experts in social networks. In: Proceedings of the 15th ACM SIGKDD International Conference on Knowledge Discovery and Data Mining, pp. 467–476 (2009)
19. Luipersbeck, M.: A new partition-based heuristic for the Steiner tree problem in large graphs. Ph.D. thesis, Technische Universität Wien (2013)
20. Pajor, T., Uchoa, E., Werneck, R.F.: A robust and scalable algorithm for the Steiner problem in graphs. Math. Program. Comput. **10**(1), 69–118 (2018)
21. Ribeiro, C.C., Uchoa, E., Werneck, R.F.: A hybrid grasp with perturbations for the Steiner problem in graphs. INFORMS J. Comput. **14**(3), 228–246 (2002)
22. Uchoa, E., Werneck, R.F.: Fast local search for the Steiner problem in graphs. J. Exp. Algorithmics (JEA) **17**, 1–2 (2012)

Analyzing the Vulnerabilities of Targets in Clean-Label Data Poisoning Attacks

Yaoyu Jin[1,2,3], Xiaochun Yang[1,4](✉), Jian Li[1], Rong Pu[1], Yujie Wang[1], and Bin Wang[1,2,3]

[1] School of Computer Science and Engineering, Northeastern University, Shenyang 110169, China
{yaoyujin,jianli,purong,yujiewang}@stumail.neu.edu.cn, {binwang,yangxc}@mail.neu.edu.cn

[2] National Frontiers Science Center for Industrial Intelligence and Systems Optimization, Northeastern University, Shenyang 110819, China

[3] Key Laboratory of Data Analytics and Optimization for Smart Industry (Northeastern University), Ministry of Education, Shenyang, China

[4] Software College, Northeastern University, Shenyang 110169, China

Abstract. The rapid evolution of artificial intelligence, propelled by large models, has led to the widespread integration of deep learning applications into everyday life. However, this convenience comes with significant security risks, especially in terms of model security. Data poisoning attacks, including targeted clean-label data poisoning attacks, pose severe threats by manipulating model outputs through covertly injected malicious poisoned samples into the training set. Despite efforts to generate more effective poisoned samples, a crucial question remains unanswered: are different target samples equally vulnerable to such attacks, and how can adversaries measure this vulnerability? To address this gap, we analyze the vulnerability of different target samples in targeted clean-label data poisoning attacks. Through experiments, we identify variations in vulnerability and demonstrate that selecting inappropriate target samples increases attack difficulty to some extent. Furthermore, we propose the **T**argeted **B**oundary **D**rift **V**elocity(**TBDV**), a novel measurement based on the velocity of the poisoned decision boundary over continuous time. Our experiments reveal that **TBDV** aids attackers in precisely identifying vulnerable samples, consequently enhancing attack success rates or diminishing the number of poisoned samples necessary in both single-target and multi-target scenarios.

Keywords: Model security · Targeted data poisoning attacks · Vulnerability assessment of targets

1 Introduction

In recent years, the field of artificial intelligence has experienced rapid growth, with the emergence of large models leading to significant advancements in deep

Y. Chen et al. (Eds.): COCOON 2024, LNCS 15162, pp. 140–151, 2025.
https://doi.org/10.1007/978-981-96-1093-8_12

learning. An increasing number of related applications are being integrated into people's daily lives. While this brings tremendous convenience, it also brings along corresponding security concerns. The outstanding performance of deep learning is attributed to its massive parameters, which rely on extensive training data. However, the diverse sources of training data also provide opportunities for malicious attackers [8]. For example, data poisoning attacks are among the primary threats to model security, where attackers manipulate the victim's model outputs by covertly injecting malicious poisoned samples into the training set.

Targeted clean-label data poisoning attack is a branch of data poisoning attacks [4]. In this type of attack, the adversary can influence the model's output for specific attack targets solely through the injection of poisoned samples with correct labels. The specificity of this attack lies in the attacker's ability to pre-determine fixed target samples before executing the attack. Attackers can conduct either single-target or multi-target attacks. The predominant focus in current research is on creating increasingly effective and inconspicuous poisoned samples for specific target samples [1,3,9,12]. However, a neglected question is: *under the same attack conditions, are different target samples equally tolerant to targeted clean-label data poisoning attacks? If not, how does the attacker measure the vulnerability level of different samples to target those that are more susceptible?* To address this challenge, we conducted data poisoning attack experiments on different target samples under identical conditions and analyzed their respective attack outcomes. We explicitly identify differences in the vulnerability of different target samples and demonstrate that selecting inappropriate (hard) targets increases the difficulty of the attack to some extent. Furthermore, we introduce a measurement based on the velocity of the poisoned decision boundary over continuous time (**TBDV**). Our experiments confirm that this approach enables attackers to more accurately hit the most vulnerable targets, thereby increasing the attack success rate or reducing the amount of poisoned samples required.

Related Works. Targeted clean-label data poisoning attackers create poisoned samples by introducing subtle perturbations to base samples. The prevailing attack methods can generally be classified into two main approaches: feature space attack and bi-level optimization attack. **Feature space attacks** [1,9,12] aim to generate poisoned samples that overlap with target in the feature space. Convex Polytope attacks (CP) [12] involve constructing a convex polytope in the feature space with the poisons, attempting to encompass the target inside the polytope. Bullseye Polytope attacks (BP) [1] are an improvement on CP, aiming to position the target precisely at the center of the polytope by assigning fixed weights to each poison. **Bi-level Optimization Attack** [3,5] formulates data poisoning attacks as a two-tier optimization problem. The Gradient Matching (GM) [3] method aligns clean-label poisoned samples and adversarially labeled targets in the gradient space.

Contribution of this Paper. This paper introduces the phenomenon that different target samples exhibit varying vulnerabilities in targeted clean-label data poisoning attacks, which has been overlooked in prior works.

- We empirically validate the diversity in vulnerability among target samples and highlight its correlation with a decrease in the attack success rate.
- We propose a novel vulnerability measurement **TBDV** from the perspective of dynamic changes in decision boundaries. We highlight that during the poisoning of target model training, the decision boundary associated with more vulnerable targets moves faster, leading to a higher rate of misclassification of targets into the adversarial class.
- Our experiments in both single-target and multi-target scenarios demonstrate the effectiveness of **TBDV** in enhancing attack efficiency and markedly reducing poison budget. In single-target experiments, target samples selected by **TBDV** exhibit a substantial reduction of 88.9% in poison ratio (i.e. the proportion of poisoned samples in the training set) compared to random targets. Similarly, this reduction exceeds 83.33% in multi-target experiments.

2 Preliminaries

We are addressing an image classification task employing a deep classifier characterized by the parameter set θ. This classifier is represented as follows: $f_\theta(x) = \arg\max_{y \in Y} g^y(x, \theta)$, where (x, y) denotes a clean input $x \in \mathcal{X}$ and its corresponding label $y \in \mathcal{Y} = \{1, 2, \cdots, K\}$. And $g^y(x, \theta)$ denotes the model's logit output for class y.

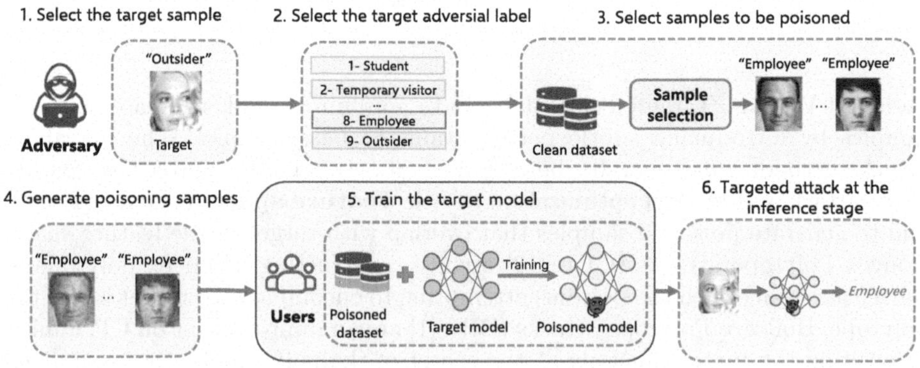

Fig. 1. The pipeline of targeted clean-label data poisoning attacks.

2.1 Targeted Clean-Label Data Poisoning Attack

In this work, our focus lies on clean-label targeted poisoning attacks, which aim to manipulate the model's output for a specified test target sample (x_t, y_t) or a set of test target samples $\mathcal{T} = \{(x_t^i, y_t^i)\}_{i=1,\cdots,n}$ by injecting a minimal amount of poisoned samples into the model's training set, causing them to be misclassified as the designated adversarial label y_{adv}, where $y_t, y_{\text{adv}} \in \mathcal{Y}$. The adversary attempts to maintain the usability of the model after poisoning.

Here, we present an illustrative scenario of clean-label targeted poisoning attacks: A facial recognition security access system can distinguish between internal employees and external visitors. Additionally, the access system is equipped with a security guard who supervises each visitor to ensure they are not wearing any special disguises. The attacker aims to manipulate this access system to classify a malicious outsider as an internal employee. However, the attacker can only introduce minimal perturbations to a small portion of the samples in the training set and cannot change the labels of the samples. This ensures that the perturbed samples are not significantly different from the original samples in the eyes of human auditors while retaining their correct labels. The access system can undergo normal training and validation and can be deployed smoothly in real-world scenarios. It operates correctly for all individuals except for the target person specified by the attacker.

As depicted in Fig. 1, clean-label targeted poisoning attacks can be delineated into six sequential steps. Initially, the attacker must identify the target sample(s) (*step 1*) and specify the label(s) to which it should be incorrectly classified (in the context of multi-class classification problems), known as the target adversarial label(*step 2*). Subsequently, the attacker selects the samples in the training dataset to be perturbed (*step 3*) and meticulously crafts poisoning perturbations tailored to the target sample(s), subsequently integrating them into the model's training set (*step 4*). Following this, the victim inadvertently employs the poisoned training set to train the model, resulting in the acquisition of a seemingly valid model during the validation phase (*step 5*). Ultimately, when the originally designated target sample is introduced to the trained model, successful misclassification to the specified target adversarial label indicates the accomplishment of the attack (*step 6*).

Most of the relevant research focuses on *step 4*, which pertains to generating more covert perturbations to be added to the original samples to create effective poisoning samples. Formally, this step is often defined as a bi-level optimization problem [3]:

$$\begin{aligned} &\min_{\Delta} \sum_{i=1}^{n} \mathcal{L}(f(x_t^i, \theta(\Delta)), y_{\text{adv}}) \\ \text{s.t.} \quad &\theta(\Delta) \in \arg\min_{\theta} \frac{1}{N} \sum_{i=1}^{N} \mathcal{L}(f(x_i + \Delta_i, \theta), y_i), \end{aligned} \tag{1}$$

where Δ represents the minute perturbations sought after by attackers. (x_i, y_i) denotes the clean samples with correct labels within the training set $\mathcal{D}$, i.e.,

$\mathcal{D} = \{(x_i, y_i)\}_{i=1}^{N}$. It's important to note that $(x_t^i, y_t^i) \notin \mathcal{D}$. Additionally, $\mathcal{L}(f(x_i+\Delta_i, \theta), y_i)$ is the loss function of the target model.

2.2 Poisoned Decision Boundary

A targeted data poisoning attack can lead to a shift in the decision boundary near the target sample [9,11]. As the adversary selects the attack target (x_t, y_t) and the adversarial targeted label y_{adv}, our focus lies primarily on the change in the decision boundary between the class y_t and y_{adv}. Therefore, we curve the poisoned decision boundary in the poisoned model.

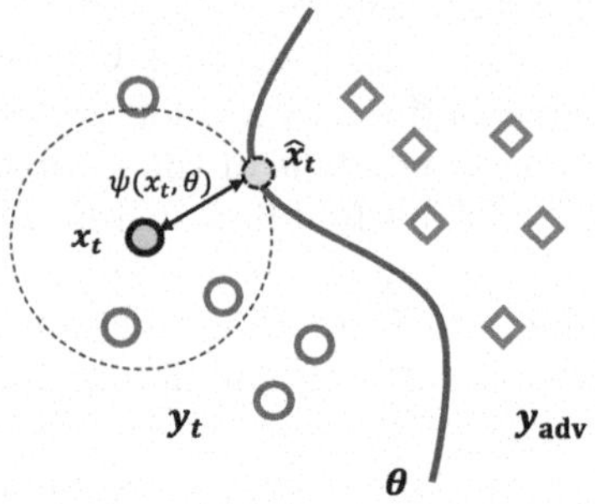

Fig. 2. The curve of the poisoned decision boundary and the poisoned margin, $\hat{x}_t$(dashed blue dots) is the closest poison boundary point of the target x_t(solid blue dots), and the samples with the adversarial label (orange diamonds) are distributed on the other side of the poisoned decision boundary. (Color figure online)

Poisoned Logit Margin. For a well-trained target model with parameters θ, when the condition of maximum logits value is met, i.e.

$$\max_{y \in \mathcal{Y}} g^{y}(x_t, \theta) \in \{g^{y_{\mathrm{adv}}}(x_t, \theta), g^{y_t}(x_t, \theta)\}. \tag{2}$$

The poisoned logit margin of the target sample x_t represents the difference in logit outputs of the model for the classes y_{adv} and y_t concerning x_t:

$$\Phi_{y_t}^{y_{\mathrm{adv}}}(x_t, \theta) = g^{y_{\mathrm{adv}}}(x_t, \theta) - g^{y_t}(x_t, \theta). \tag{3}$$

The poison margin represents the difference between the output logit values of the neural network. Observe that the target sample is successfully attacked only if $\Phi_{y_t}^{y_{\mathrm{adv}}}(x_t, \theta) > 0$. We observe that the poisoned decision boundary is determined by the zero-level set of the poisoned logit margin, denoted as

$$\Gamma_{y_t}^{y_{\mathrm{adv}}} = \{x : \Phi_{y_t}^{y_{\mathrm{adv}}}(x_t, \theta) = 0\}. \tag{4}$$

Poisoned Margin. When quantifying the distance between x_t and the poisoned decision boundary within the input space $\mathcal{X}$, we define the poisoned margin as:

$$\Psi(x_t,\theta)=\begin{cases}-\min\limits_{\hat{x}_t}||\hat{x}_t-x_t||_p, \text{s.t. } \Phi_{y_t}^{y_{\text{adv}}}(\hat{x}_t,\theta)=0, \text{ if } \max\limits_{y\in\mathcal{Y}} g^y(x_t,\theta)=g^{y_t}(x_t,\theta)\\ \min\limits_{\hat{x}_t}||\hat{x}_t-x_t||_p, \text{s.t. } \Phi_{y_t}^{y_{\text{adv}}}(\hat{x}_t,\theta)=0, \text{ if } \max\limits_{y\in\mathcal{Y}} g^y(x_t,\theta)=g^{y_{\text{adv}}}(x_t,\theta),\end{cases} \tag{5}$$

where $||\cdot||_p$ is the l_p norm with $1\leq p\leq\infty$.

The poisoned logit margin measures the gap between the logit outputs of the model. As illustrated in Fig. 2, the poison margin measures the distance from the target sample to the decision boundary in the input space. And $\hat{x}_t$ is the closest poison boundary point of the target x_t.

3 Attacking Fragile Targets

The targeted clean-label data poisoning attack is characterized by a crucial feature: the adversary has a specific target (or a set of targets) in mind. The adversary aims to tailor poison samples specifically for this target, intending to influence the model parameters during training to shift towards misclassifying the designated target. However, in prior works, the focus has predominantly been on designing optimization algorithms to generate effective perturbations for crafting poison samples [1,3,9,12], while overlooking another critical aspect: *(i) Whether the difficulty of attacking different target samples is equivalent? (ii) What type of target samples adversaries should select for the attack?*

To address the primary challenge of this paper, we first demonstrate, through exploratory experiments, the significance of target sample selection. **Not all target samples exhibit identical susceptibility during the attack process.** Subsequently, we introduce a measure termed the Targeted Boundary Drift Velocity (**TBDV**), which quantifies the rate of movement of the poisoned model's decision boundary with respect to the target samples over continuous time. Finally, we establish a targeted clean-label data poisoning attack framework that incorporates the selection of target samples.

3.1 Variability in Vulnerability Levels Across Target Samples

To address the first question, we conducted motivation experiments following the experimental setup of BP [1] on CIFAR-10. While controlling other variables, we performed data poisoning attacks on 100 different target samples from the test set across eight target models with a poison number is 40 (poison ratio is 0.08%). Figure 3 illustrates the varying attack success rates (ASR) of these 100 target samples under identical conditions.

We observe that these 100 randomly selected target samples exhibit varying vulnerabilities to the targeted clean-label data poisoning attack. Only 43% of the samples have attack success rates exceeding 0.8, with the proportion of target samples achieving attack success rates above 0.9 accounting for merely 15% of

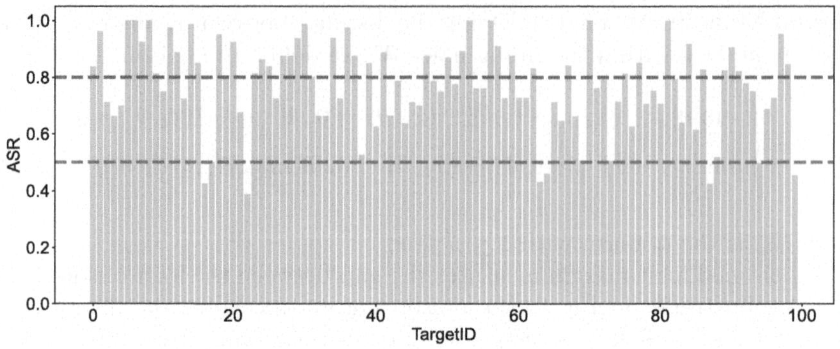

Fig. 3. The average ASR results of 100 different targets with BP-CIFAR-10 experiments.

the total target samples. Additionally, 8% of the samples exhibit attack success rates below 50%, with the most challenging sample achieving an attack success rate of 38.75%.

This set of experiments provides initial validation of our hypothesis: vulnerability varies among different target samples under identical conditions. Previous research has overlooked this aspect, where adversaries' random selection of target samples may inadvertently increase the difficulty of attacks. Conversely, strategically selecting more vulnerable target samples can significantly amplify attackers' success rates. The specificity of targeted data poisoning attacks lies in the attackers' ample freedom to specify attack targets. Therefore, the selection of target samples is a feasible and meaningful step.

3.2 Threat Model

Formally, we define the two parties involved in the attack: the adversary and the victim. The adversary can only execute the attack by injecting minute poison samples (less than 1%) into the training set, while the victim is the owner of the target model, responsible for active training, testing, and deploying the model. We consider a gray-box model setting: the attacker is aware of the framework of the target model owned by the victim but lacks detailed information about the model, including initialization details, mini-batching during training, data augmentation, and specific model parameters. Additionally, before crafting poison samples, the adversary can select their attack target samples or a set of attack target samples within a certain range. This set of samples, known as the target sample candidate pool, does not overlap with the training set samples. The adversary trains a substitute model on a dataset with the same distribution as the target training set and with the same architecture as the target model. Subsequently, the adversary crafts poison samples for the target sample(s) based on this substitute model.

3.3 Measure the Vulnerability of Targets

In this section, to address problem (ii), we draw inspiration from [10] and investigate the dynamic changes of the poisoned boundary concerning target samples over continuous time. We define the velocity of the poisoned boundary at time t and analyze its correlation with the vulnerability of target samples. Finally, we propose the Targeted Boundary Drift Velocity (**TBDV**) and integrate it with the traditional data poisoning attack process, presenting a new framework for data poisoning attacks that includes target sample selection.

The Evolution of Poisoned Decision Boundary Over Continuous Time. We illustrate the evolution of the decision boundary concerning target samples over continuous time during data poisoning attacks in Fig. 4. Figure 4a demonstrates that at time t_0, the target sample x_t (solid blue dot) is correctly classified. In Fig. 4b, at time t_1, the poisoned decision boundary moves towards the target sample, resulting in a reduction of the poisoned margin. Figure 4c indicates that at time t_2, the attacker achieves a successful attack as the target sample crosses the decision boundary to the adversarial class y_{adv}.

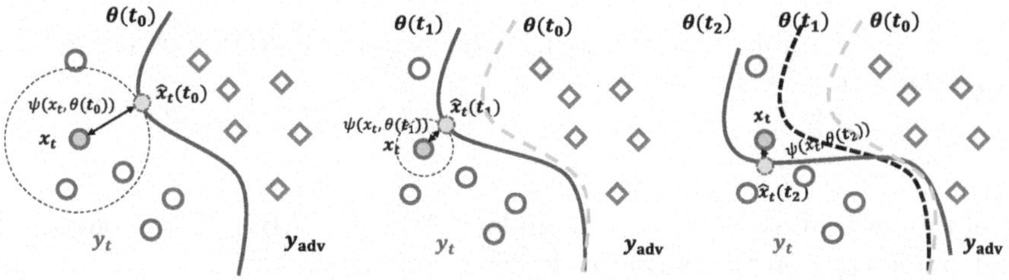

(a) The decision boundary correctly classifies the target sample with a relatively large poisoned margin.

(b) The attack causes the decision boundary near the target sample to move towards it, resulting in a smaller poisoned margin.

(c) The target sample crosses to the other side of the decision boundary, indicating a successful attack.

Fig. 4. In the continuous-time evolution of decision boundaries concerning target samples during data poisoning attacks, solid blue dot represent the target sample with a ground-truth label of y_t, while hollow blue dots represent training set samples with a label of y_t. Orange diamonds depict training set samples with a label of y_{adv}. (Color figure online)

Targeted Boundary Drift Velocity. As demonstrated in Fig. 3, at any given moment, not all target samples are successfully attacked simultaneously. We propose that during the model training phase, target samples that cross the decision boundary more quickly are more vulnerable to data poisoning attacks.

To investigate the instantaneous velocity of the poisoned decision boundary over continuous time, we substitute the model parameters θ with parameters $\theta(t)$ that are associated with time t. The update of model parameters is computed through gradient descent of the loss function i.e., $\theta'(t) = -\nabla_\theta \mathcal{L}(\theta(t))$.

To measure the vulnerability of target samples, we first define the poisoned decision boundary in continuous time.

Definition 1. *Poisoned decision boundary for specific target x_t is defined as:*

$$\Gamma_{y_t}^{y_{adv}}(t) = \{x : \Phi_{y_t}^{y_{adv}}(x_t, \theta(t)) = 0\}, \tag{6}$$

where $\Phi_{y_t}^{y_{adv}}(x_t, \theta(t))$ represents the poisoned logit margin of x_t with respect to the model parameter $\theta(t)$.

Definition 2. *we follow the [10], the closest point on poisoned decision boundary for specific target x_t is defined as:*

$$\hat{x}_t(t) = \arg\min_{\hat{x}_t} ||\hat{x}_t - x_t||_p \quad s.t. \quad \Phi_{y_t}^{y_{adv}}(\hat{x}_t, \theta(t)) = 0 \tag{7}$$

Assuming the maximum value in the logits output of the target model for x_t is either y_t or y_{adv}, we define the poisoned margin of x_t st time t as:

$$\Psi(x_t, \theta(t)) = \begin{cases} -\min\limits_{\hat{x}_t(t)} \quad ||\hat{x}_t(t) - x_t||_p, & \text{if} \quad \max\limits_{y \in \mathcal{Y}} g^y(x_t, \theta(t)) = g^{y_t}(x_t, \theta(t)) \\ \min\limits_{\hat{x}_t(t)} \quad ||\hat{x}_t(t) - x_t||_p, & \text{if} \quad \max\limits_{y \in \mathcal{Y}} g^y(x_t, \theta(t)) = g^{y_{adv}}(x_t, \theta(t)). \end{cases} \tag{8}$$

Notably, we employ the Fast Boundary Attack (FAB) [2] to compute the closest point $\hat{x}_t(t)$ on the poisoned decision boundary.

Definition 3. ***Targeted Boundary Drift Velocity (TBDV)*** *$v(x_t, t)$ is defined as the derivative of the poisoned margin:*

$$v(x_t, t) = \frac{d}{dt} \Psi(x_t, \theta(t)) \tag{9}$$

Based on the aforementioned setting, we propose a data poisoning attack framework that takes into account target sample selection. The attacker employs **TBDV** as a metric to measure and rank the vulnerability of samples in the target pool. Samples with a higher **TBDV** can traverse the poisoned decision boundary more rapidly, indicating a greater likelihood of successful attack. This allows the attacker to select the more vulnerable targets for the attack.

4 Experiments

In this section, we evaluate the effectiveness of the proposed **TBDV** on the CIFAR-10 [6] and Tiny-ImageNet [7] datasets. And compared with the random targets attack baseline.

4.1 Single Target Experiments and Results

In our initial experiments, we focused on a single-target sample setting. We employed the Bullseye Polytope (BP) [1] method on the CIFAR-10, targeting the class *frog* with an adversarial class of *ship* and a poison ratio from 0.01% to 0.1%. The target models' retraining epochs are 60. The baseline experiments utilized a random selection strategy for target samples. In each trial, one target sample was randomly selected, and the poisoning attack was conducted using eight different target models. The attack results were recorded, and the experiment was repeated ten times. The attack success rate (ASR) was calculated based on the number of successful attacks out of 80 trials. The results of the baseline experiments correspond to the first row (Random Target) of Table 1. The results of the baseline experiments correspond to the first row of Table 1.

Our proposed **TBDV** experiments involve selecting a single target and conducting the BP for eight different target models in each trial. We repeat this process for 10 trials to compute the ASR statistics. To demonstrate the effectiveness of our method, we separately computed the ASR results for the most vulnerable target samples (i.e., those with the fastest velocity) and the hardest target samples (i.e., those with the slowest velocity) obtained by the **TBDV** algorithm.

Table 1. The ASR results of BP-CIFAR-10 experiments

Poison Num	5	10	15	20	25	30	35	40	45	50
Random Target	0.50	0.55	0.54	0.64	0.70	0.74	0.76	0.83	0.84	0.88
TBDV-Vulnerable Target	0.85	0.94	0.91	0.93	0.95	0.98	0.97	0.98	1.00	1.00
TBDV-Hard Target	0.04	0.03	0.04	0.08	0.10	0.14	0.2	0.23	0.31	0.31

As we increase the poison ratio, the experimental results for three different types of targets are shown in Table 1. The results demonstrate that different target samples exhibit varying degrees of vulnerability in data poisoning attacks. **TBDV** serves as an effective measurement, enabling the selection of the most vulnerable target samples (**TBDV**-Vulnerable) and the exclusion of the hardest ones (**TBDV**-hard). For target samples selected by **TBDV**, using only a 0.01% poison ratio yields a higher attack success rate compared to using over a 0.09% poison ratio for randomly selected target samples. This significantly enhances the efficiency and stealthiness of the attack.

4.2 Multi-target Experiments and Results

We conducted similar experiments for multi-target scenarios, utilizing the Gradient Matching (GM) [3] method to execute attacks on Tiny-ImageNet. Poison ratio from 0.025% to 0.25%. The target number is 50, the target label is *otter* and the adversarial label is *Labrador Retriever*. We randomly sampled 50 targets from the candidate pool in the test set for the baseline. Subsequently, we employed three distinct models as target models to execute data poisoning attacks. The ASR was computed by aggregating the results of these 150 attacks. Similarly, we employed **TBDV** to select the most vulnerable 50 samples and the most hard 50 samples, conducting attack experiments on the same three target models. The number of successful attacks in the 150 trials was tallied for each, serving as the ASR, with results documented in Table 2.

Table 2. The ASR results of GM-TinyImageNet experiments

Poison Num	25	50	75	100	125	150	175	200	225	250
Random Targets	0.05	0.07	0.11	0.13	0.17	0.17	0.21	0.21	0.23	0.24
TBDV-Vulnerable Targets	0.19	0.19	0.21	0.2	0.21	0.26	0.36	0.37	0.37	0.39
TBDV-Hard Targets	0.00	0.00	0.02	0.02	0.06	0.06	0.05	0.13	0.13	0.15

The experimental results indicate that in multi-target attacks, our proposed **TBDV** method can significantly boost the success rate by selecting vulnerable samples. With a poison ratio of 0.25%, **TBDV** outperforms random targets selection by 15% in terms of ASR. Even with just 0.025% poison ratio, it surpasses the success rate of random targets attacks with a poison ratio of 0.15%. The **TBDV**-Hard Targets experiments underscore the risk associated with selecting hard samples, which could potentially render previously effective attack methods ineffective.

5 Conclusion

The present study addresses the issue of vulnerability in targeted clean-label data poisoning attacks across different targets. Through various experiments, we have confirmed that different target samples exhibit varying degrees of vulnerability to attacks. We propose a measurement based on the movement of the poisoned decision boundary over continuous time, termed Targeted Boundary Drift Velocity (**TBDV**), to quantify this vulnerability. Experimental results demonstrate that our method effectively identifies hard target samples, thereby enhancing the success rate of attacks by accurately selecting vulnerable target samples or significantly reducing the number of poisoned samples required.

Acknowledgements. The work is partially supported by the National Natural Science Foundation of China (No. 62072088), Liaoning Provincial Science and Technology Plan Project - Key R&D Department of Science and Technology (No.2023JH2/101300182), and 111 Project (No. B16009).

References

1. Aghakhani, H., Meng, D., Wang, Y., Kruegel, C., Vigna, G.: Bullseye polytope: A scalable clean-label poisoning attack with improved transferability. In: IEEE European Symposium on Security and Privacy, EuroS&P 2021, Vienna, Austria, September 6-10, 2021, pp. 159–178. IEEE (2021)
2. Croce, F., Hein, M.: Minimally distorted adversarial examples with a fast adaptive boundary attack. In: Proceedings of the 37th International Conference on Machine Learning, ICML 2020, 13-18 July 2020, Virtual Event. Proceedings of Machine Learning Research, vol. 119, pp. 2196–2205. PMLR (2020)
3. Geiping, J., et al.: Witches' brew: Industrial scale data poisoning via gradient matching. In: 9th International Conference on Learning Representations, ICLR 2021, Virtual Event, Austria, May 3-7, 2021. OpenReview.net (2021)
4. Goldblum, M., et al.: Dataset security for machine learning: data poisoning, backdoor attacks, and defenses. IEEE Trans. Pattern Anal. Mach. Intell. **45**(2), 1563–1580 (2023)
5. Huang, W.R., Geiping, J., Fowl, L., Taylor, G., Goldstein, T.: Metapoison: Practical general-purpose clean-label data poisoning. In: Larochelle, H., Ranzato, M., Hadsell, R., Balcan, M., Lin, H. (eds.) Advances in Neural Information Processing Systems 33: Annual Conference on Neural Information Processing Systems 2020, NeurIPS 2020, December 6-12, 2020, virtual (2020)
6. Krizhevsky, A., Hinton, G.: Learning multiple layers of features from tiny images. Master's thesis, Department of Computer Science, University of Toronto (2009)
7. Le, Y., Yang, X.: Tiny imagenet visual recognition challenge. CS **231N**, 7 (2015)
8. Schwarzschild, A., Goldblum, M., Gupta, A., Dickerson, J.P., Goldstein, T.: Just how toxic is data poisoning? A unified benchmark for backdoor and data poisoning attacks. In: Meila, M., Zhang, T. (eds.) Proceedings of the 38th International Conference on Machine Learning, ICML 2021, 18-24 July 2021, Virtual Event. Proceedings of Machine Learning Research, vol. 139, pp. 9389–9398. PMLR (2021)
9. Shafahi, A., et al.: Poison frogs! targeted clean-label poisoning attacks on neural networks. In: Bengio, S., Wallach, H.M., Larochelle, H., Grauman, K., Cesa-Bianchi, N., Garnett, R. (eds.) Advances in Neural Information Processing Systems 31: Annual Conference on Neural Information Processing Systems 2018, NeurIPS 2018, December 3-8, 2018, Montréal, Canada, pp. 6106–6116 (2018)
10. Xu, Y., Sun, Y., Goldblum, M., Goldstein, T., Huang, F.: Exploring and exploiting decision boundary dynamics for adversarial robustness. In: The Eleventh International Conference on Learning Representations, ICLR 2023, Kigali, Rwanda, May 1-5, 2023. OpenReview.net (2023)
11. Yang, Y., Liu, T.Y., Mirzasoleiman, B.: Not all poisons are created equal: Robust training against data poisoning. In: Chaudhuri, K., Jegelka, S., Song, L., Szepesvári, C., Niu, G., Sabato, S. (eds.) International Conference on Machine Learning, ICML 2022, 17-23 July 2022, Baltimore, Maryland, USA. Proceedings of Machine Learning Research, vol. 162, pp. 25154–25165. PMLR (2022)
12. Zhu, C., Huang, W.R., Li, H., Taylor, G., Studer, C., Goldstein, T.: Transferable clean-label poisoning attacks on deep neural nets. In: Chaudhuri, K., Salakhutdinov, R. (eds.) Proceedings of the 36th International Conference on Machine Learning, ICML 2019, 9-15 June 2019, Long Beach, California, USA. Proceedings of Machine Learning Research, vol. 97, pp. 7614–7623. PMLR (2019)

An Approach to Tight I/O Lower Bounds for Algorithms with Composite Procedures

Rui Xia[1,2], Ligang Cao[1,2], Huajian Zhang[1,2], Jihu Guo[1,2], Xiaowei Guo[1,2,3](✉), Jie Liu[1,2,4](✉), and Huaimin Wang[1,4]

[1] College of Computer Science and Technology, National University of Defense Technology, Changsha, China
{xiarui21,ligangcao,zhjsag,guoxiaowei,liujie}@nudt.edu.cn
[2] Laboratory of Digitizing Software for Frontier Equipment, National University of Defense Technology, Changsha, China
[3] Institute for Quantum Information and State Key Laboratory of High Performance Computing, National University of Defense Technology, Changsha, China
[4] National Key Laboratory of Parallel and Distributed Computing, National University of Defense Technology, Changsha, China

Abstract. Analyzing the I/O lower bound is crucial for enhancing dataflow performance within the memory hierarchy. Existing methods struggle with composite step programs, often underestimating I/O behavior during subcomputations. To address this, we introduce the (X_1, X_2)-Partition theory, providing tighter I/O lower bounds for statically analyzable programs. We evaluate this theory on stencil operators and show that the I/O lower bound becomes tighter compared to X-Partition theory. Furthermore, we develop a method for calculating I/O lower bounds for composite algorithms using the (X_1, X_2)-Partition theory. By analyzing the dataflow of the QR Decomposition algorithm, we demonstrate the effectiveness of our method in handling composite procedures. We also introduced this theory into the neural network architecture search task for the first time, and preliminary experimental results further demonstrated the broad application prospects of this theory.

Keywords: I/O Complexity · I/O Lower Bound · Red-blue Pebble Game · Network Architecture Search

1 Introduction

One of the major expenses in scientific computing programs is the overhead of input/output (I/O) operations across memory structures. Studying the I/O lower bound, as mentioned in [1], is crucial for achieving high-performance computing, minimizing I/O overhead, and efficiently managing I/O streams. This analytical theory is essential for designing parallel algorithms [2] and communication optimization algorithms [3], greatly benefiting scientific computing [4,5].

Y. Chen et al. (Eds.): COCOON 2024, LNCS 15162, pp. 152–163, 2025.
https://doi.org/10.1007/978-981-96-1093-8_13

The theory of I/O lower bounds has a rich history, with Hong and Kung et al. playing a pivotal role [6]. They introduced the influential Red-Blue Pebble Game model, upon which the S-Partition theory was formulated, marking the first I/O lower bound for matrix-matrix multiplication. Since then, subsequent studies have built upon this foundation. Auguste Olivry et al. [7] developed IOLB, a precise tool for calculating I/O lower bounds for affine programs. Zhang et al. [8] optimized the S-Partition theory to estimate I/O lower bounds of composite algorithms, especially for convolutional operations. Grzegorz Kwasniewski et al. [9] further advanced this field with the X-Partition theory and SOAP analysis, enhancing bounds for Polybench operators. Saachi Jain et al. [10] proposed a novel method to determine I/O bounds of computational graphs, focusing on the first few eigenvalues of the Laplacian operator. Their analysis covered algorithms like Bellman-Held-Karp and the Fast Fourier Transform, yielding precise bounds.

Previous theoretical approaches have limitations: they're specific to certain operators and algorithms, limiting their applicability to composite ones. Additionally, they neglect I/O overhead during subcomputation, making it challenging to establish tighter I/O lower bounds for write-intensive programs. This paper addresses these issues by introducing a method to establish stricter I/O lower bounds for composite process algorithms. Specifically, the contributions of this paper are as follows.

- We propose introducing the (X_1, X_2)-Partition theory to derive more stringent I/O lower bounds by incorporating potential I/O behavior from intermediate subcomputations, surpassing previous theories.
- Using (X_1, X_2)-Partition, we introduce an analysis approach for composite process programs, which not only enables a tighter I/O lower bound but also reveals previously undetectable I/O behaviors.
- Utilizing the proposed theory and method, we analyzed both the stencil operator and QR decomposition, demonstrating its efficacy in both cases.
- We first incorporated the I/O lower bound theory into neural network architecture search, and preliminary experimental results showed that the I/O performance of the architectures on CIFAR10 and CIFAR100 was improved.

The article is organized as follows: Sect. 2 covers the background. Section 3 introduces the (X_1, X_2)-Partition theory and I/O lower bound analysis for composite algorithms. Section 4 applies this theory to stencil operator dataflow. Section 5 analyzes QR decomposition dataflow using the I/O lower bound method. Section 6 presents preliminary experimental results after introducing the I/O lower bound theory into neural network architecture search. Section 7 summarizes the paper and discusses future research.

2 Background

2.1 Red and Blue Pebble Game

This section briefly introduces Red and Blue Pebble Games [11], reimagined as a directed acyclic graph (CDAG) $G = (V, E)$ with edges $(u, v) \in E$ representing computation dependencies. For execution, the program adapts to a sequential machine with a two-tiered storage system: finite-sized fast memory (red pebbles) and infinite-sized slow memory (blue pebbles). Red pebbles indicate fast memory residence, while blue pebbles represent slow memory storage. This game follows four principles:

- **Load**: placing a red pebble on a vertex which has a blue pebble.
- **Store**: placing a blue pebble on a vertex which has a red pebble.
- **Compute**: placing a red pebble on a vertex whose parents have a red pebble.
- **Discard**: removing any pebble from a vertex.

The program starts with blue pebbles on all input vertices. Program execution ends when all output vertices have blue pebbles on them. The program's I/O lower bound Q is the minimum cost of all valid pebble configurations. The pebble configuration with cost Q is called the optimal configuration.

For convenience, in addition to the above constraints, the computer model in this paper also stipulates that the computer model does not output external results before all sub-computations are completed, that is, there is a barrier between the sub-computation and the final output. How to extend it to the situation where there is no barrier will be the direction of our future research.

2.2 X-Partition

The derivation of the X-Partition theory is detailed in Grzegorz Kwasniewski et al. [9]. Within this section, we present the pertinent conclusions directly. Prior to introducing definitions and theorems, we establish a comprehensive list of symbols used throughout the text in Table 1.

We state the X-Partition as Definition 1 and we can construct an X-Partition that satisfies Theorem 1. We will give a brief proof in Appendix A[1].

Definition 1. *For any directed acyclic graph $G = (V, E)$ of any program, for any $X \geq S$, if there is a set of vertex sets $V_1, V_2, \ldots, V_h$ that satisfies the following conditions, this set is called an **X-Partition** of this program:*

- ***I**: $\forall i \neq j, V_i \cap V_j = \varnothing$, and $\cup_i V_i = V$.*
- ***II**: There is no circular data correlation between V_i.*
- ***III**: $\forall i, |In(V_i)| \leq X$.*
- ***IV**: $\forall i, |Out(V_i)| \leq X$.*

Theorem 1. *Let $\mathcal{H}(X)$ be the minimum number of subcomputations in any valid X-Partition of the CDAG $G = (V, E)$ for any $X \geq S$. The minimum number of I/O operations Q for any valid execution of a CDAG $G = (V, E)$ is bounded by inequality*(1) *and Eq.*(2)*. $\mathcal{R}(S)$ is the maximum reuse set size that satisfies*

[1] We provide the appendix of the paper at https://github.com/DongFengZero/COCOON24.

Table 1. Explanation of some symbols used in this article.

Symbol	Meaning
$W_{B,i}$	The set of vertices with blue pebbles according to the storage rules in the i-th subcomputation.
$W_{R,i}$	The set of vertices with red pebbles at the end of the i-th subcomputation and at least one child will undergo pebbling in the $i+1$-th subcomputation.
$W_{RB,i}$	The set of vertices that transition from red to blue during the i-th subcomputation, regardless of their final color.
$H_{R,i}$	Before the i-th subcomputation, the set of red nodes with at least one child will be pebbled.
$H_{BR,i}$	The blue vertices required for the i-th subcomputation.
$In(V_i)$	The input set of the i-th subcomputation.
$Out(V_i)$	The output set of the i-th subcomputation.
R^i_j	The set of red nodes (retained in fast memory) before the $j+1$-th step of the i-th subcomputation.
I^i	The set of input nodes required for the i-th subcomputation.
I^i_j	The set of input nodes required for the j-th step of the i-th subcomputation.
$\widetilde{O}^i_j$	The set of output nodes required for the j-th step of the i-th subcomputation.

$\forall i, max\{|H_{R,i}|, |W_{R,i}|\} \leq R(S)$. $\mathcal{T}(S)$ is the minimum I/O set size that satisfies $\forall i, min\{|H_{BR,i}|, |W_{RB,i}|\} \geq T(S)$.

$$Q \geq (X - \mathcal{R}(S) + \mathcal{T}(S)) \cdot (\mathcal{H}(X) - 1) \tag{1}$$

$$\mathcal{H}(X) = \frac{|V|}{|\mathcal{H}_{max}|} \tag{2}$$

Where $\mathcal{H}_{max} = arg\max_{H_i \in \mathcal{S}(X)} |\mathcal{H}_i|$ is the largest subset of vertices in the CDAG schedule $\mathcal{S}(X) = \mathcal{H}_1, \ldots, \mathcal{H}_h$.

Further, we have the following definition. **Computational intensity** is defined as $\rho = \frac{|\mathcal{H}_i|}{X-|H_{R,i}|+|W_{B,i}|}$. $\rho \leq \rho_{max} = \frac{|\mathcal{H}_{max}|}{X_0-\mathcal{R}(S)+\mathcal{T}(S)}$ is called **maximum computational intensity**. Let $\chi(X) = |\mathcal{H}_{max}|$ be the maximum size of a single subcomputation. Q is constrained by the inequality(3).

$$Q \geq |V| \frac{X_0 - \mathcal{R}(S) + \mathcal{T}(S)}{\chi(X_0)} = \frac{|V|}{\rho_{max}} \tag{3}$$

Where $X_0 = arg\min_X \frac{\chi(X)}{X-\mathcal{R}(S)+\mathcal{T}(S)}$, thus $\rho_{max} = \frac{\chi(X_0)}{X_0-\mathcal{R}(S)+\mathcal{T}(S)}$.

3 Theory and Method

3.1 (X_1, X_2)-Partition

X-Partition theory gives a way to estimate the I/O lower bound of a program. However, there is still room for further improvement of this theorem. $W_{R,i} \cap W_{RB,i}$ is often not null, giving optimization ideas for evaluating write operations. Inequality (4) is held according to **the principle of inclusion-exclusion**.

$$|Out(V_i)| \leq |W_{R,i}| + |W_{RB,i}| - |W_{R,i} \cap W_{RB,i}| \tag{4}$$

Let $\mathcal{R}_1(S) = \max_i |H_{R,i}|$,$\mathcal{R}_2(S) = \max_i |W_{R,i}|$,$\mathcal{T}_1(S) = \min_i |H_{B,i}|$,$\mathcal{T}_2(S) = \min_i |W_{RB,i}|$,$\mathcal{T}_3(S) = \min_i |W_{R,i} \cap W_{RB,i}|$. $Q(V_1, S) = \ldots = Q(V_{h-1}, S) = Y$,$Q(V_h, S) \leq Y$. Based on the above definitions, it is obvious that inequality (5)-(6) hold.

$$|In(V_i)| \leq |H_{R,i}| + |H_{B,i}| \leq \mathcal{R}_1(S) + Y - \mathcal{T}_2(S) \tag{5}$$

$$\begin{aligned} |Out(V_i)| &\leq |W_{R,i}| + |W_{RB,i}| - |W_{R,i} \cap W_{RB,i}| \\ &\leq \mathcal{R}_2(S) + Y - \mathcal{T}_1(S) - \mathcal{T}_3(S) \end{aligned} \tag{6}$$

Let $X_1 = \mathcal{R}_1(S) + Y - \mathcal{T}_2(S), X_2 = \mathcal{R}_2(S) + Y - \mathcal{T}_1(S) - \mathcal{T}_3(S)$. Building upon the existing X-Partition Definition 1 and Theorem 1, we can construct a concise definition of the (X_1, X_2)-Partition, resulting in Definition 2 and Theorem 2 for enhanced clarity and brevity.

Definition 2. *For any directed acyclic graph $G = (V, E)$ of any program, let $X_1 = \mathcal{R}_1(S) + Y - \mathcal{T}_2(S), X_2 = \mathcal{R}_2(S) + Y - \mathcal{T}_1(S) - \mathcal{T}_3(S)$, for any $\min\{X_1, X_2\} \geq S$, if there is a set of vertex sets $V_1, V_2, \ldots, V_h$ that satisfies the following conditions, this set is called an (X_1, X_2)-**Partition** of this program:*

- ***I**: $\forall i \neq j, V_i \cap V_j = \varnothing$, and $\cup_i V_i = V$.*
- ***II**: There is no circular data correlation between V_i.*
- ***III**: $\forall i, |In(V_i)| \leq \max\{X_1, X_2\}$.*
- ***IV**: $\forall i, |Out(V_i)| \leq \max\{X_1, X_2\}$.*

Theorem 2. *Let $X_1 = \mathcal{R}_1(S) + Y - \mathcal{T}_2(S), X_2 = \mathcal{R}_2(S) + Y - \mathcal{T}_1(S) - \mathcal{T}_3(S), X_{max} = \max\{X_1, X_2\}$. $\mathcal{H}(X_{max})$ be the minimum number of subcomputations in any valid (X_1, X_2)-Partition of the CDAG $G = (V, E)$. $\chi(X_{max})$ is the same as the definition in Theorem 1. For any $\min\{X_1, X_2\} \geq S$, the minimum I/O operations Q for any valid execution of a CDAG $G = (V, E)$ is bounded by inequality (7) and Eq.(8).*

$$Q \geq \mathcal{Z}(X_{max}, S) \cdot (\mathcal{H}(X_{max}) - 1) \tag{7}$$

$$\mathcal{Z}(X_{max}, S) = \begin{cases} X_2 - \mathcal{R}_2(S) + \mathcal{T}_1(S) + \mathcal{T}_3(S) & X_1 \leq X_2 \\ X_1 - \mathcal{R}_1(S) + \mathcal{T}_2(S) & X_1 > X_2 \end{cases} \tag{8}$$

Let $X_0 = arg\min_{X_{max}} \frac{\chi(X_{max})}{\mathcal{Z}(X_{max}, S)}$, then $\rho_{max} = \frac{\chi(X_0)}{\mathcal{Z}(X_0, S)}$ and Q is constrained by $Q \geq \mathcal{X}(X_0, S) \cdot (\mathcal{H}(X_0) - 1) \approx \frac{|V|}{\rho_{max}}$.

3.2 (X_1, X_2)-Partition for Subcomputations with Combined Steps

In this section, we further introduce how to use the theory in the previous section to lower the I/O lower bound of the algorithm with n combined steps.

First, we need to separate sub-calculations with compound steps according to the granularity required. We estimate the maximum sub-computation domain size $|\mathcal{H}_{max}|$ and the calculation domain size $|V|$, and calculate $\mathcal{H}(X_{max})$ according to Eq. (2). We provide a method in Appendix B.

Next, we need to calculate the five key sets. We can give the bounds of the five key sets $\mathcal{R}_1(S), \mathcal{R}_2(S), \mathcal{T}_1(S), \mathcal{T}_2(S), \mathcal{T}_3(S)$ required to solve the problem as shown in Eqs. (9)–(13) and Fig. 1.

$$\mathcal{R}_1(S) = \max_i |R^i_0| \tag{9}$$

$$\mathcal{R}_2(S) = \max_i |R^i_n| \tag{10}$$

$$\mathcal{T}_1(S) = \min_i |I^i \backslash R^i_0| \tag{11}$$

$$\mathcal{T}_2(S) = \min_i |\bigcup_{j=1}^{n} (\widetilde{O}^i_j \backslash R^i_j)| \tag{12}$$

$$\mathcal{T}_3(S) = \min_i |\bigcup_{j=1}^{n-1} (\widetilde{O}^i_j \backslash R^i_j \cap \bigcup_{r=j+1}^{n} R^i_r)| \tag{13}$$

Finally, we get X_0 according to $X_0 = arg\min_{X_{max}} \frac{\chi(X_{max})}{\mathcal{Z}(X_{max}, S)}$ and calculate the maximum calculation density ρ_{max}. We substitute it into the Theorem 2 to obtain the I/O lower bound Q.

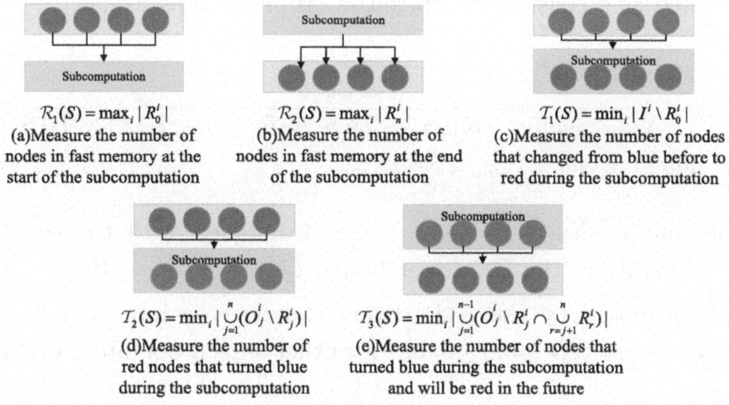

Fig. 1. Schematic diagram of the bound of the five sets

4 I/O Lower Bound Analysis for Stencil Operator

Taking *adi* as an instance, we analyze its I/O lower bound with (X_1, X_2)-Partition, utilizing a method akin to [9]. Our emphasis is on assessing Eq.(14)-Eq.(16), which is the contributor to I/O overhead Q. Since the processes of row scanning and column scanning are the same, only one of the processes is analyzed below, and is doubled when calculating Q in the end.

$$P_{j,i} = -\frac{C_3}{C_1 P_{i,j-1} + C_2} \tag{14}$$

$$Q_{i,j} = \frac{-C_4 U_{j,i-1} + (1 + 2C_4) U_{j,i} - C_5 U_{j,i+1} - C_1 Q_{i,j-1}}{C_1 P_{i,j-1} + C_2} \tag{15}$$

$$V_{j,i} = P_{i,j} V_{j+1,i} + Q_{i,j} \tag{16}$$

Based on the Loomis-Whitney inequality, $|V_{max}| = 2\sqrt{a^3b^3}$ and $\max\{R_1(S), R_2(S)\} \leq 4ab$. The required data for formula computation includes $U_{i,j}$, $P_{i,j}$, $V_{i,j}$ and $Q_{i,j}$ totaling $4ab$. Considering fast memory capacity, $\mathcal{T}_1(S) \geq \max\{4ab - S, 0\}$. Additionally, at least $12ab$ intermediate variables are generated, each operation yielding ab. Every $4ab$ initial variables can produce $12ab$ intermediate variables and $3ab$ final results. This results in $\mathcal{T}_2(S) \geq \max\{4ab - S, 0\} + \max\{12ab - \max\{S - 4ab, 0\}, 0\} + \max\{3ab - \max\{S - 16ab, 0\}, 0\}$. Assuming that starting from the calculation of the n-th group of intermediate variables will cause the initial variables that have not been used to be written back, then $S - 4(ab - n) \leq 12n$ must be satisfied, and then $8(ab - \frac{S-4ab}{8})$ variables are written back, they will still be read in future calculations, consistent with the definition of $\mathcal{T}_3(S)$. Incorporating the memory capacity constraint, we have $\mathcal{T}_3(S) \geq \max\{4ab - S, 0\} + \max\{3ab - S, 0\} + \max\{8(ab - \frac{S-4ab}{8}), 0\}$. This can be reformulated as a constrained optimization problem as Eq.(17).

$$\begin{aligned} max \quad & \rho = \frac{2\sqrt{a^3b^3}}{X_{max} + \min\{-\mathcal{R}_1(S) + \mathcal{T}_2(S), -\mathcal{R}_2(S) + \mathcal{T}_1(S) + \mathcal{T}_3(S)\}} \\ s.t. \quad & 4ab \leq S, 4ab \leq X_{max}, S \leq X_{max} \end{aligned} \tag{17}$$

The maximum density is $\rho_{max} = \frac{\sqrt{S}}{8}$. Combined with the conclusion of $Q \geq \frac{V}{\rho_{max}}$, we can get the I/O lower bound $Q = \frac{2 \cdot 2N^2T}{\rho_{max}} = \frac{32N^2T}{\sqrt{S}}$. Compared with $\frac{12N^2T}{\sqrt{S}}$ proposed by the X-Partition theory, it is improved by $\frac{8}{3}$ times. Additionally, we list I/O lower bounds for other stencil operators in Table 2.

5 I/O Lower Bound Analysis for QR Decomposition

QR decomposition refers to decomposing the matrix into a regular orthogonal matrix Q and an upper triangular matrix R. QR decomposition based on

Table 2. The leading term of the I/O lower bound of the stencil operator extracted using (X_1, X_2)-Partition is compared with previous work [9].

Kernel	I/O Lower Bound	Improv.	Kernel	I/O Lower Bound	Improv.
fdtd2d	$\frac{4\sqrt{3}N_xN_yT}{\sqrt{S}}$	2	heat3d	$\frac{6\sqrt[3]{9}N^3T}{\sqrt[3]{S}}$	$\sqrt[3]{9}$
adi	$\frac{32N^2T}{\sqrt{S}}$	$\frac{8}{3}$	jacobi1d	$\frac{4NT}{S}$	2
jacobi2d	$\frac{8N^2T}{\sqrt{S}}$	2	seidel2d	$\frac{8\sqrt{2}N^2T}{\sqrt{S}}$	$2\sqrt{2}$

Householder transformation is a commonly used implementation. For household matrices, some researchers have proposed more efficient storage schemes. In the following, we show a Householder transform-based QR decomposition algorithm for Householder matrices based on the compact WY representation [12]. The matrix Q can be obtained by $Q = I + YTY^T$. The algorithm can be divided into four main steps.

- Step 1: $v_k = \frac{\omega}{||\omega||_2}$.
- Step 2: $A'_{k:m,k:n} = -2v_kv_k^TA_{k:m,k:n}$.
- Step 3: $A_{k:m,k:n} = A_{k:m,k:n} + A'_{k:m,k:n}$.
- Step 4: $Z = -2TY^Tv_k$. $Y = [Y \quad v_k]$. $T = \begin{bmatrix} T & Z \\ 0 & -2 \end{bmatrix}$ $(k > 1)$.

We estimate the sizes of the sub-computational domain $|\mathcal{H}_{max}|$ and the global computational domain $|V|$ using a multi-variable constrained optimization approach in Appendix B. We found that with an input of $(X_{max}, 0, 0, 0)$ across four steps, the sub-computation domain achieves its maximum size. Utilizing these findings, we calculate $|V|$ and substitute them into Eq.(2) to obtain $\mathcal{H}(X_{max})$.

Next, we calculate the five key sets. If the size of the fast memory space S is large enough to store all the data needed for the next k-th step of sub-computation, then $\mathcal{R}_1(S) = \mathcal{R}_2(S) \leq \max_k\{\lceil mn - (k-1)(n - \frac{k-1}{2})\rceil\} = mn$. We consider that the size of the fast storage is S, so we can conclude that $\mathcal{R}_1(S) = \mathcal{R}_2(S) = min\{S, mn\}$.

By definition, $\mathcal{T}_1(S)$ represents the minimum number of blue nodes transitioning to red nodes during a subcomputation. The input necessitates a total of $mn - (k-1)(n - \frac{k-1}{2})$ elements. The minimal condition for $\mathcal{T}_1(S)$ occurs when fast memory usage is maximized, ensuring all data is utilized for the subsequent subcomputation, thus $\mathcal{T}_1(S) = \min_k\{\max\{\lfloor mn - (k-1)(n - \frac{k-1}{2}) - S\rfloor, 0\}\}$.

We analyze $\mathcal{T}_2(S)$ and $\mathcal{T}_3(S)$ in detail, examining each step individually. Initially, a vector of size $(m - k + 1)$ and $2(m - k)$ intermediate variables are generated. Taking inputs into account, the maximum efficient storage capacity with fast storage is limited to $mn - (k-1)(n - \frac{k-1}{2})$. Unused fast memory space is designated as $S_{re}^0 = S - (mn - (k-1)(n - \frac{k-1}{2}))$. To minimize writeback operations, data is preferentially stored in this cache space. The minimum data to be written out in the first step is $\mathcal{T}_2^1(S,k) = \max\{3(m-k)+1-\max\{\lceil S_{re}^0\rceil, 0\}, 0\}$. The minimum data to be written out in the first step is $\mathcal{T}_2^1(S,k) = \max\{3(m -$

$k)+1-\max\{\lceil S_{re}^0 \rceil, 0\}, 0\}$. Let $S_{rem}^1 = S_{re}^0 - (3m-3k+1)$ represent the residual fast memory capacity after writing intermediate variables and the final result. $S_{re}^1 = S_{re}^0 - (m-k+1)$ represents the amount of redundancy after removing data in fast memory that can be reused in the future. Reusable data amount is $\mathcal{T}_3^1(S,k) = \max\{(m-k+1) - \max\{\lceil S_{re}^0 \rceil, 0\}, 0\}$.

In the second step, a matrix of size $(m-k+1)(n-k+1)$ is generated, along with $(4m-4k+1)(n-k+1)$ intermediate variables. Taking inputs into account, the maximum available space with fast storage is $S_{re}^2 = S_{re}^1 - (m-k+1)(n-k+1)$. The residual fast memory capacity when intermediate variables and the final result are written is $S_{rem}^2 = S_{rem}^1 - (5m-5k+2)(n-k+1)$. Based on these definitions, $\mathcal{T}_2^2(S,k) = \max\{(5m-5k+2)(n-k+1) - \max\{\lceil S_{rem}^1 \rceil, 0\}, 0\}$ and $\mathcal{T}_3^2(S,k) = \max\{(m-k+1)(n-k+1) - \max\{\lceil S_{re}^1 \rceil, 0\}, 0\}$ can be derived.

In the third step, without generating new variables or intermediate variables, writeback ideally occurs in the memory space occupied by the second step. Writeback is feasible only when S is insufficient to store both $A'_{k:m,k:n}$ and $A_{k:m,k:n}$, where $A_{k:m,k:n}$ is reusable data. Following previous definitions, $S_{rem}^3 = S_{rem}^2$. Based on these definitions, $\mathcal{T}_2^3(S,k) = \max\{2(m-k+1)(n-k+1) - S, 0\}$ and $\mathcal{T}_3^3(S,k) = \max\{(m-k+1)(n-k+1) - S, 0\}$ can be derived.

The fourth step focuses on updating existing variables and producing $2mk$ intermediate variables. Given the limited capacity of fast memory S, it may not accommodate all $mn - k(n-\frac{k}{2})$ outputs. Additionally, $m+n+1$ data is written back but will not be reused. Based on these definitions, $\mathcal{T}_2^4(S,k) = \max\{\lceil mn - k(n-\frac{k}{2}) - S \rceil, 0\} + \max\{2mk - \lceil S_{rem}^3 \rceil, 0\} + (m+n+1)$,$\mathcal{T}_3^4(S,k) = \max\{\lceil mn - k(n-\frac{k}{2}) - S \rceil, 0\}$.

Summarizing the above results, we can get the formula of $\mathcal{T}_2(S)$ as $\mathcal{T}_2(S,k) = \sum_{i=1}^4 \mathcal{T}_2^i(S,k)$ and $\mathcal{T}_3(S)$ as $\mathcal{T}_3(S,k) = \sum_{i=1}^4 \mathcal{T}_3^i(S,k)$. We use matrices of different sizes and fast memories of different sizes to perform theoretical derivation and analyze the results.

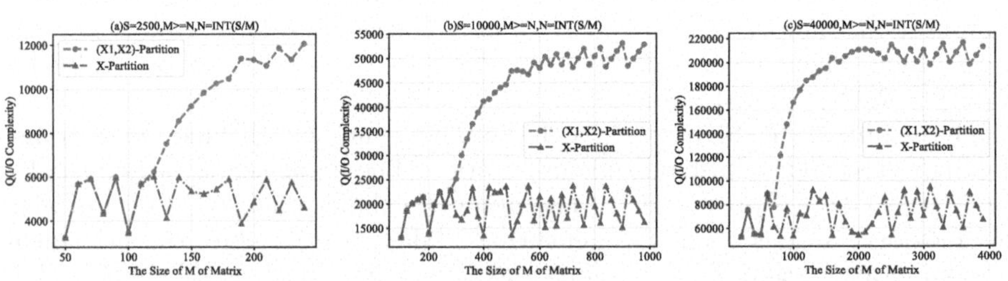

Fig. 2. The I/O lower bounds of QR decomposition in various cases are analyzed by using the method based on (X_1, X_2)-Partition we proposed. The figure shows the advantages of this approach over the method based on X-Partition. M and N are the number of rows and columns of the matrix respectively.

Figure 2(a)–(c) and Table 3 demonstrate the (X_1, X_2)-Partition's superiority over X-Partition when $N = \lfloor S/M \rfloor$ and $M \geq N$. Initially, the curve is fully con-

Table 3. This table shows the improvement in I/O lower bound for QR decomposition using the analysis method proposed in this article compared to using the method based on X-Partition theory in some cases of Fig. 2.

M	N	S	Improv.	M	N	S	Improv.	M	N	S	Improv.
100	25	2500	1.00	200	50	10000	1.00	400	100	40000	1.00
150	16	2500	1.73	300	33	10000	1.41	800	50	40000	2.26
190	**13**	**2500**	**2.93**	400	25	10000	3.05	1200	33	40000	2.60
200	12	2500	2.33	**500**	**20**	**10000**	**3.50**	1500	26	40000	2.24
220	11	2500	2.65	700	14	10000	2.96	2000	20	40000	3.87
240	10	2500	2.61	800	12	10000	2.58	**2500**	**16**	**40000**	**3.93**

sistent with the X-Partition theory within a certain range, which also indicates that the fast memory capacity is sufficient at this time. However, as M rises, the impact of I/O operations on intermediate results becomes evident, favoring the (X_1, X_2)-Partition theory. Specifically, for $S = 2500$, $S = 10000$, and $S = 40000$, the later curve increases by a factor of 2–4, peaking at 2.93, 3.50, and 3.93, respectively. This is because the length of the calculation vector in the first step gradually increases, so the I/O overhead becomes more prominent. Obviously, the larger the fast memory, the more obvious the effect.

It shows that the theoretical method we proposed can better explore the impact of potential I/O behavior on I/O performance than previous theories, thereby making the estimate of the lower bound tighter. Therefore, this theoretical method is more suitable as a criterion for evaluating program I/O performance. In the next section, we introduce it into the neural network to search for a neural network architecture with better I/O performance.

6 I/O Lower Bound Analysis for Network Architecture Search

In addition to the above work, we also introduced the I/O lower bound theory into the graph neural network and used it to search for neural network structure configurations with excellent I/O performance and accuracy. We describe the network design in Appendix C in detail. Based on this design, we conducted some simple preliminary tests on the CIFAR10 and CIFAR100 datasets.

Table 4 presents the experimental results. The data reveals that when incorporating the I/O lower bound theory as the loss term, the prediction model exhibits a slight decrease of 1.87% in accuracy on the CIFAR100 dataset but a notable increase of 4.36% on the CIFAR10 dataset, creating a balanced outcome. Furthermore, the model utilizing I/O lower bound theory demonstrates clear advantages in model parameter size, memory usage, as well as memory reading and writing capabilities on both datasets. In terms of the floating-point operations required for a single image, the improved model is only 68.53% of the original model on CIFAR100 (80.59/117.59), and 85.48% of the original model on CIFAR10 (143.27/167.61). These findings suggest that by incorporating the I/O lower bound theory as a loss term, we can devise a model that not only maintains comparable accuracy but also boasts smaller parameter size and superior I/O performance. Due to the limitation of the paper length, there is still room for further exploration, which is also the direction of future efforts.

Table 4. Performance comparison of whether to use I/O lower bound theory (IOLB) in graph convolutional networks (GCN) to search network structures on CIFAR10 and CIFAR100. The memory required for node inference is referred to as Memory. MemR+W denotes the aggregate size of data being read from and write into memory. FLOPs is floating point operations required for inference of an image. All models are trained for 100 epochs and feature four convolutional layers with the same generated structure, having output channel counts of 64, 128, 256, and 512. A linear layer is added for classification.

	Dataset	Accuracy	Params	Memory	FLOPs	MemR+W
GCN	CIFAR100	55.56%	2414720	13.13 MB	117.59MFlops	35.51 MB
GCN+IOLB	CIFAR100	53.69%	1693440	9.38 MB	80.59MFlops	25.26 MB
GCN	CIFAR10	81.37%	3266240	21.10 MB	167.61MFlops	54.69 MB
GCN+IOLB	CIFAR10	85.73%	2652480	13.13 MB	143.27MFlops	36.42 MB

7 Conclusion

This paper introduces the (X_1, X_2)-Partition theory, which offers tighter I/O lower bounds by analyzing subcomputations' potential I/O behavior. We showcase its superiority by calculating the stencil operator's I/O lower bound. Furthermore, we present an I/O lower bound analysis for algorithms with compound subcomputations, illustrated using the QR decomposition to demonstrate its advantages over prior theories. In addition, we preliminarily introduced this theory into neural network architecture search and achieved satisfactory results.

Currently, the automated application of this method remains unexplored. Our future plans involve using this theoretical framework as a performance benchmark and integrating it with machine learning and reinforcement learning to develop automated analysis tools. This integrated strategy will offer crucial insights for selecting operators, algorithms, and critical performance parameters of neural networks, ultimately facilitating the optimal design of dataflows.

Acknowledgment. The work is supported by the National Key Research and Development Program of China (No.2021YFB0300101, No.2023YFA1011704) and the National Natural Science Foundation of China (No.12102468, No.12002380) and the National University of Defense Technology Foundation (No.ZK21-02, No.ZK20-52).

References

1. Ballard, G., Demmel, J., Holtz, O., Schwartz, O.: Graph expansion and communication costs of fast matrix multiplication. J. ACM (JACM) **59**(6), 1–23 (2013)
2. Elango, V., Rastello, F., Pouchet, L.N., Ramanujam, J., Sadayappan, P.: On characterizing the data movement complexity of computational dags for parallel execution. In: Proceedings of the 26th ACM Symposium on Parallelism in Algorithms and Architectures, pp. 296–306 (2014)
3. Ballard, G., Demmel, J., Holtz, O., Lipshitz, B., Schwartz, O.: Communication-optimal parallel algorithm for Strassen's matrix multiplication. In: Proceedings of the Twenty-Fourth Annual ACM Symposium on Parallelism in Algorithms and Architectures, pp. 193–204 (2012)
4. Yan, J., Chen, X., Liu, J.: CSR&RV: an efficient value compression format for sparse matrix-vector multiplication. In: Liu, S., Wei, X. (eds.) Network and Parallel Computing, pp. 54–60. Springer, Cham (2022). https://doi.org/10.1007/978-3-031-21395-3_5
5. Chen, X., Gong, C., Liu, J., Pang, Y., Deng, L., Chi, L., Li, K.: A novel neural network approach for airfoil mesh quality evaluation. J. Parallel Distrib. Comput. **164**, 123–132 (2022)
6. Jia-Wei, H., Kung, H.T.: I/o complexity: the red-blue pebble game. In: Proceedings of the Thirteenth Annual ACM Symposium on Theory of Computing, STOC '81, pp. 326—-333. Association for Computing Machinery, New York (1981). https://doi.org/10.1145/800076.802486
7. Olivry, A., Langou, J., Pouchet, L.N., Sadayappan, P., Rastello, F.: Automated derivation of parametric data movement lower bounds for affine programs. In: Proceedings of the 41st ACM SIGPLAN Conference on Programming Language Design and Implementation, PLDI 2020, pp. 808—-822. Association for Computing Machinery, New York (2020). https://doi.org/10.1145/3385412.3385989
8. Zhang, X., Xiao, J., Tan, G.: I/o lower bounds for auto-tuning of convolutions in cnns. In: Proceedings of the 26th ACM SIGPLAN Symposium on Principles and Practice of Parallel Programming, PPoPP 2021, pp. 247—-261. Association for Computing Machinery, New York (2021). https://doi.org/10.1145/3437801.3441609
9. Kwasniewski, G., et al.: Pebbles, graphs, and a pinch of combinatorics: towards tight i/o lower bounds for statically analyzable programs. In: Proceedings of the 33rd ACM Symposium on Parallelism in Algorithms and Architectures, SPAA 2021, pp. 328–339. Association for Computing Machinery, New York (2021). https://doi.org/10.1145/3409964.3461796
10. Jain, S., Zaharia, M.: Spectral lower bounds on the i/o complexity of computation graphs. In: Proceedings of the 32nd ACM Symposium on Parallelism in Algorithms and Architectures, pp. 329–338 (2020)
11. Demaine, E.D., Liu, Q.C.: Red-blue pebble game: complexity of computing the trade-off between cache size and memory transfers. In: Proceedings of the 30th on Symposium on Parallelism in Algorithms and Architectures, pp. 195–204 (2018)
12. Schreiber, R., Van Loan, C.: A storage-efficient wy representation for products of householder transformations. SIAM J. Sci. Stat. Comput. **10**(1), 53–57 (1989). https://doi.org/10.1137/0910005

Hedonic Games for Federated Learning with Model Sharing Data

Yuqing Zhu[1(✉)], Chuanwen Luo[2], and Deying Li[3]

[1] Department of Computer Science, California State University, Los Angeles, CA 90032, USA
yuqing.zhu@calstatela.edu
[2] School of Information, Beijing Forestry University, Beijing 100107, PRC, China
chuanwenluo@bjfu.edu.cn
[3] School of Information, Renmin University of China, Beijing 100872, PRC, China
deyingli@ruc.edu.cn

Abstract. Federated Learning (FL) is a significant distributed learning setting in large scale machine learning. In FL, agents with access to their own data combine models learned from local data to build a global mode. If the agents' data are distributed differently, then FL might give a biased global model which may not be optimal for each agent. This causes a question for the agents: should they join the federation?

In this work, we assume that the agents have different amounts of data and different data distributions, and there are noises when they draw the data. Each agent's goal is to minimize the expected variance squared error (VSE) on its own distribution. Three different federating models are analyzed: In uniform federation, all agents use a single model; In coarse-grained federation, each agent has a parameter to weight the global model with its own local model; In fine-grained federation, each agent uses a vector of weights to combine other agents' contribution.

We analyze the scenario where agents may divide into coalitions where the agents in the same coalition jointly construct models under the framework of hedonic game theory. The stable partitions are thoroughly studied for uniform federation model.

Keywords: Federated Learning · Statistic Analysis · Game Theory

1 Introduction

Federated Learning (FL) is a distributed machine learning figuration. It trains statistical models over the remote devices, like mobile phones, wearable devices, and autonomous vehicles while keeping data localized [1]. A wealth of data are generated. Due to the growing computational power of edge devices, and with concerns about private information leakage, it is increasingly attractive to store data locally and push network computation to the edge. FL, which explores training statistical models directly on remote devices, has been experiencing

Y. Chen et al. (Eds.): COCOON 2024, LNCS 15162, pp. 164–175, 2025.
https://doi.org/10.1007/978-981-96-1093-8_14

rapid innovations as well as wide deployments in [2,3]. It seeks to maximize the usage of the enhanced local resources on each device, and avoids privacy issues over transmitting sensitive user data by remaining them on local device. This new machine learning paradigm significantly differs from the learning in traditional distributed environments, and has been deployed in practice by major companies [4,5]. FL has started to play a critical role in many large scale machine learning applications where training data are distributed at the edge, like predicting hospitalizations due to heart diseases on wearable health monitoring devices [6], intensive healthcare demand prediction [7], word prediction on cell phones [8], etc.

Most FL applications are motivated by the expectation to reduce the error by integrating more data. Unfortunately, adopting FL is not always a step towards more accurate modeling. For instance, there is a new algorithm proposed for small obstacle detection in autonomous driving. It is possible that some autonomous vehicles may have different models of their data, coming from the disparity of the areas they tested or variants of algorithm implementation. When two vehicles with different data federate together, they will see the reduction of the universal model's error due to data scale, and the increase for their error due to model bias. For the participating vehicles, a natural question then emerges, which other vehicles should they federate with? Or, more generally, for several agents who consider FL, which other agents should it federate together so that its error can be minimized? Are those other agents interested in federating with it? Furthermore, does there exist some stable arrangement of agents into federating groups for some special settings and general setting?

In fact, statistical heterogeneity is one of the key challenges FL is facing, and numerous solutions have been proposed for this issue [9,10]. Most of them aim at achieving each agent's equality on error rates guarantee, by involving more data or adopting transfer learning or meta learning so that the model better fits the local data. Those methods mainly use experimental methodology, and there have not been many work that consider the participating motivation for the agents to join FL. Our work introduces cooperative game theory - hedonic game to study rational agents' strategic interactions in distributed learning atmosphere. A hedonic game [11] is a game that models the formation of coalitions (groups) of players who have preferences over which group they belong to.

1.1 Our Contribution

In this work, we consider the statistical disparities of agents' data distribution, quantify the error rates of agents, and then use hedonic games to predict the agents' strategies in FL. Agents in FL are players in a hedonic game, and a coalition is a group of players who federate together. A player will prefer a coalition over another if it gets lower error in the former than in the latter.

- We assume that the true player data are normal distributed, but different players have different distribution parameters. When a player draws samples from the true distribution, it is expected to add some noises. Aiming at data variance estimation, quantitative error rate of each player is given based on the number of samples and its data distribution statistical parameters.

- Three federation methodologies are investigated, we quantify their error rates and then give stable solutions in hedonic games:
  i. *Uniform Federation*: the simplest setting where all players use a single model. We prove that generally, when the numbers of samples are large enough for all players, having them separately doing local learning is the only core stable solution. We also study two special cases: one where all players have the same number of samples and the other where the players have two different numbers of samples. For both cases, we show that if the number of samples are small enough then the only core stable solution is to have all players federating together; and for the latter case, we provide a polynomial computable procedure to obtain an individually stable solution.
  ii. *Coarse-grained Federation*: each player has a parameter to weight the global model with its own local model. We prove that in this setting all players prefer federation. We also show that when all players have the same number of samples, if the number is small enough then all players federating together is the only core stable solution, and if the number is large enough then a polynomial computable procedure for an individually stable solution is provided.
  iii. *Fine-grained Federation*: each agent uses a vector of weights to combine other agents' contribution. We give the optimal weights and show that the only core stable solution is all players federating together.
- Our results only weakly depend on the noise variance and the numbers of samples, which are not difficult to obtain.

2 Related Work

Incentive study for FL using game theory has become a hot research trend. Song et al. [12] considers using monetary incentive for the agents to participate FL. They formally define a Shapley value based concept called Contribution Index (CI) to quantify the contribution of data providers in a federating learning task, and design two approximate algorithms to calculation the CIs. [13] studies the network where agents may cooperatively perform learning tasks, and theoretically give the equilibrium with the minimum cost so the agents form stable collaborations.

In [14], Blum et el. introduce a collaborative PAC learning model, in which k players attempt to learn the same underlying concept. They study how much more information is required to learn an accurate classifier for all players simultaneously, and give an exponential improvement upon the naïve algorithm that does not share information among players. Blum et al. in [15] propose stable and envy-free equilibria in FL, to capture notions of collaboration of the agents who are interested in meeting their learning objectives while keeping their sample collection burden low. Donahue et al. in [16] investigate hedonic games for mean estimation in FL. This is the most related work, and our work significantly differs from them: First, we study the variance estimation, which is a new problem and is fundamentally much more theoretical challenging. Second, our error results are more complicated, which require more delicate techniques for stability analysis. Third, we give more rigid proofs for the stability existence and construction, and close some gaps [16] leaves.

3 Preliminary

3.1 Hedonic Game

Formally, a hedonic game is a pair $(\mathfrak{S}, (\succeq_i)_{i \in \mathfrak{S}})$ of a set $\mathfrak{S}$ of players (or agents), with a preference relation $\succeq_i$ for each player $i \in \mathfrak{S}$ over the set $\{\mathfrak{I} \subseteq \mathfrak{S}, i \in \mathfrak{I}\}$ of coalitions. A *coalition* is a subset $\mathfrak{I} \subseteq \mathfrak{S}$ of the set of players.

A *coalition structure* π is a partition of $\mathfrak{S}$. The notion of game-theoretic stability is used to evaluate coalition structures. Here in this literature, we mainly focus on following two types of stabilities:

- *Core Stable*. A coalition structure π is in the core (or is core stable) if there is no coalition $\mathfrak{I}$ whose members all prefer $\mathfrak{I}$ to π. In formal words, we say a non-empty coalition $\mathfrak{I}$ blocks π if $\forall i \in \mathfrak{I}, \mathfrak{I} \succ_i \pi(i)$. Then: π is core stable $\Leftrightarrow$ $\nexists$ blocking coalition.
- *Individually Stable*. A coalition structure π is individually stable if no players wishes to join another coalition all of whose members welcome the player. In formal words, π is individually stable $\Leftrightarrow$ $\nexists$ player $i \in \mathfrak{S}$ such that $\exists \mathfrak{I} \in \pi \cup \{\emptyset\}$, $\mathfrak{I} \cup \{i\} \succ_i \pi(i)$ and $\forall j \in \mathfrak{I}$, $\mathfrak{I} \cup \{i\} \succeq_j \mathfrak{I}$.

3.2 Data Model

Our data model is as follows: Suppose there are totally N players, and player i has ℓ_i samples. Note that N and ℓ_i are fixed. α_i is a scalar representing the true distribution of player i, $\alpha_i \sim \mathcal{N}(0, 1)$. Player i draws samples $\{X_k^i\}$ i.i.d. from its true distribution: $X \sim \mathcal{N}(0, \alpha_i)$. Then we draw samples $\{Y_k^i\}$ with noise $\{\epsilon_k^i\}$ around the samples $\{X_k^i\}$, $\epsilon_k^i \sim \mathcal{N}(0, \epsilon)$. We can use $\{Y_k^i = X_k^i + \epsilon_k^i | 1 \leq i \leq N; 1 \leq k \leq \ell_i\}$ to denote all the samples and their relationship. Also define $L = \sum_{i=1}^{N} \ell_i$ as the total number of all players' samples. Our goal of this research, is to investigate the games by varying the numbers of samples of all players.

There are three levels of randomization for each player i here,

1) To randomly choose $\alpha_i \sim \mathcal{N}(0, 1)$;
2) To randomly generate the input $X^i \sim \mathcal{N}(0, \alpha_i^2)$;
3) For each $\{X_k^i | 1 \leq k \leq \ell_i\}$ that has been generated in Step 2, to noisily add a random error $\epsilon_k^i \sim \mathcal{N}(0, \epsilon^2)$ to X_k^i to obtain Y_k^i.

Note that players have different true data distributions. Our games use variance estimation as the computational goal. Variance is very well interested in many machine learning algorithms like linear regression, decision tree, bagging, random forest, etc. We use following three methods for FL on variance: uniform federation, coarse-grained federation and fine-grained federation.

In uniform federation, we use a common model for all players in the federating coalition. In other words, $\bar{\sigma}^u$ defined in following formula is the estimation for all players sharing their samples:

$$\bar{\sigma}^u = \frac{1}{L}\sum_{i=1}^{N} \bar{\sigma}_i \cdot \ell_i.$$

In coarse-grained federation, each player uses a parameter w_i to weight the global model (within its coalition) with its own local model:

$$\bar{\sigma}_i^c = w_i \cdot \bar{\sigma}_i + (1 - w_i) \cdot \frac{1}{L}\sum_{k=1}^{N} \bar{\sigma}_k \cdot \ell_k.$$

In fine-grained federation, each player i uses a vector of weights $\boldsymbol{v_i}$ to weight every other player's contribution to their estimate:

$$\bar{\sigma}_i^f = \sum_{k=1}^{N} v_{ik} \cdot \bar{\sigma}_k$$

where $\sum_{k=1}^{N} v_{ik} = 1$.

Note that $\{\bar{\sigma}_k\}$ are the estimations of the players.

We assume that each player i knows ℓ_i: how much samples it has access to, and be aware of ϵ: the variance of their random noises. All of out stability analysis conditions depend on the relationship between $\{\ell_i\}$ are ϵ, and this is a fairly weak constraint for the players.

The purpose of this work is to provide description for stable organization for agents(players) in FL. Stability is important because players can move between coalitions. Even if they cannot change between coalitions, our analysis is informative of how optimal the coalition structure is.

4 Expected Errors

Our first contribution is to derive exact expected error for players' expected variance squared error (VSE) under different federating scenarios. These exact values allow us to precisely predict the players' incentives in later sections.

Suppose the players adopt unbiased variance estimation, then the variance square estimation (VSE) for player i with ℓ_i samples is:

$$\mathbb{E}\left[\left(\frac{1}{\ell_i - 1}\sum_{k=1}^{\ell_i}(Y_k - \overline{Y})^2 - (\alpha_i^2 + \epsilon^2)\right)^2\right]$$

where $\overline{Y} = \frac{1}{\ell_i}\sum_{k=1}^{\ell_i} Y_k$. Note that $(\alpha_i^2 + \epsilon^2)$ is the exact expected variance of $\{Y_k\}$.

Theorem 1. *The expected VSE for player i with ℓ_i examples (without federation) is $(\epsilon^4 + 2\epsilon^2 + 3)\frac{2}{\ell_i - 1}$.*

Proof of Theorem 1 Note that using Cochran's theorem, $\sum_{k=1}^{\ell_i}(Y_k - \overline{Y})^2 \sim (\alpha_i^2 + \epsilon^2)\chi^2_{(\ell_i-1)}$ since $Y_k \sim \mathcal{N}(0, \alpha_i^2 + \epsilon^2)$, where $\chi^2_{(\ell_i-1)}$ is the chi-squared distribution with degrees of freedom $(\ell_i - 1)$. The VSE error can be expanded to:

$$\left(\frac{1}{\ell_i - 1}\right)^2 \mathbb{E}\left[\left(\sum_{k=1}^{\ell_i}(Y_k - \overline{Y})^2\right)^2\right]$$
$$-2(\alpha_i^2 + \epsilon^2)\mathbb{E}\left[\frac{1}{\ell_i - 1}\sum_{k=1}^{\ell_i}(Y_k - \overline{Y})^2\right]$$
$$+(\alpha_i^2 + \epsilon^2)^2 \tag{1}$$

Since the mth moment of $X \sim \chi_k^2$ is:

$$\mathbb{E}(X^m) = k(k+2)(k+4)\cdots(k+2m-2) = 2^m \frac{\Gamma\left(m + \frac{k}{2}\right)}{\Gamma\left(\frac{k}{2}\right)}$$

The first part of (1) is the 2nd moment and it equals

$$(\alpha_i^2 + \epsilon^2)^2 \frac{1}{(\ell_i - 1)^2}(\ell_i - 1)(\ell_i + 1)$$

and

$$\mathbb{E}\left[\frac{1}{\ell_i - 1}\sum_{k=1}^{\ell_i}(Y_k - \overline{Y})^2\right] = (\alpha_i^2 + \epsilon^2)$$

Therefore,

$$\begin{aligned}(1) &= \frac{(\alpha_i^2 + \epsilon^2)^2}{(\ell_i - 1)^2}(\ell_i - 1)(\ell_i + 1) - (\alpha_i^2 + \epsilon^2)^2 \\ &= \frac{(\alpha_i^2 + \epsilon^2)^2}{(\ell_i - 1)^2}(2(\ell_i - 1)) \\ &= (\alpha_i^2 + \epsilon^2)^2 \cdot \frac{2}{\ell_i - 1}\end{aligned}$$

and

$$\begin{aligned}\mathbb{E}[(\alpha_i^2 + \epsilon^2)^2] &= \mathbb{E}[(\alpha_i^4 + 2\alpha_i^2\epsilon^2 + \epsilon^2] \\ &= (\epsilon^4 + 2\epsilon^2 + 3)\end{aligned}$$

since $\mathbb{E}[\alpha_i^4] = 3$ and $\mathbb{E}[\alpha_i^2] = 1$. At last,

$$(1) = (\epsilon^4 + 2\epsilon^2 + 3) \cdot \frac{2}{\ell_i - 1}$$

□

Proof of Theorem 2 The VSE error of player i's is:

$$\begin{aligned}
&\mathbb{E}\left[\left(\sum_{k=1}^{N} v_{ik}\bar{\sigma}_k - (\alpha_i^2 + \epsilon^2)\right)^2\right] \\
&= \mathbb{E}\left[\left(\sum_{k=1}^{N} v_{ik}\bar{\sigma}_k - \sum_{k=1}^{N} v_{ik}(\alpha_k^2 + \epsilon^2)\right)\right]^2 \\
&+\mathbb{E}\left[\left(\sum_{k=1}^{N} v_{ik}(\alpha_k^2 + \epsilon^2) - (\alpha_i^2 + \epsilon^2)\right)^2\right] \\
&+2\mathbb{E}[\left(\sum_{k=1}^{N} v_{ik}\bar{\sigma}_k - \sum_{k=1}^{N} v_{ik}(\alpha_k^2 + \epsilon^2)\right) \\
&\cdot \left(\sum_{k=1}^{N} v_{ik}(\alpha_k^2 + \epsilon^2) - (\alpha_i^2 + \epsilon^2)\right)]
\end{aligned} \tag{2}$$

The third part of (2) is 0 since from player i's perspective α_i is fixed, and $\mathbb{E}[\sum_{k=1}^{N} v_{ik}\bar{\sigma}_k - \sum_{k=1}^{N} v_{ik}(\alpha_k^2 + \epsilon^2)] = \mathbb{E}[\sum_{k=1}^{N} v_{ik}(\bar{\sigma}_k - (\alpha_k^2 + \epsilon^2))] = 0$.

The second part of (2) is:

$$\begin{aligned}
&\mathbb{E}\left[\left(\sum_{k=1}^{N} v_{ik}(\alpha_k^2 + \epsilon^2) - (\alpha_i^2 + \epsilon^2)\right)^2\right] \\
&= \mathbb{E}\left[\left(\sum_{k=1}^{N} v_{ik}(\alpha_k^2 + \epsilon^2) - \sum_{k=1}^{N} v_{ik}(\alpha_i^2 + \epsilon^2)\right)^2\right] \\
&= \mathbb{E}\left[\left(\sum_{k=1}^{N} v_{ik}(\alpha_i^2 - \alpha_k^2)\right)^2\right] \\
&= \mathbb{E}\left[\sum_{k=1,k\neq i}^{N} v_{ik}^2(\alpha_i^4 - 2\alpha_i^2\alpha_k^2 + \alpha_k^4)\right] \\
&+\mathbb{E}\left[\sum_{a\neq b,a,b\neq i} v_{ia}v_{ib}(\alpha^4 - \alpha_i^2\alpha_a^2 - \alpha_i^2\alpha_b^2 + \alpha_a^2\alpha_b^2)\right]
\end{aligned}$$

Note that α_i, α_a and α_b are mutually independent. Therefore $\mathbb{E}[\alpha_i^2\alpha_b^2] = \mathbb{E}[\alpha_a^2\alpha_b^2]$, and $\mathbb{E}[\alpha_i^4 - \alpha_i^2\alpha_a^2] = \mathbb{E}[\alpha_i^4 - \alpha_i^2\alpha_k^2] = \mathbb{E}[\alpha_i^4 - \alpha_i^2\alpha_k^2] = \mathbb{E}[\alpha_i^4] - \mathbb{E}[\alpha_i^2]\mathbb{E}[\alpha_k^2] =$

$\mathbb{E}[\alpha_i^4] - \mathbb{E}[\alpha_i^2]\mathbb{E}[\alpha_i^2]$. Then the second part of (2) is:

$$\mathbb{E}\left[\sum_{k=1,k\neq i}^{N} v_{ik}^2(2\alpha_i^4 - 2\alpha_i^2\alpha_k^2)\right]$$

$$+\mathbb{E}\left[\sum_{a\neq b,a,b\neq i} v_{ia}v_{ib}(\alpha^4 - \alpha_i^2\alpha_a^2)\right]$$

$$= \left(\mathbb{E}[\alpha_i^4] - \mathbb{E}^2[\alpha_i^2]\right)\left[2\sum_{k=1,k\neq i}^{N} v_{ik}^2 + \sum_{a\neq b,a,b\neq i} v_{ia}v_{ib}\right]$$

$\alpha_i^2 \sim \chi_1^2$, therefore $\mathbb{E}[\alpha_i^4]$ is the second moment of χ_1^2, $\mathbb{E}[\alpha_i^4] = 3$ and $\mathbb{E}[\alpha_i^2] = 1$. Also, $\sum_{a\neq b,a,b\neq i} v_{ia}v_{ib} = (\sum_{k\neq i} v_{ik})^2 - \sum_{k\neq i}^{N} v_{ik}^2 = (1 - v_{ii})^2 - \sum_{k\neq i}^{N} v_{ik}^2$. The second part of (2) is finally simplified to

$$2\left[(1 - v_{ii})^2 + \sum_{k\neq i}^{N} v_{ik}^2\right]$$

The first part of (2) is:

$$\mathbb{E}\left[\left(\sum_{k=1}^{N} v_{ik}\bar{\sigma}_k - \sum_{k=1}^{N} v_{ik}(\alpha_k^2 + \epsilon^2)\right)^2\right]$$

$$= \mathbb{E}\left[\left(\sum_{k=1}^{N} v_{ik}(\bar{\sigma}_k - (\alpha_k^2 + \epsilon^2))\right)^2\right]$$

$$+\mathbb{E}\left[\sum_{a\neq b}^{N} v_{ia}v_{ib}(\bar{\sigma}_a - (\alpha_a + \epsilon^2))(\bar{\sigma}_b - (\alpha_b + \epsilon^2))\right]$$

According to Theorem 1, $\mathbb{E}\left[\sum_{k=1}^{N} v_{ik}^2(\bar{\sigma}_k - (\alpha_k^2 + \epsilon^2))^2\right] = \sum_{k=1}^{N} v_{ik}^2(\epsilon^4 + 2\epsilon^2 + 3)\frac{2}{\ell_k - 1}$. Also, since players a and b setup their parameters independently, $\mathbb{E}\left[\sum_{a\neq b}^{N} v_{ia}v_{ib}(\bar{\sigma}_a - (\alpha_a + \epsilon^2))(\bar{\sigma}_b - (\alpha_b + \epsilon^2))\right] = \sum_{a\neq b}^{N} v_{ia}v_{ib} \cdot \mathbb{E}\left[(\bar{\sigma}_a - (\alpha_a + \epsilon^2))(\bar{\sigma}_b - (\alpha_b + \epsilon^2))\right] = \sum_{a\neq b}^{N} v_{ia}v_{ib} \cdot \mathbb{E}\left[(\bar{\sigma}_a - (\alpha_a + \epsilon^2))\right] \cdot \mathbb{E}\left[(\bar{\sigma}_b - (\alpha_b + \epsilon^2))\right] = 0$. The first part of (2) is finally simplified to

$$(\epsilon^4 + 2\epsilon^2 + 3)\sum_{k=1}^{N} v_{ik}^2 \frac{2}{\ell_k - 1}$$

Put the simplified results of three parts of (2) together, we prove this theorem.
□

Theorem 2. *For fine-grained federation, the expected VSE for player i with ℓ_i examples is*

$$(\epsilon^4 + 2\epsilon^2 + 3) \sum_{k=1}^{N} v_{ik}^2 \cdot \frac{2}{\ell_k - 1} + 2[(1 - v_{ii})^2 + \sum_{k \neq i} v_{ik}^2].$$

For the sake of conciseness, from here down in this paper, we define

$$\mathtt{E} \triangleq \epsilon^4 + 2\epsilon^2 + 3$$

to simplify our writing.

Corollary 1. *For uniform federation, the expected VSE for player i with ℓ_i examples is*

$$\mathtt{E} \sum_{k=1}^{N} \frac{\ell_k^2}{L^2} \cdot \frac{2}{\ell_k - 1} + 2 \left[\frac{(L - \ell_i)^2}{L^2} + \sum_{k \neq i} \frac{\ell_k^2}{L^2} \right].$$

Corollary 2. *For coarse-grained federation, the expected VSE for player i with ℓ_i samples is*

$$2\mathtt{E} \left[\sum_{k=1}^{N} (1 - w_i)^2 \frac{\ell_k^2}{L^2} \cdot \frac{1}{\ell_k - 1} + \left(w_i^2 + 2w_i(1 - w_i)\frac{\ell_i}{L} \right) \frac{1}{\ell_i - 1} \right]$$

$$+ 2(1 - w_i)^2 \left[\frac{(L - \ell_i)^2}{L^2} + \sum_{k \neq i} \frac{\ell_k^2}{L^2} \right].$$

5 Coalition Formation of Uniform Federation

We analyze the stability formation for uniform federation in this section. Two cases are considered: the first is where all players have the same number of samples, and the second is where the players have two numbers of samples: ℓ and 2ℓ. We will use π_g to denote the coalition structure where all players are put in a single grand set, and use π_d for the coalition structure where all players are alone.

Theorem 3. *For uniform federation, if $\ell_i \geq \frac{1}{2} \left(\mathtt{E} + 2 + \sqrt{(\mathtt{E} + 2)^2 + 4\mathtt{E}} \right)$ for all i, then the partition of individuals π_d is the unique core-stable coalition structure.*

Proof of Theorem 3: In the beginning, note that since $\ell_i > 0$, $\ell_i \geq \mathtt{E} + \sqrt{\mathtt{E}^2 + \mathtt{E}}) \Leftrightarrow \ell_i - \mathtt{E} \geq \frac{\mathtt{E}}{\ell_i} + 2$. To prove π_d is the only core-stable coalition, we aim at proving that the error is strictly minimized when every player is alone in π_d. Let us assume that S is a coalition that contains a player i, and the total number of samples in S is L_S. We prove that as long as $|S| \geq 2$, following inequality stands:

$$
\begin{aligned}
&2\texttt{E}\cdot\left(\frac{1}{L_S}+\frac{|S|}{L_S^2}\right)+2\left[\frac{(L_S-\ell_i)^2}{L_S^2}+\sum_{k\neq i}\frac{\ell_k^2}{L_S^2}\right]\\
&>2\texttt{E}\cdot\left(\frac{1}{\ell_i}+\frac{1}{\ell_i^2}\right)
\end{aligned}
\tag{3}
$$

$$
\begin{aligned}
(3)\Leftrightarrow\ &\texttt{E}\cdot\left(\frac{1}{L_S}+\frac{|S|}{L_S^2}\right)+\left[\frac{(L_S-\ell_i)^2}{L_S^2}+\sum_{k\neq i}\frac{\ell_k^2}{L_S^2}\right]\\
&-\texttt{E}\cdot\left(\frac{1}{\ell_i}+\frac{1}{\ell_i^2}\right)>0\\
\Leftrightarrow\ &\frac{1}{L_S^2\ell_i}\left[\texttt{E}(L_S\ell_i+|S|\ell_i)+(L_S-\ell_i)^2\ell_i+\sum_{k\neq i}\ell_k^2\ell_i\right]\\
&-\texttt{E}\left(L_S^2+\frac{L_S^2}{l_i}\right)>0\\
\Leftrightarrow\ &(L_S-\ell_i)^2\ell_i+\sum_{k\neq i}\ell_k^2\ell_i-\texttt{E}(L_S^2-L_S\ell_i)\\
&-\texttt{E}\frac{L_S^2}{l_i}+\texttt{E}|S|\ell_i>0\\
\Leftrightarrow\ &(L_S-\ell_i)^2(\ell_i-\texttt{E})+\ell_i\sum_{k\neq i}\ell_k(\ell_k-\texttt{E})\\
&-\texttt{E}\frac{L_S^2}{l_i}+\texttt{E}|S|\ell_i>0
\end{aligned}
\tag{4}
$$

Since $\ell_i-\texttt{E}\geq\frac{\texttt{E}}{\ell_i}+2$, $\ell_k-\texttt{E}\geq\frac{\texttt{E}}{\ell_k}+2$,

$$
\begin{aligned}
&(L_S-\ell_i)^2(\ell_i-\texttt{E})+\ell_i\sum_{k\neq i}\ell_k(\ell_k-\texttt{E})-\texttt{E}\frac{L_S^2}{l_i}+\texttt{E}|S|\ell_i\\
&\geq(L_S-\ell_i)^2\left(\frac{\texttt{E}}{\ell_i}+2\right)+\ell_i\sum_{k\neq i}\ell_k\left(\frac{\texttt{E}}{\ell_k}+2\right)-\texttt{E}\frac{L_S^2}{l_i}+\texttt{E}|S|\ell_i\\
&=L_S^2\left(\frac{\texttt{E}}{\ell_i}+2\right)-2L_S\ell_i\left(\frac{\texttt{E}}{\ell_i}+2\right)+\ell_i^2\left(\frac{\texttt{E}}{\ell_i}+2\right)\\
&+\ell_i\sum_{k\neq i}\ell_k\left(\frac{\texttt{E}}{\ell_k}+2\right)-\texttt{E}\frac{L_S^2}{l_i}+\texttt{E}|S|\ell_i\\
&=2L_S[L_S-(\texttt{E}+\ell_i)]+2\texttt{E}|S|\ell_i\\
&=2L_S\left(\sum_{k\neq i}\ell_k-\texttt{E}\right)+2\texttt{E}|S|\ell_i>0.
\end{aligned}
$$

Therefore, the inequality is strict, which means every player strictly minimizes the error when staying alone, π_d is the unique element in the core. □

★ **All players have the same number of samples**

Theorem 4. *In uniform federation where all players have the same number of samples ℓ, then:*

- *if $\ell < \mathrm{E}+1$, then π_g is the only element in the core.*
- *if $\ell > \mathrm{E}+1$, then π_d is the only element in the core.*
- *if $\ell = \mathrm{E}+1$, then all partitions of players is in the core.*

★ **Players with small and large numbers of samples**

In this part, we analyze a case where there are two types of players, one type with small number of samples, and the other type with large number of samples. To be more specific, we assume that there are N' *small* players each have ℓ samples and N' *large* players each have 2ℓ samples.

Theorem 5. *For uniform federation, if $2\ell \leq \frac{1}{2}(\mathrm{E}+\sqrt{\mathrm{E}^2+4\mathrm{E}})$, then the grand coalition π_g is core stable.*

Theorem 6. *For uniform federation, if $2\ell \geq \frac{1}{2}(\mathrm{E}+\sqrt{\mathrm{E}^2+4\mathrm{E}})$, then there exists an individually stable partition which is efficiently computable.*

6 Conclusion

As a distributed machine learning paradigm, FL has become an innovation hot topic in large scale machine learning. In this study, hedonic game framework is adopted to model the players' behaviors in FL. By using variance estimation as our case, we quantify the error rates for the players when doing local learning and FL. Three most popular federation models are investigated, and based on the error rates obtained, we give the stability analysis for uniform model.

Acknowledge. Dr. Li is supported in part by the National Natural Science Foundation of China Grant No. 12071478.

References

1. Konečný, J., Brendan McMahan, H., Yu, F.X., Richtarik, P., Suresh, A.T., Bacon, D.: Federated learning: strategies for improving communication efficiency. In: NIPS Workshop on Private Multi-Party Machine Learning (2016)
2. Li, T., Sahu, A.K., Talwalkar, A., Smith, V.: Federated learning: challenges, methods, and future directions. IEEE Sig. Process. Mag. **37**(3), 50–60 (2020)
3. Li, L., Fan, Y., Tse, M., Lin, K.-Y.: A review of applications in federated learning. Comput. Ind. Eng. **149**, 106854 (2020)
4. Bonawitz, K.A., et al.: Towards federated learning at scale: System design. In: SysML 2019 (2019). To appear

5. Sheller, M., Reina, G., Edwards, B., Martin, J., Bakas, S.: Multi-institutional Deep Learning Modeling Without Sharing Patient Data: A Feasibility Study on Brain Tumor Segmentation: 4th International Workshop, BrainLes 2018, Held in Conjunction with MICCAI 2018, Granada, Spain, September 16, 2018, Revised Selected Papers, Part I, volume 11383, pp. 92–104 (2019)
6. Brisimi, T.S., Chen, R., Mela, T., Olshevsky, A., Paschalidis, I.C., Shi, W.: Federated learning of predictive models from federated electronic health records. Int. J. Med. Inf. **112**, 59–67 (2018)
7. Huang, L., Yin, Y., Fu, Z., Zhang, S., Deng, H., Liu, D.: LoAdaBoost: loss-based AdaBoost federated machine learning on medical data (2018)
8. Hard, A., et al.: Federated learning for mobile keyboard prediction (2018)
9. Smith, V., Chiang, C.-K., Sanjabi, M., Talwalkar, A.S.: Federated multi-task learning. In: Guyon, I., et al. (eds.) Advances in Neural Information Processing Systems, vol. 30. Curran Associates, Inc., (2017)
10. Li, J., Khodak, M., Caldas, S., Talwalkar, A.: In Proceedings of International Conference on Learning Representations, volume 30 (2020)
11. Bogomolnaia, A., Jackson, M.O.: The stability of hedonic coalition structures. Games Econom. Behav. **38**, 201–230 (2002)
12. Song, T., Tong, Y., Wei, S.: Profit allocation for federated learning. In: 2019 IEEE International Conference on Big Data (Big Data), pp. 2577–2586 (2019)
13. Zhu, Y., Yan, R., Li, D.: Minimum cost stability in exchange networks. J. Comb. Optim. **38**(4), 987–1004 (2019). https://doi.org/10.1007/s10878-019-00431-z
14. Blum, A., Haghtalab, N., Procaccia, A.D., Qiao, M.: Collaborative PAC learning. In: Guyon, I., et al. (eds.) Advances in Neural Information Processing Systems, volume 30. Curran Associates, Inc. (2017)
15. Blum, A., Haghtalab, N., Phillips, R.L., Shao, H.: One for one, or all for all: Equilibria and optimality of collaboration in federated learning. In: ICML (2021)
16. Donahue, K., Kleinberg, J.: Model-sharing games: analyzing federated learning under voluntary participation. Proc. AAAI Conf. Artif. Intell. **35**(6), 5303–5311 (2021)

K-Division Framework Enhances GNNs' Expressive Power

Dun Ma, Suixiang Gao, and Wenguo Yang(✉)

University of Chinese Academy of Sciences, School of Mathematical Sciences, Beijing, China
madun21@mails.ucas.ac.cn, {sxgao,yangwg}@ucas.ac.cn

Abstract. Researches have indicated that the capabilities of MP-GNNs (Message-Passing GNN) are confined within the constraints of the Weisfeiler Lehman (WL) test for graph isomorphism. This includes the notable shortcoming of MP-GNNs being unable to effectively identify and quantify substructures within the graph, which are crucial for many applications. To address the deficiencies, we introduce K-Division Graph Neural Networks, a novel message-passing framework that is designed to be able to classify different node's neighbors. Through theoretical analysis, we have demonstrated that our architecture is strictly more powerful than WL test. Meanwhile, to understand what kinds of problem that k-div GNNs is competent to solve, we have introduced k-div GNNs' equivalent logic language and model based on model theory to investigate k-div GNNs' expressive power. Also we have proved that if the longest circle in graph is bounded, k-div WL is able to recognize all non-isomorphic graphs. We have conducted both synthetic and real-world experimental evaluations on tasks related to graph and node classification and regression. These results have validated our theory and demonstrated the effectiveness and practicality of our approach.

Keywords: Graph Neural Networks · Expressive Power · Graph Isomorphism

1 Introduction

Graph neural networks have become recently become popular and powerful methods with a wide range of applications, for graph representation learning, such as social and financial networks, molecule classification, and Web page ranking [1]. Most GNNs follow Message Passing scheme: at each iteration every node will gather feature from its neighbors to update new feature representation [2]. Since the expressive power of MP-GNNs(Message-Passing GNN) is limited by the Weisfeiler-Lehman [3] (WL) test, there are substantial tasks that MP-GNNs will fail to be competent to. In this paper, we propose a new framework as K-Division Graph Networks(k-div GNNs) to enhance the expressive power of GNNs.

Y. Chen et al. (Eds.): COCOON 2024, LNCS 15162, pp. 176–188, 2025.
https://doi.org/10.1007/978-981-96-1093-8_15

Xu [2] has revealed the theoretical connection between expressive power of GNN and the Weisfeiler-Lehman (WL) test algorithm. We found out that WL test algorithm sometimes decides some pairs of non-isomorphism graphs is isomorphism because it lacks the ability to identify different types of node's neighbors. Based on this discovery, we have proposed a new graphs isomorphism test algorithm we call it k-division Weisfeiler-Lehman algorithm, it will classify node's different types of neighbors to enhance expressive power in distinguishing non-isomorphic graphs. We have proved that (k+1)-div WL test is strictly powerful than k-div WL test, so if $k \geq 2$, then k-div WL test is strictly powerful than WL test.

Pablo [4] has researched the connection between First logic and MP-GNNs and has proved that exactly every formula that can be captured by MP-GNNs if and only if it can be equivalently expressed as modal graded logic. Based on this we have found k-div GNNs' equivalent logic language and model, having proved the equivalence between k-div GNNs' and logic language, and use this tool to research the expressive power of k-div GNNs.
The following are our main contributions:

(1) We have proposed a new framework as K-Division Graph Networks(k-div GNNs)which has higher expressive power, and proved its corresponding k-div WL test is strictly more powerful than WL test.
(2) We have introduced k-div GNNs' equivalent logic language and model and proved the equivalence. Also we demonstrate that k-div GNNs' is able to detect triangle and identify bipartite graph.
(3) We have proved that if the longest circle is bounded, the k-div WL test is able to recognize all non-isomorphic graphs. Also we have demonstrated that k-div GNNs still maintain beneficial characteristics of MP-GNNs, such as invariance, equivariance and linear complexity.
(4) We have conducted synthetic experimental evaluations on:detecting triangle, counting triangles and identifying bipartite graph, the results has validated our theory. We also have conducted real-world experimental evaluations and showed the effectiveness and practicality of our approach.

2 Related Work

Following the research Xu [2] in the field, there has been a significant push towards developing new GNN architectures that surpass the limitations of the 1-WL test in terms of expressiveness. The research in this area can be categorized into several key directions:

Higher-Order GNNs: A method to enhance the expressiveness of GNNs involves drawing inspiration from higher-order Weisfeiler-Lehman (k-WL) tests [5]. These networks go beyond traditional node feature aggregation by calculating feature vectors for each group of k nodes ($k \geq 2$) and use tensor operations to aggregate features across these node groups.

Substructure-Based GNNs: Recognizing the shortcomings of 1-WL test in detecting common substructures [6], researchers have developed GNNs that focus on incorporating substructure counting into node features. This has been further advanced through the use of homomorphism counting and the development of new aggregation schemes that factor in these substructures, enhancing the network's ability to identify and process such patterns.

Subgraph GNNs: Given that graphs indistinguishable by 1-WL often exhibit high symmetry, recent methods aim to disrupt this symmetry by processing subgraphs within an MPNN [7]. By generating subgraphs in a symmetric manner from the original graph and aggregating their outputs, these approaches seek to improve the network's ability to differentiate between structurally similar graphs.

Non-equivariant GNNs: A simple method to overcome the symmetry in 1-WL aggregation is through the use of non-equivariant GNNs. It has been shown that by assigning unique identifiers to each node [8], standard MPNNs can achieve a higher level of expressiveness. However, this comes with the trade-off of potentially increased sample complexity for training and generalization. For an in-depth exploration of these various approaches to enhancing GNN expressiveness, a range of literature is available that provides comprehensive reviews and analyses of the current state of the field.

3 Notation

Definition 1 (*k-Hop Neighbor*). *Given graph $G = (V, E)$ and positive integer k, node u is said to be node v's k-hop neighbor, if the length of the shortest path from u to v equals to k. Denote the set of v's k-hop neighbors as $N_k(v)$.*

Definition 2 (*k-Hop Adjacency Matrix*). *Given graph $G = (V, E)$ and positive integer k . If $|V| = n$ and matrix $A(i, j)$ (the diagonal elements of A equal to 0) represents the element in the i-th row and j-th column of the matrix $A^{n \times n}$. Then if $A(v, u) = 1$ if and only if $u \in N_k(v)$, $A(v, u) = 0$ if and only if $u \notin N_k(v)$, we call matrix A is graph G's k-hop adjacency matrix. Denote the k-hop adjacency matrix as A^k*

Definition 3 (*$(i_2, \cdots, i_k)^{th}$* **Neighbor**). *Given graph $G = (V, E)$ and positive integer k and $(i_2, \cdots, i_k) \in \{0, 1\}^{k-1}$. If there's a path from u to v with length equals to t if and only if $i_t = 1$, there's no path from u to v equals to t if and only if $i_t = 0$, then node u is said to be node v's $(i_2, \cdots, i_k)^{th}$ neighbor. Denote set of v's $(i_2, \cdots, i_k)^{th}$ neighbors as $N_{(i_2,\cdots,i_k)}(v)$*

Definition 4 (*$(i_2, \cdots, i_k)^{th}$* **Adjacency Matrix**). *Given graph $G = (V, E)$, positive integer k and $(i_2, \cdots, i_k) \in \{0, 1\}^{k-1}$. If $|V| = n$ and $A(i, j) \in \{0, 1\}$ represents the element in the i-th row and j-th column of the matrix $A^{n \times n}$. Then $A(v, u) = 1$ if and only if $u \in N_{(i_2,\cdots,i_k)}(v)$, $A(v, u) = 0$ if and only if $u \notin N_{(i_2,\cdots,i_k)}(v)$, we call matrix A is graph G's $(i_2, \cdots, i_k)^{th}$ adjacency matrix. Denote the $(i_2, \cdots, i_k)^{th}$ adjacency matrix as $A^{(i_2,\cdots,i_k)}$ (Fig. 1).*

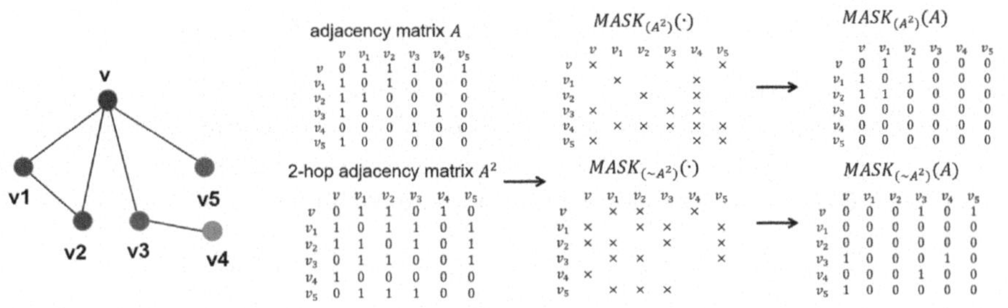

Fig. 1. In the left figure, node v_1 and v_2 are $v's$ $(1)^{th}$ neighbors while v_3 and v_5 are $v's$ $(0)^{th}$ neighbors. The right figure has demonstrated the computation process for $(1)^{th}$ and $(0)^{th}$ adjacency matrix.

Lemma 1. *Given graph* $G = (V, E)$*, if* $A^{(i_2,\cdots,i_k,1)}$ *is* $G's$ $(i_2, \cdots, i_k, 1)^{th}$ *adjacency matrix,* $A^{(i_2,\cdots,i_k,0)}$ *is* $G's$ $(i_2, \cdots, i_k, 0)^{th}$ *adjacency matrix. Then* $A^{(i_2,\cdots,i_k,1)} + A^{(i_2,\cdots,i_k,0)}$ *is* $G's$ $(i_2, \cdots, i_k)^{th}$ *adjacency matrix.*

Definition 5. *We define operator* $Mask_A(\cdot)$ *and* $Mask_{\sim A}(\cdot)$ *as follow:*

$$Mask_A(X)(i,j) = \begin{cases} 0 & if \quad A(i,j) = 0 \\ X(i,j) & if \quad A(i,j) \neq 0 \end{cases} \tag{1}$$

$$Mask_{\sim A}(X)(i,j) = \begin{cases} X(i,j) & if \quad A(i,j) = 0 \\ 0 & if \quad A(i,j) \neq 0 \end{cases} \tag{2}$$

Theorem 1. *Given graph* $G = (V, E)$ *, if* A^{k+1} *is G's (k+1)-hop adjacency matrix,* $A^{(i_2,\cdots,i_k)}$ *is G's* $(i_2, \cdots, i_k)^{th}$ *adjacency matrix, then* $Mask_{A^{k+1}}(A^{(i_2,\cdots,i_k)})$ *is G's* $(i_2, \cdots, i_k, 1)^{th}$ *adjacency matrix,* $Mask_{\sim A^{k+1}}$ $(A^{(i_2,\cdots,i_k)})$ *is G's* $(i_2, \cdots, i_k, 0)^{th}$ *adjacency matrix.*

4 Framework of K-Neighbor Division Graph Nerual Networks

The key essence of our architecture: k-div GNN is the aggregation pattern that k-div GNN aggregate the node's neighbor feature. Different from MP-GNN(message passing graph neural networks), k-div GNN assignes different parameters to different types of node neighbors. In particular, each node v updates its feature $H^t(v)$ by combining its previous state with the diiferent aggregated messages as:

$$T^t_{(i_2,\cdots,i_k)}(v) = AGG\left(\{\{H^{t-1}(u) \mid u \in N_{(i_2,\cdots,i_k)}(v)\}\}\right) \tag{3}$$

$$H^t(v) = COM\left(H^{t-1}(v), (T^t_{(i_2,\cdots,i_k)} | (i_2,\cdots,i_k) \in \{0,1\}^{k-1})\right) \tag{4}$$

where is AGG an arbitrary neighborhood aggregate function, COM is an arbitrary function approximator (MLP), $H^t(v)$ represents the feature of node v in t^{th} layer, $N_{(i_2,\cdots,i_k)}(v)$ is the set of node $v's$ $(i_2,\cdots,i_k)^{th}$ neighbors. In this paper, we will consider sum as aggregate function, the architecture of k-div GNN can be written as

$$H^t(v) = F^t\left(H^{t-1}(v)D^t + \sum_{(i_2,\cdot,i_k)\in\{0,1\}^{k-1}} \sum_{u\in N^k_{(i_2,\cdots,i_k)}} H^{t-1}(u)B^t_{(i_2,\cdot,i_k)} + c^t\right) \tag{5}$$

where D^t and $B^t_{(i_2,\cdot,i_k)}$ are learnable parameter matrixs, c^t is learnable parameter vector and F^t is an activation function. We can also derive the graph isomorphism algorithm: k-division weisfeiler-lehman test from k-neighbor division graph nerual networks architecture demonstrated in Algorithm 1.

Algorithm 1. k-Division Weisfeiler-Lehman Test

Input: Graph $G = (V, E, X)$, $G' = (V', E', X')$ $k \in N^+$.
Output: Whether Graph G and G' are isomorphic.
1: **Initialization:** $\forall$ v $\in V$, $label^0(v) \leftarrow X_v$, $\forall$ v' $\in V'$, $label^0(v') \leftarrow X'_v$, i=0.
2: **while** There's bijection $f : V \to V'$, $label^i(v) = label^i(f(v))$ **do**
3: i=i+1
4: **for** ($\forall v \in V$) **do**
5: $label^i_{(i_2,\cdots,i_k)}(v) \leftarrow Hash(\{\{label^{i-1}(u)|u \in N_{(i_2,\cdots,i_k)}(v)\}\})$
6: $label^i(v) \leftarrow Hash(label^{i-1}(v), (label^i_{(i_2,\cdots,i_k)}(v)|(i_2,\cdots,i_k) \in \{0,1\}^{k-1}))$
7: **end for**
8: **for** ($\forall v' \in V'$) **do**
9: $label^i_{(i_2,\cdots,i_k)}(v') \leftarrow Hash(\{\{label^{i-1}(u')|u' \in N_{(i_2,\cdots,i_k)}(v')\}\})$
10: $label^i(v') \leftarrow Hash(label^{i-1}(v'), (label^i_{(i_2,\cdot,i_k)}(v')|(i_2,\cdots,i_k) \in \{0,1\}^{k-1}))$
11: **end for**
12: **if** The number of labels' types does not increase **then**
13: End algorithm, **return** output Graph G and G' are isomorphic
14: **end if**
15: **end while**
16: **return** Output Graph G and G' are not isomorphic

5 How Powerful Are K-DIV GNNs

We now analyse the expressive power of k-div GNNs in comparison with MP-GNNs, in this section we will prove that k-div WL's expressive in recognizing non-isomorphic graphs will strictly increase with the increment of k, hence if integer $k \geq 2$, then k-div WL test is strictly more powerful in recognizing non-isomorphic graphs in comparison with WL test. We are also interested in what kind of problem in graph theory that k-div GNNs is able to learn.

Theorem 2. *Given a pair of non-isomorphic graph (G, G') and $k \in N^+$, then if the output of k-div WL for (G, G') is non-isomorphic, then the output of $(k+1)$-div WL for (G, G') is non-isomorphic.*

Theorem 3. *For $\forall k \in N^+$, There's a pair of non-isomorphic graph (G, G') that the output of k-div WL for (G, G') is isomorphic, and the output of $(k+1)$-div WL for (G, G') is non-isomorphic.*

Corollary 1. *If integer $k \geq 2$, then k-div WL test is strictly more powerful in distinguishing non-isomorphic graphs than WL test.*

In order to figure out what kind of problem in graph theory that k-div GNNs is able to learn. We will analyse the expressive of k-div GNNs from the perspective of mathematical logic.

The key tool for the theoretical analysis of the expressive of k-div GNNs that we will introduce is model theory. First we will introduce equivalent logic model(it's different from GNN's model) and equivalent logic language: Equivalent logic model $\Psi =< V, E, F >$ is a tool to describe graph G, it's composed of three parts:(1)V represents the node set.(2):E represents the relation function, usually $E(v, u)$ is true iff there's an edge between node v and u.(3)F is a truth function that maps node to certain characteristics.

Equivalent logic language is a set of logic formula to describe graphs or nodes. For example, in first order logic(FOL), statements can refer to specific nodes and their properties. For example, if graph G is a color graph, $G's$ logic model is defined as $\Psi =< V, E, Color >$, then logic formula $\alpha \in FOL$ can be defined as $\alpha(v) = red(v) \wedge \exists u(E(u, v) \wedge green(u))$, while $\alpha(v)$ is true iff node v is red and has at least one green neighbor node.

Definition 6. *Given a logic formula α, Σ is a class of GNNs. If there's a GNN $M \in \Sigma$, $\forall G = (V, E)$, $H(v)$ is node v's output feature in M, that $H(v) = 1$ iff $\alpha(v)$ is true, $H(v) = 0$ iff $\alpha(v)$ is false. Then we call Σ captures logic formula α.*

Definition 7. *If Σ is a class of GNNs, we will call that the set of logic formulas that Σ captures as Σ's equivalent logic language, denoted as $\mathcal{L}_\Sigma$. If every logic formulas in $\mathcal{L}_\Sigma$ can be defined by logic model $\Psi =< V, E, F >$, then we will call Ψ as Σ's equivalent logic model.*

FOC_2 is a set of formulas that replaces quantifiers $\exists$ with $\exists^{\geq n}$, meaning that at least exists n nodes satisfy the fromula and restricts the number of variables occur in formula less than two. Graded modal language is a fragment of FOC_2: instead of $\exists u \varphi(u)$ it add $E(v, u)$ restriction on $\exists^{\geq n}$ that as $\exists^{\geq n} u(E(v, u) \wedge \varphi(u))$, hence graded modal language's expressive power is less than FOC_2.

Theorem 4. *Given a graph $G = (V, E, X)$, $k \in N^+$, if node's features X can be divided into s classes, define $F_i(v)(1 \leq i \leq s)$ to be true iff node v features is divided into s^{th} class. Then k-div GNNs' equivalent logic model can be formed as $\Psi =< V, E_{(i_2,\cdots,i_k)}, F_i >$, $E_{(i_2,\cdots,i_k)}(v,u)$ is true iff $u \in N_{(i_2,\cdots,i_k)}(v)$, k-div GNNs' equivalent logic language is graded modal language, in another word, $\forall \alpha \in \mathcal{L}_{k-div\,GNN}$, α can be formed as:*

$$\neg\varphi, \varphi \wedge \psi, \exists^{\geq n} u(E_{(i_2,\cdots,i_k)}(v,u) \wedge \varphi(u)), F_i \tag{6}$$

where $\varphi, \psi \in \mathcal{L}_{k-div\,GNN}$.

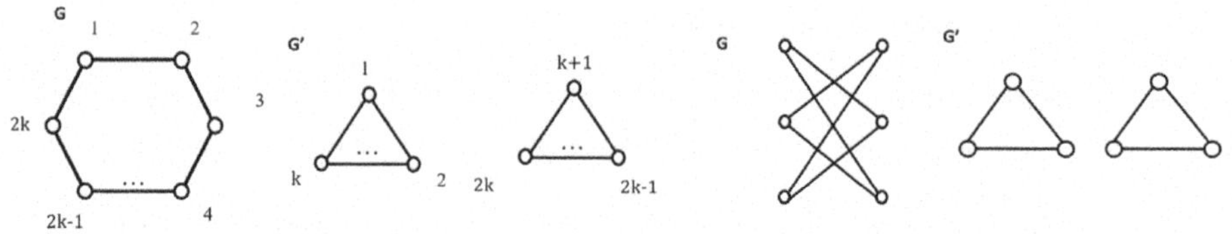

Fig. 2. In left figure, every node in graph G is not contained by a circle with length k, while in graph G' every node is contained by a circle with length k. In right figure, graph G is a bipartite graph, while graph G' is not. However the outputs of WL test for both (G, G') in left and right figures are isomorphic.

Now we are able to explore the questions that what kinds of problem that k-div GNNs are able to solve: through researching whether the problem is able to be expressed equivalently by k-div GNNs' equivalent logic language, we are able to know whether k-div GNNs are able to solve the problem. The first problem that we are interested in is that whether k-div GNNs is able to detect whether node v is contained by a circle with length k. As Fig. 2 shows, MP-GNNs is not able to detect whether node v is contained by a circle with length k. However if $l \leq k$, then node v is contained by a circle with length l+1 iff there's exists node u and $(i_2, \cdots, i_{l-2}, i_l, \cdots, i_k) \in \{0,1\}^{k-2}$ that $u \in N_{(i_2,\cdots,i_{k-l},1,i_l,\cdots,i_k)}$, we can define formula α_k^l as

$$\alpha_k^l(v) = \bigvee_{(i_2,\cdots,i_{l-2},i_l,\cdots,i_k)\in\{0,1\}^{k-2}} \exists^{\geq 1} u(E_{(i_2,\cdots,i_{l-2},1,i_l,\cdots,i_k)}(v,u) \tag{7}$$

if $l = k+1$, we can define formula α_k^l as

$$\alpha_k^l(v) = \bigvee_{(i_2,\cdots,i_{k-1})\in\{0,1\}^{k-2}} \exists^{\geq 1} u(E_{(i_2,\cdots,i_{k-1},1)}(v,u) \tag{8}$$

hence $\alpha_k^l(v)$ is true iff node v is contained by a circle with length l+1. The second problem we are interested in is that when k-div WL test will fails. If graph G and $G's$ don't have circle, the WL test has such conclusion:

Theorem 5. *[9] If graph G and G′ are both connected and don't have circle, then WL algorithm outputs G and G′ are "non-isomorphic" iff G and G′ are non-isomorphic.*

We notice that one of the reason that WL test will fails in recognizing non-isomorphic graphs because WL test is not able to recognize circle structures. Hence for k-div WL test we have following conclusion:

Theorem 6. *If graph G and G′ are both connected and the length of circle in graph G and G′ is bound by k+1(no more than k+1, $k \geq 2$), then k-div WL algorithm outputs G and G′ are "isomorphic" iff G and G′ are isomorphic.*

Bipartite graphs are a type of graphs in graph theory where the vertices can be divided into two disjoint sets such that every edge connects a vertex from the first set to one from the second set, and no edges connect vertices within the same set. This property makes bipartite graphs particularly useful in modeling relationships where one set represents one type of entity and the other set represents a different type of entity, with connections or interactions only occurring between the two different types. As Fig. 2 demonstrates, since WL outputs isomorphic for G and G', MP-GNNs is not able to recognize bipartite graphs. Meanwhile a graph is a bipartite graph iff there's no odd circle in the graph. If a graph G's length of the longest circle is bound by k+1, for k-div GNNs, we can define formula α'_k:

$$\alpha'_k(v) = \bigwedge_{t \in [1:[\frac{k}{2}]]} (\neg \alpha_k^{2t+1}(v)) \tag{9}$$

where $\alpha_k^{2t+1}(v)$ is defined in Eq. 7 and 8. Since G's length of circle is bound by k+1, then $\alpha'_k(v)$ is true iff node v is not contained by a circle with odd length, then if α'_k is true for every node in G iff G is a bipartite graph.

6 Characteristics of K-Div GNNs

6.1 Invariance and Equivariance

Definition 8. *A function $f : R^{n\times n} \times R^{n\times k} \rightarrow R^d$ is said of* ***invariance*** *iff $\forall \sigma \in Sym(n), f(\sigma(A), \sigma(X)) = f(A, X)$. A function $f : R^{n\times n} \times R^{n\times k} \rightarrow R^{n\times d}$ is said of* ***equivariance*** *iff $\forall \sigma \in Sym(n), f(\sigma(A), \sigma(X)) = \sigma(f(A, X))$.*

A consensus for GNNs is that the GNNs' outputs should not be affected by nodes' permutation [10, 11]. **Invariance** refers to the property that the output of the network should remain unchanged under specific transformations of the input data. For example, if the nodes of a graph are renumbered, the output of an invariant network should not change, as the renumbering does not affect the inherent structure or relationships within the graph. **Equivariance** refers to the property that network's output changes in a predictable and consistent manner with the transformation of the input data. In the context of graph data, an equivariant network would transform its internal representations of the nodes in the same way that the graph itself is transformed.

Theorem 7. *For $k \in N^{+}$, $\forall M \in$k-div GNNs, M is of invariance and equivariance.*

6.2 Complexity

The complexity of k-div GNNs comprises two parts: precomputation (neighbor division) and training/testing. The prominent property is that training and inference are linear to the number of edges, $O(|E|)(\leq O(n^2))$, since k-div GNNs still follows the message passing framework. For precomputation, every classification for node's neighbor can be regarded as a classification for node's adjoined edge, $O(|E|)$. Compared with other types of GNNs with higher expressive like higher-order mothods [5,12,13], with $O(n^k)$ training complexity or relational pooling [14] with $O(n!)$ training complexity, k-div GNNs has obvious-advantage dealing with large graph data.

7 Experiment

In the following section, we evaluate k-div GNNs in comparision to MP-GNNs in a variety of datasets from different domains. We have evaluated k-div GNNs for both synthesis experiments and real-world experiments. In synthesis experiments, by our theory, in graph-level test if the length of circle in graph is bound by 4, then 2-div GNNs is able to recognize all bipartite graphs. In node-level test, 2-div GNNs is able to dectect whether the node is contained by a triangle. The datasets from in real-world experiments like proteins which might requests more powerful GNNs to effectively identify and quantify substructures within the graph data, hence we have utilized GCN, GIN and their 2-div GNNs in comparision to other models and graph kernel methods to evaluate k-div GNNs. We also have synthetic experiments on the influence of differents aggregator and layers on GNN's expressive, additional implementation details can be found in the Appendix C.

7.1 Synthetic Experiments

Node-Level Experiments: To testify our theory, in node-level experiments, we utilize synthetic data to perform experiment to validate our result. We performed the experiment using common model MP-GNNs:GIN,GCN,TAG and their 2-div frameworks. Our experiments were implemented in the PyTorch Geometric library and DEEP GRAPH LIBRARY and utilize networkx to generate synthetic erdos renyi graph graphs, severally with 4000,5000,6000 nodes with 0.006,0.005,0.004 probabilities to generate edges, we have also generated random regular graph where every node's degree is 3. We set up two missions for GNNs at node-level:(1)Dectect whether the node is contained by a triangle. (2)Count the number of triangles the nodes is contained in.

Table 1 and 2 are the accuracies of detecting triangle and counting number of triangles, We can see the there is a significant gap of accuracies between MP-GNNs and their 2-div frameworks(2-div frameworks are able to detect every

Table 1. Accuracy(%) of detecting triangle

	random regular			erdos renyi		
Method	4000	5000	6000	4000	5000	6000
GCN	67 ± 0.8	51.9 ± 0.5	$63 \pm 1.$	58.5 ± 4.3	51.2 ± 1	64.7 ± 0.7
$2 - Div\,GCN$	**100**	**100**	**100**	**100**	**100**	**100**
GIN	68.1 ± 2	48.8 ± 2.1	62.3 ± 0.8	64.6 ± 0.6	54.6 ± 1.7	64.5 ± 0.4
$2 - Div\,GIN$	**100**	**100**	**100**	**100**	**100**	**100**
TAG	65 ± 2	49.5 ± 3.4	63.5 ± 0.3	61.1 ± 1	53.8 ± 0.1	65.7 ± 0.4
$2 - Div\,TAG$	**100**	**100**	**100**	**100**	**100**	**100**

Table 2. Accuracy(%) of counting number of triangles

	random regular			erdos renyi		
Method	4000	5000	6000	4000	5000	6000
GCN	66.3 ± 1.1	49.3 ± 0.1	36.2 ± 1	59.8 ± 1.7	47.5 ± 0	37.0 ± 0.5
$2 - Div\,GCN$	99.3 ± 0.6	96.7 ± 0.1	94 ± 0.8	95 ± 0.3	91.4 ± 2.4	80.9 ± 2
GIN	66.8 ± 2.4	48.9 ± 0.6	34.5 ± 2.8	60.2 ± 2.2	46.5 ± 0.9	35.4 ± 2.7
$2 - Div\,GIN$	$\mathbf{99.8 \pm 0}$	$\mathbf{99.8 \pm 0.2}$	$\mathbf{99.3 \pm 0.2}$	$\mathbf{99.2 \pm 0.3}$	$\mathbf{98.8 \pm 0.2}$	$\mathbf{98.7 \pm 0.4}$
TAG	63.7 ± 4	49.5 ± 2.3	36.1 ± 2.9	60.8 ± 4	48.6 ± 2.1	38.5 ± 4.3
$2 - Div\,TAG$	99.6 ± 0	98.6 ± 0.5	97.8 ± 1.1	96.2 ± 1.1	94.8 ± 4.6	84.6 ± 2.7

node which is contained in a triangle and get high scores in counting the number of triangle that nodes are contained in), as expected from theory.

Graph-Level Experiments: In graph-level Experiments, we utilize networkx to generate synthetic 5000 erdos renyi graphs with 20,30,40 nodes and 0.1 probabilities to generate edges, we have also generated 5000 random regular graph with 20,30,40 nodes where every node has 3°C. The length of every circle in graphs is set to no more than 4. We set up one mission for GNNs at graph-level:(1)Identify whether a graph is a bipartite graph. We randomly seperated nodes into three classes: train set with 60%, val set with 20% and test set with 20%.

Table 3 is the accuracy of identifying whether a graph is a bipartite graph. We can see the there is a significant gap of accuracies between MP-GNNs and their 2-div frameworks,(2-div framework is able to detect every bipartite graph) as expected from theory. Hence the synthetic experiments has validated our theory.

7.2 TUD Graph Classification Benchmarks

Our assessment of k-div GNNs is conducted using datasets from the real-world TUD benchmarks [15]. We have selected seven distinct datasets from the fields of

Table 3. Accuracy(%) of identifying whether a graph is a bipartite graph

	random regular			erdos renyi		
Method	20	30	40	20	30	40
GCN	55.2 ± 2.1	61.1 ± 2.1	75.3 ± 1.8	57.2 ± 1.4	54.7 ± 0.8	63.0 ± 0.4
$2 - Div\,GCN$	**100**	**100**	**100**	**100**	**100**	**100**
GIN	58.3 ± 0	67.8 ± 0.3	72.1 ± 0.4	74.6 ± 0.4	62.1 ± 0.7	59.8 ± 3.2
$2 - Div\,GIN$	**100**	**100**	**100**	**100**	**100**	**100**
TAG	59.3 ± 1.3	65.3 ± 3.4	63.5 ± 2.2	63.5 ± 1.3	64.1 ± 0.1	69.5 ± 0.7
$2 - Div\,TAG$	**100**	**100**	**100**	**100**	**100**	**100**

bioinformatics and computational social science for our analysis. These datasets are then benchmarked against a variety of GNNs and Graph Kernels. The foundational model MP-GNNs architecture we have employed are the GIN and GCN.

Table 4 provides a comprehensive overview of the different methods(Models with $*$ are graph kernels methods.) we have assessed, along with the specified split of. It is noteworthy to mention that our model 2-Div GCN and 2-Div GIN achieve high scores(get the 5 highest scores across 7 datasets) across the majority of the datasets. In certain instances, our model demonstrates a significant lead over the primary GNN baselines, showcasing its potential as a leading approach in this domain.

Table 4. Accuracy(%) of graph classification on TUD Dataset.

Dataset	MUTAG	PTC	Proteins	NCI1	Collab	IMDB-B	IMDB-M
RWK*	79.2 ± 2.1	55.9 ± 0.3	59.6 ± 0.1	>3 days	N/A	N/A	N/A
GK*	81.4 ± 1.7	55.7 ± 0.5	71.4 ± 0.31	62.5 ± 0.3	N/A	N/A	N/A
PK* [16]	76.0 ± 2.7	59.5 ± 2.4	73.7 ± 0.7	82.5 ± 0.5	N/A	N/A	N/A
WL kernel* [17]	90.4 ± 5.7	59.9 ± 4.3	75.0 ± 3.1	$\mathbf{86.0} \pm 1.8$	78.9 ± 1.9	73.8 ± 3.9	50.9 ± 3.8
GNTK* [18]	90.0 ± 8.5	67.9 ± 6.9	75.6 ± 4.2	84.2 ± 1.5	83.6 ± 1.0	76.9 ± 3.6	$\mathbf{52.8 \pm 4.6}$
DGCNN	85.8 ± 1.8	58.6 ± 2.5	75.5 ± 0.9	74.4 ± 0.5	73.8 ± 0.5	70.0 ± 0.9	47.8 ± 0.9
IGN [19]	83.9 ± 13.0	58.5 ± 6.9	76.6 ± 5.5	74.3 ± 2.7	78.3 ± 2.5	72.0 ± 5.5	48.7 ± 3.4
PPGNs [20]	90.6 ± 8.7	66.2 ± 6.6	77.2 ± 4.7	83.2 ± 1.1	81.4 ± 1.4	73.0 ± 5.8	50.5 ± 3.6
Natural GN [21]	89.4 ± 1.60	66.8 ± 1.79	71.7 ± 1.04	82.7 ± 1.35	N/A	74.8 ± 2.01	51.3 ± 1.50
WEGL [22]	N/A	67.5 ± 7.7	76.5 ± 4.2	N/A	80.6 ± 2.0	75.4 ± 5.0	52.3 ± 2.9
GCN [23]	88.6 ± 4.1	57.2 ± 6.7	70.59 ± 2.1	76.4 ± 1.0	76.1 ± 0.5	72.4 ± 3.4	49.8 ± 1.6
2-Div GCN	$\mathbf{91.6 \pm 6.5}$	66.5 ± 4.2	$\mathbf{78.92 \pm 6.1}$	84.1 ± 4.1	79.6 ± 1.7	74.6 ± 0	51.7 ± 1.8
GIN [2]	89.4 ± 5.6	64.6 ± 7.0	76.2 ± 2.8	82.7 ± 1.7	80.2 ± 1.9	75.1 ± 5.1	52.3 ± 2.8
2-Div GIN	89.4 ± 3.7	$\mathbf{69.2 \pm 8.2}$	75.7 ± 3.3	82.5 ± 4.3	$\mathbf{84.6 \pm 2.7}$	$\mathbf{79.8 \pm 5.0}$	51.2 ± 1.4

Acknowledgement. This work was supported by the National Key R&D Program of China under grant 2022YFA1003900

References

1. Ingraham, J., Garg, V., Barzilay, R., Jaakkola, T.: Generative models for graph-based protein design. Adv. Neural Inf. Process. Syst. 32 (2019)
2. Xu, K., Hu, W., Leskovec, J., Jegelka, S.: How powerful are graph neural networks? arXiv preprint arXiv:1810.00826 (2018)
3. Weisfeiler, B., Leman, A.: The reduction of a graph to canonical form and the algebra which appears therein. nti, Series **2**(9), 12–16 (1968)
4. Barceló, P., Kostylev, E.V., Monet, M., Pérez, J., Reutter, J., Silva, J.-P.: The logical expressiveness of graph neural networks. In: 8th International Conference on Learning Representations (ICLR 2020) (2020)
5. Morris, C., et al.: Weisfeiler and leman go neural: higher-order graph neural networks. In: Proceedings of the AAAI Conference on Artificial Intelligence vol. 33, pp. 4602–4609 (2019)
6. Zhao, L., Jin, W., Akoglu, L., Shah, N.: From stars to subgraphs: Uplifting any GNN with local structure awareness. arXiv preprint arXiv:2110.03753 (2021)
7. Frasca, F., Bevilacqua, B., Bronstein, M., Maron, H.: Understanding and extending subgraph GNNs by rethinking their symmetries. Adv. Neural. Inf. Process. Syst. **35**, 31376–31390 (2022)
8. Cohen, W.W., Minkov, E.: A graph-search framework for associating gene identifiers with documents. BMC Bioinform. **7**, 1–16 (2006)
9. Immerman, N., Lander, E.: Describing graphs: A first-order approach to graph canonization. Springer (1990)
10. Maron, H., Ben-Hamu, H., Shamir, N., Lipman, Y.: Invariant and equivariant graph networks. arXiv preprint arXiv:1812.09902 (2018)
11. Battaglia, P., Pascanu, R., Lai, M., Rezende, D.J., et al.: Interaction networks for learning about objects, relations and physics. Adv. Neural Inf. Process. Syst. **29** (2016)
12. Morris, C., Rattan, G., Mutzel, P.: Weisfeiler and leman go sparse: towards scalable higher-order graph embeddings. Adv. Neural. Inf. Process. Syst. **33**, 21824–21840 (2020)
13. Paliwal, A., Loos, S., Rabe, M., Bansal, K., Szegedy, C.: Graph representations for higher-order logic and theorem proving. In: Proceedings of the AAAI Conference on Artificial Intelligence, vol. 34, pp. 2967–2974 (2020)
14. Murphy, R., Srinivasan, B., Rao, V., Ribeiro, B.: Relational pooling for graph representations. In: International Conference on Machine Learning, pp. 4663–4673. PMLR (2019)
15. Dwivedi, V.P., Joshi, C.K., Luu, A.T., Laurent, T., Bengio, Y., Bresson, X.: Benchmarking graph neural networks. J. Mach. Learn. Res. **24**(43), 1–48 (2023)
16. Neumann, M., Garnett, R., Bauckhage, C., Kersting, K.: Propagation kernels: efficient graph kernels from propagated information. Mach. Learn. **102**, 209–245 (2016)
17. Shervashidze, N., Schweitzer, P., Van Leeuwen, E.J., Mehlhorn, K., Borgwardt, K.M.: Weisfeiler-lehman graph kernels. J. Mach. Learn. Res. **12**(9) (2011)

18. Du, S.S., Hou, K., Salakhutdinov, R.R., Poczos, B., Wang, R., Xu, K.: Graph neural tangent kernel: Fusing graph neural networks with graph kernels. Adv. Neural Inf. Process. Syst. **32** (2019)
19. Maron, H., Ben-Hamu, H., Shamir, N., Lipman, Y.: Invariant and equivariant graph networks. arXiv preprint arXiv:1812.09902 (2018)
20. Maron, H., Ben-Hamu, H., Serviansky, H., Lipman, Y.: Provably powerful graph networks. Adv. Neural Inf. Process. Syst. **32** (2019)
21. de Haan, P., Cohen, T.S., Welling, M.: Natural graph networks. Adv. Neural Inf. Process. Syst. **33**, 3636–3646 (2020)
22. Kolouri, S., Naderializadeh, N., Rohde, G.K., Hoffmann, H.: Wasserstein embedding for graph learning. arXiv preprint arXiv:2006.09430 (2020)
23. Kipf, T.N., Welling, M.: Semi-supervised classification with graph convolutional networks. arXiv preprint arXiv:1609.02907 (2016)

An Innovative Irregular Nesting Algorithm for Flaw Avoidance

Hongji Zhu, Liping Chen, and Shuguang Han(✉)

Zhejiang Sci-Tech University, Hangzhou, China
{2023220603110,2023210103003,dawn1024}@zstu.edu.cn

Abstract. In textile manufacturing, the presence of fabric flaws is inevitable. However, traditional nesting methods often overlook these flaws, leading to compromised product quality or delivery delays due to the need for extra fabric. This research addresses the two-dimensional irregular nesting problem for flawed fabric by innovatively proposing the Flaw-Avoidance Irregular Nesting Algorithm (FAINA). Initially, flaws on the textile fabric are conceptualized as specially prioritized pieces, with an appropriate genetic algorithm utilized to generate the sequence for arranging regular pieces. Subsequently, the strategy for generating No-Fit-Polygon is refined to achieve flaw circumvention and ascertain the pieces' adjacency positioning. Finally, an integrated sorting scheme for flaw-free pieces is obtained by calculating the tailored fitness based on the adjacency positioning relationship and flaw information, optimizing the arrangement of irregular pieces to enhance fabric utilization. Our algorithm guarantees avoidance of flaws of any shape, number and size while optimizing the layout to improve material usage. It is tested on datasets with 10, 30, 50 and 314 pieces on flawed fabric and compared to random, greedy, genetic algorithms that ignore flaws. The results demonstrate a 2% to 13% improvement in area utilization by the proposed algorithm, ensuring all pieces are uncontaminated. Our research offers significant technical support for material utilization and cost control in production, showcasing substantial practical value.

Keywords: Flawed Fabric · Two-dimensional irregular nesting · FAINA · Genetic Algorithm · No-Fit-Polygon

1 Introduction

The two-dimensional irregular nesting problem, also known as the irregular part layout problem, involves efficiently and non-overlappingly arranging irregularly shaped two-dimensional objects within a limited space to maximize material utilization or minimize cost. Due to the inevitable production of various flaws such as uneven textures, color differences and cracks in textile fabrics, nesting algorithms must flexibly and efficiently avoid flaws of various sizes, shapes and locations on the fabric. Thus, precisely navigating flaws on flawed fabrics while optimizing irregular piece layouts in two-dimensional irregular nesting research is deemed challenging. Given its NP-Hard complexity, researching relevant algorithms and strategies offers significant challenges and value.

Y. Chen et al. (Eds.): COCOON 2024, LNCS 15162, pp. 189–201, 2025.
https://doi.org/10.1007/978-981-96-1093-8_16

1.1 Related Work

For the two-dimensional irregular nesting problem with flaw-free fabric, early scholars Adamowicz and Albano [1] introduced the concept of No-Fit-Polygon to determine the positional relationship between two polygons in a plane. Jakobs [2] suggested arranging the rectangular envelopes of irregular polygons and then applying a squeezing operation to reduce the gaps between polygons. Mahadevan [3] introduced the sliding collision method to intuitively simulate the generation process of No-Fit-Polygon. Ghosh [4] proposed the Minkowski sum, a method for quickly solving the No-Fit-Polygon between two convex polygons. Jia et al. [5] used a genetic algorithm for the layout of enveloped rectangles and employed the lowest horizontal line for positioning. Wang et al. [6] proposed a hybrid algorithm that combines an edge matching degree algorithm with a variable neighborhood search algorithm to solve the 2D rectangular nesting problem. Jie et al. [7] devised a reinforcement learning algorithm, integrating Monte Carlo, Q-learning and Sarsa, to address the 2D irregular bin packing issue.

For the two-dimensional nesting problem with flawed fabric, Dong et al. [8] developed a uniparental genetic algorithm by generating rectangles abutting No-Fit-Polygon to avoid overlaps between the rectangles and flaws. Afsharian et al. [9] proposed a dynamic programming-based heuristic algorithm that redefines the expression of defect areas and the discrete point set when handling defects, reducing the space-time complexity of the algorithm. Li et al. [10] combined a dual-population genetic algorithm with a defect-constrained lowest horizontal line method for optimizing defective plate nesting.

1.2 Our Approach

Current research, mainly targeting nesting with rectangular flaws, underperforms across diverse flaw types and sample traits. This research proposes a novel irregular nesting algorithm that can flexibly and efficiently avoid flaws, meeting the application requirements of complex scenarios.

This research pioneers the exploration of the two-dimensional irregular nesting challenge while avoiding an unlimited number of flaws with unspecified positions and arbitrary shapes. To address this, a Flaw-Avoidance Irregular Nesting Algorithm (FAINA) is proposed, which constructs fabric flaw areas as specially prioritized pieces and refines the No-Fit-Polygon generation strategy. A tailored fitness evaluation method, leveraging flaw specifics and the spatial interrelation of pieces, is crafted to adeptly evade flaw zones and enhance the arrangement of the nesting of irregular pieces simultaneously. Furthermore, using different scales of piece datasets on flawed fabric, FAINA is compared with random, greedy and genetic algorithms [13] that ignore flaws, demonstrating the efficiency and practicality of FAINA.

2 Preliminaries

This research investigates the two-dimensional irregular nesting problem with flawed fabrics, where flaws can be the areas of unlimited quantity, unspecified positions and arbitrary shapes. Our objective is to meticulously circumvent these flawed regions, concurrently optimizing the nesting of irregularly shaped pieces to ensure the maximization of fabric utilization. As shown in Fig. 1 .

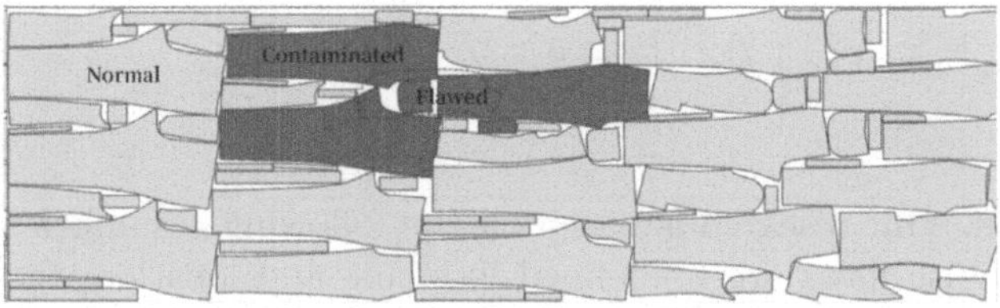

Fig. 1. Traditional Nesting Result

2.1 Problem Defnition

Let E be a band-like fabric with a width of and sufficiently large length. Fabric contains flaws whose shapes and positions are known, generally represented by several non-overlapping polygons. The parameters are as follows in Table 1:

Table 1. Explanation of Variables in the Mathematical Model

Symbol	Description
i	Index number
N	Number of positive integers
D	Number of pieces to be arranged
W	Width of the fabric
S	Sum of areas
$x_{\max}$	Maximum x-coordinate of the polygon vertices after nesting
$x_{\min}$	Minimum x-coordinate of the polygon vertices after nesting
$S_i(i = 1, 2, \ldots, D)$	Area of the i-th piece to be arranged
$S_i(i = D + 1, \ldots, D + N)$	Area of the $i - D$-th positive integer piece
P_i	Geometrical parameters of the i-th piece polygon vertices

Definition 1. Minimum Length of Rectangle Enclosed after Nesting:

$$\min L = x_{\max} - x_{\min} \tag{1}$$

Definition 2. If fabric E comprises special and normal pieces for nesting, maximizing area utilization requires considering only the area of normal pieces. (Normal pieces denote the flaw-free portion of samples):

$$\max S = \frac{\sum_{i=D+1}^{D+N} S_i}{L \times W} \tag{2}$$

Where $\{S_1, S_2, ..., S_D\}$ is the set of areas for the special pieces and $\{S_{D+1}, ..., S_{D+N}\}$ is the set of areas for the normal pieces.

Spatial Constraints. Fabric E contains $D(D \geq 0)$ special pieces $\{P_1, P_2, ..., P_D\}$ and $N(N \geq 0)$ normal pieces $\{P_{D+1}, ..., P_{D+N}\}$. Since any two pieces do not overlap and each piece is within the spatial range of fabric E.Our model is subject to the following spatial constraints:

$$\begin{cases} P_i \cap P_j = \emptyset, i \neq j \\ P_i \in E, \qquad i = 1, ..., N + D \end{cases} \tag{3}$$

Assumptions. Our research is based on the following assumptions:

Initially, fabric flaws are predefined in shape and location; these are prioritized as special pieces in the nesting sequence, ensuring their placement precedes that of regular pieces. Moreover, each nested piece corresponds to an equivalent geometric polygon and the fabric is considered a continuous strip of fixed width.

Christofdes [14] have provided an exact algorithm for two-dimensional rectangular piece optimization using a tree search approach to solve the rectangular cutting stock problem. However, the exact algorithm is computationally intensive and inefficient for large-scale problems.Moreover, considering the diverse shapes and random locations of fabric flaws further increases the complexity and limits the generalizability of this problem. Nonetheless, by prioritizing fabric flaws as special pieces, this model ensures effective flaw avoidance and significantly simplifies the theoretical model and algorithmic workflow.

3 Flaw-Avoidance Irregular Nesting Algorithm(FAINA)

Traditional nesting algorithms, primarily designed for flaw-free homogeneous materials, often lack mechanisms for avoiding flaws and are unsuitable for irregular two-dimensional layouts with flawed fabric. Existing algorithms face challenges like high computational complexity, increased computing resources, and limited generality when flaws must be avoided, making it difficult to efficiently solve practical problems. Consequently,our research introduces the Flaw-Avoidance Irregular Nesting Algorithm (FAINA), which refines the No-Fit Polygon generation strategy based on different adjacency relationships between the pieces and employs a genetic algorithm to optimize the population's sequence, integrating flaw information into the fitness evaluation for optimal arrangement of all pieces. This section will detail the FAINA's structure(see Fig. 2) and technical nuances.

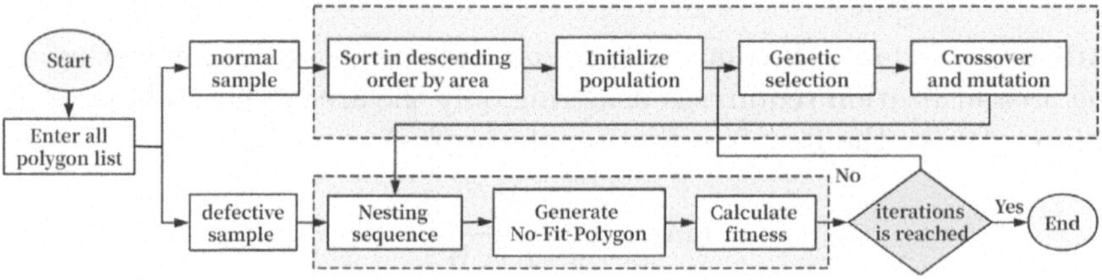

Fig. 2. FAINA Framework

3.1 Overlap Detection

In the problem of two-dimensional irregular polygon nesting, both the fabric to be cut and the pieces to be nested can be characterized by a series of geometrically equivalent polygons, hence the adjacency relationship between any two polygons can be classified as either internal or external tangency. The No-Fit Polygon [3] (NFP, as shown in Fig. 3), serves as a key tool for characterizing all possible relative positioning configurations between polygons while maintaining a non-overlapping state. It ensures that the pieces to be nested do not exceed the boundaries of the material to be cut and that there is no overlap between any two pieces. Consequently, its generation method is crucial for the nesting results.as a key tool for characterizing all possible relative positioning configurations between polygons while maintaining a non-overlapping state, plays a crucial role in the generation of nesting results,where it ensures that the pieces to be nested do not exceed the boundaries of the material to be cut and that there is no overlap between any two pieces. This research improves the No-Fit Polygon generation strategy based on different adjacency relationships between polygons, locally caching generated No-Fit Polygons as key-value pairs during the computation process to reduce redundant calculations and optimize the processing flow.

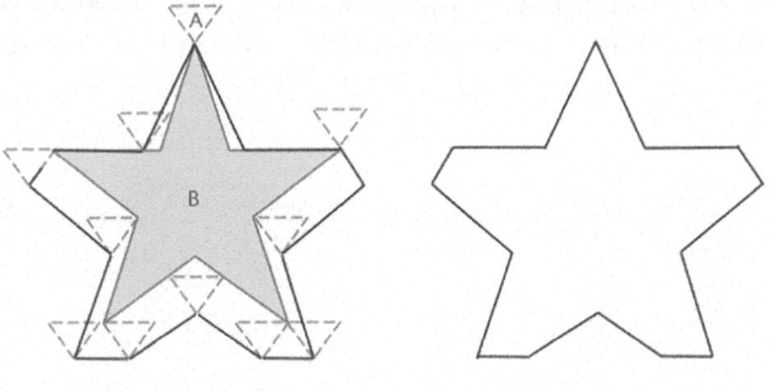

Fig. 3. Process and Results of No-Fit Polygon

When polygons have an internal tangency adjacency, implying that the piece to be nested must not exceed the fabric's spatial range, NFP calculation varies based on the shape of the fabric and the piece. For rectangular pieces and fabric, leveraging the symmetry property and simple geometric relationships allows for rapid NFP computation. However, for irregular pieces, the sliding collision method is employed. This iterative process detects contact points or edges between polygons A and B (see Fig. 3). Each contact point or edge generates a translation vector indicating feasible non-overlapping movement directions and distances of B relative to A. The largest valid translation vector from these generated vectors determines the non-overlapping position of B relative to A.

When polygons are tangentially adjacent, meaning there is non-overlapping between any two pieces to be arranged, the Minkowski Difference method [12] is used to express the adjacency relationship between polygons. In specific calculations, the Minkowski Difference method is utilized to determine the spatial range within which a geometric set can move without intersecting another set. Its mathematical definition is: $D = D_1 \ominus D_2 = \{\alpha - \beta \mid \alpha \in D_1, \beta \in D_2\}$.This research constructs a series of polygonal sets representing the maximum non-overlapping states achievable between two polygons by performing the Minkowski Difference operation on the vertex coordinates (α and β) of polygons D_1 and D_2. From these sets, the one with the smallest area is selected as the final No-Fit Polygon.

3.2 Nesting Sequencing

The FAINA addresses the problem of sequencing irregular two-dimensional nestings for flawed fabrics by using an appropriate genetic algorithm. It seeks high-quality nesting solutions through population iteration, integrating flaw information into the fitness calculation to provide a uncontaminated nesting sequencing optimization scheme. The specific process of the FAINA is as follows:

A population Z evolves over multiple generations to produce excellent individuals. The initial population Z_0 consists of pieces sorted in descending order of area size, along with their rotation sequences. During the selection process, FAINA first carries over the gene spots of the best individual from generation t into generation $t+1$, ensuring the preservation of the fittest individuals. It then randomly selects two individuals from generation t to serve as parents for the population's reproduction, which will undergo subsequent crossover and mutation.

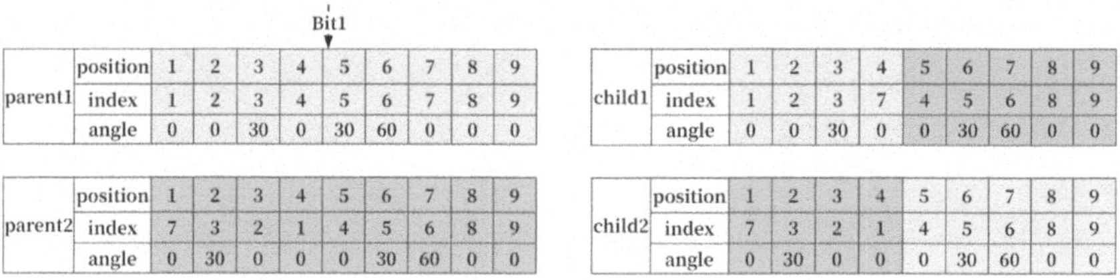

Fig. 4. Genomic Recombination

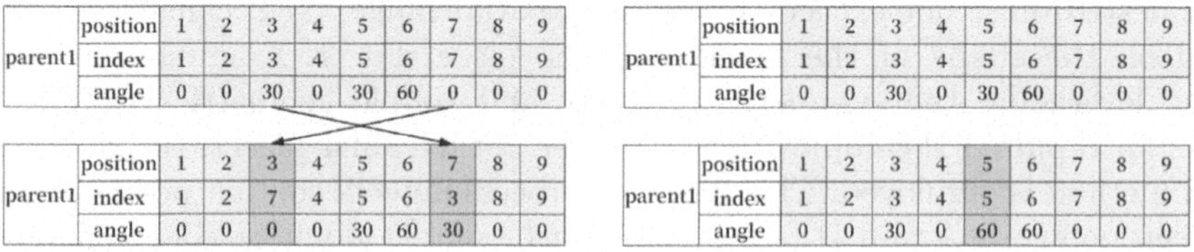

Fig. 5. Mutagenic Process

During crossover(as shown in Fig. 4), a random cut point in the gene sequence is determined, allowing offspring g_1 to inherit the gene sequence before the father's cut point and complete itself with the remaining sequence from the mother. Offspring g_2 inherits the gene sequence after the father's cut point and completes itself with the remaining sequence from the mother. The gene sequence of the offspring produced by crossover may undergo mutation with a certain probability by randomly swapping the positions of two gene spots and changing the rotation angle(see Fig. 5).

As the flaws in special pieces remain unchanged in position and angle, and normal pieces will be tangentially adjacent to them, flaw information is directly inserted at the front of the offspring's gene sequence during fitness calculation. This ensures that during the generation of the NFP and the arrangement process, special pieces are prioritized and do not overlap with normal pieces.Finally, the enclosing rectangle area after nesting is used as the evaluation criterion for the genetic algorithm solution.

Algorithm 1: FAINA Pseudocode

```
Input: object_list: List of polygon objects, flaws_list: List of flaws
Output: best_solution: Optimal nesting result
function InitObject(object_points)
    process_shapes(shapes):
        for shape in shapes: shape.area = calculate_area(shape)
    flaws = generate_flaws_list(shapes)
    return obj_information
function SortShape(object_list)
    SortShape = sorted(shapes, key=lambda shape: criteria(shape))
function LaunchWorkers(population)
    solution_list = pool.insert(config, population)
for iteration ← 1 to config['loop_time'] do
    population ← GA(population, config['offset_bin'], config)
    // Apply GA to evolve population
    process_nfp_cache ← generate_nfp(population)
    // Generate NFP cache for current population state
    for index ← 1 to population_length do
        find fitness(flaw_list + population[index])
return draw_result(best_solution)
```

4 Simulation

To visually demonstrate the performance of FAINA in arranging two-dimensional irregular nesting on flawed fabrics, this research conducts simulation experiments based on the public dataset provided by Tianchi [15]. It uses random sampling to construct datasets containing 10, 30, 50 and 314 pieces to be arranged. Randomly selected pieces from these datasets are used as fabric flaws to simulate the uncertainty of flaw information in industrial production and comprehensively evaluate the nesting capabilities of FAINA at different data scales.

4.1 Experimental Environment

- Processor: AMD Ryzen 9 7845 HX, with 12 cores and a base clock of 16 GHz.
- Operating System: Windows 11 Pro.
- RAM: 64 GB.
- Software Environment: Programming with Python 3.7, main libraries include pyclipper 1.3.0, Shapely 1.8.4 etc.

These experiments compare FAINA with random, greedy and genetic algorithms [13] that ignore flaws, testing FAINA's nesting performance in multiple dimensions. The greedy strategy orders the pieces by area from largest to smallest, according to the manual nesting habits of workers, while the random strategy nests pieces by randomly shuffling the sequence of pieces. Core parameters of the genetic algorithm are as shown in Table 2:

Table 2. Experimental Parameters

Parameter	Value
Population Size	20
Rotation Probability	0.5
Mutation Probability	0.2
Termination Generation	30

4.2 Experimental Results

These experiments show that in the nesting results obtained by the random algorithm, greedy algorithm and genetic algorithm, there are instances of sample piece contamination due to the presence of fabric flaws. The partial nesting results(as shown in Figs. 6, 7 and 8) of these three algorithms are shown in the figure below,

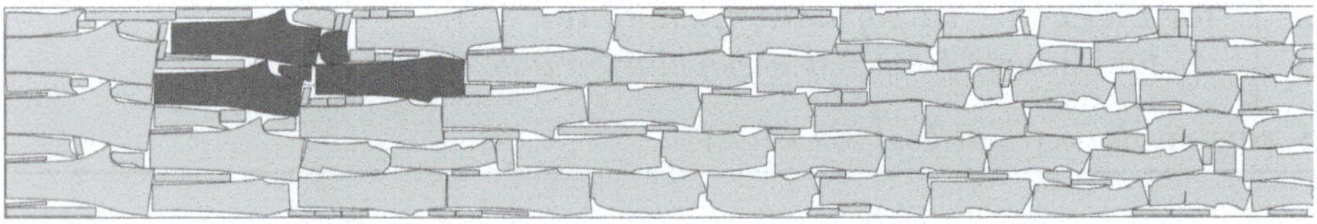

Fig. 6. Greedy Algorithm Nesting Result (314 Pieces)

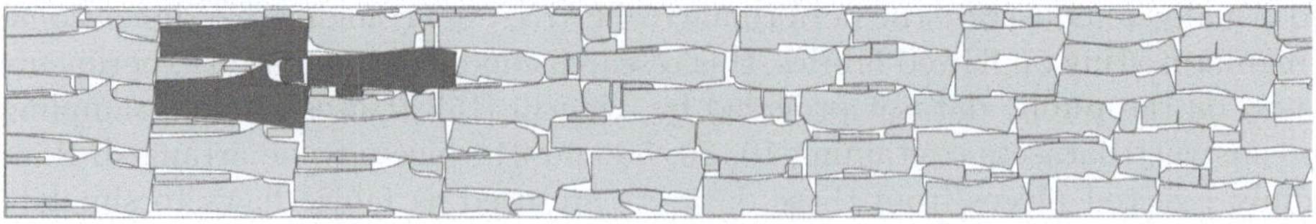

Fig. 7. GA Nesting Result (314 Pieces)

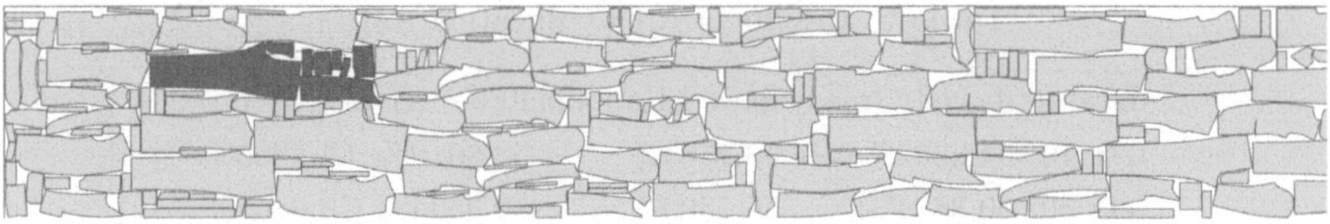

Fig. 8. Random Algorithm Nesting Result (314 Pieces)

where the yellow polygons represent normal pieces, the dashed-line area represents the flawed area and the gray polygons represent contaminated pieces.

FAINA can provide a flaw-free nesting solution for pieces, with some of the nesting results shown in Fig. 9 and Fig. 10. In these figures, the yellow polygons represent normal pieces, while the red polygons indicate areas.

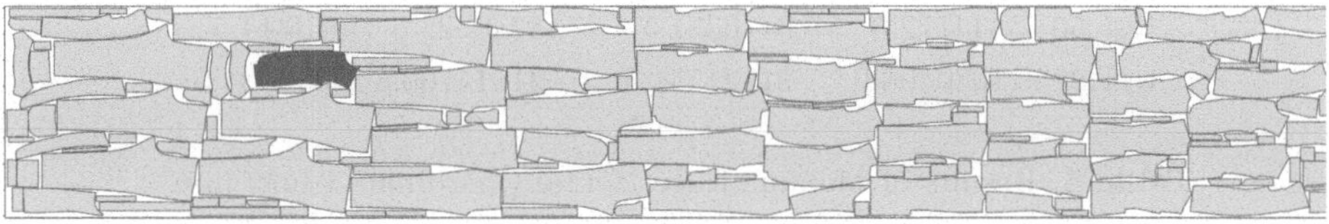

Fig. 9. FAINA Nesting Result (314 Pieces with a flaws)

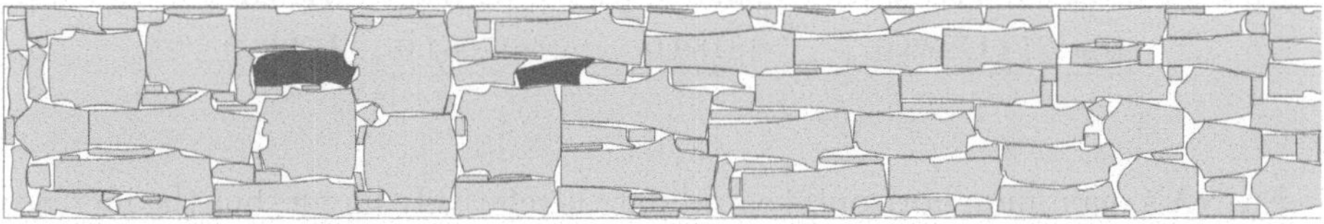

Fig. 10. FAINA Nesting Result (314 Pieces with 2 flaws)

These experiments conduct a multi-dimensional comparison of the nesting performance of four different algorithms on flawed fabric, using the following five core indicators: area utilization rate, total effective sample piece area, the maximum horizontal coordinate after nesting, the number of sample pieces contaminated by flaws, and the proportion of contaminated sample piece area.

The area utilization rate and total effective sample piece area are directly related to material efficiency and cost control, the maximum horizontal coordinate of the nesting results reflects the spatial cost required by the actual nesting plan, the number of sample pieces contaminated by flaws and the area of contaminated sample pieces affect production consumption and additional costs. The specific experimental data are as follows:

Results in Tables 3, 4 and 5 show FAINA improves area utilization by 2% to 13% and effective sample area by 0.16 to 1.04 m^2 compared to flaw-ignoring algorithms.

Table 3. Results of Area Utilization Rate

dataset	flawed_Greedy	flawed_Random	flawed_GA	FAINA
test_10	54.2382%	55.1704%	54.8019%	65.4355%
test_30	60.3874%	55.7390%	60.3524%	73.6862%
test_50	64.6429%	60.2177%	71.2045%	73.2418%
test_314	76.0828%	74.1550%	76.8803%	78.8991%

Table 4. Results of Total Effective Sample Piece Area (unit: mm^2)

dataset	flawed_Greedy	flawed_Random	flawed_GA	FAINA
test_10	687435.5	692058.0	687435.5	857114.5
test_30	1807676.5	1943283.0	1807676.5	2215302.0
test_50	3444172.5	3446381.5	3740228.5	4073414.5
test_314	21030496.0	21521551.5	21014769.5	22045034.0

Table 5. Results of Maximum Horizontal Coordinate (unit: mm)

dataset	flawed_Greedy	flawed_Random	flawed_GA	FAINA
test_10	784.0	784.0	784.0	858.0
test_30	1871.0	22179.0	1872.0	1879.0
test_50	3330.0	3577.0	3283.0	3476.0
test_314	17276.0	18139.0	17084.0	17463.0

FAINA's maximum horizontal coordinate is relatively larger, but the increase is tiny, indicating that the actual space cost required after avoiding flaws has slightly increased within a reasonable range area. This algorithm effectively avoids flaws and reduces fabric waste, producing no discarded pieces.

From Tables 6 and 7, it can be seen that when using the greedy algorithm, random algorithm, and genetic algorithm directly for nesting on flawed fabric, the nested pieces are contaminated by fabric flaws, resulting in discarded pieces. On the dataset test_314, up to eight pieces might be contaminated, while on the dataset test_10, the area of contaminated pieces can be as high as 19%. With the flaw avoidance strategy proposed in this paper, FAINA's nesting solutions have zero contaminated pieces and zero contaminated area. This algorithm effectively avoids flaws and reduces fabric waste, producing no discarded pieces.

Given the stochastic nature of genetic algorithms in their crossover and iteration processes, this study evaluates FAINA's stability on datasets of various scales based on area utilization rates. It's found that FAINA performs consistently under various conditions, the results are shown in Fig. 11.

Table 6. Results of the Number of Contaminated Sample Pieces (unit: *piece*)

dataset	flawed_Greedy	flawed_Random	flawed_GA	FAINA
test_10	1	2	2	0
test_30	2	3	2	0
test_50	5	5	3	0
test_314	5	8	5	0

Table 7. Results of Contaminated Sample Piece Area Proportion

dataset	flawed_Greedy	flawed_Random	flawed_GA	FAINA
test_10	19.7965%	19.2572%	19.7965%	0%
test_30	18.4004%	12.2791%	18.4004%	0%
test_50	15.4475%	15.3933%	8.1795%	0%
test_314	4.6021%	2.3746%	4.6735%	0%

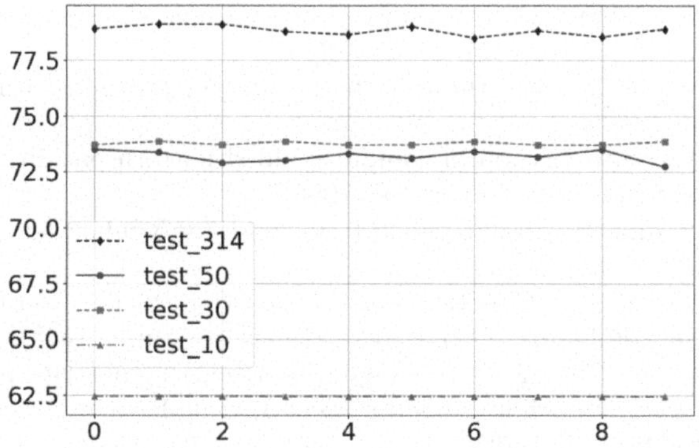

Fig. 11. Experimental Results of FAINA's Stability

5 Conclusions

This research addresses the two-dimensional irregular nesting problem with flawed fabrics by improving the generation strategy of No-Fit-Polygons and leveraging NFP for strategic flaw avoidance. It utilizes genetic algorithms to perform crossover and iteration on the population composed of normal pieces, integrating flaw information into fitness calculation for precise nesting results. Experimental results demonstrate that FAINA achieves maximum area utilization rate and effective area index. The fluctuations in area utilization rates across different datasets are tiny, indicating FAINA's ability to efficiently and stably rank irregular pieces on flawed fabrics, significantly enhancing material utilization and controlling production costs. FAINA ensures zero contaminated sample pieces and zero contaminated sample piece area, leading to reduced waste generation and energy-saving production.

FAINA is also applicable across domains such as metal processing and furniture manufacturing due to its high adaptability to different material properties and manufacturing process requirements. With the rapid progress of machine learning technology, future research will consider incorporating factors such as energy consumption reduction and min imizing cutting time into the nesting optimization process. Additionally, there will be a deeper exploration of AI-based nesting optimization algorithms to fur ther enhance nesting efficiency and material utilization.

Acknowledgments. This work was supported by the National Natural Science Foundation of China [12471304].

References

1. Adamowicz, M., Albano, A.: Nesting two-dimensional shapes in rectangular modules. Comput. Aided Des. **8**(1), 27–33 (1976)
2. Jakobs, S.: On genetic algorithm for the packing of polygons. Eur. J. Oper. Res. **88**(1), 165–181 (1996)
3. Mahadevan, A.: Optimization in computer-aided pattern packing. Ph.D. dissertation, North Carolina State University (1984)
4. Ghosh, P.K.: A unified computational framework for Minkowski operations. Comput. Graph. **17**(4), 357–378 (1993)
5. Jia, Z., Yin, G., Luo, Y.: Genetic algorithm solution to the two-dimensional irregular part nesting problem. J. Comput.-Aided Des. Graph. **14**(5), 467–470 (2002)
6. Wang, W., Ma, W., Cao, Y.: A rectangular part cutting algorithm combining edge matching and variable neighborhood search. J. Qingdao Univ. Sci. Technol. (Nat. Sci. Ed.) **44**(2), 108–115 (2023)
7. Jie, F., Yunqing, R., Xusheng, Z., et al.: A hybrid reinforcement learning algorithm for 2D irregular packing problems. Mathematics **11**(2), 327–345 (2023)
8. Dong, D., Yan, Y., Wang, Z.: Research on optimization layout problem of rectangular parts with surface defects in raw materials. J. Northeast. Univ. (Nat. Sci. Ed.) **33**(9), 1323–1326 (2012)
9. Afsharian, M., Niknejad, A., Wäscher, G.: A heuristic, dynamic programming-based approach for a two-dimensional cutting problem with defects. OR Spect. **36**(4), 971–999 (2014). https://doi.org/10.1007/s00291-014-0363-x
10. Li, Z., Yu, J., Qian, C.: Optimization research on layout of rectangular parts with defects based on dual-population genetic algorithm. Mach. Electron. **41**(3), 7–12 (2023)
11. Bennell, J.A., Oliveira, J.F.: A tutorial in irregular shape packing problems. J. Oper. Res. Soc. **60**(1), S93–S105 (2009)
12. Babu, A.R., Babu, N.R.: A generic approach for nesting of 2D parts in 2D sheets using genetic and heuristic algorithms. Comput. Aided Des. **33**(12), 879–891 (2001)
13. Lee, Y., Lu, O., Lim, W.: An application of mathematics to computer programming: connecting translation vectors, the Minkowski difference, and collision detection. Math. Teach. **112**(2), 150–155 (2018)

14. Christofides, N., Whitlock, C.: An algorithm for two-dimensional cutting problems. Oper. Res. **25**(1), 30–44 (1977)
15. Tianchi: Guangdong Industrial Intelligent Manufacturing Innovation Competition Dataset. https://tianchi.aliyun.com/competition/entrance/231749/information. Accessed 25 Oct 2023
16. Zhang, Y., Guo, B.: Optimization strategy for irregular layout based on small-habitat discrete particle swarm algorithm. J. Mech. Des. Manuf. Eng. **53**(1), 84–88 (2024)
17. Wu, S., Tang, C., Wei, L.: Research on heuristic two-dimensional irregular multi-specification sheet layout problem. Mech. Electr. Eng. Technol. **53**(1), 130–135 (2024)
18. Yang, Z., Pan, Z., Li, M., Wu, K., Gao, X.: Learning based 2D irregular shape packing. ACM Trans. Graph. **42**, 1–16 (2023)
19. Chen, Y., Tang, M., Tong, R., et al.: Irregular polygon layout based on genetic simulated annealing algorithm. J. Comput.-Aided Des. Graph. **15**(5), 598–603 (2003)
20. Li, M., Song, C., Zhou, Z.: Particle swarm algorithm solution to the two-dimensional irregular part nesting problem. J. Jiangnan Univ. **3**, 266–269 (2005)
21. Liang, L., Zhong, X.: Solving optimization problems of irregular parts layout based on immune genetic algorithm. Mech. Sci. Technol. **32**(3), 448–451 (2013)

Efficient Partitioning Algorithms for Optimizing Big Graph Computation

Baoling Ning[1(✉)] and Yupeng Gao[2]

[1] Heilongjiang University, Harbin, China
ningbaoling2009@163.com
[2] Harbin Institute of Technology, Harbin, China

Abstract. Graph partitioning is a key fundamental problem in the area of big graph computation. In this paper, motivated by optimizing the big data computing applications, two novel variants of the classical balanced graph partitioning are studied. The first one is to optimize the performance of specific workloads by graph partitioning, and the second one is to optimize the computation of motifs by graph partitioning. The formal definitions of the above two problems are introduced first, and the semidefinite programming representations are designed based on analyzing their properties. For the motif based partitioning problem, it is proved to be NP-complete even for the special case of $k = 2$ and the motif is a triangle, and its inapproximability is also shown by proving that there are no efficient algorithms with finite approximation ratio. Finally, by solving the semidefinite programs and utilizing sophisticated rounding techniques, the bi-criteria $O(\sqrt{\log n \log k})$-approximation algorithms with polynomial time cost are designed and analyzed for them.

Keywords: Graph Partitioning · Bi-criteria Approximation Algorithm · Semidefinite Programming

1 Introduction

Partitioning is a widely used strategy when processing big graph data. On one side, graph partitioning provides fundamental functions for supporting many important graph analysis operations (*e.g.*, *traversal*, *pagerank*). On the other side, it is also a key fundamental problem in big graph computing platforms, which is the core of many mechanisms such as load balancing.

In previous works, the most common problem considered is called k-balanced graph partitioning, kBGP for short. Given a graph G and an integer $k \geq 2$, the kBGP problem is to find a solution for partitioning the graph vertices, such that all vertices are partitioned into k disjoint partitions of the same size, and the partitioning cost is minimal. Here, the partitioning cost should be chosen according to the practical requirements, the most popular cost is the size (or weighted size) of the edge *cut*. The problem has been proved to be NP-complete, as one of the most classical NP-complete problems found in the 1970s, more

Y. Chen et al. (Eds.): COCOON 2024, LNCS 15162, pp. 202–213, 2025.
https://doi.org/10.1007/978-981-96-1093-8_17

precisely, even considering the special case of $k = 2$, the kBGP problem is still NP-complete [10]. Even worse, it is proved by [1] that for any $k \geq 3$ there are no polynomial-time algorithms for kBGP with finite approximation ratio unless P=NP. Due to its hardness, the following-up research interests focus on designing *bi-criteria approximation algorithms*, where the solutions are relaxed to be no larger than $\tau\lceil\frac{|V_G|}{k}\rceil$ and the cost is still compared with the optimal one of the original kBGP problem. Current works on *bi-criteria* approximation algorithms can be divided into three kinds. (1) The first kind is designed based on the algorithms for the 2BGP problem, and the best approximation ratio is $O(\log k\sqrt{\log n})$ [2,16]. (2) The second kind is designed using the mathematical programming techniques, and the best ratio is $O(\sqrt{\log n \log k})$ [5,12]. (3) The third kind is designed based on combinatorial techniques such as dynamic programming, and the best approximation ratio can be $O(\log n)$ [1,8,15].

As shown above, previous works have focused on the graph partitioning problem and produced a plenty of insightful results. However, in view of big graph processing systems such as [13] and [11], novel partitioning requirements like optimizing for specific workloads or motif computation are needed, but they do not receive enough considerations. In this paper, two typical graph partitioning problems motivated by big graph computing are studied. The first one is to optimize specific workloads by graph partitioning. Current works only focus on designing heuristic [14] or learning based methods [7], and have no performance guarantees. The second one is to optimize the motif computation by graph partitioning. As far as we know, there still lacks of research works focusing on it.

Our main contributions can be summarized as follows.

① The balanced graph partition problem for optimizing specific workloads is formally introduced, and a bi-criteria $O(\sqrt{\log n \log k})$-approximation algorithm with polynomial time cost is proposed.

② The balanced graph partition problem for optimizing motif computation is formally introduced. It is shown to be NP-complete even for the special case of $k = 2$ and the motif is a triangle, and is also proved to have no polynomial time algorithms with finite approximation ratio.

③ For the special case that the motif is a triangle, using the semidefinite programming techniques, a bi-criteria $O(\sqrt{\log n \log k})$-approximation algorithm with polynomial time cost is designed. For the general case, the proposed algorithm is extended with the same performance guarantee.

2 Preliminaries

A graph is represented by $G = (V, E)$, where V is the vertex set and $E \subseteq V \times V$ is the edge set. We also use V_G and E_G to emphasize that the graph G is referred.

A *partitioning* solution of G is defined as $\text{PAR}_G = \{V_1, \cdots, V_m\}$ where V_is are disjoint subsets of V such that $\bigcup_{i=1}^{m} V_i = V$. If $m = k$, it is also called to be a *k-partitioning* solution, denoted by $\text{PAR}_G^k = \{V_1, \cdots, V_k\}$. A k-partitioning solution PAR_G is *balanced* if we have $|V_i| \leq \lceil\frac{|V|}{k}\rceil$ for any $V_i \in \text{PAR}_G$. Moreover, for $u \in V$, we use $\text{PAR}_G(u)$ to represent the partition item which u belongs to.

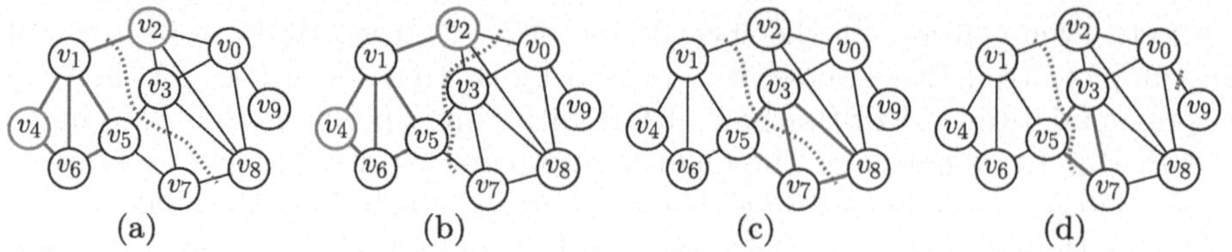

Fig. 1. An illustrated example of partitioning solutions for the workload ϕ of computing the shortest path from v_4 to v_2 and the triangle motif Δ. (a) The cost of PAR_1 on ϕ. (b) The cost of PAR_2 on ϕ. (c) The cost of PAR_1 on Δ. (d) The cost of PAR_3 on Δ.

The classical partitioning problem aims to minimize the size of edge cut.

Definition 1 (k-balanced graph partitioning, kBGP for short). *Given a graph G and an positive integer k, the kBGP problem is to find a k-balanced partitioning solution* PAR_G *such that* $\textsf{cost}(\text{PAR}_G)$ *is minimized, where* $\textsf{cost}(\text{PAR}_G) = |\{(u,v) \in E \mid \text{PAR}_G(u) \neq \text{PAR}_G(v)\}|$. □

kBGP is well known to be NP-complete [9], and there is no polynomial time algorithm with finite approximation ratio for any $k \geq 3$, unless P = NP [1].

Due to the hardness, *bi-criteria* approximation algorithms are usually considered. Given a balance factor τ, a partition solution $\text{PAR}_G = \{V_i | i \in [1,k]\}$ is (k,τ)-*balanced* if $|V_i| \leq \tau\lceil\frac{|V|}{k}\rceil$. Given a kBGP instance $\langle G,k\rangle$, a bi-criteria (k,τ) approximation algorithm $\mathcal{A}$ will return a (k,τ)-balanced partitioning solution PAR, and the approximation ratio of $\mathcal{A}$ is defined as $\rho_{\mathcal{A}} = \frac{\textsf{cost}(\text{PAR})}{\textsf{cost}(\text{PAR}^*)}$ where PAR^* is the optimal solution for the kBGP problem. Obviously, since PAR^* must be balanced, the ratio $\rho_{\mathcal{A}}$ may be less than 1.

2.1 Workload Driven Balanced Graph Partitioning Problem

In the parallel computing setting, a big graph will be partitioned and stored in different nodes, and the communication cost of a computing procedure needs to be optimized. Therefore, a key problem is to determine the optimal partitioning solution to minimize the communication cost for a specific workload.

Example 1. Consider the graph G shown in Fig. 1. Two 2-balanced partitioning solutions PAR_1 and PAR_2 are shown in Fig. 1(a) and 1(b). Here, PAR_1 includes two partitions, $\{v_1, v_4, v_5, v_6, v_7\}$ and $\{v_0, v_2, v_3, v_8, v_9\}$. For PAR_2, the partitions are $\{v_1, v_2, v_4, v_5, v_6\}$ and $\{v_0, v_3, v_7, v_8, v_9\}$. If $\textsf{cost}(\cdot)$ in Definition 1 is used, we have $\textsf{cost}(\text{PAR}_1) = 4$ and $\textsf{cost}(\text{PAR}_1) = 5$, and PAR_1 is better than PAR_2.

However, let us consider a specific workload ϕ, that is to compute the shorted path between v_4 and v_2, a commonly used algorithm will do a breath first travel starting from v_4 and the visited edges are colored with blue in Fig. 1(a). As shown in Fig. 1(a), PAR_1 destroys (v_1, v_2) and will cause once communication. As shown in Fig. 1(b), PAR_2 does not cut any blue edges, the communication cost will be 0. Therefore, PAR_2 is in fact better than PAR_1 in this case. □

A workload on graph G is represented by $\Phi = \{(\phi_1, c_1), \cdots, (\phi_w, c_w)\}$, where each ϕ_i is a specific computing job, c_i is the frequency that ϕ_i appears. The job ϕ_i is a set $\{(e_{ij}, c_{ij})\}$, where $e_{ij} \in E_G$ and c_{ij} is the times of visiting e_{ij} when computing ϕ_i. $\mathsf{cost}_\Phi(\text{PAR})$ is used to represent the expected communication cost of computing Φ on G partitioned by PAR.

$$\begin{aligned}\mathsf{cost}_\Phi(\text{PAR}) &= \frac{1}{\sum_{1\le i\le w} c_i} \sum_{1\le i\le w} \left(c_i \cdot \mathsf{cost}_{\phi_i}(\text{PAR})\right) \qquad (1)\\ &= \frac{1}{\sum_{1\le i\le w} c_i} \sum_{1\le i\le w} \left(c_i \cdot \sum_{1\le j\le |\phi_i|} \left(c_{ij} \cdot \mathbf{I}(\text{PAR}(u_{ij}) \ne \text{PAR}(v_{ij}))\right)\right)\end{aligned}$$

Here, $\mathsf{cost}_{\phi_i}(\text{PAR})$ is the communication cost needed to evaluate the job ϕ_i, the two vertices related with the edge e_{ij} are u_{ij} and v_{ij}, and $\mathbf{I}(\cdot)$ is a function whose value is 1 if the required condition is satisfied, otherwise it is 0.

Definition 2 (The WkBGP Problem). *Given a graph G, a positive integer k and a workload Φ, the WkBGP problem is to find a k-balanced partitioning solution* PAR_G *such that* $\mathsf{cost}_\Phi(\text{PAR}_G)$ *is minimized.* □

Obviously, kBGP is a special case of WkBGP, thus, WkBGP is NP-complete and admits no PTime algorithms with finite approximation ratio unless P = NP.

2.2 Motif Based Balanced Graph Partitioning Problem

Motif computation is a hot topic in the area of graph data processing, which is defined to be graphs of small limited sizes, such as 3 or 4 vertices.

Example 2. Consider the 2-balanced partitioning solutions shown in Fig. 1(c) and 1(d). Figure 1(c) is actually PAR_1, the same as Fig. 1(a). Figure 1(d) is denoted by PAR_3, where G is partitioned into $\{v_1, v_4, v_5, v_6, v_9\}$ and $\{v_0, v_2, v_3, v_7, v_8\}$. Obviously, $\mathsf{cost}(\text{PAR}_3) = 4$, and the classical measure $\mathsf{cost}(\cdot)$ can not tell the difference between PAR_1 and PAR_3. Next, let us consider a specific motif computation, that is to count the number of triangles in G. No matter what method is used, if the three vertices of one triangle are not in a single partition, there will be at least one communication. In Fig. 1(c) and 1(d), the triangles destroyed are colored with blue. Obviously, PAR_3 is better than PAR_1. □

Let c be a constant, a c-Motif is a graph of c vertices. $\mathsf{cost}_M(\text{PAR})$ is the communication cost of processing a motif M on G partitioned by PAR.

$$\mathsf{cost}_M(\text{PAR}) = \sum_{H \in G\langle M\rangle} \mathbf{I}(\exists (u,v) \in H, \text{s.t.} \text{PAR}(u) \ne \text{PAR}(v)) \qquad (2)$$

Here, $G\langle M\rangle$ means all appearances of the motif M in the graph G.

Definition 3 (The MkBGP problem). *Given a graph G, a positive interger k and a c-Motif M, the MkBGP problem is to compute a k-balanced partitioning solution* PAR_G *for G such that the cost* $\mathsf{cost}_M(\mathrm{PAR}_G)$ *is minimized.* □

Different from WkBGP, kBGP is not a special case of MkBGP, therefore, the hardness of MkBGP can not be obtained directly based on the results of kBGP.

3 Approximation Partitioning Algorithm for WkBGP

3.1 The Semidefinite Program for WkBGP

First, WkBGP is represented by Eq.(3) using semidefinite programming (SDP) [17]. A vertex v in V is represented by an m-dimensional vector $\boldsymbol{v}$ satisfying $m \geq k$, where $\boldsymbol{v}$ is expected to indicate which partition v should belong to.

$$\min \quad \sum_{(v_i,v_j)\in E} (W_{ij} \cdot \frac{1}{2}\|\boldsymbol{v}_i - \boldsymbol{v}_j\|_2^2) \tag{3}$$

$$\text{s.t.} \quad \|\boldsymbol{u} - \boldsymbol{v}\|_2^2 + \|\boldsymbol{v} - \boldsymbol{w}\|_2^2 \geq \|\boldsymbol{u} - \boldsymbol{w}\|_2^2 \qquad \forall u, v, w \in V \tag{4}$$

$$\sum_{v\in S} \frac{1}{2}\|\boldsymbol{u} - \boldsymbol{v}\|_2^2 \geq |S| - \frac{n}{k} \qquad \forall S \subseteq V, u \in S \tag{5}$$

$$W_{ij} = \sum_{1\leq x\leq w} \sum_{1\leq y\leq |\phi_x|} \frac{c_x \cdot c_{xy} \cdot \mathbf{I}(e_{xy} = (v_i, v_j))}{\sum_{1\leq x\leq w} c_x} \tag{6}$$

Here, the optimizing goal is a relaxed variant of $\mathsf{cost}_\Phi(\cdot)$ used by WkBGP, where the item $\mathbf{I}(\mathrm{PAR}(v_i) \neq \mathrm{PAR}(v_j))$ of $\mathsf{cost}_\Phi(\cdot)$ is replaced with $\frac{1}{2}\|\boldsymbol{v}_i - \boldsymbol{v}_j\|_2^2$. Intuitively, for v_i and v_j, if the distance between $\boldsymbol{v}_i$ and $\boldsymbol{v}_j$ is too large (like $\gg 1$), it means $\mathrm{PAR}(v_i) \neq \mathrm{PAR}(v_j)$, otherwise, it means $\mathrm{PAR}(v_i) = \mathrm{PAR}(v_j)$. The first constraint shown by Eq. (4) is the triangle inequality defined on ℓ_2^2, which is firstly proposed by [2] and used to express the partitioning requirements. The second constraint shown by Eq. (5) is also called the spreading constraint, which is proposed by [6] and used to control the size of each partition.

Theorem 1. *The SDP shown in Eq.*(3) *can be solved in polynomial time.* □

Proof. First, let us consider the range of the optimized goal of Eq. (3). Let $\gamma = \max\{\sum_{1\leq y\leq|\phi_x|} c_{xy} \mid x \in [1, w]\}$. W_{ij} can be limited by $\max\{\sum_{1\leq y\leq|\phi_x|}(c_{xy} \cdot \mathbf{I}(e_{xy} = (v_i, v_j)))|x \in [1, w]\}$. Then, we can let $W_{ij} = \frac{W_{ij}}{\gamma}$ without changing the optimal solution. Let the edge with the maximized distance be (u_{max}, v_{max}), then the corresponding distance will be $\frac{1}{2}\|\boldsymbol{u}_{max} - \boldsymbol{v}_{max}\|_2^2$. As a result, the final goal can be limited within the range $[0, |E_G| \cdot \frac{1}{2}\|\boldsymbol{u}_{max} - \boldsymbol{v}_{max}\|_2^2]$. Then, let us consider a trivial solution. First, randomly select $\frac{n}{k}$ vertices and represent them as $v_1, \cdots, v_{n/k}$. The corresponding vector of v_i is set to be $\boldsymbol{v}_i = [\sqrt{n}, 0, 0, \cdots, 0]^T$. Then, repeat the above operations until all vertices are selected, for each step, the value $\sqrt{n}$ is moved backward once. Since the dimension size of solutions should

Algorithm 1: WKBGPARTITION

Input: A Graph $G = \{V, E\}$, an integer k and a workload set Φ.
Output: A partition $\text{PAR}_G = \{V_1, V_2, \cdots, V_k\}$ for G.

function WKBGPARTITION(G, k, Φ)
  Initialize a $|V_G| \times |V_G|$ array W to be 0;
  Let $C = \sum_{1 \le i \le w} c_i$;
  for *each* $(\phi_x, c_x) \in \Phi$, *each* $(e_{xy}, c_{xy}) \in \phi_x$ **do**
    $W[i][j] += \frac{c_x \cdot c_{xy}}{C}$, $W[j][i] = W[i][j]$; // $e_{xy} = (v_i, v_j)$
  Build SDP (3), and let $opt = \{\boldsymbol{v}_1, \cdots, \boldsymbol{v}_n\}$ be the optimal solution of (3);
  Invoke k-PARTITION of [12] on opt to compute $\text{PAR}_G = \{V_1, \cdots, V_k\}$;
  return PAR_G;

be m, the other $m-k$ dimensions are set to be 0. It can be verified that the above trivial solution satisfy the constraints, thus, its cost is an upper bound of the optimal one. Then, according to [4], by setting the search range as $[0, |V_G| \cdot |E_G|]$ and letting $\epsilon < \frac{1}{|V_G| \cdot |E_G|}$, a $(1+\epsilon)$-approximation solution is actually an exact one. Finally, the time cost of can be bounded by $O(poly(n + |V_G| \cdot |E_G|))$. □

3.2 SDP Based Approximation Algorithm for WkBGP

This part introduces the partitioning algorithm for WkBGP, which rely on solving the SDP shown in Eq.(3) and a well designed rounding technique.

The proposed algorithm, WKBGPARTITION, is shown in Algorithm 1. The main procedure of WKBGPARTITION is to first construct the semidefinite program shown in Eq.(3) based on the inputs, then use the program solver to obtain the optimal SDP solution, and finally computes the partitioning solution using the k-PARTITION rounding method proposed by [12].

Theorem 2. *The optimal cost of Eq.* (3) *is a lower bound of the optimal partitioning cost of the input instance* $\langle G, k, \Phi \rangle$ *of WkBGP.* □

Proof. Let $\text{PAR}_G^* = \{V_1, V_2, \cdots, V_k\}$ be the optimal partitioning solution of WkBGP, and the cost is $\mathsf{cost}_\Phi(\text{PAR}_G^*)$. Assume the optimal solution of the SDP is $\{\boldsymbol{u}_1, \boldsymbol{u}_2, \cdots, \boldsymbol{u}_n\}$ and the corresponding optimal value of the goal is A^*. Then, a feasible solution $\{\boldsymbol{v}_1, \boldsymbol{v}_2, \cdots, \boldsymbol{v}_n\}$ of the SDP can be built based on PAR_G^* as follows. For each $v_i \in V_j$, let $\boldsymbol{v}_i = [0, \cdots, 1, \cdots, 0]^T$ where 1 appears in the jth position. Obviously, the solution satisfies all constraints in the SDP. Then, let the corresponding goal value of Eq.(3) be A. For any two vertices v_i and v_j, it is not hard to verify that $\mathbf{I}(\text{PAR}_G^*(v_i) \neq \text{PAR}_G^*(v_j)) \equiv \frac{1}{2}\|\boldsymbol{v}_i - \boldsymbol{v}_j\|_2^2$, that is they have the same value indeed. According to Eq. (1) and (3), we have $\mathsf{cost}_\Phi(\text{PAR}_G^*) = A$. Finally, since A^* is optimal, we have $\mathsf{cost}_\Phi(\text{PAR}_G^*) \ge A^*$. □

Lemma 1 ([12])**.** k-PARTITION *will return a solution with approximation ratio* $O(\sqrt{\log k \log n})$ *in polynomial time, and each partition size is at most* $\frac{2n}{k}$. □

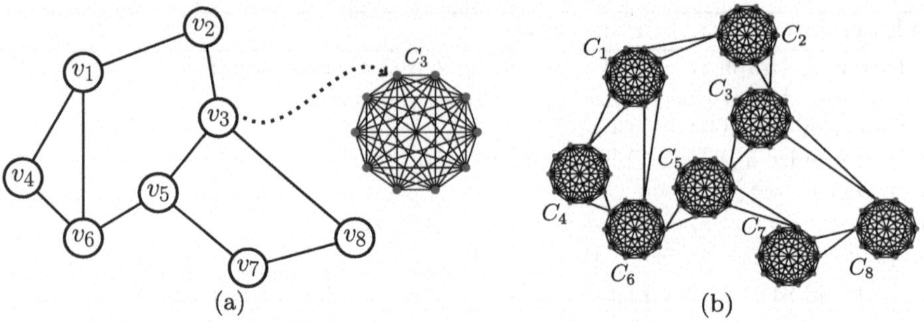

Fig. 2. An illustrated example of the reduction used by Theorem 4. (a) An instance G of the graph bisection problem. (b) An instance H of Δ-kBGP.

Combining the above two bounds, we have the following result naturally.

Theorem 3. WKBGPARTITION *has an expected polynomial time cost, and is a bi-criteria $(k, 2)$-approximation algorithm with ratio $O(\sqrt{\log k \log n})$.* □

4 Computational Hardness Results of MkBGP

Theorem 4. *If k=2 and M is a triangle, MkBGP is still NP-complete.* □

Proof. A brief proof is introduced here to explain the main idea of the reduction from the graph bisection problem, which is NP-complete [9]. Given an input (G, B) of the graph bisection problem, the reduction will output an instance (H, C) of Δ-2BGP. Here, B and C are cost limits used by the decision versions of graph bisection and Δ-2BGP. The reduction can be explained as follows. (1) First, an even integer X is selected to satisfy $X \geq 2 \cdot \mathsf{deg}(G)$ and $X \geq |V_G| + 1$. (2) Then, as shown in Fig. 2(a), for each $v_i \in G$, a clique C_i with X vertices is built, and H is initialized as the graph including all C_is. (3) Next, the following triangles are added to H, guaranteeing each vertex is utilized at most once. For each edge $(v_i, v_j) \in G$, assuming $i < j$, arbitrarily choose two unused vertices c_{i1} and c_{i2} in C_i and one unused vertex c_j in C_j, and add (c_{i1}, c_j) and (c_{i2}, c_j) to H. Given an input graph G for the bisection problem shown in Fig. 2(a), the result graph H constructed by the reduction is shown in Fig. 2(b). (4) Finally, let $C = B$. It can be verified that the above reduction is correct and can be finished within polynomial time. □

Theorem 5. *Assume that $\mathsf{P} \neq \mathsf{NP}$, even if M is a triangle, MkBGP admits no polynomial time algorithms with finite approximation ratio.* □

Proof. The proof is achieved by a reduction from 3-Partition to MkBGP. The input of 3-Partition is composed of $n = 3k$ positive integers $\{a_1, \cdots, a_n\}$ and a threshold S satisfying $S/4 < a_i < S/2$ and $\sum_{i=1}^{n} a_i = kS$. The 3-Partition problem is to determine whether there is a partitioning method to group each three integers and the corresponding sum is exactly S. According to [9], 3-partition is strongly NP-complete. Using the unary representation of 3-Partition, the following reduction will output an instance of MkBGP within polynomial time. For each integer a_i, construct a clique K_i of size a_i, the union of all K_is is just the input graph H of MkBGP. Then, it can be verified that, the answer of 3-Partition is 'yes' if and only if the reduced MkBGP problem admits a k-balanced partitioning solution with cost 0. Assume that there is a polynomial time algorithm $\mathcal{A}$ for MkBGP with finite approximation ratio, and let the output of $\mathcal{A}$ be PAR_H. If $\mathsf{cost}_{\Delta}(\text{PAR}_H) = 0$, the answer of 3-Partition must be 'yes'. Otherwise, if $\mathsf{cost}_{\Delta}(\text{PAR}_H) > 0$, since the approximation ratio of $\mathcal{A}$ is finite, we have $\mathsf{cost}_{\Delta}(\text{PAR}_{opt}) > 0$. Then, the answer of 3-Partition should be 'no'. That is, $\mathcal{A}$ can be used to solve the 3-Partition problem, which is a contradiction. □

5 Approximation Partitioning Algorithm for MkBGP

The algorithm will be introduced first for the special case that the motif is a triangle and then extended to the general case.

5.1 SDP Based Approximation Algorithm for Δ-kBGP

When the motif is a triangle, the corresponding MkBGP problem is denoted by Δ-kBGP, and the semidefinite program is defined by Eq. (7).

$$\min \sum_{\Delta_{\langle a,b,c\rangle} \in G} \|\boldsymbol{x}_{abc}\|_2^2 \tag{7}$$

$$\text{s.t.} \quad \|\boldsymbol{u}-\boldsymbol{v}\|_2^2 + \|\boldsymbol{v}-\boldsymbol{w}\|_2^2 \geq \|\boldsymbol{u}-\boldsymbol{w}\|_2^2 \quad \forall u,v,w \in V \tag{8}$$

$$\sum_{v \in S} \frac{1}{2}\|\boldsymbol{u}-\boldsymbol{v}\|_2^2 \geq |S| - \frac{n}{k} \quad \forall S \subseteq V, u \in S \tag{9}$$

$$\|\boldsymbol{x}_{abc}\|_2^2 \geq \frac{1}{2}\|\boldsymbol{m}-\boldsymbol{n}\|_2^2 \quad \forall \Delta_{\langle a,b,c\rangle} \in G, (m,n) \in \Delta_{\langle a,b,c\rangle} \tag{10}$$

The goal shown in Eq.(7) is to simulate the cost function $\mathsf{cost}_{M=\Delta}(\cdot)$ defined by Eq.(2), which can be explained as follows. Each triangle $\Delta_{\langle a,b,c\rangle}$ in G has a corresponding vector $\boldsymbol{x}_{abc}$ whose value indicates whether the triangle $\Delta_{\langle a,b,c\rangle}$ is destroyed by the partitioning solution. Ideally, if $\|\boldsymbol{x}_{abc}\|_2^2 = 1$, the triangle $\Delta_{\langle a,b,c\rangle}$ is destroyed, otherwise, all vertices of $\Delta_{\langle a,b,c\rangle}$ are placed in the same partition. Practically, the smaller $\|\boldsymbol{x}_{abc}\|_2^2$ is, the more likely the three vertices belong to the same partition. The constraints defined by Eq.(8) and (9) are the triangle inequality and the spreading constraints. Equation (10) defines the constraints between the triangles and the vectors, where each triangle have three constraints and the total size can be bounded by $O(n^3)$. According to Eq.(10), if $\frac{1}{2}\|\boldsymbol{a}-\boldsymbol{b}\|_2^2$ is large, $\|\boldsymbol{x}_{abc}\|_2^2$ will be larger. Considering the three constraints for the triangle $\Delta_{\langle a,b,c\rangle}$, $\|\boldsymbol{x}_{abc}\|_2^2$ is 0, only if the values of $\frac{1}{2}\|\boldsymbol{c}-\boldsymbol{a}\|_2^2$, $\frac{1}{2}\|\boldsymbol{a}-\boldsymbol{b}\|_2^2$ and $\frac{1}{2}\|\boldsymbol{b}-\boldsymbol{c}\|_2^2$ are all 0. That is, $\Delta_{\langle a,b,c\rangle}$ is not destroyed.

Algorithm 2: TRI-KBGPARTITION

Input: A Graph $G = \{V, E\}$ and an integer k.
Output: A partition $\text{PAR}_G = \{V_1, V_2, \cdots, V_k\}$ for G.
1 **function** TRI-KBGPARTITION(G, k)
2   Compute and maintain all triangles of G into S_Δ;
3   Solve sdp (7) and let $opt = \{\boldsymbol{v}_1, \cdots, \boldsymbol{v}_n\}$ be the optimal solution;
4   $S \leftarrow V$, $\text{PAR}_G \leftarrow \emptyset$;
5   **while** $|S| > \frac{2n}{k}$ **do**
6   Randomly choose $\boldsymbol{r} \in \mathbb{R}^m$ $s.t.$ $r_i \sim N(0, 1)$ and all r_is are $i.i.d.$;
7     $S_r \leftarrow \{v_i \mid v_i \in S, g(\boldsymbol{v}_i) \cdot \boldsymbol{r} \geq \alpha_{Ck}\}$;
8   **if** $0 < |S_r| \leq \frac{2n}{k}$ **then**
9       $\text{PAR}_G \leftarrow \text{PAR}_G \cup S_r$ and $S \leftarrow S \setminus S_r$;
10   **while** $|\text{PAR}_G| > k$ **do**
11   Find the smallest two partition items V' and V'' of PAR_G;
12   Remove V' and V'' from PAR_G and add $V' \cup V''$ to PAR_G;
13   **return** PAR_G;

Since the size of the SDP (7) is polynomial, similar with Theorem 1, it can be proved that the SDP can be solved within polynomial time.

The proposed algorithm, TRI-KBGPARTITION, is shown in Algorithm 2, which first solves the corresponding SDP and then use rounding techniques to construct the partitioning solution. First, TRI-KBGPARTITION processes all triangles in G and maintain them using S_Δ (line 2). Then, the SDP is built based on S_Δ and solved, and let the optimal solution be $\{\boldsymbol{v}_1, \cdots, \boldsymbol{v}_n\}$ (line3). Finally, using a similar rounding technique with [12], the partitioning solution can be built as follows. (1) First, TRI-KBGPARTITION builds a vector $\boldsymbol{r}$ randomly, and generates α_{Ck} based on a variable X by a normal distribution $N(0, 1)$, where the value of α_m satisfies $\Pr[X \geq \alpha_m] = \frac{1}{m}$ (line 6). (2) Then, TRI-KBGPARTITION utilizes a function $g(\cdot)$ to transform the distance from ℓ_2^2 to ℓ_2 (line 7). As shown by [3], given n vectors satisfying the triangle inequality on ℓ_2^2, there exists a constant $\delta, A > 0$ and for any $\Lambda > 0$ we can find a function $g_\Lambda : \mathbb{R}^m \to \mathbb{R}^n$ such that for any two vectors $\boldsymbol{u}$ and $\boldsymbol{v}$, we have (i) $\|g_\Lambda(\boldsymbol{u}) - g_\Lambda(\boldsymbol{v})\|_2 \leq \frac{A\sqrt{\log n}}{\Lambda}\|\boldsymbol{u} - \boldsymbol{v}\|_2^2$, (ii) $\|\boldsymbol{u} - \boldsymbol{v}\|_2^2 \geq \Lambda \Rightarrow \|g_\Lambda(\boldsymbol{u}) - g_\Lambda(\boldsymbol{v})\|_2 \geq \delta$ and (iii) $\|g_\Lambda(\boldsymbol{u})\|_2 = 1$. Here, the function $g(\cdot)$ used by TRI-KBGPARTITION is a special case of $g_\Lambda(\cdot)$ by setting $\Lambda = 2/3$. (3) Using the above two techniques, in each iteration, when the size of S is larger than $\frac{2n}{k}$, TRI-KBGPARTITION randomly selects a vector $\boldsymbol{r}$, and adds v_is satisfying the $\boldsymbol{r} \cdot g(\boldsymbol{v}_i) \geq \alpha_{Ck}$ to S_r (line 8). If the size of S_r is not larger than $\frac{2n}{k}$, it will be treated as a feasible partition, and S and PAR_G are updated (line 8–9). (4) After the rounding procedure, TRI-KBGPARTITION can not always guarantee that there are exactly k partitions. Therefore, TRI-KBGPARTITION needs a post-process (line 11–13). TRI-KBGPARTITION will merge the smallest two partitions until there are exactly k partitions. It can be verified that the sizes of the merged partitions are also not larger than $\frac{2n}{k}$.

In the worst case, the size of S_r selected may be always too large (line 6–9), the time of TRI-KBGPARTITION can not be bounded efficiently. Fortunately, the probability that S_r is too large can be verified to be no larger than 1/2, therefore, the expected time cost of TRI-KBGPARTITION is polynomial.

Similar with Theorem 2, we have the following lower bound result.

Theorem 6. *The optimal value of SDP in Eq.(7) is a lower bound of the optimal partitioning cost of the input instance $\langle G, k\rangle$ of Δ-kBGP.* □

Lemma 2 ([12]). *The probability that* TRI-KBGPARTITION *splits u and v is limited by $3AB\sqrt{2\log(2Ck)\log n}\cdot ||\boldsymbol{u}-\boldsymbol{v}||_2^2$, where A, B and C are constants.* □

For a triangle $\Delta_{\langle a,b,c\rangle}$, we use $||\Delta_{\langle a,b,c\rangle}||_2^2$ to represent $||\boldsymbol{a}-\boldsymbol{b}||_2^2+||\boldsymbol{b}-\boldsymbol{c}||_2^2+||\boldsymbol{a}-\boldsymbol{c}||_2^2$, and let $\mathcal{A}_{abc}$ be the event that $\Delta_{\langle a,b,c\rangle}$ is destroyed by TRI-KBGPARTITION.

Theorem 7. *Given $U\subseteq V$ including all vertices of $\Delta_{\langle a,b,c\rangle}$, assume that $S=U$ during an iteration of* TRI-KBGPARTITION, *then $\Pr[\mathcal{A}_{abc}|S=U]\le\frac{2||\Delta_{\langle a,b,c\rangle}||_2^2}{3}\cdot 3AB\sqrt{2\log(2Ck)\log n}$, where A, B and C are same as Lemma 2.* □

Proof. Let $\mathcal{A}_{x,y}$ be the event that x and y are separated. Assume that $||a-b||_2^2\le||b-c||_2^2\le||a-c||_2^2$. The probability of $\mathcal{A}_{abc}$ can be decomposed into two parts.

$$\Pr[\mathcal{A}_{abc}]=\Pr[\mathcal{A}_{a,b}]+\Pr[\overline{\mathcal{A}}_{a,b}\wedge\mathcal{A}_{b,c}]$$

Here, $\overline{\mathcal{A}}$ is the event that $\mathcal{A}$ does not happen. Then, we have the following result.

$$\begin{aligned}\Pr[\mathcal{A}_{abc}]&=\Pr[\mathcal{A}_{a,b}]+\Pr[\overline{\mathcal{A}}_{a,b}\wedge\mathcal{A}_{b,c}]\le\Pr[\mathcal{A}_{a,b}]+\Pr[\mathcal{A}_{b,c}]\\&\le\frac{(1+\epsilon)AB}{\epsilon}\sqrt{2\log(Ck/\epsilon)\log n}\left(||a-b||_2^2+||b-c||_2^2\right)\\&\le\frac{2}{3}\frac{(1+\epsilon)AB}{\epsilon}\sqrt{2\log(Ck/\epsilon)\log n}\left(||a-b||_2^2+||b-c||_2^2+||a-c||_2^2\right)\\&=\frac{2||\triangle_{\langle a,b,c\rangle}||_2^2}{3}\cdot\frac{(1+\epsilon)AB}{\epsilon}\sqrt{2\log(Ck/\epsilon)\log n}\end{aligned}$$

Here, the second inequality is from Lemma 2, the third one is because of the fact that the distance between a and c is the largest one in $\triangle_{\langle a,b,c\rangle}$. □

Based on Theorem 6 and 7, the approximation ratio of TRI-KBGPARTITION can be obtained by the following theorem.

Theorem 8 (Approximation Ratio of Tri-kBGPartition). *Δ-kBGP can be solved within expected polynomial time by* TRI-KBGPARTITION *with approximation ratio $O(\sqrt{\log k\log n})$, and the size of each partition is at most $\frac{2n}{k}$.* □

Proof. Given an instance $\langle G, k\rangle$ of Δ-kBGP, let PAR_G be the partitions built by Algorithm 2 and PAR_G^* be the optimal solution. Obviously, $\mathsf{cost}_\Delta(\text{PAR}_G) \geq \mathsf{cost}_\Delta(\text{PAR}_G^*)$. Then, let S^* be the optimal solution of the SDP (7), according to Theorem 6, we have $S^* \leq \mathsf{cost}_\Delta(\text{PAR}_G^*) \leq \mathsf{cost}_\Delta(\text{PAR}_G)$.

Since the rounding techniques used is randomized, we analyze the expected cost of the solution of TRI-KBGPARTITION as follows, where the first inequality is based on Theorem 7, and the last two are based on Eq.(10) and (7).

$$\begin{aligned}
\mathbf{E}[\mathsf{cost}_\Delta(\text{PAR}_G)] &= \sum_{\Delta_{\langle a,b,c\rangle} \in G} \left(\Pr[\mathcal{A}_{abc}]\right) \\
&\leq \sum_{\Delta_{\langle a,b,c\rangle} \in G} \left(\frac{2\|\Delta_{\langle a,b,c\rangle}\|_2^2}{3} \cdot 3AB\sqrt{2\log(2Ck)\log n}\right) \\
&\leq \sum_{\Delta_{\langle a,b,c\rangle} \in G} \left(4\|\boldsymbol{x}_{abc}\|_2^2 \cdot 3AB\sqrt{2\log(2Ck)\log n}\right) \\
&= \left(12AB\sqrt{2\log(2Ck)\log n}\right) \cdot S^*
\end{aligned}$$

Then, we have the following results.

$$S^* \leq \mathsf{cost}_\Delta(\text{PAR}_G^*) \leq \mathbf{E}[\mathsf{cost}_\Delta(\text{PAR}_G)] \leq \left(12AB\sqrt{2\log(2Ck)\log n}\right) \cdot S^*$$

$$\mathbf{E}\left[\frac{\mathsf{cost}_\Delta(\text{PAR}_G)}{\mathsf{cost}_\Delta(\text{PAR}_G^*)}\right] \leq 12AB\sqrt{2\log(2Ck)\log n} = O\left(\sqrt{\log k \log n}\right)$$

Thus, the expected approximation ratio is bounded by $O(\sqrt{\log k \log n})$. □

5.2 Extension to the General MkBGP Problem

For MkBGP, TRI-KBGPARTITION can be naturally extended and guarantee the same approximation ratio. Given a motif M, the corresponding SDP can be obtained by replacing Eq.(7) and (10) with the following (11) and (12) respectively.

$$\min \sum_{H \in G\langle M\rangle} \|\boldsymbol{x}_H\|_2^2 \tag{11}$$

$$\|\boldsymbol{x}_H\|_2^2 \geq \frac{1}{2}\|\boldsymbol{a} - \boldsymbol{b}\|_2^2 \quad \forall H \in G\langle M\rangle, (a,b) \in H \tag{12}$$

When the size of M is c, compared with (7), the SDP (11) needs at most $|V|^c$ constraints in addition, and it can be solved within polynomial time also. Finally, by extending the proofs of Theorem 6 and 8, we have the following result.

Theorem 9. *Given a c-motif M, MkBGP admits a bi-criteria $(k, 2)$ approximation algorithm with expected polynomial time, and the expect approximation ratio can be bounded by $O(\sqrt{\log k \log n})$.* □

6 Conclusion

In this paper, two typical problems of graph partitioning motivated by the big data computing applications are studied. For the WkBGP problem, a bi-criteria polynomial time algorithm with $O(\sqrt{\log n \log k})$ approximation ratio is designed. For the MkBGP problem, it is proved to be NP-complete and impossible to be approximated within finite ratio, then a bi-criteria polynomial time algorithm with $O(\sqrt{\log n \log k})$ approximation ratio is designed.

Acknowledgement. This work is supported by the National Key Research and Development Program of China (2021YFB1715000), the National Natural Science Foundation of China (62272137, 62022013, 61832003) and the Fundamental Research Funds for the Universities of Heilongjiang Province (2022-KYYWF-1122).

References

1. Andreev, K., Räcke, H.: Balanced graph partitioning. Theory Comput. Syst. **39**(6), 929–939 (2006)
2. Arora, S., Rao, S., et al.: Expander flows, geometric embeddings and graph partitioning. In: STOC 2004, pp. 222–231 (2004)
3. Chlamtac, E., Makarychev, K., et al.: How to play unique games using embeddings. In FOCS 2006, pp. 687–696 (2006)
4. Dingzhu, D., Ge, K., et al.: Design and Analysis of Approximation Algorithms. Springer, New York (2011)
5. Even, G., Naor, J., et al.: Fast approximate graph partitioning algorithms. SIAM J. Comput. **28**(6), 2187–2214 (1999)
6. Even, G., Naor, J., et al.: Divide-and-conquer approximation algorithms via spreading metrics. J. ACM **47**(4), 585–616 (2000)
7. Fan, W., Jin, R., et al.: Application driven graph partitioning. In: SIGMOD 2020, pp. 1765–1779 (2020)
8. Andreas Emil Feldmann and Luca Foschini: Balanced partitions of trees and applications. Algorithmica **71**(2), 354–376 (2015)
9. Garey, M.R., Johnson, D.S.: Computers and Intractability: A Guide to the Theory of NP-Completeness. W. H. Freeman, New York, USA, 1979
10. Garey, M.R., Johnson, D.S., et al.: Some simplified npcomplete graph problems. Theor. Comput. Sci. **1**(3), 237–267 (1976)
11. Gonzalez, J.E., Low, Y., et al.: Powergraph: distributed graph-parallel computation on natural graphs. In: OSDI 2012, pp. 17-30
12. Krauthgamer, R., (Seffi) Naor, J., et al.: Partitioning graphs into balanced components. In: SODA 2009, pp. 942–949 (2009)
13. Martella, C., Logothetis, D., et al.: Spinner: Scalable graph partitioning in the cloud. In: ICDE 2017, pp. 1083–1094 (2017)
14. Pacaci, A., Tamer Özsu, M.: Experimental analysis of streaming algorithms for graph partitioning. In: SIGMOD 2019, pp. 1375–1392 (2019)
15. Räcke, H.: Optimal hierarchical decompositions for congestion minimization in networks. In: STOC 2008, pp. 255–264 (2008)
16. Simon, H.D., Teng, S.-H.: How good is recursive bisection? SIAM J. Sci. Comput. **18**(5), 1436–1445 (1997)
17. Alizadeh, F.: Interior point methods in semidefinite programming with applications to combinatorial optimization. SIAM J. Optim. **5**(1), 13–51 (1995)

Distributed Generalized Deutsch-Jozsa Algorithm

Hao Li[1,2,3,4], Daowen Qiu[1,2,3,4](✉), and Le Luo[3,4,5]

[1] Institute of Quantum Computing and Software, School of Computer Science and Engineering, Sun Yat-sen University, Guangzhou 510006, China
issqdw@mail.sysu.edu.cn

[2] The Guangdong Key Laboratory of Information Security Technology, Sun Yat-sen University, Guangzhou 510006, China

[3] Shenzhen Research Institute of Sun Yat-Sen University, Nanshan Shenzhen 518057, China

[4] QUDOOR Technologies Inc., Beijing, China

[5] School of Physics and Astronomy, Sun Yat-sen University, Zhuhai 519082, China

Abstract. Generalized Deutsch-Jozsa (DJ) problem proposed by Qiu et al. in 2016 is a generalization of the DJ problem, which can be described as a symmetric partial Boolean function $\mathrm{DJ}_n^k : \{0,1\}^n \rightarrow \{0,1\}$ defined as $\mathrm{DJ}_n^k(x) = 1$ for $|x| = n/2$, $\mathrm{DJ}_n^k(x) = 0$ for $|x|$ in the set $\{0, 1, \ldots, k, n-k, n-k+1, \ldots, n\}$, and it is undefined for the rest cases, where n is even, $0 \leq k < n/2$, and $|x|$ is the Hamming weight of x. In this paper, we design a distributed quantum algorithm for computing the generalized DJ problem. A key technique in distributed generalized Deutsch-Jozsa algorithm is designing distributed quantum algorithm for the subroutine Xquery constructed by Ambainis et al. The advantage of our algorithm is that fewer qubits are required than that in the generalized DJ algorithm.

Keywords: Distributed Quantum Algorithms · Generalized Deutsch-Jozsa Problem · Xquery · Generalized Deutsch-Jozsa Algorithm

1 Introduction

Quantum algorithms have shown great power in important research areas such as information security [1], unsorted database search [2] and solving systems of linear equations [3]. However, the development of large-scale universal quantum computers is still very difficult due to the current level of physics technology. In the **Noisy Intermediate-Scale Quantum** (NISQ) era [4], one can only implement quantum algorithms with fewer qubits and shallower circuit depth. Therefore, it is fascinating to design new quantum algorithms that require fewer qubits and even shallower circuit depth compared with previous quantum algorithms for solving the same problems. To a great extend, **distributed quantum computing** (DQC) have the potential advantages [5–20].

Y. Chen et al. (Eds.): COCOON 2024, LNCS 15162, pp. 214–225, 2025.
https://doi.org/10.1007/978-981-96-1093-8_18

Distributed quantum algorithms (DQA) are a core issue of DQC. Due to limited space, we review DQA that are closely related to this paper, though there have been important contributions by many scholars. Qiu et al. [13] proposed a distributed Grover's algorithm. Tan, Xiao and Qiu et al. [14] proposed a distributed Simon's algorithm. Xiao and Qiu et al. [15,16] proposed a distributed Shor's algorithm. Li and Qiu et al. [17,18] designed distributed exact quantum algorithms for DJ problem and generalized Simon's problem. Zhou and Qiu et al. [19,20] designed a distributed exact Grover's algorithm and a distributed Bernstein-Vazirani algorithm. In general, compared to centralized quantum algorithms, DQA need smaller number of qubits and can reduce the circuit depth. Therefore, in the NISQ era, DQA are more likely to be physically realized and applied earlier.

In 2016, Qiu et al. [21,25] proposed generalized DJ problem that generalizes DJ problem [22–24]. Genaralized DJ problem [21,25] can be described as a symmetric partial Boolean function [25]. In this paper, we study a generalized DJ problem in the distributed scenario and propose a distributed generalized DJ algorithm. We focus on the decomposition of the subroutine–Xquery in the generalized DJ algorithm. The design method of Xquery was proposed by Ambainis et al. [26], and Qiu et al. [21,25] used the Xquery to design a generalized DJ algorithm. The advantage of the distributed generalized DJ algorithm proposed in this paper over the generalized DJ algorithm by Qiu et al. [21,25] is that the number of qubits is reduced.

The remainder of this paper is organized as follows. In Sect. 2, we recall the generalized DJ problem and the generalized DJ algorithm. In Sect. 3, we design the distributed generalized DJ algorithm. Then, in Sect. 4, we analyze our algorithm as well as compare our algorithm with the generalized DJ algorithm. Fianlly, we conclude with a summary in Sect. 5.

2 Preliminaries

In this section, we review the generalized DJ problem and the generalized DJ algorithm [21,25]. In this paper, we treat bit strings and their corresponding binary integers as the same.

2.1 Generalized DJ Problem

Qiu et al. [21,25] generalized the DJ problem [22] to the generalized DJ problem, and some generalizations of the DJ problem were also investigated in [29,31].

Definition 1. *The generalized DJ problem is the symmetric partial Boolean function* $DJ_n^k : \{0,1\}^n \rightarrow \{0,1\}$*, which is defined as follows:*

$$\mathrm{DJ}_n^k(x) = \begin{cases} 1, & \textit{if } |x| = n/2, \\ 0, & \textit{if } |x| \leq k \textit{ or } |x| \geq n-k, \end{cases} \tag{1}$$

where n *is even and* $0 \leq k < n/2$.

2.2 Generalized DJ Algorithm

Before introducing the generalized DJ algorithm by Qiu et al. [21,25], we recall the important subroutine, i.e., Xquery proposed by Ambainis et al. [26].

Let $x = x_1 x_2 \cdots x_m \in \{0,1\}^m$ be the input to Xquery, where m is even. Denote by $d = \lfloor \log_2 m \rfloor + 1$. The unitary operator $U_1 : \{0,1\}^d \to \{0,1\}^d$ is defined as:

$$U_1|0^d\rangle = \sum_{i=1}^{m} \frac{1}{\sqrt{m}}|i\rangle. \tag{2}$$

The query operator $O_x : \{0,1\}^d \to \{0,1\}^d$ is defined as:

$$O_x|i\rangle = (-1)^{x_i}|i\rangle, \tag{3}$$

where $1 \le i \le m$. The unitary operator $U_2 : \{0,1\}^{2d} \to \{0,1\}^{2d}$ is defined as:

$$U_2|i,0^d\rangle = \sum_{i<j} \frac{1}{\sqrt{m}}|i,j\rangle - \sum_{j<i} \frac{1}{\sqrt{m}}|j,i\rangle + \frac{1}{\sqrt{m}}|0^d,0^d\rangle, \tag{4}$$

where $1 \le i \le m$ and $1 \le j \le m$. In the following, we present Xquery by Ambainis et al. [26].

Algorithm 1. Subroutine for Xquery

procedure SX(**integer** m, **bit string** x) $\triangleright\ x \in \{0,1\}^m$
  integer $d = \lfloor \log_2 m \rfloor + 1$;
  $|\phi_0\rangle = |0^{2d}\rangle$;
  $|\phi_1\rangle = \left(U_1 \otimes I^{\otimes d}\right)|\phi_0\rangle$;
  $|\phi_2\rangle = \left(O_x \otimes I^{\otimes d}\right)|\phi_1\rangle$;
  $|\phi_3\rangle = U_2|\phi_2\rangle$;
  Measure $|\phi_3\rangle$ in the standard basis, and denote the result as Output;
  if Output$==(0^d,0^d)$ **then return** $(0,0)$;
  end if
  if Output$==(i,j)$ **then return** (i,j);
  end if
end procedure

Subsequently, we review the exact quantum algorithm, i.e. the generalized DJ algorithm proposed by Qiu et al. [21] for computing DJ_n^k.

Algorithm 2. Algorithm for DJ_n^k

```
procedure DJ(integer n, integer k, bit string x)          ▷ x ∈ {0,1}^n
    integer l = 1;
    while l ≤ k do
        Output = SX(n, x);
        if Output == (0,0) then return 0;
        end if
        if Output == (i, j) then
            x = x \ {x_i, x_j};
            l = l + 1;
            n = n − 2;
        end if
    end while
    Output = SX(n, x);
    if Output == (0,0) then return 0;
    end if
    if Output == (i, j) then return 1;
    end if
end procedure
```

3 Distributed Generalized DJ Algorithm

In this section, our main purpose is to design a distributed generalized DJ algorithm. First we describe the generalized DJ problem with a distributed scenario.

Let $x = x_1x_2 \cdots x_m \in \{0,1\}^m$, where m is even. Decompose m into positive integers p multiplied by q, i.e., $m = pq$. Divide x equally into q segments, each of which is $x_w x_{q+w} \cdots x_{(p-1)q+w}$, where $1 \leq w \leq q$. Define the function $g_w^x : \{0,1,\cdots,p-1\} \to \{0,1\}$ as $g_w^x(u) = x_i$, where $0 \leq u \leq p-1$, $1 \leq w \leq q$ and $i = u \cdot q + w$. In the following, we define these operators used in our distributed Xquery, i.e., Algorithm 3. The unitary operator $U_p : \{0,1\}^{d_1} \to \{0,1\}^{d_1}$ is defined as:

$$U_p|0^{d_1}\rangle = \frac{1}{\sqrt{p}} \sum_{0 \leq u \leq p-1} |u\rangle, \tag{5}$$

where $d_1 = \lceil \log_2 p \rceil$. The unitary operator $U_q : \{0,1\}^{d_2} \to \{0,1\}^{d_2}$ is defined as:

$$U_q|0^{d_2}\rangle = \frac{1}{\sqrt{q}} \sum_{1 \leq w \leq q} |w\rangle, \tag{6}$$

where $d_2 = \lfloor \log_2 q \rfloor + 1$. Intuitively, the effect of U_p combined with U_q can be thought of as U_1 in Algorithm 1, except that they differ in scale and scope of action. The query operators $O_{g_w^x} : \{0,1\}^{2(d_1+d_2)+w} \to \{0,1\}^{2(d_1+d_2)+w}$ are defined as:

$$O_{g_w^x}|u\rangle|b\rangle|c\rangle = |u\rangle|b\rangle|c \oplus g_w^x(u)\rangle, \tag{7}$$

where $d_1 = \lceil \log_2 p \rceil$, $d_2 = \lfloor \log_2 q \rfloor + 1$, $0 \leq u \leq p-1$, $1 \leq w \leq q$, $b \in \{0,1\}^{d_1+2d_2+w-1}$ and $c \in \{0,1\}$. The number of qubits for storage u is d_1.

Essentially, $O_{g_w^x}$ is a decomposition of O_x in Algorithm 1 into a number of smaller query operators. The unitary operator $A : \{0,1\}^{2d_2+q+1} \to \{0,1\}^{2d_2+q+1}$ can be regarded as a more general query operator, defined as:

$$A|i\rangle|a\rangle\left(\bigotimes_{1\leq j\leq q}|c_j\rangle\right)|e\rangle = |i\rangle|a\rangle\left(\bigotimes_{1\leq j\leq q}|c_j\rangle\right)|e\oplus c_i\rangle, \tag{8}$$

where $d_2 = \lfloor\log_2 q\rfloor + 1$, $1 \leq i \leq q$, $a \in \{0,1\}^{d_2}$, $c_j \in \{0,1\}$, $\bigotimes_{1\leq j\leq q}|c_j\rangle \triangleq |c_1\rangle|c_2\rangle\cdots|c_q\rangle$ and $e \in \{0,1\}$. The unitary operator $V_p : \{0,1\}^{2d_1} \to \{0,1\}^{2d_1}$ is defined as:

$$V_p|u,0^{d_1}\rangle = \frac{1}{\sqrt{p}}\left(\sum_{0\leq u<v\leq p-1}|u,v\rangle - \sum_{0\leq v<u\leq p-1}|v,u\rangle + |0^{d_1},0^{d_1}\rangle\right), \tag{9}$$

where $d_1 = \lceil\log_2 p\rceil$. The unitary operator $V_q : \{0,1\}^{2d_2} \to \{0,1\}^{2d_2}$ is defined as:

$$V_q|w,0^{d_2}\rangle = \frac{1}{\sqrt{q}}\left(\sum_{1\leq w<z\leq q}|w,z\rangle - \sum_{1\leq z<w\leq q}|z,w\rangle + |0^{d_2},0^{d_2}\rangle\right), \tag{10}$$

where $d_2 = \lfloor\log_2 q\rfloor + 1$. As U_p and U_q, the effect of V_p and V_q is similar to U_2 in Algorithm 1. Clearly, the unitarity of U_p, U_q, V_p, V_q, $O_{g_w^x}$ and A is easy to be verified.

For the sake of convenience, we need two notations: For any operator T_w with $1 \leq w \leq q$, denote by $\prod_{1\leq w\leq q} T_w \triangleq T_q\cdots T_2T_1$, and $\prod'_{1\leq w\leq q} T_w \triangleq T_1T_2\cdots T_q$.

We call the qubits actually work as *active qubits*. In Algorithm 1, the number of active qubits required for U_2 is $2\lfloor\log_2 m\rfloor + 2$, and it is the largest in these operators. In Algorithm 3, the numbers of active qubits required for A, V_p and V_q are $\lfloor\log_2(m/p)\rfloor + m/p + 2$, $2\lceil\log_2 p\rceil$ and $2\lceil\log_2 q\rceil + 2$ respectively. In fact, we can verify that the operator requiring the most active qubits in Algorithm 3 is either A or V_p. In order to make the single operator in Algorithm 3 that requires the most qubits demand fewer qubits than the operator in Algorithm 1, it can be proved that $2d - (d_2 + q + 1) = 2(\lfloor\log_2 m\rfloor + 1) - (\lfloor\log_2(m/p)\rfloor + m/p + 2) > 0$, and $2d - 2d_1 = 2(\lfloor\log_2 m\rfloor + 1) - 2\lceil\log_2 p\rceil > 0$. Denote the set D_m as:

$$D_m = \left\{r \,\middle|\, r|m, \lfloor\log_2(m/r)\rfloor + m/r < 2\lfloor\log_2 m\rfloor, \lceil\log_2 r\rceil < \lfloor\log_2 m\rfloor + 1\right\}. \tag{11}$$

To minimize the number of qubits required for the operator requiring the most active qubits in Algorithm 3, i.e., A or V_p, we can prove that $p \in D_m$ and p satisfies the function $\max\{\lfloor\log_2(m/p)\rfloor + m/p + 2, 2\lceil\log_2 p\rceil\}$ attaining the smallest value. Define the function $E_m : D_m \to \mathbb{Z}$ as:

$$E_m(r) = \max\{\lfloor\log_2(m/r)\rfloor + m/r + 2, 2\lceil\log_2 r\rceil\}, \tag{12}$$

where $r \in D_m$.

In the following, we describe a distributed Xquery.

Algorithm 3. Distributed subroutine for Xquery

---

procedure DSX(**integer** m, **bit string** x) $\triangleright\ x \in \{0,1\}^m$

  $p = \min\left\{ s \middle| s \in \arg\min_r E_m(r) \right\}$, where E_m is defined in Eq. (12);

  $q = m/p$;

  $d_1 = \lceil \log_2 p \rceil$;

  $d_2 = \lfloor \log_2 q \rfloor + 1$;

  for $u = 0$ to $p - 1$ **do**

    for $w = 1$ to q **do**

      $i = u \cdot q + w$;

      $g_w^x(u) = x_i$;

    end for

  end for

  $|\psi_0\rangle = |0^{2(d_1+d_2)+q+1}\rangle$;

  $|\psi_1\rangle = \left(U_p \otimes I^{d_1} \otimes U_q \otimes I^{\otimes(d_2+q+1)}\right)|\psi_0\rangle$;

  $|\psi_2\rangle = \prod_{1\le w\le q}\left(O_{g_w}^x \otimes I^{\otimes(q+1-w)}\right)|\psi_1\rangle$;

  $|\psi_3\rangle = (I^{\otimes 2d_1} \otimes A)\,|\psi_2\rangle$;

  $|\psi_4\rangle = \left(I^{\otimes(2(d_1+d_2)+q)} \otimes Z\right)|\psi_3\rangle$;

  $|\psi_5\rangle = \left({\prod}'_{1\le w\le q}\left(O_{g_w}^x \otimes I^{\otimes(q+1-w)}\right)\right)(I^{\otimes 2d_1} \otimes A)\,|\psi_4\rangle$;

  $|\psi_6\rangle = \left(V_p \otimes V_q \otimes I^{\otimes(q+1)}\right)|\psi_5\rangle$;

  Measure the first $2(d_1+d_2)$-th qubits of $|\psi_6\rangle$, and denote the result as Output;

  if Output$==(0^{2d_1}, 0^{2d_2})$ **then return** $(0,0)$;

  end if

  if (Output $== (uv, 0^{2d_2})) \wedge (u < v)$ **then**

    Find i and j in the set $\{u \cdot q + w | 1 \le w \le q\} \cup \{v \cdot q + z | 1 \le z \le q\}$ that satisfy $i \ne j$ and $x_i \ne x_j$.

    return (i,j);

  end if

  if (Output $== (0^{2d_1}, wz)) \wedge (w < z)$ **then**

    Find i and j in the set $\{u \cdot q + w | 0 \le u \le p-1\} \cup \{v \cdot q + z | 0 \le v \le p-1\}$ that satisfy $i \ne j$ and $x_i \ne x_j$.

    return (i,j);

  end if

  if (Output $== (uv, wz)) \wedge (u < v) \wedge (w < z)$ **then**

    Find i and j in the set $\{u \cdot q + w, v \cdot q + w, u \cdot q + z, v \cdot q + z\}$ that satisfy $i \ne j$ and $x_i \ne x_j$.

    return (i,j);

  end if

end procedure

Now we are ready to present a distributed generalized DJ algorithm, i.e., Algorithm 4, which uses Algorithm 3.

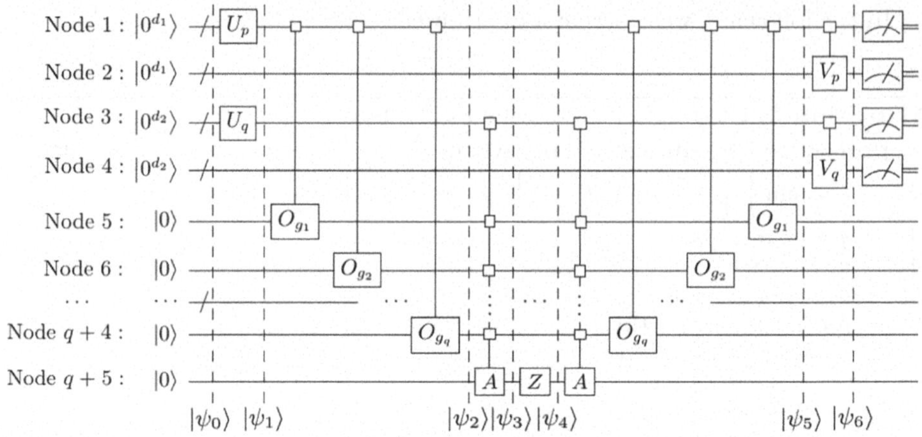

Fig. 1. Circuit for Algorithm 3.

Algorithm 4. Distributed algorithm for DJ_n^k

---

procedure DDJ(**integer** n, **integer** k, **bit string** x) $\triangleright x \in \{0,1\}^n$
    integer $l = 1$;
    while $l \leq k$ **do**
        Output = DSX(n, x);
        if Output == $(0, 0)$ **then return** 0;
        end if
        if Output == (i, j) **then**
            $x = x \setminus \{x_i, x_j\}$;
            $l = l + 1$;
            $n = n - 2$;
        end if
    end while
    Output = DSX(n, x);
    if Output == $(0, 0)$ **then return** 0;
    end if
    if Output == (i, j) **then return** 1;
    end if
end procedure

---

4 Analysis and Comparison of Algorithms

In this section, we analyze the correctness of our algorithm and compare our algorithm with the generalized DJ algorithm by Qiu et al. [21,25]. First, we give Lemma 1 related to Algorithm 3.

Lemma 1. *Let* $x = x_1 x_2 \cdots x_m \in \{0,1\}^m$ *be an input to Algorithm 3, where* m *is even. In line 16 of Algorithm 3, after the first* $2(d_1 + d_2)$*-th qubits of* $|\psi_6\rangle$ *are measured, we have:*

- *If the result is* $\left(0^{2d_1}, 0^{2d_2}\right)$, *then* $\sum_{i=1}^{m}(-1)^{x_i} \neq 0$ *and* $|x| \neq m/2$.
- *If the result is* $\left(uv, 0^{2d_2}\right)$ *and* $u < v$, *then there exist* $i \neq j$ *and* $i, j \in \{u \cdot q + w | 1 \leq w \leq q\} \cup \{v \cdot q + z | 1 \leq z \leq q\}$ *such that* $x_i \neq x_j$.
- *If the result is* $\left(0^{2d_1}, wz\right)$ *and* $w < z$, *then there exist* $i \neq j$ *and* $i, j \in \{u \cdot q + w | 0 \leq u \leq p-1\} \cup \{v \cdot q + z | 0 \leq v \leq p-1\}$ *such that* $x_i \neq x_j$.
- *If the result is* (uv, wz), $u < v$ *and* $w < z$, *then there exist* $i \neq j$ *and* $i, j \in \{u \cdot q + w, v \cdot q + w, u \cdot q + z, v \cdot q + z\}$ *such that* $x_i \neq x_j$.

Proof. We verify the evolution of quantum states in Algorithm 3. The state after step 12 is:

$$
\begin{aligned}
|\psi_1\rangle &= \left(U_p \otimes I^{d_1} \otimes U_q \otimes I^{\otimes(d_2+q+1)}\right)|\psi_0\rangle \\
&= \frac{1}{\sqrt{pq}} \sum_{\substack{0 \leq u \leq p-1 \\ 1 \leq w \leq q}} |u\rangle|0^{d_1}\rangle|w\rangle|0^{d_2}\rangle|0^{q+1}\rangle.
\end{aligned} \tag{13}
$$

Applying operators $O_{g_w}^{x}$ $(1 \leq w \leq q)$ and A results in the following state:

$$
\begin{aligned}
|\psi_3\rangle &= (I^{\otimes 2d_1} \otimes A)\left(\prod\nolimits'_{1 \leq w \leq q}\left(O_{g_w}^{x} \otimes I^{\otimes(q+1-w)}\right)\right)|\psi_1\rangle \\
&= \frac{1}{\sqrt{pq}} \sum_{\substack{0 \leq u \leq p-1 \\ 1 \leq w \leq q}} |u\rangle|0^{d_1}\rangle|w\rangle|0^{d_2}\rangle|g_1^x(u)\rangle|g_2^x(u)\rangle \cdots |g_q^x(u)\rangle|g_w^x(u)\rangle.
\end{aligned} \tag{14}
$$

After that, by means of operator Z, the following state is obtained:

$$
\begin{aligned}
|\psi_4\rangle &= \left(I^{\otimes(2(d_1+d_2)+q)} \otimes Z\right)|\psi_3\rangle \\
&= \frac{1}{\sqrt{pq}} \sum_{\substack{0 \leq u \leq p-1 \\ 1 \leq w \leq q}} (-1)^{g_w^x(u)}|u\rangle|0^{d_1}\rangle|w\rangle|0^{d_2}\rangle|g_1^x(u)\rangle|g_2^x(u)\rangle \cdots |g_q^x(u)\rangle|g_w^x(u)\rangle.
\end{aligned} \tag{15}
$$

Furthermore, by combining operators A with $O_{g_w}^{x}$ $(1 \leq w \leq q)$, it leads to the following state:

$$
\begin{aligned}
|\psi_5\rangle &= \left(\prod\nolimits'_{1 \leq w \leq q}\left(O_{g_w}^{x} \otimes I^{\otimes(q+1-w)}\right)\right)(I^{\otimes 2d_1} \otimes A)|\psi_4\rangle \\
&= \frac{1}{\sqrt{pq}} \sum_{\substack{0 \leq u \leq p-1 \\ 1 \leq w \leq q}} (-1)^{g_w^x(u)}|u\rangle|0^{d_1}\rangle|w\rangle|0^{d_2}\rangle|0^{q+1}\rangle.
\end{aligned} \tag{16}
$$

Proceeding to applying operators V_p and V_q, we get the state:

$$
\begin{aligned}
|\psi_6\rangle &= \left(V_p \otimes V_q \otimes I^{\otimes(q+1)}\right)|\psi_5\rangle \\
&= \frac{1}{pq}\left\{\sum_{\substack{0\le u\le p-1\\1\le w\le q}}(-1)^{g_w^x(u)}|0^{2d_1},0^{2d_2}\rangle + \sum_{\substack{0\le u<v\le p-1\\1\le w\le q}}\left[(-1)^{g_w^x(u)}-(-1)^{g_w^x(v)}\right]\right. \\
&|u,v,0^{2d_2}\rangle + \sum_{\substack{0\le u\le p-1\\1\le w<z\le q}}\left[(-1)^{g_w^x(u)}-(-1)^{g_z^x(u)}\right]|0^{2d_1},w,z\rangle \\
&\left.+\sum_{\substack{0\le u<v\le p-1\\1\le w<z\le q}}\left[(-1)^{g_w^x(u)}-(-1)^{g_w^x(v)}+(-1)^{g_z^x(v)}-(-1)^{g_z^x(u)}\right]|u,v,w,z\rangle\right\} \\
&|0^{q+1}\rangle.
\end{aligned}
\tag{17}
$$

Measure the first $2(d_1+d_2)$-th qubits of $|\psi_6\rangle$. Then, according to Eq. (17), the following cases result:

If the result is $\left(0^{2d_1},0^{2d_2}\right)$, then $\sum_{\substack{0\le u\le p-1\\1\le w\le q}}(-1)^{g_w^x(u)}\neq 0$, i.e., $\sum_{i=1}^{m}(-1)^{x_i}\neq 0$ and $|x|\neq m/2$.

If the result is $\left(uv,0^{2d_2}\right)$ and $u<v$, then there exist $i\neq j$ and $i,j\in\{u\cdot q+w|1\le w\le q\}\cup\{v\cdot q+z|1\le z\le q\}$ such that $x_i\neq x_j$.

If the result is $\left(0^{2d_1},wz\right)$ and $w<z$, then there exist $i\neq j$ and $i,j\in\{u\cdot q+w|0\le u\le p-1\}\cup\{v\cdot q+z|0\le v\le p-1\}$ such that $x_i\neq x_j$.

If the result is (uv,wz), $u<v$ and $w<z$, then there exist $i\neq j$ and $i,j\in\{u\cdot q+w,v\cdot q+w,u\cdot q+z,v\cdot q+z\}$ such that $x_i\neq x_j$.

Now we give Theorem 1 to show the correctness of Algorithm 4.

Theorem 1. *Let $x=x_1x_2\cdots x_n\in\{0,1\}^n$ be the input to the function $DJ_n^k:\{0,1\}^n\rightarrow\{0,1\}$ in the generalized DJ problem and the input to Algorithm 4, where n is even and $0\le k<n/2$. If Algorithm 4 returns 0, then $DJ_n^k(x)=0$. If Algorithm 4 returns 1, then $DJ_n^k(x)=1$.*

Proof. According to lines 20 to 21 in Algorithm 3, and lines 2 to 15 in Algorithm 4, as well as Lemma 1, it can be concluded that if Algorithm 4 returns 0, then there exists an even number m, $2\le m\le n$, such that there are $\sum_{i=1}^{m}(-1)^{x_i}\neq 0$ and $|x|\neq m/2$ for $x=x_1x_2\cdots x_m\in\{0,1\}^m$. Since there is $\mathrm{DJ}_n^k(x)=\mathrm{DJ}_{n-2}^{k-1}(x\setminus\{x_i,x_j\})$ for $x_i\neq x_j$, it follows that if Algorithm 4 returns 0, then $\mathrm{DJ}_n^k(x)=0$.

According to lines 22 to 33 in Algorithm 3, and lines 2 to 17 in Algorithm 4, as well as Lemma 1, it can be obtained that if Algorithm 4 returns 1, then there is $|x|=(n-2k)/2$, where $x=x_1x_2\cdots x_{n-2k}\in\{0,1\}^{n-2k}$. Since there is $\mathrm{DJ}_n^k(x)=\mathrm{DJ}_{n-2}^{k-1}(x\setminus\{x_i,x_j\})$ for $x_i\neq x_j$, it follows that if Algorithm 4 returns 1, then $\mathrm{DJ}_n^k(x)=1$.

In light of the parameter p set by Algorithm 3, it can be concluded that the operator requiring the most active qubits in Algorithm 3 demands fewer active qubits than Algorithm 1 and therefore achieves relatively better separation.

In Fig. 1, the number of qubits required at Node $q+5$ is equal to the number of active qubits required to implement the operator A. The number of qubits required for Node 2 is equal to the number of active qubits to realize the operator V_P. Therefore, the number of qubits required for a single computing node in Algorithm 3 is smaller than that of Algorithm 1 (Table 1).

Table 1. Comparison of the quantum part of Algorithm 2 and Algorithm 4.

Algorithm	Number of qubits required for the largest single computing node
Algorithm 2	$2\lfloor \log_2 m \rfloor + 2$
Algorithm 4	$\max\{\lfloor \log_2(m/p) \rfloor + m/p + 2, 2\lceil \log_2 p \rceil\}$

5 Conclusion

In this paper, we have designed a distributed generalized DJ algorithm. The key to our algorithm is to study the subroutine–Xquery proposed by Ambainis et al. [26] that was used in the generalized DJ algorithm designed by Qiu et al. [21,25]. We have designed the decomposition of the unitary operator in Xquery into a number of unitary operators that require fewer qubits. Compared to the generalized DJ algorithm, the implementation of the circuit of a single computing node in the distributed generalized DJ algorithm requires fewer qubits, with shallower circuit depth, and thus is more resistant to noise interference. As a result, our algorithm proposed is likely more feasible to be implemented physically in the NISQ era.

Acknowledgments. We are grateful to the three anonymous referees for important comments and suggestions that help us improve the quality of the paper. This work is supported in part by the National Natural Science Foundation of China (Grant No. 61876195), the Natural Science Foundation of Guangdong Province of China (Grant No. 2017B030311011), the Shenzhen Science and Technology Program (Grant No. JCYJ20220818102003006), and the Guangdong Provincial Quantum Science Strategic Initiative (No. GDZX2200001).

Disclosure of Interests. The authors have no competing interests to declare that are relevant to the content of this article.

References

1. Shor, P.W.: Polynomial-time algorithms for prime factorization and discrete logarithms on a quantum computer. SIAM J. Comput. **26**(5), 1484–1509 (1997)
2. Grover, L.K.: A fast quantum mechanical algorithm for database search. In: Proceedings of the 28th Annual ACM Symposium on Theory of Computing, STOC'96, New York, USA, pp. 212–219. ACM (1996)
3. Harrow, A.W., Hassidim, A., Lloyd, S.: Quantum algorithm for linear systems of equations. Phys. Rev. Lett. **103**, 150502 (2009)
4. Preskill, J.: Quantum computing in the NISQ era and beyond. Quantum **2**, 79 (2018)
5. de Wolf, R.: Quantum communication and complexity. Theor. Comput. Sci. **287**(1), 337–353 (2002)
6. Buhrman, H., Röhrig, H.: Distributed quantum computing. In: Rovan, B., Vojtáš, P. (eds.) Mathematical Foundations of Computer Science 2003, pp. 1–20. Springer, Heidelberg (2003)
7. Beals, R., et al.: Efficient distributed quantum computing. Proc. Math. Phys. Eng. Sci. **469** (2153) (2013)
8. Li, K., Qiu, D., Li, L., Zheng, S., Rong, Z.: Application of distributed semiquantum computing model in phase estimation. Inf. Process. Lett. **120**, 23–29 (2017)
9. Caleffi, M., Cacciapuoti, A.S., Bianchi, G.: Quantum internet: from communication to distributed computing! In: Proceedings of the 5th ACM International Conference on Nanoscale Computing and Communication, pp. 1-4. ACM, New York, USA (2018)
10. Le Gall, F., Magniez, F.: Sublinear-time quantum computation of the diameter in CONGEST networks. In: Proceedings of the 2018 ACM Symposium on Principles of Distributed Computing. PODC'18, New York, USA, pp. 337–346. ACM (2018)
11. Le Gall, F., Nishimura, H., Rosmanis, A.: Quantum advantage for the LOCAL model in distributed computing. In: Niedermeier, R., Paul, C. (eds.) 36th International Symposium on Theoretical Aspects of Computer Science (STACS 2019). Leibniz International Proceedings in Informatics (LIPIcs), Dagstuhl, Germany, vol. 126, pp. 49:1–49:14. Schloss Dagstuhl-Leibniz-Zentrum für Informatik (2019)
12. Avron, J., Casper, O., Rozen, I.: Quantum advantage and noise reduction in distributed quantum computing. Phys. Rev. A **104**, 052404 (2021)
13. Qiu, D., Luo, L., Xiao, L.: Distributed Grover's algorithm. Theor. Comput. Sci. **993**, 114461 (2024). arXiv:2204.10487
14. Tan, J., Xiao, L., Qiu, D., Luo, L., Mateus, P.: Distributed quantum algorithm for Simon's problem. Phys. Rev. A **106**, 032417 (2022). arXiv:2204.11388
15. Xiao, L., Qiu, D., Luo, L., Mateus, P.: Distributed Shor's algorithm. Quantum Inf. Comput. **23**, 27–44 (2022). arXiv:2207.05976
16. Xiao, L., Qiu, D., Luo, L., Mateus, P.: Distributed quantum-classical hybrid Shor's algorithm (2023). arXiv:2304.12100
17. Li, H., Qiu, D., Luo, L.: Distributed exact quantum algorithms for Deutsch-Jozsa problem (2023). arXiv:2303.10663
18. Li, H., Qiu, D., Luo, L.: Exact distributed quantum algorithm for generalized Simon's problem. Acta Inform. **61**(1) (2024). arXiv:2307.14315
19. Zhou, X., Qiu, D., Luo, L.: Distributed exact Grover's algorithm. Front. Phys. **18**, 51305 (2023)
20. Zhou, X., Qiu, D., Luo, L.: Distributed Bernstein-Vazirani algorithm. Phys. A **629**, 129209 (2023)

21. Qiu, D., Zheng, S.: Generalized Deutsch-Jozsa problem and the optimal quantum algorithm. Phys. Rev. A **97**, 062331 (2018)
22. Deutsch, D., Jozsa, R.: Rapid solution of problems by quantum computation. Proc. R. Soc. Lond. A **439**, 553–558 (1992)
23. Qiu, D., Zheng, S.: Revisiting Deutsch-Jozsa algorithm. Inf. Comput. **275**, 104605 (2020)
24. Qiu, D., Xu, G.: Exact Quantum 1-Query Algorithms and Complexity. SPIN **11**(3), 1–10 (2021)
25. Qiu, D., Zheng, S.: Characterizations of symmetrically partial Boolean functions with exact quantum query complexity (2016). arXiv:1603.06505
26. Ambainis, A., Iraids, J., Smotrovs, J.: Exact quantum query complexity of EXACT and THRESHOLD. In: Severini, S., Brandao, F. (eds.) 8th Conference on the Theory of Quantum Computation, Communication and Cryptography (TQC 2013). Leibniz International Proceedings in Informatics (LIPIcs), Dagstuhl, Germany, vol. 22, pp. 263–269. Schloss Dagstuhl - Leibniz-Zentrum für Informatik (2013)
27. Even, S., Selman, A.L., Yacobi, Y.: The complexity of promise problems with applications to public-key cryptography. Inf. Control **61**(2), 159–173 (1984)
28. Goldreich, O.: In: Goldreich, O., Rosenberg, A.L., Selman, A.L. (eds.) On Promise Problems: A Survey, pp. 254-290. Springer, Heidelberg (2006)
29. Montanaro, A., Jozsa, R., Mitchison, G.: On exact quantum query complexity. Algorithmica **71**, 775–796 (2015)
30. Chi, D. P., Kim, J., Lee, S.: Quantum algorithm for generalized Deutsch-Jozsa problem (2000). arXiv:0005059
31. Gruska, J., Qiu, D., Zheng, S.: Generalizations of the distributed Deutsch-Jozsa promise problem. Math. Struct. Comput. Sci. **27**(3), 311–331 (2017)

Average AoI Optimization at Wireless-Powered Network Edge with Stochastic Arrivals

Quan Chen[1], Jungen Xia[1], Jing Li[2(✉)], Yuan Chai[1(✉)], Hong Gao[3], and Zhipeng Cai[4]

[1] School of Computer Science and Technology, Guangdong University of Technology, Guangzhou, China
{quan.c,y.chai}@gdut.edu.cn, 2112205340@mail2.gdut.edu.cn

[2] Department of Computer Science, City University of Hong Kong, Hong Kong, China
jing.li@cityu.edu.hk

[3] School of Computer Science and Technology, Zhejiang Normal University, Jinhua, China
honggao@zjnu.edu.cn

[4] Department of Computer Science, Georgia State University, Atlanta, USA
zcai@gsu.edu

Abstract. Age of Information (AoI) has been introduced as a new metric for characterizing data freshness in wireless networks. To optimize the system AoI at wireless-powered networks, we investigate the first work to optimize the weighted sum of average AoI with stochastic arrivals in this paper. Firstly, the average AoI optimization problem with stochastic arrivals is formulated, which tries to minimize the weighted sum of average AoI from the point of directional energy transferring. The lower bound of the achievable AoI by any network operating is derived. Then, to assist in the energy harvesting process of the source nodes, the stationary semi-randomized scheduling policy is introduced and the closed-form expression under the space of such a policy is derived. The optimal stationary semi-randomized scheduling policy is also proposed. Finally, based on the obtained closed-form expression of average AoI, an age-based max-weight charging scheduling policy is proposed with Lyapunov optimization, which is proved to have a bounded approximation ratio.

1 Introduction

Recently, significant attention has been paid to Age of Information (AoI), which is a new metric to characterize the freshness of information [1,2]. This is mainly because of a growing demand for maintaining timely status update in many network controlling and monitoring applications, such as the autonomous driving systems and the smart grid systems [3–7]. For these applications, usually only the freshest status update is valuable for the information receiver, while the old data may own little value. Considering the importance of maintaining data freshness in these applications, it has motivated numerous researches for optimizing AoI, which is defined as the elapsed time of the most recent packet received by the destination since it is first generated at the source.

In this paper, we are interested in the problem of optimizing average AoI for data collection at wireless-powered network edge. Considering a typical scenario at the network edge, where there exist multiple source nodes targeting for different applications,

Y. Chen et al. (Eds.): COCOON 2024, LNCS 15162, pp. 226–237, 2025.
https://doi.org/10.1007/978-981-96-1093-8_19

try to collect samples of information from the ambient environment and send the time-sensitive data to the edge server. Considering the limited network bandwidth and energy capacity, in order to reduce the AoI, it is critical to allocate the channel and energy resources efficiently among the source nodes [8–12]. Furthermore, the Wireless Power Transfer (WPT) technology has been widely adopted to charge devices via electromagnetic waves, to provide a controllable and reliable power supply [13–15]. At such networks, to optimize the system AoI, one not only needs to design the efficient scheduling policy of communications, but also take the schedule of energy transferring into account, which makes the AoI optimization problem much more challenging.

Early studies in [16–23] try to optimize the system AoI under omnidirectional WPT, where the electromagnetic waves are broadcasted evenly in all directions regardless of the locations of the rechargeable devices. When there exists a single source node in the network, the authors in [16] first studied the achievable age-energy region and the trade-off between energy and the achieved AoI. In [17], the joint optimization problem of the sampling process, update transmission and the energy transferring process is studied. A threshold-based structure of the optimal policy is derived. The average AoI optimization problem with the selfish charging nodes and penalty function are studied by [18, 19]. In [20], the authors consider a wireless-powered relay aided network, where there exists a rechargeable relay node responsible for forwarding the information from a sensor to the destination. The average AoI optimization problem with multiple source nodes were studied by [21, 22], and several Lyapunov optimization-based approaches are proposed. In [23], the authors investigate the max AoI minimization problem with energy harvesting for single source and multiple sources, respectively.

Besides omnidirectional WPT, the directional WPT has been proposed to further improve the energy transfer efficiency recently, by concentrating radiated energy in a fixed direction with energy beamforming. The AoI optimization problem under directional WPT are studied by [24–27]. In [24], the problem of optimizing the maximum peak AoI with the directional chargers is studied. The theoretical bound of the maximum peak AoI is derived, and an approximation ratio of at most 1.5 is proposed. After then, the authors in [25] try to optimize the weighted sum of average peak AoI with the directional chargers. An age-based scheduling algorithm is proposed by considering the charging process and communication process simultaneously. It is proved to have an approximation ratio of $(1+\varphi)$, where φ is related to the weight of the source nodes. The average AoI optimization problem at wireless-powered network edge is studied by [26] recently. The authors first derived the theoretical bound of the weighted sum of average AoI, and then, an AoI-aware periodical charging scheduling algorithm is proposed, which has an approximation ratio of up to 1.5625. In [27], the authors study the average AoI optimization problem at a two-hop wireless-powered network, which tries to obtain the optimal time allocation among the source nodes. However, all these methods assume a simple data arriving model, *i.e.*, the generate-at-will model, where the source node can always transmit the freshest packets which are also generated currently.

Thus, to make the work more practical and general, we investigate how to optimize the average AoI at wireless-powered network edge with stochastic arrivals. Firstly, the average AoI optimization problem with stochastic arrivals is formulated, which tries to minimize the weighted sum of average AoI from the point of directional energy transferring. The lower bound of the achievable AoI by any network operating is also derived. Then, to assist in the energy harvesting process of the source nodes, the stationary semi-randomized scheduling policy is introduced. The closed-form expression under the space of such a policy is derived, and the optimal stationary semi-randomized scheduling policy is also proposed. Finally, based on the obtained closed-form expression of average AoI, an age-based max-weight charging scheduling policy is proposed with Lyapunov optimization, which is proved to have a bounded approximation ratio.

2 Problem Definition

2.1 Network Model

In this paper, we study a multi-source monitoring system in the wireless-powered network as in [24,27], in which the source nodes, denoted by $V = \{s_1, s_2, ..., s_n\}$, report the time-sensitive data to a Hybrid Access Point (HAP), denoted as s_0. In the system, we employ the discrete-time model, where the network time is slotted with a fixed length τ. For each source node s_i, the packets arrive at the beginning of each time slot with a probability $p_i \in (0, 1]$. To be noted that, for each source node, it only maintains the freshest update, which means the old packet will be dropped when a new packet arrives. The arriving process is i.i.d. over time and independent across the source nodes with $\mathcal{P}(\lambda_i(t) = 1) = p_i, \forall i, t$, where $\lambda_i(t) \in \{0, 1\}$ denotes whether a packet arrives at source node s_i at time slot t $(t \geq 1)$.

To support the transmission of the source nodes in the uplink, the HAP employs the directional energy transferring technology to focus the energy in narrow beams toward the source nodes in the downlink. As in [24,25,27], we assume the downlink directional energy transferring and the uplink data transmission can perform simultaneously. Actually, as the prototype shown in [27], the HAP is equipped with L $(L > 1)$ high-gain antennas for directional energy transfer and a single antenna for data transmission. For each source node, it is also equipped with two antennas, which are used for energy harvesting and data transmission, respectively. In addition, since it takes a few milliseconds to transmit a small packet, which is much less than the one of harvesting enough energy for packet transmission, *i.e.*, several seconds. Thus, we mainly consider the energy transferring decision for AoI optimization in this paper. To be noted that, at each time slot t, the HAP can only choose one source node to transfer the energy.

At each time slot t, let $e_i(t)$ denote whether the source node s_i is harvesting energy from the HAP. Obviously, one can have $\sum_{i=1}^{N} e_i(t) \leq 1$. Assume the HAP transmits the RF signals with a power of $\mathcal{P}_0$, then the harvested energy of source node s_i from the HAP, *i.e.*, $H_i(t)$, can be computed as

$$H_i(t) = \begin{cases} \mathcal{P}_0 \cdot h_i \cdot \tau, & if\ e_i(t) = 1, \\ 0, & otherwise. \end{cases}$$

where $h_i \propto \|s_i, s_0\|^{-\beta}$ is the downlink channel power gain between the HAP and the source node, $\|s_i, s_0\|$ is the distance between the source node and HAP, and β is the channel path loss exponent.

Denote $E_i(t)$ as the residual energy of source node s_i at the beginning of time slot t. Then, for any time slot t $(t \geq 1)$, $E_i(t)$ can be updated as $E_i(t) = E_i(t-1) + H_i(t -$

$1) - \chi_i(t-1) \cdot \varepsilon_i$, where $\chi_i(t) \in \{0, 1\}$ denotes whether the source node s_i transmits a packet to the HAP at time t, and ε_i denotes the required energy for transmitting the packet. To be noted that, when it tries to transmit a packet, *i.e.*, $\chi_i(t) = 1$, it must satisfy that $E_i(t) \geq \varepsilon_i$. For ease of implementation, as the setting in [27], we assume the source node transmits the packet to the HAP directly when it has harvested enough energy.

2.2 AoI Model

In order to measure how closely the update is collected by the HAP, we adopt the concept of Age of Information (AoI), which can measure the freshness of data from the destination quantitatively. Formally, the AoI of source node s_i is defined as the time elapsed since the generation time of the latest update received by the HAP. At each time slot t, let $\hbar_i(t)$ denote the generation time of the freshest update stored at the source node s_i, and let $\hat{\hbar}_i(t)$ denote the generation time of the freshest update received by the HAP from source node s_i. To be noted that, the packet stored at the source node is usually fresher than the one received by the HAP, *i.e.*, $\hbar_i(t) \geq \hat{\hbar}_i(t)$. According to the definition, the AoI of source node s_i, denoted as $\Delta_i(t)$, can be computed as

$$\Delta_i(t) = t - \hat{\hbar}_i(t). \tag{1}$$

Note that, at time slot $t-1$ ($t \geq 1$), if source node s_i transmits an update to the HAP, *i.e.*, $\chi_i(t-1) = 1$, then one can have $\hat{\hbar}_i(t) = \hbar_i(t-1)$. In this case, at the next time slot t, its AoI can be computed as $t - \hbar_i(t-1)$. Otherwise, if $\chi_i(t-1) = 0$, its AoI will just increased by 1, *i.e.*, $\Delta_i(t-1)+1$. Therefore, the AoI of source node s_i, *i.e.*, $\Delta_i(t)$ ($t \geq 1$), can be recomputed as

$$\Delta_i(t) = \begin{cases} t - \hbar_i(t-1), & if\ \chi_i(t-1) = 1 \\ \Delta_i(t-1) + 1, & otherwise \end{cases} \tag{2}$$

Note that, for each source node s_i, $\Delta_i(0)$ is assumed to be 0 for simplicity. One can find that $\Delta_i(t)$ is actually a zigzag-like function which only drops when the source node transmits a new update and grows linearly otherwise. As the example shown in Fig. 1, where the blue arrow denotes a new packet arrives at source node s_i, while a red arrow denotes it transmits a packet to the HAP. With respect to the packet arrival process, different scheduling policies result in much different AoI performance.

2.3 Problem Formulation

In this paper, due to the randomness in the packet arrival process, we aim to optimize the expected weighted sum of average AoI of the entire network in the limit as the time-horizon grows to infinity.

Let $\Pi := \{e_i(t) | \sum_{i=1}^{N} e_i(t) \leq 1, \forall t\}$ be the set of all the feasible scheduling policies. For a scheduling policy $\pi \in \Pi$, the expected weighted sum of average AoI of the entire network, *i.e.*, $\overline{\Delta_\pi}$, can be computed as $\overline{\Delta_\pi} \triangleq \lim_{T \to \infty} \frac{1}{TN} \mathbb{E}[\sum_{i=1}^{N} \sum_{t=1}^{T} \omega_i \Delta_i(t)]$, where ω_i is the weight of source node s_i.

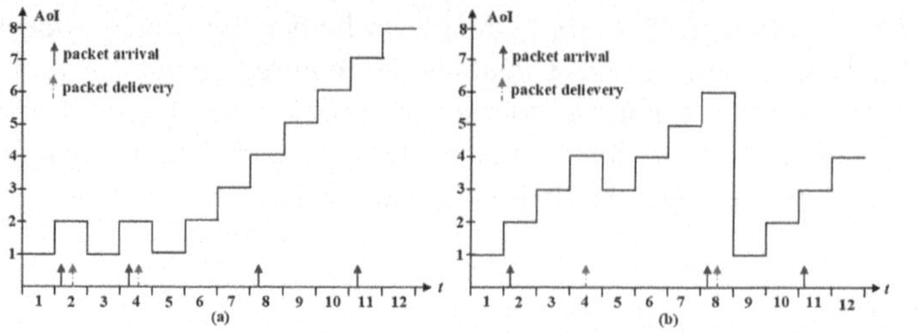

Fig. 1. An example of source node s_i's AoI.

For a given network setup (N, p_i, ε_i, ω_i), we try to find a scheduling policy $\pi \in \Pi$ to achieve the minimum expected weighted sum of average AoI, denoted as $\overline{\Delta_{opt}}$. It can be formulated as:

$$\mathbf{P1}: \quad \overline{\Delta_{opt}} \triangleq \min_{\pi \in \Pi} \lim_{T \to \infty} \frac{1}{TN} \mathbb{E}[\sum_{i=1}^{N} \sum_{t=1}^{T} \omega_i \Delta_i(t)] \tag{3}$$

subject to:

$$\sum_{i=1}^{N} e_i(t) \leq 1, \; t = 1, 2, ..., T. \tag{4}$$

$$e_i(t) \in \{0, 1\}, \; 1 \leq i \leq N, \; t = 1, 2, ..., T. \tag{5}$$

3 The Age-Based Charging Scheduling Algorithm

3.1 Lower Bound Analysis

Before introducing the proposed scheduling policy, we first try to derive a lower bound to the above AoI optimization problem for a given network setup.

Considering a scheduling policy π under a given network setup, the total harvest energy of source node s_i during the period of T, *i.e.*, $\mathcal{H}_i^\pi$, can be computed as $\mathcal{H}_i^\pi = \sum_{t=1}^{T} H_i(t) = \sum_{t=1}^{T} \mathcal{P}_0 \cdot h_i \cdot \tau \cdot e_i(t)$. Then, the total number of packets delivered by the source node s_i during the period of T, denoted as $\mathcal{N}_i(T)$, can be computed as $\mathcal{N}_i^\pi = \lfloor \frac{\mathcal{H}_i^\pi}{\varepsilon_i} \rfloor$. Let r_i^π be the transmission rate of source node s_i under the scheduling policy π, which can be computed as $r_i^\pi = \mathcal{N}_i^\pi / T = \lfloor \frac{\sum_{t=1}^{T} H_i(t)}{\varepsilon_i} \rfloor / T$.

Let Ω be sample space under the network setup and $o \in \Omega$ be a sample path. For source node s_i, let $\varphi_i(k)$ be the time slot that source node s_i transmits its k ($k = 1, 2, ..., \mathcal{N}_i^\pi$)-th packet. To be noted that, after source node s_i transmit the packet at time slot $\varphi_i(k)$, its AoI at the next time slot $\varphi_i(k) + 1$ can be computed as $\Delta_i(\varphi_i(k) + 1) = \varphi_i(k) + 1 - \hbar_i(\varphi_i(k))$, where $\hbar_i(\varphi_i(k))$ is the generation time of the freshest update received by the HAP from source node s_i at time slot $\varphi_i(k)$. Let $\Gamma_i(k) = \varphi_i(k) - \hbar_i(\varphi_i(k))$ denote the waiting latency of the k-th packet being delivered. Then, the AoI at the time slot $\varphi_i(k) + 1$ would be $\Delta_i(\varphi_i(k) + 1) = \Gamma_i(k) + 1$.

Then, from time slot $\varphi_i(k-1)$ to $\varphi_i(k)$, the AoI of source node s_i would be $\Gamma_i(k-1)+1$, $\Gamma_i(k-1)+2,...,\Gamma_i(k-1)+\varphi_i(k)-\varphi_i(k-1)$. Let $\mathcal{I}_i(k)=\varphi_i(k)-\varphi_i(k-1)$ be the time interval between the (k-1)-th packet and the k-th packet. Then, the sum of AoI of source node s_i from time $\varphi_i(k-1)$ to $\varphi_i(k)$ can be computed as $\Gamma_i(k-1)+1+\Gamma_i(k-1)+2+\ldots+\Gamma_i(k-1)+\mathcal{I}_i(k)=\Gamma_i(k-1)\cdot\mathcal{I}_i(k)+\frac{\mathcal{I}_i(k)\cdot(\mathcal{I}_i(k)+1)}{2}$.

In the following, we will prove the lower bound of the achievable AoI for a given network setup, which is denoted by $\overline{\Delta^{LB}}$.

Theorem 1. *For a given network setup* $(N, p_i, \varepsilon_i, \omega_i)$*, one can have* $\overline{\Delta^{LB}}=\frac{1}{2N}\sum_{i=1}^{N}\omega_i\cdot(\frac{1}{r_i^*}+1)$*, where* $r_i^*=min\{p_i, \frac{\mathcal{P}_0\cdot h_i\cdot\tau\sqrt{\omega_i}}{\varepsilon_i\sum_{j=1}^{N}\sqrt{\omega_j}}, \frac{\sqrt{\omega_i}}{\sum_{j=1}^{N}\sqrt{\omega_j}}\}$.

Proof. Given a network setup $(N, p_i, \varepsilon_i, \omega_i)$ and a scheduling policy π, let $\overline{\Delta_i^\pi}$ be the average AoI of source node s_i during the period of T. According to the above analysis, it can be computed as

$$\overline{\Delta_i^\pi}=\frac{1}{T}\sum_{t=1}^{T}\Delta_i(t)=\frac{1}{T}[\sum_{k=1}^{N_i^\pi}(\Gamma_i(k-1)\cdot\mathcal{I}_i(k)+\frac{\mathcal{I}_i(k)\cdot(\mathcal{I}_i(k)+1)}{2})+\Gamma_i(N_i^\pi)\cdot\mathcal{R}_i+\frac{\mathcal{R}_i\cdot(\mathcal{R}_i+1)}{2}],$$

where $\mathcal{R}_i$ is the left time since the delivery of the last packet of source node s_i.

For each source node s_i, since the waiting latency of its k ($k=1,2,...,N_i^\pi$)-th packet is at least 0, which means $\Gamma_i(k)\ge 0$ for $k=1,2,...,N_i^\pi$. Then, one can have $\overline{\Delta_i^\pi}=\frac{1}{T}\sum_{t=1}^{T}\Delta_i(t)\ge\frac{1}{T}[\sum_{k=1}^{N_i^\pi}(\frac{\mathcal{I}_i(k)^2+\mathcal{I}_i(k)}{2})+\frac{\mathcal{R}_i^2+\mathcal{R}_i}{2}]$. In addition, as for the entire period, one can also have $T=\sum_{k=1}^{N_i^\pi}\mathcal{I}_i(k)+\mathcal{R}_i$. According to the inequality that $\frac{\sum_{i=1}^{n}x_i}{n}\le\sqrt{\frac{\sum_{i=1}^{n}x_i^2}{n}}$, where $\{x_i, 1\le i\le n\}$ are n non-negative numbers, then one can have $\sum_{k=1}^{N_i^\pi}\mathcal{I}_i(k)^2\ge\frac{(T-\mathcal{R}_i)^2}{N_i^\pi}$. As a result, the average AoI of source node s_i during the period of T can be re-computed as $\overline{\Delta_i^\pi}=\frac{1}{T}\sum_{t=1}^{T}\Delta_i(t)\ge\frac{1}{T}\cdot(\frac{(T-\mathcal{R}_i)^2}{2N_i^\pi}+\frac{\mathcal{R}_i^2}{2}+\frac{T}{2})$. By setting $\mathcal{R}_i=\frac{T}{N_i^\pi+1}$, according to the Jensen inequality, one can achieve the minimum value $\frac{1}{2}(\frac{T}{N_i^\pi+1}+1)$. Then, taking the expectation and applying the limit $T\to\infty$ to the above equation, one can have

$$\overline{\Delta_i^\pi}\triangleq\lim_{T\to\infty}\frac{1}{T}\sum_{t=1}^{T}\mathbb{E}[\Delta_i(t)]\ge\frac{1}{2}[\frac{1}{\mathbb{E}[\frac{N_i^\pi}{T}]+\frac{1}{T}}+1]\ge\frac{1}{2}(\frac{1}{r_i^\pi}+1),$$

where $r_i^\pi=\frac{N_i^\pi}{T}$ is the transmission rate of s_i. As for the entire network, one can have

$$\overline{\Delta^\pi}\triangleq\sum_{i=1}^{N}\omega_i\cdot\overline{\Delta_i^\pi}\ge\frac{1}{2N}\sum_{i=1}^{N}\omega_i(\frac{1}{r_i^\pi}+1).$$

To be noted that, for source node s_i, since the packets arrive with a rate p_i, then one can have $r_i^\pi\le p_i$. In addition, let κ_i denote the average rate of source node s_i being charged under the scheduling policy π. Obviously, one can have $\sum_{i=1}^{N}\kappa_i\le 1$. Since the source node can only transmit its packet after it has been charged ready, one can have $r_i^\pi\le\kappa_i\cdot\mathcal{P}_0\cdot h_i\cdot\tau/\varepsilon_i$, where ε_i is the required energy for one transmission. To minimize $\frac{1}{2N}\sum_{i=1}^{N}\omega_i(\frac{1}{r_i^\pi}+1)$, the optimal value of r_i^π can be computed by the KKT condition, which can be computed as $r_i^*=min\{p_i, \frac{\mathcal{P}_0\cdot h_i\cdot\tau\sqrt{\omega_i}}{\varepsilon_i\sum_{j=1}^{N}\sqrt{\omega_j}}, \frac{\sqrt{\omega_i}}{\sum_{j=1}^{N}\sqrt{\omega_j}}\}$. The theorem is proved. □

3.2 Optimal Stationary Semi-randomized Scheduling Algorithm

In this subsection, we first consider a stationary semi-randomized scheduling algorithm, which is quite simple to implement in practical. To be noted that, in the stationary randomized scheduling policy, the HAP selects the source node to transfer the energy according to a computed probability, without considering the AoI value $\Delta_i(t)$ and the waiting latency of the packet, *i.e.*, $\Gamma_i(k)$.

Denote the set of stationary randomized scheduling policies as Π_R. For any stationary randomized scheduling policy that $\pi \in \Pi_R$, at each time slot t, source node s_i $(1 \leq i \leq N)$ is selected with a probability $\mu_i \in (0, 1]$. That is, for each source node s_i, $\mathcal{P}(e_i(t) = 1) = \mu_i$. Obviously, for the entire network, it should satisfy the constraint that $\sum_{i=1}^{N} \mu_i \leq 1$. At wireless-powered networks, since a source node s_i can harvest at most $\mathcal{P}_0 \cdot h_i \cdot \tau$ unit of energy at a time slot. Then, the number of time slots that node s_i required to transmit a packet can be computed as $B_i = \lceil \frac{\varepsilon_i}{\mathcal{P}_0 \cdot h_i \cdot \tau} \rceil$. For simplicity, we use $X_i(t)$ to denote the energy status of source node s_i, which belongs to the state space of $\{0, 1, 2, ..., B_i\}$. To be noted that, for each source node s_i, it will just transmit its packet after it has been charged enough energy.

To assist in the charging process, we consider a stationary semi-randomized scheduling policy in this paper. That is, if a source node has been chosen to transfer the energy, then it will continue to be chosen to harvest the energy at the following time slots, until it is charged ready for transmission. For example, if the HAP select the source node s_i to transfer the energy at time slot t, *i.e.*, $e_i(t) = 1$. Then, at the next time slot $t + 1$, the HAP will continue to charge source node s_i, until it is charged ready. Different from the stationary randomized scheduling policy, which selects the source nodes at each time slot randomly, we try to accelerate the charging process for a chosen source node in the stationary semi-randomized scheduling policy.

Let $\mathcal{P}(X_i(t) = B_i)$ be the probability that source node s_i's energy status at time slot t is B_i. Then, we can have the following lemma.

Lemma 1. *For each source node s_i, one can have* $\lim_{t\to\infty} \mathcal{P}(X_i(t) = B_i) = \frac{\mu_i}{1-\mu_i+B_i\mu_i}$, *and* $\lim_{t\to\infty} \mathcal{P}(0 \leq X_i(t) < B_i) = 1 - \frac{\mu_i}{1-\mu_i+B_i\mu_i}$.

Proof. For each source node s_i, let $\mathcal{P}(X_i(t) = n|X_i(t-1) = m)$ $(1 \leq n, m \leq B_i)$ be the state transmission probability. According to the stationary semi-randomized scheduling policy, when $X_i(t) = 0$, the HAP will select the source node s_i to transfer the energy with a probability μ_i. Thus, one can have $\mathcal{P}(X_i(t) = 1|X_i(t-1) = 0) = \mu_i$ and $\mathcal{P}(X_i(t) = 0|X_i(t-1) = 0) = 1 - \mu_i$. If source node s_i is selected at time slot t, *i.e.*, $e_i(t) = 1$, at the next time slot t, the HAP will continue to charge it until it is charged ready. It means $\mathcal{P}(X_i(t) = j|X_i(t-1) = j-1) = 1$ for $j = 2, ..., B_i - 1$. To be noted that, when $X_i(t) = B_i$, it means it has harvested enough energy for transmitting the packet at the beginning of time slot t. As a result, it can continue to be chosen to harvest the energy after it transmits the packet, *i.e.*, $\mathcal{P}(X_i(t) = 0|X_i(t-1) = B_i) = 1 - \mu_i$ and $\mathcal{P}(X_i(t) = 1|X_i(t-1) = B_i) = \mu_i$. Then, the Markov chain of the energy status can be constructed as in Fig. 2, which is an irreducible ergodic chain.

Let $\Upsilon = \{\gamma_1, \gamma_2, ..., \gamma_{B_i}\}$ be the steady-state vector, where γ_j denotes the probability of stabilization in state j, *i.e.*, $\gamma_j = \lim_{t\to\infty} \mathcal{P}(X_i(t) = j)$. Then, according to the Ergodic

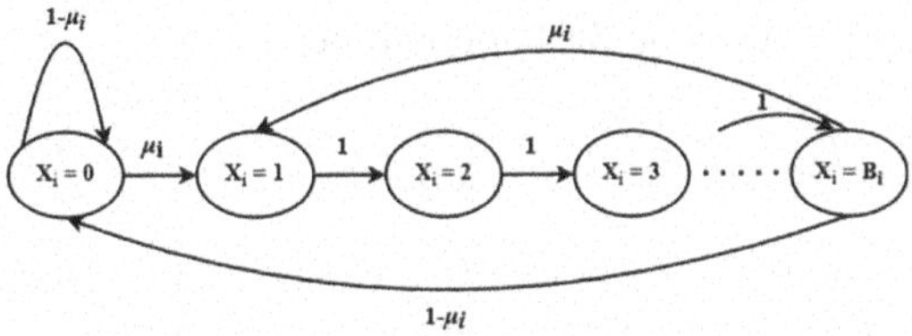

Fig. 2. The Markov chain of the energy status of source node s_i.

Theorem [22], one can have the following equation

$$\begin{cases} \gamma_0 + \gamma_1 + \gamma_2 + \cdots + \gamma_{B_i} = 1. \\ \Upsilon P = \Upsilon. \end{cases}$$

By solving the above equation, one can get $\gamma_{B_i} = \frac{\mu_i}{(1-\mu_i+B_i \cdot \mu_i)}$. It means $\lim_{t\to\infty} \mathcal{P}(X_i(t) = B_i) = \frac{\mu_i}{1-\mu_i+B_i\mu_i}$, and $\lim_{t\to\infty} \mathcal{P}(0 \leq X_i(t) < B_i) = 1 - \frac{\mu_i}{1-\mu_i+B_i\mu_i}$. The lemma is proved. □

Let $\overline{\Delta_{opt}^{SR}}$ denote the optimal expected sum of average AoI under the stationary semi-randomized scheduling policy, and let Π_{SR} denote the set of the stationary semi-randomized scheduling policies. Based on Lemma 1, one can have the following Theorem 2, where the proof is omitted here.

Theorem 2. *Given a network setup (N, p_i, ε_i, ω_i), the closed-form expression of average AoI under the stationary semi-randomized scheduling policy is:*

$$\overline{\Delta_{opt}^{SR}} \triangleq \min_{\pi_r \in \Pi_{SR}} \{\frac{1}{N} \sum_{i=1}^{N} \omega_i(\frac{1}{p_i} - 1 + \frac{1}{\mu_i} + B_i - 1)\}.$$

According to the above Theorem 2, to obtain the optimal $\overline{\Delta_{opt}^{SR}}$, it is equivalent to solve the optimization problem (p_i and B_i are given as the parameters).

$$\textbf{P2}: \quad \min_{\pi_r \in \Pi_{SR}} \frac{1}{N} \sum_{i=1}^{N} \frac{\omega_i}{\mu_i} \tag{6}$$

$$s.t. \sum_{i=1}^{N} \mu_i \leq 1; \tag{7}$$

$$0 < \mu_i \leq 1, 1 \leq i \leq N; \tag{8}$$

Such a problem can be solved by adopting the KKT conditions. Let ϱ_i ($1 \leq i \leq N$) be the Lagrange multiplier for the constraint $0 < \mu_i$, and σ_i ($1 \leq i \leq N$) be the Lagrange multiplier for the constraint $\mu_i \leq 1$. In addition, let ς be the Lagrange multiplier for the constraint $\sum_{i=1}^{N} \mu_i \leq 1$. Then, the Lagrange function $\mathcal{L}(\mu_i, \varrho_i, \sigma_i, \varsigma)$ can be computed as

$$\mathcal{L}(\mu_i, \varrho_i, \sigma_i, \varsigma) = \frac{1}{N} \sum_{i=1}^{N} \frac{\omega_i}{\mu_i} - \sum_{i=1}^{N} \varrho_i\mu_i - \sum_{i=1}^{N} \sigma_i(\mu_i - 1) - \varsigma(\sum_{i=1}^{N} \mu_i - 1),$$

where $\nabla_{\mu_i}\mathcal{L}(\mu_i, \varrho_i, \sigma_i, \varsigma) = 0$ can be further computed as:

$$\nabla_{\mu_i}\mathcal{L}(\mu_i, \varrho_i, \sigma_i, \varsigma) = -\frac{\omega_i}{N{\mu_i}^2} + \varrho_i + \sigma_i + \varsigma = 0.$$

According to the condition $\varrho_i(\mu_i - 1) = 0$, it can be first derived that $\varrho_i = 0$ $(1 \le i \le N)$ since $\mu_i > 0$ for $1 \le i \le N$. Similarly, one can also have that σ_i $(1 \le i \le N) = 0$ for the condition $\sigma_i(\mu_i - 1) = 0$ since $\sum_{i=1}^{N} \mu_i \le 1$.

Then, according to the condition $\nabla_{\mu_i}\mathcal{L}(\mu_i, \varrho_i, \sigma_i, \varsigma) = 0$, one can have $\varsigma = \frac{\omega_i}{N{\mu_i}^2}$. Since $0 < \mu_i \le 1$, then one can have $\varsigma > 0$. As for the condition $\varsigma\left(\sum_{i=1}^{N} \mu_i - 1\right) = 0$, then one can have $\sum_{i=1}^{N} \mu_i - 1 = 0$. Since $\varsigma = \frac{\omega_i}{N{\mu_i}^2}$, which means $\mu_i = \sqrt{\frac{\omega_i}{\varsigma N}}$, then one can also have $\sum_{i=1}^{N} \sqrt{\frac{\omega_i}{\varsigma N}} = 1$.

Then, the optimal solution ς^* and μ_i^* can be computed as

$$\varsigma^* = \Big(\sum_{i=1}^{N} \frac{\sqrt{\omega_i}}{\sqrt{N}}\Big)^2, \mu_i^* = \frac{\sqrt{\omega_i}}{\sum_{j=1}^{N} \sqrt{\omega_j}}.$$

With the optimal μ_i^*, then one can have the optimal AoI under the stationary semi-randomized scheduling policy be

$$\overline{\Delta_{opt}^{SR}} \triangleq \frac{1}{N} \sum_{i=1}^{N} \omega_i\big(\frac{1}{p_i} - 1 + B_i - 1\big) + \frac{1}{N}\Big(\sum_{i=1}^{N} \sqrt{\omega_i}\Big)^2.$$

3.3 The Age-Based Charging Scheduling Algorithm

Although the stationary semi-randomized scheduling policy is easy to be implemented, it ignores the AoI and waiting latency of the source nodes during scheduling. To further reduce the AoI, in this subsection, we will introduce an age-based max-weight charging scheduling algorithm, which takes the AoI and waiting latency of each source node at each time slot t, *i.e.*, $\Delta_i(t)$ and $\Gamma_i(t)$, into account.

The main idea is to employ the Lyapunov Optimization method to design the age-based max-weight charging scheduling policy based on the derived closed-form properties of the average AoI in the aforementioned subsection, by reducing the expected drift of the Lyapunov Function at each time slot t. To develop the age-based max-weight charging scheduling policy, we first construct the Lyapunov Function as follows

$$\mathcal{L}(t) = \mathcal{L}\{\Delta_i(t)_{i=1}^N\} = \frac{1}{N} \sum_{i=1}^{N} \alpha_i \Delta_i(t),$$

where α_i is the parameter which is used to tune the max-weight policy for different network settings [9]. The Lyapunov Drift is defined as

$$\mathbb{D}(\mathcal{L}(t)) = \mathbb{E}[\mathcal{L}(t+1) - \mathcal{L}(t)|\mathbb{S}(t)],$$

where $\mathbb{S}(t) = \{\Delta_i(t)_{i=1}^N, \Gamma_i(t)_{i=1}^N\}$ is the AoI and waiting latency of each source node at time slot t. Let $\chi_i^{\pi}(t)$ be the binary variable which denotes whether s_i transmits packet to the HAP at time t under the scheduling policy π. According to the definition of AoI in Eq. (2), the AoI value of $\Delta_i(t)$ can be represented with $\chi_i^{\pi}(t)$ as follows

$$\Delta_i(t+1) = \chi_i^{\pi}(t) \cdot (\Gamma_i(t)+1) + (1-\chi_i^{\pi}(t))(\Delta_i(t)+1).$$

To be noted that, $t+1-\hbar_i(t)$ in Eq. (2) equals to $\Gamma_i(t)+1$. Then, the Lyapunov Drift $\mathbb{D}(\mathcal{L}(t))$ can be recomputed as

$$\mathbb{D}(\mathcal{L}(t)) = \frac{1}{N}\sum_{i=1}^{N}\alpha_i - \frac{1}{N}\sum_{i=1}^{N}\alpha_i(\Delta_i(t)-\Gamma_i(t))\mathbb{E}[\chi_i^{\pi}(t)|\mathbb{S}(t)].$$

One can find that, the scheduling decision at each time slot t only impacts the second term in the above equation. To minimize the Lyapunov Drift, the proposed age-based max-weight scheduling policy chooses the source node with the maximum value of $\alpha_i(\Delta_i(t)-\Gamma_i(t))$ at each time slot t, where α_i is set $\alpha_i = \frac{\omega_i}{\gamma_{B_i}} = \omega_i(\frac{1}{\mu_i^*}+(B_i-1))$ and $\mu_i^* = \frac{\sqrt{\omega_i}}{\sum_{j=1}^{N}\sqrt{\omega_j}}$.

Now, the age-based max-weight charging scheduling algorithm is introduced. In the following, we will show the proposed algorithm has a bounded approximation ratio.

Theorem 3. *The proposed age-based max-weight charging scheduling algorithm has an approximation ratio of* $max\{4, 2B_{max}\}$*, where* $B_{max} = max\{B_i, 1 \le i \le N\}$.

Proof. Let $\overline{\Delta^{MW}}$ denote the achieved expected sum of average AoI by the proposed age-based max-weight charging scheduling algorithm. First, consider the transmission decision of source node s_i under the stationary semi-randomized scheduling policy π at each time slot t, *i.e.*, $\chi_i^{\pi}(t)$. In this case, the transmission decision of s_i actually is determined by whether its energy status is full, *i.e.*, $X_i(t) = B_i$. Thus, one can have $\mathbb{E}[\chi_i^{\pi}(t)|\mathbb{S}(t)] = \gamma_{B_i}$. Since the age-based max-weight charging scheduling policy minimizes the Lyapunov Drift $\mathbb{D}(\mathcal{L}(t))$, then, one can have

$$\mathbb{D}(\mathcal{L}(t)) \le \frac{1}{N}\sum_{i=1}^{N}\alpha_i - \frac{1}{TN}\sum_{t=1}^{T}\sum_{i=1}^{N}\alpha_i\mathbb{E}\left[\Delta_i(t)-\Gamma_i(t)\right]\gamma_{B_i}. \tag{9}$$

Then, consider the time-average on the interval $[1, .., T]$ and take the expectation with respect to $\mathcal{L}(t)$, one can have

$$\frac{\mathbb{E}(\mathcal{L}(T))-\mathbb{E}(\mathcal{L}(1))}{T} \le \sum_{i=1}^{N}\alpha_i - \frac{1}{TN}\sum_{t=1}^{T}\sum_{i=1}^{N}\alpha_i\mathbb{E}[\Delta_i(t)-\Gamma_i(t)]\gamma_{B_i}. \tag{10}$$

By substituting $\alpha_i = \frac{\omega_i}{\gamma_{B_i}} = \omega_i(\frac{1}{\mu_i^*}+(B_i-1))$ and taking the limit as $T \to \infty$, then one can have

$$\overline{\Delta^{MW}} \le \frac{1}{N}\sum_{i=1}^{N}\omega_i\left(\frac{1}{\mu_i^*}+(B_i-1)\right) + \lim_{T\to\infty}\frac{1}{TN}\sum_{t=1}^{T}\sum_{i=1}^{N}\omega_i\mathbb{E}[\Gamma_i(t)]. \tag{11}$$

Note that, for source node s_i, the packets arrive with a probability p_i. It means the expected waiting latency after transmitting a packet is $\frac{1}{p_i}-1$. Then, one can have $\lim_{T\to\infty}\frac{1}{TN}\sum_{t=1}^{T}\sum_{i=1}^{N}\omega_i\mathbb{E}[\Gamma_i(t] = \frac{1}{N}\sum_{i=1}^{N}\omega_i\left(\frac{1}{p_i}-1\right)$. As a result, one can have

$$\overline{\Delta^{MW}} \le \frac{1}{N}\sum_{i=1}^{N}\omega_i\left(\frac{1}{\mu_i^*}+(B_i-1)\right) + \frac{1}{N}\sum_{i=1}^{N}\omega_i\left(\frac{1}{p_i}-1\right).$$

According to Theorem 1, when given a network setup (N, p_i, ε_i, ω_i), the lower bound can be computed as $\overline{\Delta^{LB}} = \frac{1}{2N}\sum_{i=1}^{N}\omega_i\left(\frac{1}{r_i^{LB}}+1\right)$, where r_i^{LB} is the transmission rate of the source nodes in the lower bound. Thus, one can have

$$\frac{\overline{\Delta^{MW}}}{\overline{\Delta^{LB}}} \le \frac{\frac{1}{N}\sum_{i=1}^{N}\omega_i\left(\frac{1}{p_i}-1+\frac{1}{\mu_i^*}+(B_i-1)\right)}{\frac{1}{2N}\sum_{i=1}^{N}\omega_i\left(\frac{1}{r_i^{LB}}+1\right)} \le \frac{\sum_{i=1}^{N}\omega_i\left(\frac{4}{r_i^M}+2(B_i-2)\right)}{\sum_{i=1}^{N}\omega_i\left(\frac{1}{r_i^{LB}}+1\right)} \le \frac{\sum_{i=1}^{N}\omega_i(4+2r_i^{LB}(B_i-2))}{\sum_{i=1}^{N}\omega_i(1+r_i^{LB})}.$$

Note that, r_i^M is actually the transition rate of source node s_i under the age-based max-weight scheduling policy. Obviously, one can have $r_i^M \le p_i$ and $r_i^M \le \mu_i^*$. Additionally, to derive the approximation ratio, there exists an energy transfer rate $\mu_i \in [0, 1]$ such that $r_i^M = r_i^{LB}$. Substituting r_i^M into the objective function of the stationary randomized scheduling policy yields a large AoI. Based on the aforementioned equation, we can formulate the following conclusion:

- If $B_i - 2 \le 0$ for each source node s_i $(1 \le i \le N)$, then one can have $\frac{\overline{\Delta^{MW}}}{\overline{\Delta^{LB}}} \le 4$.
- Otherwise, if $\exists s_i$ that $B_i - 2 > 0$, then one can have $\frac{\overline{\Delta^{MW}}}{\overline{\Delta^{LB}}} \le 2B_{max}$.

Therefore, the proposed algorithm has an approximation ratio of $max\{4, 2B_{max}\}$. □

4 Conclusions

In this paper, we investigate the first work to optimize the weighted sum of average AoI at wireless-powered networks with stochastic arrivals. Firstly, the lower bound of the achievable AoI with directional energy transferring is derived. Then, to assist in the energy harvesting process of the source nodes, the stationary semi-randomized scheduling policy is introduced. The closed-form expression under the space of such a policy is derived, and the optimal stationary semi-randomized scheduling policy is also proposed. Finally, based on the obtained closed-form expression of average AoI, an age-based max-weight charging scheduling policy is proposed with Lyapunov optimization, which is proved to have a bounded approximation ratio.

References

1. Yates, R.D., Sun, Y., Brown, D.R., et al.: Age of information: an introduction and survey. In: IEEE Journal on Selected Areas in Communications, vol. 39, no. 5, pp. 1183–1210 (2021)
2. Maatouk, A., Kriouile, S., et al.: The age of incorrect information: a new performance metric for status updates. IEEE/ACM Trans. Netw. **28**(5), 2215–2228 (2020)
3. Cai, Z., et al.: Battery-free wireless sensor networks: a comprehensive survey. IEEE Internet Things J. **10**(6), 5543–5570 (2023)
4. Li, J., Guo, S., Liang, W., et al.: Digital twin-enabled service provisioning in edge computing via continual learning. IEEE Trans. Mobile Comput. (2023)
5. Zhang, J., et al.: Task-oriented energy scheduling in wireless rechargeable sensor networks. ACM Trans. Sens. Netw. **19**(4), 1–32 (2023)
6. Li, J., Guo, S., Liang, W., et al.: Mobility-Aware Utility Maximization in Digital Twin-Enabled Serverless Edge Computing. IEEE Transactions on Computers (2024)

7. Cai, Z., et al.: Latency-and-Coverage Aware Data Aggregation Scheduling for Multihop Battery-Free Wireless Networks. IEEE Trans. on Wirel. Commun. (2021)
8. Kadota, I., Modiano, E.: Minimizing the age of information in wireless networks with stochastic arrivals. In: Proceedings of ACM Mobihoc, pp. 221–230. July 2-5 (2019)
9. Kadota, I., Modiano, E.: Age of information in random access networks with stochastic arrivals. In: Proceedings IEEE INFOCOM, Vancouver, BC, Canada, pp. 1–10 (2021)
10. Wang, Q., Chen, H.: Age of information in reservation multi-access networks with stochastic arrivals: analysis and optimization. IEEE Trans. Commun. **71**(8), 4707–4720 (2023)
11. Tripathi, V., Talak, R., Modiano, E.: Information freshness in multihop wireless networks. IEEE/ACM Trans. Netw. **31**(2), 784–799 (2023)
12. Liu, Q., Li, C., Hou, Y.T., et al.: Ao2I: minimizing age of outdated information to improve freshness in data collection. In: Proceedings of IEEE INFOCOM (2022)
13. Dai, H., et al.: Charging task scheduling for directional wireless charger networks. In: IEEE Trans. Mob. Comput. **20**(11), 3163–3180 (2021)
14. Chen, Q., et al.: Joint near-optimal age-based data transmission and energy replenishment scheduling at wireless-powered network edge. In: Proceedings of IEEE INFOCOM (2022)
15. Chen, K., et al.: Distributed energy-adaptive aggregation scheduling with coverage guarantee for battery-free wireless sensor networks. In: Proceedings of INFOCOM, Paris, France (2019)
16. Zheng, H., Xiong, K., Fan, P., et al.: Age-energy region in wireless powered communication networks. In: Proceedings of IEEE INFOCOM WKSHPS, pp. 334–339 (2020)
17. Abd-Elmagid, M.A., Dhillon, H.S., Pappas, N.: AoI-optimal joint sampling and updating for wireless powered communication systems. IEEE Trans. Veh. Technol. **69**(11), 14110–14115 (2020)
18. Hu, H., Xiong, K., Lu, Y., et al.: $\alpha - \beta$ AoI penalty in wireless-powered status update networks. In: IEEE Internet of Things Journal, vol. 9, no. 1, pp. 474–484 (2022)
19. Zheng, H., Xiong, K., Fan, P., et al.: Age of information-based wireless powered communication networks with selfish charging nodes. IEEE J. Sel. Areas Commun. **39**(5), 1393–1411 (2021)
20. Zheng, Y., Hu, J., Yang, K.: Average age of information in wireless powered relay aided communication network. IEEE IoT J. **9**(13), 11311–11323 (2022)
21. Hu, H., et al.: AoI-minimal online scheduling for wireless-powered IoT: a Lyapunov optimization-based approach. IEEE Trans. Green Commun. Netw. **7**(4), 2081–2092 (2023)
22. Sun, S., Wu, W., Fu, C., Qiu, X., Luo, J.: Energy-aware age optimization: AoI analysis in multi-source update network systems powered by energy harvesting. In: Proceedings of IEEE INFOCOM, New York City, NY, USA, pp. 1–10 (2023)
23. He, T., Chin, K.-W., Zhang, Z., Liu, T., Wen, J.: Optimizing information freshness in RF-powered multi-hop wireless networks. IEEE Trans. Wireless Commun. **21**(9), 7135–7147 (2022)
24. Chen, Q., et al.: Peak AoI minimization at wireless-powered network edge: from the perspective of both charging and transmitting. In: IEEE/ACM Transactions on Networking (2023)
25. Chen, Q., et al.: Peak AoI minimization with directional charging for data collection at wireless-powered network edge. In: IEEE Transactions on Services Computing (2023)
26. Chen, Q., Guo, S., Xu, W., et al.: Optimizing average AoI with directional charging for wireless-powered network edge. In: IEEE/ACM 31st International Symposium on Quality of Service (IWQoS), Orlando, FL, USA, pp. 1–10 (2023)
27. Zhou, Z., Yun, Z., Fu, C., Xue, C.J., Han, S.: Maintaining real-time data freshness in wireless powered communication networks. In: IEEE Real-Time Systems Symposium (RTSS), Houston, TX, USA, pp. 166–177 (2020)

Proactive Bi-objective Multi-path Planning for Wireless Sensor Networks

Yamin Wang[1(✉)], Xiaoping Li[2], Shuai Ma[1], Yuan Li[1], Qianfan Jia[1], Yizheng Li[1], Zhiqiang Wu[3], and Qing Feng[3]

[1] China Mobile Research Institute, Beijing, China
261371052@qq.com
[2] China Southeast University, Nanjing, Beijing, China
[3] PKU-Wuhan Institute for Artificial, Wuhan, Beijing, China

Abstract. Optimizing routing paths in IoT Wireless Sensor Networks (WSNs) is crucial for efficient data transmission. The fault tolerance of data transmission is intricately linked to various uncertain variables, encompassing dynamic environmental conditions, equipment malfunctions, and intricate network topologies. Therefore, pre-constructing multiple optimized paths becomes essential to effectively manage uncertainties in real-time scenarios. Nonetheless, prevailing optimization techniques for multi-path routing in wireless sensor networks often neglect the consideration of sensor device reliability and the possible inconsistency between transmission reliability and energy consumption. Moreover, identifying all non-dominated paths for multi-objective optimization problems consumes excessive time. In this study, addressing the optimization objectives of both reliability and energy consumption, we introduce a Fast Bi-objective Multi-path Search (FBMS) algorithm tailored to swiftly generate multiple paths. These paths are strategically disjointed within the decision space to achieve a balanced distribution of transmission loads. The proposed method significantly reduces the time required to identify multiple non-dominated solution paths compared to conventional approaches. The experimental results validate the efficiency and effectiveness of the proposed algorithm.

Keywords: Multi-path Planning · Bi-Objective · Wireless Sensor Networks

1 Introduction

In WSNs, sensor nodes are typically deployed across challenging environments, leading to difficulties in system maintenance and fault detection [1]. These nodes possess constrained resources such as power, computational capabilities, storage, and communication abilities. Upon manual deployment or dispersion via aircraft, these intelligent nodes autonomously create a network, connecting to a base station or sink sensor node, thereby constructing a Directed Acyclic Graph (DAG)

This work is supported by the National Key Research and Development Program of China (No. 2022YFB3305500), the National Natural Science Foundation of China (Nos. U21B2029, 62273089).

Y. Chen et al. (Eds.): COCOON 2024, LNCS 15162, pp. 238–251, 2025.
https://doi.org/10.1007/978-981-96-1093-8_20

comprising sensors and communication paths [2]. However, the dynamic changes in reliability and the energy consumption of sensors always result in changes in the DAG over time, and the optimal path and its cost could change dynamically. Relying solely on one transmission path may result in incomplete and delayed data transmission. Therefore, optimizing multiple routing paths is critical for the reliability of data transmission in WSNs.

Generally, there are two methods for routing paths: reactive routing and proactive routing, respectively. In reactive routing, it takes time to establish the optimal path only when data packets need to be sent. However, due to the complexity of the DAG construction in WSNs, it is challenging to ensure reactive routing to meet real-time requirements. In contrast, proactive routing establishes the optimal path in advance, which can realize fast data transmission. Proactive routing is a fast way to plan one optimal path, but it is difficult to guarantee the effectiveness of one planned path because of the sensor equipment failures, complex topologies, and other uncertain factors. Therefore, considering uncertain real-time transmissions in WSNs, it is necessary to develop multiple optimal paths in advance.

For real-time WSN traffic, it is required to maximize the sensor path reliability and minimize energy consumption in a short time because they greatly influence WSN communications in harsh and complex environments with limited sensor power. As such, a bi-objective optimization issue is considered because the path with the minimum energy consumption is always not the path with the maximum reliability, and vice versa. However, finding all non-dominated solution paths is much more time-consuming for the multi-path planning problem using traditional exact multi-objective search algorithms. Moreover, due to a large number of candidate paths, it is not necessary to search all non-dominated solutions in real-time applications.

Above all, there are two fundamental issues to be considered: (i) For any node intersected with selected candidate paths, the path number should be as few as possible since the failure of the intersected node leads to the involved candidate paths failing, which could greatly decrease the system's reliability. (ii) All selected candidate paths should be diverse as much as possible to avoid the search process being trapped into areas highlighting only one objective. In other words, independent and evenly distributed paths are desirable for the considered situation.

This paper considers the proactive bi-objective multi-path planning problem for real-time transmission in WSNs with unreliability environments and various uncertainties, a FBMS scheme based on decomposition and iterative searching is proposed. The objective is to find multiple optimal paths quickly to maximize reliability and minimize energy consumption. The main contributions are as follows:

- The structure of a WSN is constructed as a DAG. The reliability of sensor equipment and the energy consumption are modeled as weights of a path in the DAG. The objective of the considered problem is to find multiple optimal paths in the DAG.
- Diversities in both decision and objective spaces are considered. The diversity in the decision space means that data transmissions are balanced to avoid

overloading on only a few nodes, which guarantees that the multiple optimal paths are disjointed as much as possible. The diversity in the objective space avoids preferring only optimizing energy consumption objective or reliability objective.
- A FBMS scheme is proposed. The bi-objective search in DAGs is decomposed into several iterative single-objective searches based on the maximum distance between adjacent found solutions in the objective space. Multiple exact optimal paths can be obtained in a shorter time by FBMS compared to traditional accurate bi-objective search algorithms.

The rest of the paper is organized as follows: Sect. 2 reviews the related work. The considered problem is described in Sect. 3. Section 4 details the proposed fast bi-objective multi-path search algorithm for the considered problem. Experimental results of the compared algorithms are shown in Sect. 5, followed by conclusions and future research in Sect. 6.

2 Related Work

WSNs are utilized in military and civilian applications, and their performance is significantly influenced by routing manners, objectives, and the number of routing paths [3,4]. Path planning in WSNs involves either single-objective routing, focusing on a sole objective [5,6], or multi-objective routing, which seeks optimal paths to simultaneously optimize multiple objectives [5]. Existing solutions for multi-objective optimization problems fall into three categories: (i) Calculating the extreme value of each objective independently. (ii) Combining multiple objectives into a single objective through weighting, which poses challenges due to changing user preferences [6]. (iii) Pareto-dominated methods are most studied [7], like NSGAIII [8], SPEA2, MOA* and NAMOA* [9], which provide either time-consuming approximate optimal paths or accurate solutions with a rapid increase in computation time for large-scale wireless sensor networks. In summary, these methods are not suitable for multi-objective path routing problems.

In WSNs, routing paths can be single or multiple. Single path routing is straightforward but lacks fault tolerance and reliability. On the other hand, multi-path routing establishes multiple transmission paths, offering advantages such as reduced latency, improved network reliability, load balancing, and bandwidth aggregation [10]. Multi-path routing methods including Split Multipath Routing (SMR), Ad hoc On-demand Multipath Distance Vector (AOMDV) protocols and others are mostly single objective optimization or changing multi-objective to single objective optimization by weighted sum method.

3 Problem Formulation

Firstly, dominance is defined as follows:

Definition 1. *A vector* $\vec{u}= (u_1, u_2, \cdots, u_k)$ *dominates* $\vec{v} = (v_1, v_2, \cdots, v_k)$ *(denoted by* $\vec{u} \prec \vec{v}$ *) if* $\vec{u}$ *is partially less than* $\vec{v}$, *i.e.*, $\forall i \in \{1, 2, \cdots, k\}$, $u_i \leq v_i \bigwedge \exists j \in \{1, 2, \cdots, k\}$, $u_j < v_j$.

3.1 Problem Description

In this study, we assume that number of sensors $N = \{n_0, n_1, \cdots, n_{|N|-1}\}$ are deployed in a field. This assumption is based on the premise that the information detected by these sensors can be effectively communicated to the base station or sink sensor node. The WSN is regarded as a directed graph composed of sensor nodes. Each sensor n_i has position (x_i, y_i). Each sensor node n_i has a sensing range d_i and uses time t_i. A node is covered by n_i if it is within the radius of node n_i. The considered proactive bi-objective multi-path planning is finding multiple accurate optimal paths from the starting sensor node to the sink node in the directed graph. The objective is to minimize energy consumption and maximize reliability of path from the start sensor to the goal sensor.

Let G be a finite directed graph $G =< N, A, \vec{c} >$ with a set of $|N|$ sensor nodes and a set of $|A|$ arcs. Node n_i is a wireless sensor with usage time t_i. Node $\zeta = n_0$ and γ denote the starting sensor and the sink node, respectively. If node n_j is in the transmission range of sensor node n_i, arc (n_i, n_j) will exist. Arc (n_i, n_j) is labeled with a positive cost $\vec{c}(n_i, n_j) = (c_1, c_2)$ in which $c_1 \in R^+$ and $c_2 \in R^+$. c_1 is the energy consumption for direct transmission from sensor node n_i to sensor node n_j. It is composed of the energy consumption of sensor node n_i and n_j. c_2 is the reliability related cost of arc (n_i, n_j). It is calculated based on the reliability of sensor node n_j. Let P be a path that is a sequence of nodes going from the start node $s = n_0 \in N$ to some other node $n_k \in N$, i.e., $n_0 \rightarrow n_{[1]} \rightarrow n_{[2]} \rightarrow \ldots \rightarrow n_{[j]} \rightarrow n_k$. $n_{[i]} \in N$ is the $(i+1)^{th}$ node on the path P and $(n_{[i]}, n_{[i+1]}) \in A$ for all $0 < i < j$, $(n_0, n_{[1]}) \in A$, and $(n_{[j]}, n_k) \in A$. $\vec{g}(P) = (g_1(P), g_2(P))$ is the cost of path P. The cost $g_1(P)$ is the sum of the energy consumption costs of its arcs, i.e., $g_1(P) = \sum_{i=0}^{k-1} c_1(n_{[i]}, n_{[i+1]})$. $g_2(P)$ is the sum of the reliability costs of its nodes, i.e., $g_2(P) = \sum_{i=0}^{k} c_2(n_{[i]}, n_{[i+1]})$.

The considered multiple paths planning problem consists of finding the set of nondominated cost paths from the sensor node $\zeta = n_0$ to the sink node γ, and these paths have the following two characteristics: (i) They are diverse in the objective space to balance the two objectives. In other words, these solutions are uniformly distributed on the Pareto front. (ii) The solution paths are disjointed as much as possible for the reliable communication and load balance in WSNs, i.e., they are diverse in the decision space. Paths satisfying the diversity in solution space may not necessarily satisfy the diversity in objective space, and vice versa. Therefore, paths satisfying both of these conditions simultaneously need to be found.

For the problem in this study, we consider the following assumptions: (1) The WSN (graph G) is locally finite due to the finite number of sensor nodes. (2) The cost vector of each arc is dominated by vector $(0,0)$, that is, $(0,0) \prec \vec{c}(n_i, n_j)$ where $(n_i, n_j) \in A$ and $\vec{c}(n_i, n_j) = (c_1, c_2)$. (3) The heuristic function $H(n_i)$ is **admissible**. That is, for each nondominated path $P^*_{(n_i,\gamma)} = (n_i, \cdots, \gamma)$ from node n_i $(n_i \in N)$ to the sink node γ, there always exists an estimated cost vector $\vec{h}(n_i)$ in $H(n_i)$ satisfies $\vec{h}(n_i) \preceq \vec{h}^*(n_i)$, assuming $\vec{h}^*(n_i)$ $(\vec{h}^*(n_i) \in H^*(n_i))$ to be the cost vector of path $P^*_{(n_i,\gamma)}$. (4) Sensor nodes are statistically independent and identically distributed.

3.2 Energy Consumption Calculation

In this work, the WSN energy consumption model consists of three parts: sending data Tx, receiving data Rx, and processing unit. To calculate the energy consumption for communication, we applied the following model [11]:

$$E_{TX}(u, d) = u \times E_{trans} + u \times \epsilon_{amp} \cdot d^{\tau} \tag{1}$$

Where E_{TX} represents the energy required to transmit a packet of size u bits over distance d, E_{trans} is the energy required for the communication unit to process a single bit, τ is the path loss exponent, ϵ_{amp} is the transceiver's energy dissipation described by:

$$\epsilon_{amp} = \frac{\frac{SN}{N_d} \cdot NF_{Rx} \cdot N_0 \cdot BW(\frac{4\pi}{\lambda})^{\tau}}{G_{ant} \cdot \eta_{amp} \cdot R_{bit}} \tag{2}$$

Where$\frac{SN}{N_d}$ is the signal to noise ratio at the receiver, NF_{RX} is the receiver noise figure, N_0 is the noise power spectra density, BW is the channel noise bandwidth, λ is the wavelength in meters, G_{ant} is the antenna gain, η_{amp} is the transmitter efficiency and R_{bit} is the channel data rate in bits per second. Here, we assume the transmitting and receiving processes require the same amount of energy; The cost $c_1(n_i, n_j)$ is set for each arc (n_i, n_j) by Eq. (3) if there exists arc (n_i, n_j) in the directed graph G.

$$c_1(n_i, n_j) = \begin{cases} 2PW_p \times t_p + 2E_{TX}(u, d), if \ j = |N| - 1 \\ PW_p \times t_p + 2E_{TX}(u, d), \quad\quad otherwise \end{cases} \tag{3}$$

where PW_p is the power of the process unit and t_p is the processing time.

3.3 Reliability Calculation

The reliability of data transmission is related not only to the reliability of a wireless sensor device node but also to the reliability of the path constructed by wireless sensor devices. In this paper, a 2-parameter Weibull distribution model for each sensor node is applied and constructed paths in the process of optimal path searching. The probability density function of Weibull distribution is given by $f(t) = \frac{m}{\eta}(\frac{t}{\eta})^{m-1} e^{-(\frac{t}{\eta})^m}$, where $t > 0$, m is the shape parameter, and η is the scale parameter. The reliability $r(t)$ of sensor equipment with the using time t can be given by $r(t) = e^{-(\frac{t}{\eta})^m}$.

The details of sensor equipment path reliability calculation is shown as follows: Assumed that the path P is composed of wireless sensor nodes n_0, n_1, $\cdots$, $n_i = \gamma$, and the equipment failure rate of each sensor node is independent and identically distributed. Then, the reliability of path P, denoted by R_P, is determined by the sensor equipment reliability of each node on the path. R_P is calculated as:

$$R_P = \prod_{n_j \in \{n_0, n_1, \cdots, n_i\}} r(t_j) = \prod_{n_j \in \{n_0, n_1, \cdots, n_i\}} e^{-(\frac{t_j}{\eta})^{m_j}} \tag{4}$$

where t_j, m_j and $r(t_j)$ are the usage time, the shape parameter, and the reliability of sensor node n_j, respectively.

The reliability cost $c_2(n_i, n_j)$ is set for each arc (n_i, n_j) by Eq. (5) if there exists arc (n_i, n_j) in the directed graph G. According to whether the arc connects the head node or tail node, $c_2(n_i, n_j)$ is described in the following four situations.

$$c_2(n_i, n_j) = \begin{cases} rc(n_i) + \frac{1}{2}rc(n_j) & i = 0 \ and \ j \neq |N| - 1 \\ \frac{1}{2}rc(n_i) + rc(n_j) & i \neq 0 \ and \ j = |N| - 1 \\ rc(n_i) + rc(n_j) & i = 0 \ and \ j = |N| - 1 \\ \frac{1}{2}rc(n_i) + \frac{1}{2}rc(n_j) & otherwise \end{cases} \tag{5}$$

Where $rc(n_j)$ is regarded as the reliability-related cost of sensor node $n_j \in N$, $rc(n_j) = -ln(\frac{t_j}{\eta})^{m_j}$.

Therefore, the objective of minimizing energy consumption and maximizing the reliability of the path from the start sensor to the goal sensor is given:

$$\min_{P(\varsigma,\gamma)} \sum_{(n_i,n_j) \ on \ P_{\varsigma,\gamma}} (c_1(n_i, n_j), c_2(n_i, n_j)) \tag{6}$$

4 Proposed FBMS Algorithm

When the number of sensor nodes increases, the number of nondominated solution paths also increases. However, WSN route planning does not need many candidate solution paths in practice, and several optimal paths with diversity are sufficient for nondominated solution pathfinding. In this paper, we propose FBMS algorithm for fast constructing multiple paths considering optimization of both reliability and energy consumption.

4.1 Framework of the FBMS Algorithm

Initially, for convenient description, $\lambda-$*shortest* path, $\lambda-$*minimum* and $\lambda-$*cost* are defined as follows:

Definition 2. *Let P be a path from node n_i to node n_j and the cost of path P be $\vec{c}$. $C_{i,j}$ is the set of path costs from node n_i to node n_j. When $\vec{c} = \arg\min_{c' \in C_{i,j}} (\lambda c_1' + (1-\lambda)c_2')$, P is $\lambda -$* **shortest** *path from node n_i to node n_j, $\vec{c}$ is $\lambda -$* **minimum** *cost vector within $C_{i,j}$ and $c_\lambda = \lambda c_1 + (1-\lambda)c_2$ is the minimum $\lambda-$***cost** *of path from node n_i to node n_j.*

The proposed FBMS algorithm divides the bi-objective path optimization problem into multiple iterated single-objective path optimization problems. The objective is to minimize the f_1 and f_2. In each search, the objective is obtained by calculating the weighted sum of all objectives, and each search is executed by a single objective search algorithm. One or more minimum cost paths are found during each iterated search process. The steps of the FBMS method shown in Algorithm 1 are as follows: Firstly, two or more solutions are obtained by Algorithm 2 to minimize cost f_1 and f_2. Other solutions are searched using the information of previously found solutions, then repeat the above process. Next, two adjacent found solutions with maximum distance are selected to compute weight λ. Subsequently, a new single objective $f = \lambda f_1 + (1 - \lambda) f_2$ is obtained. Then, single objective path searches are implemented repeatedly based on previous search results. In this process, to obtain multiple paths that are disjointed as much as possible, the intersection degree is set for each node. The above process is repeated until no new nondominated solution path is found or the running time is out of range.

The detail of computing weight λ is followed: As seen from Fig. 1, a line L with a fixed normal vector is to intersect with the objective space Ω. When the intercept on the axis is minimum, the line L intersects with the objective space Ω at one or more points known as nondominated solution cost vectors. let $\vec{c}^1$ and $\vec{c}^2$ be two nondominated solution cost vectors, where $\vec{c}^1 = (c_1^1, c_2^1)$, $\vec{c}^2 = (c_1^2, c_2^2)$ and $c_1^1 < c_1^2$. Let k ($k \leq 0$) be the slope in two dimensions of f_1 and f_2 (i.e., the objective space Ω). k is equal to the slope of the line obtained by connecting point $\vec{c}^1$ and point $\vec{c}^2$, $k = (c_2^2 - c_2^1)/(c_1^2 - c_1^1)$. Denote line L with $f_2 = k \times f_1 + b$ where $b > 0$. Let line L intersect with the objective space Ω. When the intercept reaches the minimum b_{min}, the line L and the objective space Ω intersect at the point $\vec{c}$. Obviously, $\vec{c}$ is one nondominated solution cost vector between $\vec{c}^1$ and $\vec{c}^2$.

Considering to obtain minimum b_{min}, it is given by

$$\min_{(f_1,f_2)\in\Omega} \quad b = -k \times f_1 + f_2 \tag{7}$$

the objective is also seen as Eq. (8).

$$\min_{(f_1,f_2)\in\Omega} \quad f = \lambda f_1 + (1 - \lambda) f_2 \tag{8}$$

According to Eq. (8) and $0 < \lambda < 1$, the objective can be described as Eq. (9).

$$\min_{(f_1,f_2)\in\Omega\Omega} \quad f/(1 - \lambda) = \lambda f_1/(1 - \lambda) + f_2 \tag{9}$$

Because the objective of both Eq. (7) and Eq. (9) are finding the nondominated solution cost vectors, b can be considered as $f/(1 - \lambda)$, $-k$ can be considered as $\lambda/(1 - \lambda)$. That is, $b = f/(1 - \lambda)$ and $-k = \lambda/(1 - \lambda)$. It is known that $\lambda = k/(k - 1)$ because of $-k = \lambda/(1 - \lambda)$. Eq. (10) is gotten for $k = (c_2^2 - c_2^1)/(c_1^2 - c_1^1)$.

$$\lambda = (c_2^2 - c_2^1)/(c_2^2 - c_2^1 - c_1^2 + c_1^1) \tag{10}$$

Algorithm 1 shows the steps of the FBMS algorithm, in which the bi-objective minimum cost paths search is transformed into multiple single objective searches

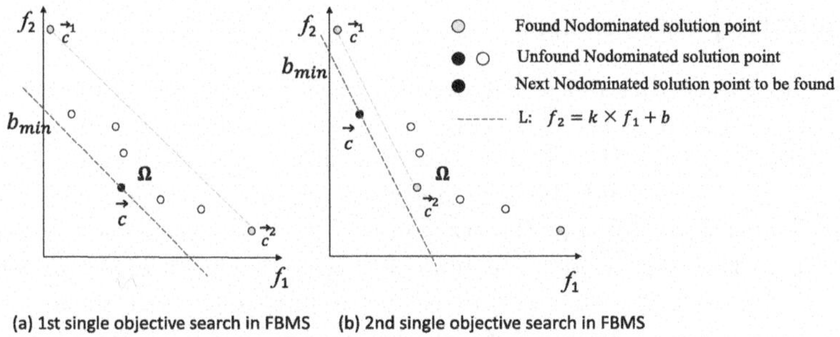

Fig. 1. Diagrammatic sketch of computing weight λ

according to lines 10–12. The single objective search is executed by the SS algorithm shown as Algorithm 2. The heuristic value $h(n, \lambda)$ is calculated by the bi-objective heuristic cost vector $\vec{h}(n)$. $\vec{c}_1$ and $\vec{c}_2$ are two adjacent found solution costs with maximum distance. The objective space between $\vec{c}_1$ and $\vec{c}_2$ has not been searched so far. Based on $\vec{c}_1$ and $\vec{c}_2$, the weight λ is updated by the equations on line 10 of Algorithm 1 to ensure that the obtained optimal path costs are uniformly distributed in the objective space. At the same time, the constraint of path intersection degree is considered on line 12 of Algorithm 1. This makes the obtained solution paths disjointed as much as possible.

Algorithm 1: FBMS

1: **for** $n \in N \setminus \{\zeta, \gamma\}$ **do**
2: $b(n) \leftarrow 0$;
3: **end for**
4: Initialize path intersection threshold b^*; $P \leftarrow \emptyset$; $flag \leftarrow true$.
5: $\vec{c}^1 \leftarrow$ **SS**$(1,+\infty,b^*,\vec{b})$; $\vec{c}^2 \leftarrow$ **SS**$(0,+\infty,b^*,\vec{b})$; $C \leftarrow \{\vec{c}^1, \vec{c}^2\}$.
6: **while** $flag = true$ **do**
7: $\lambda \leftarrow (c_2^1 - c_2^2)/(c_2^1 - c_2^2 - c_1^1 + c_1^2)$.
8: $c_\lambda \leftarrow \lambda \times c_1^1 + (1 - \lambda) \times c_2^1$.
9: $\vec{c}^*$, **P** $\leftarrow$ **SS**$(\lambda, c_\lambda, b^*, \vec{b})$.
10: $C \leftarrow \{\vec{c}^*\} \bigcup C$.
11: **for** node n on P (except for node ζ and γ) **do**
12: $b(n) \leftarrow b(n) + 1$.
13: **end for**
14: **if** each objective space part between adjacent vectors in set C has been searched **or** $\forall n \in Successors(\zeta)$, $b(n) \geq b^*$ **then**
15: $flag \leftarrow false$.
16: **else**
17: In set C, set two adjacent solution costs with the biggest distance in which the objective space has not been searched as $\vec{c}^1$ and $\vec{c}^2$
18: **end if**
19: **end while**

4.2 Single Objective Search

Different from the traditional A* algorithm, in the proposed Algorithm 2, there are three data structures for each node n_i in the search subgraph SG: i) intersection degree $b(n_i)(n_i \in N \setminus \{\zeta, \gamma\})$, i.e., the number of paths passing node n_i (note, the start node ζ and the sink node γ have no limitation of intersection degree), ii) λ-cost $g(n_i, \lambda)$ from the start node ζ to node n_i, iii) cost vector $\vec{g}(P)$ of path P from the start node ζ to node n_i. $b(n_i)$ is the intersection degree of node n_i, $g_{min}(n_i, \lambda)$ is $\lambda - minimum$ cost of the path $P_{(\zeta,n_i)}$ from the start node ζ to the current node n_i in graph G. Additionally, list $OPEN$ contains partial solution paths that can be further expanded in the search of the Algorithm 2. In $OPEN$, each element is a vector $(n_i, \vec{g}(n_i)$,$g(n_i, \lambda), f(n_i, \lambda))$. For fast selection, $OPEN$ is stored in a heap structure of the path with minimum $f(n_i, \lambda)$.

The Algorithm 2 keeps selecting (line 3–4) and expanding subpaths (line 5–22) until $OPEN$ list is empty, which means the optimal path(s) is found. In each round of the cycle, firstly, the path P with the minimum $f(n_i, \lambda)$ is selected to expand; secondly, $h(n_i, \lambda)$, $g(n, \lambda)$ and $f(n_i, \lambda)$ of paths constituted by path P and its successive nodes are calculated in the expansion process. Function $g(n_i, \lambda)$ denotes the $\lambda-$cost of the path from the start node ζ to node n_i, the heuristic function $h(n_i, \lambda)$ estimates the $\lambda-$cost of the path from node n_i to the goal node γ. Heuristic functions can be devised for particular problems depending on their structure. When the subpaths do not meet the constraints or cannot be expanded to the optimal solution path, it is pruned to reduce the search time. Let $P_{(\zeta,m)}$ be a path from node ζ to node m constructed through the subpath expansion. λ-cost $g(m, \lambda)$, the intersection degree $b(m)$, and an estimated cost $f(m, \lambda)$ obtained by $f(m, \lambda) = g(m, \lambda) + h(m, \lambda)$, are calculated for the path $P_{(\zeta,m)}$ during the expansion. As shown in line 9 to 10, when $g(m, \lambda) > g_{min}(m, \lambda)$, path $P_{(\zeta,m)}$ is not the minimum λ-cost path to node n_i; therefore, it cannot be the subpath of λ-minimum solution path. $b(m) > b^*$ indicates that the number of paths going through node m exceeds threshold b^*, and node m can no longer be used.

4.3 Diversity of Nondominated Solution Paths

The diversity of solution paths in this paper includes two meanings. 1) The diversity of solutions in the objective space: To take simultaneous reliability maximization and energy consumption minimization into consideration fairly, the optimal solutions are expected to be uniformly distributed on the Pareto front. To ensure the diversity of solution paths in the objective space, two adjacent solution vectors with the maximum distance are selected in these nondominated path solutions found so far. Based on two selected vectors, the objective weight λ is calculated for next time search. 2) The diversity in the decision space: In order to load balance during data transmission in WSNs, candidate transmission paths are hoped to be disjointed as much as possible.

Algorithm 2: SS$(\lambda, c_\lambda, b^*, \vec{b})$

Input: Weight λ, minimum λ-cost found so far c_λ, path intersection threshold b^*, the number vector of times sensor node on the optimal path $\vec{b} = (b(n_1), \cdots, b(n_{|N|-2}))$.
Output: Path starting from node n_i to the sink node n_j, One cost vector of path P.

Initialize the OPEN list.
while $OPEN \neq \emptyset$ **do**
  Select the path $(n, \vec{g}(n), g(n,\lambda), f(n,\lambda))$ in $OPEN$ with the lowest $f(n,\lambda)$.
  Delete $(n, \vec{g}(n), g(n,\lambda), f(n,\lambda))$ from $OPEN$.
  if $n \neq \gamma$ **then**
    Generate all successor sensor nodes set M by expanding node n and calculate $g(m,\lambda)$, $h(m,\lambda)$ and $f(m,\lambda)$.
    for $m \in M$ **do**
      $f(m,\lambda) \leftarrow g(m,\lambda) + h(m,\lambda)$.
      if $g(m,\lambda) < g_{min}(m,\lambda)$ **and** $b(m) < b^*$ **then**
        Eliminate the element $(m, \vec{g}'(m), g'(m,\lambda), f'(m,\lambda))$ from $OPEN$ with $g'(m,\lambda) < g(m,\lambda)$ and remove the arc (m', m) $(m' \in Predecessor(m))$ from the path with $g'(m,\lambda)$ in SG.
        $g_{min}(m,\lambda) \leftarrow g(m,\lambda)$.
        Set the point from node m to node n with cost $\vec{g}(m)$ in search graph SG.
        Insert tuple $(m, \vec{g}(m), g(m,\lambda), f(m,\lambda))$ into list $OPEN$.
        if $m = \gamma$ **and** $g(m,\lambda) < c_\lambda$ **then**
          $c_\lambda \leftarrow g(m,\lambda)$.
        end if
      end if
    end for
  else
    $\vec{c^*} \leftarrow \vec{g}(n)$; insert $\vec{g}(n)$ into set C.
  end if
end while
Construct the subgraph of SG by backtracking from node γ to start node ζ according to λ-cost c_λ.
return Path $P_{(\zeta,\gamma)}$ and cost vector $\vec{c}$;

5 Experimental Results

Given the lack of prior research on proactive bi-objective multi-path planning in WSNs, we compare the efficiency of our proposed FBMS algorithm with the commonly used NAMOA* algorithm. Similar to some variants of A* [9], we compare the algorithms on bi-dimensional grids with $s \times s$ nodes. Each node denotes a sensor node. Node n_i (identified by its coordinate (x, y)) has several neighbors as successors. Each node on the angle points has two neighbors, each node on the edges has three, and each inside node has four. Therefore, there are $4s^2 - 4s$ arcs in the grid. Each arc cost c_1, representing the energy consumption, is randomly generated with a uniform distribution in the range [1,10]. Each sensor node n_i has a reliability value $r(n_i)$ in the range [0.01, 1). When $r(n_i) = 0.01$, the reliability related cost $rc(n_i)$ is about 4.6. When $r(n_i) = 1$, the reliability related cost $rc(n_i)$ is 0. Therefore, each cost c_2 of arcs from the start node ζ or to the sink node γ, meaning the reliability factor, is randomly created with a uniform distribution in the range (0, 6.9) according to Eq. (5). The cost of other acrs c_2 is randomly generated with a uniform distribution in the range (0, 4.6). The involved algorithms search nondominated solution paths from the start node (0,0) to the sink node $(s-1, s-1)$. An extreme case is that the longest path has s^2 nodes. The proposed FBMS algorithm is compared to the NAMOA*

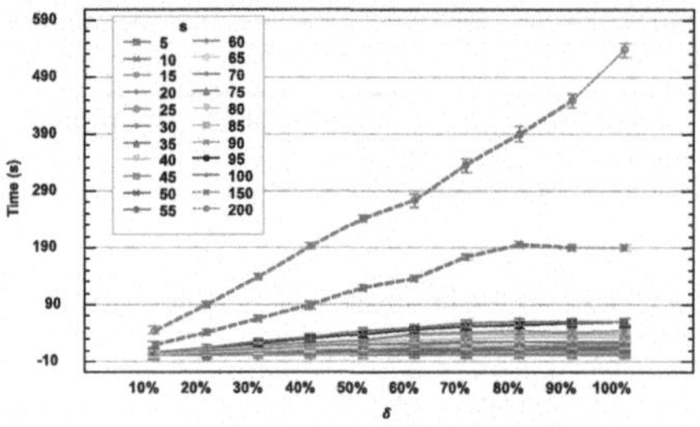

Fig. 2. Interactions between parameter δ and *Time* 95.0% confidence level Tukey HSD intervals on random instances.

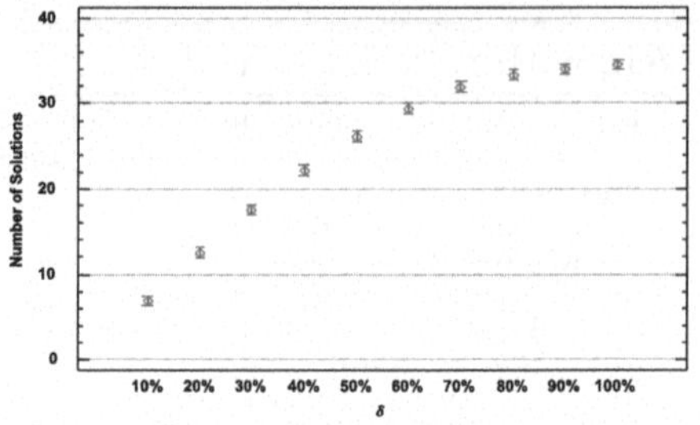

Fig. 3. Means plot and 95.0% confidence level Tukey HSD intervals for parameter δ on random instances

algorithm with $s \in \{5,\ 10,\ \cdots, 100,\ 150,\ 200\}$. The problem parameter s has 22 values. Five instances are generated for each $s \times s$. Therefore, 110 instances in total are tested. For each node, the heuristic value $\vec{h}$ is set as $(0, 0)$ for a fair comparison. All algorithms are coded in Visual C++ Community 2019 and run on a computer with Windows 10 (64 bits), 8 GBytes RAM and Intel(R) Core(TM) i5-8250U running at CPU 1.60 GHz.

5.1 Parameter Analysis

The proposed FBMS algorithm has only one parameter, the maximum intersection degree of solution paths b^*. Because the range of parameter b^* is greatly related to the problem size parameter s, b^* is set as $\delta \times s$ where δ is the intersection degree parameter. δ is set as $10\% \leq \delta \leq 100\%$ and the step is 10%. That is to say, the parameter δ has ten values $\{10\%, 20\%, \cdots, 100\%\}$. The FBMS algorithm runs 10 times for each instance. Therefore, there are $10 \times 110 = 1100$ experimental results about the FBMS algorithm. After carrying out the statistical analyses, we show the means plot and 95.0% confidence level Tukey Honest Significant Difference (HSD) intervals in Fig. 2~5. Overlapping confidence intervals imply statistical insignificance in the corresponding averages, in another word, the observed differences in the overlapped means is not statistically significant at the indicated confidence level.

Figure 2 shows that a larger intersection degree δ implies longer CPU time for different parameters δ. When δ is the same, the running time increases quickly along with the problem size $s \times s$. Especially when the problem size is greater than 100×100, the running time increases more rapidly. It is seen from Fig. 3 and Fig. 4 that the changes in parameters s and δ affect the number of nondominated solution paths. It is obtained from Fig. 3 that a larger parameter δ implies more nondominated solutions when $\delta < 80\%$. When $\delta \geq 80\%$, the number of nondominated solutions is stable. Therefore, the intersection degree parameter δ of the FBMS algorithm can be set to the range $[10\%, 80\%]$. Figure 4 shows that a larger problem size s implies more nondominated solutions when the parameter δ is the same. And a bigger parameter δ implies more nondominated solutions

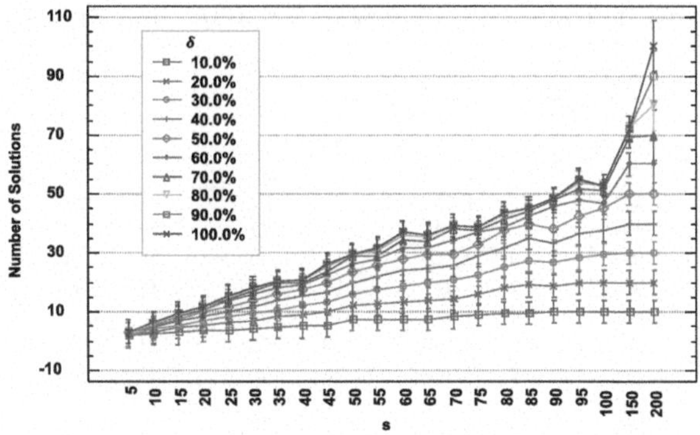

Fig. 4. Interactions between parameter δ and s 95.0% confidence level Tukey HSD intervals on random instances

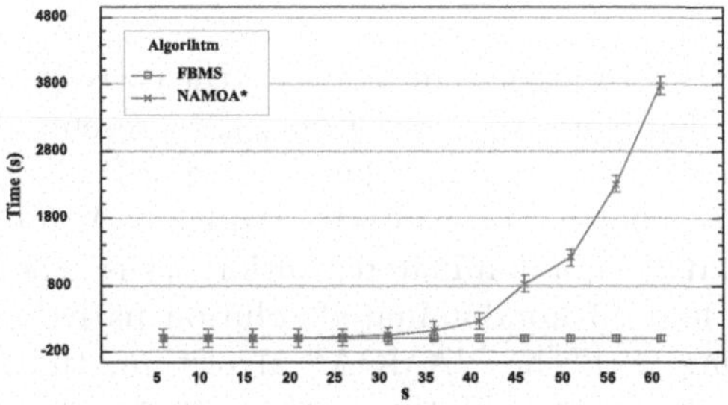

Fig. 5. Interactions between the compared algorithms with $s \leq 60$ and 95.0% confidence level Tukey HSD intervals.

for the same size problems. A larger problem size means a greater change of nondominated solution number for different δ.

5.2 Algorithm Comparison

Because the NAMOA* algorithm cannot be finished in one hour when $s \geq 60$, we analyze the time of the NAMOA* algorithm and the FBMS algorithm in Fig. 5. It shows the interactions between the compared algorithms with $s \leq 60$ and 95.0% confidence level Tukey HSD intervals. Compared with the NAMOA* algorithm, the FBMSA algorithm obtains multiple nondominated solution paths with diversity in a much shorter time especially when the problem size is greater than 30 × 30. The FBMS algorithm demonstrates a more substantial advantage over the $NAMOA^*$ algorithm as the problem scale increases.

Table 1 shows the comparison of CPU time and the number of solutions between the NAMOA* algorithm and the FBMSA algorithm for all 12 cases of $s \times s$. k is the number of solutions found by the NAMOA* algorithm. k_δ is the solution number found by the FBMSA algorithm where δ is the intersection degree parameter and the maximum intersection degree of solution paths b^* is

Table 1. Comparison between NAMOA* and FBMSA

s×s	NAMOA*		FBMSA									
	k	$Time(s)$	$k_{20\%}$	$Time_{20\%}(s)$	$k_{40\%}$	$Time_{40\%}(s)$	$k_{60\%}$	$Time_{60\%}(s)$	$k_{80\%}$	$Time_{80\%}(s)$	k_{agv}	$Time_{agv}(s)$
5×5	5	**0.02**	2	0.50	3	0.78	3	0.76	3	0.77	3	0.70
10×10	17	**0.26**	3	0.77	5	1.18	6	1.43	6	1.50	5	1.22
15×15	24	**1.10**	5	1.24	7	1.71	8	2.07	9	2.24	7	1.82
20×20	47	7.19	5	1.45	8	2.10	11	2.71	12	3.10	9	**2.34**
25×25	61	17.40	6	1.87	10	2.42	14	3.29	15	3.84	11	**2.86**
30×30	65	39.10	7	1.96	12	3.07	16	4.02	18	4.57	13	**3.40**
35×35	97	112.99	8	2.43	14	3.53	18	4.66	20	5.55	15	**4.04**
40×40	122	259.09	9	2.66	16	4.29	19	5.58	21	6.29	16	**4.71**
45×45	161	824.30	10	3.17	17	5.08	23	7.25	25	8.90	19	**6.10**
50×50	176	1212.29	12	4.22	20	6.68	26	9.24	29	10.95	22	**7.77**
55×55	235	2309.76	13	5.07	22	8.47	28	10.71	30	13.08	23	**9.34**
60×60	235	3497.81	13	6.56	24	9.50	32	13.46	36	16.06	26	**11.40**

− : The algorithm did not finish in one hour.

equal to $\delta \times s$ in the FBMSA algorithm. $Time_{\delta}$ is the corresponding CPU running time of the FBMSA algorithm. k_{agv} and $Time_{agv}$ represents the average number of solutions and the average running time of the FBSP algorithm, respectively. For example, for the problems in grid 60 × 60, the NAMOA* algorithm costs about 3497.81 s to find 235 nondominated solution paths. The FBMSA algorithm only costs 6.56 s to find 13 nondominated solution paths with diversity when $\delta = 20\%$. Comparing with the NAMOA* algorithm, the FBMSA algorithm obtains multiple nondominated solution paths with diversity in a much shorter time. The solution number found by the FBMSA algorithm generally meets the demand of wireless sensor network path quantity.

6 Conclusions

In this paper, the proactive bi-objective multi-path planning problem is studied for real-time traffic in WSNs. Addressing the uncertainty inherent in real-time scenarios, this study models sensor equipment reliability and energy consumption as path costs within a directed graph, thereby constructing multiple optimization paths. Diversity control strategies are designed in both the decision space and the objective space. By decomposing the multi-objective search into multiple iterative searches, the proposed FBMS algorithm obtains distinct non-dominated paths within a short time, which is almost impossible by traditional exact multi-objective search algorithms. The search efficiency is improved by reducing the search scope using the pruning strategy to the iterative search process. Experimental results demonstrate efficiency and effectiveness of the proposed FBMS algorithm compared to the classical NAMOA* algorithm.

References

1. Al-Tous, H., Barhumi, I.: Differential game for resource allocation in energy harvesting wireless sensor networks. IEEE Trans. Green Commun. Netw. **4**(4), 1165–1173 (2020)
2. Revanesh, M., Acken, J.M., Sridhar, V.: DAG block: trust aware load balanced routing and lightweight authentication encryption in WSN. Future Gener. Comput. Syst. **140**, 402–421 (2023)
3. Kumar, A., Jain, V.K., Bhattacharya, P.P.: Shortest path algorithms for sensor node localization for internet of things. In: IEEE 2020 International Conference on Emerging Trends in Communication, Control and Computing, Lakshmangarh, Sikar, India, pp. 1–5 (2020)
4. Judith, G.M.C., Geetha, V.: Energy optimization with higher information quality for SHM application in wireless sensor networks. IEEE Sens. J. **19**(9), 3513–3520 (2019)
5. Wang, Y.-C., Chen, K.-C.: Efficient path planning for a mobile sink to reliably gather data from sensors with diverse sensing rates and limited buffers. IEEE Trans. Mob. Comput. **18**(7), 1527–1540 (2019)
6. Wang, Y., Li, X., Ruiz, R.: A fast algorithm for finding the bi-objective shortest path in complicated networks. In: 2018 IEEE 22nd International Conference on Computer Supported Cooperative Work in Design (CSCWD), Nanjing, China, pp. 104–109 (2018)
7. Peng, H., Wang, C., Han, Y., Xiao, W., Zhou, X., Wu, Z.: Micro multi-strategy multi-objective artificial bee colony algorithm for microgrid energy optimization. Future Gener. Comput. Syst. FGCS **131**, 59–74 (2022)
8. Deb, K., Jain, H.: An evolutionary many-objective optimization algorithm using reference-point-based nondominated sorting approach, part i: Solving problems with box constraints. IEEE Trans. Evol. Comput. **18**(4), 577–601 (2014)
9. Mandow, L., Pérez de la Cruz, J.-L.: Multiobjective A* search with consistent heuristics. J. ACM (JACM) **57**(5), 1–25 (2010)
10. Hossain, M.S., You, X., Xiao, W., Lu, J., Song, E.: QoS-oriented multimedia transmission using multipath routing. Future Gener. Comput. Syst. (2019). https://doi.org/10.1016/j.future.2019.04.006
11. Salah-Ddine, K., Benaddy, M., Habil, B.E., Ouali, M.E., Meslouhi, O.E.: A mutlipath routing algorithm for wireless sensor networks under distance and energy consumption constraints for reliable data transmission, pp. 1–4 (2017). http://www.iares.net/Conference/ICEMIS2017

TCFNet: Temporal-Correlated Feature Fused Network for Multivariate Time Series Classification

Wenlong Liang, Chenghao Li, Yahui Zhao, and Zhenguo Zhang(✉)

Department of Computer Science and Technology, Yanbian University,
977 Gongyuan Road, Yanji 133002, China
{2022010102,2022050064,yhzhao,zgzhang}@ybu.edu.cn

Abstract. Multivariate Time Series Classification (MTSC), one of the most essential tasks for time series data mining, has continuously been attracting great interest over the past decades. For MTSC, previous methods usually extracted temporal features directly from the entire time series, which neglects the inherent multi-periodic and correlated characteristics of multivariate time series. In this paper, we propose a novel network, called Temporal feature Flip Fusion Network(TCFNet), to extract abundant temporal and correlated features from the multi-scale subsequences of multivariate time series, which are decomposed by different periods. To obtain the appropriate multi-scale subsequences, we first transform the time series from the time domain to the frequency domain by utilizing the Fast Fourier Transform(FFT). The original time series is decomposed into subsequences of different scales based on the multiple periods according to top-k frequencies. Then, all the subsequences of each scale are mapped onto a two-dimensional space and we design a two-branch feature extraction and fusion block to extract the features. After handling k scales, all the features are concatenated to complete the MTSC task. To verify the performance of our model, we do experiments on 26 UEA benchmark multivariate time series datasets. The results show that our model gains better accuracy compared to the baselines.

Keywords: multivariate time series classification · period · feature flipping · feature fusion

1 Introduction

Multivariate time series classification (MTSC), an essential task in time series analysis, usually employs a machine learning model to classify the time series data based on their collected attributes. The task of MTSC has garnered widespread attention from the academic and industrial communities due to its practical value.

Shapelet-based methods [17], which take the most discriminative subsequences as features to retain the temporal dependency, have attracted a lot of

Y. Chen et al. (Eds.): COCOON 2024, LNCS 15162, pp. 252–264, 2025.
https://doi.org/10.1007/978-981-96-1093-8_21

attention in the last decade. In recent years, an increasing number of studies try to utilize the powerful deep learning techniques to learn the effective features from MTS data. Among these, CNN-based methods unaffected by sequence length constraints, contribute to high computational efficiency and superior capabilities in extracting temporal features. Besides, several works take the transformer-based encoder-decoder architecture to learn the essential features of MTS such as Autoformer [16] and FEDformer [20]. However, these methods have not been thoroughly explored in MTSC tasks and tend to experience rapid performance degradation on datasets with less pronounced periodicity within MTS.

In this paper, we propose a novel Temporal-Correlated feature Fused Network(TCFNet) to learn the rich information embedded in MTS from two aspects: temporal dependence features and correlated features. Concretely, we find the main k periods from the frequency domain representation of data, where the Fast Fourier Transform(FFT) method is employed to complete the transformation. Then, a two-branch structure is designed to learn temporal and correlated features from the two-dimensional representation of time series. These two kinds of features are then fused to represent the integral features. Finally, we use a full connection work to get the class label of each sample. The experimental results on all 26 UEA benchmark MTS datasets show that our model has good performance. In summary, the main contributions of our work are as follows:

- An innovative two-branch feature flipping extraction architecture is proposed to effectively learn the temporal dependency and correlation of different univariate time series.
- A comprehensive two-stage feature fusion architecture is introduced to integrate different types of features.
- Extensive experiments on all UEA benchmark datasets demonstrate that the proposed model exhibits superior performance compared to several state-of-the-art methods.

2 Related Work

The earliest MTSC methods are the extension of the 1-nearest neighbor with Dynamic Time Warping (1NN-DTW) [7], which combined the k-nearest neighbor (KNN) classifier and the Dynamic Time Warping (DTW) distance metric. These methods are simple but rely heavily on the similarity metric of time series. When the number of training samples is large, they usually have high computational complexity and, more importantly, they lack interpretability. To mitigate these limitations, several works focus on individually handling each variable of each time series, including DTW_D and DTW_I, both of which have similar performance [11]. Compared to the above distance-based methods, feature-based methods that employ the on-the-shelf classifier can enhance efficiency by relying on static features manually crafted by domain experts. The Generalized Random

Shapelets Forests (gRSF) [6] stand out as a cutting-edge methodology, showcasing superior results when compared to its direct contender, Ultra Fast Shapelets (UFS) [13]. In the field of bag-of-words, WEASEL+MUSE [10] utilizes Symbolic Fourier Approximation to form "words" and takes these "words" as features to complete the task. It initially transforms time series data into symbolic sequences and then extracts relevant features from these symbolic sequences using Fourier approximation. Although these methods have been successfully applied to MTSC tasks, capturing time-related multidimensional interaction information is still a challenge.

In recent years, with the tremendous success of deep learning techniques in the fields of NLP and CV, an increasing number of time series tasks are adopting deep neural network architecture. CNN-based methods have proven that they are effective in extracting temporal latent features. Two types of deep neural networks, Fully Convolutional Network (FCN) [14] and Residual Network (ResNet) [4], have been employed for end-to-end time series classification. In pursuit of better performance, InceptionTime [5] explores larger filters than previous models, which comprises a collection of five randomly initialized network models and has good feature learning ability for time series. In addition to adjusting the network structure, some studies focus on modifying the convolutional kernels to better adapt to time series classification tasks. Recently, Disjoint-CNN [3] demonstrate that decomposing 1D convolutional kernels into disjoint spatio-temporal components can improve accuracy with minimal additional computational cost. And *Zhang et al.* devise a semi-supervised MTSC model named TapNet [19], which employs a multi-layer convolutional network with random group permutation to learn low-dimensional features from MTS and constructs an attention prototype network, which mitigates the limitations of sparse training labels. The Transformer [15] has demonstrated remarkable performance across various natural language processing tasks and has shown exceptional proficiency in capturing long-term dependencies. TST [18], utilizing only the encoder part of the Transformer, underwent unsupervised pretraining with proportionally masked data to alleviate the issues of data sparsity.

3 Model

3.1 Model Architecture Overview

Figure 1 depicts the overall architecture of our proposed TCFNet. It consists of four parts: (a) Times series decomposition. In this module, the original MTS are partitioned into subsequences according to the different periods of each univariate time series. (b) 1-D to 2-D mapping. All the subsequences with the same length are mapped to a series of 2-D structures (k) which are the multiple forms of the original MTS. (c) Feature extraction and Fusion. It is the core of our model. For each 2-D representation of the original MTS, a feature extraction and fusion block, which consists of a two-branch structure, is designed to separately extract the temporal dependency and correlated features. Then, a feature fusion operation is employed to combine these two features, where an attention-based technique is utilized to find an effective way of fusion. (d) Feature fusion again and classification. After all k 2-D representations are handled, k fusion features are further fused to form the final feature. A fully connected layer with *softmax* is used to complete the classification.

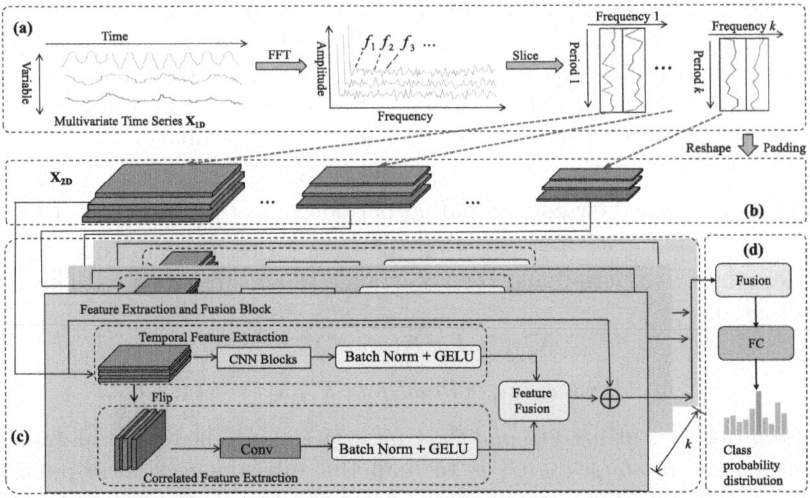

Fig. 1. Model Architecture Overview. By exploring periodicity, MTS can be decomposed into subsequences with periodic patterns. By 2D mapping, temporal dependency and correlated features are extracted through feature extraction and fusion block. After going through two stages of fusion, we can obtain the class probability distribution by an FC layer with *softmax*.

3.2 Two-Dimensional Representation of MTS

Considering that the one-dimensional structure of a time series can only depict changes between adjacent data observations, we try to form a two-dimensional representation for a time series. We propose this transformation based on the following ideas: firstly, the temporal dependence of time series is contained in subsequences, which has been widely recognized; secondly, the two-dimensional structure allows for convenient utilization of mature feature extraction modules to achieve effective time series features. Actually, the number of subsequences for a time series dataset is huge, as it can range from several observations to the entire time series. However, the vast majority of these subsequences are worthless for the classification task. We partition the original time series according to its period and then reshape all obtained subsequences to get the 2-D representation.

Specifically, the Fast Fourier Transform (FFT) is employed to get the frequency domain representation of time series, where we can find the main periods.

The top-k periods are calculated by computing the frequency and corresponding amplitude values of the time series as follows:

$$Period\left(X_{1D}\right) = argtopk\left(Avg\left(Amp\left(FFT\left(X_{1D}\right)\right)\right)\right). \quad (1)$$

Where $X_{1D} \in R^{m \times C}$ represents an MTS with a length of m and the number of univariate time series C. The functions $FFT\left(\cdot\right)$ and $Amp\left(\cdot\right)$ are utilized to get the frequency and the amplitude of the corresponding period for a univariate time series, respectively. $Avg\left(\cdot\right)$ is employed to calculate the average amplitude of periods calculated for multiple sets of time series, portraying their collective periodic amplitudes. Given the sparsity of frequencies and in order to mitigate the noise impact caused by nonsensical high-frequency information, we utilize $argtopk\left(\cdot\right)$ to acquire the top-k significant frequencies $f_i \in \left\{1, \ldots \left[\frac{m}{2}\right]\right\}, i \in \{1, \ldots, k\}$ along with their corresponding periods $p_i = \left\lceil \frac{m}{f_i} \right\rceil, i \in \{1, \ldots, k\}$.

For each period i, we first partition the time series into subsequences based on the length of i-th period and then rearrange these subsequences. The process is represented as

$$X_{1D}^{i} = Partition_i\left(X_{1D}\right) \quad (2)$$

$$X_{2D}^{i} = Reshape_{f_i, p_i}\left(Padding\left(X_{1D}^{i}\right)\right), i \in \{1, \ldots k\}. \quad (3)$$

Here, $Padding\left(\cdot\right)$ is utilized to pad the time series along the temporal dimension. The purpose of $Reshape_{f_i, p_i}\left(\cdot\right)$ is to map the subsequences corresponding to period p_i into a two-dimensional space. After the transformation is completed, p_i and f_i can represent the rows and columns of the transformed two-dimensional tensor, respectively. The process is illustrated in Fig. 1(a).

3.3 Temporal and Correlated Feature Extraction

After transforming the 1-D representation of the time series to an image-like structure, we get k 2-D representations of MTS where each one is stacked by the subsequences with the same period. Due to the Overlapping of different periods, the time dependency of time series between intermittent observations isn't lost. Therefore, we can make full use of the 2-D off-the-shelf feature extraction module to mine the essential time series features. We employ CNN Blocks to extract the temporal features. Considering that the correlation between different univariate time series at the same time is also an issue that cannot be ignored, we use convolution to extract correlated features. According to the extraction ideas of the above two features of MTS, we design a two-branch module to learn two aspects' features. The structure is shown in Fig. 1(c), where the upper branch aims to extract the temporal dependency while the below branch tries to find the correlated features.

For the branch of temporal feature extraction, the CNN block, which is considered a powerful module in the time series community, is employed first to extract the periodic features X_p^i. The process is formalized as follows:

$$X_p^i = GELU\left(BatchNorm\left(CNNBlocks\left(X_{2D}^{i}\right)\right)\right), i \in \{1, \ldots k\}. \quad (4)$$

where $X_{2D}^{i} \in R^{p_i \times f_i \times C}$ is the 2-D representation obtained corresponding to period k. After extracting the features by powerful CNN block, we do the batch normalization operation to standardize data and take $GELU$ active function to complete the nonlinear transform. X_p^i is the extracted temporal features from the subsequences divided by period k.

On the other hand, in order to find the correlated features between different univariate time series at the same time, another branch is designed for this purpose in parallel. Specifically, we flip X_{2D}^{i} (denoted as $X_{2D}^{i,T}$. Figure 1(c) illustrates the process) and employ 1-D convolution to find the correlation of each time hidden in MTS. Again, Batch normalization and $GELU$ active function are utilized to obtain the correlated features corresponding to period i. Equations 5 and 6 denotes the process.

$$X_{2D}^{i,T} = Flip\left(X_{2D}^{i}\right), \tag{5}$$

$$X_f^i = GELU\left(BatchNorm\left(Conv1D\left(X_{2D}^{i,T}\right)\right)\right), i \in \{1, ...k\}. \tag{6}$$

where X_f^i represents the extracted correlated features.

3.4 Features Fusion

When obtaining temporal dependency X_p^i and correlated features X_f^i by the above two-branch structure, we introduce a comprehensive feature fusion module to get the integral features X^i. Figure 2 details the architecture. To fuse features from different scales, we integrate them from two aspects: One is to extract local relationships between features, and the other is to fuse features from a global perspective. Specifically, features X_p^i and X_f^i are first added by elements to complete preliminary integration (denoted as $X_{p,f}^i$). Then, it is used to extract local and global information by points-wise convolution separately. What's different is that a global average pooling operation is used for global information integration. The process can be formulated as follows:

$$Local\left(X_{p,f}^i\right) = BN\left(PWConv\left(\delta\left(BN\left(PWConv\left(X_{p,f}^i\right)\right)\right)\right)\right). \tag{7}$$

$$Global\left(X_{p,f}^i\right) = BN\left(PWConv\left(\delta\left(BN\left(PWConv\left(GAP\left(X_{p,f}^i\right)\right)\right)\right)\right)\right). \tag{8}$$

where the feature denoted by $X_{p,f}^i \in R^{C \times H \times W}$ corresponds to a channel number of C and a feature map size of $p_i \times f_i$. $PWConv(\cdot)$ represents a point-wise convolution, while $BN(\cdot)$ and $\delta(\cdot)$ refer to the BatchNorm layer and ReLU activation function, respectively. Unlike the local information $Local(\cdot)$, the $Global(\cdot)$ involves an initial global average pooling operation $GAP(\cdot)$ applied to the input features.

As depicted in Fig. 2, the process of Eq. 7 and Eq. 8 enable the calculation of weights, which are used to perform a weighted fusion of the input features, thereby obtaining a novel feature representation, that is,

$$W = Sigmoid\left(BC\left(Global\left(X_{p,f}^i\right) \oplus \left(Local\left(X_{p,f}^i\right)\right)\right)\right). \tag{9}$$

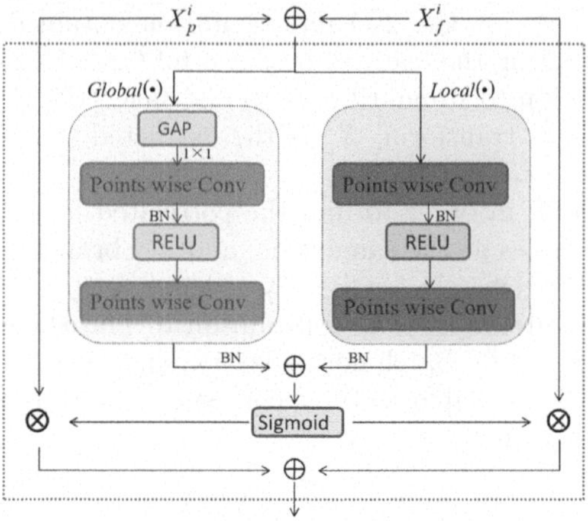

Fig. 2. Feature Fusion Module.

$$X^i = W \otimes X_p^i + (1 - W) \otimes X_f^i. \tag{10}$$

Where $X_i, W \in R^{C \times H \times W}$ represent the integral new features and the generated weights, respectively. $BC(\cdot)$ denotes the broadcast operation that matches the input feature sizes so that the addition can be performed, $\oplus$ represents element-wise addition, and $\otimes$ signifies element-wise multiplication. Equation 9 performs an addition operation on the global channel attention and local channel attention using broadcasting. The resulting values are then passed through a $Sigmoid(\cdot)$ activation function to constrain the output weight values within the range of 0 to 1. Equation 10 encapsulates the entire calculation process, starting with the fusion of the initial features, which are subsequently fed into the network. Through training, specific weights are determined to perform a weighted fusion of the two distinct input features. As a result, our fully integrated feature fusion module facilitates weighted fusion of features at different scales, thereby further enhancing the capability of feature representation.

When completing the fusion of temporal and correlated features for all periods, we concatenate these integral features to obtain the final features X_F, i.e.,

$$X_F = Concat\left(X^1, ..., X^i, ...X^k\right). \tag{11}$$

Finally, we utilize a fully connected layer with *softmax* to get the label of input MTS samples.

4 Experiments

4.1 Experimental Setting

Datasets. The UEA archive [1] is widely recognized as the standard dataset source in the field of multivariate time series classification, garnering substantial attention and widespread application. It comprises a series of real-world multivariate time series datasets encompassing various domains, including Human Activity Recognition (HAR), Motion Classification, Electrocardiography (ECG), Electroencephalography (EEG), and others.

Baselines. To highlight the effectiveness of our TCFNet, we compare it with the following baselines, including conventional methods: ED-1NN [12], DTW [9], DTW-1NN-I [12] and DTW-1NN-D [12]; deep learning-based models: FCN [14], Disjoint-CNN [3], Inception-Time [5], ResNet [4], TapNet [19] and TST [18].

4.2 Classification Performance Evaluation

To validate the performance of our model, we compared it with the baselines for all datasets listed in Table 1. The results are presented in Table 1, with baseline classification accuracy values as references. The best and second-best results are highlighted in **bold** and underlined, respectively. The results in Table 1 demonstrate that our model achieved the highest classification accuracy on 15 datasets. In terms of average accuracy, our model exhibited the best performance, with an accuracy of 0.6983, which represents a 3.19% improvement over the second-best method(Disjoint-CNN). For one-versus-one (1v1) comparisons, our model outperforms all the baselines. Furthermore, our model shows significant improvements in dataset types related to human activity recognition (HAR), electrocardiograms (ECG), and electroencephalograms (EEG). Excitingly, these datasets all pertain to biological signals, featuring prominent periodic characteristics. Our model effectively captures these features, greatly enhancing classification performance.

Additionally, we assess the performance of all methods in Table 1 using the Wilcoxon signed-rank test [2]. Except for comparisons based on average rankings, we generated critical difference diagrams with a significance level of $\alpha = 0.05$ based on the results. As depicted in Fig. 3, our model achieved an average ranking of 3.7308, which is lower than the rankings of the currently most popular methods. Combining this information with Table 1, it becomes evident that our model exhibits the highest average accuracy among all MTSC methods across 26 datasets.

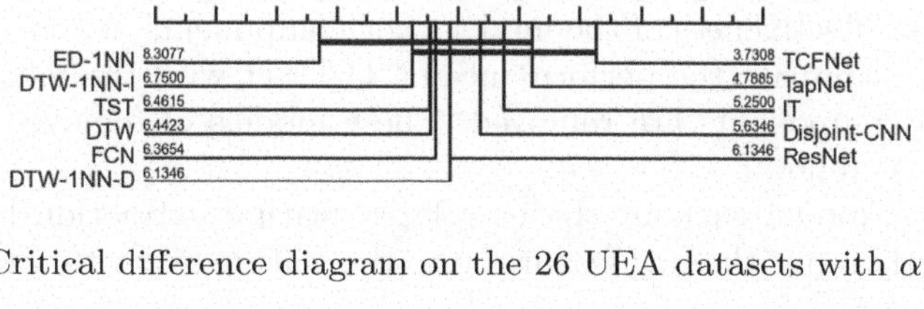

Fig. 3. Critical difference diagram on the 26 UEA datasets with $\alpha = 0.05$.

Table 1. The comparison of classification accuracy on uea benchmark datasets between tcfnet and baselines: Accuracy (%), Average Accuracy and Win/Loss/Tie Ratio. The Best Results Are highlighted In **Bold** and the second best is Underlined.

Datasets	ED-1NN	DTW	DTW -1NN-I	DTW -1NN-D	FCN	Disjoint -CNN	IT	ResNet	TapNet	TST	TCFNet
ArticularyWord	0.97	**0.987**	0.98	**0.987**	0.6353	0.7422	0.76	0.687	**0.987**	0.7655	0.9867
AtrialFibrillation	0.267	0.2	0.267	0.2	0.3333	**0.4**	0.2	0.3333	0.333	0.2	**0.4**
BasicMotions	0.675	0.975	**1**	0.975	**1**	**1**	**1**	**1**	**1**	0.975	**1**
Cricket	0.944	**1**	0.986	**1**	0.9306	0.9772	0.9861	0.9722	0.958	**1**	0.9583
DuckDuckGeese	0.275	**0.6**	0.55	**0.6**	0.36	0.5	0.36	0.24	0.575	0.5	**0.6**
EigenWorms	0.55	0.618	0.603	0.618	0.4198	0.5934	0.53	0.4198	0.489	0.4503	**0.6556**
Epilepsy	0.667	0.964	0.978	0.964	**0.9928**	0.8898	0.9	**0.9928**	0.971	0.9492	0.8986
EthanolConcen	0.293	0.323	0.304	0.323	0.3232	0.2775	**0.3489**	0.3155	0.323	0.151	0.3156
ERing	0.133	0.133	0.133	0.133	0.9037	0.9111	0.93	0.9296	0.133	0.9296	**0.9333**
FaceDetection	0.519	0.529	0.513	0.529	0.5037	0.5665	0.59	0.5948	0.556	0.6542	**0.693**
FingerMovements	0.55	0.53	0.52	0.53	0.53	0.54	0.56	0.54	0.53	0.58	**0.59**
HandMovement	0.279	0.231	0.306	0.231	0.2973	0.5405	0.38	0.2838	0.378	0.5405	**0.6081**
Handwriting	0.371	0.286	0.509	**0.607**	0.376	0.2372	0.3	0.18	0.357	0.2752	0.3435
Heartbeat	0.62	0.717	0.659	0.717	0.678	0.717	0.62	0.7268	0.751	0.6975	**0.7756**
Libras	0.833	0.87	**0.894**	0.872	0.85	0.8577	0.87	0.8389	0.85	0.8222	0.7833
LSST	0.456	0.551	**0.575**	0.551	0.5616	0.5559	0.45	0.5725	0.568	0.2846	0.3986
MotorImagery	0.51	0.5	0.39	0.5	0.55	0.49	0.53	0.52	0.59	0.48	**0.65**
NATOPS	0.86	0.883	0.85	0.883	0.8778	0.9277	0.92	0.8944	0.939	**0.95**	**0.95**
PenDigits	0.973	0.977	0.939	0.977	0.9857	0.9708	0.98	0.9771	0.98	0.9694	**0.9874**
PEMS-SF	0.705	0.711	0.734	0.711	0.8324	**0.8901**	**0.8901**	0.7399	0.751	0.7572	0.8728
PhonemeSpectra	0.104	0.151	0.151	0.151	0.1599	**0.2821**	0.16	0.1596	0.175	0.089	0.1297
RacketSports	0.868	0.803	0.842	0.803	0.8223	0.8355	0.82	0.8223	0.868	0.8815	**0.8947**
SelfRegSCP1	0.771	0.775	0.765	0.775	0.7816	0.8839	0.86	0.8362	0.652	0.86	**0.9044**
SelfRegSCP2	0.483	0.539	0.533	0.539	0.4667	0.5166	0.47	0.5	0.55	0.5333	**0.5556**
StandWalkJump	0.2	0.2	0.333	0.2	**0.4**	0.3333	**0.4**	**0.4**	**0.4**	0.3333	**0.4**
UWaveGesture	0.881	0.903	0.869	0.903	0.85	0.8906	**0.9093**	0.85	0.894	0.8906	0.8719
Average accuracy	0.5676	0.614	0.6224	0.6261	0.6316	0.6664	0.65	0.6279	0.6368	0.6354	**0.6983**
Total best accuracy	0	1	2	2	3	3	5	3	2	1	**16**
Ours 1-to-1-Wins	22	19	19	18	19	18	17	20	17	21	–
Ours 1-to-1-Draws	0	0	1	0	1	2	1	1	1	1	–
Ours 1-to-1-Losses	4	7	6	8	6	6	8	5	8	4	–

4.3 Ablation Study

To validate the effectiveness of the model's key components, we conduct an ablation study. We compare the performance of TCFNet with the following models where special components are removed. These models of removed components are described as follows:

Wo/FE: A model without the correlated feature extraction block. In this case, we directly input the raw features into the feature extraction block without flipping.

Wo/FF: A model without the feature fusion module. Here, we fused the temporal and correlated features using a standard averaging method.

Figure 4 illustrates the impact of removing different components on the model's classification results. The horizontal and vertical axes represent classification accuracy, and points within the light blue region indicate that TCFNet outperforms the corresponding dataset.

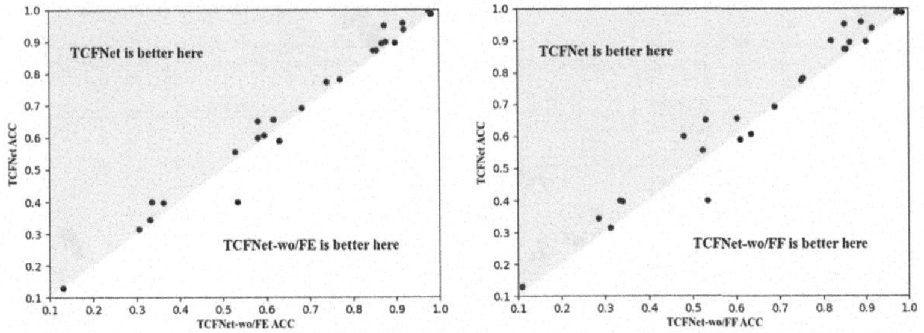

Fig. 4. Scatter plot for TCFNet, TCFNet-wo/FE and TCFNet-wo/FF on the UEA MTSC problems.

As can be seen, when removing the correlated feature extraction block(called TCF-wo/FE), the accuracy on most datasets decreases and TCF-wo/FE wins only 3 times. The result gives us two insights: (1) the correlation between different univariate time series of MTS indeed exists in most scenarios; (2) the designed correlated feature extraction block is effective in uncovering this correlation. When removing the feature fusion module(called TCF-wo/FF), there is a significant decrease in performance on the vast majority of datasets. It demonstrates that the fusion module is a key part of extracting effective MTS features and the proposed fusion model can efficiently fuse temporal dependency features and correlated features of MTS.

4.4 Visualization of the Extracted Features

To intuitively illustrate the features learned by TCFNet, we randomly select examples from the *ERing* and *PEMS-SF* datasets and employ the t-SNE algorithm [8] to visualize the representations embedded in two-dimensional images. Figure 5 displays the results. Each point in the graph represents an example and points with the same color indicate that they have the same label. Clearly, TCFNet is capable of projecting the data into a separable space, where samples from the same class are close while samples from the different classes are separated. The results reveal notable differences in the embeddings learned from different categories. It demonstrates that TCFNet can distinctly differentiate the classes of different samples in MTS by acquiring effective representations.

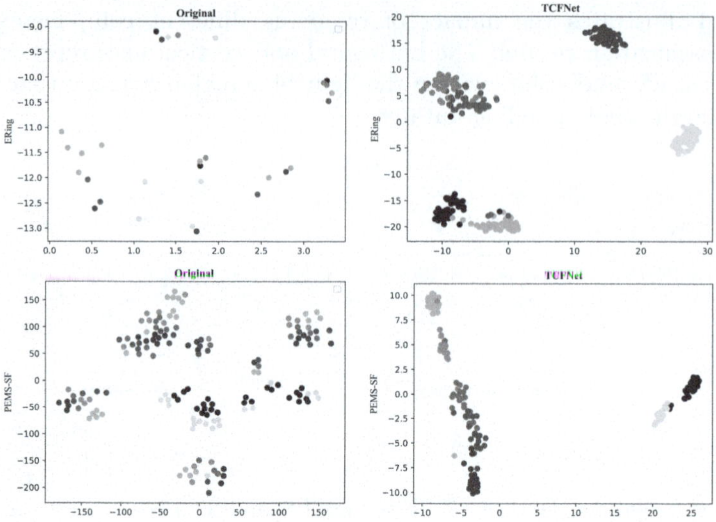

Fig. 5. The t-SNE visualization of the representation space for the datasets of *ERing* and *PEMS-SF*.

5 Conclusion

In this study, we analyze the characteristics of MTS from two aspects, temporal dependency and correlation, and propose a TCFNet to complete the MTSC task. Based on the multiple periodicity, we partition the MTS into subsequences and reorganize them to form a series of 2-D representations. Two types of features are extracted by a specially designed branch structure and a comprehensive feature fusion module is also proposed to merge features of different scales. After fusing all integral features of different 2-D representations of the original MTS, we complete the classification by an FC layer. Compared to the current state-of-the-art methods, TCFNet shows outstanding performance on the UEA archive datasets. From the experimental results on the UEA archive, we conclude that the consideration of multiple periodicity during feature extraction of multivariate time series is necessary, especially for data with clear periodicity. Additionally, the relationships between different univariate time series are also an important factor in MTS that should not be overlooked.

Acknowledgement. This work is supported by the National Natural Science Foundation of China [grant number 62162062] and the Science and Technology Development Plan Project of Jilin Province [20220203127SF].

References

1. Bagnall, A., et al.: The UEA multivariate time series classification archive (2018). arXiv:1811.00075 (2018)
2. Benavoli, A., Corani, G., Mangili, F.: Should we really use post-hoc tests based on mean-ranks? J. Mach. Learn. Res. **17**(1), 152–161 (2016)
3. Foumani, S.N.M., Tan, C.W., Salehi, M.: Disjoint-CNN for multivariate time series classification. In: 2021 International Conference on Data Mining Workshops (ICDMW), pp. 760–769. IEEE (2021)
4. He, K., Zhang, X., Ren, S., Sun, J.: Deep residual learning for image recognition. IEEE (2016)
5. Ismail Fawaz, H., et al.: Inceptiontime: finding alexnet for time series classification. Data Min. Knowl. Discov. **34**(6) (2020)
6. Karlsson, I., Papapetrou, P., Boström, H.: Generalized random shapelet forests. Data Min. Knowl. Disc. **30**(5), 1053–1085 (2016). https://doi.org/10.1007/s10618-016-0473-y
7. Keogh, E., Pazzani, M.: Derivative dynamic time warping. First SIAM International Conference on Data Mining **1** (2002). https://doi.org/10.1137/1.9781611972719.1
8. Van der Maaten, L., Hinton, G.: Visualizing data using T-SNE. Journal of machine learning research **9**(11) (2008)
9. Rakthanmanon, T., Campana, B., Mueen, A., Batista, G., Westover, B., Zhu, Q., Zakaria, J., Keogh, E.: Searching and mining trillions of time series subsequences under dynamic time warping. In: Proceedings of the 18th ACM SIGKDD International Conference on Knowledge Discovery and Data Mining, pp. 262–270 (2012)
10. Schäfer, P., Leser, U.: Multivariate time series classification with weasel+ muse. arXiv preprint arXiv:1711.11343 (2017)
11. Shokoohi-Yekta, M., Hu, B., Jin, H., Wang, J., Keogh, E.: Generalizing DTW to the multi-dimensional case requires an adaptive approach. Data Min. Knowl. Disc. **31**(1), 1–31 (2017)
12. Tang, W., Long, G., Liu, L., Zhou, T., Blumenstein, M., Jiang, J.: Omni-scale CNNs: a simple and effective kernel size configuration for time series classification. arXiv preprint arXiv:2002.10061 (2020)
13. Vlachos, M., Hadjieleftheriou, M., Gunopulos, D., Keogh, E.: Indexing multi-dimensional time-series with support for multiple distance measures. In: Proceedings of the Ninth ACM SIGKDD International Conference on Knowledge Discovery and Data Mining, pp. 216–225 (2003)
14. Wang, Z., Yan, W., Oates, T.: Time series classification from scratch with deep neural networks: a strong baseline. In: 2017 International Joint Conference on Neural Networks (IJCNN), pp. 1578–1585. IEEE (2017)
15. Wen, Q., et al.: Transformers in time series: a survey. arXiv preprint arXiv:2202.07125 (2022)
16. Wu, H., Xu, J., Wang, J., Long, M.: Autoformer: decomposition transformers with auto-correlation for long-term series forecasting. Adv. Neural. Inf. Process. Syst. **34**, 22419–22430 (2021)
17. Ye, L., Keogh, E.: Time series shapelets: a new primitive for data mining. In: Proceedings of the 15th ACM SIGKDD International Conference on Knowledge Discovery and Data Mining, pp. 947–956 (2009)

18. Zerveas, G., Jayaraman, S., Patel, D., Bhamidipaty, A., Eickhoff, C.: A transformer-based framework for multivariate time series representation learning. In: Proceedings of the 27th ACM SIGKDD Conference on Knowledge Discovery and Data Mining, pp. 2114–2124 (2021)
19. Zhang, X., Gao, Y., Lin, J., Lu, C.T.: Tapnet: multivariate time series classification with attentional prototypical network. In: Proceedings of the AAAI Conference on Artificial Intelligence, vol. 34, pp. 6845–6852 (2020)
20. Zhou, T., Ma, Z., Wen, Q., Wang, X., Sun, L., Jin, R.: Fedformer: frequency enhanced decomposed transformer for long-term series forecasting. In: International Conference on Machine Learning, pp. 27268–27286. PMLR (2022)

Strong Chromatic Index of Graphs with Small Girth

Yuehua Bu[1,2], Hongrui Zheng[1], and Hongguo Zhu[1(✉)]

[1] Department of Mathematics, Zhejiang Normal University, Zhejiang, China
zhuhongguo@zjnu.edu.cn

[2] Department of Basics, Zhejiang Guangsha Vocational and Technical University of Construction, Zhejiang, China

Abstract. A strong k-edge-coloring of a graph G is a mapping φ: $E(G) \longrightarrow \{1, \ldots, k\}$, such that $\varphi(e) \neq \varphi(e^{'})$ for every pair of distinct edges at distance at most two. In this paper, we prove $\chi_s^{'}(G) \leq 3\Delta + 1$ for a planar graph with $g(G) \geq 5$ and 5-cycles are disjoint with 6^{-}-cycles by analyzing the structural properties of minimal counterexample and using the discharging method.

Keywords: strong edge-coloring · girth · discharge · planar graph

1 Introduction

All graphs mentioned in this paper are finite simple graphs. For a graph G, we denote its vertex set, edge set, face set, maximum degree, girth and the distance of edges e and $e^{'}$ in G by $V(G)$, $E(G)$, $F(G)$, $\Delta(G)$, $g(G)$ and $d_G(e, e^{'})$ respectively. The maximum average degree of graph G is defined as $mad(G)$, and $mad(G) = max\{\frac{2|E(H)|}{|V(H)|}, H \subseteq G\}$. A strong k-edge-coloring of G is a mapping φ: $E(G) \longrightarrow \{1, \ldots, k\}$, such that $\varphi(e) \neq \varphi(e^{'})$ if $d_G(e, e^{'}) \leq 2$. The strong chromatical index of a graph G is the smallest integer k such that G has a strong k-edge-coloring, denoted by $\chi_s^{'}(G)$. We define $\varphi(E^{'}) = \{\varphi(e) | e \in E^{'}\}$ for an edge set $E^{'}$ and a strong k-edge-coloring φ. The notion of strong edge-coloring was introduced by Fouquet and Jolivet [8].

In 1989, Erdős and Nešetřil [7] posed the classical conjecture with respect to strong chromatical index.

Conjecture 1. Let G be a graph with maximum degree Δ, then
(1) $\chi_s^{'}(G) \leq \frac{5}{4}\Delta^2$, if Δ is even;
(2) $\chi_s^{'}(G) \leq \frac{1}{4}(5\Delta^2 - 2\Delta + 1)$, if Δ is odd.

By now, Conjecture 1 is still open. Andersen [1] proved $\chi_s^{'}(G) \leq 10$ when $\Delta(G) = 3$. In 1997, Molloy and Reed [11] proved $\chi_s^{'}(G) \leq 1.998\Delta^2$ when Δ is large enough. Bruhn and Joos [2] improved this upper bound to $1.93\Delta^2$. Very recently, this was further improved to $1.85\Delta^2$ by Bonamy et al. [5]. For more information and some related topics, we refer the reader to [4,6,10,12].

Y. Chen et al. (Eds.): COCOON 2024, LNCS 15162, pp. 265–271, 2025.
https://doi.org/10.1007/978-981-96-1093-8_22

There have been numerous investigations about strong edge-coloring of planar graphs. In 2013, Hocquard et al. [9] gave an upper bound $\chi_s'(G) \le 4\Delta + 4$ for planar graphs with $\Delta \ge 3$. In 2014, Bensmail et al. [3] gave a better bound by proving that $\chi_s'(G) \le 4\Delta$ for planar graphs with $g \ge 5$. In the same paper, Bensmail et al. gave the other conclusion: if G is a planar graph with $g(G) \ge 6$, then $\chi_s'(G) \le 3\Delta(G) + 1$. Our first result improves this result by showing.

Theorem 1. *If G is a planar graph with $g(G) \ge 5$ and 5-cycles are disjoint with 6^--cycles, then $\chi_s'(G) \le 3\Delta(G) + 1$.*

This paper is organized as follows. In the remainder of this section, we describe notation and terminology used in our proofs. In Sect. 2, we give some reducible configurations of a minimal counterexample. Using the discharging method, we prove Theorem 1 in Sect. 3.

Let G be a planar graph with a fixed embedding in the plane. For $v \in V(G)$, let $N(v)$ denote the set of vertices adjacent to v, and let $d(v) = |N(v)|$ denote the *degree* of v. For $f \in F(G)$, the *degree* of f, denoted by $d(f)$, is the number of edges in its boundary walk. A *k-face*(resp. *k^+-face*, *k^--face*) is a face of $d(f) = k$(resp. $d(f) \ge k$, $d(f) \le k$). A *k-vertex*(resp. *k^+-vertex*, *k^--vertex*) is a vertex of $d(v) = k$(resp. $d(v) \ge k$, $d(v) \le k$). For an edge e in G, let $N_2(e) = \{e' | e' \in E(G) - e, d(e, e') \le 2\}$. A *$k_i$-vertex* is a k-vertex adjacent to i 2-vertices. A *$k_{(i)}$-vertex* is a k-vertex adjacent to i 3-vertices. A *good 4-vertex* is a 4-vertex not adjacent to 3^--vertices. A *k^--cycle* is the cycle of length at most k. A *special face* is a $(l + 2\alpha)$-face which is structured with a l-cycle and α 1-vertices ($\alpha \ge 1$).

2 Proof of Theorem 1

2.1 Structure of Minimal Counterexample

Let G be a counterexample of Theorem 1 such that $|V(G)| + |E(G)|$ is smallest, namely G is a planar graph with $g(G) \ge 5$ and 5-cycles are disjoint with 6^--cycles, $\chi_s'(G) > 3\Delta + 1$. For any edge e in G, $\chi_s'(G - e) \le 3\Delta + 1$.

At first, G is a connected planar graph. Then we consider the structural properties of G by using the minimality of G. We can obtain $\chi_s'(G) \le 10$ when $\Delta \le 3$ in [1]. Thus set $\Delta \ge 4$ in this paper. Let $C = \{1, 2, \ldots, 3\Delta + 1\}$ be the color set for convenience. The following two lemmas can be found in [3].

Lemma 1. *G does not contain a 1-vertex adjacent to a 4^--vertex.*

Lemma 2. *G does not contain a 2-vertex adjacent to two 3^--vertices, or adjacent to a 3^--vertex and a 4_i-vertex ($i = 2, 3$), or adjacent to a 4_3-vertex and a 4_i-vertex ($i = 2, 3$).*

Lemma 3. *Let v be a 3-vertex adjacent to a 2-vertex in G, then the other two neighbors of v are not two 3^--vertices, or two 4_3-vertices, or a 3^--vertex and a 4_3-vertex.*

Proof. Let v be a 3-vertex adjacent to a 2-vertex v_1, suppose that v_2 and v_3 are another neighbors of v.

(1) Suppose that v_2 and v_3 are 3^--vertices, and $G' = G - v$. According to the minimality of G, there is a strong $(3\Delta + 1)$-edge-coloring φ of G'. Since $|\varphi(N_2(vv_2))| \leq 2\Delta + 3 \leq 3\Delta - 1$, $|\varphi(N_2(vv_3))| \leq 2\Delta + 3 \leq 3\Delta - 1$, and $|\varphi(N_2(vv_1))| \leq \Delta + 4 \leq 2\Delta$. Let $\varphi(vv_2) \in C - \varphi(N_2(vv_2))$, $\varphi(vv_3) \in C - \varphi(N_2(vv_3)) \cup \{\varphi(vv_2)\}$, and $\varphi(vv_1) \in C - \varphi(N_2(vv_1)) \cup \{\varphi(vv_2), \varphi(vv_3)\}$. Hence, φ can be extended to a strong $(3\Delta + 1)$-edge-coloring of G, a contradiction.

(2) Suppose that v_2 and v_3 are 4_3-vertices, and $G' = G - v$. According to the minimality of G, there is a strong $(3\Delta + 1)$-edge-coloring φ of G'. Since $|\varphi(N_2(vv_2))| \leq 10 \leq 3\Delta - 2$, $|\varphi(N_2(vv_3))| \leq 10 \leq 3\Delta - 2$, and $|\varphi(N_2(vv_1))| \leq \Delta + 6 \leq 2\Delta + 2$. Then there are at least three available colors for vv_2, vv_3 and vv_1, we can color vv_2, vv_3 and vv_1 independently. Hence, φ can be extended to a strong $(3\Delta + 1)$-edge-coloring of G, a contradiction.

(3) Suppose that v_2 be a 3^--vertex and v_3 be a 4_3-vertex, and $G' = G - v$. According to the minimality of G, there is a strong $(3\Delta + 1)$-edge-coloring φ of G'. Since $|\varphi(N_2(vv_2))| \leq 2\Delta + 4 \leq 3\Delta$, $|\varphi(N_2(vv_3))| \leq 9 \leq 3\Delta - 3$, and $|\varphi(N_2(vv_1))| \leq \Delta + 5 \leq 2\Delta + 1$, then we can color vv_2, vv_3 and vv_1 independently. Therefore, φ can be extended to a strong $(3\Delta + 1)$-edge-coloring of G, a contradiction.

Lemma 4. *G does not contain a 4_3-vertex adjacent to a 3^--vertex or a 4_3-vertex.*

Proof. Let v be a 4_3-vertex, we assume that $N(v) = \{v_1, v_2, v_3, v_4\}$, where v_1, v_2 and v_3 are 2-vertices. Then take $G' = G - v$. According to the minimality of G, there is a strong $(3\Delta + 1)$-edge-coloring φ of G'.

(1) If v_4 is a 3-vertex, then $|\varphi(N_2(vv_4))| \leq 2\Delta + 3 \leq 3\Delta - 1$ for vv_4, and $|\varphi(N_2(vv_i))| \leq \Delta + 4 \leq 2\Delta$ for vv_i $(i \in \{1, 2, 3\})$. Hence, there are at least two available colors for vv_4, and there are at least five available colors for vv_1, vv_2 and vv_3. Thus we can color vv_4, vv_3, vv_2 and vv_1 in turn and then φ can be extended to a strong $(3\Delta + 1)$-edge-coloring of G, a contradiction.

(2) If v_4 is a 4_3-vertex, then $|\varphi(N_2(vv_4))| \leq 9 \leq 2\Delta + 1$ for vv_4 and $|\varphi(N_2(vv_i))| \leq \Delta + 5 \leq 2\Delta + 1$ for vv_i $(i \in \{1, 2, 3\})$. Therefore, there are at least four available colors for $vv_i (i \in \{1, 2, 3, 4\})$ and we can color vv_i independently. Hence, φ can be extended to a strong $(3\Delta + 1)$-edge-coloring of G, a contradiction.

Lemma 5. *Let v be a k-vertex of G.*

(1) If $k \geq 4$, then v is adjacent to at most $(k - 1)$ 2^--vertices;

(2) If $k \geq 5$, then v is adjacent to at most $(k - 3)$ 1-vertices. If v is just adjacent to $(k - 3)$ 1-vertices, then v is not adjacent to other 2-vertices;

(3) If $k \geq 5$ and v is adjacent to α 1-vertices and $(k-2-\alpha)$ 2-vertices, then $0 \leq \alpha \leq k-4$, and there are at most $(k-4-\alpha)$ 2-vertices adjacent to 3^--vertices or 4_3-vertices;

(4) If $k \geq 4$, and v is just adjacent to $(k-1)$ 2^--vertices, then the $(k-1)$ 2^--vertices are all 2-vertices, and the other neighbor of 2-vertex is a 4^+-vertex distinct from 4_3-vertex.

Proof. The properties (1), (2), (4) are proved in [3], it is still work for our result. Similarly, Now we prove property (3).

Suppose that v is adjacent to α 1-vertices and $(k-2-\alpha)$ 2-vertices $v_1, v_2, \cdots, v_{k-2-\alpha}$, then $0 \leq \alpha \leq k-4$. Otherwise, suppose that v is just adjacent to $(k-3)$ 1-vertices, then v is not adjacent to 2-vertex according to (2). Hence, there is a contradiction since the number of 2-vertices in $N(v)$ is $k-2-(k-3)=1$. Assume there are $(k-3-\alpha)$ 2-vertices $v_1, v_2, \cdots, v_{k-3-\alpha}$ adjacent to 3^--vertices or 4_3-vertices $u_1, u_2, \ldots, u_{k-3-\alpha}$. Let $G' = G - \{v_1, v_2, \cdots, v_{k-3-\alpha}\}$, by the minimality of G, there is a strong $(3\Delta+1)$-edge-coloring φ of G'. At first, we color $vv_i (i=1,2,\ldots,k-4-\alpha)$ one by one. Since $|\varphi(N_2(vv_i)| \leq 2\Delta + (i-1) + \alpha + 5 \leq 2\Delta + k \leq 3\Delta$, then we can color $vv_i (i=1,2,\ldots,k-4-\alpha)$ independently.

If $u_{k-3-\alpha}$ is a 3^--vertex, then for $vv_{k-3-\alpha}$ we have $|\varphi(N_2(vv_{k-3-\alpha}))| \leq 2\Delta + k \leq 3\Delta$, and we can color $vv_{k-3-\alpha}$ independently. If $u_{k-3-\alpha}$ is a 4_3-vertex, let r be a 2-vertex adjacent to $u_{k-3-\alpha}$ and distinct from $v_{k-3-\alpha}$. We uncolor the edge $ru_{k-3-\alpha}$. Then for $vv_{k-3-\alpha}$ we have $|\varphi(N_2(vv_{k-3-\alpha}))| \leq 2\Delta + k \leq 3\Delta$, and we can color $vv_{k-3-\alpha}$ independently. Next, for $ru_{k-3-\alpha}$, we can color $ru_{k-3-\alpha}$ independently since $|\varphi(N_2(ru_{k-3-\alpha}))| \leq 2\Delta + 3 \leq 3\Delta - 1$. At last, we color $u_i v_i (i=1,2,\ldots,k-3-\alpha)$ one by one. If u_i is not adjacent to u_j $(1 \leq j \leq k-2-\alpha)$, then $|\varphi(N_2(u_i v_i))| \leq 2\Delta + k \leq 3\Delta$ or $|\varphi(N_2(u_i v_i))| \leq \Delta + k + 4 \leq 3\Delta$, and we can color $u_i v_i$ independently. If there is an vertex u_i adjacent to u_j, then u_i is only adjacent to one vertex in $\{u_1, u_2, \ldots, u_{i-1}, u_{i+1}, \ldots, u_{k-2-\alpha}\}$ since 5-cycles are disjoint with 6^--cycles in G. According to Lemmas 3 and 4, u_i and u_j are 3^--vertices when $j \neq k-2-\alpha$, then $|\varphi(N_2(u_i v_i))| \leq \Delta + k + 3 \leq 3\Delta - 1$. For $j = k-2-\alpha$, since u_i is a 3^--vertex or a 4_3-vertex, we have $|\varphi(N_2(u_i v_i))| = \max\{\Delta + k + 4, 2\Delta + k\} \leq 3\Delta$. Then we can color $u_i v_i$ independently. Hence, φ can be extended to a strong $(3\Delta+1)$-edge-coloring of G, a contradiction.

2.2 Discharging

In this section, we prove Theorem 1 by contradiction. Let G be a minimal counterexample of Theorem 1. Note that G is connected. By Euler's formula $|V(G)| + |F(G)| - |E(G)| = 2$, we have

$$\sum_{v \in V(G)} (2d(v) - 6) + \sum_{f \in F(G)} (d(f) - 6) = -12. \quad (1)$$

First, we define the weight function μ by $\mu(v) = 2d(x) - 6$ for each $v \in V(G)$, and $\mu(f) = d(f) - 6$ for each $f \in F(G)$. Then we will design several discharging rules and reassign weight accordingly. Once the discharging is finished, a new weight function μ' is produced and we will show that for each $x \in V(G) \cup F(G)$, $\mu'(x) \geq 0$. However, the total sum of weights is unchanged. This leads to the following contradiction:

$$0 \leq \sum_{x \in V(G) \cup F(G)} \mu'(x) = \sum_{x \in V(G) \cup F(G)} \mu(x) = -12. \quad (2)$$

We define the following discharging rules:

- **R1** Every special face sends 2 to each incident 1-vertex; Every 5^+-vertex sends 2 to each adjacent 1-vertex.
- **R2** Let v be a 2-vertex.
  R2.1 If v is adjacent to a 3^--vertex and either a 4_1-vertex or a 5^+-vertex, then each 4_1-vertex or 5^+-vertex sends 2 to v;
  R2.2 If v is adjacent to a 4_3-vertex and either a 4_1-vertex or a 5^+-vertex, then each 4_1-vertex or 5^+-vertex sends $\frac{4}{3}$ to v, and each 4_3-vertex sends $\frac{2}{3}$ to v;
  R2.3 If v is adjacent to two 4^+-vertices different from 4_3-vertices, then each 4^+-vertex sends 1 to v.
- **R3** Let f be a 5-face, then each 7^+-face sends $\frac{1}{5}$ to f through every edge adjacent to f.

At first, we check $\mu'(v) \geq 0$ for each $v \in V(G)$.

(1) $d(v) = 1$ and then $\mu(v) = -4$.

According to Lemma 1, 1-vertex is adjacent to 5^+-vertex. By R1, v receives $2+2$ from its adjacent 5^+-vertex and incident face, then $\mu'(v) = -4+2+2 = 0$.

(2) $d(v) = 2$ and then $\mu(v) = -2$.

Supposer that u and w are two neighbors of v. According to Lemma 2, G does not contain a 2-vertex adjacent to two 3^--vertices. W.l.o.g., suppose $d(w) \geq 4$.

If u is a 3^--vertex, then w is not a 4_i-vertex ($i = 2, 3$) by Lemma 2. Note that v receives 2 from w by R2.1, and then $\mu'(v) = -2 + 2 = 0$. If u is a 4_3-vertex, then w is not a 4_i-vertex ($i = 2, 3$) by Lemma 2. According to R2.2, v receives $\frac{2}{3} + \frac{4}{3}$ from u and w. Therefore, $\mu'(v) = -2 + \frac{2}{3} + \frac{4}{3} = 0$. Otherwise, suppose that u is a 4^+-vertex distinct from 4_3-vertex, then v receives $1 + 1$ from u and w when w is not a 4_3-vertex by R2.3, and receives $\frac{2}{3} + \frac{4}{3}$ from u and w when w is a 4_3-vertex by R2.2. So $\mu'(v) = -2 + 2 = 0$.

(3) $d(v) = 3$ and then $\mu'(v) = \mu(v) = 0$.

(4) $d(v) = 4$ and then $\mu(v) = 2$.

Note that v is adjacent to at most three 2-vertices by Lemma 5. If v is adjacent to three 2-vertices, then $\mu'(v) = 2 - 3 \times \frac{2}{3} = 0$ by R2.2. If v is adjacent to two 2-vertices, then $\mu'(v) \geq 2 - 2 \times 1 = 0$ by R2.3. If v is adjacent to at most one 2-vertex, then v sends at most $\max\{2 \times 1, \frac{4}{3} \times 1\} = 2$ to adjacent 2-vertex by R2. So, $\mu'(v) \geq 2 - 2 = 0$.

(5) $d(v) = k \geq 5$ and then $\mu(v) = 2k - 6$.

According to Lemma 5, there are at most $(k-1)$ 2^--vertices adjacent to v, and there are at most $(k-3)$ 1-vertices adjacent to v. If v is adjacent to $(k-1)$ 2^--vertices, then these 2^--vertices are all 2-vertices, and there are not

2-vertices adjacent to 3^--vertices or 4_3-vertices by Lemma 5. By R2.3, $\mu'(v) = 2k-6-1\times(k-1) = k-5 \geq 0$. If v is adjacent to $(k-2)$ 2^--vertices, and the number of 1-vertices adjacent to v is α, then $0 \leq \alpha \leq k-4$ by Lemma 5. According to R1.2,R2, $\mu'(v) \geq 2k-6-2\times(k-4-\alpha)-2\alpha-1\times 2 = 0$. If v is adjacent to at most $(k-3)$ 2^--vertices, then by R1.2 and R2, we have $\mu'(v) \geq 2k-6-2\times(k-3) = 0$.

Now we check $\mu'(f) \geq 0, \forall f \in F(G)$. At first, suppose that f is not a special face.

(1) $d(f) = 5$ and then $\mu(f) = -1$.

Obviously, f has five common edges with adjacent 7^+-faces. According to R3, f receives $5 \times \frac{1}{5}$ from its adjacent 7^+-faces. Hence, $\mu'(f) = -1 + 5 \times \frac{1}{5} = 0$.

(2) $d(f) = 6$ and then $\mu'(f) = \mu(f) = 6-6 = 0$.

(3) $d(f) = 7$ and then $\mu(f) = 1$.

Since 5-cycles are disjoint with 6^--cycles according to the condition of Theorem 1, f has at most three common edges with adjacent 5-faces. By R3, we have $\mu'(f) \geq 1 - 3 \times \frac{1}{5} = \frac{2}{5}$.

(4) $d(f) = k \geq 8$ and then $\mu(f) = k-6$.

Observe that f has at most k common edges with adjacent 5-faces. By R3, we have $\mu'(f) \geq k - 6 - \frac{1}{5}k = \frac{4}{5}k - 6 \geq \frac{2}{5}$.

Next, suppose that f is a special face.

(5) $d(f) = l + 2\alpha$ and then $\mu(f) = l + 2\alpha - 6$.

It is known that $l \geq 5$ and $\alpha \geq 1$ since f is a special face. If $l = 5$, since 5-cycles are disjoint with 6^--cycles according to the condition of Theorem 1, then a 5-face has exactly 5 common edges with adjacent 7^+-faces. By R1 and R3, $\mu'(f) = 5+2\alpha-6+5\times\frac{1}{5}-2\alpha = 0$. If $l = 6$, then $\mu'(f) = 6+2\alpha-6-2\alpha = 0$ by R1. If $l = 7$, then $\mu'(f) \geq 7+2\alpha-6-5\times\frac{1}{5}-2\alpha = 0$ by R1 and R3. If $l \geq 8$, then f has at most l common edges with adjacent 5-faces. By R1 and R3, we have $\mu'(f) \geq l + 2\alpha - 6 - \frac{1}{5}l - 2\alpha = \frac{4}{5}l - 6 \geq \frac{2}{5}$.

Thus, we have $\mu'(x) \geq 0$ for each $x \in V(G) \cup F(G)$, which is a contradiction. This completes the proof of Theorem 1.

Acknowledgments. This study was supported by by the National Natural Science Foundation of China(12031018, 12201569).

Disclosure of Interests. The authors declare that they have no known competing financial interests or personal relationships that could have appeared to influence the work reported in this paper.

References

1. Andersen, L.D.: The strong chromatic index of a cubic graph is at most 10. Discret. Math. **108**, 31–252 (1992)
2. Bruhn, H., Joos, F.: A stronger bound for the strong chromatic index. Comb. Probab. Comput. **27**, 21–43 (2018)
3. J. Bensmail, A. Harutyunyan, H. Hocquard. P. Valicov. Strong edge-coloring of sparse planar graphs. Discrete Applied Mathematics, 2014, 179:229–234

4. Brualdi, R.A., Quinn Massey, J.J.: Incidence and strong edge colorings of graphs. Discrete Math. **122**, 51–58 (1993)
5. Bonamy, M., Perrett, T., Postle, L.: Colouring graphs with sparse neighbourhoods: bounds and applications. arXiv:1810.06704
6. Cranston, D.W.: Strong edge-coloring of graphs with maximum degree 4 using 22 colors. Discret. Math. **306**, 2772–2778 (2006)
7. Erdős, P., Nešetřil, J.: Problem. In: Halász, G., Sós, V.T. (eds.) Irregularities of Partitions, pp. 162–163. Springer, Berlin (1989)
8. Fouquet, J.L., Jolivet, J.L.: Strong edge-colorings of graphs and applications to multi-K-GONs. Ars Combin. **16A**, 141–150 (1983)
9. Hocquard, H., Montassier, M., Raspaud, A., Valicov, P.: On strong edge-colouring of subcubic graphs. Discret. Appl. Math. **161**, 2467–2479 (2013)
10. Huang, M., Santana, M., Yu, G.: Strong chromatic index of graphs with maximum degree four. Electron. J. Combin. **25**, P3.31 (2018)
11. Molloy, M., Reed, B.: A bound on the strong chromatic index of a graph. J. Combin. Theory Ser. B **69**, 103–109 (1997)
12. Steger, A., Yu, M.-L.: On induced matchings. Discret. Math. **120**, 291–295 (1993)

A Space Efficient Algorithm for Multiset Multicover with Multiplicity Constraints Problem via Algebraic Method

Pu Wu[1(✉)], Huiqin Jiang[2], Zehui Shao[2], and Jin Xu[1,3]

[1] School of Computer Science, Peking University, Beijing 100871, China
puwu1997@126.com, jxu@pku.edu.cn

[2] Institute of Computing Science And Technology, Guangzhou University, Guangzhou 510006, China
{hqjiang,zshao}@gzhu.edu.cn

[3] Key Laboratory of High Confidence Software Technologies (Peking University), Ministry of Education, School of Computer Science, Peking University, Beijing 100871, China

Abstract. Given a set S with n elements and a collection $\mathcal{F}$ of multisets over S, the multiset multicover with multiplicity constraints problem (MSMC-MCP) entails finding the minimum number of multisets from $\mathcal{F}$ required to cover each element $s_i \in S$ at least specified b_i times, while ensuring that each multiset $F_i \in \mathcal{F}$ can be selected no more than specified d_i times. Here, $b = \max_{1 \leq i \leq n} b_i$. MSMC-MCP is a generalization of classic set cover problem.

In this paper, we employ a novel algebraic approach that combines generating functions and the Discrete Fourier Transform over a ring. Through this method, we introduce an exact algorithm for MSMC-MCP, which exhibits $O^*(b^4)$ space complexity and $O^*(b^6(b+1)^n|\mathcal{F}|)$ time complexity (where the O^* notation suppresses polynomial factors involving n and $\ln b$). Compared to the results presented in Hua et al.'s works (TCS, 2009) and (TCS, 2010), our approach significantly reduces the space complexity.

Keywords: Multiset multicover · Discrete Fourier Transform over a ring · Exact algorithm

1 Introduction

Formally, we are provided with a set $S = \{s_1, s_2, \ldots, s_n\}$, a collection of multisets $\mathcal{F} = \{F_1, F_2, \ldots, F_m\}$ over S, a positive integer coverage requirement vector $B = (b_i)_{1\times n}$ and a positive integer multiplicity constraints vector $D = (d_i)_{1\times|\mathcal{F}|}$. Define $d = \max\{d_i | 1 \leq i \leq |\mathcal{F}|\}$ and $b = \max\{b_i | 1 \leq i \leq n\}$. The goal of MSMC-MCP, an NP-hard combinatorial optimization problem [9], is to determine the minimum number of multisets from $\mathcal{F}$ required to cover each $s_i \in S$ at least b_i times, with the condition that $F_j \in \mathcal{F}$ can be selected no more than d_j times.

Y. Chen et al. (Eds.): COCOON 2024, LNCS 15162, pp. 272–286, 2025.
https://doi.org/10.1007/978-981-96-1093-8_23

MSMC-MCP is found to have applications in various real-world scenarios. For example, the special case known as the set cover problem (SCP), involves selecting a minimum number of sets from a given collection to cover each element in S. The work cited in [17] emphasizes the significance of fields like view planning for unmanned aerial vehicles, where the objective is to identify near-optimal viewpoints to cover a target geometric model, a problem naturally framed as a set cover problem. When the objective is to cover each element s_i in S at least a specified b_i times, SCP transforms into the set multicover problem (SMCP) [8]. SMCP addresses scenarios where objects require multiple coverings, such as k-coverage problem in wireless sensor networks [16], where the objective is to select the minimum number of sensors to cover each given area at least k times. Furthermore, if the elements in $\mathcal{F}$ are multisets rather than sets, SMCP transitions to multiset multicover problem (MSMCP) [9]. MSMCP is pertinent in situations like marker selection for gene expression [7], where the task involves selecting the minimum features to cover each phenotype at specified times. Here, a feature covering each phenotype with different times can be regarded as a multiset of phenotypes. When F_i in $\mathcal{F}$ has a selection limit d_i, SMCP and MSMCP respectively become set multicover with multiplicity constraints problem (SMC-MCP) and multiset multicover with multiplicity constraints problem (MSMC-MCP) [9], presenting real-world constraints like capacity limits, security goals, and fault tolerance [10,15].

The time and space complexities of exact algorithms for MSMC-MCP and its special cases have been extensively studied. As shown in Table 1, the mentioned special cases, except SMC-MCP, feature algorithms with $O^*((b+1)^n|\mathcal{F}|)$ time complexity and polynomial space. Following the work of Hua et al. [9] in 2010, there has been no progress in exact algorithm for MSMC-MCP for a significant duration. As of now, there are currently no algorithms with such complexities for SMC-MCP and MSMC-MCP. Thus, the challenge remains to develop an exact algorithm for these cases with $O^*((b+1)^n|\mathcal{F}|)$ time complexity and polynomial space.

In this paper, we propose an algorithm for MSMC-MCP with a time complexity of $O^*(b^6(b+1)^n|\mathcal{F}|)$ and a space complexity of $O^*(b^4)$. Our approach, as depicted in Table 1, introduces novel techniques such as generating functions and Discrete Fourier Transform over a ring, setting it apart from previous research. Notably, our method eliminates the need for dynamic programming, a key component in the approach by Hua et al. [9]. Consequently, compared to their findings, we achieve a substantial reduction in space complexity, particularly evident when $b = O^*(n)$, rendering our algorithm a polynomial space algorithm.

Main Idea of Our Algorithm. We utilize algebraic methods to design our algorithm. Algebraic methods were utilized in algorithm design previously, for example, Björklund et al. [2] use inclusion-exclusion principle and fast zeta transform to design algorithms for set partition problem, and Lokshtanov et al. [11] use convolution and Discrete Fourier Transform to save space for some dynamic programs. In this paper, different with them, we find generating function and

Table 1. Summary of previous works and our result

Problem	Main technologies	Complexity		Reference
		Time	Space	
SCP	inclusion-exclusion principle	$O^*(2^n\|\mathcal{F}\|)$	Polynomial	[2]
SCP	inclusion-exclusion principle and fast zeta transform	$O^*(2^n)$	$O^*(2^n)$	[2]
SMCP	inclusion-exclusion principle and dynamic program	$O^*((2b)^n)$	$O^*((b+1)^n)$	[8]
SMCP	inclusion-exclusion principle	$O^*(2^{O(bn^2)})$	Polynomial	[8]
SMCP		$O^*((b+1)^n\|\mathcal{F}\|)$		[12]
MSMCP		$O^*((b+1)^n\|\mathcal{F}\|)$		[9]
SMC-MCP MSMC-MCP	inclusion-exclusion principle and dynamic program	$O^*((b+1)^n\|\mathcal{F}\|)$	$O^*((b+1)^n)$	[9]
MSMC-MCP	generating function and Discrete Fourier Transform over a ring	$O^*(b^6(b+1)^n\|\mathcal{F}\|)$	$O^*(b^4)$	This Paper

Multidimensional Discrete Fourier Transform over a ring can be combined to design algorithm, and it is powerful for MSMC-MCP.

To address MSMC-MCP, we initially reframe it as determining whether a coefficient of a term in the generating function is zero. We then employ the Discrete Fourier Transform to compute this coefficient. We introduce the classic Discrete Fourier Transform in Definition 1 for clarity and utilize Example 1 to illustrate the process of obtaining the coefficient of a specified term in an univariate polynomial. However, since classic Discrete Fourier Transform is working with floating-point numbers, there is a potential for accuracy errors. To mitigate this, we employ Discrete Fourier Transform over a ring instead of the classic Discrete Fourier Transform. Equation (4) is formulated to compute the desired coefficient, leveraging its ability to calculate a single component of Multidimensional Discrete Fourier Transform over a ring. Readers unfamiliar with Multidimensional Discrete Fourier Transform over a ring need not be concerned, as we provide a comprehensive proof in Lemma 5 to establish correctness.

Definition 1. *(N-point Discrete Fourier Transform) [14]. The Discrete Fourier Transform transforms a sequence of N complex numbers $\{x_r\} := x_0, x_1, \ldots, x_{N-1}$ into another sequence of N complex numbers $\{X_r\} := X_0, X_1, \ldots, X_{N-1}$ which is defined by*

$$X_k = \sum_{r=0}^{N-1} x_r e^{-\frac{2\pi i}{N} kr}$$

where i is the imaginary unit. And we call X_k as the single component of Discrete Fourier Transform for $0 \le k \le N-1$.

Example 1. [5] Let N be a positive integer. Consider a polynomial $f(x) = a_N x^N + a_{N-1}x^{N-1} + \cdots + a_0$, and a sequence of $N+1$ complex numbers $\{x_r\} := x_0, x_1, \ldots, x_N$, where $x_r = f(e^{\frac{2\pi i}{N+1}r})$ for $0 \leq r \leq N$ and i is the imaginary unit. Perform the $(N+1)$-point Discrete Fourier Transform on $\{x_r\}$ to get the another sequence of $N+1$ complex numbers $\{X_k\} := X_0, X_1, \ldots, X_N$, where $X_k = \sum_{r=0}^{N} x_r e^{-\frac{2\pi i}{N+1}kr}$ for $0 \leq k \leq N$. It can be verified that $X_k = (N+1)a_k$ for $0 \leq k \leq N$.

2 Preliminary

For a multiset $F \in \mathcal{F}$, we can denote $F = \{(s_{i_1}, t_1), (s_{i_2}, t_2), \ldots, (s_{i_p}, t_p)\}$ where t_j is the number of occurrences of s_{i_j} in F. Denote $s \in F$ if the number of occurrences of s in F is no less than 1. If the number of occurrences of s in F_1 is no more than the number of occurrences of s in F_2 for any $s \in F_1$, we refer to F_1 as a submultiset of F_2. In the case where F_1 is a submultiset of F_2, it is denoted as $F_1 \subseteq F_2$. For example, $\{(s_1, 1), (s_2, 2)\} \subseteq \{(s_1, 2), (s_2, 2)\}$.

Call $Solution = \{(F_{i_1}, t_{i_1}), (F_{i_2}, t_{i_2}), \ldots, (F_{i_k}, t_{i_k})\}$, where (F_{i_j}, t_{i_j}) denotes selecting F_{i_j} t_{i_j} times, as a solution of MSMC-MCP, if $Solution$ covers $s_i \in S$ at least b_i times and $t_{i_j} \leq d_{i_j}$ for $1 \leq j \leq k$. Additionally, call $Solution$ as a t-solution if $\sum_{j=1}^{k} t_{i_j} = t$. For a given integer t, if we can test whether there is a t-solution of a given MSMC-MCP, we can find the minimum t by using binary search. Therefore, we only need to test whether there is a t-solution of a given MSMC-MCP.

In this paper, O^* suppresses polynomial factors involving n and $\ln b$. For a vector ℓ with length n, let ℓ_i be the i-th element of ℓ for $1 \leq i \leq n$. For positive integers a and b, denote $\gcd(a, b)$ as the greatest common divisor of a and b. For a positive integer a and a nonnegative integer b, $\binom{a}{b}$ is the combinatorial number. For a real number r, $\lceil r \rceil$ is the smallest integer greater than or equal to r.

For a polynomial $f(x_1, x_2, \ldots, x_n)$, if all coefficients are nonnegative integers, we refer to f as a nonnegative integer polynomial. Let $\deg(f)$ denote the degree of the polynomial f. An univariate polynomial f is defined as a polynomial with a single variable. The leading coefficient of an univariate polynomial f corresponds to the coefficient of the term with the highest degree in f. A monic polynomial is a non-zero univariate polynomial with a leading coefficient of 1.

Some Basic Notes in Ring Theory.

Definition 2. *(Ring) [4]. A ring is a set R equipped with two binary operations $+$ (addition) and $\cdot$ (multiplication) satisfying the following laws:*

- *For the operation "+ ": (1) $a+b = b+a$ for $a, b \in R$; (2) $(a+b)+c = a+(b+c)$ for $a, b, c \in R$; (3) there is a $0 \in R$ such that $a + 0 = a$; (4) for each $a \in R$, there is a $-a \in R$ with $a + (-a) = 0$.*

- *For the operation "$\cdot$": (1) $a(bc) = (ab)c$ for $a, b, c \in R$; (2) there is a $1 \in R$ such that $1a = a1 = a$.*
- *For operations "$\cdot$" and "$+$": $(a + b)c = ac + bc, c(a + b) = ca + cb$ for $a, b, c \in R$.*

In a ring R, we denote an element e in R as $e \in R$. $\mathbb{Z}$ represents the ring of integers. The univariate polynomial ring over a ring R is denoted as $R[x]$. The ideal generated by $a_1, a_2, \ldots, a_t$ in a ring R is symbolized by $\langle a_1, a_2, \ldots, a_t \rangle$. If I is an ideal of a ring R, the quotient ring of R by I is represented as R/I. We introduce the ring mainly used in this paper as follows: in the ring $R[x]/\langle p \rangle$, where p is a monic polynomial in $R[x]$ with $\deg(p) \geq 1$, the set is defined as $\{a \in R[x] \mid \deg(a) < \deg(p)\}$, with addition and multiplication performed modulo p. For example, $x+1$ and $x-1$ belong to $\mathbb{Z}[x]/\langle x^2+1 \rangle$, where $(x+1)(x-1) = -2$. In $R'[y]/\langle y^2+1 \rangle$, with $R' = \mathbb{Z}[x]/\langle x^2+1 \rangle$, $(x+1)y+(x-1)y+x-1 = 2xy+x-1$ and $(x+1)y((x-1)y+x-1) = -2y+2$. Readers interested in a deeper understanding of ring theory and the concepts discussed are encouraged to refer to [4].

Lemma 1. *(Exponentiation By Squaring Method) [6] For an integer $k > 0$ and $x \in R$ for some ring R, computing x^k can be accomplished within $O(\ln k)$ multiplications.*

Lemma 2. *[3] Multiplying two univariate polynomials of degree no more than n with coefficients from an arbitrary (possibly non-commutative) ring R can be accomplished within $O(n \ln n \ln \ln n)$ additions and subtractions and $O(n \ln n)$ multiplications over R.*

Lemma 3. *[13] (1) The multiplication of two integers, each represented with a bit length no greater than n, can be achieved within $O(n \ln n)$ time complexity.*

(2) The division of two integers, each represented with a bit length no greater than n, can be achieved within $O(n \ln n)$ time complexity.

(3) Dividing two integer coefficient polynomials, each with a degree no greater than n, can be executed in $O(n \ln n)$ operations, including additions, subtractions, and divisions on their coefficients.

For a positive integer r, the Cyclotomic polynomial [1] of order r, denoted as $\Phi_r(x)$, is defined as $\prod_{1 \leq k \leq r, \gcd(k,r)=1}(x - e^{2\pi i \frac{k}{r}})$, where i is the imaginary unit.

Recall $b = \max\{b_i \mid 1 \leq i \leq n\}$, where $B = (b_i)_{1 \times n}$ represents the positive integer coverage requirement vectors. Let $\mathbb{Z}[x]/\langle \Phi_{b+1}(x) \rangle$ be the quotient ring of $\mathbb{Z}[x]$ by $\langle \Phi_{b+1}(x) \rangle$, i.e., $\mathbb{Z}[x]/\langle \Phi_{b+1}(x) \rangle$ is the ring obtained by taking the modulo $\Phi_{b+1}(x)$ operation on $\mathbb{Z}[x]$.

Observation 1 *[1] (1) $x^{b+1} = 1$ over $\mathbb{Z}[x]/\langle \Phi_{b+1}(x) \rangle$.*

(2) $\sum_{j=0}^{b} x^{jr} = 0$ over $\mathbb{Z}[x]/\langle \Phi_{b+1}(x) \rangle$, if $r \not\equiv 0 \pmod{b+1}$.

3 Design of Algorithm

3.1 Transform MSMC-MCP to Decide Whether $\alpha_{(t,b_1,b_2,\ldots,b_n)}$ is Zero

In this section, we transform MSMC-MCP to a new problem.

For any $F = \{(s_{i_1}, t_1), (s_{i_2}, t_2), \ldots, (s_{i_p}, t_p)\}$ in $\mathcal{F}$, we define the generating function of F as

$$h(F; s_1, s_2, \ldots, s_n) = \prod_{j=1}^{p} (1 + s_{i_j})^{t_j} \tag{1}$$

Note that $h(F; s_1, s_2, \ldots, s_n)$ is a polynomial in $s_1, s_2, \ldots, s_n$. Upon expanding $h(F; s_1, s_2, \ldots, s_n)$, each monomial corresponds to a submultiset of F. This observation is formally captured in Observation 2. For convenience, define

$$h'(F; s_1, s_2, \ldots, s_n) = \prod_{j=1}^{p} s_{i_j}^{t_j},$$

for any $F = \{(s_{i_1}, t_1), (s_{i_2}, t_2), \ldots, (s_{i_p}, t_p)\}$, and specially $h'(\emptyset; s_1, s_2, \ldots, s_n) = 1$.

Observation 2 *For any $F = \{(s_{i_1}, t_1), (s_{i_2}, t_2), \ldots, (s_{i_p}, t_p)\}$ in $\mathcal{F}$,*

$$h(F; s_1, s_2, \ldots, s_n) = \sum_{F' \subseteq F} C_{F'} h'(F'; s_1, s_2, \ldots, s_n),$$

where $C_{F'}$ is a positive integer for any $F' \subseteq F$.

Now, we define the generating function of $\mathcal{F}$ and study the relationship between the generating function of $\mathcal{F}$ and MSMC-MCP. Let $\mathcal{F} = \{F_1, F_2, \ldots, F_m\}$, and recall that each $F_i \in \mathcal{F}$ can be selected no more than d_i times. With introducing a more variable y, we define the generating function of $\mathcal{F}$ as follows:

$$g(\mathcal{F}; s_1, s_2, \ldots, s_n, y) = \prod_{i=1}^{m} (1 + y h(F_i; s_1, s_2, \ldots, s_n))^{d_i}. \tag{2}$$

Observation 3 *$g(\mathcal{F}; s_1, s_2, \ldots, s_n, y)$ is a nonnegative integer polynomial.*

By Observation 3, we assume

$$g(\mathcal{F}; s_1, s_2, \ldots, s_n, y) = \sum_{\ell \in \mathbb{N}^{n+1}} \alpha_\ell y^{\ell_1} \prod_{i=1}^{n} s_i^{\ell_{i+1}} \tag{3}$$

where α_ℓ are nonnegative integers for any $\ell \in \mathbb{N}^{n+1}$.

For convenience, when there is no ambiguity, denote $h'(F; s_1, s_2, \ldots, s_n)$, $h(F; s_1, s_2, \ldots, s_n)$, and $g(\mathcal{F}; s_1, s_2, \ldots, s_n, y)$, as $h'(F)$, $h(F)$, and $g(\mathcal{F})$ respectively. If we expand $g(\mathcal{F}; s_1, s_2, \ldots, s_n, y)$, we will find that MSMC-MCP has a t-solution if and only if the coefficient of $y^t \prod_{i=1}^{n} s_i^{b_i}$ in the expansion of $g(\mathcal{F}; s_1, s_2, \ldots, s_n, y)$ is not zero. Formally, we prove the following lemma.

Lemma 4. *There is a t-solution of MSMC-MCP if and only if* $\alpha_{(t,b_1,b_2,\ldots,b_n)} \neq 0$.

Proof. **Proof of sufficiency**. Assume $\{(F_{i_1}, t_{i_1}), (F_{i_2}, t_{i_2}), \ldots, (F_{i_k}, t_{i_k})\}$ is a t-solution of MSMC-MCP, where (F_{i_j}, t_{i_j}) denotes selecting F_{i_j} t_{i_j} times. Since this multiset is a t-solution, there exist multisets $F_{i_j,z} \subseteq F_{i_j}$ for $1 \leq j \leq k$ and $1 \leq z \leq t_{i_j}$ such that the multiset $\{F_{i_j,z} \mid 1 \leq j \leq k, 1 \leq z \leq t_{i_j}\}$ covers each $s_i \in S$ exactly b_i times. In other words, $\prod_{j=1}^{k}\prod_{z=1}^{t_{i_j}} h'(F_{i_j,z}) = \prod_{i=1}^{n} s_i^{b_i}$.

By Observation 2, the coefficient of $h'(F_{i_j,z})$ in the expansion of $h(F_{i_j})$ is nonzero for $1 \leq j \leq k$ and $1 \leq z \leq t_{i_j}$. Consequently, the coefficient of $\prod_{z=1}^{t_{i_j}} h'(F_{i_j,z})$ in the expansion of $h^{t_{i_j}}(F_{i_j})$ is nonzero for $1 \leq j \leq k$.

Since $\{(F_{i_1}, t_{i_1}), (F_{i_2}, t_{i_2}), \ldots, (F_{i_k}, t_{i_k})\}$ is a t-solution, we have $t_{i_j} \leq d_{i_j}$ for $1 \leq j \leq k$. Thus, the coefficient of $y^{t_{i_j}} \prod_{z=1}^{t_{i_j}} h'(F_{i_j,z})$ in the expansion of $(1 + yh(F_{i_j}))^{d_{i_j}}$ is nonzero for $1 \leq j \leq k$.

Since $\sum_{j=1}^{k} t_j = t$, by Equation (2), the coefficient of $y^t \prod_{j=1}^{k}\prod_{z=1}^{t_{i_j}} h'(F_{i_j,z}) = y^t \prod_{i=1}^{n} s_i^{b_i}$ in the expansion of $g(\mathcal{F})$ is nonzero, i.e., $\alpha_{(t,b_1,b_2,\ldots,b_n)} \neq 0$.

Proof of Necessity. Assume $\alpha_{(t,b_1,b_2,\ldots,b_n)} \neq 0$. By the definition of $g(\mathcal{F})$, there exist k with $1 \leq k \leq t$ and $t_{i_j} \leq d_{i_j}$ for $1 \leq j \leq k$ such that the coefficient of $y^t \prod_{i=1}^{n} s_i^{b_i}$ in the expansion of $\prod_{j=1}^{k} y^{t_{i_j}} h^{t_{i_j}}(F_{i_j})$ is not zero.

Firstly, we have $\sum_{j=1}^{k} t_{i_j} = t$. Since the coefficient of $\prod_{i=1}^{n} s_i^{b_i}$ in the expansion of $\prod_{j=1}^{k} h^{t_{i_j}}(F_{i_j})$ is not zero, then by Observation 2, there exist $F_{i_j,z} \subseteq F_{i_j}$ for $1 \leq j \leq k, 1 \leq z \leq t_{i_j}$ such that $\prod_{j=1}^{k}\prod_{z=1}^{t_{i_j}} h'(F_{i_j,z}) = \prod_{i=1}^{n} s_i^{b_i}$. This implies that $\{(F_{i_1}, t_{i_1}), (F_{i_2}, t_{i_1}), \ldots, (F_{i_k}, t_{i_k})\}$ covers each s_i at least b_i times for $1 \leq i \leq n$. Since $\sum_{j=1}^{k} t_{i_j} = t$ and $t_{i_j} \leq d_{i_j}$ for $1 \leq j \leq k$, $\{(F_{i_1}, t_{i_1}), (F_{i_2}, t_{i_1}), \ldots, (F_{i_k}, t_{i_k})\}$ is a t-solution. □

Now, MSMC-MCP transforms into deciding whether $\alpha_{(t,b_1,b_2,\ldots,b_n)}$ is zero. A natural approach to determine this is by utilizing Discrete Fourier Transform.

3.2 Solving MSMC-MCP by Discrete Fourier Transform over a Ring

Discrete Fourier Transform over a ring. $\mathbb{Z}[x]/\langle \Phi_{b+1}(x)\rangle$ is the quotient ring of $\mathbb{Z}$ by $\langle \Phi_{b+1}(x)\rangle$, i.e., $\mathbb{Z}[x]/\langle \Phi_{b+1}(x)\rangle$ is the ring obtained by taking the modulo $\Phi_{b+1}(x)$ operation on $\mathbb{Z}[x]$. Construct $D(\mathcal{F}; y)$ as following:

$$D(\mathcal{F}; y) = \sum_{j_1=0}^{b} \cdots \sum_{j_n=0}^{b} g(\mathcal{F}; x^{j_1}, x^{j_2}, \ldots, x^{j_n}, y) \prod_{z=1}^{n} x^{(b+1-b_z)j_z} \tag{4}$$

Now, we employ Lemma 5 to elucidate the relationship between Equation (4) and the original problem MSMC-MCP. Notably, $D(\mathcal{F}; y)$ is a polynomial in

y with coefficients in $\mathbb{Z}[x]/\langle \Phi_{b+1}(x)\rangle$. Thus, w.l.o.g, assume

$$D(\mathcal{F};y) = \sum_{j\in\mathbb{N}} \beta_j y^j \tag{5}$$

where $\beta_j \in \mathbb{Z}[x]/\langle \Phi_{b+1}(x)\rangle$.

Lemma 5. *For $t \in \mathbb{N}$, there is a t-solution of MSMC-MCP if and only if $\beta_t \neq 0$.*

Proof. Denote $A_q = \{\ell \in \mathbb{N}^{n+1} \mid \ell_1 = q, \ell_{i+1} \equiv b_i \pmod{b+1} \text{ for } 1 \le i \le n\}$ for $q \in \mathbb{N}$.

Claim. $\beta_j = \sum_{\ell\in A_j} \alpha_\ell (b+1)^n$ for $j \in \mathbb{N}$.

Proof. By Equations (3) and (4), we have

$$\begin{aligned} D(\mathcal{F};y) &= \sum_{j_1=0}^{b}\cdots\sum_{j_n=0}^{b} g(\mathcal{F};x^{j_1},x^{j_2},\ldots,x^{j_n},y)\prod_{z=1}^{n} x^{(b+1-b_z)j_z} \\ &= \sum_{\ell\in\mathbb{N}^{n+1}}\sum_{j_1=0}^{b}\sum_{j_2=0}^{b}\cdots\sum_{j_n=0}^{b} \alpha_\ell y^{\ell_1}\prod_{z=1}^{n} x^{(\ell_{z+1}-b_z+b+1)j_z}. \end{aligned} \tag{6}$$

Notably, $\mathbb{N}^{n+1} = (\bigcup_{q\in\mathbb{N}} A_q) \cup \{\ell \in \mathbb{N}^{n+1} \mid$ there exists a p with $1 \le p \le n$ such that $\ell_{p+1} \not\equiv b_p \pmod{b+1}\}$. First we consider a ℓ in $\{\ell \in \mathbb{N}^{n+1} \mid$ there exists a p with $1 \le p \le n$ such that $\ell_{p+1} \not\equiv b_p \pmod{b+1}\}$.

By (2) of Observation 1, for a ℓ in $\{\ell \in \mathbb{N}^{n+1} \mid$ there exists a p with $1 \le p \le n$ such that $\ell_{p+1} \not\equiv b_p \pmod{b+1}\}$, we have

$$\begin{aligned} &\sum_{j_1=0}^{b}\sum_{j_2=0}^{b}\cdots\sum_{j_n=0}^{b} \alpha_\ell y^{\ell_1}\prod_{z=1}^{n} x^{(\ell_{z+1}-b_z+b+1)j_z} \\ =&\sum_{j_1=0}^{b}\cdots\sum_{j_{p-1}=0}^{b}\sum_{j_{p+1}=0}^{b}\cdots\sum_{j_n=0}^{b} \alpha_\ell y^{\ell_1}\prod_{1\le z\le n, z\neq p} x^{(\ell_{z+1}-b_z+b+1)j_z}\sum_{j_p=0}^{b} x^{(\ell_{p+1}-b_p+b+1)j_p} \\ =&0 \quad \text{since} \sum_{j_p=0}^{b} x^{(\ell_{p+1}-b_p+b+1)j_p} = 0 \text{ by (2) of Observation 1.} \end{aligned}$$

Thus by Equation (6) and the definition of A_t,

$$\begin{aligned} D(\mathcal{F};y) &= \sum_{\ell\in\mathbb{N}^{n+1}}\sum_{j_1=0}^{b}\cdots\sum_{j_n=0}^{b} \alpha_\ell y^{\ell_1}\prod_{z=1}^{n} x^{(\ell_{z+1}-b_z+b+1)j_z} \\ &= \sum_{q\in\mathbb{N}}\sum_{\ell\in A_q}\sum_{j_1=0}^{b}\cdots\sum_{j_n=0}^{b} \alpha_\ell y^{\ell_1}\prod_{z=1}^{n} x^{(\ell_{z+1}-b_z+b+1)j_z} \end{aligned} \tag{7}$$

For $q \in \mathbb{N}$ and $\ell \in A_q$, by the definition of A_q, w.l.o.g, assume $\ell_{z+1} = r_z(b+1) + b_z$ for $1 \leq z \leq n$. Thus, for $q \in \mathbb{N}$ and $\ell \in A_q$, we obtain

$$
\begin{aligned}
&\sum_{j_1=0}^{b}\sum_{j_2=0}^{b}\cdots\sum_{j_n=0}^{b} \alpha_\ell y^{\ell_1} \prod_{z=1}^{n} x^{(\ell_{z+1}-b_z+b+1)j_z} \\
=&\sum_{j_1=0}^{b}\sum_{j_2=0}^{b}\cdots\sum_{j_n=0}^{b} \alpha_\ell y^{q} \prod_{z=1}^{n} x^{(b_z+r_z(b+1)-b_z+b+1)j_z} \\
=&\sum_{j_1=0}^{b}\sum_{j_2=0}^{b}\cdots\sum_{j_n=0}^{b} \alpha_\ell y^{q} \quad \text{since} x^{b+1} = 1 \text{ by (1) of Observation1} \\
=&(b+1)^n \alpha_\ell y^q.
\end{aligned}
$$

Thus by Equation (7),

$$
\begin{aligned}
D(\mathcal{F}; y) &= \sum_{q\in\mathbb{N}}\sum_{\ell\in A_q}\sum_{j_1=0}^{b}\sum_{j_2=0}^{b}\cdots\sum_{j_n=0}^{b} \alpha_\ell y^{\ell_1} \prod_{z=1}^{n} x^{(\ell_{z+1}-b_z+b+1)j_z} \\
&= \sum_{q\in\mathbb{N}}\sum_{\ell\in A_q} (b+1)^n \alpha_\ell y^q.
\end{aligned}
$$

Since we assume that $D(\mathcal{F}; y) = \sum_{j\in\mathbb{N}} \beta_j y^j$, then $\beta_j = \sum_{\ell\in A_j} (b+1)^n \alpha_\ell$.

Thus, this claim holds. □

Proof of Sufficiency. By Lemma 4, existing a t-solution implies $\alpha_{(t,b_1,b_2,\ldots,b_n)} \neq 0$. Since $g(\mathcal{F}; s_1, s_2, \ldots, s_n, y)$ is a nonnegative integer polynomial by Observation 3 and $(t, b_1, b_2, \ldots, b_n) \in A_t$ by the definition of A_t, we have $\beta_t = \sum_{\ell\in A_t} (b+1)^n \alpha_\ell \neq 0$.

Proof of Necessity. If $\beta_t = \sum_{\ell\in A_t} (b+1)^n \alpha_\ell \neq 0$, then considering that $g(\mathcal{F}; s_1, s_2, \ldots, s_n, y)$ is a nonnegative integer polynomial by Observation 3, there is an $\ell \in A_t$ with $\alpha_\ell \neq 0$. W.l.o.g, let $\ell' = (t, \ell'_2, \ldots, \ell'_{n+1}) \in A_t$ with $\alpha_{\ell'} \neq 0$. By the definition of A_t, $\ell'_{z+1} \equiv b_z \pmod{b+1}$ for $1 \leq z \leq n$. As $\ell' \in \mathbb{N}^{n+1}$ by the definition of A_t, it follows that $\ell'_{z+1} \geq b_z$ for $1 \leq z \leq n$. Let P be the original MSMC-MCP which requests finding minimum number of multisets from $\mathcal{F}$ required to cover $s_i \in S$ at least b_i times where each multiset $F_i \in \mathcal{F}$ can be only selected no more than d_i times. Consider a new MSMC-MCP P' which requests finding minimum number of multisets from $\mathcal{F}$ required to cover $s_i \in S$ at least ℓ'_{i+1} times where each multiset $F_i \in \mathcal{F}$ can be only selected no more than d_i times. Since $\ell' = (t, \ell'_2, \ldots, \ell'_{n+1})$ and $\alpha_{\ell'} \neq 0$, then by Lemma 4, there is a t-solution for P'. In other words, there are t multisets in $\mathcal{F}$ that cover each $s_i \in S$ at least ℓ'_{i+1} times. Since $\ell'_{i+1} \geq b_i$ for $1 \leq i \leq n$, it's worth noting that a t-solution for P' also constitutes a t-solution for the original problem P. Thus there is a t-solution of P. □

Construct a Ring to Obtain the Coefficient β_t of $D(\mathcal{F}; y)$. Denote R_{b+1} as $\mathbb{Z}[x]/\langle \Phi_{b+1}(x) \rangle$. Recall that for a ring R, $R[x]$ is the univariate polynomial ring over ring R. Let k be a positive integer. Consider the quotient ring $R_{b+1}[w]/\langle w^{k+1} \rangle$, i.e., $R_{b+1}[w]/\langle w^{k+1} \rangle$ is the ring obtained by taking the modulo w^{k+1} operation on $R_{b+1}[w]$. Thus, replacing the variable y in $D(\mathcal{F}; y)$ with the element w, we have

$$D(\mathcal{F}; w) = \sum_{j=0}^{k} \beta_j w^j, \tag{8}$$

over $R_{b+1}[w]/\langle w^{k+1} \rangle$.

Theorem 1. *For a positive integer k, there is a t-solution of MSMC-MCP with $t \le k$ if and only if $D(\mathcal{F}; w) \ne 0$ over $R_{b+1}[w]/\langle w^{k+1} \rangle$.*

Proof. By definition of $R_{b+1}[w]/\langle w^{k+1} \rangle$, $w^p = 0$ for $p \ge k+1$. By Lemma 5, there is a t-solution of MSMC-MCP with $t \le k$ if and only if there is a $\beta_j \ne 0$ for $1 \le j \le k$. Moreover by Equation (8), there is a t-solution of MSMC-MCP with $t \le k$ if and only if $D(\mathcal{F}; w) \ne 0$ over $R_{b+1}[w]/\langle w^{k+1} \rangle$. □

By Theorem 1, we design the following Algorithm 1 to test the existence of t-solution of P with $t \le k$.

Algorithm 1. TTS: Test if there is a t-solution of a MSMC-MCP P with $t \le k$.

Input: A MSMC-MCP P and an integer k.
Output: Existence of t-solution of P with $t \le k$.

```
1: Compute f = D(F; w) over R_{b+1}[w]/⟨w^{k+1}⟩ by Equations (4), (2), (1);
2: if f ≠ 0 then
3:     return Yes;
4: else
5:     return No;
6: end if
```

Based on Algorithm 1, we drive a binary search to find the minimum t such that there is a t-solution of the given MSMC-MCP P. Specifically, the binary search is implemented as the following Algorithm 2 (Algorithm *Bsearch*). And $Bseach(P, 0, bn+1)$ returns the minimum t such that there is a t-solution of P.

Algorithm 2. BSearch:Find the minimum t with $left \leq t < right$ such that there is a t-solution of the given MSMC-MCP P.

Input: A MSMC-MCP P and two integers $left, right$.
Output: The minimum t if there is a t-solution of P such that $left \leq t < right$, and $right$ otherwise.

if $right = left + 1$ **then**
  if $TTS(P, left) = Yes$ **then**
    return $left$
  else
    return $right$
  end if
else
  if $TTS(P, \lfloor \frac{left+right}{2} \rfloor - 1) = Yes$ **then**
    return $BSearch(P, left, \lfloor \frac{left+right}{2} \rfloor)$
  else
    return $BSearch(P, \lfloor \frac{left+right}{2} \rfloor, right)$
  end if
end if

4 Complexity of Solving MSMC-MCP

4.1 Arithmetic Complexity

Arithmetic complexity is the time complexity with regard to the time complexity of each addition and multiplication as a constant.

Lemma 6. *For any $F \in \mathcal{F}$ and $s_1, s_2, \ldots, s_n$, the arithmetic complexity of computing $h(F; s_1, s_2, \ldots, s_n)$ is $O(n \ln b)$.*

Proof. Referring to Equation (1), $(1 + s_{i_j})^{t_j}$ requires at most $O(\ln t_j)$ multiplications by Lemma 1(Exponentiation By Squaring Method). Considering that we only aim to cover s_{i_j} with b_{i_j} times, and t_j denotes the occurrences of s_{i_j} in F, we can assume, w.l.o.g, that $t_j \leq b_{i_j} \leq b$. Since p in Equation (1) is no more than n, the arithmetic complexity of computing $h(F; s_1, s_2, \ldots, s_n)$ is $O(n \ln b)$. □

Lemma 7. *For any $s_1, s_2, \ldots, s_n, y$, the arithmetic complexity of computing $g(\mathcal{F}; s_1, s_2, \ldots, s_n, y)$ is $O(|\mathcal{F}| n \ln b)$.*

Proof. Referring to Equation (2), by Lemma 6, the arithmetic complexity of computing $1 + yh(F_i; s_1, s_2, \ldots, s_n)$ is $O(n \ln b)$ for any $F_i \in \mathcal{F}$ and $s_1, s_2, \ldots, s_n$, y. Additionally, $(1 + yh(F_i; s_1, s_2, \ldots, s_n))^{d_i}$ requires at most $O(\ln d_i)$ multiplications by Lemma 1(Exponentiation By Squaring Method). Thus, the arithmetic complexity of computing $(1 + yh(F_i; s_1, s_2, \ldots, s_n))^{d_i}$ is $O(n \ln b + \ln d_i)$. Since we only need to cover s_i with b_i times and F_i can be selected at most d_i times, w.l.o.g, we can assume that $d_i \leq b$(Recall that $b = \max\{b_1, b_2, \ldots, b_n\}$). Consequently, the arithmetic complexity of computing $g(\mathcal{F}; s_1, s_2, \ldots, s_n, y)$ is $O(|\mathcal{F}| n \ln b)$. □

Lemma 8. *The arithmetic complexity of computing $D(\mathcal{F}; y)$ in Equation (4) is $O((b+1)^n|\mathcal{F}|n \ln b)$.*

Proof. For any integers $j_1, j_2, \ldots, j_n$ where $0 \leq j_p \leq b$ for $1 \leq p \leq n$, and a fixed integer z with $1 \leq z \leq n$, the computation of $x^{(b+1-b_z)j_z}$ can be accomplished within $O(\ln(b+1-b_z)j_z) = O(\ln b)$ arithmetic complexity by Lemma 1(Exponentiation By Squaring Method). Consequently, for any integers $j_1, j_2, \ldots, j_n$ where $0 \leq j_p \leq b$ for $1 \leq p \leq n$, $\prod_{z=1}^{n} x^{(b+1-b_z)j_z}$ in Equation (4) can be accomplished within $O(n \ln b)$ arithmetic complexity. For any $j_1, j_2, \ldots, j_n$ where $0 \leq j_p \leq b$ for $1 \leq p \leq n$, the computations $x^{j_1}, x^{j_2}, \ldots, x^{j_n}$ can be performed within $O(n \ln b)$ arithmetic complexity by Lemma 1(Exponentiation By Squaring Method). Consequently, by Lemma 7, $g(\mathcal{F}; x^{j_1}, x^{j_2}, \ldots, x^{j_n}, y)$ can be computed within $O(n \ln b + |\mathcal{F}|n \ln b) = O(|\mathcal{F}|n \ln b)$ arithmetic complexity. Thus, for each $j_1, j_2, \ldots, j_n$ where $0 \leq j_p \leq b$ for $1 \leq p \leq n$, we have that $g(\mathcal{F}; x^{j_1}, x^{j_2}, \ldots, x^{j_n}, y) \prod_{z=1}^{n} x^{(b+1-b_z)j_z}$ can be accomplished within $O(|\mathcal{F}|n \ln b + n \ln b) = O(|\mathcal{F}|n \ln b)$ arithmetic complexity. Thus the arithmetic complexity of computing $D(\mathcal{F}; y)$ in Equation (4) is $O((b+1)^n|\mathcal{F}|n \ln b)$. □

4.2 Time Complexity and Space Complexity

Recall that R_{b+1} is $\mathbb{Z}[x]/\langle \Phi_{b+1}(x) \rangle$. Let's consider the representation of an element e in $R_{b+1}[w]/\langle w^{k+1} \rangle$. Since $\deg(\Phi_{b+1}(x)) \leq b+1$, for $a \in R_{b+1}$, it can be expressed as $\sum_{j=0}^{b} a_j x^j$ with $a_j \in \mathbb{Z}$ for $0 \leq j \leq b$. Furthermore, for $e \in R_{b+1}[w]/\langle w^{k+1} \rangle$, its representation can be given as $\sum_{z=0}^{k}(\sum_{j=0}^{b} e_{j,z} x^j) w^z$ where $e_{j,z} \in \mathbb{Z}$ for $0 \leq j \leq b, 0 \leq z \leq k$.

Theorem 2. *MSMC-MCP can be solved within $O^*(b^4)$ space complexity and $O^*(b^6(b+1)^n)$ time complexity.*

Proof. For a given MSMC-MCP P and an integer k, Algorithm 1 tests if there is a t-solution for P with $t \leq k$. Since every multiset F in $\mathcal{F}$ covers at least one element s at least once, it is feasible to choose nb multisets from $\mathcal{F}$ in a way that ensures the coverage of each s_i at least b times, where $1 \leq i \leq n$. Thus w.l.o.g, we consider $k \leq nb$. And since each element need to cover at most b times, w.l.o.g, we assume $d_j \leq b$ for $1 \leq j \leq m$.

Space Complexity. Referring to representation of an element $e \in R_{b+1}[w]/\langle w^{k+1} \rangle$, it can be expressed as $\sum_{z=0}^{k}(\sum_{j=0}^{b} e_{j,z} x^j) w^z$, where $e_{j,z} \in \mathbb{N}$. Consider an element $e \in R_{b+1}[w]/\langle w^{k+1} \rangle$ during the computation of f in line 1 of Algorithm 1. Let $N_z = \{\ell \in \mathbb{N}^{n+1} \mid \ell_1 = z\}$ for $z \in \mathbb{N}$. By Equation (4), we have $\sum_{j=0}^{b} e_{j,z} \leq \sum_{j_1=0}^{b} \cdots \sum_{j_n=0}^{b} \sum_{\ell \in N_z} \alpha_\ell = (b+1)^n \sum_{\ell \in N_z} \alpha_\ell$ for $1 \leq z \leq k$. Recall that $b = \max\{b_1, b_2, \ldots, b_n\}$ and $|\mathcal{F}| = m$. Since we aim to cover s_{i_j} with b_{i_j} times, and t_j in Equation (1) represents the occurrences of s_{i_j} in F, we can assume, w.l.o.g, that $t_j \leq b_{i_j} \leq b$. Thus for $F_i \in \mathcal{F}$, by Equation (1), the sum of coefficients in $g(F_i; s_1, s_2, \ldots, s_n)$ is no more than $2^{bp} \leq 2^{bn}$. And since $|\mathcal{F}| = m$

and $d_j \leq b$ for $1 \leq j \leq m$, we have $mb \leq b(b+1)^n$. Therefore by Equation (1) and (2), we have $\sum_{\ell \in N_z} \alpha_\ell \leq \binom{mb}{z} 2^{bnz} \leq b^z(b+1)^{nz}2^{bnz}$ for $1 \leq z \leq k$. Since we consider $k \leq nb$, we have $\sum_{z=0}^{k}\sum_{j=0}^{b} e_{j,z} \leq \sum_{z=0}^{k}(b+1)^n \sum_{\ell \in N_z} \alpha_\ell \leq (nb+1)(b+1)^n b^{nb}(b+1)^{n^2b}2^{n^2b^2}$. Since element $e \in R_{b+1}[w]/\langle w^{k+1}\rangle$ can be presented a nk-length integer vector, the needed space is no more than $b^2 n \ln\left((nb+1)(b+1)^n b^{nb}(b+1)^{n^2b}2^{n^2b^2}\right) = O(n^3b^4 \ln nb)$. Then the space complexity is $O^*(b^4)$ where O^* suppresses a polynomial about $\ln b$ and n.

Time Complexity. Recall that R_{b+1} is $\mathbb{Z}[x]/\langle \Phi_{b+1}(x)\rangle$. When computing f in line 1 of Algorithm 1, we perform the computation over $R_{b+1}[w]/\langle w^{k+1}\rangle$. Since the needed space for element $e \in R_{b+1}[w]/\langle w^{k+1}\rangle$, when computing f, is no more than $O(n^3b^4\ln nb)$, all additions, subtractions, and multiplications for $e_{i,j} \in \mathbb{Z}$ can be accomplished within $O(n^3b^4\ln^2 nb)$ time complexity by Lemma 3. Thus Lemma 2 implies that additions, subtractions, and multiplications over $\mathbb{Z}[x]$ can be accomplished within $O(\mathrm{d}^*\ln \mathrm{d}^* \ln\ln \mathrm{d}^* n^3b^4\ln^2 nb)$ time complexity, where d^* is the degree of the input polynomials. Recall $\Phi_r(x) = \prod_{1\leq k\leq r, \gcd(k,r)=1}(x - e^{2\pi i\frac{k}{r}})$, implying $\deg(\Phi_{b+1}(x)) \leq b+1$. Considering the representation of $e \in \mathbb{Z}[x]/\langle \Phi_{b+1}(x)\rangle$, we express e as $\sum_{i=0}^{b} a_i x^i$ with $a_i \in \mathbb{Z}$ for $0 \leq j \leq b$. The maximum degree of element e in $\mathbb{Z}[x]/\langle \Phi_{b+1}(x)\rangle$ is at most b. Therefore, the additions, subtractions, and multiplications over $\mathbb{Z}(x)/\langle \Phi_{b+1}(x)\rangle$ for $1 \leq j \leq r$ can be accomplished within

$$O(b\ln b \ln\ln b\; n^3b^4\ln^2 nb + (b\ln b)n^3b^4\ln^2 nb) = O(n^3b^5\ln^2 nb \ln b\ln\ln b).$$

The $(b\ln b)n^3b^4\ln^2 nb$ part in the equation arises from the division by $\Phi_{b+1}(x)$, by (2) and (3) of Lemma 3. Similarly, the additions, subtractions and multiplications over $R_{b+1}[w]/\langle w^{k+1}\rangle$ can be accomplished within

$$O(k\ln k\ln\ln k\; n^3b^5\ln^2 nb\ln b\ln\ln b)$$

time complexity. Since $k \leq nb$, the additions, subtractions, and multiplications over $R_{b+1}[w]/\langle w^{k+1}\rangle$ can be accomplished within

$$O(nb\ln nb\ln\ln nb\; n^3b^5\ln^2 nb\ln b\ln\ln b) = O^*(b^6)$$

time complexity.

By the definition of arithmetic complexity, the time complexity of computing Equation (4) is the product of the arithmetic complexity of computing Equation (4) and the time complexity of each multiplication over $R_{b+1}[w]/\langle w^{k+1}\rangle$. Since the arithmetic complexity of computing Equation (4) is $O(n(b+1)^n|\mathcal{F}|\ln b)$ by Lemma 8, the time complexity of computing Equation (4) is $O^*(b^6(b+1)^n|\mathcal{F}|)$. Finally, by binary search (Algorithm 2), MSMC-MCP can be solved within $O^*(b^6(b+1)^n|\mathcal{F}|\ln bn) = O^*(b^6(b+1)^n|\mathcal{F}|)$ time complexity. □

5 Conclusions

We present a novel algorithm for MSMC-MCP, significantly improving space complexity. Our method may offer insights for algorithm development in related problems, including the cover integer problem. Our algorithm's reduced space requirements enhance practical applicability. However, for large b (e.g., $b = O(2^n)$), it may exhibit high space consumption. Hence, the open question remains: is there an exact solution for MSMC-MCP with time complexity $O^*(|\mathcal{F}|(b+1)^n)$ and space complexity $f(n, \ln b)$, where $f(n, \ln b)$ represents a polynomial in n and $\ln b$, thus constituting a genuine polynomial space algorithm?

Acknowledgments. This work is supported by National key R&D Program of China under Grant 2019YFA0706401 and National Natural Science Foundation of China under Grants 62172116, 62172014, 62172015, 62272009, 62372008, 62332006.

References

1. Al-Kateeb, A.Q.M.: Structures and properties of cyclotomic polynomials. Master Thesis, North Carolina State University, The USA (2016)
2. Björklund, A., Husfeldt, T., Koivisto, M.: Set partitioning via inclusion-exclusion. SIAM J. Comput. **39**(2), 546–563 (2009)
3. Cantor, D.G., Kaltofen, E.: Fast multiplication of polynomials over arbitrary rings. Acta Informatica **28** (1987)
4. Cohn, P.M.: Introduction to Ring Theory, 3th ed. Springer (2004)
5. Cormen, T.H., Leiserson, C.E., Rivest, R.L., Stein, C.: Introduction to Algorithms, 4th ed. MIT press (2022)
6. Gordon, D.M.: A survey of fast exponentiation methods. J. Algorithms **27**(1), 129–146 (1998)
7. Hasanaj, E., Alavi, A., Gupta, A., Póczos, B., Bar-Joseph, Z.: Multiset multicover methods for discriminative marker selection. Cell Rep. Meth. **2**(11) (2022)
8. Hua, Q.S., Wang, Y., Yu, D., Lau, F.C.: Set multi-covering via inclusion-exclusion. Theoret. Comput. Sci. **410**(38–40), 3882–3892 (2009)
9. Hua, Q.S., Wang, Y., Yu, D., Lau, F.C.: Dynamic programming based algorithms for set multicover and multiset multicover problems. Theoret. Comput. Sci. **411**(26–28), 2467–2474 (2010)
10. Kolliopoulos, S.G., Young, N.E.: Approximation algorithms for covering/packing integer programs. J. Comput. Syst. Sci. **71**(4), 495–505 (2005)
11. Lokshtanov, D., Nederlof, J.: Saving space by algebraization. In: Proceedings of the Forty-second ACM Symposium on Theory of Computing, pp. 321–330 (2010)
12. Nederlof, J.: Inclusion exclusion for hard problems. Master Thesis, Utrecht University, The Netherlands (2008)
13. Pan, V.: Complexity of computations with matrices and polynomials. SIAM Rev. **34**(2), 225–262 (1992)
14. Proakis, J.G.: Digital signal processing: principles, algorithms, and applications, 4th ed. Pearson Education India (2007)
15. Srinivasan, A.: Improved approximation guarantees for packing and covering integer programs. SIAM J. Comput. **29**(2), 648–670 (1999)

16. Yeasmin, N.: k-coverage problems and solutions in wireless sensor networks: a survey. Int. J. Comput. Appl. **100**(17), 1–6 (2014)
17. Zhou, X., Yi, Z., Liu, Y., Huang, K., Huang, H.: Survey on path and view planning for UAVs. Virtual Reality Intell. Hardware **2**(1), 56–69 (2020)

Generative Flow Networks for Influence Maximization in Social Networks

Zizhen Zhang[1], Deying Li[1(✉)], Yongcai Wang[1], Wenping Chen[1], and Yuqing Zhu[2]

[1] Renmin University of China, Beijing 100872, China
{zizhenzhang,deyingli,ycw,chenwenping}@ruc.edu.cn
[2] California State University, Los Angeles, USA
yuqing.zhu@calstatela.edu

Abstract. Influence maximization (IM) involves selecting a set of initial users from a social network to maximize the expected number of influenced users. In recent years, learning-based combinatorial optimization (CO) methods have emerged to develop generalized policies for specific CO problems on graphs. However, these algorithms struggle to manage diversified diffusion patterns, which directly limits their generalization ability. In this paper, we use a reverse influence sampling technique to simplify influence maximization to stochastic maximum coverage on hyperedges. We subsequently formulated the stochastic maximum coverage problem as a generative process, called GFlowIM. This model generates various samples through sequential actions, with probabilities precisely proportional to a predefined reward function. By training on multiple graphs, we learn a transferable seed selection policy that can generalize to unseen test graphs. Extensive experiments demonstrate that our method outperforms recent learning-based and traditional methods on both real and synthetic datasets for the IM problem.

Keywords: Influence maximization · Hypergraph · Generative flow networks

1 Introduction

Social influence among individuals is pivotal in decision-making and information acquisition. The advent of online social networks has magnified this influence, enabling it to spread on an unprecedented scale. The study of understanding, predicting and controlling social influence has evolved into a significant field known as computational social influence [5]. One of the most actively researched algorithmic problems within this field is influence maximization (IM) [10,11], which was originally inspired by viral marketing [7]. Conceptually, IM involves identifying a small number of key individuals within a network who can maximize

Dr. Li is supported in part by the National Natural Science Foundation of China Grant No. 12071478.

Y. Chen et al. (Eds.): COCOON 2024, LNCS 15162, pp. 287–299, 2025.
https://doi.org/10.1007/978-981-96-1093-8_24

the spread of influence. Kempe et al. [10] formulated IM as a combinatorial optimization problem on graphs, and their framework has gained broad acceptance within the research community [3,5,15,22,23,25,26].

Identifying a (near) optimal seed set to maximize influence in a network is a classic challenge in combinatorial optimization, primarily due to the hardness of the problem. Traditional methods for solving the IM problem can be categorized into heuristic algorithms and general greedy algorithms. The former approach reduces the time required for seed set generation. However, it may also lead to a significantly lower number of influenced nodes generated by the seed set. If the objective function exhibits submodularity, the latter approach can offer a $(1-1/e)$-approximation guarantee. These graphs are often dynamic in nature and the greedy algorithm needs to be run repeatedly at regular intervals. Recently, studies have focused on using learning-based methods to develop generalized policies for combinatorial optimization problems on graphs, including influence maximization [4,12]. The key idea is to use a Graph Neural Network (GNN) to embed nodes of the input graph. By integrating GNNs with deep reinforcement learning (RL), a greedy node selection policy is learned from a set of training graphs, with the aim of generalizing this policy to unseen test graphs.

Despite numerous recent attempts to apply deep RL to the IM problem, these approaches inherently face fundamental limitations. (1) Existing methods usually adopt GNN specially designed according to the information diffusion model. Devising GNN architectures that generalize and simultaneously capture essential patterns remains an challenge. However, increased expressiveness might negatively impact generalization. (2) Standard RL algorithms are designed to maximize cumulative rewards, which often leads to a lack of diversity in solutions. Consequently, the agent may become stuck in solutions accessible by many trajectories, limiting its ability to generate a wide range of candidate solutions.

To tackle these limitations, we propose GFlowIM, a **G**enerative **Flow** Networks for **I**nfluence **M**aximization (GFlowIM). Instead of designing a neural network to model the influence diffusion process, we employ the reverse influence sampling technique to reduce the influence maximization problem into a stochastic maximum coverage problem. This approach enhances our method's generalization capability. Unlike previous RL methods, we transform the sampling procedure in maximum coverage into a generation problem, progressively adding nodes to the seed set until the complete set is formed. Our method generates diverse samples through sequential actions with probabilities precisely aligned with a predefined reward function, enabling it to learn the optimal node selection policy more effectively and avoid local optima. Additionally, GFlowIM exhibits superior exploration capabilities compared to RL-based methods, resulting in faster convergence and enhanced performance.

2 Related Work

Influence Maximization was first introduced as an algorithmic problem by Domingos et al. [7], sparking extensive research and applications over the following decade. Kempe et al. [10] later formulated it as a discrete optimization problem on graphs and proposed a greedy algorithm with a (1-1/e) approximation guarantee. Subsequent improvements to the greedy algorithm, such as CELF [15], RIS [3], and IMM [26], have enhanced efficiency through more effective spread estimation techniques. The above works can be categorised as traditional IM methods which achieve near optimal solutions.

Learning-based approaches leverage deep learning to overcome the limitations of traditional IM methods, particularly their lack of generalization capability. Early research, such as the works by [1,18], were the first to integrate reinforcement learning with IM. These pioneering studies sparked a wide range of subsequent research that utilized deep reinforcement learning to address the IM problem. Existing state-of-the-art solutions [4,17,19] follow a similar paradigm. They learn latent embeddings of nodes or networks, using the current node embedding as the agent's state to select the next seed node as an action, with the reward being its marginal influence gain. In addition to reinforcement learning-based IM methods, there are also approaches [14,21] that solely utilize graph neural networks to encode social influence into node embeddings and guide the node selection process. The main challenge with current learning-based IM methods is their model complexity and adaptability, which still fall short compared to traditional methods. Specifically, current ML-based algorithms struggle to handle diverse diffusion patterns and cannot ensure the same quality of solutions or scalability as traditional models.

3 Preliminaries

3.1 Problem Statement

In this study, we concentrate on the Independent Cascade (IC) [10] to tackle the influence maximization problem. The social network is represented by a directed graph $G = (V, E, p)$, where V denotes the set of nodes and $E \subseteq V \times V (|E| = m)$ denotes the directed edges. Each directed edge $(u, v) \in E$ is associated with a probability $p_{u,v}$. During the information diffusion process, an activated node u gets a single opportunity to activate each of its outgoing neighbors v, with the success occurring independently with probability $p_{u,v}$. The live-edge graph $L = (V, L(E))$ is a stochastic subgraph of G, where each edge $(u, v) \in E$ is included in $L(E)$ based on the independent probability $p_{u,v}$. The diffusion process can be described as follows: At time $\tau = 0$, a live-edge graph L is sampled, and the nodes in the seed set $S \subseteq V$ are activated. In each subsequent discrete time step $\tau = 1, 2, \ldots$, if a node u was activated at time $\tau - 1$, its outgoing neighbors in L are activated at time τ. This propagation continues until no new nodes are activated in a given time step.

Definition 1 (Influence Maximization Problem). *Given a seed set S, the influence spread of S, denoted as $\sigma(S)$, represents the expected number of nodes activated from the seed set. This is expressed as $\sigma(S) = \mathbb{E}_{L \sim G}[|\Gamma(S, L)|]$, where $\Gamma(S, L)$ denotes the set of nodes reachable from S in the graph L. The objective* *in the influence maximization problem is to identify a seed set S of size at most k that maximizes the expected influence. Formally, this involves finding $S^* = \arg\max_{|S|\leq k} \sigma(S)$.*

Learning for IM (LIM). The objective of LIM is to learn an algorithm (or policy) $\pi_\theta(\cdot)$: $G \rightarrow \{0,1\}^n$, say a neural network (NN) parameterized by θ to solve IM problems. Given the IM problem over graphs drawn from distribution $\mathcal{D}$, we expect π_θ to generate a valid solution $\hat{S} = \pi_\theta(G)$ and maximize $\sigma(\hat{S})$ on an unseen graph G generated from $\mathcal{D}$.

3.2 Reverse Influence Sampling

Since computing $\sigma(S)$ is #P-hard [6], most methods for estimating the expected influence $\sigma(S)$ employ a sampling technique known as Reverse Influence Sampling (RIS) [3]. This approach is based on the concept of a *random reverse reachable (RR) set*, defined as follows:

Definition 2 (Reverse Reachable Set[3]). *Under the Independent Cascade (IC) model, a reverse reachable (RR) set with a root node v, randomly selected from the set of nodes V, is denoted as R_v. This RR set represents the random set of nodes that can reach node R_v during one reverse propagation process. Specifically, R_v is determined by randomly sampling a live-edge graph $L = (V, L(E))$ and including all nodes in V that can reach v within this live-edge graph L.*

Lemma 1 ([3]). *Let R be a random RR set, then $\sigma(S) = n \cdot \Pr(S \cap R \neq \emptyset)$.*

Lemma 1 demonstrates that the expected influence of any seed set S can be estimated using random RR sets. We say that S covers an RR set R if $S \cap R \neq \emptyset$. Assume we generate a collection $\mathcal{R}$ of random RR sets. The coverage $\Lambda_{\mathcal{R}}(S)$ of a seed set S with respect to $\mathcal{R}$ is defined as the number of RR sets in $\mathcal{R}$ that are covered by S. Then, $n \cdot \Lambda_{\mathcal{R}}(S)/|\mathcal{R}|$ offers an unbiased estimate of the expected influence of S. Likewise, if S^* covers the majority of the RR sets, it is likely to maximize $\sigma(S^*)$.

In this paper, we treat the reverse reachable set as a hyperedge, and we first introduce some preliminary knowledge of hypergraphs. Consider a hypergraph $\mathcal{H} = (V, \mathcal{E})$ comprising a set of nodes $V = \{v_i\}_{i=1}^N$ and a set of hyperedges $\mathcal{E} = \{\epsilon_j\}_{j=1}^M$. Each hyperedge $\epsilon_j \in \mathcal{E}$ is a subset of nodes, $\epsilon_j \subseteq V$. The (i, j)-th entry of the incidence matrix $\mathbf{H} \in \mathbb{R}^{N \times M}$ of $\mathcal{H}$ is 1 if $v_i \in \epsilon_j$, and 0 otherwise. Then $\mathbf{D} \in \mathbb{R}^{N \times N}$ and $\mathbf{B} \in \mathbb{R}^{M \times M}$ be two diagonal matrices, which denote the degrees of nodes and hyperedges, respectively.

3.3 GFlowNets

Generative Flow Networks (GFlowNets) [2] were initially introduced to aid in discovering diverse modes within an unnormalized distribution. These networks belong to a class of generative models that operate over a discrete and structured sample space $\mathcal{X}$. Here, we provide a brief summary of the formulation and key training algorithms for GFlowNets. For a more detailed explanation, please refer to [2]. A GFlowNet's structure is defined by a directed acyclic graph (DAG) over a set of states $s \in \mathcal{S}$. The sample space, where we aim to define a distribution, is a subset of this state space: $\mathcal{X} \subseteq \mathcal{S}$. Samples $s \in \mathcal{X}$ are constructed sequentially by traversing the edges of the DAG, starting from a fixed initial state s_0. We also define a special absorbing state s_f, known as the terminal state, which marks the end of the sequential construction. Certain states $s \in \mathcal{X}$ are connected to s_f and are referred to as complete states. For instance, Bengio et al. [2] use a GFlowNet to define a distribution over molecules, where $\mathcal{X}$ represents the space of all complete molecules. These molecules are incrementally constructed by attaching new blocks to atoms within a potentially partially constructed molecule, corresponding to states in $\mathcal{S} \backslash \mathcal{X}$. A GFlowNet is structurally analogous to a Markov Decision Process (MDP) with deterministic dynamics. In addition to the DAG structure over states, each complete state $s \in \mathcal{X}$ is assigned a reward $R(s) \geq 0$, indicating a "preference" for certain states. By convention, $R(s) = 0$ for any incomplete state $s \in \mathcal{S} \backslash \mathcal{X}$. The goal of a GFlowNet is to find a flow that satisfies the following flow-matching condition for all states $s' \in \mathcal{S}$:

$$\sum_{s \in \text{Parent}(s')} F_\theta(s \to s') - \sum_{s'' \in \text{Child}(s')} F_\theta(s' \to s'') = R(s'), \tag{1}$$

where $F_\theta(s \to s') \geq 0$ is a scalar representing the flow from state s to s', typically parameterized by a neural network. In other words, the total flow entering state s' must be equal to the flow exiting s', plus an additional residual $R(s')$. To learn the parameters θ of the flow using Stochastic Gradient Descent (SGD), we can reformulate the flow-matching condition into a regression problem. This can be achieved by applying a least squares objective over sampled states, i.e.,

$$\mathcal{L}_{FM}(s;\theta) = \left(\log \frac{\sum_{s' \in \text{Parent}(s)} F(s' \to s)}{\sum_{s'' \in \text{Child}(s)} F(s \to s'')} \right)^2 \tag{2}$$

If the conditions in (1) are satisfied for all states s', a GFlowNet generates complete states $s \in \mathcal{X}$ with a probability proportional to $R(s)$. Starting from the initial state s_0, a complete trajectory $(s_0, s_1, \ldots, s_T, s, s_f)$ is sampled using the transition probability defined as the normalized outgoing flow $P(s_{t+1}|s_t) \propto F_\theta(s_t \to s_{t+1})$. With the conventions $s_{T+1} = s$ and $s_{T+2} = s_f$, the state s is then sampled with probability $P(s) \propto R(s)$.

4 Method

In this section, our proposed model GFlowIM comprises two stages: (1) **Hypergraph Construction**. This phase iteratively generates random hyperedges on a graph G, until a certain stopping condition is met. (2) **Building seed set and training GFlowIM.** Use the GFlowIM for the Max-coverage problem to

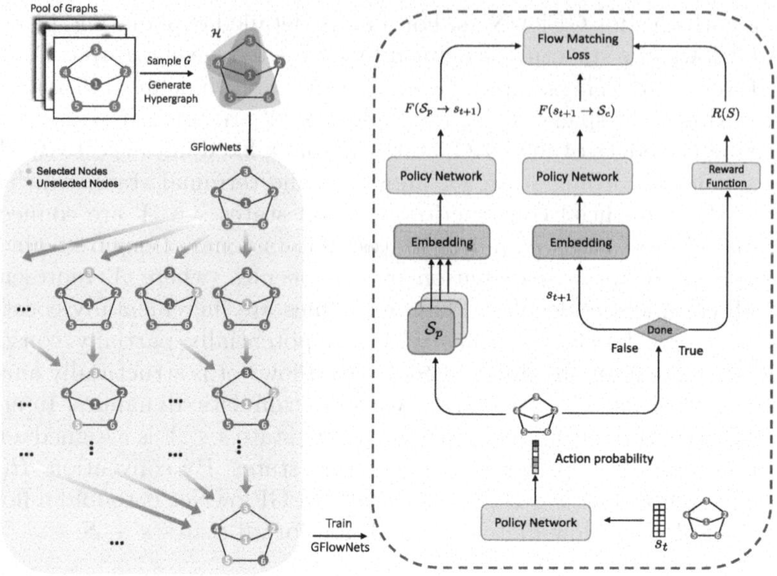

Fig. 1. The framework of GFlowIM.

find S, which covers the maximum number of hyperedges based on the forward policy. After obtaining a feasible solution, the policy network is trained based on the FM objective. Figure 1 shows the pipeline of our method. We will elaborate on how we construct hypergraph and train GFlowIM later in following sections.

4.1 Hypergraph Construction

As detailed in Sect. 3.2, the RIS method comprehends the influence landscape of a given graph $G(V, E, p)$ by creating a corresponding hypergraph $\mathcal{H} = (V, \{\mathcal{E}_1, \mathcal{E}_2, \dots\})$. In this hypergraph, each hyperedge $\mathcal{E}_j$ is a subset of nodes from V. The construction of each hyperedge $\mathcal{E}_j$ follows a specific process: (1) a random node $v \in V$ is selected, (2) a sample subgraph $g \subseteq G$ is generated, and (3) $\mathcal{E}_j$ is defined as the set of nodes that can reach the node v within the sample graph g. By generating multiple such random hyperedges, nodes with significant influence tend to appear more frequently within these hyperedges. Consequently, a seed set S that intersects with most hyperedges is likely to maximize the spread of influence. A hyperedge $\mathcal{E}_j$ is covered by the seed set S if there is an intersection between them, i.e., $S \cap \mathcal{E}_j \neq \emptyset$.

4.2 GFlowIM

Overall Framework. An overview of the policy training framework is illustrated in Fig. 1. We start by initializing a new empty set, represented by the initial graph with all blue nodes above layer 1 on the left side. At each step, GFlowIM computes the forward policy as unnormalized action probabilities based on the current input state. It then samples an unselected node and adds it to the seed set, corresponding to the nodes highlighted in orange in the graph. The layer i represents the i-th state transition in a trajectory, and this process continues until the budget k is exhausted. When a set S is generated, we count the number of hyperedges it covers and use this count as a reward. In GFlowIM (shown on the right side), the unselected node v_t is sampled based on the action probabilities calculated by the neural networks. After completing the trajectories, the policy network is trained using the FM objective, which is calculated from the inflow, outflow, or rewards of the states.

Designing MDP for GFlowIM. After hyperedges generation, IM is equivalent to the maximum coverage problem on hyperedges. This problem can naturally be formalized as a Markov Decision Process (MDP).

State. The state in graph G at step t is denoted as a matrix s_G^t, where each row s_v^t represents the state of node v. The state representation of each node is defined based on heuristic criteria as follows: (1) We use an indicator variable to represent whether a node has been selected, i.e., $s_v^t(1) = \mathbb{1}\{v \in S^{t-1}\}$. (2) We use the remaining budget to measure the length of the trajectory after step t, i.e., $s_v^t(2) = k - |S^{t-1}|$. These features are concatenated to form the state representation of each node s_v^t. The graph state matrix s_G^t is then passed into the GFlowIMπ to generate action probabilities.

Action. At time step t, the action involves selecting a node v from $V \backslash S^{t-1}$ to add to the seed set S^{t-1}, based on the action probability $\pi(\cdot|s_G^t)$ provided by the policy network at step t.

Reward. The reward is set to be the size of the covered hyperedges, i.e., $R(S) = |\{e \in \mathcal{E}, e \cap S \neq \emptyset\}|$, where $|\{e \in \mathcal{E}, e \cap S \neq \emptyset\}|$ denotes the size of the set of hyperedges e in $\mathcal{E}$ that intersect with S.

4.3 Policy Network Architecture and Training

A key characteristic of maximum coverage in a hypergraph is the strong correlation between nodes dictated by the hypergraph topology. To automatically extract this information and model the influence of hypergraph structures on the forward policy, we parameterize the policy network as a Hypergraph Convolution Network (HGCN) [8], which iteratively aggregates neighborhood information via hyperedges to update the state representation for each node. Then we define the hypergraph convolution operator as:

$$\mathbf{L} = \mathbf{D}^{-1/2}\mathbf{H}\mathbf{B}^{-1}\mathbf{H}^{\top}\mathbf{D}^{-1/2}, \tag{3}$$

$$\mathbf{X}^{(l+1)} = \text{LeakyReLU}(\mathbf{L}\mathbf{X}^{(l)}\mathbf{W}^{(l+1)}). \tag{4}$$

Algorithm 1. GFlowIM training algorithm.

Input: graph pool; k: seed set size; π_θ: GFlowIM model; $R(\cdot)$: reward function.
repeat
    Sample G from graph pool.
    Construct Hypergraph $\mathcal{H}$.
    repeat
        Sample a valid action a_t: $v_{t+1} \sim \pi(a_t|s_t)$ s.t. $v_{t+1} \in V \backslash S$
        Update the seed set $S^{t+1} = S^t \cup \{v_{t+1}\}$
        Make a state transition $s_{t+1} = T(s_t, a_t)$
    until $|S| = k$
    Calculate $R(S)$.
    Update the network parameter θ based on $\nabla \mathcal{L}_\theta(\tau)$ according to (6).
until Some convergence condition
Output: GFlowIM model with parameters θ.

where $\mathbf{L}$ is the Laplacian matrix for the hypergraph $\mathcal{H}$, and $\mathbf{X}^{(l+1)}$ represents the representations from the l-th layer in the hypergraph module. $\mathbf{W}^{(l+1)} \in \mathbb{R}^{d^{(l)} \times d^{(l+1)}}$ is the parameter matrix in the $(l+1)$-th layer. Notably, our neural network is designed to approximate the flow, specifically to obtain $\hat{F}(s \rightarrow s')$. Therefore, to derive the predicted probability, a softmax operator is applied for normalization.

$$\pi_\theta(a_t|s_t) = P(\cdot|s_t) = \text{Softmax}(\mathbf{X}^{(L)}). \tag{5}$$

We predict the next seed a_t, by sampling or choosing greedily from $\pi_\theta(a_t|s_t)$.

For simplicity, we outline our GFlowIM algorithm in Algorithm 1. Beginning with an empty seed set $S = \emptyset$, GFlowIM iteratively samples a valid action and selects a new node to transition the state from $s_t \rightarrow s_{t+1}$. Subsequently, we update the seed set to $S^{t+1} = S^t \cup \{v_{t+1}\}$. Next, we calculate the dynamic node vector feature to update s_{t+1}. This process continues until the budget k is exhausted. The complete trajectory $(s_0, s_1, \ldots, s_f)$ is collected. After that we calculate $R(S)$ according to the number of hyperedges covered by S. The policy network is then updated with stochastic gradient

$$\mathbb{E}_{(s_0 \rightarrow \cdots \rightarrow s_f) \sim \pi_\theta} \nabla_\theta \left[\sum_{t=1}^{s_f} \mathcal{L}_{\text{FM}}(s_t) + \mathcal{L}'_{\text{FM}}(s_f) \right]. \tag{6}$$

5 Experiments

All tests are conducted on a system equipped with an Intel Xeon E5-2698v4 CPU containing 64 cores, 6 Nvidia 2080 Ti GPUs featuring 12GB of GPU memory, and 256GB of RAM, operating on Ubuntu 20.04.

Datasets. We assess the performance of our approach using both real and synthetic datasets. The real datasets include Cora [13], Citeseer [13], Facebook

[24], and GrQc [16]. For the synthetic dataset, we adopt Erdos-Renyi (ER) and Barabasi-Albert (BA) graphs following [12]. The IM problems of different scales are named BA/ER-50/100/1000, respectively, where BA/ER-n indicates that the IM instance has n nodes on BA/ER graph.

Evaluating the Generalizability. Another concern within the combinatorial optimization community is whether policies trained on small instances can generalize to larger instances. As shown in Table 1, we demonstrate that machine learning methods, particularly the GFlowIM model, can effectively generalize to significantly larger instances. We train a GFlowIM model with 10,000 instances on BA/ER-2000, and use this model to generate IM solutions on BA/ER-2000/3000/4000/5000 of 100 instances to evaluate generalization performance. Furthermore, the GCOMB, LeNSE, RL4IM, and HGCN-DQN models were also trained using BA/ER-1000 data to assess their transferability to larger-scale problems. The results, averaged over 100 IM instances, demonstrate that our GFlowIM model surpasses all learning-based methods when trained with $n = 2000$ and applied to larger-scale problems.

Table 1. Our results w.r.t. existing baselines, tested on 100 instances respectively with $n = 2000, 3000, 4000, 5000$.

Method	Barabasi Albert Graph				Erdős-Rényi Graph			
	Test	Generalization			Test	Generalization		
	$n = 2000$	$n = 3000$	$n = 4000$	$n = 5000$	$n = 2000$	$n = 3000$	$n = 4000$	$n = 5000$
Degree	381.0	553.6	755.4	924.7	101.6	249.3	1156.6	2764.6
CELF [15]	442.3	665.1	885.6	1088.5	93.7	242.7	1167.1	2808.8
IMM [25]	447.0	659.7	890.3	1081.1	106.0	252.2	1172.8	2818.6
GCRL [20]	454.4	540.5	773.0	875.5	68.3	181.9	832.4	2038.0
RL4IM [4]	391.2	542.8	743.7	836.4	45.0	166.1	925.5	2323.2
GCOMB [19]	461.3	657.7	889.9	1066.8	87.2	236.5	1169.8	2803.6
LeNSE [9]	420.2	629.4	836.4	1033.4	112.9	222.0	1119.6	2679.9
HGCN-DQN [12]	469.1	698	940.8	1168.6	**118.8**	261.4	1252.2	2998.6
GFlowIM	**489.3**	**711.1**	**983.3**	**1171.6**	117.7	**284.5**	**1295.1**	**2998.9**

Evaluation on Various other Problem Settings. We are also keen to explore the effectiveness of different methods across various problem settings, including: (a) Stochastic Propagation: We define the range of $p(e)$ as $[0.1, 0.2, 0.3, 0.4, 0.5]$. In this stochastic context, we run Greedy and DPIM algorithms 20 times and report the average results. (b) Seed Budget k: We set the range of $[10, 20, 30, 40, 50]$.

The results are presented in Fig. 2, where consistent trends are observed across different networks. Specifically, Figs. 2(a)(d) display the outcomes on

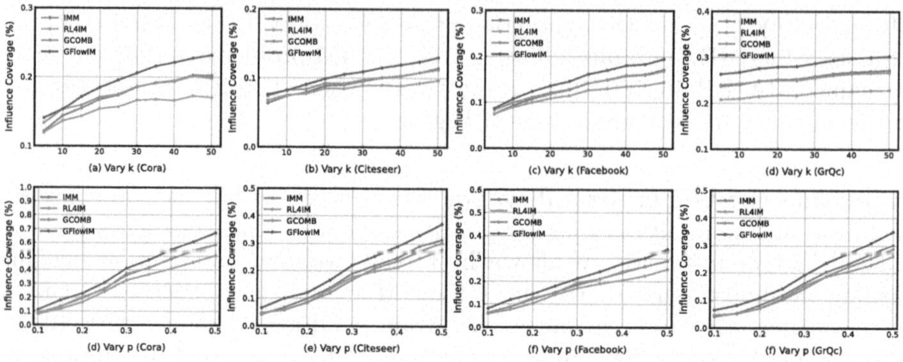

Fig. 2. Performance of different methods under various settings, on networks Cora, Citeseer, Facebook and GrQc. The x-axis is the value of the underlying parameter being evaluated, and the y-axis is the normalized influence.

the Cora network. The key observations are: (i) GFlowIM consistently outperforms all baselines across various settings, demonstrating its robustness. (ii) As expected, the influence value increases with a higher propagation probability p and a larger seed budget k.

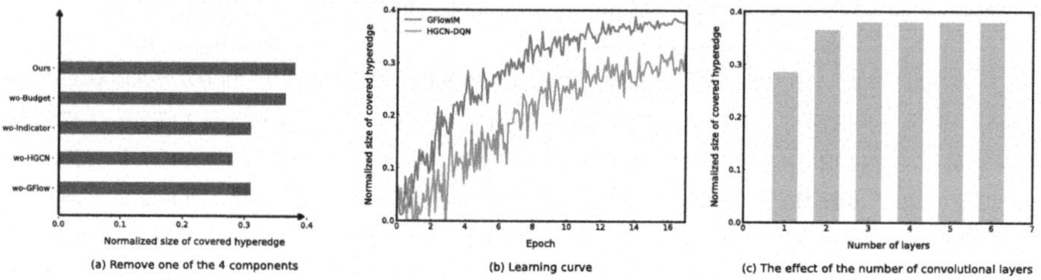

Fig. 3. Ablation study on GFlowIM training.

Ablation Study. To assess the effectiveness of the various customized components, we conducted an ablation study on the BA200, utilizing the basic problem setting. In this study, we created four versions of GFlowIM by removing one of the components each time: 1) wo-GFlow (without GFlowNets, using deep reinforcement learning instead), 2) wo-HGCN (without HGCN, using as a multi-layer perceptron instead), 3) wo-Indicator (without indicator feature), 4) wo-Budget (without budget feature). The comparison is illustrated in Fig. 3(a). It is evident that removing each component leads to a significant decrease in the normalized influence value: 7.3% (wo-GFlow), 7.2% (wo-Indicator), and 1.5% (wo-Budget). Our hypergraph-based policy network specifically models node informativeness by considering hypergraph structures. To evaluate its effectiveness, we compared it with a policy network that does not take hypergraph structures into account. We parameterized the policy network as a multi-layer perceptron (MLP) that only utilizes single-node information. To ensure a fair comparison, we used a 4-layer MLP with hidden layers of size 256, matching the number of parameters in the HGCN policy network. After training both policy networks, the experimental results demonstrated that HGCN outperformed the MLP by a significant margin of 10.4%, highlighting the importance of modeling node interactions in hypergraphs.

Figure 3(b) illustrates a comparison of learning efficiency between GFlowIM and RL method implementations. Despite two methods sharing similar speeds for one epoch of training, the figure demonstrates that GFlowIM achieves the most stable and best training performance. This suggests that GFlowIM may be more sample-efficient than traditional RL method. This is because GFlowIM focuses on directly learning the target distribution, rather than learning a policy through maximizing cumulative rewards. In the third ablation experiment, we tested the effect of the number of hypergraph convolution layers on the number of hyperedge covers. In Fig. 3(c), we find that even 1 layer of hypergraph convolution already achieves the effect of multilayer mlp. There is no significant change on the number of hyperedges covered when the convolution layer exceeds 3 layers. The GFlowIM is also robust to this hyperparameter.

6 Conclusion

In this work, we leveraged the powerful generative capabilities of GFlowNets to propose a novel approach called GFlowIM, which models the influence maximization problem as a generation problem. We equivalently transform the influence maximization problem with complex information diffusion properties into a stochastic maximum coverage problem. GFlowIM learns the optimal policy to select seed set that covers the largest number of hyperedges, considered a compositional object, through sequential actions. We conducted extensive experiments to demonstrate the superiority of GFlowIM in terms of performance and generalization.

References

1. Ali, K., Wang, C.Y., Chen, Y.S.: Boosting reinforcement learning in competitive influence maximization with transfer learning. In: 2018 IEEE/WIC/ACM International Conference on Web Intelligence (WI), pp. 395–400. IEEE (2018)
2. Bengio, Y., Lahlou, S., Deleu, T., Hu, E.J., Tiwari, M., Bengio, E.: GFlowNet foundations. J. Mach. Learn. Res. **24**(210), 1–55 (2023)
3. Borgs, C., Brautbar, M., Chayes, J., Lucier, B.: Maximizing social influence in nearly optimal time. In: Proceedings of the Twenty-Fifth Annual ACM-SIAM Symposium on Discrete Algorithms, pp. 946–957. SIAM (2014)
4. Chen, H., Qiu, W., Ou, H.C., An, B., Tambe, M.: Contingency-aware influence maximization: a reinforcement learning approach. In: Uncertainty in Artificial Intelligence, pp. 1535–1545. PMLR (2021)

5. Chen, W.: Computational social influence. In: Proceedings of the 1st International Conference on Social Influence Analysis-Volume 1398, pp. 1–1 (2015)
6. Chen, W., Wang, C., Wang, Y.: Scalable influence maximization for prevalent viral marketing in large-scale social networks. In: Proceedings of the 16th ACM SIGKDD International Conference on Knowledge Discovery and Data Mining, pp. 1029–1038 (2010)
7. Domingos, P., Richardson, M.: Mining the network value of customers. In: Proceedings of the Seventh ACM SIGKDD International Conference on Knowledge Discovery and Data Mining, pp. 57–66 (2001)
8. Feng, Y., You, H., Zhang, Z., Ji, R., Gao, Y.: Hypergraph neural networks. In: Proceedings of the AAAI Conference on Artificial Intelligence. vol. 33, pp. 3558–3565 (2019)
9. Ireland, D., Montana, G.: LeNSE: learning to navigate subgraph embeddings for large-scale combinatorial optimisation. In: International Conference on Machine Learning, pp. 9622–9638. PMLR (2022)
10. Kempe, D., Kleinberg, J., Tardos, É.: Maximizing the spread of influence through a social network. In: Proceedings of the ninth ACM SIGKDD International Conference on Knowledge Discovery and Data Mining, pp. 137–146 (2003)
11. Kempe, D., Kleinberg, J., Tardos, É.: Influential nodes in a diffusion model for social networks. In: Caires, L., Italiano, G.F., Monteiro, L., Palamidessi, C., Yung, M. (eds.) ICALP 2005. LNCS, vol. 3580, pp. 1127–1138. Springer, Heidelberg (2005). https://doi.org/10.1007/11523468_91
12. Khalil, E., Dai, H., Zhang, Y., Dilkina, B., Song, L.: Learning combinatorial optimization algorithms over graphs. Adv. Neural Inf. Process. Syst. **30** (2017)
13. Kipf, T.N., Welling, M.: Semi-supervised classification with graph convolutional networks. arXiv preprint arXiv:1609.02907 (2016)
14. Kumar, S., Mallik, A., Khetarpal, A., Panda, B.: Influence maximization in social networks using graph embedding and graph neural network. Inf. Sci. **607**, 1617–1636 (2022)
15. Leskovec, J., Krause, A., Guestrin, C., Faloutsos, C., VanBriesen, J., Glance, N.: Cost-effective outbreak detection in networks. In: Proceedings of the 13th ACM SIGKDD International Conference on Knowledge Discovery and Data Mining, pp. 420–429 (2007)
16. Leskovec, J., Mcauley, J.: Learning to discover social circles in ego networks. Adv. Neural Inf. Process. Syst. **25** (2012)
17. Li, H., Xu, M., Bhowmick, S.S., Rayhan, J.S., Sun, C., Cui, J.: PIANO: Influence maximization meets deep reinforcement learning. IEEE Transactions on Computational Social Systems (2022)
18. Lin, S.C., Lin, S.D., Chen, M.S.: A learning-based framework to handle multi-round multi-party influence maximization on social networks. In: Proceedings of the 21th ACM SIGKDD International Conference on Knowledge Discovery and Data Mining, pp. 695–704 (2015)
19. Manchanda, S., Mittal, A., Dhawan, A., Medya, S., Ranu, S., Singh, A.: GCOMB: learning budget-constrained combinatorial algorithms over billion-sized graphs. Adv. Neural. Inf. Process. Syst. **33**, 20000–20011 (2020)
20. Meirom, E., Maron, H., Mannor, S., Chechik, G.: Controlling graph dynamics with reinforcement learning and graph neural networks. In: International Conference on Machine Learning, pp. 7565–7577. PMLR (2021)
21. Panagopoulos, G., Malliaros, F.D., Vazirgiannis, M.: Multi-task learning for influence estimation and maximization. IEEE Trans. Knowl. Data Eng. **34**(9), 4398–4409 (2020)

22. Rao, G., Li, D., Wang, Y., Chen, W., Zhou, C., Zhu, Y.: Maximizing the influence with κ-grouping constraint. Inf. Sci. **629**, 204–221 (2023)
23. Rao, G., Li, D., Wang, Y., Chen, W., Zhou, C., Zhu, Y.: Online conflict resolution: algorithm design and analysis. Inf. Sci. **651**, 119718 (2023)
24. Rossi, R., Ahmed, N.: The network data repository with interactive graph analytics and visualization. In: Proceedings of the AAAI Conference on Artificial Intelligence. vol. 29 (2015)
25. Tang, Y., Shi, Y., Xiao, X.: Influence maximization in near-linear time: a martingale approach. In: Proceedings of the 2015 ACM SIGMOD International Conference on Management of Data, pp. 1539–1554 (2015)
26. Tang, Y., Xiao, X., Shi, Y.: Influence maximization: near-optimal time complexity meets practical efficiency. In: Proceedings of the 2014 ACM SIGMOD International Conference on Management of Data, pp. 75–86 (2014)

Construction Algorithm of Vertex-Disjoint Paths in Circulant-Based Recursive Networks

Xiaoqing Liu[1], Hai Liu[2], Baolei Cheng[1], Yan Wang[1], and Jianxi Fan[1](✉)

[1] School of Computer Science and Technology, Soochow University, Suzhou 215006, China
jxfan@suda.edu.cn

[2] Department of Computing, Hang Seng University of Hong Kong, Siu Lek Yuen, Hong Kong

Abstract. Parallel and distributed systems have seen significant development and application in areas such as big data processing, cloud computing, edge computing, deep learning, and artificial intelligence. The network architecture serves as a crucial component in parallel and distributed systems. Circulant-based recursive networks (CRNs), one kind of networks, include well-known networks such as the interconnection network hypercube, the k-ary n-cube network, and the data center network BCube, as well as some less-known networks. In this paper, we design an $O(lk)$ algorithm, named DJ-Path1, to construct $l(k+1)$ disjoint paths between any two adjacent vertices in k-dimensional n-order CRN with the maximum length being $l+n-2$. Additionally, we propose another $O(kl(k+l))$ algorithm, called DJ-Path2, to construct $l(k+1)$ disjoint paths between any two nonadjacent vertices with the maximum length being $\langle\lceil\frac{n-1}{l}\rangle\rceil k+(n-l+2)$, where l denotes the degree of the 0-dimensional n-order CRN.

Keywords: Circulant-based recursive networks · Vertex-disjoint paths · Algorithms

1 Introduction

With the continuous development of computer software and hardware technology, parallel and distributed systems are continuously evolving and have been widely applied in many fields including scientific computing, big data processing, cloud computing, and artificial intelligence. Meanwhile, emerging technologies such as quantum computing and biological computing will also bring new development opportunities to parallel and distributed systems. As the scale of parallel and distributed systems continues to increase, failures in processors/severs or communication links are inevitable. Network architectures, as an important part of parallel and distributed systems, largely determine the system's fault

Y. Chen et al. (Eds.): COCOON 2024, LNCS 15162, pp. 300–312, 2025.
https://doi.org/10.1007/978-981-96-1093-8_25

tolerance, diagnosability, robustness and so on. Data center networks (DCNs) and Interconnection Networks (INs) are both fundamental network architectures in parallel and distributed systems.

The k-ary n-cube network Q_n^k is one of the well-known INs. Based on the desirable properties of Q_n^k, such as recursiveness, low latency, high bandwidth, vertex-symmetry and edge-symmetry, it has been applied to build parallel and distributed systems. For example, CamCube [1] employs a k-ary 3-cube network as its topology structure, enabling its internal communication without the need for switches or routers. Furthermore, WaveCube [4] and IBM Gene/L [2] all adopted k-ary n-cube networks as their underlying topologies. DCN BCube, proposed by Guo *et al.*. [7], is one of the important server-centric DCNs. It places intelligence on servers in a modular data center and significantly reduces network costs by using inexpensive commodity switches. Circulant-based recursive networks (CRNs), a new class of networks proposed by [11], include well-known networks such as the interconnection network hypercube, the k-ary n-cube network, and the data center network BCube, as well as some less-known networks. In addition, CRNs possess some desirable properties such as recursivity, regularity, small diameter, and low cost. Therefore, we explore the vertex-disjoint paths of circulant-based recursive networks (CRNs) in this paper.

The path serves as the channel for transmitting data between processors/servers, which is crucial for ensuring the efficiency of information transmission. Finding vertex-disjoint paths between any two processors/servers in parallel and distributed systems can establish secure communication paths that do not interfere with each other. This implies that constructing as many vertex-disjoint paths as possible between any two processors/servers can make it easier to find alternative propagation paths in the event of server/processor or communication link failures, thereby increasing the likelihood of finding a fault-free path. Simultaneously, vertex-disjoint paths can also enhance the robustness and load balance of the network. Recently, Lv et al. [10] constructed $2n$ vertex-disjoint paths between any two vertices in k-ary n-cube networks Q_n^k, where $2n$ is equal to the connectivity of Q_n^k. Furthermore, they proposed the path construction algorithms of k-ary n-cube networks. An algorithm for constructing disjoint paths in H-free graphs was introduced by Kern et al. [9]. They also designed an algorithm to find disjoint connected subgraphs in graphs with n vertices and m edges in H-free graphs. In fact, the problem of disjoint paths in networks is also a type of path embedding problem, which has received extensive attention in recent years. For instance, Fan et al. [5] considered fault-tolerant embedding of paths in twisted cube TN_n. They showed that a path of length l can be embedded between any two distinct vertices with dilation 1 for any faulty set F with $|F| \leq n-3$. Furthermore, Wang et al. [15] explored embedding of meshes in TN_n. Other issues related to disjoint paths can be referred to [6,8,12–14,16].

In this paper, we construct $l(k+1)$ disjoint paths between any two vertices in a k-dimensional n-order CRN, where $l(k+1)$ equals the regular degree of the k-dimensional n-order CRN and l denotes the degree of the 0-dimensional n-order CRN. And the maximum length of these $l(k+1)$ disjoint paths is no more than

$\langle\lceil \frac{n-1}{l} \rangle\rceil k + (n - l + 2)$. Furthermore, we provide two algorithms DJ-Path1 and DJ-Path2 to find $l(k+1)$ disjoint paths between any two adjacent vertices and any two nonadjacent vertices in the k-dimensional n-order CRN, respectively.

The outline of the rest of this paper is organized as follows: Sect. 2 introduces the definition of circulant-based recursive networks (CRNs), and reviews some notations. Section 3 shows the path construction algorithms between any two adjacent vertices and two nonadjacent vertices in the CRN, respectively. Finally, conclusions of this paper and suggestions for future work are given in Sect. 4.

2 Preliminaries

2.1 Terminologies and Notions

A network can be modeled as a simple undirected graph $\mathcal{G} = (V(\mathcal{G}), E(\mathcal{G}))$, where a vertex $u \in V(\mathcal{G})$ represents a processor/server and an edge $(u, v) \in E(\mathcal{G})$ represents a link between vertices u and v. For any vertex $w \in V(\mathcal{G})$, The degree of vertex w refers to the number of edges incident with v, denoted by $d_{\mathcal{G}}(w)$. For any vertex $x \in V(\mathcal{G})$, If $d(v) = k$, then call the graph $\mathcal{G}$ a k-regular graph. The open neighborhood of x, denoted by $N_{\mathcal{G}}(x)$, refers to the vertex-set in $\mathcal{G}$ that are adjacent to x. In the absence of confusion, we usually ignore the subscript $\mathcal{G}$. Furthermore, let $\mathcal{F}$ be the vertex subset of graph $\mathcal{G}$, we use $\mathcal{G}\backslash\mathcal{F}$ to denote the subgraph of $\mathcal{G}$ obtained by removing all the vertices of $\mathcal{F}$ from G. For any vertex $u \in \mathcal{F}$, the neighbors of u in $\mathcal{G}\backslash\mathcal{F}$ refer to the set of vertices adjacent to u in $\mathcal{G}\backslash\mathcal{F}$, denoted as $N_{\mathcal{G}\backslash\mathcal{F}}(u)$. A path in graph $\mathcal{G}$ is a sequence of distinct vertices connected by edges between consecutive vertices, denoted by $\mathcal{P} = \langle m_1, m_2, \ldots, m_n \rangle$, where $m_1, \ldots, m_n \in V(\mathcal{G})$. The length of the path $\mathcal{P}$ is the number of edges in $\mathcal{P}$. For any two paths $\mathcal{P}_1$ and $\mathcal{P}_2$ from x to y in $\mathcal{G}$, $\mathcal{P}_1$ and $\mathcal{P}_2$ are vertex-disjoint paths if $V(\mathcal{P}_1) \cap V(\mathcal{P}_2) = \{x, y\}$. For convenience, the vertex-disjoint path is abbreviated as the disjoint path. We use $\mathcal{P}[n]$ and $\mathcal{P}[-n]$ to represent the nth vertex from the left and the right in the path $\mathcal{P}$, respectively.

A graph $\mathcal{G}$ with n vertices is called a complete graph if any two vertices in the graph are adjacent, denoted as K_n. Graphs $\mathcal{G}_1$ and $\mathcal{G}_2$ are isomorphic, denoted by $\mathcal{G}_1 \cong \mathcal{G}_2$, if there exists a bijection $\psi : V(\mathcal{G}_1)\rangle \to V(\mathcal{G}_2)$ such that $(x, y) \in E(\mathcal{G}_1)$ if and only if $(\psi(x), \psi(y)) \in E(\mathcal{G}_2)$. Graph $\mathcal{H}$ is said to be a subgraph of graph $\mathcal{G}$ if $V(\mathcal{H}) \subseteq V(\mathcal{G})$ and $E(\mathcal{H}) \subseteq E(\mathcal{G})$. Let $[n] = \{1, 2, \ldots, n\}$. For terminologies and notations not defined here, we follow [3].

2.2 Circulant-Based Recursive Betworks

For convenience, let $I = \{2i \mid 1 \le i \le \langle\lfloor \frac{n-1}{2} \rangle\rfloor\} \cup \{1, n-1\}$. The definitions of circulant-based recursive network [11] are listed as follows.

Definition 1. *For* $n \ge 2$, $k \ge 0$ *and* $l \in I$, *a circulant-based recursive network* (CRN) *is denoted as* $C_{l,n,k}$, *where* $V(C_{l,n,k}) = \{(u^0, u^1, \ldots, u^k) \mid u^i = 0, 1, \ldots, n-1$ and $i = 0, 1, \ldots, k\}$. *Two vertices* $u = (u^0, u^1, \ldots, u^k), v =$

$(v^0, v^1, \ldots, v^k) \in V(C_{l,n,k})$ *are adjacent if and only if* $v^i = u^i \pm s$ *for* $i \in [k] \cup \{0\}$ *and* $v^j = u^j$ *for each* $j \in ([k] \cup \{0\}) \setminus \{i\}$, *where the symbol "±" is under (mod n),* $s = 1, 2, \ldots, \langle \lceil \frac{l}{2} \rangle \rceil$.

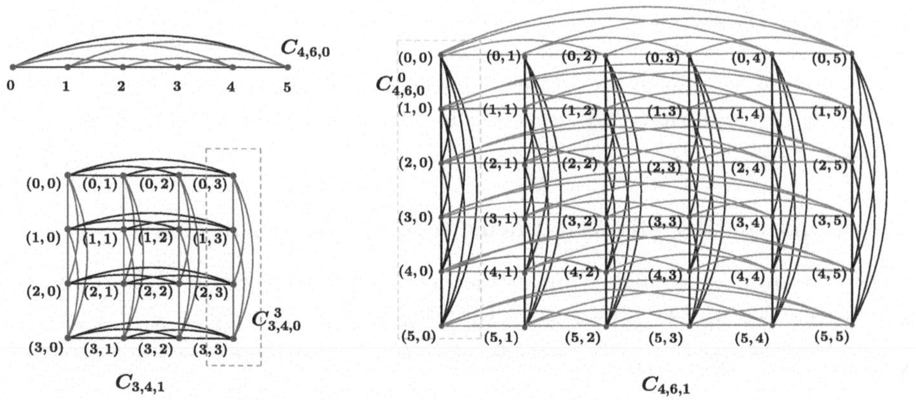

Fig. 1. Illustration of $C_{4,6,0}$, $C_{3,4,1}$ and $C_{4,6,1}$.

Definition 2. *For two positive integers* $n \geq 2$ *and* $l \in I$, *the k-dimensional* $C_{l,n,k}$ *is defined recursively as follows:*

(1) *0-dimensional* $C_{l,n,0}$ *is a undirected graph with* $V(C_{l,n,0}) = \{0, 1, 2, \ldots, n-1\}$ *and* $E(C_{l,n,0}) = \{ij \mid i = j \pm s, j = 0, 1, \ldots, n-1, s = 1, 2, \ldots, \langle \lceil \frac{l}{2} \rangle \rceil\}$.

(2) *For any integer* $k \geq 1$, *k-dimensional* $C_{l,n,k}$ *is composed of n* $(k-1)$*-dimensional* $C_{l,n,k-1}$*s, denoted by* $C^0_{l,n,k-1}, C^1_{l,n,k-1}, \ldots, C^{n-1}_{l,n,k-1}$. *It is clear that* $V(C^i_{l,n,k-1}) = \{(a^0, \ldots, a^{k-1}, i) \mid (a^0, \ldots, a^{k-1}, i) \in V(C_{l,n,k})\}$ *for each* $i \in [n-1] \cup \{0\}$. *For any two vertices* $u = (u^0, u^1, \ldots, u^{k-1}, i)$, $v = (v^0, u^1, \ldots, u^{k-1}, i) \in V(C^i_{l,n,k-1})$ $(i \in [n-1] \cup \{0\})$, *if u is adjacent to v, then the vertex* $(i, u^1, \ldots, u^{k-1}, u^0)$ *is adjacent to the vertex* $(i, u^1, \ldots, u^{k-1}, v^0)$ *in* $C_{l,n,k}$.

Clearly, when $n = 2$ and $k \geq 0$, then $l = 1$. Thus, $C_{1,2,k} \cong Q_{k+1}$; when $n \geq 3$, $l = 2$ and $k \geq 0$, then $C_{2,n,k} \cong Q^n_{k+1}$; when $n \geq 3$, $l = n-1$ and $k \geq 0$, then $C_{n-1,n,k} \cong BC_{n,k}$. Figure 1 shows the illustration of $C_{4,6,0}$, $C_{4,6,1}$ and $C_{3,4,1}$. $C_{4,6,0}$ and $C_{4,6,1}$ are two distinct graphs from hypercubes, k-ary n-cube networks and BCubes.

According to Definition 1, it is clear that $C_{l,n,k}$ is $l(k+1)$-regular. By Definition 2, we know that $C_{l,n,k}$ can be divided into n^r disjoint $(k-r)$-dimensional n-order $C_{l,n,k-r}$s and n^{r+1} $(k-r)$-dimensional n-order distinct $C_{l,n,k-r}$s $(r \geq 1)$, respectively. Hence, we put all n^{r+1} $(k-r)$-dimensional n-order subgraphs of $C_{l,n,k}$ in order, denoted as $C^0_{l,n,k-r}, C^1_{l,n,k-r}, \ldots, C^{n^{r+1}-1}_{l,n,k-r}$ $(r \geq 1)$. For any set

$J \subseteq [n^{r+1}-1] \cup \{0\}$, let $C_{l,n,k-r}^{J} = \cup_{j \in J} C_{l,n,k-r}^{j}$. For any vertex $u \in V(C_{l,n,k-1}^{i})$ ($i \in [n-1] \cup \{0\}$), we call the neighbors of u in $C_{l,n,k} \setminus C_{l,n,k-1}^{i}$ the external neighbors of u, denoted by $N_{\text{out}}(u)$. Furthermore, for any two vertices $u, v \in V(C_{l,n,k-1}^{i})$ ($i \in [n-1] \cup \{0\}$), we have $N_{\text{out}}(u) \cap N_{\text{out}}(v) = \emptyset$.

3 Main Results

In this section, we consider the construction of disjoint path algorithms in CRNs, analyzing the following two aspects: (1) vertex x is adjacent to vertex y; (2) vertex x is not adjacent to vertex y.

3.1 Constructing Disjoint Paths of Situation (1) in $C_{l,n,k}$

Firstly, we propose an algorithm, named DJ-Path1, to construct $l(k+1)$ disjoint paths between any two adjacent vertices in $C_{l,n,k}$.

Algorithm 1. DJ-Path1($C_{l,n,k}; x, y$)

Input: four positive integers $i \in [k] \cup \{0\}$, $j \in ([k] \cup \{0\}) \setminus \{i\}$, $s \in [\frac{l}{2}]$ and $l \in \{2i \mid 1 \le i \le \lfloor \frac{n-1}{2} \rfloor\}$; a $C_{l,n,k}$; two distinct vertices $x = (x^0, \ldots, x^j, \ldots, x^k)$ and $y = (y^0, \ldots, y^j, \ldots, y^k)$ satisfy $xy \in E(C_{l,n,k})$.

Output: $l(k+1)$ disjoint paths $\mathcal{P}_{0,s}, \mathcal{P}'_{0,s}, \ldots, \mathcal{P}_{k,s}, \mathcal{P}'_{k,s}$.

- **for** $s = 1$ to $\frac{l}{2}$ **do**
  - **for** $i = k$ to 0 **do**
    - **if** $x^i = y^i$ **then**
      - $\mathcal{P}_{i,s} \leftarrow \langle x, (x^0, \ldots, x^i+s, \ldots, x^j, \ldots, x^k), (x^0, \ldots, x^i+s, \ldots, y^j, \ldots, x^k), y \rangle$;
      - $\mathcal{P}'_{i,s} \leftarrow \langle x, (x^0, \ldots, x^i-s, \ldots, x^j, \ldots, x^k), (x^0, \ldots, x^i-s, \ldots, y^j, \ldots, x^k), y \rangle$;
    - **else**
      - $\mathcal{A} \leftarrow |x^i - y^i|$;
      - $m_s \leftarrow$ the smallest nonnegative integer such that $0 < |(x^i + s + \mathcal{A} \cdot m_s) - y^i| \le \frac{l}{2}$ (the "+" after x^i is under "mod n");
      - $w_s \leftarrow$ the smallest nonnegative integer such that $0 < |(x^i - s - \mathcal{A} \cdot w_s) - y^i| \le \frac{l}{2}$ (the "−" after x^i is under "mod n");
      - $\mathcal{P}_{i,s} \leftarrow \langle x, (x^0, \ldots, x^i + s, \ldots, x^k), (x^0, \ldots, x^i + s + \mathcal{A}, \ldots, x^k), \ldots, (x^0, \ldots, x^i + s + \mathcal{A} \cdot m_s, \ldots, x^k), y \rangle$;
      - $\mathcal{P}'_{i,s} \leftarrow \langle x, (x^0, \ldots, x^i - s, \ldots, x^k), (x^0, \ldots, x^i - s - \mathcal{A}, \ldots, x^k), \ldots, (x^0, \ldots, x^i - s - \mathcal{A} \cdot w_s, \ldots, x^k), y \rangle$;
    - **end if**
  - **end for**
- **end for**
- **return** $\mathcal{P}_{0,s}, \mathcal{P}'_{0,s}, \ldots, \mathcal{P}_{k,s}, \mathcal{P}'_{k,s}$.

Theorem 1. *Let $n \ge 3$, $k \ge 0$ and $l \in \{2i \mid 1 \le i \le \langle \lfloor \frac{n-1}{2} \rangle \rfloor\}$. For any two vertices $x, y \in V(C_{l,n,k})$ satisfy $xy \in E(C_{l,n,k})$, the algorithm* DJ-Path1 *can find $l(k+1)$ disjoint paths from x to y in $C_{l,n,k}$ with the maximum length being $n - l + 1$. Furthermore, it's time complexity is $O(lk)$.*

Proof. Note that the vertex $x = (x^0, \ldots, x^k)$ is adjacent to the vertex $y = (y^0, \ldots, y^k)$, then there exactly exists one integer $j \in [k] \cup \{0\}$ such that $|x^j - y^j| = s$ for $s \in [\frac{l}{2}]$ and $x^i = y^i$ for each $i \in ([k] \cup \{0\}) \backslash \{j\}$. Next, we discuss the following two situations.

If $x^i = y^i$, then the vertex $(x^0, \ldots, x^i + s, \ldots, x^j, \ldots, x^k)$ is adjacent to the vertex x and the vertex $(x^0, \ldots, x^i + s, \ldots, y^j, \ldots, x^k)$ is adjacent to the vertex y. Since $|x^j - y^j| = s$ for $j \in [k] \cup \{0\}$ and $i \neq j$, the vertex $(x^0, \ldots, x^i + s, \ldots, x^j, \ldots, x^k)$ is adjacent to the vertex $(x^0, \ldots, x^i + s, \ldots, y^j, \ldots, x^k)$. Based on the arbitrariness of i and s, there are $\frac{lk}{2}$ disjoint paths are obtained in line 4. Similar to line 4, there are $\frac{lk}{2}$ disjoint paths are constructed in line 5.

If $x^i \neq y^i$, then $i = j$. Let $\mathcal{A} = |x^i - y^i|$. It is clear that vertices $(x^0, \ldots, x^i + s, \ldots, x^k)$ and $(x^0, \ldots, x^i - s, \ldots, x^k)$ are both adjacent to the vertex x. We add (resp., subtract) $\mathcal{A}$ to $x^i + s$ (resp., $x^i - s$) at a time until there exists a nonnegative integer m_s (resp., w_s) such that $0 < |(x^i + s + \mathcal{A} \cdot m_s) - y^i| \leq \frac{l}{2}$ (resp., $0 < |(x^i - s - \mathcal{A} \cdot w_s) - y^i| \leq \frac{l}{2}$). Then $x^i + s + \mathcal{A} \cdot m_s$ (resp., $x^i - s - \mathcal{A} \cdot w_s$) can reach y^i in one step. By the arbitrariness of s, we have l disjoint paths in line 10.

Table 1. An example illustrating the flow of Algorithm DJ-Path1.

$C_{6,14,2}$ ($l = 6$, $n = 14$, $k = 2$)		
$x = (3,4,5)$, $y = (3,6,5)$		
$s = 1$	$i = 2$, $x^2 = y^2$	$\mathcal{P}_{1,2} = \langle(3,4,5),(3,4,\mathbf{6}),(3,\mathbf{6},6),(3,6,5)\rangle$
		$\mathcal{P}'_{1,2} = \langle(3,4,5),(3,4,\mathbf{4}),(3,\mathbf{6},4),(3,6,5)\rangle$
	$i = 1$, $x^1 \neq y^1$	$\mathcal{P}_{1,1} = \langle(3,4,5),(3,\mathbf{3},5),(3,6,5)\rangle$
		$\mathcal{P}'_{1,1} = \langle(3,4,5),(3,\mathbf{5},5),(3,6,5)\rangle$
	$i = 0$, $x^0 = y^0$	$\mathcal{P}_{1,0} = \langle(3,4,5),(\mathbf{4},4,5),(4,\mathbf{6},5),(3,6,5)\rangle$
		$\mathcal{P}'_{1,0} = \langle(3,4,5),(\mathbf{2},4,5),(2,\mathbf{6},5),(3,6,5)\rangle$
$s = 2$	$i = 2$, $x^2 = y^2$	$\mathcal{P}_{2,2} = \langle(3,4,5),(3,4,\mathbf{7}),(3,\mathbf{6},7),(3,6,5)\rangle$
		$\mathcal{P}'_{2,2} = \langle(3,4,5),(3,4,\mathbf{3}),(3,\mathbf{6},3),(3,6,5)\rangle$
	$i = 1$, $x^1 \neq y^1$	$\mathcal{P}_{2,1} = \langle(3,4,5),(3,\mathbf{2},5),(3,\mathbf{0},5),(3,\mathbf{12},5),(3,\mathbf{10},5),(3,\mathbf{8},5),\rangle.$ $\langle.(3,6,5)\rangle$
		$\mathcal{P}'_{2,1} = \langle(3,4,5),(3,\mathbf{6},5)\rangle$
	$i = 0$, $x^0 = y^0$	$\mathcal{P}_{2,0} = \langle(3,4,5),(\mathbf{5},4,5),(5,\mathbf{6},5),(3,6,5)\rangle$
		$\mathcal{P}'_{2,0} = \langle(3,4,5),(\mathbf{1},4,5),(1,\mathbf{6},5),(3,6,5)\rangle$
$s = 3$	$i = 2$, $x^2 = y^2$	$\mathcal{P}_{3,2} = \langle(3,4,5),(3,4,\mathbf{8}),(3,\mathbf{6},8),(3,6,5)\rangle$
		$\mathcal{P}'_{3,2} = \langle(3,4,5),(3,4,\mathbf{2}),(3,\mathbf{6},2),(3,6,5)\rangle$
	$i = 1$, $x^1 \neq y^1$	$\mathcal{P}_{3,1} = \langle(3,4,5),(3,\mathbf{1},5),(3,\mathbf{13},5),(3,\mathbf{11},5),(3,\mathbf{9},5),(3,6,5)\rangle$
		$\mathcal{P}'_{3,1} = \langle(3,4,5),(3,\mathbf{7},5),(3,6,5)\rangle$
	$i = 0$, $x^0 = y^0$	$\mathcal{P}_{3,0} = \langle(3,4,5),(\mathbf{6},4,5),(6,\mathbf{6},5),(3,6,5)\rangle$
		$\mathcal{P}'_{3,0} = \langle(3,4,5),(\mathbf{0},4,5),(0,\mathbf{6},5),(3,6,5)\rangle$

To sum up, there are $l(k+1)$ disjoint paths from x to y obtained in Algorithm DJ-Path1. In particular, the length of paths obtained in lines 4–5 is 4. In lines 10–11, the maximum length is $n-l+1$ in extreme case. Hence, the maximum length of the paths obtained in Algorithm DJ-Path1 is $n-(l-1)$.

In Algorithm DJ-Path1, lines 1–14 execute two nested loops, where the outer loop iterates once and the inner loop completes all iterations before moving on to the next iteration of the outer loop. When s is held fixed, lines 2–12 form a loop that runs a total of $k+1$ times to find $k+1$ disjoint paths from x to y. Furthermore, the time spent in each step of the loop is consistent. Thus, lines 2–12 take $O(k+1)$ times. Furthermore, the outer loop executes $\frac{l}{2}$ times, which implies that the time complexity of DJ-Path1 is $O(\frac{l}{2}(k+1)) = O(lk)$.

To further demonstrate the path construction process of Algorithm DJ-Path1, Table 1 takes $x=(3,4,5)$ and $y=(3,6,5)$ in $C_{6,14,2}$ as an example. There are $l(k+1)=18$ disjoint paths from x to y in $C_{6,14,2}$.

3.2 Constructing Disjoint Paths of Situation (2) in $C_{l,n,k}$

Before giving an algorithm for constructing $l(k+1)$ disjoint paths between any two nonadjacent vertices, we consider that, for any two vertices $x=(x^0,\ldots,x^{i-1}, x^i,x^{i+1},\ldots,x^k)$ and $y=(x^0,\ldots,x^{i-1},y^i,x^{i+1},\ldots,x^k)\in V(C_{l,n,k})$ $(i\in[k]\cup\{0\})$, how to find l disjoint paths from x to y by changing only the bit x^i. Algorithm P-DJPath solves the above problem.

Algorithm 2. P-DJPath$(C_{l,n,k};x,y)$

Input: three positive integers $s\in[\frac{l}{2}]$, $l\in\{2i \mid 1\le i\le \lfloor\frac{n-1}{2}\rfloor\}$ and $i\in[k]\cup\{0\}$; a $C_{l,n,k}$; two distinct vertices $x=(x^0,\ldots,x^{i-1},x^i,x^{i+1},\ldots,x^k)$ and $y=(x^0,\ldots,x^{i-1}, y^i,x^{i+1},\ldots,y^k)$.
Output: l disjoint paths $\mathcal{P}_1,\mathcal{P}'_1,\ldots,\mathcal{P}_{\frac{l}{2}},\mathcal{P}'_{\frac{l}{2}}$.

for $s=1$ to $\frac{l}{2}$ **do**
  if $1\le|x^i-y^i|\le\frac{l}{2}$ **then**
    $\mathcal{A}\leftarrow|x^i-y^i|$;
    $m_s\leftarrow$ the smallest nonnegative integer such that $0<|(x^i+s+\mathcal{A}\cdot m_s)-y^i|\le\frac{l}{2}$ (the "+" after x^i is under "mod n");
    $w_s\leftarrow$ the smallest nonnegative integer such that $0<|(x^i-s-\mathcal{A}\cdot w_s)-y^i|\le\frac{l}{2}$ (the "−" after x^i is under "mod n");
    $\mathcal{P}_s\leftarrow\langle x,(x^0,\ldots,x^i+s,\ldots,x^k),(x^0,\ldots,x^i+s+\mathcal{A},\ldots,x^k),\ldots,(x^0,\ldots,x^i+s+\mathcal{A}\cdot m_s,\ldots,x^k),y\rangle$;
    $\mathcal{P}'_s\leftarrow\langle x,(x^0,\ldots,x^i-s,\ldots,x^k),(x^0,\ldots,x^i-s-\mathcal{A},\ldots,x^k),\ldots,(x^0,\ldots,x^i-s-\mathcal{A}\cdot w_s,\ldots,x^k),y\rangle$;
  else if $|x^i-y^i|>\frac{l}{2}$ **then**
    $m_s\leftarrow$ the smallest nonnegative integer such that $0<|(x^i+(s+\frac{l}{2}\cdot m_s))-y^i|\le\frac{l}{2}$ (the "+" after x^i is under "mod n");
    $w_s\leftarrow$ the smallest nonnegative integer such that $0<|(x^i-(s+\frac{l}{2}\cdot w_s))-y^i|\le\frac{l}{2}$ (the "−" after x^i is under "mod n");
    $\mathcal{P}_s\leftarrow\langle x,(x^0,\ldots,x^i+s,\ldots,x^k),(x^0,\ldots,x^i+s+\frac{l}{2},\ldots,x^k),\ldots,(x^0,\ldots,x^i+s+\frac{l}{2}\cdot m_s,\ldots,x^k),y\rangle$;

12: $\mathcal{P}_s \leftarrow \langle x, (x^0, \ldots, x^i - s, \ldots, x^k), (x^0, \ldots, x^i - s - \frac{l}{2}, \ldots, x^k), \ldots, (x^0, \ldots, x^i - s - \frac{l}{2} \cdot w_s, \ldots, x^k), y \rangle$;
13: **end if**
14: **end for**
15: **return** $\mathcal{P}_1, \mathcal{P}'_1, \ldots, \mathcal{P}_{\frac{l}{2}}, \mathcal{P}'_{\frac{l}{2}}$.

Theorem 2. *Let $n \geq 3$, $k \geq 0$ and $l \in \{2i \mid 1 \leq i \leq \langle\lfloor\frac{n-1}{2}\rangle\rfloor\}$. For any two vertices $x = (x^0, \ldots, x^i, \ldots, x^k), y = (x^0, \ldots, y^i, \ldots, x^k) \in V(C_{l,n,k})$, the algorithm* P-DJPath *can construct l disjoint paths from x to y with the maximum length being $n - l + 1$. Furthermore, it's time complexity is $O(l)$.*

Proof. By Definition 1, it is obvious that both vertices x and y are in the same 0-dimensional subgraph. If $1 \leq |x^i - y^i| \leq \frac{l}{2}$, then $xy \in E(C_{l,n,k})$ and the method of constructing paths is the same as Algorithm DJ-Path1.

If $|x^i - y^i| > \frac{l}{2}$, then $xy \notin E(C_{l,n,k})$. For $s \in [\frac{l}{2}]$, we know that the vertex $(x^0, \ldots, x^i + s, \ldots, x^k)$ and $(x^0, \ldots, x^i - s, \ldots, x^k)$ are adjacent to the vertex x. Then adding $\frac{l}{2}$ to $x^i + s$ at a time until it exists a nonnegative integer m_s such that $0 < |(x^i + s + \frac{l}{2} \cdot m_s) - y^i| \leq \frac{l}{2}$. Hence, the path $\mathcal{P}_s$ is constructed in line 11; Similarly, we subtracting $\frac{l}{2}$ to $x^i - s$ at a time until it exists a nonnegative integer w_s such that $0 < |(x^i - s - \frac{l}{2} \cdot w_s) - y^i| \leq \frac{l}{2}$, which implies that the path $\mathcal{P}'_s$ is constructed in line 12. Due to the arbitrariness of s, we construct l disjoint paths in Algorithm P-DJPath. Obviously, the paths constructed in lines 6–7 are longer than those constructed in lines 11–12. Thus, the maximum length of the paths obtained in Algorithm P-DJPath is $n - l + 1$.

Furthermore, lines 1–14 form a loop that runs a total of $\frac{l}{2}$ times to find l disjoint paths from x to y, and the time required for each step of the loop remains constant. Hence, the time complexity of this algorithm is $O(\frac{l}{2}) = O(l)$.

In the following, we design an algorithm, called DJ-Path2, to find $l(k+1)$ disjoint paths between any two nonadjacent vertices $x = (x^0, \ldots, x^k)$ and $y = (y^0, \ldots, y^k)$ in $C_{l,n,k}$.

Algorithm 3. DJ-Path2($C_{l,n,k}; x, y$)

Input: four positive integers $s \in [\frac{l}{2}]$, $l \in \{2i \mid 1 \leq i \leq \lfloor\frac{n-1}{2}\rfloor\}$, $i \in [k] \cup \{0\}$ and $j \in ([k] \cup \{0\}) \backslash \{i\}$; a $C_{l,n,k}$; two distinct vertices $x = (x^0, \cdots, x^k)$ and $y = (y^0, \ldots, y^k)$ satisfy $xy \notin E(C_{l,n,k})$.
Output: $l(k+1)$ disjoint paths $\mathcal{P}_{0,s}, \mathcal{P}'_{0,s}, \ldots, \mathcal{P}_{k-1,s}, \mathcal{P}'_{k-1,s}, \mathcal{P}_{k,s}, \mathcal{P}'_{k,s}$.

for $s = 1$ to $\frac{l}{2}$ **do**
  for $i = 0$ to $k - 1$ **do**
    $\mathcal{P}_{i,s} \leftarrow \langle x, (x^0, \ldots, x^i + s, \ldots, x^k) \rangle$;
    $\mathcal{P}'_{i,s} \leftarrow \langle x, (x^0, \ldots, x^i - s, \ldots, x^k) \rangle$;
    for $j = k$ to 0 except i **do**
      let $(x^0, \ldots, x^i + s, \ldots, x^j, y^{j+1} \ldots, y^k)$ be the last vertex of $\mathcal{P}_{i,s}$, and $(x^0, \ldots, x^i - s, \ldots, x^j, y^{j+1} \ldots, y^k)$ be the last vertex of $\mathcal{P}_{i,s}$;
      if $x^j = y^j$ **then**
        $j \leftarrow j - 1$;
      else

$m_s^j \leftarrow$ the smallest nonnegative integer such that $0 < |(x^j + \frac{l}{2} \cdot m_s^j) - y^j| \leq \frac{l}{2}$;
$w_s^j \leftarrow$ the smallest nonnegative integer such that $0 < |(x^j - \frac{l}{2} \cdot w_s^j) - y^j| \leq \frac{l}{2}$;
$\mathcal{P}_{i,s} \leftarrow \Big\langle \mathcal{P}_{i,s}, (x^0, \ldots, x^i + s, \ldots, x^j + \frac{l}{2}, y^{j+1}, \ldots, y^k), \ldots, (x^0, \ldots, x^i + s, \ldots, x^j + \frac{l}{2} \cdot m_s^j, y^{j+1}, \ldots, y^k), (x^0, \ldots, x^i + s, \ldots, y^j, \ldots, y^k) \Big\rangle$;
$\mathcal{P}'_{i,s} \leftarrow \Big\langle \mathcal{P}'_{i,s}, (x^0, \ldots, x^i - s, \ldots, x^j - \frac{l}{2}, y^{j+1}, \ldots, y^k), \ldots, (x^0, \ldots, x^i - s, \ldots, x^j - \frac{l}{2} \cdot w_s^j, y^{j+1}, \ldots, y^k), (x^0, \ldots, x^i - s, \ldots, y^j, \ldots, y^k) \Big\rangle$;
end if
end for
$z_{i,s} \leftarrow \mathcal{P}_{i,s}[-1]$;
$z'_{i,s} \leftarrow \mathcal{P}'_{i,s}[-1]$;
$z_i \leftarrow (y^0, \ldots, y^{i-1}, x^i, y^{i+1}, \ldots, y^k)$;
$\mathcal{P}^1_{i,s} \leftarrow$ P−DJPath$(C_{l,n,k}; z_i, y)$;
$\mathcal{P}^2_{i,s} \leftarrow$ P−DJPath$(C_{l,n,k}; z_i, y)$;
$\mathcal{P}_{i,s} \leftarrow \Big\langle \mathcal{P}_{i,s}, \mathcal{P}^1_{i,s} \backslash \{z_{i,s}, z_i\} \Big\rangle$;
$\mathcal{P}'_{i,s} \leftarrow \Big\langle \mathcal{P}'_{i,s}, \mathcal{P}^2_{i,s} \backslash \{z'_{i,s}, z_i\} \Big\rangle$;
end for
$f_s^k \leftarrow$ the smallest nonnegative integer such that $0 < |(x^k + s + \frac{l}{2} \cdot f_s^k) - y^k| \leq \frac{l}{2}$;
if $x^k = y^k$ **then**
$\mathcal{P}_{k,s} \leftarrow \Big\langle x, (x^0, \ldots, x^{k-1}, x^k + s), (x^0, \ldots, x^{k-1} + \frac{l}{2}, x^k + s), \ldots, (x^0, \ldots, x^{k-1} + \frac{l}{2} \cdot m_s^{k-1}, x^k + s), (x^0, \ldots, y^{k-1}, x^k + s), (x^0, \ldots, x^{k-2} + \frac{l}{2}, y^{k-1}, x^k + s), \ldots, (x^0, \ldots, x^{k-2} + \frac{l}{2} \cdot m_s^{k-2}, y^{k-1}, x^k + s), (x^0, \ldots, y^{k-2}, y^{k-1}, x^k + s), \ldots, (y^0, \ldots, y^{k-1}, x^k + s), y \rangle$;
$\mathcal{P}'_{k,s} \leftarrow \Big\langle x, (x^0, \ldots, x^{k-1}, x^k - s), (x^0, \ldots, x^{k-1} - \frac{l}{2}, x^k - s), \ldots, (x^0, \ldots, x^{k-1} - \frac{l}{2} \cdot m_s^{k-1}, x^k - s), (x^0, \ldots, y^{k-1}, x^k - s), \ldots, (x^0, \ldots, x^{k-2} - \frac{l}{2}, y^{k-1}, x^k - s), (x^0, \ldots, x^{k-2} - \frac{l}{2} \cdot m_s^{k-2}, y^{k-1}, x^k - s), (x^0, \ldots, y^{k-2}, y^{k-1}, x^k - s), \ldots, (y^0, \ldots, y^{k-1}, x^k - s), y \rangle$;
else
$\mathcal{P}_{k,s} \leftarrow \Big\langle x, (x^0, \ldots, x^{k-1}, x^k + s), (x^0, \ldots, x^{k-1}, x^k + s + \frac{l}{2}), \ldots, (x^0, \ldots, x^{k-1}, x^k + s + \frac{l}{2} \cdot f_s^k), (x^0, \ldots, x^{k-1} + \frac{l}{2}, x^k + s + \frac{l}{2} \cdot f_s^k), \ldots, (x^0, \ldots, x^{k-1} + \frac{l}{2} \cdot m_s^{k-1}, x^k + s + \frac{l}{2} \cdot f_s^k), (x^0, \ldots, x^{k-2} + \frac{l}{2}, x^{k-1} + \frac{l}{2} \cdot m_s^{k-1}, x^k + s + \frac{l}{2} \cdot f_s^k), \ldots, (x^0 + \frac{l}{2} \cdot m_s^0, \ldots, x^k + s + \frac{l}{2} \cdot f_s^k), (y^0, x^1 + \frac{l}{2} \cdot m_s^1, \ldots, x^k + s + \frac{l}{2} \cdot f_s^k), \ldots, y \Big\rangle$;
$\mathcal{P}'_{k,s} \leftarrow \Big\langle x, (x^0, \ldots, x^{k-1}, x^k - s), (x^0, \ldots, x^{k-1}, x^k - s - \frac{l}{2}), \ldots, (x^0, \ldots, x^{k-1}, x^k - s - \frac{l}{2} \cdot f_s^k), (x^0, \ldots, x^{k-1} - \frac{l}{2}, x^k - s - \frac{l}{2} \cdot f_s^k), \ldots, (x^0, \ldots, x^{k-1} - \frac{l}{2} \cdot m_s^{k-1}, x^k - s - \frac{l}{2} \cdot f_s^k), (x^0, \ldots, x^{k-2} - \frac{l}{2}, x^{k-1} - \frac{l}{2} \cdot m_s^{k-1}, x^k - s - \frac{l}{2} \cdot f_s^k), \ldots, (x^0 - \frac{l}{2} \cdot m_s^0, \ldots, x^k - s - \frac{l}{2} \cdot f_s^k), (y^0, x^1 - \frac{l}{2} \cdot m_s^1, \ldots, x^k - s - \frac{l}{2} \cdot f_s^k), \ldots, y \Big\rangle$;
end if
end for
return $\mathcal{P}_{0,s}, \mathcal{P}'_{0,s}, \ldots, \mathcal{P}_{k-1,s}, \mathcal{P}'_{k-1,s}, \mathcal{P}_{k,s}, \mathcal{P}'_{k,s}$.

Theorem 3. *Let $n \geq 3$, $k \geq 0$ and $l \in \{2i \mid 1 \leq i \leq \langle \lfloor \frac{n-1}{2} \rangle \rfloor\}$. For any two vertices $x = (x^0, \ldots, x^k), y = (x^0, \ldots, x^k) \in V(C_{l,n,k})$ satisfy $xy \notin E(C_{l,n,k})$, the algorithm* DJ-Path2 *can construct $l(k+1)$ disjoint paths from x to y with* *the maximum length is* $\langle\lceil\frac{n-1}{l}\rangle\rceil k+(n-l+2)$. *Furthermore, it's time complexity is* $O(kl(k+l))$.

Proof. For $s \in [\frac{l}{2}]$, vertices $(x^0,\ldots,x^i+s,\ldots,x^k)$ and $(x^0,\ldots,x^i-s,\ldots,x^k)$ are adjacent to vertex x. Next, we consider the following two situations.

Case 1. $0 \leq i \leq k-1$.

For integer j from k to 0 except for i, if $x^j = y^j$, then we consider whether x^{j-1} is equal to y^{j-1}. If $x^j \neq y^j$, we add $\frac{l}{2}$ to x^j at a time until there exists a nonnegative integer m_s^j such that $0 < |(x^j+\frac{l}{2}\cdot m_s^j)-y^j| \leq \frac{l}{2}$. Then $x^j+\frac{l}{2}\cdot m_s^j$ can reach y^j in one step. By the arbitrariness of j, we construct the path $\mathcal{P}_{i,s}^1$ from $(x^0,\ldots,x^i+s,\ldots,x^k)$ to $z_{i,s}=(y^0,\ldots,y^{i-1},x^i+s,y^{i+1},\ldots,u^k)$ in line 12; Furthermore, we subtract $\frac{l}{2}$ to x^j at a time until there exists a nonnegative integer w_s^j such that $0<|(x^j-\frac{l}{2}\cdot w_s^j)-y^j| \leq \frac{l}{2}$. Then $x^j-\frac{l}{2}\cdot w_s^j$ can reach y^j in one step. By the arbitrariness of j, we obtain the path $\mathcal{P}_{i,s}^{1'}$ from $(x^0,\ldots,x^i-s,\ldots,x^k)$ to $z'_{i,s}=(y^0,\ldots,y^{i-1},x^i-s,y^{i+1},\ldots,u^k)$ for $s=1,2,\ldots,\frac{l}{2}$ in line 13.

It is obviously that $z_{i,s}=\mathcal{P}_{i,s}[-1]$ and $z'_{i,s}=\mathcal{P}'_{i,s}[-1]$. Simultaneously, they are all adjacent to $z_i=(y^0,\ldots,y^{i-1},\ x^i,y^{i+1},\ldots,y^k)$. By invoking Algorithm P-DJPath, there are l disjoint paths from z_i to y in lines 18–19, denoted by $\mathcal{P}_{i,s}^2$ and $\mathcal{P}_{i,s}^{2'}$, such that $z_{i,s}=\mathcal{P}_{i,s}^2[2]$ and $z'_{i,s}=\mathcal{P}_{i,s}^{2'}[2]$. Hence, the path

$$\mathcal{P}_{i,s}=\langle x,\mathcal{P}_{i,s}^1,\mathcal{P}_{i,s}^2\backslash\{z_{i,s},z_i\}\rangle$$

and

$$\mathcal{P}'_{i,s}=\langle x,\mathcal{P}_{i,s}^{1'},\mathcal{P}_{i,s}^{2'}\backslash\{z'_{i,s},z_i\}\rangle$$

are obtained in lines 20 and 21, respectively. By the arbitrariness of s and i, there are lk disjoint paths.

Case 2. $i=k$.

First, we add (resp., subtract) s to x^k. If $x^k=y^k$, for j from $k-1$ to 0, we add (resp., subtract) $\frac{l}{2}$ to x^j at a time until there exists a nonnegative integer m_s^j (resp., w_s^j) such that $0<|(x^j+\frac{l}{2}\cdot m_s^j)-y^j| \leq \frac{l}{2}$ (resp., $0<|(x^j-\frac{l}{2}\cdot w_s^j)-y^j| \leq \frac{l}{2}$). Then $x^j+\frac{l}{2}\cdot m_s^j$ (resp., $x^j-\frac{l}{2}\cdot w_s^j$) can reach y^j in one step. By the arbitrariness of j, the path from x to $(y^0,\ldots,y^{k-1},x^k+s)$ (resp., the path from x to $(y^0,\ldots,y^{k-1},x^k-s)$) is obtained for $s=1,2,\ldots,\frac{l}{2}$. Since $x^k=y^k$, $(y^0,\ldots,y^{k-1},x^k-s))$ is adjacent to y. Hence, there are l disjoint path from x to y in lines 26–27.

If $x^k \neq y^k$, we add (resp., subtract) $\frac{l}{2}$ to x^k+s at a time until there exists a nonnegative integer f_s (resp., f'_s) such that $0<|(x^k+s+\frac{l}{2}\cdot f_s)-y^k| \leq \frac{l}{2}$ (resp., $0<|(x^k-s-\frac{l}{2}\cdot f_{s'})-y^k| \leq \frac{l}{2}$). Furthermore, for j from $k-1$ to 0, by adding (resp., subtracting) $\frac{l}{2}$ to x^j at a time until it reaches $x^j+s+\frac{l}{2}\cdot m_s^j$, where m_s^j (resp., w_s^j) is a nonnegative integer such that $0<|(x^j+\frac{l}{2}\cdot m_{s'}^j)-y^j| \leq \frac{l}{2}$ (resp., $0<|(x^j-\frac{l}{2}\cdot w_{s'}^j)-y^j| \leq \frac{l}{2}$). Finally, due to $x^j+\frac{l}{2}\cdot m_s^j$ (resp., $x^j-\frac{l}{2}\cdot w_s^j$) being able to reach y^j in one step, we change them sequentially from $j=0$ to $k-1$. Therefore, there are l disjoint paths obtained in lines 29–30.

Table 2. An example illustrating the flow of Algorithm DJ-Path2.

$C_{6,13,2}$ ($l=6$, $n=13$, $k=2$)		
$x=(2,5,1)$, $y=(10,7,5)$ ($\lvert x^1-y^1\rvert=8, \lvert x^2-y^2\rvert=2, \lvert x^3-y^3\rvert=4$)		
$s=1$	$i=0$	$\mathcal{P}_{0,1}=\langle(2,5,1),(\mathbf{3},5,1),(3,5,\mathbf{4}),(3,5,\mathbf{5}),(3,\mathbf{7},5),(\mathbf{6},7,5),(\mathbf{9},7,5),\rangle.$ $\langle.(10,7,5)\rangle$
	$j\neq 0$	$\mathcal{P}'_{0,1}=\langle(2,5,1),(\mathbf{1},5,1),(1,5,\mathbf{4}),(1,5,\mathbf{5}),(1,\mathbf{7},5),(\mathbf{11},7,5),(10,7,5)\rangle.$
	$i=1$	$\mathcal{P}_{1,1}=\langle(2,5,1),(2,\mathbf{6},1),(2,6,\mathbf{4}),(2,6,\mathbf{5}),(\mathbf{5},6,5),(\mathbf{8},6,5),(\mathbf{10},6,5),\rangle.$ $\langle.(10,7,5)\rangle$
	$j\neq 1$	$\mathcal{P}'_{1,1}=\langle(2,5,1),(2,\mathbf{4},1),(2,4,\mathbf{4}),(2,4,\mathbf{5}),(\mathbf{5},4,5),(\mathbf{8},4,5),(\mathbf{10},4,5),\rangle.$ $\langle.(10,7,5)\rangle$
	$i=2$	$\mathcal{P}_{1,2}=\langle(2,5,1),(2,5,\mathbf{2}),(\mathbf{5},5,2),(\mathbf{8},5,2),(\mathbf{10},5,2),(10,\mathbf{7},2),(10,7,5)\rangle$
	$j\neq 2$	$\mathcal{P}'_{1,2}=\langle(2,5,1),(2,5,\mathbf{0}),(2,5,\mathbf{10}),(2,5,\mathbf{7}),(2,\mathbf{2},7),(2,\mathbf{12},7),(2,\mathbf{9},7),\rangle.$ $\langle.(\mathbf{12},9,7),(\mathbf{10},9,7),(10,\mathbf{7},7),(10,7,5)\rangle$
$s=2$	$i=0$	$\mathcal{P}_{2,0}=\langle(2,5,1),(\mathbf{4},5,1),(4,5,\mathbf{4}),(4,5,\mathbf{5}),(4,\mathbf{7},5),(\mathbf{7},7,5),(10,7,5)\rangle$
	$j\neq 0$	$\mathcal{P}'_{2,0}=\langle(2,5,1),(\mathbf{0},5,1),(0,5,\mathbf{4}),(0,5,\mathbf{5}),(0,\mathbf{7},5),(10,7,5)\rangle$
	$i=1$	$\mathcal{P}_{2,1}=\langle(2,5,1),(2,\mathbf{7},1),(2,7,\mathbf{4}),(2,7,\mathbf{5}),(\mathbf{5},7,5),(\mathbf{8},7,5),(10,7,5)\rangle$
	$j\neq 1$	$\mathcal{P}'_{2,1}=\langle(2,5,1),(2,\mathbf{3},1),(2,3,\mathbf{4}),(2,3,\mathbf{5}),(\mathbf{5},3,5),(\mathbf{8},3,5),(10,\mathbf{3},5),\rangle.$ $\langle.(10,\mathbf{1},5),(10,\mathbf{12},5),(10,\mathbf{10},5),(10,7,5)\rangle$
	$i=2$	$\mathcal{P}_{2,2}=\langle(2,5,1),(2,5,\mathbf{3}),(\mathbf{5},5,3),(\mathbf{8},5,3),(\mathbf{10},5,3),(10,\mathbf{7},3),(10,7,5)\rangle$
	$y\neq 2$	$\mathcal{P}'_{2,2}=\langle(2,5,1),(2,5,\mathbf{12}),(2,5,\mathbf{9}),(2,5,\mathbf{6}),(\mathbf{5},5,6),(\mathbf{8},5,6),(\mathbf{10},5,6),\rangle.$ $\langle.(10,\mathbf{7},6),(10,7,5)\rangle$
$s=3$	$i=0$	$\mathcal{P}_{3,0}=\langle(2,5,1),(\mathbf{5},5,1),(5,5,\mathbf{4}),(5,5,\mathbf{5}),(5,\mathbf{7},5),(\mathbf{8},7,5),(10,7,5)\rangle$
	$j\neq 0$	$\mathcal{P}'_{3,0}=\langle(2,5,1),(\mathbf{12},5,1),(12,5,\mathbf{4}),(12,5,\mathbf{5}),(12,\mathbf{7},5),(10,7,5)\rangle$
	$i=1$	$\mathcal{P}_{3,1}=\langle(2,5,1),(2,\mathbf{8},1),(2,8,\mathbf{4}),(2,8,\mathbf{5}),(\mathbf{5},8,5),(\mathbf{8},8,5),(\mathbf{10},8,5),\rangle.$ $\langle.(10,7,5)\rangle$
	$y\neq 1$	$\mathcal{P}'_{3,1}=\langle(2,5,1),(2,\mathbf{2},1),(2,2,\mathbf{4}),(2,2,\mathbf{5}),(\mathbf{5},2,5),(\mathbf{8},2,5),(\mathbf{10},2,5),\rangle.$ $\langle.(10,\mathbf{0},5),(10,\mathbf{11},5),(10,\mathbf{9},5),(10,7,5)\rangle$
	$i=2$	$\mathcal{P}_{3,2}=\langle(2,5,1),(2,5,\mathbf{4}),(\mathbf{5},5,4),(\mathbf{8},5,4),(\mathbf{10},5,4),(10,\mathbf{7},4),(10,7,5)\rangle$
	$j\neq 2$	$\mathcal{P}'_{3,2}=\langle(2,5,1),(2,5,\mathbf{11}),(2,5,\mathbf{8}),(2,\mathbf{2},8),(2,\mathbf{12},8),(2,\mathbf{9},8),(\mathbf{12},9,8),\rangle.$ $\langle.(\mathbf{10},9,8),(10,\mathbf{7},8),(10,7,5)\rangle$

Next, we consider the maximum length of $l(k+1)$ paths obtained in the algorithm DJ-Path2. In lines 20–21, the maximum length is $1+\langle\lceil\frac{n-1}{l}\rangle\rceil k+(n-l+1)=\langle\lceil\frac{n-1}{l}\rangle\rceil k+(n-l+2)$; In lines 25–26, the maximum length is $\langle\lceil\frac{n-1}{l}\rangle\rceil k+2$; And the maximum length is $\langle\lceil\frac{n-1}{l}\rangle\rceil(k+1)+2$ in lines 28–29. Thus, the maximum length of Algorithm DJ-Path2 is $\max\{\langle\lceil\frac{n-1}{l}\rangle\rceil k+n-l+2, \langle\lceil\frac{n-1}{l}\rangle\rceil k+2, \langle\lceil\frac{n-1}{l}\rangle\rceil(k+1)+2\}=\langle\lceil\frac{n-1}{l}\rangle\rceil k+(n-l+2)$.

In Algorithm DJ-Path2, lines 5–15 form a loop that runs a total of k times to find two disjoint paths from x to $(y^0,\ldots,y^{i-1},x^i,y^{i+1},\ldots,y^k)$, and the time required for each step of the loop remains constant. Hence, lines 5–15 take $O(k)$ times. In lines 16–17, it takes $O(1)$ time. By Theorem 3, it takes $O(l)$ time to call Algorithm P-DJPath in line 18 and line 19, respectively. In lines 20–21, it takes

$O(1)$ time. In lines 2–22, a loop nested outside of the lines 5–15 executes k times. Thus, the time complexity of lines 2–22 is $O(k(k+1+l+1)) = O(k(k+l))$. It takes $O(1+1+1) = O(1)$ time in lines 23–30 to obtain two disjoint paths from x to y. Furthermore, lines 1–31 constitute a loop nested outside the other loops, which executes a total of $\frac{l}{2}$ times. Therefore, the time complexity of Algorithm DJ-Path2 is $O(\frac{l}{2}(k^2+kl)) = O(kl(k+l))$.

To further illustrate the path construction process of Algorithm DJ-Path2, Table 2 takes $x = (2,5,1)$ and $y = (10,7,5)$ in $C_{6,13,2}$ as an example. There are $l(k+1) = 18$ disjoint paths from x to y in $C_{6,13,2}$.

4 Conclusions

In this paper, we design two algorithms to construct the $l(k+1)$ disjoint paths between any two vertices in a k-dimensional n-order CRN, respectively. The maximum length of those paths is no more than $\langle\lceil\frac{n-1}{l}\rangle\rceil k+(n-l+2)$. Given the benefits of disjoint paths in enhancing network robustness and load balancing, our future research will delve into exploring vertex-disjoint path covers within circulant-based recursive networks.

Acknowledgments. This work was partly supported by the National Natural Science Foundation of China (Nos. 62172291 and 62272333), and the Postgraduate Research & Practice Innovation Program of Jiangsu Province (No. KYCX24-3319).

Disclosure of Interests. The authors have no competing interests to declare that are relevant to the content of this article.

References

1. Abu-Libdeh H., Costa P., Rowstron A., O'Shea G., Donnelly A.: Symbiotic routing in future data centers. In: Proceedings of the 10th ACM SIGCOMM conference on Applications, Technologies, Architectures, and Protocols for Computer Communications, New Delhi, India, pp. 51–62 (2010). https://doi.org/10.1145/1851182.1851191.
2. Adiga, N.R., et al.: Blue Cene/L torus interconnection networks. IBM J. Res. Devlop. **49**(2/3), 265–276 (2005). https://doi.org/10.1147/rd.492.0265
3. Bondy, J.A., Murty, U.S.R.: Graph Theory. Springer, New York (2008)
4. Chen K., Wen X., Ma X., Chen Y., Hu C., Dong Q.: WaveCube: a scalable, fault-tolerant, high-performance optical data center architecture. In: Proceedings of the 15th ACM SIGCOMM conference on Applications, Technologies, Architectures, and Protocols for Computer Communications, Kowloon, Hong Kong, pp. 1903–1911 (2015). https://doi.org/10.1109/INFOCOM.2015.7218573.
5. Fan, J., Lin, X., Pan, Y., Jia, X.: Optimal fault-tolerant embedding of paths in twisted cubes. J. Parallel Distrib. Comput. **67**(2), 205–214 (2007). https://doi.org/10.1016/j.jpdc.2006.04.004

6. Fan, W., Xiao, F., Cai, H., Chen, X., Yu, S.: Disjoint paths construction and fault-tolerant routing in BCube of data center networks. IEEE Trans. Comput. **72**(9), 2467–2481 (2023). https://doi.org/10.1109/TC.2023.3251849
7. Guo C., et al.: BCube: a high performance, server-centric network architecture for modular data centers. In: Proceedings of the 9th ACM SIGCOMM conference on Applications, Technologies, Architectures, and Protocols for Computer Communications, Barcelona, Spain, pp. 63–74 (2009). https://doi.org/10.1145/1594977.1592577.
8. He, Y., Zhang, S., Chen, L., Yang, W.: The disjoint path cover in the data center network HSDC with prescribed vertices in each path. Appl. Math. Comput. **459**, 128262 (2023). https://doi.org/10.1016/j.amc.2023.128262
9. Kern, W., Martin, B., Paulusma, D., Smith, S., Leeuwen, E.J.: Disjoint paths and connected subgraphs for H-free graphs. Theor. Comput. Sci. **898**, 59–68 (2022). https://doi.org/10.1016/j.tcs.2021.10.019
10. Lv, M., Fan, J., Cheng, B., Yu, J., Jia, X.: Construction algorithms of fault-tolerant paths and disjoint paths in k-ary n-cube networks. J. Parallel Distrib. Comput. **183**, 104761 (2024). https://doi.org/10.1016/j.jpdc.2023.104761
11. Liu X., Fan J., Cheng B., Wang Y., Yin B., Jia X.: A family of general architectures towards interconnection networks and dater center networks. IEEE/ACM Trans. Netw. https://doi.org/10.1109/TNET.2024.3411021.
12. Ma, M., Guo, C., Li, X.-J.: Disjoint paths in the enhanced hypercube with a faulty subgraph. J. Appl. Math. Comput. **69**(2), 1343–1354 (2023). https://doi.org/10.1007/s12190-022-01794-z
13. Park, J.-H.: Torus-like graphs and their paired many-to-many disjoint path covers. Discret. Appl. Math. **289**, 64–77 (2021). https://doi.org/10.1016/j.dam.2020.09.008
14. Wu, S., Fan, J., Cheng, B., Yu, J., Wang, Y.: Connectivity and constructive algorithms of disjoint paths in dragonfly networks. Theor. Comput. Sci. **922**, 257–270 (2022). https://doi.org/10.1016/j.tcs.2022.04.028
15. Wang, X., Fan, J., Jia, X., Zhang, S., Yu, J.: Embedding meshes into twisted-cubes. Inf. Sci. **181**(14), 3085–3099 (2011). https://doi.org/10.1016/j.ins.2011.02.019
16. Wang, Q., et al.: On searching multiple disjoint shortest paths in scale-free networks with hyperbolic geometry. IEEE Trans. Netw. Sci. Eng. **9**(4), 2772–2785 (2022). https://doi.org/10.1109/TNSE.2022.3169691

Enumerating Floorplans with Any Set of Columns

Yosuke Kikuchi[1] and Shin-ichi Nakano[2(✉)]

[1] National Institute of Technology, Gunma College, Maebashi 371-8530, Japan
[2] Gunma University, Maebashi 371-8510, Japan
nakano@gunma-u.ac.jp

Abstract. A floorplan is a partition of an axis-aligned rectangle into a set of smaller rectangles. Given an axis-aligned rectangle R and a set P of points in R we wish to partition R into a set S of rectangles so that each point in P is on the boundary of rectangles in S. We call such a partition of R a floorplan covering P. Intuitively P is the locations of columns and a floorplan covering P is a floorplan in which no column is in the proper inside of a room.

If no two points in P have the same x-coordinate and no two points in P have the same y-coordinate, then one can generate all floorplans covering P in $O(|P|)$ time for each.

In this paper we design an algorithm to generate all floorplans covering P, when P is any set of points, which is more practical setting in architecture. The running time of the algorithm is $O(|P|)$ time for each.

Keywords: enumeration · floorplan · algorithm

1 Introduction

Given an axis-aligned rectangle R and a set P of points in R we wish to partition R into a set S of smaller rectangles so that each point in P is on the boundary of rectangles in S. We call such a partition of R a floorplan *covering* P. See an example in Fig. 1. Intuitively, P is the locations of columns and a floorplan covering P is a floorplan in which no column is in the proper inside of a room, i.e., each column is located in a wall.

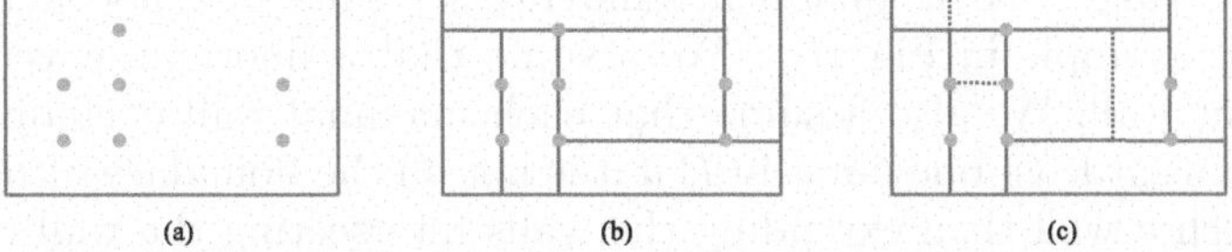

Fig. 1. (a) A set P of points. (b) A floorplan covering P. (c) Redundant edges

Y. Chen et al. (Eds.): COCOON 2024, LNCS 15162, pp. 313–323, 2025.
https://doi.org/10.1007/978-981-96-1093-8_26

Some efficient algorithms to enumerate all floorplans covering P is known [1,2,10,12]. Let S_P be the set of the floorplans covering P. Ackerman et al. [1,2] gave an algorithm to enumerate all floorplans covering P. The algorithm is based on the reverse search method [3,4] and enumerates all such floorplans in either $O(n|S_P|)$ time using $O(n)$ space or $O(\log n|S_P|)$ time using $O(n^3)$ space, where $n = |P|$. Yamanaka et al. [10,12] gave a faster algorithm, which is also based on the reverse search method. The algorithm uses $O(n)$ space, and enumerates all such floorplans in $O(|S_P|)$ time.

Several variants of floorplan enumeration algorithms are also known [5–7,9, 11]. See the brief survey in Sect. 1 of [5].

In those papers P is in a general position, that is, no two points in P have the same x-coordinate, and no two points in P have the same y-coordinate.

In this paper we design an enumeration algorithm when points may share the same x-coordinate or the same y-coordinate. See an example in Fig. 1(a). Our algorithm generates all floorplans covering P in $O(|P|)$ time for each, where P may not be in a general position, which is more practical setting in architecture.

The rest of the paper is organized as follows. Section 2 gives some definitions. Section 3 defines a tree structure among the floorplans covering P. Section 4 gives our enumeration algorithm using the tree structure. Finally Sect. 5 is a conclusion.

2 Preliminaries

In this section we give some definitions.

A *floorplan* is a partition of an axis-aligned rectangle R into a set of smaller rectangles. Each line segment in a floorplan is called a *wall*. A wall is either horizontal or vertical. Each of the four line segments of the boundary of R is called *the outer wall*. Each smaller rectangle in the set is called a *room*. The two horizontal line segments on the boundary of a room is *the upper wall* and *the lower wall* of the room. Similarly the two vertical line segments on the boundary of a room is *the left wall* and *the right wall* of the room.

Let P be a set of points located in the proper inside of R. See an example in Fig. 1(a). A floorplan *covers* P if every point in P is in a wall. See an example in Fig. 1(b).

We regard a floorplan as a graph, in which each corner of a room, each corner of R, or each point in P is a vertex, and each minimal line segment connecting two vertices is an edge. Each edge is called a *basic* wall. A basic wall s in a floorplan covering P is *redundant* if removing s results in a floorplan covering P. See some example in Fig. 1(c). We assume that a floorplan covering P has no redundant wall. We also assume that each maximal wall excluding the end points contains at least one point in P if it is not on the boundary of R. (Because removing such a wall then extending the walls intersecting the wall results in a floorplan covering P.)

3 Family Tree

In this section we define a tree structure among the set of floorplans covering P.

Let f_r be the floorplan covering P having only horizontal walls except the left and the right outer walls. See an example in Fig. 5, where the top floorplan is f_r.

Given a floorplan $f \neq f_r$ covering P, we define the parent floorplan $p(f)$ of f, which also covers P, as follows. We need some definitions.

Let s be the lowest maximal vertical wall among the leftmost maximal vertical walls in f except the left outer wall. (Since $f \neq f_r$ there exists at least one such vertical wall.) We call such s the *critical* wall of f.

Note that, s contains at least two basic vertical walls since otherwise it is redundant and never appears, and by assumption s contains at least one point in P not at the ends.

We have the following two cases.

Case 1: s contains three or more basic vertical walls.

We have the following two subcases. Let v_D be the lower end vertex of s, and v be the upper neighbor vertex of v_D. (See Fig. 2(a).)

Case 1(a): s has at least one point in P which is not at v.

Note that, since the face below v_D is either a rectangle room or the outer face, v_D has horizontal walls on the both sides. If v has horizontal wall, say t, only on the left then the left end vertex of t is on the left outer wall and t contains at least one point in P, since otherwise t is redundant.

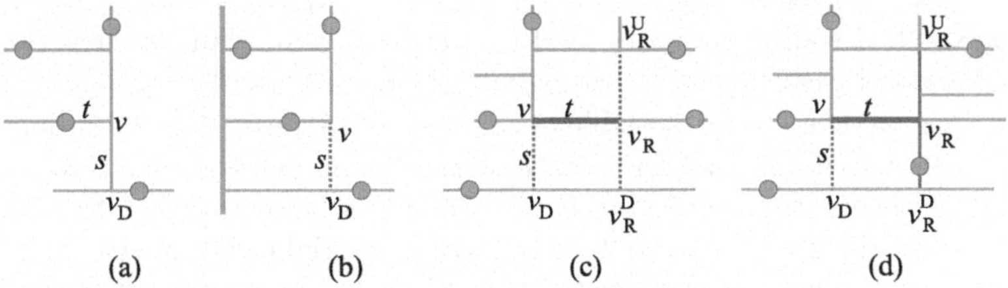

Fig. 2. (a) An illustration for v and v_D. (b) The horizontal wall containing v_D is not redundant after the removal. (c) (d) Illustrations for subroutine Right.

Then $p(f)$ is derived from f by removing the lowest basic vertical wall from s with some modification around v explained below.

After the removal of the lowest basic vertical wall between v and v_D, the horizontal wall containing v_D is not redundant.

Also after the removal of the lowest basic vertical wall between v and v_D, we need basic horizontal walls on the both sides of v since the face below v is a rectangle room in the resulting floorplan. So we may append a basic horizontal wall on each side of v if it is missing.

Appending such basic horizontal walls may cause some new redundant walls. We have the following two subroutines to care this.

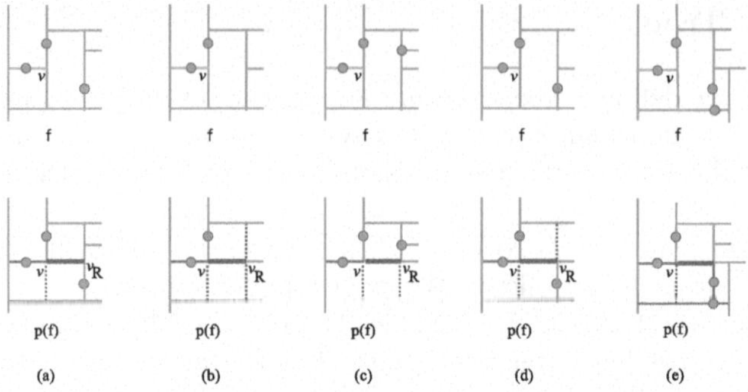

Fig. 3. Illustration for the parent floorplan in Case 1.

Subroutine Right (Appending a basic horizontal wall on the right of v.)

If v has no basic horizontal wall on the right then append the basic horizontal wall t on the right of v. Let v_R be the right end vertex of t, v_R^D be the lower neighbor vertex of v_R, and v_R^U be the upper neighbor vertex of v_R. See Fig. 2(c).

If (1) v_R has a basic horizontal wall on the right, (2) v_R^D has basic horizontal walls on the both sides, and (3) the right wall of the room below t has exactly one basic vertical wall, then after appending t the basic vertical wall between v_R and v_R^D becomes redundant, so it is removed. See Fig. 2(c) and Figs. 3(b)(c). (See Figs. 4(b)(c) for Case 2.) Similarly, if (1) v_R has a basic horizontal wall on the right, (2) v_R^U has basic horizontal walls on the both sides, and (3) the right wall of the room above t has exactly one basic vertical wall, then after appending t the basic vertical wall between v_R and v_R^U becomes redundant, so it is removed. See Fig. 2(c) and Figs. 3(b)(d). (See Figs. 4(b)(d) for Case 2.)

Note that we care those possible occurrence of at most two redundant basic vertical walls at v_R. If the right wall of the room below t has two or more basic vertical walls, then no redundant wall occurs below v_R. See Fig. 2(d) and Fig. 3(d). (See Fig. 4(d) for Case 2.) Similarly if the right wall of the room above t has two or more basic vertical walls, then no redundant wall occurs above v_R. See Fig. 2(d).

Subroutine Left (Appending a basic horizontal wall on the left of v.)

If v has no basic horizontal wall on the left then append the basic horizontal wall t on the left of v. No redundant wall occurs on the left side of v since s is the leftmost vertical wall. The left end vertex of t is always on the left outer wall.

Now we explain the detail. We have the following four subsubcases.

Case 1(a)1: v has a horizontal wall on neither side. (So $v \in P$.)

Execute **Right** and **Left** above.

Case 1(a)2: v has a horizontal wall on the right only.

Execute **Left** above.

Case 1(a)3: v has a horizontal wall on the left only.

Execute **Right** above.

Case 1(a)4: v has horizontal walls on the both sides.

Then the basic vertical wall below v is redundant so it has been removed, a contradiction. Thus this case never occur.

Case 1(b): s has exactly one point in P and it is at v.

Let v_U be the upper end vertex of s, and v' be the lower neighbor vertex of v_U. Note that, since the face above v_U is either a rectangle room or the outer face, v_U has horizontal walls on the both sides.

Then $p(f)$ is derived from f by removing the highest basic vertical wall from s with some modification around v'.

This is an upside-down symmetrical case to **Case 1(a)**. Omitted.

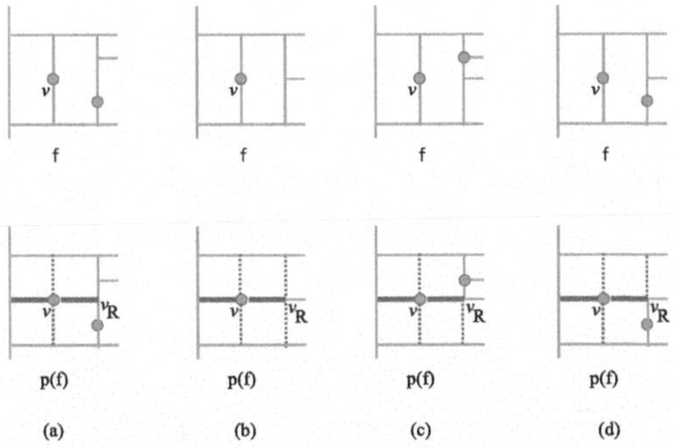

Fig. 4. Illustration for the parent floorplan in Case 2.

Case 2: s contains exactly two basic vertical walls. (So $v \in P$ holds.)

Let v_D be the lower end vertex of s, v be the upper neighbor vertex of v_D, v_U be the upper neighbor vertex of v. Note that, since the face below v_D is either a rectangle room or the outer face, v_D has horizontal walls on the both sides. In Case 2, c_U also has horizontal walls on the both sides. (For example see Fig. 4.)

Then $p(f)$ is derived from f by removing the lowest basic vertical wall from s with some modification around v explained below.

After the removal of the lowest basic vertical wall between v and v_D, the horizontal wall containing v_D is not redundant.

Also after the removal of the lowest basic vertical wall between v and v_D, we need horizontal edges on the both sides of v since the face below v is a rectangle room in the resulting floorplan. Additionally, we always need to remove the basic vertical wall above v, since it is redundant. Thus we remove whole s.

We have the following four subcases.

Case 2–1: v has a horizontal wall on neither side.

Execute **Right** and **Left** above. Then remove the basic vertical wall above v.

Case 2–2: v has a horizontal wall on the right only.

Execute **Left** above. Then remove the basic vertical wall above v.

Case 2–3: v has a horizontal wall on the left only.

Execute **Right** above. Then remove the basic vertical wall above v.

Case 2–4: v has horizontal walls on the both sides.

Then the basic vertical wall below v is redundant in f so it has been removed, a contradiction. Thus this case never occur.

Now we have defined $p(f)$.

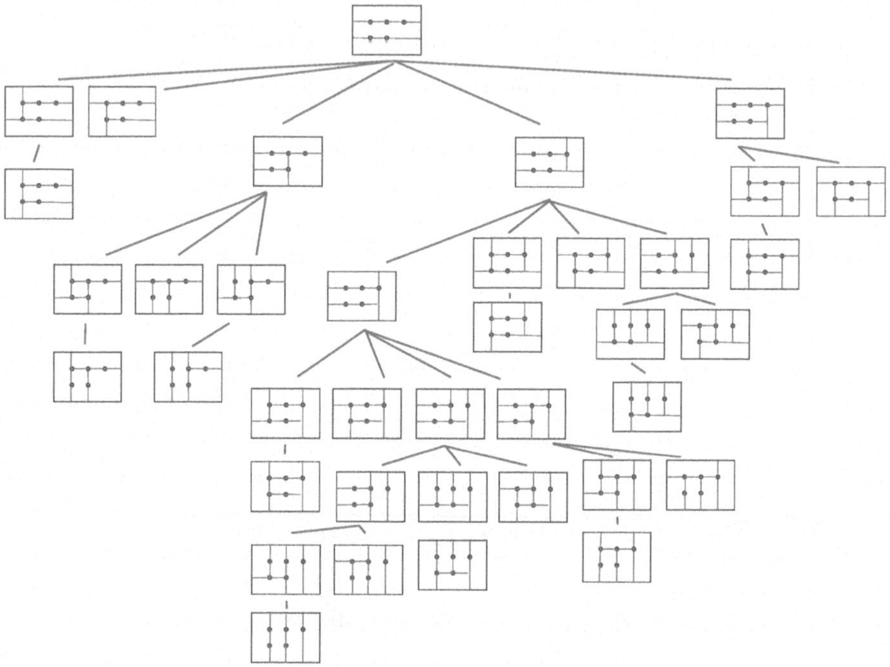

Fig. 5. An example of the family tree.

In those cases the total length of the vertical walls in $p(f)$ is always less than that of f. Given a floorplan covering P, by repeatedly computing the parent of the derived floorplan, say $f, p(f), p(p(f)), \cdots$, we have the unique sequence of floorplans, called *the removing sequence* of f, and it always ends with f_r, which is the floorplan covering P with only horizontal walls.

Given the set S_P of floorplans covering P, by merging the removing sequences of the floorplans in S_P we have the tree structure T_P of floorplans in S_P, in which the root corresponds to f_r, each vertex corresponds to a floorplan in S_P, each edge corresponds to each parent-child relation among the floorplans. See an example in Fig. 5. The tree structure is called *the family tree* of S_P.

If $p(f)$ is the parent of f then we say f is a *child* of $p(f)$. If $p(f)$ is derived from f by Case x above then we say f is a child floorplan of $p(f)$ with Type x.

We have the following lemma.

Lemma 1. *In Case 1, the critical wall s of f remains the critical wall of* $p(f)$ *(after removing either the lowest or highest basic vertical wall).*

In Case 2, the critical wall s of f is completely removed in $p(f)$ *and the critical wall of* $p(f)$ *is located either on the right of (removed) s of f or on the vertical line containing (removed) s but above s.*

Proof. By case analysis. See Fig. 3 and Fig. 4. □

4 Algorithm

In this section we design an algorithm to generate all floorplans covering P.

Given a floorplan f covering P, we design an algorithm to generate all child floorplans of f, as a basic routine. Then, by recursively executing the all child floorplans generation algorithm, we generate all floorplans covering P. We construct the family tree in the depth first order.

Recall that each child floorplan is either Type 1(a), Type 1(b), or Type 2.

Let f be a floorplan covering P and s the critical wall of f. Especially if $f = f_r$ holds then we regard the right outer wall as s.

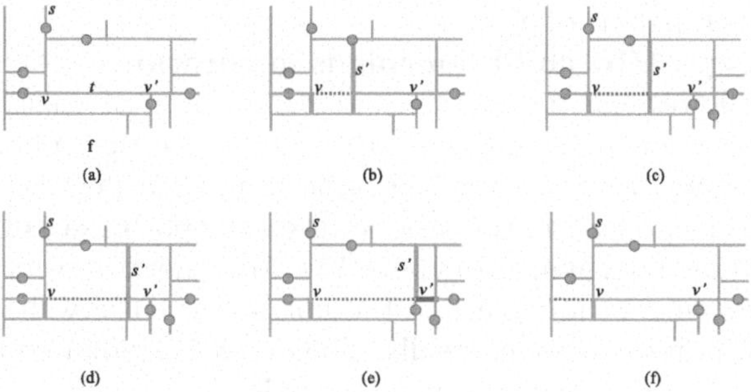

Fig. 6. Illustrations for child floorplans with Type 1(a). (a) A floorplan f. (b)-(d) Child floorplans of f with a part of the horizontal wall on the right of v is removed. (e) A possible child floorplan of f but not a child floorplan of f. (f) A child floorplan of f with the horizontal wall on the lftt of v is removed if the part has no point in P.

STEP 1(a) (Type 1(a) child floorplans generation)

Let v be the lower end vertex of s. Now v has horizontal walls on the both sides. See Fig. 6(a).

If v is not on the outer wall then we extend s downward by one vertical basic wall, then perform every possible modification of the two horizontal basic walls on the both sides of v to generate every possible child floorplan of Type 1(a). We generate every possible child floorplan by removing various part of the horizontal wall on the right of v then reinstall at most two removed redundant walls s' so that the room on the right of v becomes rectangle. See examples in Figs. 6(b)-(d). Note that if a redundant horizontal wall appears then it is not a child floorplan, since we need to remove it. See an example in Fig. 6(e). On the other hand we may remove whole part of the horizontal basic wall on the left of v if the part has no point in P. See an example in Fig. 6(f). Also we may remove the both sides.

Otherwise v is on the outer wall then f has no child floorplan with Type 1(a).

The detail is as follows. Let t be the common horizontal wall of the room having v at the lower left corner and the room having v on the upper wall, v' be the right end point of t. Let t_u be the upper horizontal wall of the room above t, and t'_u be the part of t_u located on the left of v'. Let t_d be the lower horizontal wall of the room below t, and t'_d be the part of t_d located on the left of v' and the right of v.

For each point in P on either t, t'_u and t'_d, each lower end vertex of a vertical wall on t'_u, and each upper end vertex of a vertical wall on t'_d, we may append a vertical wall s' consisting of at most two vertical basic walls and remove the part of t located on the left of s' so that the room on the right of v (and the left of s') becomes a rectangle. If the parent of the resulting floorplan f_c is f then f_c is a child floorplan of f. See examples in Figs. 6(b)-(d). Especially, the floorplan f_c in Fig. 6(e) is not a child floorplan of f since when we remove whole t, a redundant horizontal basic wall appears on the right of t, so we remove it. Because of this the parent of f_c is not f.

STEP 1(b) (Type 1(b) child floorplans generation)

Let v be the lower end vertex of s and v' be the upper neighbour of v and s has exactly one point in P and it is at v'. Let v'' be the upper end vertex of s. We extend s upward by one vertical basic wall, then perform every possible modification of the two horizontal basic walls on the both sides of v'' to generate every possible child floorplan of Type 1(b). We generate every possible child floorplan by removing various part of the horizontal basic wall on the right of v'' then reinstall removed dummy walls. The detail is similar to Step 1(a).

STEP 2 (Type 2 child floorplans generation)

For each p in P located either (1) on the left of s, or (2) on the line containing s but below s, we append a vertical basic wall above p and a vertical basic wall below p with some modification on the horizontal wall t containning p so that the appended two vertical basic walls becomes the critical wall of the resulting floorplan f_c. If the parent of the resulting floorplan f_c is f then f_c is a child floorplan of f. We generate every possible child floorplan with Type 2 by performing every possible modification. We generate every possible child floorplan by removing various part of the horizontal wall on the right of p then reinstall removed dummy walls as in Type 1(a).

We remove either (1) the part of t located on the left of p if the removed part has no point in P (See Fig. 7(b)), (2) a part of t located on the right of p (See Figs. 7(c)-(g)), or (3) both (1) and (2) (See Figs. 7(h)-(k)). Additionally when (2) or (3) we reinstall at most two removed redundant walls of f so that the room on the right of p becomes a rectangle.

The detail is similar to Case 1(a).

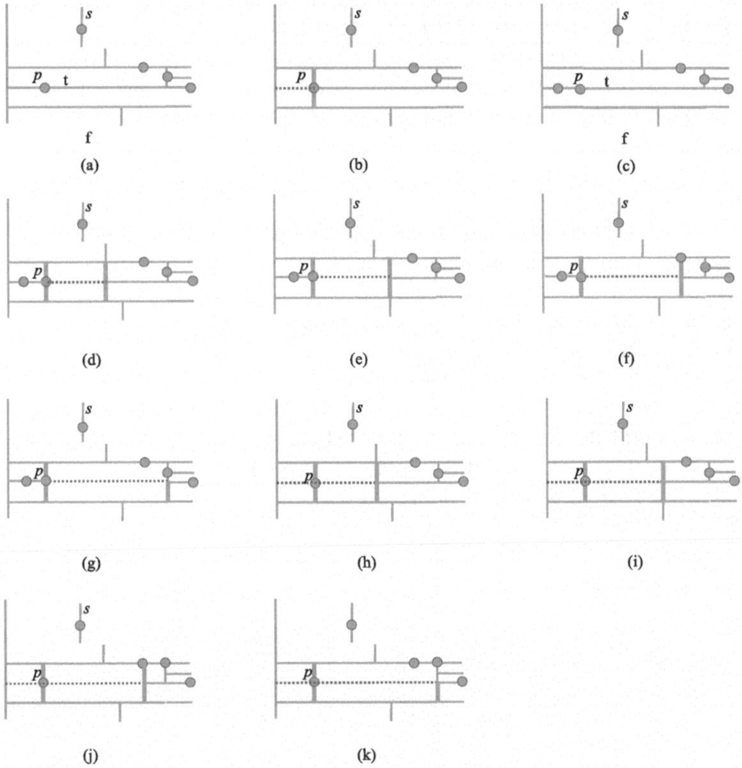

Fig. 7. Illustration for child floorplans with Type 2. (a) A floorplan f. (b) A child floorplan of f in (a). (c) A floorplan f. (d)-(g) Child floorplans of f in (c). (h)-(k) Child floorplans of f in (a).

Based on the cases above we have the following algorithm to generate all child floorplans of a given floorplan.

When we append a vertical basic wall at a vertex, we need to compute the other end vertex of the basic vertical wall to update the graph corresponding to the floorplan. For efficient update we need the following data structure.

For the current floorplan f, for each horizontal wall having the left end vertex on the left outer wall, we maintain the rightmost point w in P on the horizontal wall for which we have already generated child floorplans with Type 2. If we append a vertical basic wall t below a vertex v we need to compute the lower end vertex u of t on the horizontal wall h below v. The lower end vertex u of t is inserted at the right of the rightmost point w in P on h, which we maintain (See Fig. 8), so we can update the graph corresponding to the current floorplan in $O(1)$ time. On the other hand if we compute u without the location of w we may perform a linear search on h and it takes $O(|P|)$ time.

Algorithm 1. Generate all child floorplans(f, s)

```
for p ∈ P located on the left of s in the increasing order of x(p) do
    Generate each possible child floorplan f_c of Type 2 with respect to p
    and output f_c if f_c is a child floorplan of f
end for
for p ∈ P located on the line containing s and below s in the increasing order of
  y(p) do
    Generate each child floorplan f_c of Type 2 with respect to p
    and output f_c if f_c is a child floorplan of f
end for
if f has the child floorplan of Type 1(a) then
    Generate each possible child floorplan f_c of Type 1(a)
    and output f_c if f_c is a child floorplan of f
end if
if f has the child floorplan of Type 1(b) then
    Generate each possible child floorplan of Type 1(b)
    and output f_c if f_c is a child floorplan of f
end if
```

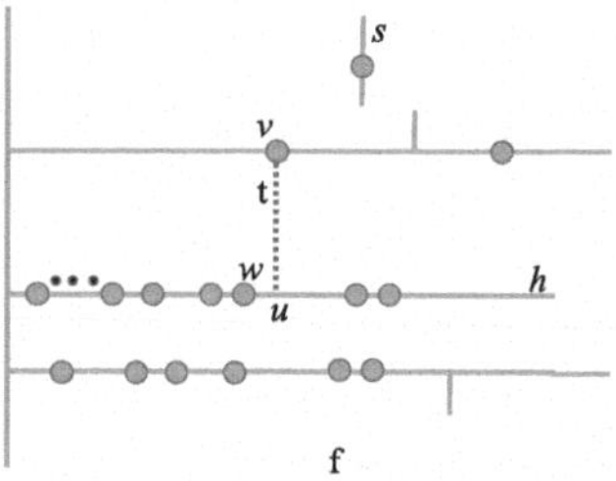

Fig. 8. Computing the crossing point u of h and t.

In this way we can generate all child floorplans of f in $O(|P|)$ time. And so we can enumerate all floorplans covering P in $O(|P||S_P|)$ time, where S_P is the set of floorplans covering P.

If we generate each floorplan in the depth first order in the family tree we may need much time to generate the next floorplan of the last floorplan in a large subtree. However, by the prepost order method [8], which generates each floorplan (Case ODD) before its child floorplans if the depth of recursive call is odd and (Case EVEN) after its child floorplans otherwise, one can generate each floorplan in $O(1)$ time, as the difference from the preceding floorplan. Note that by tracing at most three edges of the family tree we can output the next floorplan.

We have the following theorem.

Theorem 1. *After sorting P in $O(|P|\log|P|)$ time, one can generate all floorplans covering P in $O(|P|)$ time for each, where P is any set of points.*

5 Conclusion

In this paper we have designed an algorithm to generate all floorplans covering P when P is any set of points. After sorting P in $O(|P| \log |P|)$ time the algorithm generate all floorplans covering P in $O(|P|)$ time for each.

Can we design an algorithm to generate all floorplans covering P, for any set of points P, when each point of P has a type, either vertical or horizontal, and each vertical type point must be contained in a vertical wall and each horizontal type point must be contained in a horizontal wall. If no two points in P have the same x-coordinate and no two points in P have the same y-coordinate, then one can generate all such floorplan in $O(|P|)$ time for each [11].

References

1. Ackerman, E., Barequet, G., Pinter, R.Y.: On the number of rectangulations. Proc. SODA **2004**, 729–738 (2004)
2. Ackerman, E., Barequet, G., Pinter, R.Y.: On the number of rectangulations of a planar point set. J. Comb. Theory Ser. A **113**, 1072–1091 (2006)
3. Avis, D.: Generating rooted triangulations without repetitions. Algorithmica **16**, 618–632 (1996)
4. Avis, D., Fukuda, K.: Reverse search for enumeration. Discret. Appl. Math. **65**, 21–46 (1996)
5. Merino, A.I., Mütze, T.: Efficient Generation of Rectangulations via permutation languages. Proc. SoCG **2021**(54), 1–18 (2021)
6. Nakano, S.: Enumerating Floorplans with n Rooms. In: Eades, P., Takaoka, T. (eds.) ISAAC 2001. LNCS, vol. 2223, pp. 107–115. Springer, Heidelberg (2001). https://doi.org/10.1007/3-540-45678-3_10
7. Nakano, S.: Enumerating Floorplans with n Rooms. IEICE Trans. Fundam. Electron. Commun. Comput. Sci. E85-A, pp. 1746–1750 (2002)
8. Nakano, S., Uno, T.: Constant time generation of trees with specified diameter. In: Proceedings of WG04, LNCS 3353, pp. 33–45 (2004)
9. Takagi, M., Nakano, S.: Listing all rectangular drawings with certain properties. Syst. Comput. Jpn. **35**, 1–8 (2004)
10. Yamanaka, K., Rahman, M.S., Nakano, S.-I.: Floorplans with columns. In: Gao, X., Du, H., Han, M. (eds.) COCOA 2017. LNCS, vol. 10627, pp. 33–40. Springer, Cham (2017). https://doi.org/10.1007/978-3-319-71150-8_3
11. Yamanaka, K., Nakano, S.: Floorplans with Walls. In: Chen, J., Feng, Q., Xu, J. (eds.) TAMC 2020. LNCS, vol. 12337, pp. 50–59. Springer, Cham (2020). https://doi.org/10.1007/978-3-030-59267-7_5
12. Yamanaka, K., Rahman, S., Nakano, S.: Enumerating floorplans with columns. IEICE Trans. Fundam. Electron. Commun. Comput. Sci. **E101-A**, 1392–1397 (2018)

Streaming Algorithm for Balance Gain and Cost with Cardinality Constraint on the Integer Lattice

Jingjing Tan[1], Cuiping Ge[1](✉), Fengmin Wang[2], and Ziyang Li[1]

[1] School of Mathematics and Statistics, Weifang University, Weifang 261061, Shandong, China
wfgecuiping@163.com

[2] Beijing Jinghang Research Institute of Computing and Communication, Beijing 100074, People's Republic of China

Abstract. Team formation problem is a very important problem in the labor market, and it is proved to be NP-hard. This paper proposes an efficient bicriteria streaming algorithm aimed at striking a balance between gain and cost in team formation problems with cardinality constraints on the integer lattice. In addressing this, we utilize a optimized model of maximizing the difference between a nonnegative normalized monotone submodule function and a nonnegative linear function. Combining the lattice binary search with the threshold method, we present an online algorithm called bicriteria streaming algorithms.

Keywords: Knapsack constraint · Integer lattice · Cardinality function · Streaming algorithm

1 Introduction

Team formation considerations occur in a diverse range of scenarios, including but not limited to project management, workforce allocation, sports team composition, and academic collaboration. These problems have been the pivot point of intense attention in recent research efforts [1–4]. Nikolakaki et al. [5] formulated the team formation issue as a maximization problem, which can be characterized as follows:

$$\max_{V \subset E} g(V) - c(V),$$

where the function $g(V)$ is monotone and non-negative submodular while $c(V)$ represents a non-negative linear function that sums up the costs of selected elements. Du et al. [6] proposed a two-criterion approximation algorithm. Feldman [7] designed bicriteria approximation algorithm.

The concepts of lattice theory and optimization was first proposed by Lovasz [8]. For a finite set E of size n, where each element e_i belongs to E, we set χ_{e_i} as the i-th unit vector. Then, we use $\mathbb{N}^E$ to represent the integer grid defined over E. We denote g as a DR-submodular function if the following condition is satisfied: for any $\boldsymbol{x}, \boldsymbol{y}$, given that $\boldsymbol{x} \leq \boldsymbol{y}$, the inequality

$$g(\boldsymbol{y} + \chi_{e_i}) - g(\boldsymbol{y}) \leq f(\boldsymbol{x} + \chi_{e_i}) - f(\boldsymbol{x})$$

Y. Chen et al. (Eds.): COCOON 2024, LNCS 15162, pp. 324–331, 2025.
https://doi.org/10.1007/978-981-96-1093-8_27

holds. Soma et al. [9] conducted research on the DR-submodular maximization problem, taking into account constraints, such as cardinality, knapsack, and matroid. Gottschalk et al. [10] cleverly used the bidirectional greed technique offers a 1/3-approximation ratio, tackling the issue of maximising non-monotonic modular functions on the integer lattices. Nong et al. [11] further improved the approximate ratio to 1/2. Soma et al. [12] established a monotone lattice submodular maximum on the integer lattices with knapsack constraints. Tan et al. [13] proposed an online threshold streaming algorithm with 1/3-approximate ratios for the lattice submodular maximization problems with knapsack constraints in a flow model. There are many excellent results, see [9, 12, 14–19]

Inspired by the above results, we design an efficient bicriteria streaming algorithms to construct a balance between gain and cost in a team formation problem with cardinality constraint on the integer lattice. The problem is formulated as follows: Let E be a ground set of size n , $\boldsymbol{x}$ be a n-dimensional vector from $\mathbb{N}^E$, g be a normalized submodular function defined on $\mathbb{N}^E$ satisfying nonnegativity and monotonicity, and c be a linear function also defined on $\mathbb{N}^E$,

$$\begin{aligned} \text{maximize } & g(\boldsymbol{x}) - c(\boldsymbol{x}), \\ \text{s.t. } \quad & \boldsymbol{x} \leq \boldsymbol{b}, \\ & \boldsymbol{x}(E) \leq k. \end{aligned} \tag{1}$$

where the total budget is $k \in \mathbb{N}$, and $\boldsymbol{B} = \{\boldsymbol{x} \in \mathbb{N}^E : \boldsymbol{x} \leq \boldsymbol{b}\}$ is a box in $\{\mathbb{N} \bigcup \{+\infty\}\}^E$.

2 Preliminaries

In this section, we elucidate numerous symbols, conceptual definitions and axioms.

We use $[k]$ to denote all positive integers between 1 and k. We establish a base set $E = \{e_1, e_2, \cdots, e_n\}$ and $\boldsymbol{x}$ as an n-dimensional vector in $\mathbb{N}^E$. The segment of coordinate $e_i \in E$ for $\boldsymbol{x}$ is represented as $\boldsymbol{x}(e_i)$. We employ $\mathbf{0}$ to depict the null vector and χ_{e_i} as the standard base vector. Thus, all components are at zero, except for the i-th component that stands at unity. For any subset $X \subseteq E$, its characteristic vector is χ_X and $\boldsymbol{x}(X)$ amounts to the summation of $\boldsymbol{x}(e_i)$ over all elements e_i belonging to X. Within the set $\boldsymbol{x} \in \mathbb{N}^E$, $supp^+(\boldsymbol{x}) = \{e \in E | \boldsymbol{x}(e) > 0\}$ signifies the support set of $\boldsymbol{x}$. We define a multi-set $\{\boldsymbol{x}\}$, where e appears $\boldsymbol{x}(e)$ times, and use $|\{\boldsymbol{x}\}| := \boldsymbol{x}(E)$ to express the total amount of $\boldsymbol{x}(E)$. $\boldsymbol{x} \vee \boldsymbol{y}$ and $\boldsymbol{x} \wedge \boldsymbol{y}$ stand for the coordinate-wise maximum and minimum of vectors $\boldsymbol{x}$ and $\boldsymbol{y}$ respectively. Multi-sets $\{\boldsymbol{x}\}$ and $\{\boldsymbol{y}\}$ have their difference denoted as: $\{\boldsymbol{x}\} \setminus \{\boldsymbol{y}\} := \{(\boldsymbol{x} \setminus \boldsymbol{y}) \vee \mathbf{0}\}$.

A function $f : \mathbb{N}^E \rightarrow \mathbb{R}_+$ is known as monotone non-decreasing if and only if $f(\boldsymbol{x}) \leq f(\boldsymbol{y})$ can be satisfied for any $\boldsymbol{x} \leq \boldsymbol{y}$. A function f is deemed nonnegative and normalized if it invariably fulfills the conditions: $f(\boldsymbol{x}) \geq 0$ (for all $\boldsymbol{x} \in \mathbb{N}^E$) and $f(\mathbf{0}) = 0$.

The value denoted by $f(\boldsymbol{x}|\boldsymbol{y})$ presents the marginal increase of a vector $\boldsymbol{x}$ concerning $\boldsymbol{y}$ expressed as:

$$f(\boldsymbol{x}|\boldsymbol{y}) = f(\boldsymbol{x} + \boldsymbol{y}) - f(\boldsymbol{y}).$$

.

Definition 1. *For any $\boldsymbol{s}, \boldsymbol{t} \in \mathcal{B}$ with $\boldsymbol{s} \leq \boldsymbol{t}$ and $e \in E$, the α-weakly submodular function g in $\mathcal{G}_b$ is the maximum scalar α_g such that*

$$\alpha_g g(\chi_e|\boldsymbol{t}) \leq g(\chi_e|\boldsymbol{s})$$

where $\boldsymbol{t} + \chi_e \in \mathcal{B}$.

3 Bicriteria Streaming Algorithms

We will design two combinatorial algorithms and analyze the performance of the two respectively in this section.

Denote $\boldsymbol{x}^*$ be the optimal solution, $\boldsymbol{x}$ is the final solution vector returned by Algorithm 1. $p = \boldsymbol{x}^*(E)$, $q = \boldsymbol{x}(E)$. Assume that $\boldsymbol{x}_{i-1}$ is the set when the element e_i arrives, $i = 1, 2, \cdots, q$, $\boldsymbol{x}_0 = \boldsymbol{0}$, $\boldsymbol{x}_q = \boldsymbol{x}$. There are two cases $\boldsymbol{x}(E) = k$(Lemma1) and $\boldsymbol{x}(E) < k$(Lemma2). By construct an intermediary function $g(\boldsymbol{x}) - tc(\boldsymbol{x})$, where $t \geq 1$, we can select element e whose marginal gain is maximal the threshold τ. The detailed presentation is shown in Algorithm 1.

Algorithm 1. Streaming algorithm for G-C on the integer lattice

Input: Data stream E, $g \in \mathcal{G}_b$, $c \in \mathcal{C}_b$, cardinality constraint $k \in \mathbb{N}_+$, threshold value τ, $t \geq 1$.
Output: a vector $\boldsymbol{x} \in \mathbb{N}^E$.

```
1: x ← 0;
2: for e ∈ E do
3:    if x(E) < k, then
4:       find the level l with BS Algorithm(g, c, x, b, e, k, τ);
5:       if x + lχ_{e_i}(E) ≤ k, then
6:          update x ← x + lχ_{e_i};
7:       end if
8:    end if
9: end for
10: return x
```

Lemma 1. *Let $\boldsymbol{x}$ represent the result acquired from Algorithm 1. Provided $\boldsymbol{x}(E) = k$, it follows that the computation of $g(\boldsymbol{x}) - c(\boldsymbol{x})$ yields a value greater than or equal to $k\tau$.*

Proof. According to Algorithm 2, for any element $e_i \in \boldsymbol{x}$, we get that the following inequality is satisfied

$$g(l_i\chi_{e_i}|\boldsymbol{x}_{i-1}) - tc(l_i\chi_{e_i}) \geq l_i\tau,$$

summing up all the inequalities above, we get

$$\sum_{e_i \in \{\boldsymbol{x}\}} [g(l_i\chi_{e_i}|\boldsymbol{x}_{i-1}) - tc(l_i\chi_{e_i})] \geq \sum_{e_i \in \{\boldsymbol{x}\}} l_i\tau = \tau\boldsymbol{x}(E).$$

Algorithm 2. BS Algorithm $(g, c, \boldsymbol{x}, \boldsymbol{b}, e, k, \tau)$

Input: Data stream E, $e \in E$, submodular function $g \in \mathcal{G}_b$, $c \in \mathcal{C}_b$, $\boldsymbol{x} \in \mathbb{N}^E$, and $\tau \in \mathbb{R}_+$.
Output: l

$l_m \leftarrow 1$;
$l_n \leftarrow \min \left\{ \boldsymbol{b}(e) - \boldsymbol{x}(e), k - \boldsymbol{x}(e) \right\}$;
if $g(\chi_e|\boldsymbol{x}) - tc(\chi_e) < \tau$ **then**
  return 0.
end if
if $\frac{g(l_n\chi_e|\boldsymbol{x}) - tc(l_n\chi_e)}{l_n} \geq \tau$ **then**
  return l_n
end if
while $l_m < l_n + 1$, **do**
  $a = \lfloor \frac{l_n + l_m}{2} \rfloor$
  if $\frac{g(a\chi_e|\boldsymbol{x}) - tc(a\chi_e)}{a} \geq \tau$ **then**
    $l_m = a$,
  else
    $l_n = a$,
  end if
end while
return l_m

With the aid of the submodularity of the function g and the linearity of function c, we can rewrite the above inequality as following

$$g(\boldsymbol{x}) - tc(\boldsymbol{x}) \geq \tau \boldsymbol{x}(E) = \tau k.$$

Since $t \geq 1$, we get

$$g(\boldsymbol{x}) - c(\boldsymbol{x}) \geq g(\boldsymbol{x}) - tc(\boldsymbol{x}) \geq \tau k.$$

thus finalizing the demonstration.

Lemma 2. *Suppose $\boldsymbol{x}$ is the outcome solution yielded from Algorithm 1, and it adheres to the condition $\boldsymbol{x}(E) < k$, under such supposition, the function $g(\boldsymbol{x}) - c(\boldsymbol{x})$ presents a value that conforms to a specific inequality*

$$g(\boldsymbol{x}) - c(\boldsymbol{x}) \geq (1 - \frac{1}{t})g(\boldsymbol{x}^*) - (t-1)c(\boldsymbol{x}^*) + k\tau[\frac{1}{t}\nu - (1 - \frac{1}{t})\mu],$$

where $\mu, \nu : 0 < \mu, \nu \leq 1, t \geq 1$.

Proof. First, we denote $\{\bar{\boldsymbol{x}}\} = \{\boldsymbol{x}^*\} \setminus \{\boldsymbol{x}\}$, $s = \bar{\boldsymbol{x}}(E)$. let $\boldsymbol{x}_{i-1}$ be the corresponding output solution for each element e_i $(i \in \{1, \cdots, s\})$ that χ_{e_i} has arrived but not been added to $\boldsymbol{x}$.

Next, we will prove the lower bound of function $g(\boldsymbol{x})$ and the upper bound of $c(\boldsymbol{x})$ respectively. Now, We proof the the lower bound of $g(\boldsymbol{x})$. According to Algorithm 2, for each element $e_i \in \{\bar{\boldsymbol{x}}\}$, we have

$$g(\chi_{e_i} \mid \boldsymbol{x}_{i-1}) - tc(\chi_{e_i}) < \tau.$$

Summing up all the above inequalities of the elements in $\bar{\boldsymbol{x}}$, we obtain

$$\sum_{e_i \in \{\boldsymbol{x}\}} g(\chi_{e_i}|\boldsymbol{x}_{i-1}) - t \sum_{e_i \in \{\boldsymbol{x}\}} c(\chi_{e_i}) < \sum_{e_i \in \{\boldsymbol{x}\}} \tau = \tau|\{\bar{\boldsymbol{x}}\}| \tag{2}$$

Using the submodularity of the function g, we have

$$\begin{aligned} &\sum_{e_i \in \{\bar{\boldsymbol{x}}\}} g(\chi_{e_i}|\boldsymbol{x}_{i-1}) \\ \geq &\sum_{e_i \in \{\bar{\boldsymbol{x}}\}} g(\chi_{e_i}|\boldsymbol{x}) \\ \geq &\, g(\bar{\boldsymbol{x}}|\boldsymbol{x}) \\ \geq &\, g(\boldsymbol{x}^* + \boldsymbol{x}) - g(\boldsymbol{x}). \end{aligned}$$

Given that non-negative function c is linear, we derive

$$\sum_{e_i \in \{\bar{\boldsymbol{x}}\}} c(\chi_{e_i}) = c(\bar{\boldsymbol{x}}) \leq c(\boldsymbol{x}^*).$$

By combining the above two inequalities with the monotonicity of g, we can derive the following inequality

$$g(\boldsymbol{x}^*) - g(\boldsymbol{x}) - tc(\boldsymbol{x}^*) \leq \tau|\bar{\boldsymbol{x}}(E)|,$$

that is,

$$g(\boldsymbol{x}) \geq g(\boldsymbol{x}^*) - tc(\boldsymbol{x}^*) - \tau|\bar{\boldsymbol{x}}(E)|. \tag{3}$$

Furthermore, we can determine the upper bound of function $c(\boldsymbol{x})$ that is related to the function $g(\boldsymbol{x})$. From the running process of the algorithm, we can see that all the elements selected into the output solution $\boldsymbol{x}$ satisfy the following inequality

$$g(l_i\chi_{e_i}|\boldsymbol{x}_{i-1}) - tc(l_i\chi_{e_i}) \geq l_i\tau.$$

Summing up above all the inequality which the element $e_i \in \{\boldsymbol{x}\}$, we get

$$\sum_{e_i \in \{\boldsymbol{x}\}} g(l_i\chi_{e_i}|\boldsymbol{x}_{i-1}) - t \sum_{e_i \in \{\boldsymbol{x}\}} c(l_i\chi_{e_i}) \geq \sum_{e_i \in \{\boldsymbol{x}\}} l_i\tau = \tau\boldsymbol{x}(E).$$

Assuming that $g(\boldsymbol{x})$ is a monotonic submodular function and c represents a linear function, it enables us to approximate the term of the $c(\boldsymbol{x})$, that is,

$$c(\boldsymbol{x}) \leq \frac{1}{t} g(\boldsymbol{x}) - \frac{\tau}{t}\boldsymbol{x}(E). \tag{4}$$

Then, by combining inequality (3) and inequality (4), we get the lower bound of the value of objective function $g(\boldsymbol{x}) - c(\boldsymbol{x})$ as follows

$$\begin{aligned} &g(\boldsymbol{x}) - c(\boldsymbol{x}) \\ \geq &g(\boldsymbol{x}) - \frac{1}{t} g(\boldsymbol{x}) + \frac{\tau}{t}\boldsymbol{x}(E) \\ \geq &\frac{t-1}{t} g(\boldsymbol{x}^*) - (t-1)c(\boldsymbol{x}^*) - \frac{t-1}{t}\bar{\boldsymbol{x}}(E)\tau + \frac{\tau}{t}\boldsymbol{x}(E) \end{aligned} \tag{5}$$

For the remaining part of the proof, we approach this by introducing two parameters, μ (where $0 \leq \mu \leq 1$) and ν (where $0 \leq \nu \leq 1$), that are reliant on either the ultimate output set or the present set. We can assume without loss of generality that $\bar{\boldsymbol{x}}(E)$ equals to ν_k, and $\boldsymbol{x}(E)$ equals to ν_k. Throughout the algorithmic analysis, we only employ μ and ν to denote multiple factors, so they are need to consistent with other parameters in the current analysis. Then we can rewrite the inequality (5)

$$
\begin{aligned}
& g(\boldsymbol{x}) - c(\boldsymbol{x}) \\
& \geq (1 - \frac{1}{t}) g(\boldsymbol{x}^*) - (t-1) c(\boldsymbol{x}^*) - (1 - \frac{1}{t}) \mu k \tau + \frac{\tau}{t} \boldsymbol{x}(E) \\
& = (1 - \frac{1}{t}) g(\boldsymbol{x}^*) - (t-1) c(\boldsymbol{x}^*) + k\tau [\frac{1}{t} \nu - (1 - \frac{1}{t}) \mu]
\end{aligned}
$$

So, we can come to the conclusion that

$$
g(\boldsymbol{x}) - c(\boldsymbol{x}) \geq (1 - \frac{1}{t}) g(\boldsymbol{x}^*) - (t-1) c(\boldsymbol{x}^*) + k\tau [\frac{1}{t} \nu - (1 - \frac{1}{t}) \mu,
$$

for $0 \leq \mu, \nu \leq 1, t \geq 1$.

which completes the proof.

Theorem 1. *Suppose* $t = \frac{2+\mu+\sqrt{\delta}}{2}$, $\delta = \mu^2 + 4 - 4\nu$, $0 \leq \mu, \nu \leq 1$, $\rho = \frac{(t-1)}{t+\mu(t-1)-\nu}$, *then Algorithm 1's bicriteria ratio is* $(\rho, \rho \cdot t)$.

Proof. Lemma 1 and Lemma 2 show that for any $0 \leq \mu, \nu \leq 1, t \geq 1$, we have

$$
g(\boldsymbol{x}) - c(\boldsymbol{x}) \geq k\tau, \boldsymbol{x}(E) = k
$$

and

$$
g(\boldsymbol{x}) - c(\boldsymbol{x}) \geq (1 - \frac{1}{t}) g(\boldsymbol{x}^*) - (t-1) c(\boldsymbol{x}^*) + k\tau [\frac{1}{t} \nu - (1 - \frac{1}{t}) \mu], \boldsymbol{x}(E) < k
$$

for $0 \leq \mu, \nu \leq 1, t \geq 1$. Therefore, the value of the final objective function is minimum of $k\tau$ and $(1 - \frac{1}{t}) g(\boldsymbol{x}^*) - (t-1) c(\boldsymbol{x}^*) + k\tau [\frac{1}{t} \nu - (1 - \frac{1}{t}) \mu]\}$. Let

$$
k\tau = (1 - \frac{1}{t}) g(\boldsymbol{x}^*) - (t-1) c(\boldsymbol{x}^*) + k\tau [\frac{1}{t} \nu - (1 - \frac{1}{t}) \mu],
$$

we obtain

$$
k\tau = \rho g(\boldsymbol{x}^*) - \rho \cdot t \cdot c(\boldsymbol{x}^*).
$$

We calculate the value of t such that

$$
\rho \cdot t = 1,
$$

that is, make the coefficient of $c(\boldsymbol{x}^*)$ becomes 1. We can get values for t from the above equality

$$
t_1 = \frac{2 + \mu + \sqrt{\delta}}{2} > 1 + \mu,
$$

and

$$t_2 = \frac{2 + \mu - \sqrt{\delta}}{2} < 1 + \mu,$$

where $\delta = (2+\mu)^2 - 4(\mu+\nu) = 4+\mu^2 - 4\nu$ and $\delta > 0$, $\sqrt{\delta} > \mu$ under the assumption $0 \leq \mu \leq 1$ and $0 \leq \nu \leq 1$.

Depending on the range of t, we take $t = t_1$. Then

$$k\tau - \rho g(\boldsymbol{x}^*) \quad \rho \quad t \quad c(\boldsymbol{x}^*).$$

and the threshold value

$$\tau = \frac{1}{k}\rho g(\boldsymbol{x}^*) - \frac{1}{k}\rho \cdot t \cdot c(\boldsymbol{x}^*).$$

where

$$t = \frac{2 + \mu + \sqrt{\delta}}{2} > 1 + \mu,$$
$$\delta = 4 + \mu^2 - 4\nu,$$
$$0 \leq \mu \leq 1,$$
$$0 \leq \nu \leq 1.$$

Therefore, we can get the conclusion of the Theorem 1 that the bicriteria ratio is $(\rho, \rho{\cdot}t)$.

The amount of $e \in E$ is k. For any element, the binary search algorithm takes $O(\log k)$ time per e. So, we obtain the total memory is $O(\log k)$.

We also discuss the case where the first function g in the objective function is α-weakly submodularity.

4 Conclusions

In this paper, we present the development of combinatorial approximation algorithms designed to address the maximization problem of monotone non-negative submodular function minus a nonnegative linear function in integer lattice. Simultaneously, we present comprehensive analyses for this models. Features of these models include that the query number per is $O(\log K)$.

Acknowledgements. The first author is supported by Natural Science Foundation of Shandong Province (No. ZR2022MA034) and National Natural Science Foundation of China (No. 12301417). The third author is supported by Beijing Municipal Natural Science Foundation (No.Z2 20004)

References

1. Anagnostopoulos, A., Becchetti, L., Castillo, C., Gionis, A., Leonardi, S.: Online team formation in social networks. In: Proceedings of the 21st International Conference on World Wide Web, pp. 839–848 (2012)
2. Anagnostopoulos, A., Castillo, C., Fazzone, A., Leonardi, S., Terzi, E.: Algorithms for hiring and outsourcing in the online labor market. In: Proceedings of the 24th ACM SIGKDD International Conference on Knowledge Discovery and Data Mining, pp. 1109–1118 (2018)
3. Lappas, T., Liu, K., Terzi, E.: Finding a team of experts in social networks. In: Proceedings of the 15th ACM SIGKDD International Conference on Knowledge Discovery and Data Mining, pp. 467–476 (2009)
4. Wang, Y., Xu, D., Du, D., Jiang, Y.: Bicriteria streaming algorithms to balance gain and cost with cardinality constraint. J. Comb. Optim. **44**, 2946–2962 (2022)
5. Nikolakaki, S., Ene, A, Terzi, E.: An Efficient Framework for Balancing Submodularity and Cost. ArXiv preprint arXiv: 2002.07782
6. Du, D., Li, Y., Xiu, N., Xu, D.: Simultaneous approximation of multi-criteria submodular functions maximization. J. Oper. Res. Soc. China **2**, 271–290 (2014)
7. Feldman, M.: Guess free maximization of submodular and linear sums. In: Proceedings of the 16th International Conference Workshop on Algorithms and Data Structures, pp. 380–394 (2019)
8. Lovasz, L.: On the Shannon capacity of a graph. IEEE Trans. Inf. Theory **25**(1), 1–7 (1979)
9. Soma, T., Yoshida, Y.: Maximization monotone submodular functions over the integer lattice. Math. Program. **172**, 539–563 (2018)
10. Gottschalk, C., Peis, B.: Submodualr function maximization on the bounded integer lattice. In: Proceedings of WAOA, pp. 133–144 (2015)
11. Nong, Q., Fang, J., Gong, S., Du, D., Feng, Y., Qu, X.: A 1/2-approximation algorithm for maximizing a non-monotone weak-submodular function on a bounded integer lattice. J. Comb. Optim. **39**(4), 1208–1220 (2020). https://doi.org/10.1007/s10878-020-00558-4
12. Soma, T., Kakimura, N., Inaba, K., Kawarabayashi, K.: Optimal budget allocation: theoretical guarantee and efficient algorithm. In: Proceedings of ICML, pp. 351–359 (2014)
13. Tan, J., Xu, Y., Zhang, D., Zhang, X.: On streaming algorithms for maximizing a supermodular function plus a MDR-submodular function on the integer lattice. J. Comb. Optim. **45**(2), 1–19 (2023)
14. Sviridenko, M., Vondrak, J., Ward, J.: Optimal approximation for submodular and supermodular optimization with bounded curvature. In: Proceedings of SODA, pp. 1134–1148 (2015)
15. Sarpatwar, K.K., Schieber, B., Shachnai, H.: Constrained submodular maximization via greedy local search. Oper. Res. Lett. **47**, 1–6 (2019)
16. Krause, A., Singh, A., Guestrin, C.: Near-optimal sensor placements in gaussian processes: theory, efficient algorithms and empirical studies. J. Mach. Learn. Res. **9**, 235–284 (2008)
17. Wolsey, L.: Maximising real-valued submodular set function: primal and dual heuristics for location problems. Math. Oper. Res. **7**, 410–425 (1982)
18. Kapralov, M., Post, I., Vondrák, J.: Online submodular welfare maximization: greedy is optimal. In: Proceedings of SODA, pp. 1216–1225 (2012)
19. Das, A., Kempe, D.: Submodular meets spectral: greedy algorithms for subset selection, sparse approximation and dictionary selection. In: Proceedings of ICML, pp. 1057–1064 (2011)

Enhancing Crowding Event Detection on Campus with Multidimensional Logs: A Meta-Heuristic Search Approach

Maoyi Wang[1,2], Jiajie Shen[3], Jack Mao[4], Jihan Dai[1,2], Bochun Wu[3], Yun Xiong[1,2], and Xin Wang[1,2](✉)

[1] School of Computer Science, Fudan University, Shanghai, China
xinw@fudan.edu.cn
[2] The Shanghai Key Lab of Intelligent Information Processing, Fudan University, Shanghai, China
[3] Infomatization Office, Fudan University, Shanghai, China
[4] Software School, Fudan University, Shanghai, China

Abstract. This research addresses the pressing need for rapid detection of emergencies in campus crowding events by introducing wireless log-based crowding detection (WLCD). The WLCD employs a novel meta-heuristic search algorithm, coupled with the utilization of wireless fidelity (Wi-Fi) network logs, hierarchical attribute relationships, and information entropy theory. As the first attempt to transform a crowding event detection problem into a combinatorial optimization problem, we validated our approach using a real-world data set of 500 million records from a campus Wi-Fi network with over 27,000 access points. Our study demonstrates the superior performance of the WLCD, with increased accuracy by up to 61% and reduced execution time by up to 82% compared to state-of-the-art methods.

Keywords: crowding event detection · campus management · meta-heuristic search · log analysis · time series analysis

1 Introduction

Crowding event detection is crucial for public safety in smart campus environments with the avoidance of potential disasters like human stampedes [1]. Various methods have been proposed for crowding event detection, e.g., modeling travel patterns [2], analyzing human mobility behavior [3], and studying crowd behaviors [4]. However, existing approaches rely mainly on offline data sets and suffer limited real-time performance [5].

Since the wireless fidelity (Wi-Fi) plays a crucial role in connecting devices across various locations [6], we utilize Wi-Fi network log records for real-time crowding event detection. Log recording is operational all day and unaffected by dense crowds, consisting of a series of attributes. Crowding events generate

Y. Chen et al. (Eds.): COCOON 2024, LNCS 15162, pp. 332–342, 2025.
https://doi.org/10.1007/978-981-96-1093-8_28

multiple log records for specific attribute combinations, termed *effective combinations* [7]. By identifying effective combinations, we formulate the crowding event detection problem as a combinatorial optimization process.

Due to the multiple attributes and values in logs, the search space of attribute combinations is huge. To approach an optimal solution to effective combinations, there are two challenging targets: i) precise detection and ii) efficient detection. To address these, we propose a wireless log-based crowding detection (WLCD) scheme, which reduces the search space by combining the information entropy and the hierarchical relationships among attributes [8]. The main contributions of this paper are summarized below.

- We point out the correlation between the number of network logs and the pedestrian traffic. Based on this, to the best of our knowledge, we are the first to transform the crowding event detection problem into a combinatorial optimization problem.
- We propose a meta-heuristic search algorithm for effective combinations identification to solve the combinatorial optimization problem. To reduce the search space, we let an adjacency matrix denote all of the hierarchical relationships among attributes. Furthermore, detection results of our method can be effectively used to analyze the reasons for crowding event due to utilizing the relationships between different locations.
- We conduct experiments on synthetic and real-world data sets. The WLCD efficiently identifies crowding events from Wi-Fi logs, improving F1-Score by 15% in synthetic data sets and detecting up to 216% more events in real-world data sets while reducing execution time by 82% compared to state-of-the-art algorithms [7]. Deployed in a campus Wi-Fi network, our scheme effectively detects crowding events in various cases.

2 Related Work

Researchers have explored various approaches for modeling individual and collective human mobility, with a focus on detecting crowding events. Nouvellet et al. [9], investigated the impact of the COVID-19 pandemic on crowd mobility. Redondo et al. [10] utilized local-based social network (LBSN) data for early anomaly detection in crowd dynamics. Lin et al. [11] proposed a video-based method to detect anomaly events in crowd scenes. Another approach by Behera et al. [12] involved analyzing crowd movements using an active Langevin equation-based force model.

Existing methods for crowding event detection can be categorized into global navigation satellite systems (GNSS)-based, short-range wireless network-based, and vision-based approaches. GNSS-based methods, primarily suitable for outdoor use due to limited indoor performance, rely on global navigation satellite systems. Short-range wireless network-based methods perform crowding event detection by estimating crowd density or number of people. Taking the Wi-Fi based method as an example, Determe et al. measured the rate of Wi-Fi control packets sent by mobile devices to obtain the number of devices in the area and infer the number of people [13]. Yuan estimates crowd density based on Wi-Fi signal strength and Bluetooth query protocol [14]. They incur high deployment costs and achieve limited performance in densely crowded scenarios [5]. Vision-based methods, reliant on camera surveillance, struggle with inconsistent camera configurations and unfavorable lighting conditions [5].

In campuses with vast areas, complex building distributions, and diverse indoor layouts, there exist indoor and outdoor scenarios. Existing crowding event detection methods are not well-suited to campus scenarios, which calls for a method with wide coverage, scene versatility, and low cost.

3 Problem Formulation

3.1 Search Space and Effective Combination

Let $\mathcal{L}$ denote a set of log records. When log records are captured, timestamps are associated. Suppose there are n attributes in set $\mathcal{L}$. We can collect n attributes in the set

$$\mathcal{A} = \{a_1, a_2, \cdots, a_n\}, \tag{1}$$

where a_i for $\forall i = 1, \cdots, n$ is the i-th attribute in $\mathcal{L}$.

For attribute $\forall a_i \in \mathcal{A}$, there are m_i possible values as collected in a value set

$$\mathcal{V}_i = \{v_i^1, v_i^2, \cdots, v_i^{m_i}\}, \quad \forall i = 1, \cdots, n, \tag{2}$$

where v_i^j for $\forall j = 1, \cdots, m_i$ is the j-th possible value for attribute a_i.

An attribute a_i and a possible value of its v_i^j can construct tuple $u = (a_i, v_i^j)$. For each attribute a_i, we can construct a set of tuples $\mathcal{U}_i = a_i \times \mathcal{V}_i$, which contains all combinations of attribute a_i and its possible value $v_i^j \in \mathcal{V}_i$. Accordingly, we can define the set of all possible tuples as $\mathcal{U} = \cup_{i=1}^n \mathcal{U}_i$. An effective combination is the subset of $\mathcal{U}$, and contains at most one tuple from $\mathcal{U}$.

Thus, we can define the search space as

$$\mathbb{S} = \{s \in P(\mathcal{U}) | \forall U_i, G(s \cap U_i) \leq 1\}, \tag{3}$$

where $P(\mathcal{U})$ represents the power set[1] of $\mathcal{U}$, and $G(s \cap \mathcal{U}_i)) \leq 1$ represents the cardinality constraint that guarantees that each attribute can be involved at most once in the attribute combination.

3.2 Change Point Detection for Time Series

A change point refers to a point in a time series that undergoes significant changes. We should determine the change point of the time series, so that the crowding event detection algorithms can identify the effective combination within a given time window t. According to the detected change points, we can assign an object value to the attribute combination C_i.

[1] https://en.wikipedia.org/wiki/Power_set.

Let N_i denote the number of log records at time slot t_i.

For a given attribute combination $C_i \in \mathbb{S}$, we can define the time series $\mathcal{T}_i$ as a sequence that contains the numbers of log records as

$$\mathcal{T}_i = \{N_1, N_2, \cdots, N_{\max}\}, \tag{4}$$

where $N_{\max}$ represents the number of log records in the last item in time series $\mathcal{T}_i$.

Inspired by *Fisher distance* [15], we measure the quality of the combination by the following objective function:

$$R(x) = p_{b(x)} \ln \frac{p_{b(x)}}{p_{a(x)}}, \tag{5}$$

where x represents the change point, $a(x)$ and $b(x)$ represent the time periods before and after change point x, respectively, and p_t can be given by

$$p_t = \frac{n_t}{v_t}, \tag{6}$$

where n_t denotes the total number of log records under the current attribute combination during period t, and v_t denotes the total volume of logging records at t.

In this paper, we name $R(x)$ in Eq. (5) as the *Fisher value*. The Fisher value not only considers the change ratio of logging records $\frac{p_{b(x)}}{p_{a(x)}}$, but also emphasizes the global impact of the attribute combination on $p_{b(x)}$.

To calculate the Fisher value, we search for the change point x for a given time series. We apply the bottom-up segmentation algorithm [16] to fast detect change points, within the computational complexity of $O(n \ln n)$.

Since the button-up segmentation algorithm [16] may output multiple change points, we need to determine the Fisher value of the given effective combination. Let $\mathcal{X}$ denote the set of all change points. The maximum Fisher value can be given by

$$f = \max_{x \in \mathcal{X}} R(x). \tag{7}$$

Accordingly, we can calculate the Fisher value for each effective combination. With the computational complexity of $O(n \ln n)$, it meets the efficient requirements of the WLCD scheme. According to the object function f in Eq. (7), the WLCD can efficiently perform the search operators and identify the effective combinations.

3.3 Search Operators

Before discussing the search process for log records, we define the operations during the search process. Suppose there is an attribute combination C, we can perform the following search operators to modify the tuple in combination C.

- **Adding Tuple (AT):** This operator adds a tuple $(a', v') \in \{(a_a, v_a) \in T | \forall (a, v) \in C : a_a \neq a\}$ to attribute combination C.

- **Swapping Value (SV):** This operator swaps a tuple $(a_s, v_s) \in C$ with another tuple with the same attribute, *i.e.*, $(a', v') \in \{(a_s, v_s') \in T_s | v_s \neq v_s'\}, a_s \neq a$.
- **Swapping Tuple (ST):** This operator swaps a tuple $(a_s, v_s) \in C$ and another tuple $(a', v') \in \{(a_s', v_s') \in T | \forall (a, v) \in C : a_s' \neq a_s\}$.
- **Deleting Tuple (DT):** This operator deletes a tuple from attribute combination C.

3.4 Scoring Function

The scoring function is crucial in guiding the meta-heuristic search. Since there are no explicit scoring functions to identify crowding events, we use the *information entropy theory* [17] to construct the scoring function. For a given attribute a_i and its value set $\mathcal{V}_i$, the proportion of log records with this attribute value $v_i^j \in \mathcal{V}_i$ to all log records is denoted as $P(v_i^j)$. It can be calculated by

$$P(v_i^j) = \frac{\text{count}(v_i^j)}{\sum_{v_i^l \in \mathcal{V}_i} \text{count}(v_i^l)}, \tag{8}$$

where $\text{count}(v_i^j)$ denotes the number of log records containing attribute a_i with the value v_i^j.

For given $v_i^j \in \mathcal{V}_i$, the entropy function can be calculated by

$$E(v_i^j) = -P(v_i^j) \ln P(v_i^j). \tag{9}$$

We define the scoring function for attribute a_i based on the ideas of information theory. According to Eq. (9), we can calculate the entropy value of a_i by

$$E(a_i) = \sum_{v_i^j \in \mathcal{V}_i} E(v_i^j) = - \sum_{v_i^j \in \mathcal{V}_i} P(v_i^j) \ln P(v_i^j). \tag{10}$$

According to Eq. (9) and Eq. (10), the possibility of select the value v_i^j for attribute a_i can be calculated by

$$\Pr((a_i, v_i^j)) = \frac{E(v_i^j)}{E(a_i)}. \tag{11}$$

It is widely recognized that a larger entropy value of a tuple typically indicates more information involved. Inspired by the great success of information theory research areas, the WLCD tends to select the tuples with larger values of entropy.

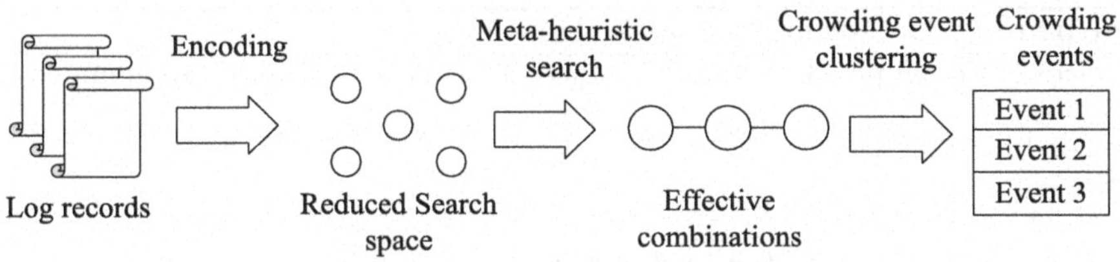

Fig. 1. Crowding event detection process via meta-search, including encoding, meta-heuristic search, and crowding event detecting.

4 Wireless Log-Based Crowding Detection

4.1 Overview of Detection Process

Motivated by the great demand for identifying crowding events, we developed the WLCD framework that automatically detects these events from multidimensional log records. While traditional methods may lead to an exponential computational complexity, the WLCD utilizes an effective and efficient meta-heuristic search approach.

As Fig. 1 shows, the design of the WLCD consists of three parts. First, we encode the search space and mark the relationship between the attributes. Second, the WLCD identifies the effective combinations according to log records by using the meta-heuristic search. Finally, we can apply a hierarchical clustering algorithm on the effective combinations to identify the crowding events.

4.2 Encoding

To reduce the search space of attributes as defined in Sect. 3, we innovatively take into account the relationships among attributes.

We define the *attribute chain* to describe the relationships among the attributes.

Suppose an attribute chain contains n attributes, we express a chain as

$$H_i = \{a_1^i \rightarrow a_2^i \rightarrow \cdots \rightarrow a_n^i\}. \tag{12}$$

According to Eq. (12), we can define an adjacency matrix M, with each element regarded as a distance between a pair of attributes in chain H_l, *i.e.*,

$$e_{i,j} = \begin{cases} d(a_i, a_j), & a_i, a_j \in H_i, a_i \neq a_j, \\ 0, & \text{else}, \end{cases} \tag{13}$$

where $d(a_i, a_j)$ represents the distance between attributes a_i and a_j. It is the number of edges between a_i and a_j on the chain.

Algorithm 1. Meta-heuristic search algorithm

Require: Set of all possible tuples U, Adjacency matrix A, Current attribute combination C
Ensure: Result set RS

```
RS ← ∅;
C ← ∅;
while termination conditions not met do
    if with probability p then
        C ← Random search mode(U, A, C);
    else
        C ← Greedy search mode(U, A, C);
    end if
    if C is effective combination then
        RS ← RS ∪ C;
    end if
end while
return RS
```

4.3 Meta-Heuristic Search

To efficiently find the appropriate combination from the huge search space, the WLCD adopts the meta-heuristic search. In particular, there are two types of meta-heuristic search strategies: i) the greedy mode which tends to select tuples with the largest information entropy values and ii) the random mode which randomly modifies attribute combinations.

By integrating both modes above, our algorithm trades off exploitation and exploration. This combination draws inspiration from multidimensional incident detection (MID) and is designed to enhance the efficiency of identifying effective attribute combinations in a reduced search space [7]. **Algorithm 1** shows the process of meta-heuristic search. The algorithm starts with empty attribute combination. In each iteration, it selects random search mode with probability p and greedy search mode with probability $1 - p$ to modify the attribute combination and verify whether it can be added to the result set.

4.4 Crowding Event Clustering

To identify redundant effective combinations implying the same crowding event, we analyze their corresponding time series and shared characteristics. For two combinations c_1 and c_2, represented by time series t_1 and t_2, respectively, we estimate their similarity. Similar attribute values and comparable time series indicate the same crowding event. We use the Jaccard distance and the cosine distance for attribute combinations and time series, respectively. Hierarchical clustering is then employed for event clustering [7].

Specifically, the similarity between combinations c_1 and c_2 can be quantified by the similarity distance between c_1 and c_2, denoted by $d(c_1, c_2)$:

$$d(c_1, c_2) = \frac{J(c_1, c_2) + \cos < t_1, t_2 >}{2}, \tag{14}$$

where $J(c_1, c_2)$ calculate Jaccard distance between attribute combinations c_1 and c_2, and $\cos < t_1, t_2 >$ donates the cosine distance of the cumulative time series t_1 and t_2. The smaller $d(c_1, c_2)$, the greater the similarity.

5 Experiments

5.1 Experimental Setup

We used two diverse types of data sets to evaluate the performance of detection methods.

Synthetic Data Set: We use synthetic data sets to assess the performance of the WLCD. Each data set spans 30 days, with 100 log records daily, excluding timestamps. Each record comprises 20 attributes, each with 8 possible values, resulting in 10 data sets of 3000 log records, each with various effective combinations adhering to adjacency matrix constraints. Data set names denote effective combination count and attribute count, such as 'one-4' indicating one effective combination of four attributes.

Real-World Data Set: All the log data are collected from Wi-Fi authorization in a campus over three years (2019–2021), and divided into months. Labeled as 'year-i' where 'i' represents the month.

As for the experiment metrics, *Precision* (P), *Recall* (R), and *F1-Score* (F), these metrics can be calculated by the following equations.

$$P = \frac{N^{\text{TP}}}{N^{\text{TP}} + N^{\text{FP}}}, R = \frac{N^{\text{TP}}}{N^{\text{TP}} + N^{\text{FN}}}, F = \frac{2 * P * R}{P + R}, \tag{15}$$

where N^{TP} is the number of correctly reported attribute combinations, N^{FP} is the number of wrongly reported attribute combinations, and N^{FN} is the number of effective combinations not reported.

We calculate three metrics, *i.e.*, precision (P), recall (R), and F1-Score (F), to evaluate the accuracy, detection rate, and overall effectiveness, respectively. We also measure the execution time (T) to evaluate the efficiency. To improve accuracy, we count the user accounts instead of log records.

We employ *Pandas* to calculate the effective combination of different detection approaches. In terms of efficiency, we compare our scheme to four state-of-the-art alternatives:

i) **MID** from Microsoft Azure [7],
ii) **NSGA-II** based on Genetic Algorithm [18],
iii) **MOEA/D** based on decomposition and evolutionary computation [19], and
iv) **Random search (RS)** as a baseline approach.

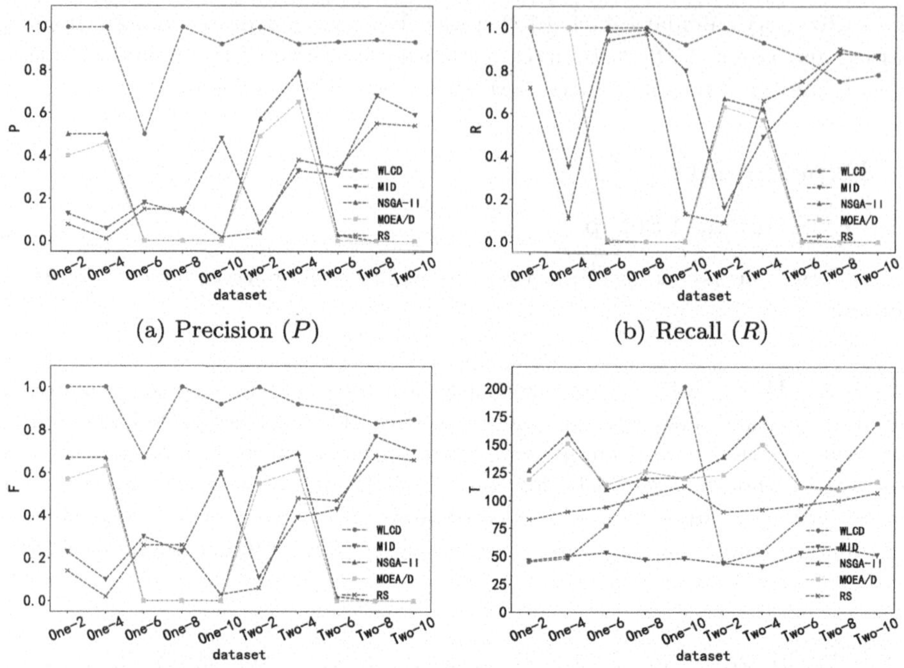

(a) Precision (P) (b) Recall (R)

(c) F1-score (F) (d) Execution Time (T)

Fig. 2. Experimental results on synthetic data sets under parameters $p = 0.3$ and $d = 0.1$.

5.2 Synthetic Evaluation

We evaluate the performance of the WLCD and its competitors on synthetic data sets. Each experiment is performed 100 times, with the averaged results shown in Fig. 2. Based on the relationships among attributes, the WLCD improves Precision, Recall, and F1-Score by up to 61%, 20%, and 52%, respectivley, compared to the state-of-the-art alternatives. Notably, the WLCD's Precision is only 0.50 on the One-6 data set, due to an unexpected attribute combination, identified as effective by WLCD but judged as a false positive.

5.3 Real-World Test

We also evaluate the performance under different detection algorithms on the real-world data set. The number of detected crowding events N and search time T are illustrated in Fig. 3. Specifically, the WLCD can raise the number of detected crowding events by up to 216% and reduce the time of the searching process by up to 82% compared to the alternatives.

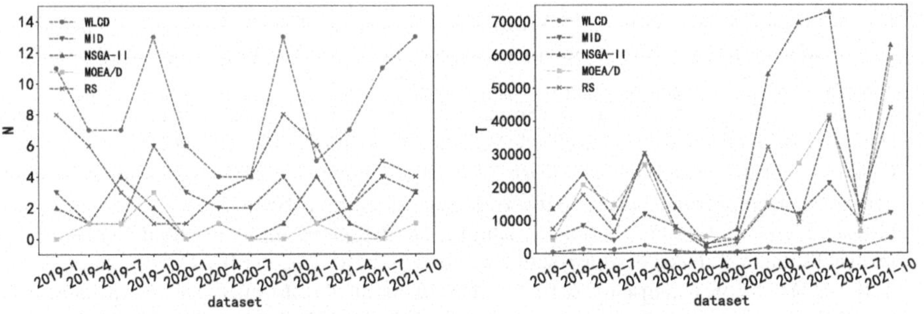

(a) Number of detected events (N) (b) Search Time (T)

Fig. 3. Experimental results on real-world data sets under parameters $p = 0.3$ and $d = 0.3$.

6 Conclusion

In this paper, we proposed the WLCD scheme. We formulated the crowding event detection problem as a combinatorial optimization problem by depicting the relationship between different attributes with an adjacency matrix. By employing the information entropy, the proposed WLCD constructs an efficient meta-heuristic search scheme, which can integrate both random search and greedy strategy and obtain the effective combinations. Based on synthetic and real-world data sets, we conducted extensive experiments to verify the effectiveness of the proposed approach. The experimental results showed that the WLCD can improve the F1-Score by up to 52%, and reduce the time of searching process by up to 82%, compared to the state-of-the-art approaches. Furthermore, we applied the WLCD to a practical Wi-Fi network at a campus. In several cases, We studied real-world crowding events in campus management.

Acknowledgments. This work was funded in part by the University-Industry Collaborative Education Program (No. 231002842310022), Open Fund of PDL (Project No. WDZC20245250106), and the National Natural Science Foundation of China Projects No. U1936213.

References

1. Zhiren, H., Pu, W., Fan, Z., Jianxi, G., Maximilian, S.: A mobility network approach to identify and anticipate large crowd gatherings. Transp. Res. Part B: Methodol. **114**, 147–170 (2018)
2. Haghani, M., Sarvi, M.: Crowd behaviour and motion: Empirical methods. Transp. Res. Part B Methodol. 253–294 (2018)
3. Riccardo, G., Armando, B., Sandro, R., Marc, B.: A stochastic model of randomly accelerated walkers for human mobility. Nat. Commun. **7**(1), 1–6 (2016)

4. Lu, L., Chan, C.Y., Wang, J., Wang, W.: A study of pedestrian group behaviors in crowd evacuation based on an extended floor field cellular automaton model. Transp. Res. Part C Emerg. Technol. 317–329 (2016)
5. Weng, W., Wang, J., Shen, L., Song, Y.: Review of analyses on crowd-gathering risk and its evaluation methods. J. Saf. Sci. Resilience 93–107 (2023)
6. Cisco: Annual internet report (2020). www.cisco.com/c/en/us/solutions/executive-perspectives/annual-internet-report/index.html
7. Gu, J., Luo, C., Qin, S., Qiao, B., Zhang, D.: Efficient incident identification from multi-dimensional issue reports via meta-heuristic search. In: Proceedings of the 28th ACM Joint European Software Engineering Conference and Symposium on the Foundations of Software Engineering (ESEC/FSE 20), pp. 292–303 (2020)
8. Al-Shaery, A.M., Khozium, M.O., Farooqi, N.S., Alshehri, S.S., Al-Kawa, M.A.M.: Problem solving in crowd management using heuristic approach. IEEE Access 25422–25434 (2022)
9. Nouvellet, P., et al.: Reduction in mobility and COVID-19 transmission. Nat. Commun. **12**(1), 1–9 (2021)
10. Redondo, R.P.D., Garcia-Rubio, C., Vilas, A.F., Campo, C., Rodriguez-Carrion, A.: A hybrid analysis of LBSN data to early detect anomalies in crowd dynamics. Futur. Gener. Comput. Syst. **109**, 83–94 (2020)
11. Lin, W., Gao, J., Wang, Q., Li, X.: Learning to detect anomaly events in crowd scenes from synthetic data. Neurocomputing **436**, 248–259 (2021)
12. Behera, S., Dogra, D.P., Bandyopadhyay, M.K., Roy, P.P.: Understanding crowd flow patterns using active-langevin model. Pattern Recogn. **119**, 108037 (2021)
13. Determe, J.F., Azzagnuni, S., Singh, U., Horlin, F., De Doncker, P.: Monitoring large crowds with WiFi: a privacy-preserving approach. IEEE Syst. J. **16**(2), 2148–2159 (2022)
14. Yuan, Y.: Crowd monitoring using mobile phones. In: 2014 Sixth International Conference on Intelligent Human-Machine Systems and Cybernetics. vol. 1, pp. 261–264 (2014)
15. Han, J., Kamber, M., Pei, J.: Data Mining Concept and Techniques (3rd edition). Springer (2011)
16. Aminikhanghahi, S., Cook, D.J.: A survey of methods for time series change point detection. Knowl. Inf. Syst. **51**(2), 339–367 (2017)
17. Thomas, M.C., Thomas, J.: Elements of Information Theory (2 ed.). Wiley (2006)
18. Deb, K., Pratap, A., Agarwal, S., Meyarivan, T.: A fast and elitist multiobjective genetic algorithm: NSGA-II. IEEE Trans. Evol. Comput. **6**(2), 182–197 (2002). https://doi.org/10.1109/4235.996017
19. Zhang, Q., Li, H.: MOEA/D: a multiobjective evolutionary algorithm based on decomposition. IEEE Trans. Evol. Comput. **11**(6), 712–731 (2007)

Construction of Binary Cooperative MSR Codes with Multiple Repair Degrees

Lei Li[1(✉)], Xinchun Yu[2], Yaqian Zhang[1], Liang Chen[3], Yuanyuan Dong[3], Chenhao Ying[1], and Yuan Luo[1]

[1] School of Electronic Information and Electrical Engineering, Shanghai Jiao Tong University, Shanghai, China
{staroverseas,zhangyq9,yingchenhao,yuanluo}@sjtu.edu.cn

[2] Institute of Data and Information, Tsinghua University Shenzhen International Graduate School, Shenzhen, China
yuxinchun@sz.tsinghua.edu.cn

[3] Alibaba Group, Hangzhou, China
{cl304641,yuanyuan.dyy}@alibaba-inc.com

Abstract. Modern large-scale distributed storage systems usually deploy erasure codes to provide high data reliability at a small storage overhead. Among them, maximum distance separable codes are the most common choice as they can achieve the optimal tradeoff between fault tolerance and storage overhead. In this paper, we focus on the repair problem of MDS codes and present constructions of binary MDS array codes which achieve the optimal repair bandwidth for multiple node failures. Specifically, by stacking multiple Blaum-Roth code instances whose "evaluation points" are judiciously designed, we obtain two families of binary MDS array codes with optimal cooperative repair bandwidth. The first family of codes of length n and dimension k can achieve the optimal cooperative repair bandwidth for $2 \leq h \leq n-k$ and $k+1 \leq d \leq n-h$ where h and d are the number of failed nodes and helper nodes, respectively. The second family of codes are optimal for multiple values of d simultaneously. Both of the codes are constructed on a special polynomial ring over binary field, thereby only XORs and cyclic shifts are involved during nodes repair and file reconstruction. Moreover, due to the inherent parallel structure of the codes, both the encoding and decoding procedures can be implemented in parallel, which can speed up the computing process.

Keywords: Binary MSR Codes · Repair Bandwidth · Distributed Storage Systems

1 Introduction

As an alternative redundancy scheme to replication, erasure codes (EC) have been widely used in large-scale distributed storage systems (DSS) to provide high data reliability at a low storage overhead. Among them, maximum distance

Y. Chen et al. (Eds.): COCOON 2024, LNCS 15162, pp. 343–354, 2025.
https://doi.org/10.1007/978-981-96-1093-8_29

separable (MDS) codes, such as Reed-Solomon (RS) code [11], are most implemented in DSS as they achieve the optimal trade-off between fault tolerance and storage overhead. For a file of size $\mathcal{M}$, the storage system using an (n, k) code first divides the file into k packets of equal size, then encodes them into n packets which are dispersed over n distinct storage nodes. With the MDS property, the system can reconstruct the file from any k out of n packets, meaning that the system can tolerate any $r := n - k$ node failures. Therefore, in face of node failures, the system can recover the lost data by accessing and communicating the data stored on any k out of the n surviving nodes to recover the whole file as long as there are no more than r failures. However, the real-world DSS usually only need to repair one or a few nodes ($< r$), instead of the whole file, in an efficient manner. In [3], the authors took the repair bandwidth, defined as the amount of data communicated within the storage network, as a measure for the efficiency of the repair procedure and derived a trade-off between storage and repair bandwidth. Codes that achieve the trade-off are called regenerating codes in the literature. There are two extremal points on the storage repair bandwidth trade-off, which are called minimum storage regenerating (MSR) codes and minimum bandwidth regenerating (MBR) codes, respectively. MSR codes have drawn much attention due to their optimal storage efficiency and the reader can refer to [2,4,7,10,12–16,18] for details. The common way to constructing MSR codes is through sub-packetization, i.e., each of the k packets is further divided into l sub-packets and the encoding procedure is implemented on these sub-packets with a finer granularity. Thus, an MSR code is actually an MDS array code which can be denoted by a triple (n, k, l), where n, k and l are the length, dimension and sub-packetization level of the code, respectively.

There are few works that construct binary MDS array codes with optimal repair bandwidth (i.e., binary MSR codes). Existing constructions of binary MSR codes for single-node failures can be found in [5,6]. One limitation of the codes constructed in [5,6] is that, when the number of helper nodes $d < n - 1$, a fixed set of nodes are forced to participate in the repair of the failed node. Some improvements have been made by the authors in [8,9] where they construct binary MDS array codes with optimal repair/access bandwidth, and any set of d nodes among the $n - 1$ surviving nodes can help finish the repair procedure. Moreover, binary MDS array codes with optimal repair/access bandwidth for multiple values of d simultaneously are also presented in [8,9]. Studies on MSR codes in the literature originally are focused on repairing single-node failures and turn to multiple-node failures recently [17,19,20], where the codes are usually constructed over some nonbinary finite field. In the present paper, we focus on repairing multiple-node failures and construct binary MDS array codes with optimal cooperative repair bandwidth. Specifically, we present explicit constructions of two families of binary cooperative MSR codes with the first family of codes having single repair degree (fixed d) and the second having multiple repair degrees (multiple values of d simultaneously). Both of the codes are constructed on a special polynomial ring over binary field, thereby only XORs and cyclic shifts are involved during nodes repair and file reconstruction. Moreover, due to the inherent parallel structure of the codes, both the encoding and decoding procedures can be implemented in parallel, which can speed up the computing process.

The rest of this paper is organized as follows. Section 2 provides some preliminaries that are needed in the paper. In Sect. 3, we present explicit constructions of two families of binary cooperative MSR codes over the binary polynomial ring $\mathcal{R} = \mathbb{F}_2[x]/(1+x+\cdots+x^{p-1})$ where p is a prime. Section 4 concludes the paper by pointing out some future research directions.

2 Preliminaries

Given two integers i and j with $i < j$, denote by $[i]$ and $[i,j]$ two ordered sets $\{1,2,\cdots,i\}$ and $\{i,i+1,\cdots,j\}$, respectively. For an (n,k) code, denote by $r := n-k$ the number of parity nodes. Following the literature of codes for distributed storage, we use the two words "coordinate" and "node" interchangeably. As a result, repairing failed nodes in a distributed storage system can be viewed as correcting erasures of a codeword.

2.1 Blaum-Roth Codes

An (n,k,l) binary array code can be viewed as a set of matrices of size $l \times n$ over $\mathbb{F}_2$. Let p be a prime, a Blaum-Roth code $\mathcal{C}_{BR}$ is a code over the polynomial ring $\mathcal{R} = \mathbb{F}_2[x]/(1+x+\cdots+x^{p-1})$. Each codeword in $\mathcal{C}_{BR}$ is a $(p-1)\times n$ matrix $c[i,a]$ where $i \in [n]$ and $a \in [0,p-2]$. The $p-1$ bits $c_{i,0}, c_{i,1}, \ldots, c_{i,p-2}$ in the i-th column of the matrix represent a polynomial in $\mathcal{R}$, which can be written as $c_i(x) = c_{i,0}+c_{i,1}\cdot x+c_{i,2}\cdot x^2+\cdots+c_{i,p-2}\cdot x^{p-2}$. For $r < n < p$, an $(n, n-r, p-1)$ Blaum-Roth code is defined by its parity-check matrix

$$H_{BR} = \begin{bmatrix} 1 & 1 & 1 & \cdots & 1 \\ 1 & x & x^2 & \vdots & x^{n-1} \\ \vdots & \vdots & \vdots & \vdots & \vdots \\ 1 & x^{r-1} & x^{2(r-1)} & \cdots & x^{(n-1)(r-1)} \end{bmatrix},$$

where $1, x, x^2, \ldots, x^{n-1}$ are n distinct nonzero elements that have multiplicative inverse in $\mathcal{R}$. Note that the matrix H_{BR} can be defined by $1, x, x^2, \ldots, x^{n-1}$ and we call these n elements the "evaluation points" of the $(n, k, l = p-1)$ Blaum-Roth code. Blaum-Roth codes are MDS since any r columns of H_{BR} are linearly independent over $\mathcal{R}$. The reader can refer to [1] for more details about Blaum-Roth codes.

2.2 Lower Bounds on the Repair Bandwidth of MDS Codes for Multiple Failures

While most of the studies on MSR codes focus on repairing single-node failure, it is also important to construct MSR codes handling multiple-node failures.

There are mainly two repair models for multi-node failures, i.e., the centralized repair model and the cooperative repair model. The latter is more suitable for DSS and we focus on cooperative-repair MDS array codes in this paper. Let $\mathcal{C}$ be an (n, k, l) MDS array code over some finite field $\mathbb{F}$ where l is called sub-packetization level in the literature. For each codeword $C = (C_1, C_2, \ldots, C_n) \in \mathcal{C}$, coordinate C_i is stored on node i where $i \in [n]$. Let $\mathcal{F} \subset [n]$ with $|\mathcal{F}| = h \in [2, r]$ be the set of indices of the failure nodes and $\mathcal{H} \subseteq [n]\backslash\mathcal{F}$ with $|\mathcal{H}| = d$ be the set of indices of the helper nodes, where $d \in [k, n-h]$, usually called repair degree, is the number of helper nodes. Under the cooperative repair model, the erased coordinates $\{C_i\}_{i\in\mathcal{F}}$ can be repaired usually through the two phases, i.e., the download phase and the collaboration phase.

- **Download phase.** For each $i \in \mathcal{F}$ and $j \in \mathcal{H}$, node i downloads β_1 symbols from node j by accessing at most l symbols stored on node j.
- **Collaboration phase.** For any i and $i' \in \mathcal{F}$, nodes i and i' exchange β_2 symbols, where $\beta_2 \le \beta_1$.

Obviously, $\beta_2 \le \beta_1 \le l$ and the total amount of data communicated through the storage network during the repair of the h failed nodes is $\gamma = h[d\beta_1 + (h-1)\beta_2]$. It has been proved in [19] that for MDS array codes, it has

$$\beta_1 \ge \beta_2 \ge \frac{l}{d-k+h}. \tag{1}$$

Therefore, the the total amount of data γ communicated through these two phases satisfies

$$\gamma \ge \frac{h(d+h-1)l}{d-k+h}. \tag{2}$$

We say a code has optimal cooperative repair bandwidth when γ meets the lower bound with equality.

3 Code Construction

In this section, we present the explicit constructions of two families of MDS array codes with optimal cooperative repair bandwidth over the binary polynomial ring $\mathcal{R} = \mathbb{F}_2[x]/(1 + x + \cdots + x^{p-1})$. The codes are named $\mathcal{C}_1$ and $\mathcal{C}_2$ respectively in the sequel. Code $\mathcal{C}_1$ achieves the optimal cooperative repair bandwidth for any fixed $d \in [k+1, n-h]$ where $h \in [2, r]$, while code $\mathcal{C}_2$ is optimal for multiple values of $d \in [k+1, n-h]$ simultaneously.

3.1 Construction of Binary Cooperative MSR Codes for $h \in [2, r]$ and fixed $d \in [k+1, n-h]$

In this subsection, we present the codes with single repair degree $d \in [k+1, n-h]$ where $h \in [2, r]$. The code constructed in this subsection is named $\mathcal{C}_1$ whose sub-packetization level is $l = (p-1)(d-k+h)s^n$ where $s = d-k+1$.

Each codeword $C = (C_1, C_2, \ldots, C_n) \in \mathcal{C}_1$ can be viewed as a matrix of size $(d-k+h)s^n \times n$ over the ring $\mathcal{R}$, i.e., for any $i \in [n]$, the coordinate C_i is a column vector of length $(d-k+h)s^n$. Obviously, the codeword C can also be viewed as a matrix of size $l \times n$ over the binary field $\mathbb{F}_2$. For any $i \in [n]$, we write the coordinate $C_i = \{c_{i,(a,b)}\}_{a\in[0,s^n-1],b\in[0,d-k+h-1]}$, where $c_{i,(a,b)}$ can be viewed as a polynomial in the ring $\mathcal{R}$. For any $a \in [0, s^n-1]$, we write its n-digit s-ary expansion as $a = (a_1, a_2, \ldots, a_n)$, i.e., $a = \sum_{i=1}^{n} a_i \cdot s^{i-1}$ where $a_i \in [0, s-1]$ for all $i \in [n]$. We also define $a(i, v) := (a_1, \ldots, a_{i-1}, v, a_{i+1}, a_n)$ where $v \in [0, s-1]$, i.e., $a(i, v)$ is obtained by substituting the i-th digit of a with v.

Construction 1. *Let n and k, respectively, be the length and dimension of the array code. Define $r := n-k$, $s := d-k+1$, and let $\{x_{i,\tau} = x^{\tau n+i}\}_{i\in[n],\tau\in[0,s-1]}$ be a set of sn distinct nonzero elements in the polynomial ring $\mathcal{R} = \mathbb{F}_2[x]/(1+x+\cdots+x^{p-1})$ where p is a prime and $p-2 \geq sn$. Code $\mathcal{C}_1$ is defined by the following parity-check equations:*

$$\sum_{i=1}^{n} x_{i,a_i}^{t} c_{i,(a,b)} = 0, \tag{3}$$

where $t \in [0, r-1]$, $a \in [0, s^n-1]$, $b \in [0, d-k+h-1]$ and $x_{i,a_i}^{t} = x^{t(a_i \cdot n+i)} \in \mathcal{R}$.

The MDS property of code $\mathcal{C}_1$ is straightforward as the set $\{c_{i,(a,b)}\}_{i\in[n]}$ forms an (n, k) Blaum-Roth code for any $a \in [0, s^n-1]$ and $b \in [0, d-k+h-1]$. We now present the repair bandwidth of $\mathcal{C}_1$ in the following Theorem 1.

Theorem 1. *Code $\mathcal{C}_1$ achieves the optimal cooperative repair bandwidth for fixed $h \in [2, n-k]$ and $d \in [k+1, n-h]$.*

Proof. Without loss of generality, we assume the first h coordinates are erased and prove the theorem by reconstructing $C_1, C_2, \ldots, C_h$ with optimal cooperative repair bandwidth. For any $e \in [h]$, according to Eq. (3), we have

$$x_{e,a_e}^{t} c_{e,(a,b)} + \sum_{j\in[n]\setminus\{e\}} x_{j,a_j}^{t} c_{j,(a,b)} = 0.$$

Substituting (a, b) with $(a(e, a_e \oplus v), v)$, where $\oplus$ denotes the operation of addition modulo s and $v \in [0, s-2]$, we obtain

$$x_{e,a_e\oplus v}^{t} c_{e,(a(e,a_e\oplus v),v)} + \sum_{j\in[n]\setminus\{e\}} x_{j,a_j}^{t} c_{j,(a(e,a_e\oplus v),v)} = 0 \tag{4}$$

and similarly

$$\begin{aligned} &x_{e,a_e\oplus(s-1)}^{t} c_{e,(a(e,a_e\oplus(s-1)),s-2+e)} + \\ &\sum_{j\in[n]\setminus\{e\}} x_{j,a_j}^{t} c_{j,(a(e,a_e\oplus(s-1)),s-2+e)} = 0 \end{aligned} \tag{5}$$

for any $a \in [0, s^n - 1]$ and $t \in [0, r-1]$. Summing up the s equations defined by (4) and (5), we have

$$\sum_{v=0}^{s-2} x^t_{e,a_e \oplus v} c_{e,(a(e,a_e\oplus v),v)} + x^t_{e,a_e\oplus(s-1)} c_{e,(a(e,a_e\oplus(s-1)),s-2+e)} +$$

$$\sum_{j\in[n]\setminus\{e\}} x^t_{j,a_j} \left(\sum_{v=0}^{s-2} c_{j,(a(e,a_e\oplus v),v)} + c_{j,(a(e,a_e\oplus(s-1)),s-2+e)} \right) = 0,$$

which defines an $(n' = s+n-1, k' = n'-r) = (d+r, d)$ Blaum-Roth code for any $a \in [0, s^n - 1]$. Thus, C_e can obtain

$$\{c_{e,(a(e,a_e\oplus v),v)}\}_{v\in[0,s-2]} \cup \{c_{e,(a(e,a_e\oplus(s-1)),s-2+e)}\}$$

and

$$\{\sum_{v=0}^{s-2} c_{j,(a(e,a_e\oplus v),v)} + c_{j,(a(e,a_e\oplus(s-1)),s-2+e)}\}_{j\in[h]\setminus\{e\}}$$

by downloading any d out of the $n-h$ symbols

$$\{\sum_{v=0}^{s-2} c_{j,(a(e,a_e\oplus v),v)} + c_{j,(a(e,a_e\oplus(s-1)),s-2+e)}\}.$$

for any $a \in [0, s^n - 1]$. As

$$\{c_{e,(a(e,a_e\oplus v),v)}\}_{a\in[0,s^n-1],v\in[0,s-2]} = \{c_{e,(a,v)}\}_{a\in[0,s^n-1],v\in[0,s-2]}$$

and

$$\{c_{e,(a(e,a_e\oplus(s-1)),s-2+e)}\}_{a\in[0,s^n-1]} = \{c_{e,(a,s-2+e)}\}_{a\in[0,s^n-1]},$$

one can easily find that for any $e \in [h]$, C_e can obtain

$$\{c_{e,(a,b)}\}_{a\in[0,s^n-1],b\in[0,s-2]} \cup \{c_{e,(a,s-2+e)}\}_{a\in[0,s^n-1]}$$

and

$$\{\sum_{v=0}^{s-2} c_{j,(a(e,a_e\oplus v),v)} + c_{j,(a(e,a_e\oplus(s-1)),s-2+e)}\}_{j\in[h]\setminus\{e\},a\in[0,s^n-1]}$$

by downloading $(p-1)s^n$ bits from each of the d helper nodes, resulting in an amount of $hd(p-1)s^n$ bits network bandwidth consumption in the download phase for all the h erased coordinates.

In the collaboration phase, C_e, $e \in [h]$ downloads

$$\{\sum_{v=0}^{s-2} c_{e,(a(j,a_j\oplus v),v)} + c_{e,(a(j,a_j\oplus(s-1)),s-2+j)}\}_{a\in[0,s^n-1]}$$

from $C_j, j \in [h]\backslash\{e\}$ to further obtain

$$\{c_{e,(a(j,a_j\oplus(s-1)),s-2+j)}\}_{a\in[0,s^n-1]} = \{c_{e,(a,s-2+j)}\}_{a\in[0,s^n-1]}$$

since $\{c_{e,(a,b)}\}_{a\in[0,s^n-1],b\in[0,s-2]}$ are already known for C_e. Thus, C_e can recover the remaining data

$$\{c_{e,(a,s-2+j)}\}_{a\in[0,s^n-1],j\in[h]\backslash\{e\}},$$

completing the overall repair of C_e. It is not hard to find that the total amount of data (in bit) communicated among the h erased coordinates in the collaboration phase is $h(h-1)(p-1)s^n$. Together with the data communicated in the download phase, there are $h(d+h-1)(p-1)s^n$ bits communicated during the repair of the h erased coordinates, achieving the lower bound (2) on the repair bandwidth under the cooperative repair model. This completes the proof.

3.2 Construction of Binary Cooperative MSR Codes for $h \in [2, r]$ and Multiple $d \in [k+1, n-2]$ Simultaneously

In the previous subsection, we constructed binary MSR code for some fixed d, meaning that the repair degree d has only one value in $[k+1, n-h]$. In the present subsection, we construct codes with multiple repair degrees simultaneously by using the previous codes as building blocks. The code constructed in this subsection is named $\mathcal{C}_2$.

Construction 2. *For integers n, k and m distinct integers $d_1, d_2, \ldots, d_m \in [k+1, n-h]$ where $h \in [2, n-k]$. Denote $s_i := d_i - k + 1$ for $i \in [m]$ and $s := \mathrm{lcm}(d_1 - k + h, d_2 - k + h, \ldots, d_m - k + h)$ where $\mathrm{lcm}()$ is the least common multiple of some integers. Let $\{x_{i,\tau} = x^{\tau n+i}\}_{i\in[n],\tau\in[0,s_m-1]}$ be a set of $s_m \cdot n$ distinct nonzero elements in the polynomial ring $\mathcal{R} = \mathbb{F}_2[x]/(1 + x + \cdots + x^{p-1})$ where p is a prime and $p - 2 \geq s_m \cdot n$. Code $\mathcal{C}_2$ is defined by the following parity-check equations:*

$$\sum_{i=1}^{n} x_{i,a_i}^t c_{i,(a,b)} = 0 \tag{6}$$

where $t \in [0, r-1]$, $a \in [0, s_m^n - 1]$, $b \in [0, s-1]$ and $x_{i,a_i}^t = x^{t(a_i \cdot n+i)} \in \mathcal{R}$ with $a_i \in [0, s_m - 1]$ for any $i \in [n]$.

The MDS property of code $\mathcal{C}_2$ is also straightforward as the set $\{c_{i,(a,b)}\}_{i\in[n]}$ forms an (n,k) Blaum-Roth code for any $a \in [0, s_m^n - 1]$ and $b \in [0, s-1]$. We now present the repair bandwidth of $\mathcal{C}_2$ in the following Theorem 2.

Theorem 2. *Code $\mathcal{C}_2$ achieves the lower bound on cooperative repair bandwidth for fixed $h \in [2, n-k]$ and multiple $d \in [k+1, n-h]$.*

Proof. For any $i \in [m]$, let $u_i = \frac{s}{d_i-k+h} = \frac{s}{s_i+h-1}$, and the set $[0, s-1]$ is partitioned into u_i disjoint sets of equal size, i.e., $[0, s-1] = \cup\Gamma_u, u \in [u_i]$ where

$$\Gamma_u = \{(u-1)(s_i+h-1), (u-1)(s_i+h-1)+1, \cdots, (u-1)(s_i+h-1)+s_i+h-2\}.$$

Obviously, $|\Gamma_u| = s_i + h - 1$ for any $u \in [u_i]$. For some $a \in [0, s_m^n - 1]$ and $b \in \Gamma_u$, we have the corresponding parity-check equations

$$\sum_{j=1}^{n} x_{i,a_j}^t c_{j,(a,b)} = 0,$$

where $t \in [0, r-1]$. Without loss of generality, we assume the first h nodes are failed, i.e., $\mathcal{F} = [h]$. Divide the set Γ_u into h overlapping sets of equal size, i.e.,

$$\Gamma_u = \cup \Gamma_{u,e}, \quad e \in [h]$$

where

$$\begin{aligned}\Gamma_{u,e} = \{&(u-1)(s_i+h-1), (u-1)(s_i+h-1)+1, \cdots, \\ &(u-1)(s_i+h-1)+s_i-2, (u-1)(s_i+h-1)+s_i-2+e\}\end{aligned}$$

and for $e_1 \neq e_2 \in [h]$,

$$\begin{aligned}&|\Gamma_{u,e_1}| = |\Gamma_{u,e_2}| = s_i, \\ &\Gamma_{u,e_1} \cap \Gamma_{u,e_2} = [(u-1)(s_i+h-1), (u-1)(s_i+h-1)+s_i-2], \\ &\Gamma_{u,e_1} - \Gamma_{u,e_2} = \{e_1\}, \\ &\Gamma_{u,e_2} - \Gamma_{u,e_1} = \{e_2\}.\end{aligned}$$

For some fixed $a \in [0, s_m^n - 1]$ and $b \in \Gamma_{u,e}$, we have

$$x_{e,a_e}^t c_{e,(a,b)} + \sum_{j \in [n] \setminus \{e\}} x_{j,a_j}^t c_{j,(a,b)} = 0 \tag{7}$$

where $t \in [0, r-1]$. Replacing a with $a(e, a_e \oplus v)$ where $\oplus$ denotes the operation of addition modulo s_m and $v \in [0, s_i - 2]$, we can rewrite Eq. (7) as

$$\begin{aligned}&x_{e,a_e \oplus}^t c_{e,(a(e,a_e \oplus v),(u-1)(s_i+h-1)+v)} + \\ &\sum_{j \in [n] \setminus \{e\}} x_{j,a_j}^t c_{j,(a(e,a_e \oplus v),(u-1)(s_i+h-1)+v)} = 0\end{aligned} \tag{8}$$

and similarly

$$\begin{aligned}&x_{e,a_e \oplus (s_i-1)}^t c_{e,(a_{e,a_e \oplus (s_i-1)},(u-1)(s_i+h-1)+s_i-2+e)} + \\ &\sum_{j \in [n] \setminus \{e\}} x_{j,a_j}^t c_{j,(a(e,a_e \oplus (s_i-1)),(u-1)(s_i+h-1)+s_i-2+e)} = 0\end{aligned} \tag{9}$$

where $t \in [0, r-1]$. Summing up the s_i equations defined by (8) and (9), we obtain

$$\sum_{v=0}^{s_i-2} x^t_{e,a_e\oplus v} c_{e,(a(e,a_e\oplus v),(u-1)(s_i+h-1)+v)} +$$
$$x^t_{e,a_e\oplus(s_i-1)} c_{e,(a(e,a_e\oplus(s_i-1)),(u-1)(s_i+h-1)+s_i-2+e)} + \sum_{j\in[n]\setminus\{e\}} x^t_{j,a_j} \cdot$$
$$\left(\sum_{v=0}^{s_i-2} c_{j,(a(e,a_e\oplus v),(u-1)(s_i+h-1)+v)} + c_{j,(a(e,a_e\oplus(s_i-1)),(u-1)(s_i+h-1)+s_i-2+e)}\right)$$
$$= 0,$$

which defines an $(n' = s_i+n-1, k' = n'-r) = (d_i+r, d_i)$ Blaum-Roth code whose evaluation points are $\{x_{e,a_e\oplus v}\}_{v\in[0,s_i-1]} \cup \{x_{j,a_j}\}_{j\in[n]\setminus\{e\}}$. Thus, the symbols

$$\{c_{e,(a(e,a_e\oplus v),(u-1)(s_i+h-1)+v)}\}_{v\in[0,s-2]}\cup$$
$$\{c_{e,(a(e,a_e\oplus(s_i-1)),(u-1)(s_i+h-1)+s_i-2+e)}\}\cup$$
$$\{\sum_{v=0}^{s_i-2} c_{j,(a(e,a_e\oplus v),(u-1)(s_i+h-1)+v)}+c_{j,(a(e,a_e\oplus(s_i-1)),(u-1)(s_i+h-1)+s_i-2+e)}\}_{j\in[h]\setminus\{e\}}$$

can be recovered by downloading any d_i out of the $n-h$ symbols

$$\{\sum_{v=0}^{s_i-2} c_{j,(a(e,a_e\oplus v),(u-1)(s_i+h-1)+v)}+c_{j,(a(e,a_e\oplus(s_i-1)),(u-1)(s_i+h-1)+s_i-2+e)}\}_{j\in[h+1,n]}$$

for any $a \in [0, s_m^n - 1]$. Note that

$$\begin{aligned}
&\{(a(e,a_e\oplus v),(u-1)(s_i+h-1)+v)\}_{a\in[0,s_m^n-1],v\in[0,s_i-2]}\\
&\cup\{(a(e,a_e\oplus(s_i-1)),(u-1)(s_i+h-1)+e)\}_{a\in[0,s_m^n-1]}\\
&=\{(a,b)\}_{a\in[0,s_m^n-1],b\in\Gamma_{u,e}},
\end{aligned}$$

we conclude that the symbols

$$\{c_{e,(a,b)}\}_{a\in[0,s_m^n-1],b\in\Gamma_{u,e}}$$

and

$$\{\sum_{v=0}^{s_i-2} c_{j,(a(e,a_e\oplus v),(u-1)(s_i+h-1)+v)}+c_{j,(a(e,a_e\oplus(s_i-1)),(u-1)(s_i+h-1)+e)}\}_{j\in[h]\setminus\{e\},a\in[0,s_m^n-1]}$$

can be recovered. It is not hard to verify that there are $d_i s_m^n h(p-1)$ bits are downloaded from the d_i helper nodes to the h failed nodes in the download phase.

In the collaboration phase, the failed node $e \in [h]$ downloads the symbols

$$\{\sum_{v=0}^{s_i-2} c_{e,(a(j,a_j\oplus v),(u-1)(s_i+h-1)+v)}+c_{e,(a(j,a_j\oplus(s_i-1)),(u-1)(s_i+h-1)+s_i-2+j)}\}_{a\in[0,s_m^n-1]}$$

from the failed node $j \in [h]\backslash\{e\}$ to further obtain

$$\{c_{e,(a(j,a_j\oplus(s_i-1)),(u-1)(s_i+h-1)+s_i-2+j)}\}_{a\in[0,s_m^n-1]} = \{c_{e,(a,(u-1)(s_i+h-1)+s_i-2+j)}\}_{a\in[0,s_m^n-1]}$$

since $\{c_{e,(a,b)}\}_{a\in[0,s_m^n-1],b\in\Gamma_{u,e}}$ are already known. In the collaboration phase, there are $h(h-1)(p-1)s_m^n$ bits in total communicated between the h failed nodes.

Through the download phase and collaboration phase, the symbols

$$\{c_{j,(a,b)}\}_{j\in[h],a\in[0,s_m^n-1],b\in\Gamma_u}$$

can be recovered and the amount of data (in bit) communicated through the storage network is

$$\gamma_u = d_i s_m^n h(p-1) + h(h-1)(p-1)s_m^n.$$

Thus, to complete the overall repair of the symbols $\{c_{j,(a,b)}\}_{j\in[h],a\in[0,s_m^n-1],b\in[s]}$, a total number of

$$\begin{aligned}\gamma &= \frac{s}{|\Gamma_u|}\gamma_u \\ &= \frac{s}{d_i-k+h}(d_i s_m^n h(p-1) + h(h-1)(p-1)s_m^n) \\ &= \frac{h(d+(h-1))l}{d_i-k+h}\end{aligned}$$

bits of data are communicated through the storage network, achieving the lower bound (2) on repair bandwidth under the cooperative repair model.

4 Conclusion

In this paper, we propose two families of binary MDS array codes with optimal cooperative repair bandwidth, i.e., binary cooperative MSR codes, for multiple erasures. The first family of codes can repair $h \in [2, n-k]$ failures with fixed repair degree $d \in [k+1, n-1]$ while the second family of codes have multiple repair degrees simultaneously. Both of the codes are constructed by stacking multiple Blaum-Roth code instances whose "evaluation points" are judiciously selected from the special ring $\mathcal{R} = \mathbb{F}_2[x]/(1+x+\cdots+x^{p-1})$. As a result, only XORs and cyclic shifts are involved during the file reconstruction and nodes repair procedures, avoiding complex multiplications and divisions over large finite fields. Moreover, due to the inherent parallel structure of the codes, advanced encoding/decoding algorithms of Blaum-Roth code can be directly implemented in parallel, which can speed up the computing process.

We conclude the paper by pointing out two further research directions. Note that during the repair procedure, although the amount of data communicated through the storage network are optimal, all the data need to be accessed (read) on the helper nodes, which will consume significant disk I/Os. Thus, we are working on the constructions of binary MDS array codes with optimal cooperative repair and access bandwidth. Also, the sub-packetization level of the codes constructed in this paper is very large, which is not friendly to practical use. It is of great significance to construct binary MDS array codes with small sub-packetization level and low repair bandwidth for repairing single-node or multiple-node failures.

Acknowledgments. This work is supported in part by National Natural Science Foundation of China under Grant 62171279.

Disclosure of Interests. The authors have no competing interests to declare that are relevant to the content of this article.

References

1. Blaum, M., Roth, R.M.: New array codes for multiple phased burst correction. IEEE Trans. Inf. Theory **39**(1), 66–77 (1993)
2. Cadambe, V.R., Huang, C., Li, J.: Permutation code: optimal exact-repair of a single failed node in MDS code based distributed storage systems. In: 2011 IEEE International Symposium on Information Theory Proceedings (ISIT), pp. 1225–1229 (2011)
3. Dimakis, A.G., Godfrey, P.B., Wu, Y., Wainwright, M.J., Ramchandran, K.: Network coding for distributed storage systems. IEEE Trans. Inf. Theory **56**(9), 4539–4551 (2010)
4. Goparaju, S., Fazeli, A., Vardy, A.: Minimum storage regenerating codes for all parameters. IEEE Trans. Inf. Theory **63**(10), 6318–6328 (2017)
5. Hou, H., Lee, P.P.: Binary MDS array codes with optimal repair. IEEE Trans. Inf. Theory **66**(3), 1405–1422 (2019)
6. Li, J., Tang, X., Hollanti, C.: A generic transformation for optimal node repair in MDS array codes over f2. IEEE Trans. Commun. **70**(2), 727–738 (2021)
7. Li, J., Tang, X., Tian, C.: A generic transformation to enable optimal repair in MDS codes for distributed storage systems. IEEE Trans. Inf. Theory **64**(9), 6257–6267 (2018)
8. Li, L., Ying, C., Chen, L., Dong, Y., Luo, Y.: New constructions of binary MDS array codes with optimal repair bandwidth. In: 2023 IEEE International Symposium on Information Theory (ISIT), pp. 2045–2050 (2023)
9. Li, L., Yu, X., Chen, L., Dong, Y., Luo, Y.: Constructions of binary MDS array codes with optimal repair/access bandwidth. IEEE Trans. Commun. (2023)
10. Papailiopoulos, D.S., Dimakis, A.G., Cadambe, V.R.: Repair optimal erasure codes through Hadamard designs. IEEE Trans. Inf. Theory **59**(5), 3021–3037 (2013)
11. Reed, I.S., Solomon, G.: Polynomial codes over certain finite fields. J. Soc. Ind. Appl. Math. **8**(2), 300–304 (1960)
12. Suh, C., Ramchandran, K.: Exact-repair MDS codes for distributed storage using interference alignment. In: 2010 IEEE International Symposium on Information Theory (ISIT), pp. 161–165 (2010)
13. Tamo, I., Wang, Z., Bruck, J.: Zigzag codes: MDS array codes with optimal rebuilding. IEEE Trans. Inf. Theory **59**(3), 1597–1616 (2012)

14. Vajha, M., Balaji, S., Kumar, P.V.: Small-d MSR codes with optimal access, optimal sub-packetization and linear field size. IEEE Trans. Inf. Theory (2023)
15. Vajha, M., et al.: Clay codes: moulding MDS codes to yield an MSR code. In: 16th USENIX Conference on File and Storage Technologies (FAST 18), pp. 139–154 (2018)
16. Wang, N., Li, G., Hu, S., Ye, M.: Constructing MSR codes with subpacketization $2n/3$ for $k+1$ helper nodes. IEEE Trans. Inf. Theory (2023)
17. Ye, M.: New constructions of cooperative MSR codes: reducing node size to exp $(o(n))$. IEEE Trans. Inf. Theory **66**(12), 7457–7464 (2020)
18. Ye, M., Barg, A.: Explicit constructions of optimal-access MDS codes with nearly optimal sub-packetization. IEEE Trans. Inf. Theory **63**(10), 6307–6317 (2017)
19. Ye, M., Barg, A.: Cooperative repair: constructions of optimal MDS codes for all admissible parameters. IEEE Trans. Inf. Theory **65**(3), 1639–1656 (2018)
20. Zhang, Y.: New centralized MSR codes with small sub-packetization. arXiv preprint arXiv:2401.16866 (2024)

Opportunistic Routing Using Q-Learning with Context Information

Xiaojing Liu[1,2], Jiayu Cui[1,2], Winston K. G. Seah[3], Xiaodong Xu[1,2], Gang Xu[1,2(✉)], and Celimuge Wu[4]

[1] College of Computer Science, Inner Mongolia University, Hohhot 010021, China
[2] Engineering Research Center of Ecological Big Data, Ministry of Education, Inner Mongolia University, Hohhot 010021, China
csxugang@imu.edu.cn
[3] School of Engineering and Computer Science, Victoria University of Wellington, Wellington 6140, New Zealand
[4] The University of Electro-Communications, 1-5-1 Chofugaoka, Chofu, Tokyo 182-8585, Japan

Abstract. Due to the dynamic nature of opportunistic networks, the network environment undergoes frequent changes where efficient forwarding is crucial to alleviate the occurrences of situations such as node buffer overflow or inefficient utilization of network resources. In this research, we apply reinforcement learning by treating each node in the network as a reinforcement learning agent. Firstly, we utilize the encounter probability between nodes and the remaining cache ratio to construct a forwarding utility learning model based on node cache optimization. Then, integrating the contextual information of nodes, we propose the Opportunistic Routing using Q-Learning with Context Information (ORQLCI) algorithm. We validated the ORQLCI algorithm proposed in this paper using simulations and the results demonstrate that compared to opportunistic network routing algorithms assisted by node information, our scheme effectively improves message delivery ratio, reduces routing overhead, and decreases network latency.

Keywords: Opportunistic networks · Q-learning · Context information

1 Introduction

In recent years, spurred by the rapid advancements in mobile communication and wireless network technologies, opportunistic networks have garnered increasing attention from researchers. These networks represent a distinct category within wireless communication networks, characterized by transient and spontaneous connections between nodes, rather than relying on enduring end-to-end communication paths as seen in conventional networks.

In this context, reinforcement learning (RL) has been increasingly applied to solve networking problems.

Y. Chen et al. (Eds.): COCOON 2024, LNCS 15162, pp. 355–367, 2025.
https://doi.org/10.1007/978-981-96-1093-8_30

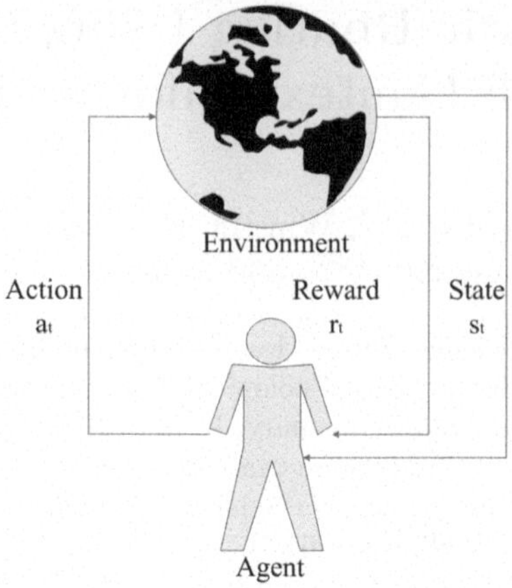

Fig. 1. Basic working procedure of the RL mechanism [1].

RL is an important branch of machine learning (ML) that is often used to solve complex time-varying problems. As shown in Fig. 1, the agent receives a reward or penalty for every action based on its impact on the environment. The agent learns which action should be performed to maximize the rewards and to avoid penalties. In these problems, the intelligent agent often do not have access to global information about the behaviour of the system, but need to actively learn the solution to the problem based on feedback obtained from the environment. The agent receives a reward or penalty for every action based on its impact on the environment.

Meanwhile, in opportunistic networks, individual nodes only have access to local contextual information within the scope of their relationship, but they can collaborate with each other in order to solve one or more discrete optimisation problems. Hence, the relevant variants of RL (e.g. Q-Learning) can be applied to opportunistic networks.

Successful delivery of messages in opportunistic networks can be improved with knowledge of the expected network topology, the mobility pattern and encounter probability of nodes, as well as, nodes' resource availability. Such information is often referred to as the *context information* of the networks, which is information that can be used to achieve efficient and effective forwarding of messages. Despite the obvious value of context information and its adoption in opportunistic routing, existing RL-based routing algorithms for opportunistic networks often under utilize contextual information carried by network nodes.

Hence, this paper proposes an opportunistic network routing algorithm that leverages Q-learning alongside a blend of contextual node information, which we refer to as Opportunistic Routing using Q-Learning with Context Information (ORQLCI). The algorithm updates the Q-value based on interactions during node encounters, while considering factors such as the likelihood of future node connections, node cache management, and dynamic changes in the network. By incorporating node encounter probability and cache utilization, the learning process adapts to the time-varying nature of opportunistic networks. Given that message-carrying nodes in opportunistic networks may not always encounter optimal relay nodes, this paper confines message delivery directions to regions where optimal relay nodes are likely to be encountered, aiming to enhance algorithm performance.

2 Related Work

The use of RL in the design of routing algorithms for opportunistic networks allows nodes to adaptively adjust their routing strategies, thereby improving the performance of the algorithms.

Yuan *et al.* proposed an opportunistic network routing algorithm called Double Q-Learning Routing (DQLR) [2], which employs two value functions to address the issue of overestimation in Q-value estimation that may occur when using the greedy strategy to update state-action values in the Q-learning algorithm. WU *et al.* proposed a Q-learning-based traffic-aware routing protocol called QTAR [3]. The algorithm takes advantage of the benefits of the geographic routing paradigm and also successfully utilizes the road side unit (RSU) to deliver the routing packet to the destination; however QTAR does not estimate the vehicle's direction, which will impair the performance of the protocol in real life. The Context Adaptive Reinforcement Learning-based Routing Algorithm in Delay Tolerant Network (CARL-DTN) [4] algorithm proposed by Yesuf *et al.* is based on the Q-Learning algorithm to replicate messages to the node with the best expected total discount return. However, Q-Learning utilizes the maximum Q-value as the criterion for selecting the best action, which may suffer from the problem of "overestimation".

While various RL-based algorithms have been proposed for routing messages in opportunistic networks, these fail to take full advantage of the contextual information carried by the nodes in the network, such as the energy, cache, movement direction and location information of the nodes, which reflect the node's movement patterns and message forwarding capabilities.

3 Opportunistic Routing Using Q-Learning with Context Information

In opportunistic networks, frequent disconnections and partitioning of the network are the norm rather than the exception. Hence, message forwarding schemes can benefit from context information, like node encounter probability and cache utilization, and incorporating such information into the learning model can significantly enhance the message delivery process.

3.1 Context Information

Node Encounter Probability: Node encounter probability serves as a common metric for depicting the potential connectivity among nodes. The calculation of node encounter probability typically involves three distinct stages and their associated probabilities [5]:

(1) Encounter update phase: When nodes A and B meet, the encounter probability between them is updated according to Eq. (1),

$$P_{(a,b)} = P_{(a,b)old} + (1 - P_{(a,b)old}) \times P_{init} \tag{1}$$

where $P_{(a,b)old}$ represents the probability of the most recent encounter between node A and node B, and $P_{init} \in (0, 1]$ is an initialization constant.

(2) Encounter probability decay phase over time: The encounter probability diminishes over time due to the weakening connection between the reflecting nodes, as illustrated in Eq. (2),

$$P_{(a,b)} = P_{(a,b)old} \times \eta^t \tag{2}$$

where η represents the decay factor, indicating the rate at which the probability of encounters between nodes decreases over time. The larger η is, the faster the probability of encounters between nodes decays, and vice versa. t denotes the number of time units since the last time the metric was decayed.

(3) Encounter probability transfer phase: If node A frequently encounters node B and node B frequently encounters node C, the encounter probability between node A and node C is computed using Eq. (3),

$$P_{(a,c)} = P_{(a,c)old} + (1 - P_{(a,c)old}) \times P_{(a,b)} \times P_{(b,c)} \times \beta \tag{3}$$

where β represents the transitivity factor, quantifying the influence of a node's transitivity on the probability of node encounters. The larger β is, the more pronounced the impact of a node's transitivity on the probability of node encounters, and conversely.

Node Cache Utilization: In opportunistic networks, a multicopy transmission mechanism is commonly employed to enhance the likelihood of successful message delivery. However, the limited cache space of a node poses a challenge; when the cache reaches its capacity, the node is unable to accept additional messages, leading to a degradation in network performance. Hence, when devising routing algorithms for opportunistic networks, the ratio of the current remaining cache to the maximum cache of a node can be leveraged to calculate the node's cache utilization [6], as follows,

$$BF = 1 - \frac{\sum_{m=1}^{n} N_m \times B_m}{C_{init}} \tag{4}$$

where B_m represents the node's current cache occupancy, N_m denotes the number of message m, and C_{init} denotes the maximum cache capacity of the node.

3.2 Q-Learning Model

Each node is an *agent* and each possible selection of the next node to forward a message (i.e. the relay node) is considered a *state* of the agent. The set of all possible candidate nodes for the relay node is the state space. The learning task is to find the best relay node for the corresponding network environment in relation to the feedback. An *action* aims to select the relay node that would be used for storing the data.

Each node maintains and updates a Q-Table, where each element $Q_d(s, x)$ of the table represents the Q-value of node s carrying a message to node d utilizing node x as a relay node to transmit the message. To accommodate the dynamically changing network structure of the opportunistic network, this paper employs a dynamic Q-Table.

Learning Algorithm and Reward Process: In the forwarding utility learning model, the Q-value update process is given by Eq. (5).

$$Q_d(s,x) \leftarrow (1-\alpha) \times Q_d(s,x) + \alpha \times \left\{ R_d(s,x) + \gamma_d(s,x) \times \max_{y \in N_x} Q'_d(x,y) \right\} \quad (5)$$

$Q_d(s, x)$ is the expected reward value for the current node s to forward a message to the destination node d with node x as the relay node. During the process of updating the Q-value table, the corresponding Q-value needs to be extracted and substituted into Eq. (5) for computation. $\alpha \in (0, 1]$ is the learning coefficient used to control the learning speed of the intelligent agent.

$R_d(s, x)$ is referred to as the immediate reward value, computed as shown in Eq. (6); when node s and node x meet, if there is a message in node s with node x as the destination node, then the message delivery is done directly and the immediate reward value is 1. If the two nodes do not meet, then the immediate reward value is 0.

$$R_d(s,x) = \begin{cases} 1, & \text{if } x = d \\ 0, & \text{others} \end{cases} \quad (6)$$

The dynamic discount factor $\gamma_d(s, x)$ is as shown in Eq. (7) where γ is the discount factor constant, $0 \leq \gamma \leq 1$.

$$\gamma_d(s,x) = \gamma \times BF_x \quad (7)$$

In Eq. (7), BF_x denotes the cache utilization of the node x and is calculated as in Eq. (4). It can be observed that a smaller cache utilization implies that the node has sufficient space to receive messages and is more suitable to act as a relay node, resulting in a larger Q-value. The term $\max_{y \in N_x} Q'_d(x, y)$ denotes the maximum Q-value for each node location information and activity interest area, defined in Eq. (8) as follows,

$$\max_{y \in N_x} Q'_d(x,y) = \max_{y \in N_x} Q_d(x,y) \times P(x,y) \quad (8)$$

where N_x denotes the one-hop neighbor set of x and $P(x, y)$ denotes the encounter probability between node x and node y. Sometimes, although node y is the best relay node for a message in the Q-table recorded by node x, it is not certain whether the two nodes can establish a communication connection at some time in the future, so the encounter probability between the nodes is introduced to further adapt to the dynamically changing network structure such as opportunistic networks.

After substituting Eqs. (6), (7) and (8) into Eq. (5), the following results are obtained: if the encounter node is not the destination node, then Eq. (9) is used to update the Q-value, and if the encounter node is the destination node, then Eq. (10) is used to update the Q-value.

$$Q_d(s,x) \leftarrow (1-\alpha) \times Q_d(s,x) + \alpha \times \gamma \times BF_x \times \max_{y \in N_x} Q_d(x,y) \times P(x,y) \tag{9}$$

$$Q_d(s,x) \leftarrow (1-\alpha) \times Q_d(s,x) + \alpha \times R_d(s,x) \tag{10}$$

***Q*-Value Ageing and Forwarding Strategy:** Like the encounter probability decay (cf: Eq. (2)), the Q-value also ages over time whereby nodes need to gradually "forget" the previously learnt Q-values as time elapsed. If a pair of nodes did not exchange gossip over a certain period of time, $Q_d(s, x)$ decreases, resulting in lost learned knowledge. The aging effect on $Q_d(s, x)$ is computed as shown in Eq. (11),

$$Q_d(s,x) = Q_d(s,x)_{old}\, \omega^t \tag{11}$$

where $\omega \in [0, 1)$ is the aging constant and t is the number of time units that have elapsed since the last time d exchanged gossip about x.

3.3 Learning and Forwarding Design

When an arbitrary node *sn* encounters another node *rn* during its movement, it gathers the contextual information of node *rn*, and then proceeds to maintain and update its Q-table based on this information, as shown in Algorithm 1. Upon receiving the context information from its neighbour *rn*, node *sn* first determines whether there is any record of its neighbour *rn* in its Q-Table. If not, it adds the information of *rn* into its Q-Table. Finally, it updates *Q_rn* (*sn*,*rn*) (the action-state value of node *sn* sending the message whose destination node is *rn* to *rn*) according to Eq. (10) and *Q_dn* (*sn*,*rn*) (the action-state value of node *sn* sending the message whose destination node is *dn* to *rn*) according to Eq. (9).

Next, node *sn* executes Algorithm 2 to forward the message(s) it is carrying. For all messages it is carrying, it determines the message's destination *dn* and examines its own Q-table to find a match. If the corresponding destination *dn* is recorded in the Q-table, it adopts a greedy strategy to select a message relay node. For a message whose destination is not found in *sn*'s Q-Table, the message delivery direction is adopted to find the optimal relay node. The node *sn* repeats the process for all messages it is carrying and the algorithm is repeated each time a node encounters new neighbour nodes.

Algorithm 1: Q-Table Update

```
if thisNode.connections.size ≠ 0 then
    foreach con ∈ connections do
        rn = con.getHost; // Select the encounter node to process;
        sn updates its Q-value using Eq. (11);
        thisNode sn gets Context Information from rn;
        if rn ≠ dn then
            // Encountered node is NOT the destination;
            Compute Q-value using Eq. (9);
        else
            // Encountered node is the destination node;
            Compute Q-value using Eq.(10);
        end
        Observe the next state s';
        Compute (new) Q-value using Eq. (11);
        Update thisNode Q-Table;
    end
end
```

Algorithm 2: Message Forwarding

```
if thisNode.connections.size ≠ 0 then
    foreach con ∈ connections do
        rn = con.getHost; // Select the encounter node to process;
        thisNode sn get Context Information from rn;
        Update thisNode sn's Q-Table;
        foreach m ∈ message do
            m.destination = d; // Determine message's destination;
            if d ∈ thisNode Q-Table then
                a* = max Q_d(s, x); // Select the optimal policy;
                execute a*;
            else
                Predict the destination node direction V_des (vector);
                if cos(V_des, V_m) ≥ cos δ then
                    // Within the range of the angle threshold;
                    thisNode.m ⟶ rn; // Forward message m to rn
                end
            end
        end
    end
end
```

4 Simulation and Analysis

We use the ONE [7] simulation package to validate our proposed scheme. ONE is a simulation platform developed specifically for delay/disruption tolerant networks (DTN) and opportunistic network environments, which can ideally simulate the parameters of opportunistic networks in a variety of operating environments. The ORQLCI algorithm is compared with three other opportunistic network routing algorithms, namely, Epidemic [8], Prophet [5] and EC-CW [9], in terms of message delivery success rate, transmission delay, and routing overhead to evaluate algorithm performance. To allow the simulation to stabilize, the initial 1000 s of data are discarded at the beginning of the simulation.

Epidemic [8] stands as one of the most representative opportunistic network routing algorithms, employing a flooding transmission strategy. The algorithm's core concept involves exchanging unstored messages when two nodes encounter each other. In theory, this approach facilitates rapid message forwarding through the network, provided there are ample resources like cache and bandwidth available. However, in practical application environments where resources of opportunistic network nodes are limited, this algorithm can easily result in message congestion due to the excessive generation of message copies [10].

Prophet's algorithm [5] defines a node's historical encounter probability as a utility value. It assumes that nodes frequently encountering each other in the past are likely to meet again in the future. Based on this historical encounter probability, the algorithm decides whether to perform data forwarding, thus achieving the routing function of opportunistic networks.

EC-CW [9] is an opportunistic network routing algorithm based on the degree of a node's willingness to communicate. It optimizes Prophet's algorithm by determining a node's willingness to communicate using information such as energy and cache during the message forwarding phase.

4.1 Optimal Learning Coefficient and Discount Factor

To determine suitable values to use in the learning algorithm, we conduct multiple experimental simulation runs and a comparative analysis. The specific parameter settings are illustrated in Table 1.

In Fig. 2, the graph illustrates the variation of the delivery rate concerning the learning coefficient α and discount factor γ in the algorithm proposed in this paper. It is evident from the graph that when the discount factor γ remains constant, the trend of the delivery rate initially increases and then decreases with the rise of the learning coefficient α. This phenomenon occurs because α governs the extent to which new information supersedes old information. A larger α implies a greater emphasis on new information acquisition by the node, leading to enhanced routing effectiveness. However, excessively large α results in the node solely focusing on the latest information while disregarding older information, ultimately degrading routing performance.

Table 1. Parameter values for coefficient determination.

Parameter	Value
Simulation time	43200 s
Simulation area size	4500 $*$3400 (width $*$ height) m^2
Number of nodes	200
Message lifetime	300 s
Message size	500 kB– 1MB
Node cache size	50 MB
Node transfer rate	250 kB/s
Node communication radius	10 m
Node movement model	Shortest Path Map Based Movement(SPMBM)
Message generation interval	15 s–25 s
Angle threshold setting	45°

Considering the impact of the learning coefficient α and discount factor γ on the message delivery rate, another key observation from Fig. 2 is that when the learning coefficient is 0.8 and the discount factor is 0.6, the delivery rate reaches its peak value of 0.658.

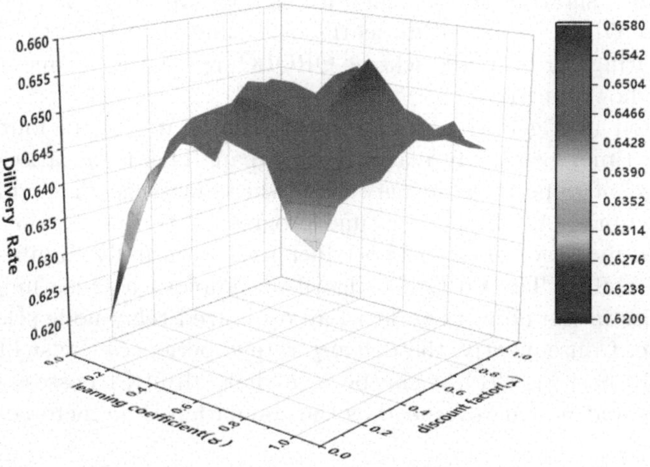

Fig. 2. Delivery rate change with variation of learning coefficient and discount factor.

In our performance evaluation of ORQLCI, we use the optimal learning coefficient α identified in Sect. 4.1 taking the value of 0.8 and the discount factor γ taking the value of 0.6. For our Q-value aging constant, as shown in Eq. (11), we use the value of the decay constant in Prophet's algorithm [5], viz. $\omega = 0.98$. We measure performance based on the packet *delivery probability* (proportion

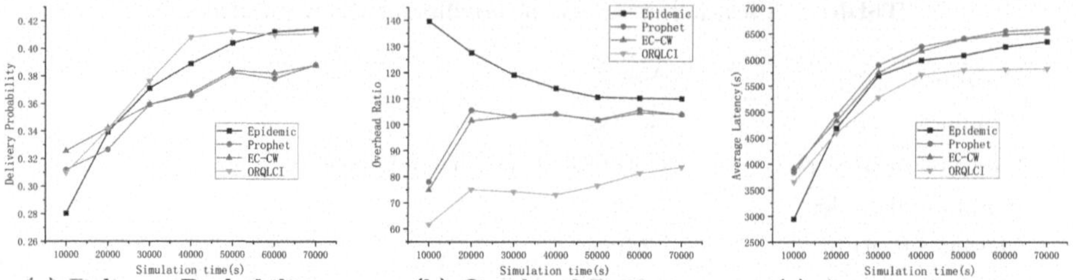

(a) Delivery Probability (b) Overhead Ratio (c) Average Latency

Fig. 3. Performance over elapsed time

of messages delivered to their destinations out of all generated messages), the *overhead ratio* (total number of replications divided by the total number of generated messages) and *average latency* (delay in delivering messages from sources to destinations.) The simulation study is divided into two distinct sets.

4.2 Performance over Elapsed Time

The first set focuses on setting the node cache to a fixed value of 20MB and the simulation time ranging from 10,000s to 70,000s. This segment primarily aims to evaluate the routing performance concerning the trend with the increase in simulation time, showing how the algorithms fluctuate over time, and in particular, how our ORQLCI algorithm learns over time. The first set of simulation results comparing our proposed scheme ORQLCI against Epidemic, Prophet and EC-CW are shown in Fig. 3.

As depicted in Fig. 3(a), the message delivery rate for all four algorithms improve over time, eventually reaching a plateau. This trend arises due to the salient nature of opportunistic networks, where message dissemination relies on multiple copies. As simulation time progresses, the proportion of successfully delivered messages in the network increases, eventually stabilizing the final delivery probability. The ORQLCI algorithm employs a Q-learning approach, enabling nodes in the network to learn an optimized relay node selection strategy over time. Consequently, the delivery rate experiences the swiftest growth. From Fig. 3(b) and Fig. 3(c), it can be seen that ORQLCI possesses the lowest average delay and routing overhead as the simulation time increases.

4.3 Performance Under Different Cache Sizes

In the second set of simulations, the simulation time is 43200s and the node cache varies from 10MB to 50MB in increments of 10MB, which is mainly used to study the routing performance with node cache variation. Evident from Fig. 4, as the cache size of nodes increases, ORQLCI has the highest overall delivery rate and the lowest average delay as well as routing overhead compared to the other three algorithms.

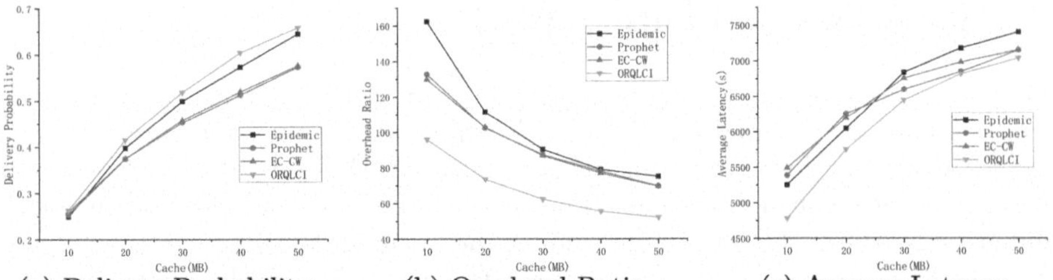

(a) Delivery Probability (b) Overhead Ratio (c) Average Latency

Fig. 4. Performance under different cache sizes

In scenarios where node resources are constrained, e.g., 10MB, the Epidemic algorithm's approach of flooding resulted in node cache overflow, high routing overhead, and a diminished message delivery rate. Conversely, Prophet alleviates network congestion by selecting nodes with the highest likelihood of encountering the target node for message forwarding, thus enhancing message delivery rates while curbing routing overhead. However, since the message-carrying nodes lack the ability to select optimal relay nodes, this method tends to incur higher average latency. Building upon the Prophet algorithm, the EC-CW algorithm refines its approach by additionally considering factors such as node energy levels and cache availability. By prioritizing nodes with higher remaining energy and more abundant cache resources for message delivery, EC-CW marginally outperforms Prophet in overall performance. Leveraging Q-Learning, nodes employing ORQLCI gather contextual information from neighboring nodes and dynamically adjust transmission strategies during network operation, thereby optimizing performance.

When the node cache exceeds 15MB, ORQLCI exhibits superior performance compared to the other three algorithms, achieving the highest overall delivery rate alongside the lowest average latency and routing overhead. Epidemic, leveraging flooding mechanisms for message delivery, exhibits a commendable message delivery rate when resources are ample. However, this comes at the expense of elevated routing overhead and average latency. In contrast, Prophet and EC-CW algorithms sacrifice message delivery rates to mitigate routing overhead. While these algorithms formulate relay node selection strategies, their lack of adaptability sometimes results in suboptimal relay node choices, leading to missed message forwarding opportunities compared to the Epidemic algorithm. The ORQLCI algorithm's adaptive transmission strategy adjustment capability enables it to consistently achieve the highest message delivery rate. By dynamically regulating forwarding strategies, ORQLCI minimizes average delay and routing overhead, thereby outperforming alternative algorithms.

5 Conclusions and Future Work

This paper introduces a forwarding utility learning model that optimizes node cache through a combination of the Q-learning algorithm. It culminates in a routing algorithm for opportunistic networks, which integrates node context information into routing decisions. Through simulations conducted on the ONE simulation platform, the study determines the optimal values of two key parameters: the learning coefficient and discount factor. Subsequently, further simulations are conducted varying two network scenario parameters: simulation time and node cache. Results confirm that, compared to the Epidemic, Prophet, and EC-CW algorithms, the proposed ORQLCI algorithm exhibits superior adaptability to changes in network conditions and offers enhanced network forwarding performance.

Nevertheless, there are opportunities for future work. Specifically, nodes with high forwarding loads are susceptible to failure due to energy depletion, given the limited energy resources in opportunistic networks. Hence, designing energy-efficient routing algorithms with dormant mechanisms emerges as a critical area of focus for future research endeavors in the field of opportunistic networks.

Acknowledgments. This work was supported by the National Natural Science Foundation of China under Grants 62061036; the University Youth Science and Technology Talent Development Project (Innovation Group Development Plan) of Inner Mongolia A. R. of China under Grants NMGIRT2318; the Self-Open Project of Engineering Research Center of Ecological Big Data, Ministry of Education; and the China Scholarship Council.

References

1. Sutton, R.S., Barto, A.G.: Reinforcement Learning: An Introduction. MIT press, Cambridge (2018)
2. Yuan, F., Wu, J., Zhou, H., Liu, L.: A double Q-learning routing in delay tolerant networks. In: Proceedings of the IEEE International Conference on Communications (ICC), pp. 1–6 (2019)
3. Wu, J., Fang, M., Li, H., Li, X.: RSU-assisted traffic-aware routing based on reinforcement learning for urban vanets. IEEE Access **8**, 5733–5748 (2020)
4. Yesuf, F.Y., Prathap, M.: CARL-DTN: context adaptive reinforcement learning based routing algorithm in delay tolerant network. arXiv preprint arXiv:2105.00544
5. Lindgren, A., Doria, A., Schelén, O.: Probabilistic routing in intermittently connected networks. ACM SIGMOBILE Mob. Comput. Commun. Rev. **7**(3), 19–20 (2003)
6. Zhang, F., Wang, X., Zhang, S.: Buffer aware prophet routing algorithm for opportunistic networks. Comput. Eng. Des. **36**(5), 1145–1150 (2015)
7. Keränen, A., Ott, J., Kärkkäinen, T.: The ONE simulator for DTN protocol evaluation. In: SIMUTools '09: Proceedings of the 2nd International Conference on Simulation Tools and Techniques, ICST, New York, NY, USA (2009)
8. Vahdat, A., Becker, D.: Epidemic routing for partially connected ad hoc networks. Technical report. CS-200006. Duke University (2000)

9. Chen, J., Xu, G., Wu, X., Wei, F., He, L.: Energy balance and cache optimization routing algorithm based on communication willingness. In: Proceedings of the IEEE Wireless Communications and Networking Conference (WCNC), pp. 1–6 (2021)
10. Rashidi, L., Towsley, D., Mohseni-Kabir, A., Movaghar, A.: On the performance analysis of epidemic routing in non-sparse delay tolerant networks. IEEE Trans. Mob. Comput. **22**(7), 4134–4149 (2023). https://doi.org/10.1109/TMC.2022.3144683

A Combinatorial View of Holant Problems on Higher Domains

Yin Liu(✉)

Google, Mountain View, CA 94043, USA
hughliu@google.com

Abstract. On the Boolean domain, there is a class of symmetric signatures called "Fibonacci gates" for which a beautiful P-time combinatorial algorithm has been designed for the corresponding Holant problems.

In this work, I give a combinatorial view for Holant($\mathcal{F}$) problems on a domain of size 3 where $\mathcal{F}$ is a set of arity 3 functions with inputs taking values on the domain of size 3 and the functions share some common properties. The combinatorial view can also be extended to the domain of size 4.

Specifically, I extend the definition of "Fibonacci gates" to the domain of size 3 and the domain of size 4. Moreover, I give the corresponding combinatorial algorithms.

Keywords: Holant problem · Combinatorial algorithm · Higher domain

1 Introduction and Background

In a lot of fields in computer science, machine learning and statistical physics, counting problems play a role. Holant problems encompass a broad class of counting problems [1,2,7,8,10,12,16,17,19,23–25]. This framework extends edge-coloring models [21,22] while the latter is when the constraint functions are symmetric. These problems also extend counting constraint satisfaction problems (a.k.a. CSP). It was proved that some prototypical Holant problems including counting perfect matchings, cannot be expressed as vertex-coloring models known as graph homomorphisms [14,18]. The complexity classification program of counting problems is to classify the computational complexity of these problems.

A Holant problem on a domain of size D is defined on a graph $G = (V, E)$ where V, E represent the set of vertices and edges respectively. In our context, edges are variables and vertices are constraints. Given a set of constraint functions $\mathcal{F}$ defined on D, a *signature grid* $\Omega = (G, \pi)$ assigns to each vertex $v \in V$ a function (a.k.a. signature) $f_v \in \mathcal{F}$. The goal is to compute the following *partition* function

$$\text{Holant}_\Omega = \sum_{\sigma:E\to D} \prod_{v\in V} f_v\left(\sigma|_{E(v)}\right).$$

Y. Chen et al. (Eds.): COCOON 2024, LNCS 15162, pp. 368–380, 2025.
https://doi.org/10.1007/978-981-96-1093-8_31

σ is each edge assignment and the summation contains in total $|E|^D$ many assignments. $E(v)$ are edges incident to vertex v. The computational problem is denoted by Holant($\mathcal{F}$). Specifically, on the Boolean domain, it is over all $\{0,1\}$-edge assignments. On the domain of size 3, it is over all $\{R,G,B\}$-edge assignments, signifying three colors Red, Green and Blue. And for size 4, it is over all $\{R,G,B,W\}$-edge assignments. On the Boolean domain, if every vertex has the EXACT-ONE function (which evaluates to 1 if exactly one incident edge is 1, and evaluates to 0 otherwise), then the partition function gives the number of perfect matchings. As another example, on domain size k, if every vertex has the ALL-DISTINCT function, then the partition function calculates the number of valid k-edge colorings.

A Holant* problem is a class of Holant problems where we assume all the unary signatures are freely available. In other words, Holant* considers function sets containing all unary functions. Holant$^*(\mathcal{F})$ = Holant($\mathcal{F} \cup \mathcal{U}$) where we denote by $\mathcal{U}$ the set of all unary functions.

A *symmetric* signature is a function that is invariant under any permutation of its variables. The value of such a signature depends only on the numbers of each domain assigned to its input variables (i.e., edges). The number of variables is its arity; unary, binary, ternary signatures have arities 1, 2, 3 respectively. We denote a symmetric ternary signature g on a domain of size 3 by a "triangle" consisting of 10 numbers:

$$\begin{array}{ccccccc} & & & g_{3,0,0} & & & \\ & & g_{2,1,0} & & g_{2,0,1} & & \\ & g_{1,2,0} & & g_{1,1,1} & & g_{1,0,2} & \\ g_{0,3,0} & & g_{0,2,1} & & g_{0,1,2} & & g_{0,0,3} \end{array}$$

where $g_{i,j,k}$ is the value on inputs having i Red, j Green and k Blue. Similarly, we denote a symmetric ternary signature g on a domain of size 4 by a "tetrahedron":

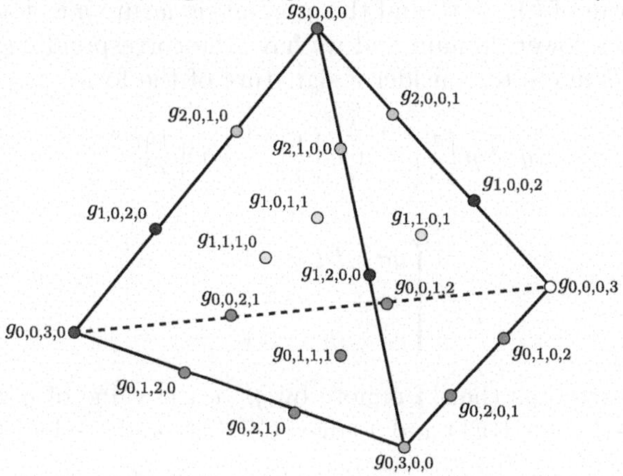

For the classification of counting CSP, researchers have made a lot of progress [3–5,13]. Also, for Holant problems, [6,11,15] are some of the milestone works. Specifically, in [9], "Fibonacci gates" is firstly proposed and studied on the Boolean domain. However, studying Holant problems on higher domains is particularly challenging. There are few existing works on higher domains. Among them, one is [11], in which a dichotomy for Holant$^*(f)$ is proved where f is a ternary complex symmetric function on domain size 3. Another piece of work is [6] which investigates some higher domain problems including κ-edge coloring and derives some dichotomies. Another one is [20] which proves some Holant dichotomies on domains 3 and 4 in restricted settings.

Essentially, people only have scratched a little of higher domain classification programs and there are still many unknowns to explore.

My Contribution. In this paper, I focus on problems on domain sizes 3 and 4 while this time I am more interested in problems with the "Fibonacci" feature. I explore the quintessential nature of such specific signatures. I define " *Generalized* Fibonacci Gates" on higher domains (Definitions 1 and 2), show that those problems are tractable and deliver our combinatorial algorithms (Theorems 1 and 2). With these new exploration and discovery, I conjecture that there are Fibonacci Gates and the associated P-time algorithms on all higher domains.

2 Fibonacci Gates on a Domain of Size 3

By an orthogonal holographic transformation [24], a domain 3 ternary symmetric signature is tractable when it is of the form $g = \alpha^{\otimes 3} + \beta^{\otimes 3} + \gamma^{\otimes 3}$ where $\alpha, \beta, \gamma \in \mathbb{C}^3$ are mutually orthogonal to each other (specifically, $Tg = ae_1^{\otimes 3} + be_2^{\otimes 3} + ce_3^{\otimes 3}$ is tractable, for some $a, b, c \in \mathbb{R}$ and T is an orthogonal matrix). Viewing $\alpha = (\alpha_1, \alpha_2, \alpha_3)^T, \beta = (\beta_1, \beta_2, \beta_3)^T, \gamma = (\gamma_1, \gamma_2, \gamma_3)^T$, I show that we can w.l.o.g. find $i \in \{1, 2, 3\}$ such that $\alpha_i\beta_i\gamma_i \neq 0$. Otherwise, it's easy to prove that either one of α, β, γ is a 0 vector in which case the problem is reducible to a lower domain, or, with some permutation of the three domains, $\alpha = c(1, 0, 0)^T, c \neq 0$ and β, γ both have the form of $(0, *, *)^T$ and the original signature g is domain separable, also reducible to a lower domain and we have the corresponding dichotomies.

Therefore, it suffices to consider a signature of the form

$$g = p \begin{bmatrix} 1 \\ a \\ b \end{bmatrix}^{\otimes 3} + q \begin{bmatrix} 1 \\ c \\ d \end{bmatrix}^{\otimes 3} + r \begin{bmatrix} 1 \\ e \\ f \end{bmatrix}^{\otimes 3}$$

where

$$\begin{cases} ac + bd = -1 \\ ae + bf = -1 \\ ce + df = -1 \end{cases} \tag{1}$$

And g can be written as (here I denote by $g_{i,j,k}$ the value of g when i incident edges are assigned color RED and j edges GREEN and k edges BLUE.)

$$\begin{array}{ccccccc} & & & g_{3,0,0} & & & \\ & & g_{2,1,0} & & g_{2,0,1} & & \\ & g_{1,2,0} & & g_{1,1,1} & & g_{1,0,2} & \\ g_{0,3,0} & & g_{0,2,1} & & g_{0,1,2} & & g_{0,0,3} \end{array}$$

where

$$\begin{cases} g_{3,0,0} = p+q+r & g_{2,1,0} = pa+qc+re \\ g_{2,0,1} = pb+qd+rf & g_{1,2,0} = pa^2+qc^2+re^2 \\ g_{1,1,1} = pab+qcd+ref & g_{1,0,2} = pb^2+qd^2+rf^2 \\ g_{0,3,0} = pa^3+qc^3+rf^3 & g_{0,2,1} = pa^2b+qc^2d+re^2f \\ g_{0,1,2} = pab^2+qcd^2+ref^2 & g_{0,0,3} = pb^3+qd^3+rf^3 \end{cases} \tag{2}$$

Consider each **"medium-sized" triangle** with depth 2 consisting of 6 numbers over the 10 signature values of g, i.e.,

$$\begin{array}{ccccc} & & g_{i,j,k} & & \\ & g_{i-1,j+1,k} & & g_{i-1,j,k+1} & \\ g_{i-2,j+2,k} & & g_{i-2,j+1,k+1} & & g_{i-2,j,k+2} \end{array}$$

There are in total 3 such triangles. It can be proved directly that there exist four parameters s, x, y, t such that

$$\begin{cases} g_{i-2,j+2,k} = g_{i,j,k} + sg_{i-1.j+1,k} + xg_{i-1,j,k+1} \\ g_{i-2,j+1,k+1} = xg_{i-1,j+1,k} + yg_{i-1,j,k+1} \\ g_{i-2,j,k+2} = g_{i,j,k} + yg_{i-1,j+1,k} + tg_{i-1,j,k+1} \end{cases} \tag{3}$$

and s, x, y, t are independent of p, q, r. Moreover, they satisfy $sy+xt+1 = x^2+y^2$. The coefficients could be written in a "**triangular**" way below:

$$\begin{array}{ccccccccc} & 1 & & & 0 & & & 1 & \\ s & & x & x & & y & y & & t \end{array}$$

Simply speaking, for each **"medium-sized" triangle** of the shape $\begin{array}{ccccc} & & * & & \\ & * & & * & \\ * & & * & & * \end{array}$ (6 signature values among the total of 10), the three signature values of the bottom line have a linear relationship with other three values on the top through s, x, y, t. In fact, in the settings above,

$$\begin{cases} x = -bdf \\ y = -ace \\ s = ace + a + c + e \\ t = bdf + b + d + f \end{cases} \tag{4}$$

It's easy to see that

$$\begin{aligned}
sy + xt + 1 &= -ace(ace + a + c + e) - bdf(bdf + b + d + f) + 1 \\
&= -ace(ace - a(ce + df) - c(ae + bf) - e(ac + bd)) \\
&\quad -(ac + bd)(ae + bf)(ce + df) \\
&= ace(2ace + adf + cbf + ebd) + bdf(2bdf + bce + dae + fac) \\
&\quad -(ac + bd)(ae + bf)(ce + df) \\
&= (ace)^2 + (bdf)^2 \\
&= x^2 + y^2.
\end{aligned}$$

Similarly, we can verify that the equations of x, y, s, t represented by a, b, c, d, e, f satisfy the linear recurrence relation 3.

Another important thing is that if we find such s, x, y, t, we can then recover the orthogonal matrix $\begin{bmatrix} 1 & 1 & 1 \\ a & c & e \\ b & d & f \end{bmatrix}$ by calculating the corresponding a, c, e. In fact, from Eq. 1 and Eq. 4, we get that

$$\begin{aligned}
x^2 = (bdf)^2 &= (-1 - ac)(-1 - ae)(-1 - ce) \\
&= -((ace)^2 + (a + c + e)ace + (ac + ae + ce) + 1) \\
&= -y^2 + (s + y)y - (ac + ae + ce) - 1
\end{aligned}$$

Let $X = -y = ace, Y = s+y = a+c+e, Z = -x^2-y^2+(s+y)y-1 = ac+ae+ce$, we know that a, c, e are the three roots of equation

$$t^3 - Yt^2 + Zt - X = 0.$$

Definition 1. *We call a domain 3 symmetric signature g (arity ≥ 2) a* generalized Fibonacci gate *(with parameter s, x, y, t where $sy + xt + 1 = x^2 + y^2$) if*

$$\begin{cases}
g_{i-2,j+2,k} = g_{i,j,k} + sg_{i-1.j+1,k} + xg_{i-1,j,k+1} \\
g_{i-2,j+1,k+1} = xg_{i-1,j+1,k} + yg_{i-1,j,k+1} \\
g_{i-2,j,k+2} = g_{i,j,k} + yg_{i-1,j+1,k} + tg_{i-1,j,k+1}
\end{cases}$$

$\forall i,\ 2 \leq i \leq \text{arity}(g)$. *Specifically, any unary signature is a Fibonacci gate.*

A set of signatures $\mathcal{F}$ is called generalized Fibonacci *if for some $s, x, y, t \in \mathbb{C}$, $sy + xt + 1 = x^2 + y^2$, every signature in $\mathcal{F}$ is a generalized Fibonacci gate with parameters s, x, y, t.*

Theorem 1. *On a domain of size 3, for any finite set of generalized Fibonacci gates $\mathcal{F}$, the Holant problem* Holant($\mathcal{F}$) *is computable in polynomial time.*

Proof. If $\Gamma_1, \Gamma_2, ..., \Gamma_k$ are the connected components of a graph Γ, then

$$\text{Holant}_\Gamma = \prod_{j=1}^{k} \text{Holant}_{\Gamma_j}.$$

So we only need to consider connected graphs as inputs.

Suppose Γ has n nodes and m edges. First we cut all the edges in Γ. A node with degree d can be viewed as an $\mathcal{F}$-gate with d dangling edges. Now step by step we merge two dangling edges into one regular edge in the original graph, until we recover Γ after m steps. We prove that all the intermediate $\mathcal{F}$-gates still have generalized Fibonacci signatures with the same parameters s, x, y, t, and at every step we can compute the intermediate signature in polynomial time. After m steps we get Γ as an $\mathcal{F}$-gate with no dangling edge; the only value of its signature is the Holant value we want. To carry this out, we only need to prove that it is true for one single step. There are two cases, depending on whether the two dangling edges to be merged are in the same component or not.

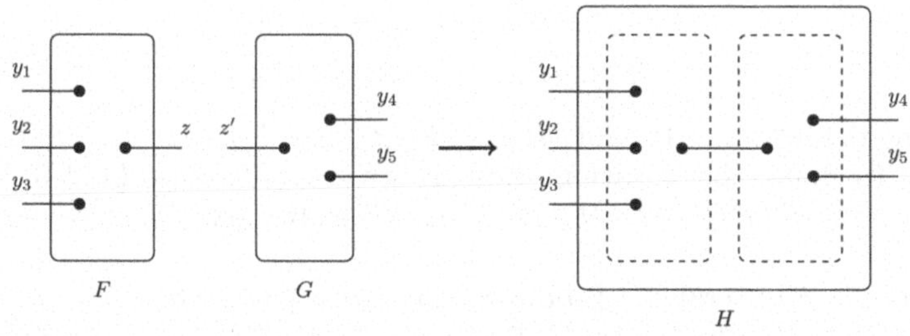

Fig. 1. First operation

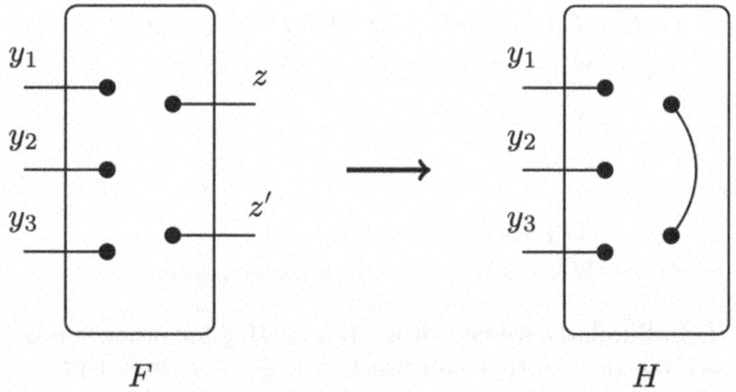

Fig. 2. Second operation

These two operations are illustrated in Figs. 1 and 2.

In the first case, the two dangling edges belong to two components before their merging (Fig. 1). Let F have dangling edges $y_1, y_2, ..., y_r, z$ and G have dangling edges $y_{r+1}, ..., y_{r+w}, z'$. After merging z and z', we have a new $\mathcal{F}$-gate H with dangling edges $y_1, ..., y_{r+w}$. Inductively the signatures of F and G are both generalized Fibonacci gates with the same parameters s, x, y, t. We show that this remains so for the resulting $\mathcal{F}$-gate H.

We first prove that H is symmetric. We only need to show that the value of H is unchanged if the values of two inputs are exchanged. Because F and G are symmetric, if both inputs are from $\{y_1, ..., y_r\}$ or from $\{y_{r+1}, ..., y_{r+w}\}$, the value of H is clearly unchanged. Suppose one input is from $\{y_1, ..., y_r\}$ and the other is from $\{y_{r+1}, ..., y_{r+w}\}$. By the symmetry of F and G we may assume these two inputs are y_1 and y_{r+1}. Thus we will fix an arbitrary assignment for $y_2, ..., y_r, y_{r+2}, ..., y_{r+w}$, and we want to show $H(u, y_2, ..., y_r, v, y_{r+2}, ..., y_{r+w}) = H(v, y_2, ..., y_r, u, y_{r+2}, ..., y_{r+w})$ where $u, v \in \{R, G, B\}$.

We will suppress the fixed values $y_2, .., y_r, y_{r+2}, ..., y_{r+w}$ and denote

$$F_{uz} = F(u, y_2, ..., y_r, z)$$

$$G_{vz} = G(v, y_{r+2}, ..., y_{r+w}, z)$$

$$H_{uv} = H(u, y_2, ..., y_r, v, y_{r+2}, ..., y_{r+w}).$$

Then by the definition of Holant, $H_{uv} = F_{uR}G_{vR} + F_{uG}G_{vG} + F_{uB}G_{vB}$, $u, v \in \{R, G, B\}$. To make the notation simpler, we use subscript $\{1, 2, 3, 4, 5, 6\}$ to represent $\{RR, RG, RB, GG, GB, BB\}$ respectively now. We also use M_{ij} to denote F_iG_j.

Because F is a generalized Fibonacci gate with parameters s, x, y, t, we have $F_4 = F_1 + sF_2 + xF_3$, $F_5 = xF_2 + yF_3$, and $F_6 = F_1 + yF_2 + tF_3$. Similar for G. Then we have the following

$$\begin{aligned} H_{RG} &= F_{RR}G_{GR} + F_{RG}G_{GG} + F_{RB}G_{GB} = F_1G_2 + F_2G_4 + F_3G_5 \\ &= F_1G_2 + F_2(G_1 + sG_2 + xG_3) + F_3(xG_2 + yG_3) \\ &= M_{12} + M_{21} + sM_{22} + xM_{23} + xM_{32} + yM_{33} \end{aligned}$$

$$\begin{aligned} H_{GR} &= F_{GR}G_{RR} + F_{GG}G_{RG} + F_{GB}G_{RB} = F_2G_1 + F_4G_2 + F_5G_3 \\ &= F_2G_1 + (F_1 + sF_2 + xF_3)G_2 + (xF_2 + yF_3)G_3 \\ &= M_{12} + M_{21} + sM_{22} + xM_{23} + xM_{32} + yM_{33} \end{aligned}$$

So $H_{RG} = H_{GR}$. Similarly we can prove $H_{uv} = H_{vu}$ for other $u, v \in \{R, G, B\}$. Hence H is symmetric. We thus can use $1, 2, 3, 4, 5, 6$ as subscripts for H too.

Now we show that $H(y_1, ..., y_{r+w})$ is also a generalized Fibonacci gate with parameters s, x, y, t. Since we have proved that H is symmetric, we can choose any two input variables to prove it being Fibonacci. Again, we choose y_1 and y_{r+1}. (This assumes that y_1 and y_{r+1} exist, i.e., F and G are not unary functions. If either one of them is unary, the proof is just as easy.) For any fixed values of all other variables, we have

$$\begin{aligned} H_1 = H_{RR} &= F_{RR}G_{RR} + F_{RG}G_{RG} + F_{RB}G_{RB} = F_1G_1 + F_2G_2 + F_3G_3 \\ &= M_{11} + M_{22} + M_{33} \end{aligned}$$

$$\begin{aligned}H_2 = H_{RG} &= F_{RR}G_{GR} + F_{RG}G_{GG} + F_{RB}G_{GB} = F_1G_2 + F_2G_4 + F_3G_5\\ &= F_1G_2 + F_2(G_1 + sG_2 + xG_3) + F_3(xG_2 + yG_3)\\ &= sM_{22} + yM_{33} + (M_{12} + M_{21}) + x(M_{23} + M_{32})\end{aligned}$$

$$\begin{aligned}H_3 = H_{RB} &= F_{RR}G_{BR} + F_{RG}G_{BG} + F_{RB}G_{BB} = F_1G_3 + F_2G_5 + F_3G_6\\ &= F_1G_3 + F_2(xG_2 + yG_3) + F_3(G_1 + yG_2 + tG_3)\\ &= xM_{22} + tM_{33} + (M_{13} + M_{31}) + y(M_{23} + M_{32})\end{aligned}$$

$$\begin{aligned}H_4 = H_{GG} &= F_{GR}G_{GR} + F_{GG}G_{GG} + F_{GB}G_{GB} = F_2G_2 + F_4G_4 + F_5G_5\\ &= F_2G_2 + (F_1 + sF_2 + xF_3)(G_1 + sG_2 + xG_3) + (xF_2 + yF_3)(xG_2 + yG_3)\\ &= M_{11} + (s^2 + x^2 + 1)M_{22} + (x^2 + y^2)M_{33}\\ &\quad + s(M_{12} + M_{21}) + x(M_{13} + M_{31}) + x(s + y)(M_{23} + M_{32})\end{aligned}$$

$$\begin{aligned}H_5 = H_{GB} &= F_{GR}G_{BR} + F_{GG}G_{BG} + F_{GB}G_{BB} = F_2G_3 + F_4G_5 + F_5G_6\\ &= F_2G_3 + (F_1 + sF_2 + xF_3)(xG_2 + yG_3) + (xF_2 + yF_3)(G_1 + yG_2 + tG_3)\\ &= x(s + y)M_{22} + y(x + t)M_{33} + x(M_{12} + M_{21})\\ &\quad + y(M_{13} + M_{31}) + (1 + sy + xt)M_{23} + (x^2 + y^2)M_{32}\\ &= x(s + y)M_{22} + y(x + t)M_{33} + x(M_{12} + M_{21}) + y(M_{13} + M_{31})\\ &\quad + (x^2 + y^2)(M_{23} + M_{32})\end{aligned}$$

(Here we use the condition that $x^2 + y^2 = sy + xt + 1$.)

$$\begin{aligned}H_6 = H_{BB} &= F_{BR}G_{BR} + F_{BG}G_{BG} + F_{BB}G_{BB} = F_3G_3 + F_5G_5 + F_6G_6\\ &= F_3G_3 + (xF_2 + yF_3)(xG_2 + yG_3) + (F_1 + yF_2 + tF_3)(G_1 + yG_2 + tG_3)\\ &= M_{11} + (x^2 + y^2)M_{22} + (y^2 + t^2 + 1)M_{33}\\ &\quad + y(M_{12} + M_{21}) + t(M_{13} + M_{31}) + y(t + x)(M_{23} + M_{32})\end{aligned}$$

Using the relation of s, x, y, t (i.e., $sy + xt + 1 = x^2 + y^2$), we can easily prove that $H_4 = H_1 + sH_2 + xH_3$, $H_5 = xH_2 + yH_3$, $H_6 = H_1 + yH_2 + tH_3$.

Next we consider the second case, where the two dangling edges to be merged are in the same component (Fig. 2). Obviously, the signature for the new gate H is symmetric. If F below is the signature before the merging operation,

$$\begin{array}{ccccccccc} & & & & F_1 & & & & \\ & & & F_2 & & F_3 & & & \\ & & F_4 & & F_5 & & F_6 & & \\ & F_7 & & F_8 & & F_9 & & F_{10} & \\ F_{11} & & F_{12} & & F_{13} & & F_{14} & & F_{15} \\ & & & & \cdots\cdots & & & & \end{array}$$

then the signature after the merging operation is H below

$$\begin{array}{ccccc} & & F_1+F_4+F_6 & & \\ & F_2+F_7+F_9 & & F_3+F_8+F_{10} & \\ F_4+F_{11}+F_{13} & & F_5+F_{12}+F_{14} & & F_6+F_{13}+F_{15} \\ & & \cdots\cdots & & \end{array}$$

Such an operation preserves the linear recurrence. It follows that H is also a generalized Fibonacci gate with parameters s, x, y, t, and the signature has already been computed.

3 Fibonacci Gates on a Domain of Size 4

We now move on to domain 4. Similarly, through an orthogonal transformation, we know that a ternary domain 4 symmetric signature is tractable when it is of the form $g = \alpha^{\otimes 3} + \beta^{\otimes 3} + \gamma^{\otimes 3} + \delta^{\otimes 3}$ where $\alpha, \beta, \gamma, \delta \in \mathbb{C}^3$ are mutually orthogonal to each other. Let $\alpha = (\alpha_1, \alpha_2, \alpha_3, \alpha_4)^T$, $\beta = (\beta_1, \beta_2, \beta_3, \beta_4)^T$, $\gamma = (\gamma_1, \gamma_2, \gamma_3, \gamma_4)^T$, $\delta = (\delta_1, \delta_2, \delta_3, \delta_4)^T$. Like on domain 3, we focus on the case that there is some $i \in \{1, 2, 3, 4\}$ such that $\alpha_i\beta_i\gamma_i\delta_i \neq 0$. By renaming the i-th domain to be the first domain, we can assume $\alpha_1\beta_1\gamma_1\delta_1 \neq 0$. Then it has the form of

$$g = q\begin{bmatrix}1\\t_1\\t_2\\t_3\end{bmatrix}^{\otimes 3} + r\begin{bmatrix}1\\t_4\\t_5\\t_6\end{bmatrix}^{\otimes 3} + s\begin{bmatrix}1\\t_7\\t_8\\t_9\end{bmatrix}^{\otimes 3} + t\begin{bmatrix}1\\t_{10}\\t_{11}\\t_{12}\end{bmatrix}^{\otimes 3} \quad (5)$$

where the four vectors are orthogonal to each other and $qrst \neq 0$.

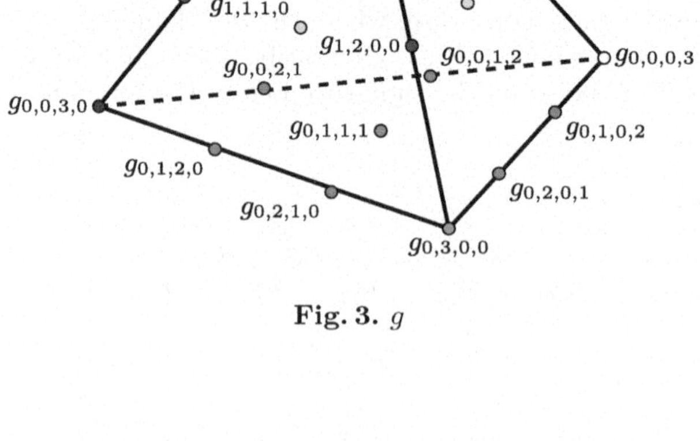

Fig. 3. g

Figure 3 shows a domain 4 signature whose four domains are $\{R, G, B, W\}$ respectively. Here we denote by $g_{w,x,y,z}$ the value of g when w incident edges are assigned color RED, x edges GREEN, y edges BLUE and z edges WHITE. Similar to domain 3 signatures, if it is defined by Eq. 5, then there exist 10 parameters $a, b, c, d, e, f, h, i, j, p$ such that in each **"medium-sized" tetrahedron** (i.e., depth-2 tetrahedron) consisting of 10 numbers in "3 layers",

$$\begin{cases} g_{w-2,x+2,y,z} = g_{w,x,y,z} + a g_{w-1,x+1,y,z} + b g_{w-1,x,y+1,z} + c g_{w-1,x,y,z+1} \\ g_{w-2,x,y+2,z} = g_{w,x,y,z} + d g_{w-1,x+1,y,z} + e g_{w-1,x,y+1,z} + f g_{w-1,x,y,z+1} \\ g_{w-2,x,y,z+2} = g_{w,x,y,z} + h g_{w-1,x+1,y,z} + i g_{w-1,x,y+1,z} + j g_{w-1,x,y,z+1} \\ g_{w-2,x+1,y+1,z} = b g_{w-1,x+1,y,z} + d g_{w-1,x,y+1,z} + p g_{w-1,x,y,z+1} \\ g_{w-2,x+1,y,z+1} = c g_{w-1,x+1,y,z} + p g_{w-1,x,y+1,z} + h g_{w-1,x,y,z+1} \\ g_{w-2,x,y+1,z+1} = p g_{w-1,x+1,y,z} + f g_{w-1,x,y+1,z} + i g_{w-1,x,y,z+1} \end{cases} \tag{6}$$

and the 10 parameters also satisfy

$$\begin{cases} ad + be + cf + 1 = b^2 + d^2 + p^2 \\ dh + ei + fj + 1 = f^2 + i^2 + p^2 \\ ha + ib + jc + 1 = h^2 + c^2 + p^2 \\ p^3 - (bi + cf + dh + 1)p + bfh + cdi = 0 \end{cases} \tag{7}$$

The coefficients could be written as in Fig. 4. It's **a top view of six tetrahedra** each consisting of 4 values. Three of them have a 1 on top and three of them have a 0 on top. Each of these tetrahedra corresponds to a linear relationship mentioned in Eq. 6. For example, the bottom tetrahedron is the one with a "1" on top and a, b, c at the bottom. This corresponds to the linear relation $g_{w-2,x+2,y,z} = g_{w,x,y,z} + a g_{w-1,x+1,y,z} + b g_{w-1,x,y+1,z} + c g_{w-1,x,y,z+1}$ in Eq. 6. Moreover, the "dot product" (i.e., sum of element-wise product) of every two different tetrahedra with a 1 on top is equal to the "dot product" of the tetrahedron (with a 0 on top) between the two with itself. For example, in Fig. 4, for tetrahedron indicated by $1, d, e, f$ and tetrahedron indicated by $1, h, i, j$, the tetrahedron between them is the one indicated by $0, p, f, i$, and $\langle \begin{bmatrix} 1 \\ d \\ e \\ f \end{bmatrix}, \begin{bmatrix} 1 \\ h \\ y \\ j \end{bmatrix} \rangle = \langle \begin{bmatrix} 0 \\ p \\ f \\ i \end{bmatrix}, \begin{bmatrix} 0 \\ p \\ f \\ i \end{bmatrix} \rangle$ is one of the first three equations in 7. Last but not least, there is a **cubic equation** regarding the center parameter p in Eq. 7, $p^3 - (bi + cf + dh + 1)p + bfh + cdi = 0$.

It's worth mentioning that there are also another 3 quadratic relationships (Eq. 8) among the parameters, however, with some calculation, they are incorporated in Eq. 7.

$$\begin{cases} ap + bf + ci = bc + dp + ph \\ bh + di + pj = cp + pf + hi \\ cd + pe + hf = bp + df + pi \end{cases} \tag{8}$$

In Fig. 4, basically Eq. 8 says that the "dot product" of each "corner tetrahedron" and the one in the opposite direction is equal to the "dot product" of the two "side tetrahedra". For example, the first equation shows that tetrahedron $1, a, b, c$ dot

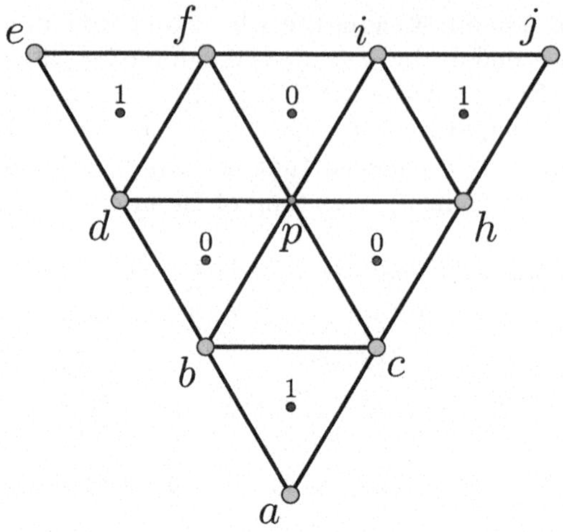

Fig. 4. Parameters of domain 4 fibonacci gate

producting with tetrahedron $0, p, f, i$ produces the same result as tetrahedron $0, b, d, p$ dot producting with $0, c, p, h$. These 3 relations will be useful when we prove the combinatorial algorithm later.

In this way, the 10 parameters have in total four constraints 7 and so the degree of freedom is 6 which is the same as that of a matrix whose four columns are orthogonal to each other written in $\begin{bmatrix} 1 & 1 & 1 & 1 \\ t_1 & t_4 & t_7 & t_{10} \\ t_2 & t_5 & t_8 & t_{11} \\ t_3 & t_6 & t_9 & t_{12} \end{bmatrix}$. With the 10 parameters satisfying Eq. 7, we can form uniquely a n-arity domain 4 signature from the "top" 4 signature values, i.e., $g_{n,0,0,0}$, $g_{n-1,1,0,0}$, $g_{n-1,0,1,0}$, and $g_{n-1,0,0,1}$.

Similar to domain 3, we have the following definition of Fibonacci gates and a corresponding tractable theorem. We leave the verbose proof in Appendix section of the full version paper.

Definition 2. *We call a domain 4 symmetric signature g (arity ≥ 2) a* generalized Fibonacci gate *(with parameters $a, b, c, d, e, f, h, i, j, p$ which satisfy Eq. 7) if it satisfies Eq. 6 $\forall w, 2 \leq w \leq \text{arity}(g)$. Any unary signature is also a Fibonacci gate.*

A set of signatures $\mathcal{F}$ is called generalized Fibonacci *if for some $a, b, c, d, e, f, h, i, j, p \in \mathbb{C}$ which satisfy Eq. 7, every signature in $\mathcal{F}$ is a generalized Fibonacci gate with parameters $a, b, c, d, e, f, h, i, j, p$.*

Theorem 2. *On a domain of size 4, for any finite set of generalized Fibonacci gates $\mathcal{F}$, the Holant problem* Holant$(\mathcal{F})$ *is computable in polynomial time.*

Acknowledgments. I sincerely thank my PhD advisor Jin-Yi Cai. He really ought to be a coauthor for this paper for his contribution, but graciously declined. I am greatly indebted to his guidance and technical help. I also would like to thank Austen Z. Fan for his insightful discussion.

References

1. Backens, M.: A full dichotomy for Holant^c, inspired by quantum computation. SIAM J. Comput. **50**(6), 1739–1799 (2021). https://doi.org/10.1137/20M1311557
2. Backens, M., Goldberg, L.A.: Holant clones and the approximability of conservative Holant problems. ACM Trans. Algorithms **16**(2), 23:1–23:55 (2020). https://doi.org/10.1145/3381425
3. Bulatov, A.A.: A dichotomy theorem for constraint satisfaction problems on a 3-element set. J. ACM (JACM) **53**(1), 66–120 (2006). https://doi.org/10.1145/1120582.1120584
4. Cai, J.Y., Chen, X.: Complexity of counting CSP with complex weights. JACM **64**(3), 19:1–19:39 (2017). https://doi.org/10.1145/2822891
5. Cai, J.Y., Chen, X., Lu, P.: Nonnegative weighted #CSP: an effective complexity dichotomy. SIAM J. Comput. **45**(6), 2177–2198 (2016). https://doi.org/10.1137/15M1032314
6. Cai, J.Y., Guo, H., Williams, T.: The complexity of counting edge colorings and a dichotomy for some higher domain Holant problems. In: 55th IEEE Annual Symposium on Foundations of Computer Science, FOCS 2014, Philadelphia, PA, USA, 18–21 October 2014, pp. 601–610. IEEE Computer Society (2014). https://doi.org/10.1109/FOCS.2014.70
7. Cai, J.Y., Guo, H., Williams, T.: A complete dichotomy rises from the capture of vanishing signatures. SIAM J. Comput. **45**(5), 1671–1728 (2016). https://doi.org/10.1137/15M1049798
8. Cai, J.Y., Lu, P.: Holographic algorithms: From art to science. J. Comput. Syst. Sci. **77**(1), 41–61 (2011). https://doi.org/10.1016/j.jcss.2010.06.005
9. Cai, J.Y., Lu, P., Xia, M.: Holographic algorithms by Fibonacci gates and Holographic reductions for hardness. In: 2008 49th Annual IEEE Symposium on Foundations of Computer Science, pp. 644–653 (2008). https://doi.org/10.1109/FOCS.2008.34
10. Cai, J.Y., Lu, P., Xia, M.: Holant problems and counting CSP. In: Proceedings of the Forty-First Annual ACM Symposium on Theory of Computing, pp. 715–724 (2009). https://doi.org/10.1145/1536414.1536511
11. Cai, J.Y., Lu, P., Xia, M.: Dichotomy for Holant* problems with domain size 3. In: Proceedings of the Twenty-Fourth Annual ACM-SIAM Symposium on Discrete Algorithms, SODA 2013, New Orleans, Louisiana, USA, 6–8 January 2013, pp. 1278–1295. SIAM (2013). https://doi.org/10.1137/1.9781611973105.93
12. Cai, J.Y., Lu, P., Xia, M.: Holographic algorithms with matchgates capture precisely tractable planar #CSP. SIAM J. Comput. **46**(3), 853–889 (2017). https://doi.org/10.1137/16M1073984
13. Dyer, M., Richerby, D.: An effective dichotomy for the counting constraint satisfaction problem. SIAM J. Comput. **42**(3), 1245–1274 (2013). https://doi.org/10.1137/100811258
14. Freedman, M., Lovász, L., Schrijver, A.: Reflection positivity, rank connectivity, and homomorphism of graphs. J. Amer. Math. Soc. **20**(1), 37–51 (2007)

15. Fu, Z., Yang, F., Yin, M.: On blockwise symmetric matchgate signatures and higher domain #CSP. Inf. Comput. **264**, 1–11 (2019). https://doi.org/10.1016/j.ic.2018.09.012
16. Guo, H., Huang, S., Lu, P., Xia, M.: The complexity of weighted Boolean #CSP modulo k. In: 28th International Symposium on Theoretical Aspects of Computer Science, STACS, pp. 249–260 (2011). https://doi.org/10.4230/LIPIcs.STACS.2011.249
17. Guo, H., Lu, P., Valiant, L.G.: The complexity of symmetric Boolean parity Holant problems. SIAM J. Comput. **42**(1), 324–356 (2013). https://doi.org/10.1137/100815530
18. Hell, P., Nešetřil, J.: Graphs and Homomorphisms. Oxford Lecture Series in Mathematics and Its Applications, vol. 28. Oxford University Press (2004)
19. Kowalczyk, M., Cai, J.Y.: Holant problems for 3-regular graphs with complex edge functions. Theory Comput. Syst. **59**(1), 133–158 (2016). https://doi.org/10.1007/s00224-016-9671-7
20. Liu, Y., Fan, A.Z., Cai, J.Y.: Restricted Holant dichotomy on domains 3 and 4. In: COCOA 2023, Proceedings, Part II. Lecture Notes in Computer Science, vol. 14462, pp. 83–96. Springer (2023)
21. Szegedy, B.: Edge coloring models and reflection positivity. J. Am. Math. Soc. **20**(4), 969–988 (2007)
22. Szegedy, B.: Edge coloring models as singular vertex coloring models, pp. 327–336. Bolyai Society Mathematical Studies. Springer, Heidelberg (2010)
23. Valiant, L.G.: Accidental algorithms. In: 47th Annual IEEE Symposium on Foundations of Computer Science, FOCS, pp. 509–517. IEEE (2006). https://doi.org/10.1109/FOCS.2006.7
24. Valiant, L.G.: Holographic algorithms. SIAM J. Comput. **37**(5), 1565–1594 (2008). https://doi.org/10.1137/070682575
25. Xia, M.: Holographic reduction: a domain changed application and its partial converse theorems. Int. J. Softw. Informatics **5**(4), 567–577 (2011). http://www.ijsi.org/ch/reader/view_abstract.aspx?file_no=i109

Labor: Adaptive Lazy Compaction for Learned Index in LSM-Tree

Chunpu Huang, Yukai Huang, Lulu Chen, Rui Zhang, Ming Yan, and Jie Wu(✉)

The School of Computer Science, Fudan University, Shanghai, China
{cphuang20,ykhuang20,llchen18,zhangrui21,myan,jwu}@fudan.edu.cn

Abstract. The log-structured merge tree (LSM-tree) is widely employed in Key-Value (KV) storage to serve data-intensive applications. In order to improve read performance, the LSM-tree has been enhanced by integrating the learned index. However, the index model invalidation resulting from frequent data writes remains unresolved. To tackle this problem, we present a novel LSM-tree named Labor, which simultaneously optimizes both read and write performance. Labor primarily involves a sheet-based lazy compaction mechanism that modifies the compaction granularity from SSTable to segments. The fine-grained compaction not only improves the efficiency of the learned index but also reduces the write amplification. Furthermore, we introduce an adaptive compaction algorithm to achieve a balanced performance for varying read-write patterns. Experimental results on different datasets show that, compared to the advanced LSM-tree (Bourbon), Labor can reduce the latency by 29.6%-37.3% in the mixed read-write case and 39.4%-46.0% in the write-only case, while maintaining read-only performance.

Keywords: Database · Data structure · LSM-tree · Learned index

1 Introduction

Key-Value (KV) storage has been extensively deployed to serve data-intensive applications, such as OLTP, cloud storage, social graphs, etc. Due to the efficiency in writing, the log-structured merge tree (LSM-tree [4,9]) has become a prevalent data structure in KV systems (e.g., LevelDB, Cassandra, RocksDB). It improves write performance by converting random writes to sequential writes and achieves high storage resource utilization by the multi-level structures.

Indexing improves KV system performance by precisely locating KV pairs. The rise of machine learning inspires a prospective work of *learned index* [2,6]. Unfortunately, current researches on the learned index are primarily concentrated on the B-tree [2,6,15]. Only a few studies on integrating the learned index into the LSM tree mainly focus on static scenarios [3], with limited consideration for the update problem of the learned index model in dynamic scenarios.

Nevertheless, even in read-dominated OLTP applications, write operations account for 30% [13], so dynamic scenarios must be considered. The most demanding task is maintaining the learned index model, which needs retraining for the

Y. Chen et al. (Eds.): COCOON 2024, LNCS 15162, pp. 381–393, 2025.
https://doi.org/10.1007/978-981-96-1093-8_32

latest index when new data writing invalidates it. Frequent write operations can lead to excessive and inefficient retraining. And using a neural network to address the update problem is not recommended due to inevitable overhead [8].

Upon thorough analysis, we identify that the primary factor for the inefficient retraining is the compaction operation in LSM-tree. Compaction involves data changes in several SSTables, causing the corresponding models to be updated. Even worse, compaction between multiple levels may happen simultaneously [5]. The frequent compation operations make learned models inaccurate and introduce model retraining overhead. Furthermore, the compaction operation itself can also be a bottleneck as it requires rewriting data to the disk, bringing serious write amplification, especially for the write-intensive workloads. In certain cases, the write amplification may reach 50×, seriously degrading the performance [7].

In this paper, we have abandoned the approach of modifying the learned index model but modifying the underlying LSM-tree system. We propose **Labor**, a new LSM-tree with lazy compaction mechanism based on Bourbon [3] to improve the efficiency of the learned index. Labor employs the lazy compaction approach to minimize redundant read and write I/O in mixed read-write and write scenarios, thereby extending the validity period of the learned index model.

However, it introduces two challenges: 1) Unsatisfactory SLA (Service Level Agreement): lazy compaction may increase the compaction granularity and affect client latency; 2) Dynamic Workloads: although Labor is friendly for the write requests, it decreases read performance for reading some additional space. So we design a *sheet-based* compaction method and an adaptive compaction algorithm in Labor. Firstly, it splits large compaction tasks into small ones based on segments for user experience. Secondly, it can dynamically adjust for workload-specific balance between read and write performance. We implement Labor based on Bourbon which improves levelDB by using the learned index. Experimental results on both real-world and synthetic datasets show Labor outperforms Bourbon by 29.6%-37.3% in mixed read-write and 39.4%-46.0% in write-only cases.

The main contributions are summarized as follows:

- We propose a novel LSM-tree named **Labor** to efficiently address the issue of retraining the learned index model in dynamic scenarios.
- We design a *sheet-based* lazy compaction triggered by the *larger level* for Labor in Sect. 3.2, which can obviously improve the write performance.
- We propose an adaptive compaction algorithm for Labor in Sect. 3.3, along with the learned index to achieve a balance of read and write performance.
- Experiments demonstrate that Labor outperforms other state-of-the-art LSMs (Wisckey and Bourbon) on both real-world and synthetic datasets.

2 Background and Motivation

2.1 Log-Structured Merge Tree

Storage Structure. As Fig. 1(a) shows, the LSM-tree consists of two components: memory and disk. The disk part has multiple $(n + 1)$ levels , denoted

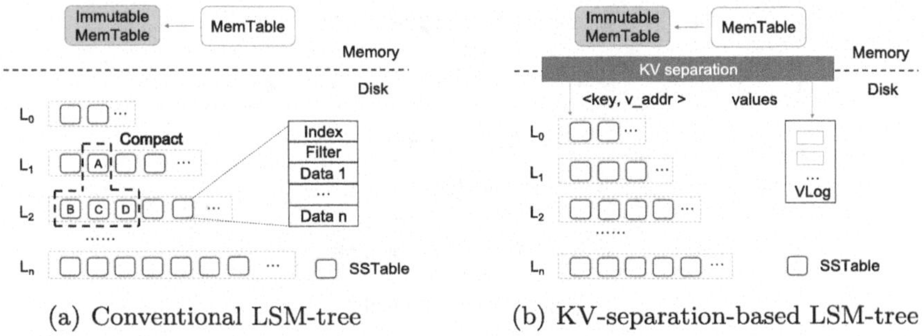

Fig. 1. LSM-tree KV storage solutions.

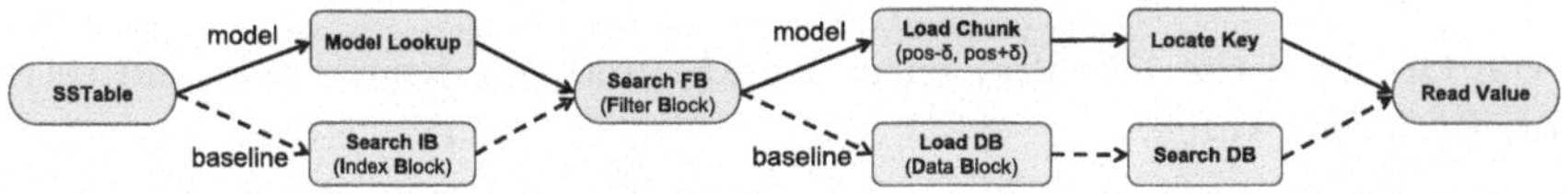

Fig. 2. The detailed lookup paths whether the learned index model for an SSTable exists or not. Solid line: model search path. Dashed line: baseline path.

by $L_0, L_1, \ldots, L_n$. And a higher level L_i usually has 10 times the capacity of a lower level L_{i-1} in LevelDB. Each level holds multiple SSTables (containing index block (IB), filter block (FB), and data blocks (DB)) to store KVs. In memory, there are two buffers: MemTable, where new KVs are first written, and Immutable MemTable, used when MemTable is full and flushed to disk level L_0.

Compaction. LSM-tree merges excessive SSTables through compaction operations. Compaction is triggered to merge the SSTables from L_i to L_{i+1} when the data in L_i is full (where $i \geq 0$). As shown in Fig. 1(a), supposing the key range of SSTable A in L_1 has overlap with SSTable B, C, and D. The compaction will first read SSTable A from L_1 and B, C, D from L_2, then sort the KVs, creates new SSTables, and write them back to L_2. So compaction incurs the write amplification problem. The lower level is called the *larger level*, as its capacity is larger than the upper level. For example, L_2 is the *larger level* compared to L_1.

KV Separation. To mitigate I/O amplification, KV separation is introduced (e.g., Wisckey [10] in Fig. 1(b)), storing keys in the LSM-tree and values in an append-only log. This reduces the size of LSM-tree and alleviates read/write amplification. Our method, Labor, is also based on this concept.

Learned Index for LSM-Tree. The learned index is a machine learning-based index structure reflecting data patterns [6]. Figure 2 illustrates the read process from an SSTable that may contain a learned index model. First, the IB and FB of the target SSTable are loaded into memory. If this SSTable is learned, it will look up the model, then use FB to confirm the key's existence and locate its value. Otherwise, it rolls back to the baseline: Perform a binary search in IB to

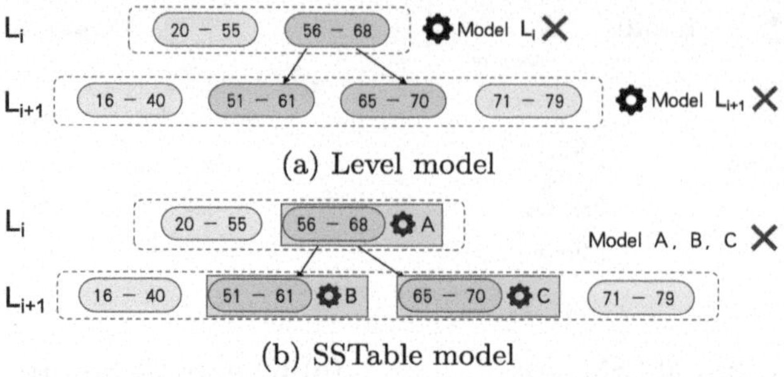

Fig. 3. The compation between two levels.

lock the DB, load the DB after checking FB, find the desired key, and retrieve its corresponding value.

The Reason of Learned Index for LSM-Tree. In-memory data indexing significantly impacts lookup time, sometimes equaling data access time. A learned index can halve this latency, even in non-memory scenarios like Optane SSDs where it can improve performance by 1.8 times [3]. Hence, it is useful for LSM-tree read optimization. Although write operations prompt a retraining of the model due to data changes, these changes mainly impact memory and the upper levels. The large, unchanging SSTables at the bottom are ideal for the learned index due to their immutability.

2.2 Motivation

However, simply choosing relatively static SSTable files to construct the learned index is insufficient [2,3,6], as they are typically cold data that is infrequently accessed. We should select SSTables that are often accessed and exhibit a moderate data change frequency. This requires minimizing the model failures and frequent rebuilds caused by data changes, to maximize the benefits of the learned index.

Upon analyzing, we find the main reason for index model failure is tied to the compaction operation. When compaction alters data in SSTables, previous models no longer accurately locate the value for a given key in a new query.

Figure 3 details how compaction affects the model availability. First, the file whose key range is $56-68$ in level L_i is selected and merged with $51-61$ and $65-70$ in level L_{i+1}. The compaction operation will load these files, sort all KVs via merge, and finally build new SSTables to write into L_{i+1}. When compaction is finished, data of levels L_i and L_{i+1} undergoes modifications, reducing the accuracy of the previous index model. Note that non-overlapping ranges $51-56$ and $68-70$ are rewritten to the L_{i+1}, causing write amplification overhead.

The learning granularity of learned index model can be on the entire level (e.g., Model L_i, L_{i+1} in Fig. 3(a)) or a specific SSTable (e.g., Model A, B, C in Fig. 3(b)). ① For level-based models, the compaction from $56-68$ to $51-61$ and

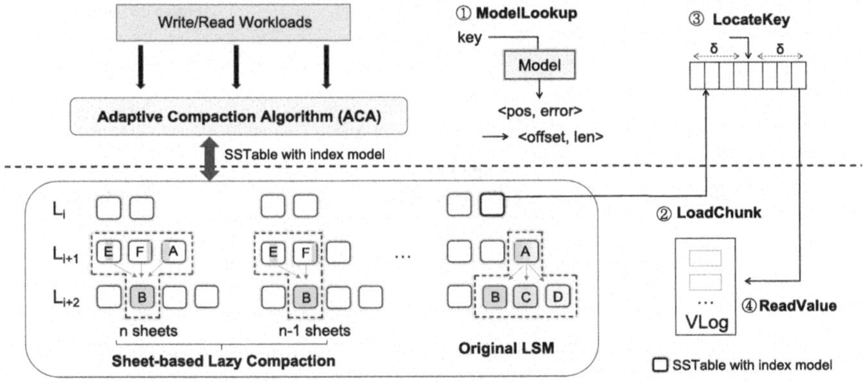

Fig. 4. The system architecture of Labor.

65 – 70 invalidates Model L_i and L_{i+1}. Before compaction, according to Model L_i, the position of 56 – 68 should be in the L_i level. But after compaction, 56 – 68 is merged into the L_{i+1} level, rendering the old Model L_i ineffective, and the same applies to Model L_{i+1}. ② For SSTable-based models, the compaction from 56 – 68 to 51 – 61 and 65 – 70 only invalidates Model A, B, and C, and not the entire Model of L_i, L_{i+1}. Models based on unaffected SSTables can still work.

Therefore, we must consider the effect of compaction on the learned index model in LSM-tree. This motivates us to design Labor, a *larger-level-driven* compaction strategy that delays data changes and lowers write amplification. We also designed an adaptive algorithm to mitigate read performance impacts.

3 System Architecture

Figure 4 shows the system architecture of Labor. It consists of three modules: 1) The learned index based on SSTables for reading steps (①②③④); 2) A *sheet-based* lazy compaction mechanism; 3) An adaptive compaction algorithm (**ACA**) based on real workloads. The learned index is the foundation of improving read performance. The lazy compaction mechanism is the key module, which reduces the compaction granularity, alleviates write amplification and slows down the model update frequency, thereby improving the efficiency of the learned index. **ACA** adjusts the number of *sheets* according to the real-time read/write ratio and throughput, thereby weakening the negative impact on reading. In the read-only case, Labor falls back to the original LSM-tree compaction mechanism.

3.1 Learned Index

First, we introduce the learned index structure for reading, using a lightweight linear regression model, learning only at the relatively static SSTable granularity.

Model. To minimize the influence of front-end and back-end traffic competition, the model of the learned index should be lightweight, memory-friendly, and fast

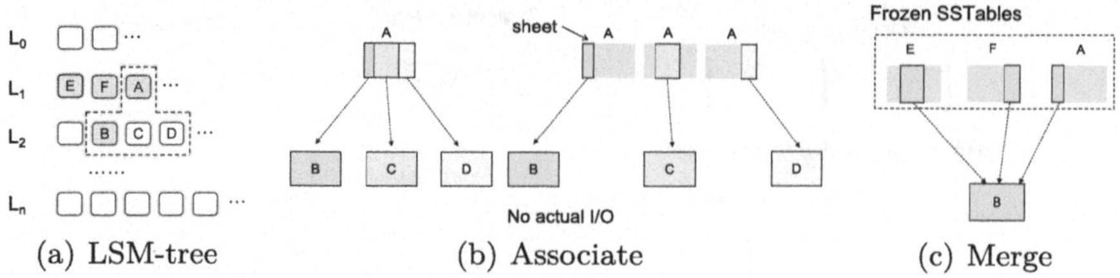

Fig. 5. The Sheet-based Lazy Compaction.

training. Thus, we use the Greedy-PLR (piecewise linear regression) algorithm [14] as the model for the learned index.

Learning Granularity. We can train the model in each level of the LSM-tree or each SSTable file of the level. However, learning at the level is not suitable, since the compaction operations will involve adjacent levels. Even a tiny compaction operation will lead to the models of at least two levels to be retrained. Therefore, learning at the SSTable granularity is optimal. Compared with the entire level, the average survival time of SSTables is longer. In addition, to maximize the efficiency of the learned index, the CBA algorithm [3] is used here to selectively build the model on SSTable files with longer lifetimes.

Reading Steps. The reading steps are shown in Fig. 4. ① ModelLookup: Labor looks up the model of the candidate SSTable to obtain the position (*pos*) and error boundary (*error*) of the target key. ② LoadChunk: Load the byte range containing the key. ③ LocateKey: The key may be in the middle of the loaded chunk, which is the predicted position of the model. ④ ReadValue: Read the value associated with the key from Value log.

3.2 Sheet-Based Lazy Compaction

To minimize the model retraining overhead caused by compaction, we design the *sheet-based* lazy compaction in Labor. To prevent excessive lazy compaction operations affecting read/write traffic, we break down compaction granularity from SSTable to small segments. As shown in Fig. 5, Labor can be divided into two phases: 1) Associate, link upper small segments with SSTables of the *larger level*; 2) Merge, execute the actual compaction I/O.

Associate. When the compaction starts, it first selects an SSTable of the level which is almost full as the compaction target file (S_u) and then get the set of SSTables of next level (S_l, generally two or more) that overlap with the S_u. For each in S_l, according to its overlapping range with S_u, divide S_u into *segments*, and *associate* these segments with the corresponding SSTables in S_l. Then, we call such a segment that has an association with the SSTable of the next level a *sheet*. The upper level SSTable that is about to be merged but delayed is called *frozen SSTable*, which is removed from LSM-tree management, but still accessible by the model. Note that there is no actual I/O performed, just a kind of link between a segment and an SSTable at the *larger level*.

Merge. Merge is the next phase of associate. The key indicators for deciding whether to merge are the size and number of *sheets*. We set two thresholds: ① *TSheetNum*, the number of *sheets* (default 10); ② *TMergeSizeRatio*(R_s), for SSTable of a *larger level*, calculate the ratio of the total *sheet* size pointing to it, compared to its own size (default 1.0). The two thresholds prevent a large number of *sheets* from affecting reading. The actual compaction I/O will execute when *sheets* hit the threshold. The compaction is triggered by the SSTable in the *larger level*, sorts and merges the SSTable with the associated *sheets* in memory, then writes the newly generated SSTable back to the *larger level*.

An Example. Fig. 5 depicts the *sheet-based* lazy compaction details. Figure 5(a) shows the structure in the LSM-tree, A, B, C, and D are the four SSTables involved in the original compaction operation. A is in the upper level, and B, C, and D are the *larger-level* files with overlapping key ranges with A. Figure 5(b) shows the operation of the first associate phase: divide A into segments according to the key ranges of the three files in the *larger level*, then link these segments with B, C, and D respectively, and freeze A. In this way, three *sheets* are generated that point to B, C, and D from A. The next phase is merge, as shown in Fig. 5(c), E, F, and A represent frozen SSTables with *sheets* in the upper level, and their *sheets* all point to the same file B. After meeting the threshold conditions, the merge operation from E, F, A to B will be triggered. By combining the three merge operations of E, F, A to B into one operation, the write amplification is greatly reduced.

Analysis. The design of *sheet-based* lazy compaction avoids additional I/O, mitigates write amplification, extends the lifetime of the SSTable, and reduces the cost of model retraining. At the same time, the extra overhead of this design is controllable. In the first association phase, there is no actual read or write I/O is performed, only fast metadata processing in memory, so the association phase is lightweight and has minimal overhead. And during the next merge phase, only the segments related to the merge in the upper level, not the entire SSTable, are loaded into memory, so the design will not cause excessive space utilization.

3.3 Adaptive Compaction Algorithm

In reality, read/write rate often fluctuates, but the *sheet-based* lazy compaction is more suitable for mixed read/write or a large proportion of writes. However, read-only cases suffer lower performance as Labor requires read more space. Therefore, we propose the adaptive compaction algorithm (**ACA**), dynamically adjusting the compaction strategy for real workloads.

It is crucial to understand that both software and hardware design contribute to unbalanced reading and writing. The write amplification in LSM-tree will cause the write throughput to be lower than the read throughput; the inherent characteristics of hard disks also lead to unbalanced reading and writing. Labor can reduce write amplification, but cannot change the characteristics of the underlying hardware. Therefore, the actual read and write throughput should be adjusted to achieve the highest total throughput with a fixed read/write ratio.

TSheetNum. Its value range is: $0 \leq TSheetNum \leq n$, where n is the maximum number of *sheets* set by the system. The larger n is, the deeper the degree of delay compaction is. In a unique scenario, the system encounters a situation where the number of *sheets* reaches the maximum value of n, with each *sheet* pointing to a distinct SSTable file. In such cases, even if the *TMergeSizeRatio* does not meet the compaction threshold, it becomes necessary to trigger compaction. Therefore this value of *TSheetNum* is used for global control and is not used as the primary adjustment parameter in ACA.

Algorithm 1: Adaptive Compaction Algorithm (ACA)

Require: ① increase R_s: $R_{s1} = k_1 \cdot R_s$; ② reduce R_s: $R_{s2} = k_2 \cdot R_s$. $\Delta T, \Delta w, \Delta r$ represents changes of I/O consumption time, the write generated by compaction, the read brought by the Labor design(sheet) and compaction, and th_w, th_r represent the read and write throughput.

$k_1 = (R_s + \Delta R_s)/R_s$
$k_2 = (R_s - \Delta R_s)/R_s$
$tmp = 0, k' = 1$
for each k in k_1, k_2 **do**
  $\Delta T = \Delta w/th_w + \Delta r/th_r$
  if $\Delta T < tmp$ **then**
    $tmp = \Delta T, k' = k$ /* chose a minimum ΔT below 0 */
  end if
end for
$R_s = R_s \cdot k'$

***TMergeSizeRatio*(R_s).** A key point of *sheet-based* lazy compaction is merge threshold *TMergeSizeRatio*(R_s), which is the ratio between the size of *sheets* and the associated SSTable at the *larger level*. When the compaction I/O increases, the smaller the R_s value, the more frequently the compaction operation is performed. At this time, the read amplification is lower and the write amplification is higher, so the read performance is better. On the contrary, if we set larger R_s, the presence of more *sheets* affects the read performance, although the merge of more segments is delayed making the write amplification smaller.

ACA. Therefore, a fixed R_s value is not suitable for practical read/write workloads. We propose ACA to tune read and write performance in Labor. The key idea is to adjust the value of R_s based on the real-time read/write rate and performance of the underlying device, aiming to enhance the overall system performance. When the value of R_s is small enough, we can consider that the compaction falls back to the original not-delayed compaction mechanism.

As shown in Algorithm 1, ACA makes three adjustment strategies for R_s according to the information of the previous period: 1) maintain the same R_s;

2) appropriately increase R_s, i.e., $R_{s1} = R_s + \Delta R_s$; 3) appropriately reduce R_s, i.e., $R_{s2} = R_s - \Delta R_s$ (line 1-2). For R_{s1} and R_{s2}, calculate ΔT respectively, i.e., $\Delta T = \Delta w/th_w + \Delta r/th_r$, where $\Delta T, \Delta w, \Delta r$ represents changes of the read and write I/O consumption time caused by the change of R_s, the write generated by compaction, the read brought by the Labor design (*sheet*) and compaction, and th_w, th_r represent the read and write throughput respectively (line 3-5). If ΔT is negative, then choose the value of R_s (R_{s1} or R_{s2}) that minimizes ΔT (line 6-9). We update R_s in this way to minimize read and write I/O consumption, so as to achieve a balance between read and write (line 10).

4 Evaluation

4.1 Experimental Setup

We conduct the experiments with several datasets and perform 10 millon operations per experiment. The keys and values are set as 16B and 64B, respectively.

Hardware Settings. The machine is with a 64-core Intel(R) Xeon(R) Gold 6242 CPU @ 2.80GHz processor, 128GB DRAM, 1.8TB SATA HDD, running Ubuntu 18.04.6 LTS with Linux 4.15.0-196-generic kernel and ext4 file system.

Datasets. We mainly use two real-world ones: AR [1] with 33.5M KVs and OSM [11] with 21.9M KVs. We also use four synthetic datasets: Linear, Seg1%, Normal, Seg10%, each containing 64M KVs. Linear keys are consecutive, while Seg1% and Seg10% have gaps inserted every 100 and 10 keys respectively. Normal dataset keys follow the standard normal distribution.

KV Systems. We compare Labor with two LSM-tree systems. ① Wisckey [10], an LSM-tree system with KV separation based on levelDB. ② Bourbon [3], integrating the static learned index into Wisckey to improve read performance.

Learned Index Model. PLR model (Sect. 3.1) as the learned index.

4.2 Performance of Labor Without ACA

We assess the performance of the *sheet-based* lazy compaction, namely **Labor w/o ACA**, using the average latency of all read/write operations under various workloads including read-only, write-only, and mixed read-write.

Lazy compaction Mechanism can Effectively Reduce Write Latency. Fig. 6(a) shows results on AR dataset. The abscissa Read and Write represent read-only and write-only, respectively, and Mix-50% represents that the mixed ratio of read/write operations is 1:1. Bourbon has 27.9% less latency than Wisckey in the read-only case, but higher latency than Wisckey in mixed read-write and write-only cases due to the internal learned index. In comparison, Labor without ACA reduces latency by 33.2% and 42.9% compared to Bourbon in mixed read-write and write-only cases, respectively. Because the *sheet-based* lazy compaction mechanism can reduce write amplification and model learning overhead. But since Labor without ACA introduces an additional read overhead, its latency is higher than Bourbon in the read-only case, which we will optimize by ACA. And the performance of Labor without ACA is better than Wisckey because of the improvement of the learned index in read-only case. As shown in Fig. 6(b), we also observed a similar phenomenon on the OSM dataset, where Labor without ACA saves 27.4% and 47.7% latency respectively in mixed read-write and write-only compared to Bourbon. And since the OSM dataset is smaller, its overall performance is a bit better than AR. Meanwhile, as shown in Fig. 6(c-f), similar experimental results on the synthetic datasets (Linear, Seg1%, Normal, Seg10%) further validate the effects of our method.

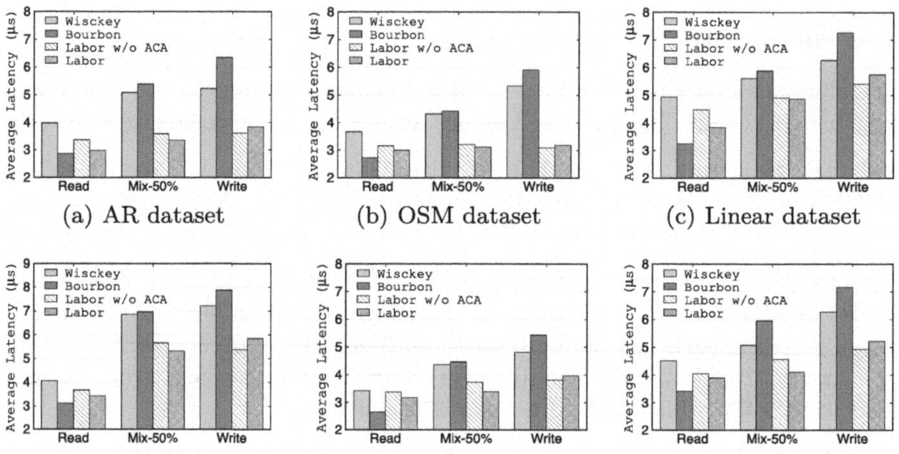

(a) AR dataset (b) OSM dataset (c) Linear dataset

(d) Seg1% dataset (e) Normal dataset (f) Seg10% dataset

Fig. 6. The average latency on real-world and synthetic datasets.

4.3 Performance of Labor

This section evaluates the performance of Labor under different workloads. The experimental conditions are consistent with Sect. 4.2.

ACA Efficiently Achieves a Balance Between Read and Write Performance. The main goal of ACA is to balance read/write performance by adjusting the number of *sheets* based on specific read/write ratios, etc. Figure 6 also depicts the comparison of Labor with the basic Labor without ACA. Taking Fig. 6(a) as an example, under the read-only workload, the performance of Labor is 11.6% higher than Labor without ACA, 25.2% higher than Wisckey, and comparable to the performance of Bourbon. In the case of mixed read-write, Labor outperforms Labor without ACA and saves 37.3% latency than Bourbon. Although in the write-only case, the performance of Labor is lower than Labor without ACA due to the additional overhead of the adaptive algorithm, it still reduces the average latency by 39.4% compared to Bourbon. Similarly, as shown in Fig. 6(b), on OSM dataset, Labor outperforms Labor without ACA and saves 18.5% latency than Wisckey in read-only workload, and also comparable to Bourbon. In mixed read-write workload, Labor also outperforms Labor without ACA, and saves 29.6% latency compared to Bourbon. And in the write-only situation, the average latency is reduced by 46.0% compared to Bourbon. The experimental results of the synthetic datasets in Fig. 6(c-f) further demonstrate the ability of Labor to balance the read and write performance.

4.4 Lifetime Analysis

We analyze SSTable lifetime in level 2, 3 and 4 on AR dataset, setting the read/write ratio is 1:1 (Mix-50%). 500 iterative experiments are conducted.

Labor can extend SSTable Lifetime on Different Levels Through Adaptive Lazy Compaction. The results are shown in Fig. 7. Compared to Bourbon, Labor extended the SSTable lifetime in different levels. Taking Fig. 7(a) as an example, the average lifetime of SSTables in level 2 is 27.6 s for Bourbon. After using Labor's adaptive lazy compaction mechanism, the average lifetime of SSTables at level 2 has increased to 45.1 s. Similarly, as shown in Fig. 7(b) and Fig. 7(c), the average SSTable lifetimes in levels 3 and 4 have increased from 134.5 to 229.1 s, and from 1213.3 to 2262.3 s, respectively. That indicates our compaction mechanism effectively extends SSTable lifetime, thus prolonging the learned index model's validity based on SSTable.

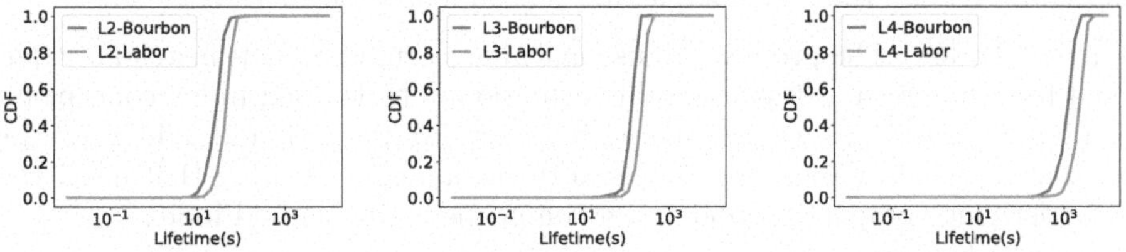

(a) Level 2 (b) Level 3 (c) Level 4

Fig. 7. The lifetime changes of SSTables in level 2, 3 and 4 on AR dataset.

Table 1. The percentage (%) decrease over Bourbon in learning and compaction time.

dataset	AR	OSM	Linear	Seg1%	Normal	Seg10%
learning	19.2	15.8	13.6	16.1	17.8	20.0
compaction	42.2	31.5	21.0	31.1	29.1	38.3

4.5 Time Saving Analysis

We compare the percentage improvement of Labor over Bourbon in model learning and compaction time, i.e., the percentage of time decrease, under a read/write ratio of 1:1. Here, 70% of the total time is spent on compaction and learning.

Labor can Significantly Reduce the Compaction Time and has a Positive Effect on the Learned Index Model. As shown in Table 1, compared to Bourbon, Labor reduces model learning time and compaction time by 19.6% and 42.2% on the AR dataset, and 15.8% and 31.5% on the OSM dataset, respectively. Therefore, we affirm that Labor's lazy compaction design notably minimizes compaction-induced latency and positively impacts the learned index model. Table 1 also depicts the results on synthetic datasets. Compared with Bourbon, the model learning time is reduced by 13.6%-20.0% and the compaction time is reduced by 21.0%-38.3%, respectively. We also note that lower dataset continuity leads to more time savings, for the higher cost of learned index and compaction overhead with complex data, highlighting Labor's improvements.

5 Related Work

Compaction. Research has been conducted to decrease compaction overhead in LSM-tree. Spooky [5] uses a combination of partial and full preemptive merges, while dCompaction [12] constructs a virtual SSTable to delay compaction. Dostoevskii [4] uses lazy leveling to reduce merge operations. In contrast, our method refines and adapts the lazy compaction scheme based on real workload.

Learned Index. The proposal to use machine learning models instead of traditional indexing [6] has drawn notable attention. The learned index concept has been implemented in LSM-tree-based KVs like Bourbon [3] and TridentKV [9], etc. Our approach, Labor, is orthogonal to them, demonstrating that modifying the underlying compaction policy can complement the learned index.

6 Conclusion

This paper introduces Labor, an adaptive compaction mechanism for the learned index in LSM-trees, designed to reduce model invalidation issues due to data updates or inserts. Specifically, we propose a *sheet-based* lazy compaction scheme to improve the efficiency of the learned index in mixed read-write and write scenarios. Furthermore, to balance read-write performance, we design an adaptive compaction algorithm to adaptively adjust the *sheet* number based on the read/write rate, etc. Experimental results show that Labor outperforms the state-of-the-art works (Bourbon and Wisckey) with different read-write patterns.

Acknowledgements. The work of this paper is supported by National Key Research and Development Program of China (2021YFC3300600).

References

1. Amazon: Reviews dataset. https://registry.opendata.aws/amazon-reviews/
2. Beutel, A., Kraska, T., Chi, E., Dean, J., Polyzotis, N.: A machine learning approach to databases indexes. In: ML Systems Workshop. NIPS (2017)
3. Dai, Y., Xu, Y., Ganesan, A., Alagappan, R., et al.: From {WiscKey} to bourbon: A learned index for {Log-Structured} merge trees. In: 14th USENIX Symposium on Operating Systems Design and Implementation (OSDI 20), pp. 155–171 (2020)
4. Dayan, N., Idreos, S.: Dostoevsky: Better space-time trade-offs for lsm-tree based key-value stores via adaptive removal of superfluous merging. In: Proceedings of the 2018 International Conference on Management of Data, pp. 505–520 (2018)
5. Dayan, N., Weiss, T., Dashevsky, S., et al.: Spooky: granulating LSM-tree compactions correctly. Proc. VLDB Endow. **15**(11), 3071–3084 (2022)
6. Kraska, T., Beutel, A., et al.: The case for learned index structures. In: Proceedings of the 2018 International Conference On Management of Data, pp. 489–504 (2018)
7. Li, J., Yue, Y., Wang, W.: GHStore: a high performance global hash based key-value store. In: Bhattacharya, A., Lee Mong Li, J., Agrawal, D., Reddy, P.K., Mohania, M., Mondal, A., Goyal, V., Uday Kiran, R. (eds.) Database Systems for Advanced Applications: 27th International Conference, DASFAA 2022, Virtual Event, April 11–14, 2022, Proceedings, Part I, pp. 493–508. Springer International Publishing, Cham (2022). https://doi.org/10.1007/978-3-031-00123-9_39
8. Livni, R., Shalev-Shwartz, S., Shamir, O.: On the computational efficiency of training neural networks. Adv. Neural Inform. Process. Syst. **27** (2014)
9. Lu, K., Zhao, N., Wan, J., Fei, C., Zhao, W., Deng, T.: Tridentkv: A read-optimized lsm-tree based kv store via adaptive indexing and space-efficient partitioning. IEEE Trans. Parallel Distrib. Syst. **33**(8), 1953–1966 (2021)
10. Lu, L., Pillai, T.S., et al.: Wisckey: separating keys from values in ssd-conscious storage. ACM Trans. Storage (TOS) **13**(1), 1–28 (2017)
11. openstreetmap: https://www.openstreetmap.org/#map=4/38.01/-95.84
12. Pan, F., Yue, Y., Xiong, J.: dcompaction: Delayed compaction for the lsm-tree. Int. J. Parallel Prog. **45**(6), 1310–1325 (2017)
13. Rehrmann, R., et al.: Sharing opportunities for oltp workloads in different isolation levels. Proc. VLDB Endowment **13**(10), 1696–1708 (2020)
14. Xie, Q., Pang, C., Zhou, X.: Maximum error-bounded piecewise linear representation for online stream approximation. VLDB J. **23**, 915–937 (2014)
15. Zhang, Z., Jin, P.Q., Wang, X.L., Lv, Y.Q., Wan, S.H., Xie, X.K.: Colin: a cache-conscious dynamic learned index with high read/write performance. J. Comput. Sci. Technol. **36**(4), 721–740 (2021)

A Distributed Algorithm for Rumor Blocking on Social Networks

Ruidong Yan[1], Zhenhua Guo[1](✉), Yaqian Zhao[1], Rengang Li[1], and Xingjian Ding[2]

[1] Shandong Massive Information Technology Research Institute, Jinan 205101, China
guozhenhua_hl@163.com
[2] College of Computer Science, Beijing University of Technology, Beijing 100124, China
dxj@bjut.edu.cn

Abstract. The optimization of influence on social networks has received significant attention, yet there has been limited research on blocking rumors targeted at specific individuals. In this paper, we introduce the *Rumor Blocking for Target Users* (RBTU) problem, which aims to identify a set of influential seed nodes to minimize the number of target users affected by the rumor. To address scalability challenges, we propose a two-stage distributed algorithm that utilizes community partitioning and reverse sampling techniques. In the first stage, the original network is partitioned into non-overlapping communities. This division serves to reduce the complexity of the problem and allows for parallel processing. In the second stage, reverse sampling is performed independently for each community. This process involves the strategic selection of seed nodes to restrict the spread of the rumor. Additionally, we provide both theoretical and empirical analyses of our proposed algorithm. Theoretically, we prove that the objective function is submodular, which enables the greedy algorithm to achieve an approximate ratio of $(1 - 1/e)$. Through numerical simulations, we demonstrate that our algorithm outperforms other comparison methods.

Keywords: social network · distributed optimization algorithm · rumor blocking · reverse sampling · combinatorial optimization

1 Introduction

Social Networks (SNs) play a crucial role in information dissemination. Users share new inventions, opinions, and knowledge with friends via social networks. However, misinformation like rumors can spread on SNs, causing economic loss and social panic if uncontrolled. For instance, during COVID-19, rumors about

This work is supported by project ZR2021QF073 from the Shandong Provincial Natural Science Foundation.

Y. Chen et al. (Eds.): COCOON 2024, LNCS 15162, pp. 394–406, 2025.
https://doi.org/10.1007/978-981-96-1093-8_33

certain vegetables and medicines preventing infection caused their prices to skyrocket. When they proved ineffective against COVID-19, it brought attention to the rumor blocking problem on social networks [1].

Currently, rumor blocking has been extensively studied by academia and industry. These studies can be roughly classified into two categories. The first category modifies the network structure by removing key nodes or edges to limit rumor spread [11]. These methods partially destroy the network structure, rendering them less ideal for practical applications. By contrast, methods in the second category spread positive information, such as truth from governments or official organizations, without modifying the network structure [3].

While existing studies focus on rumor blocking, they face two challenges. The **first challenge** is poor scalability due to centralized algorithms. Most centralized methods search for the optimal solution across the entire network, resulting in an exponential increase in computational overhead and memory footprint as the network size grows [4]. The **second challenge** is that rumor control comes with a delay. The centralized algorithm iteratively selects seed nodes in the network from the global optimal perspective, ignoring local features. This results in slow seed selection and delay in rumor control. These challenges limit the application of centralized methods in practice.

In this paper, **we consider a novel rumor blocking problem for specific target users**. Specifically, assume each user has a potential value in marketing. One may pay more attention to users with higher potential values (i.e., *target users*). These target users are highly influential and authoritative, with high probability of activating other users. Once target users are activated (infected) by the rumor, they will cause more users to accept the rumor. Therefore, it is necessary to prevent rumor seeds (i.e., rumor sources) from activating target users by selecting a set of seeds containing a fixed number of users to spread positive information and limit further rumor propagation. In summary, our goal is to minimize the number of target users activated by the rumor in a social network when selecting a fixed number of positive information seed nodes.

To address the challenges of centralized algorithms, we focus on a distributed algorithm which leverages the "divide-and-conquer" principle to reduce computational complexity and improve algorithm performance. Specifically, our algorithm comprises two stages. In the first stage, accounting for the community property of social networks, it divides the original network into several subnetworks (i.e., communities) and limits the spread of influence (rumor or truth) within each community. In the second stage, based on community partition results and target user as well as rumor seeds distribution, it designs a distributed algorithm for parallelly selecting positive information seeds via reverse sampling technique in communities to minimize the number of target users activated by the rumor. The main contributions are as follows.

- We propose a novel rumor blocking problem for target users on social network, which is NP-hard.
- We design a two-stage approach including *community partition* and *seed selection*. In the community partition stage, a primitive network is divided into

some subnetworks. In the seed selection stage, a distributed seed selection strategy based on reverse sampling is proposed to improve the algorithm's performance. Furthermore, we provide a theoretical analysis of the approach.
- We evaluate the algorithm's effectiveness on real datasets. Experimental results show the proposed algorithm outperforms the comparison methods.

Roadmap: The related work is shown in Sect. 2. The influence propagation model and the problem statement are in Sect. 3. The algorithms and the theoretical analysis are in Sect. 4. The experiments are in Sect. 5. Finally, the conclusions and future work are in Sect. 6.

2 Related Work

Distributed Influence Diffusion. Kempe et al. [4] propose the *Influence Maximization* (IM) problem which selects k seed nodes from a network to maximize the influence spread (the number of nodes accepting the influence). They present the *Independent Cascade* (IC) model and the *Linear Threshold* (LT) model. Kim et al. [5] develop an *Independent Path Algorithm* (IPA) that approximates the influence spread by multiple independent influence paths. These valid paths are independent of each other, so IPA can be executed in parallel. Song et al. [8] solve the IM by mining top-k influential nodes based on community detection. To this end, they propose a *Community-based Greedy Algorithm* (CGA). CGA first divides a network into some communities, and then finds top-k influential nodes within communities. These algorithms focus on evaluating influential nodes instead of rumor blocking.

Reverse Sampling Technique. The greedy algorithms rely on Monte Carlo simulation to calculate influence spread, which is time-consuming. To improve computational efficiency, Borges et al. [2] propose a reverse sampling technique including two steps. The first step is to randomly select a node on the transposition network (i.e., if there is an edge (u, v) in original network, then an edge (v, u) exists in transposition network.) and traverse other nodes. As a result, the *Reverse Reachable* (RR) set of the node is obtained. Repeat the above operations sufficient times. The second step is to pick the node that can cover the most RR sets in each iteration as a seed node, according to the greedy strategy. They theoretically prove that if a node appears in multiple RR sets at the same time, it should become a seed node with high probability. However, they do not provide an optimal solution for the number of RR sets.

Distributed Rumor Blocking. To our knowledge, centralized rumor blocking has been well studied while distributed versions have received less attention. There are some attempts to explore distributed algorithms [6,9]. Tong et al. [9] abstract rumor blocking as a 2-player non-cooperative game. They define influence propagation as a valid utility function that is submodular. Theoretical analysis reveals social the utility of any Nash equilibrium in the game is at least half of the optimal social utility. Ni et al. [6] propose a two-step approach with a 1/2 performance guarantee based on community detection. They detect communities and allocate seed nodes within communities in the first stage. They select seed nodes to maximize the expected number of nodes infected by seeds in the second stage. However, these approaches do not consider the target users. Therefore, they cannot be directly applied to this paper.

3 Preliminaries and Problem Statement

A directed social network is denoted by $G = (V, E, p)$, where V is the node set, E is the edge set and p is the propagation probability set. $v \in V$ represents a user, $(u, v) \in E$ denotes the relationship between u and v, $p_{uv} \in p$ indicates the probability that u activates v. $\mathcal{T}$ is a target set. b is a positive integer budget.

Influence Propagation Model. We utilize the *Competitive Independent Cascade* (CIC) [3] model for influence propagation. In the model, two competitive cascades, rumor $\mathcal{R}$ and positive truth $\mathcal{P}$, disseminate simultaneously in a network. $S_{\mathcal{R}}$ and $S_{\mathcal{P}}$ are initially *active* nodes (seeds) in $\mathcal{R}$ and $\mathcal{P}$, respectively. Each node u in $\mathcal{R}$ or $\mathcal{P}$ spreads rumor or positive truth similarly to the IC model. Influence propagation unfolds in discrete time steps. Specifically, if u becomes *active* in $\mathcal{R}$ or $\mathcal{P}$ at time step t, it has a single chance to activate each currently *inactive* neighbor v through edge (u, v) with probability p_{uv}. The state of v is determined at time step $t + 1$. The process continues until no newly activated nodes in either cascade. Furthermore, a node's state can only be switched from the *inactive* to *active*. Note that the positive cascade $\mathcal{P}$ takes effect when a node is activated by *active* nodes in both cascades $\mathcal{R}$ and $\mathcal{P}$ simultaneously.

Problem 1. **Rumor Blocking for Target Users (RBTU)**
RBTU aims to identify a seed set $S_{\mathcal{P}} \subseteq V \backslash \{S_{\mathcal{R}} \cup \mathcal{T}\}$ with a budget constraint b that spreads positive truth to minimize the number of nodes in $\mathcal{T}$ activated by $S_{\mathcal{R}}$ in a social network G.

Let $\sigma(S_{\mathcal{R}}, \emptyset)$ be the number of nodes in $\mathcal{T}$ activated by $S_{\mathcal{R}}$ without $S_{\mathcal{P}}$. Let $\sigma(S_{\mathcal{P}})$ be the number of nodes in $\mathcal{T}$ activated by $S_{\mathcal{P}}$. Let $\sigma(S_{\mathcal{R}})$ be the final number of nodes in $\mathcal{T}$ activated by $S_{\mathcal{R}}$. The problem 1 is formalized as follows

$$\sigma(S_{\mathcal{R}}) = \sigma(S_{\mathcal{R}}, \emptyset) - \sigma(S_{\mathcal{P}}). \tag{1}$$

Inspired by [3,12], minimizing $\sigma(S_{\mathcal{R}})$ is equivalent to selecting a set $S_{\mathcal{P}}$ that maximizes $\sigma(S_{\mathcal{P}})$. Therefore, we convert the Problem 1 into the Problem 2.

Problem 2. **Positive Truth Maximization (PTM)**
PTM selects a seed set $S_{\mathcal{P}}$ with b nodes such that $\sigma(S_{\mathcal{P}})$ is maximized, i.e.,

$$S^* = \arg \max_{S_{\mathcal{P}} \subseteq V \backslash \{S_{\mathcal{R}} \cup \mathcal{T}\}, |S_{\mathcal{P}}| = b} \sigma(S_{\mathcal{P}}). \tag{2}$$

Theorem 1. ***Problem** 1 and **Problem** 2 are NP-hard under the CIC model.*

4 Algorithm

We propose a distributed strategy including: 1) **community partition**, i.e., a network is divided into m communities and the influence is limited within each community, and 2) **seed selection**, i.e., we select b positive seed nodes by reverse sampling technique from each community for maximizing $\sigma(S_{\mathcal{P}})$ (See Fig. 1).

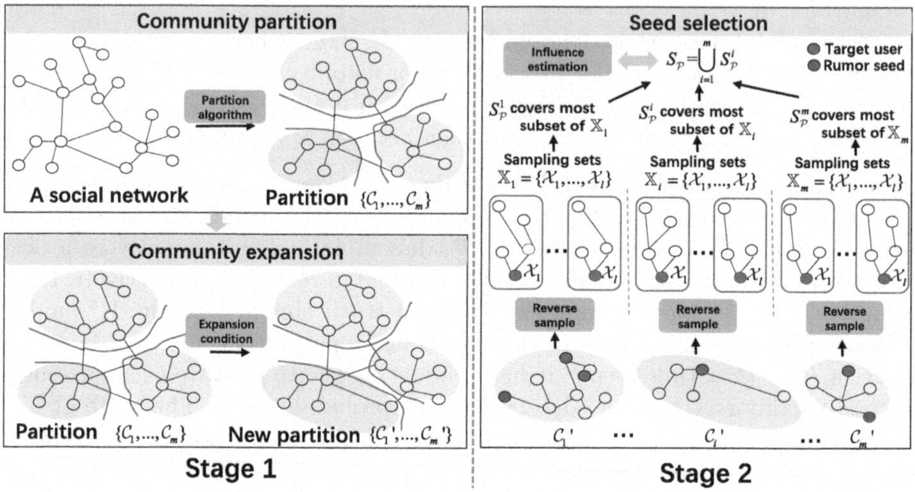

Fig. 1. The overview of our distributed strategy for rumor blocking.

4.1 Community Partition Stage

The community structure is the essential feature of SNs. Users within a community have frequent contact and thus are more likely to influence each other, while users across communities have less contact and are less likely to influence each other [6,8]. Therefore, this suggests selecting seed nodes in parallel within communities may be a feasible approximation to the whole network.

Community Partition. We employ [7] to identify non-overlapping communities due to low time complexity and suitability for large-scale networks.

Definition 1. *(Non-overlapping Community) Given a network $G = (V, E, p)$, the non-overlapping community partition $\mathcal{C} = \{\mathcal{C}_1, \mathcal{C}_2, ..., \mathcal{C}_m\}$ satisfies the conditions: $\bigcup_{1\leq j\leq m} \mathcal{C}_j = V$ and $\forall\ i \neq j \in [m]$, $\mathcal{C}_i \cap \mathcal{C}_j = \emptyset$, where m is the number of communities and $[m] = \{1, 2, ..., m\}$.*

It is worth mentioning that the quality of communities generated by [7] is not perfect, and these communities are often small and dispersed. This is because it only considers the connections between nodes (network structure) and ignores

Algorithm 1: Community Expansion

Input: $G = (V, E, p)$, $\mathcal{C} = \{\mathcal{C}_1, \mathcal{C}_2, ..., \mathcal{C}_m\}$, θ, η.
Output: $\mathcal{C}' = \{\mathcal{C}_1', \mathcal{C}_2', ..., \mathcal{C}_m'\}$.
$\hat{c} \leftarrow |V|/m$;
for *each* $\mathcal{C}_i$ **do**
  if $|\mathcal{C}_i| \leq \hat{c}$ **then**
    for *each* $u \in \mathcal{C}_i$ **do**
      compute $maxPath(u, v)$ based on θ where $v \notin \mathcal{C}_i$;
      obtain K nodes v_1, ..., $v_K \in \mathcal{C} \backslash \mathcal{C}_i$ and $maxPath(u, v_1) \geq \theta$, ..., $maxPath(u, v_K) \geq \theta$;
      if $\sum_{k=1}^{K} maxPath(u, v_k) \geq \eta$ add $\{v_1, ..., v_k, ..., v_K\}$ into $\mathcal{C}_i$;
      $\mathcal{C}_i' \leftarrow \mathcal{C}_i \cup \{v_1, ..., v_k, ..., v_K\}$;
      remove $\{v_1, ..., v_K\}$ from $\mathcal{C}_j$ where $j \neq i$;
return $\mathcal{C}' = \{\mathcal{C}_1', \mathcal{C}_2', ..., \mathcal{C}_m'\}$

the information diffusion between them. To address this issue, we propose a community expansion strategy and obtain high-quality communities.

Community Expansion. We attempt to achieve community expansion by an influence threshold-based approach. Based on the influence propagation model, the exact probability that node v is activated by u depends on all paths $Path_G(u, v)$ between u and v in network G. However, identifying all paths between two nodes in a directed graph is #P-hard. Therefore, it is not feasible to calculate precise activation probability. Instead, inspired by [13], we approximate the exact probability by computing the *Maximum Influence Path* (MIP) $maxPath(u, v)$ from u to v. Furthermore, $maxPath(u, v)$ is equivalent to the shortest path between them if we set $-\log p_{uv}$ for edge (u, v).

Algorithm 1 shows the steps of community expansion. The input requires a network $G = (V, E, p)$, a community partition $\mathcal{C} = \{\mathcal{C}_1, \mathcal{C}_2, ..., \mathcal{C}_m\}$, parameters θ and η. It first calculates the average size of the community size $\hat{c} = |V|/m$[1], and then extends those small communities (e.g., $|\mathcal{C}_i| \leq \hat{c}$, $i \in [m]$). Specifically, for each $u \in \mathcal{C}_i$, we compute the shortest path $maxPath(u, v)$ which $maxPath(u, v) \geq \theta$ from u to v based on influence threshold θ where $v \notin \mathcal{C}_i$ to filter out many MIP with weak influence. If there are K nodes ($v_k \notin \mathcal{C}_i$) satisfy the above condition and $\sum_{k=1}^{K} maxPath(u, v_k) \geq \eta$, thus we add these K nodes into $\mathcal{C}_i$. At the same time, we remove these K nodes from the original communities $\mathcal{C}_j$ where $j \neq i$. Finally, we obtain the new community partition $\mathcal{C}' = \{\mathcal{C}_1', \mathcal{C}_2', ..., \mathcal{C}_m'\}$. Each $\mathcal{C}_i'$ is called newly expanded community.

Discussion: The benefits of community partition are twofold. Firstly, it divides the entire network into independent subnetworks, known as communities, and subsequent seed selection is performed within each community rather than the

[1] $|V|$ denotes the cardinality of set V.

entire network. Furthermore, influence computation and seed selection can be performed concurrently by exploiting the independence of each community.

4.2 Seed Selection Stage

In this part, we introduce the definition of *Sampling Set* based on *Reverse Sampling Technique*, and then select the seeds effectively from sampling sets in each community. We also provide theoretical analysis of proposed algorithms.

Sampling Set. We introduce a definition of *sampling set* which is generated by *reverse sampling technique* in a community $C_i^{'}$ where $i \in [m]$.

Definition 2. *(Sampling set). Given a directed subnetwork* $C_i^{'} = (V^{(i)}, E^{(i)}, p^{(i)})$, *a* ***sampling set*** *is generated according to the following random rules: (1) One sampling set* $\mathcal{X} = (V(\mathcal{X}), E(\mathcal{X}), p(\mathcal{X}))$ *is a special subgraph of* $C_i^{'}$, *where* $V(\mathcal{X}) \subseteq V^{(i)}$, $E(\mathcal{X}) \subseteq E^{(i)}$. *(2)* $E(\mathcal{X})$ *is generated as follows. For each edge* $e \in E^{(i)}$ *in* $C_i^{'}$, *we generate a random number* $\xi_e \in [0, 1]$. *If* $\xi_e \leq p_e$ *where* $e \in E^{(i)}$, *then we add this edge* e *to the* $E(\mathcal{X})$ *and let* $p_e = 1$ *in the sampling set* $\mathcal{X}$. *(3) Let* Ω *be all possible samples space of* $C_i^{'}$. *Obviously, there are* $2^{|E^{(i)}|}$ *samples in* Ω.

We now define the probability of a sampling set $\mathcal{X}$ as follow: $Pro[\mathcal{X}] = \prod_{e \in E(\mathcal{X})} p_e \times \prod_{e \in E^{(i)} \setminus E(\mathcal{X})} (1 - p_e)$. Notice that one sampling set $\mathcal{X}$ is a deterministic IC network since all the $p_e = 1$ for all $e \in E(\mathcal{X})$. However, the traditional IC network is randomized. In other words, the randomness of the IC network is reflected in the process of generating sampling sets. Once a sampling set has been generated, the information propagating over it is deterministic. This is precisely the key to the success of the reverse sampling technique [2]. Existing studies such as [4] have shown that an influence performs a random propagation process in $C_i^{'}$ is equivalent to performing a deterministic propagation process in a sample set $\mathcal{X}$.

Reverse Sampling in Communities. We present the sampling set generation algorithm (Algorithm 2). Unlike reverse sampling [2], it conducts distributed sampling by considering the target user and rumor seed distributions within individual communities, rather than sampling the entire network. Instead of selecting a random node, it utilizes a target user as the initial sampling node.

The algorithm takes as input a community partition $C^{'} = \{C_1^{'}, ..., C_m^{'}\}$ with edge weight set p, a target user set T, a queue $\mathcal{Q}$ for recording reverse sampling nodes, and a random variable $\zeta \in (0, 1)$. It outputs a sample set $\mathbb{X}$ on communities. The algorithm mainly comprises three steps: 1) it first initializes $V(\mathcal{X}) = \emptyset$, $E(\mathcal{X}) = \emptyset$, $\mathcal{Q} = \emptyset$, and $\mathbb{X} = \emptyset$ in line 1; 2) it executes loops in parallel from lines 2-19 to generate sampling results; 3) it finally outputs $\mathbb{X}$ in m communities $C^{'}$.

Let's discuss the second step in detail. Since we have divided $G = (V, E, p)$ and got $C^{'} = \{C_1^{'}, ..., C_m^{'}\}$. Therefore, the algorithm performs concurrently reverse sampling on these communities. First, to reduce unnecessary computational overhead, it checks if the target user v belongs to the current community $C_i^{'}$. If yes, sampling is performed on $C_i^{'}$; otherwise, nothing is done for the community. When

Algorithm 2: Reverse Sampling in Communities

Input: $\mathcal{C}' = \{\mathcal{C}_1', ..., \mathcal{C}_m'\}$, $\mathcal{T}$, $\mathcal{Q}$, ζ.
Output: $\mathbb{X}$.
1 $V(\mathcal{X}) = \emptyset$, $E(\mathcal{X}) = \emptyset$, $\mathcal{Q} = \emptyset$, $\mathbb{X} = \emptyset$;
2 **for** *each* $\mathcal{C}_i' \subseteq \mathcal{C}'$ *and* $i \in [m]$ ***parallel*** **do**
3   **if** $\mathcal{T} \cap \mathcal{C}_i' = M$ *and* $M \neq \emptyset$ **then**
4   Generate the transposition graph of $\mathcal{C}_i'$, i.e., $Trans(\mathcal{C}_i')$;
5   **for** *each* $v \in M$ **do**
6   **for** *each* $(v, u) \in Trans(\mathcal{C}_i')$ *and* $u \notin S_{\mathcal{R}}$ **do**
7   **if** *a random variable* $\zeta \leq p_{uv}$ **then**
8                     $V(\mathcal{X}) \leftarrow V(\mathcal{X}) \cup \{u\}$, $E(\mathcal{X}) \leftarrow E(\mathcal{X}) \cup \{(u, v)\}$;
9                     $\mathcal{X} = (V(\mathcal{X}), E(\mathcal{X}))$;
10                     $\mathcal{Q} \leftarrow \mathcal{Q} \cup \{u\}$;
11   **while** $\mathcal{Q} \neq \emptyset$ **do**
12   **for** *each node* $u \in V(\mathcal{X})$ **do**
13   **if** $u \notin S_{\mathcal{R}}$ *and* u *can traverse in* $Trans(\mathcal{C}_i')$ **then**
14   Repeat line 7-10 to get a sample set of u, i.e., $u[\mathcal{X}] = (V_u(\mathcal{X}), E_u(\mathcal{X}))$;
15   Update $V(\mathcal{X}) \leftarrow V(\mathcal{X}) \cup V_u(\mathcal{X})$;
16   Update $E(\mathcal{X}) \leftarrow E(\mathcal{X}) \cup E_u(\mathcal{X})$;
17   Update $\mathcal{Q} \leftarrow \mathcal{Q} \cup V_u(\mathcal{X})$;
18   Delete node u from $\mathcal{Q}$, i.e., $\mathcal{Q} \leftarrow \mathcal{Q} \backslash u$;
19   Generate l reverse samples of community $\mathcal{C}_i^i$, *i.e.*, $\mathbb{X}_i = \{\mathcal{X}_1, ..., \mathcal{X}_l\}$;
20 Generate the reverse samples of a network, i.e., $\mathbb{X} = \{\mathbb{X}_1, ..., \mathbb{X}_m\}$;
21 **return** $\mathbb{X}$.

the conditions $\mathcal{T} \cap \mathcal{C}_i' = M$ and $M \neq \emptyset$ are true, indicating target users belong to $\mathcal{C}_i'$, the algorithm generates the transposition graph $Trans(\mathcal{C}_i')$ for $\mathcal{C}_i'$ in line 4. For each $v \in M$ in $\mathcal{C}_i'$, a reverse sampling set starting at node v is generated by maintaining a queue $\mathcal{Q}$ from line 5 to 18. For each edge (v, u) in $Trans(\mathcal{C}_i')$, if a random variable $\zeta \leq p_{uv}$, node u is added to $V(\mathcal{X})$, edge (u, v) to $E(\mathcal{X})$, and u to queue $\mathcal{Q}$. Note that $\mathcal{Q}$ is the set of nodes reachable from v. If $\mathcal{Q}$ is not empty, the algorithm continues sampling by using similar operations, and updating $V(\mathcal{X})$, $E(\mathcal{X})$, and $\mathcal{Q}$. If $\mathcal{Q}$ is empty or a rumor seed node is encountered, sampling cannot continue and the algorithm stops.

Based on the sampling set $\mathcal{X}$, we define the probability that a node v in target set $\mathcal{T}$ can be activated by the seed set $S_{\mathcal{P}}$ on a community $\mathcal{C}_i'$ as follow

$$F_{(\mathcal{X}, \mathcal{C}_i')}(v, S_{\mathcal{P}}) = \begin{cases} 1, \text{ if } v \text{ is activated successfully by } S_{\mathcal{P}} \text{ in } \mathcal{X} \\ 0, \text{ otherwise.} \end{cases} \tag{3}$$

Algorithm 3: Seed Nodes Selection in Communities

Input: Community partition $\mathcal{C}' = \{\mathcal{C}_1', ..., \mathcal{C}_m'\}$, a sample set $\{\mathbb{X}_1, ..., \mathbb{X}_m\}$ and $\mathbb{X}_i = \{\mathcal{X}_1, ..., \mathcal{X}_l\}$ for $i \in [m]$, a budget set $\{b_1, ..., b_m\}$.
Output: Seed set $S_{\mathcal{P}}$.

```
for each S_P^i do
    S_P^i ← ∅;
for each C_i parallel do
    for 1 ≤ z ≤ b_i do
        Let v be the node covering the most subsets of X_i;
        S_P^i ← S_P^i ∪ {v};
        Remove the v from each X_j and j ∈ [l] ;
        z = z + 1;
S_P = ∪_{i=1}^m S_P^i;
return S_P
```

Now we rewrite the **Problem** 2 as follow

$$\sigma(S_{\mathcal{P}}) = \sum_{i=1}^{m} \sum_{\mathcal{X} \in \Omega} \sum_{v \in T} Pro[\mathcal{X}] \cdot F_{(\mathcal{X}, \mathcal{C}_i')}(v, S_{\mathcal{P}}),$$

where $Pro[\mathcal{X}]$ indicates the probability of sample $\mathcal{X}$. Our goal is to maximize $\sigma(S_{\mathcal{P}})$ with restriction of $|S_{\mathcal{P}}| \leq b$, i.e.,

$$\max\{\sigma(S_{\mathcal{P}}) : |S_{\mathcal{P}}| \leq b\}. \tag{4}$$

For the function $\sigma(S_{\mathcal{P}})$, we have the following theoretical result.

Theorem 2. *The function $\sigma(S_{\mathcal{P}})$ is monotone non-decreasing and submodular with respect to $S_{\mathcal{P}}$ in a sample $\mathcal{X}$ on each community $\mathcal{C}_i'$, where $i \in [m]$.*

According to the properties of the reverse sampling set $\mathcal{X}$, for a set $S_{\mathcal{P}} \subseteq V \backslash \{S_{\mathcal{R}} \cup T\}$ in a community $\mathcal{C}_i'$ where $i \in [m]$, we define

$$\mathbb{I}_{(\mathcal{X}, \mathcal{C}_i')}(S_{\mathcal{P}}, V(\mathcal{X})) = \begin{cases} 1, \text{ if } S_{\mathcal{P}} \cap V(\mathcal{X}) \neq \emptyset \text{ in } \mathcal{C}_i' \\ 0, \text{ otherwise.} \end{cases} \tag{5}$$

Based on the equation (2), we have following theorem.

Theorem 3. *For a seed set $S_{\mathcal{P}} \subseteq V \backslash \{S_{\mathcal{R}} \cup T\}$ and a sampling set $\mathcal{X}$, $\sigma(S_{\mathcal{P}})$ in a community $\mathcal{C}_i'$ is $|T_i| \cdot \mathbb{E}[\mathbb{I}_{(\mathcal{X}, \mathcal{C}_i')}(S_{\mathcal{P}}, V(\mathcal{X}))]$, where $|T_i|$ denote the number of target nodes in $\mathcal{C}_i'$ and $i \in [m]$.*

Seeds Selection and Estimating $\sigma(S_{\mathcal{P}})$. Assume that we have a RR set $\mathbb{X}_i = \{\mathcal{X}_1, ..., \mathcal{X}_l\}$ including l samples returned by Algorithm 2 in $\mathcal{C}_i'$. To facilitate seed selection, we allocate a budget b_i to each community $\mathcal{C}_i'$ based on

---

Algorithm 4: Estimation $\sigma(S_{\mathcal{P}})$ in Communities

---

Input: Community partition $\mathcal{C}' = \{\mathcal{C}_1', ..., \mathcal{C}_m'\}$, a sample set $\{\mathbb{X}_1, ..., \mathbb{X}_m\}$ and $\mathbb{X}_i = \{\mathcal{X}_1, ..., \mathcal{X}_l\}$ for $i \in [m]$, a parameter ρ.

Output: Influence spread $\sigma(S_{\mathcal{P}})$.

for *each* $\mathcal{C}_i'$ *and* $\mathcal{C}_i' \cap \mathcal{T} \neq \emptyset$ ***parallel*** **do**
- $H_i = \log_2 |\mathcal{T}_i|$;
- **for** $1 \le t \le H_i$ **do**
  - $Cent_t \leftarrow \frac{|\mathcal{T}_i|}{2^t}$;
  - $S_{\mathcal{P}}^i \leftarrow$ Call Algorithm 3;
  - **if** $|\mathcal{T}_i| \cdot \frac{g(S_{\mathcal{P}}^i, \mathbb{X}_i)}{l} \ge (1+\rho) \cdot Cent_t$ **then**
    - $\sigma_i(S_{\mathcal{P}}) \leftarrow \frac{|\mathcal{T}_i| \cdot g(S_{\mathcal{P}}^i, \mathbb{X}_i)}{l \cdot (1+\rho)}$;
    - Break;
  - **else**
    - $|\mathcal{T}_i| \leftarrow Cent_t$;
  - $t = t + 1$;

$\sigma(S_{\mathcal{P}}) = \sum_{i=1}^{m} \sigma_i(S_{\mathcal{P}})$;

return $\sigma(S_{\mathcal{P}})$

---

the distribution of the target users $\mathcal{T}$ in communities such that $b = \sum_{i=1}^{m} b_i$ and $b_i = (|\mathcal{T}_i|/|\mathcal{T}|)b$. For a set $S_{\mathcal{P}} \subseteq V \backslash \{S_{\mathcal{R}} \cup \mathcal{T}\}$ and $\mathbb{X}_i$, let $g(S_{\mathcal{P}}^i, \mathbb{X}_i)$ be the number of target users activated by $S_{\mathcal{P}}^i$ (seed nodes are in $\mathcal{C}_i'$) in $\mathcal{C}_i'$, i.e., $g(S_{\mathcal{P}}^i, \mathbb{X}_i) = \sum_{j=1}^{l} \mathbb{I}_{(\mathcal{X}_j, \mathcal{C}_i')}(S_{\mathcal{P}}^i, V(\mathcal{X}_j))$. We consider the following problem.

Problem 3. Choose a seed set $S_{\mathcal{P}}^i$ and $|S_{\mathcal{P}}^i| \le b_i$ in $\mathcal{C}_i'$ that maximizes $g(S_{\mathcal{P}}^i, \mathbb{X}_i)$.

Problem 3 is a classic set cover set problem and using greedy *Hill Climbing* algorithm that can provide a $(1 - 1/e)$-approximation. We propose the seed selection algorithm (Algorithm 3).

In the algorithm, it first initializes the seed set $S_{\mathcal{P}}^i \leftarrow \emptyset$ for each community $\mathcal{C}_i'$. Secondly, b_i nodes (covering the most subsets of $\mathbb{X}_i$) are selected as seed nodes in each community in parallel. Finally, m seed subsets are aggregated into a global seed set $S_{\mathcal{P}}$. We have the following theory.

Theorem 4. *Let $S_{\mathcal{P}}^i$ and $S_{\mathcal{P}}^{i,*}$ be the seed set returned by algorithm 3 and the optimal seed set in community $\mathcal{C}_i'$. For a given a sample set $\mathbb{X}_i$, the following approximate ratio holds: $g(S_{\mathcal{P}}^i, \mathbb{X}_i) \ge (1 - 1/e) \cdot g(S_{\mathcal{P}}^{i,*}, \mathbb{X}_i)$ where $i \in [m]$.*

Estimating $\sigma(S_{\mathcal{P}})$. According to theorem 3, $g(S_{\mathcal{P}}^i, \mathbb{X}_i)$ can be calculated by $|\mathcal{T}_i| \cdot \mathbb{E}[\mathbb{I}_{(\mathcal{X}, \mathcal{C}_i')}(S_{\mathcal{P}}, V(\mathcal{X}))]$ in community $\mathcal{C}_i'$. Furthermore, We can utilize theorem 4 to estimate $g(S_{\mathcal{P}}^i, \mathbb{X}_i)$ with a constant factor guarantee in community $\mathcal{C}_i'$ when we select $S_{\mathcal{P}}^i$ as the seed set. By cumulatively adding each community's $g(S_{\mathcal{P}}^i, \mathbb{X}_i)$, we can achieve the estimation of $\sigma(S_{\mathcal{P}})$ using theorem 3. That is to say,

$$\sigma(S_{\mathcal{P}}) = \sum_{i=1}^{m} g(S_{\mathcal{P}}^i, \mathbb{X}_i) = \sum_{i=1}^{m} |\mathcal{T}_i| \cdot \mathbb{E}[\mathbb{I}_{(\mathcal{X}, \mathcal{C}_i^{'})}(S_{\mathcal{P}}^i, V(\mathcal{X})]. \tag{6}$$

On the basis of the above analysis, our objective is to search for $g(S_{\mathcal{P}}^i, \mathbb{X}_i)$ in each community $\mathcal{C}_i^{'}$ from the interval $[0, |\mathcal{T}_i|]$ (The number of target users in community $\mathcal{C}_i^{'}$ is between 0 and $|\mathcal{T}_i|$). Specifically, we adopt the half-interval search. The centre of the interval Cen_t is compared with $g(S_{\mathcal{P}}^i, \mathbb{X}_i)$ by an approximation parameter ρ during the t-iteration search. We find that $g(S_{\mathcal{P}}^i, \mathbb{X}_i)$ can be effectively estimated if $g(S_{\mathcal{P}}^i, \mathbb{X}_i)$ and Cen_t are close enough. The estimation of $\sigma(S_{\mathcal{P}})$ is shown in Algorithm 4.

In Algorithm 4, we perform the following operations in parallel. First, we determine the number of execution iterations H_i of the half-interval search based on the number of target users $\mathcal{T}_i$ in $\mathcal{C}_i^{'}$. Second, we determine the centre Cen_t of the half-interval search. Then we employ Algorithm 3 to select a seed set $S_{\mathcal{P}}^i$. We compare $|\mathcal{T}_i| \cdot \frac{g(S_{\mathcal{P}}^i, \mathbb{X}_i)}{l}$ and $(1 + \rho) \cdot Cent_t$ w.r.t the parameter ρ. If the

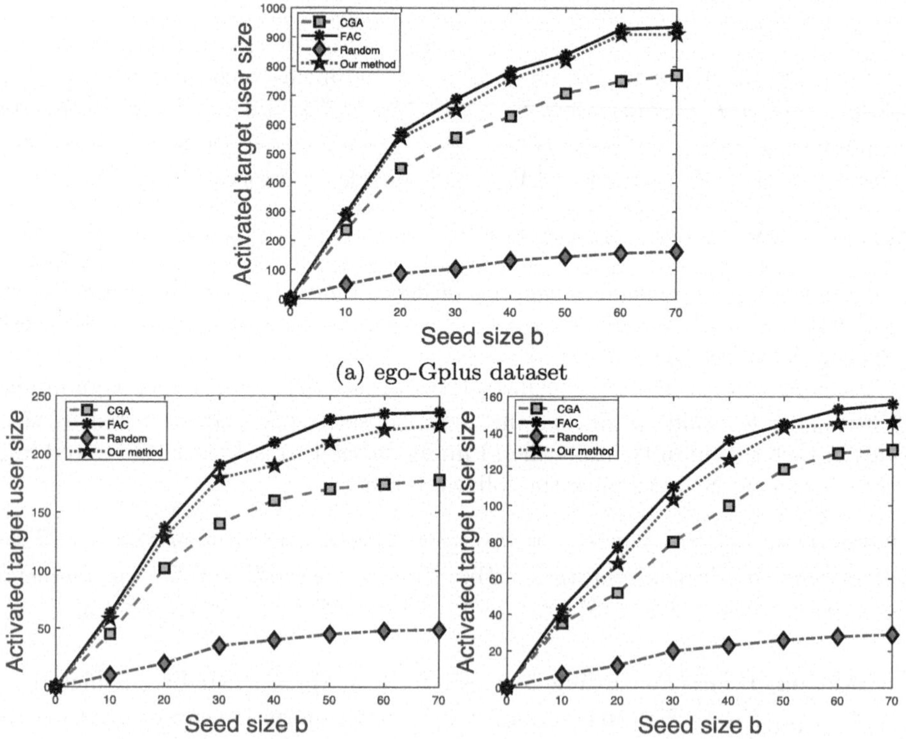

(a) ego-Gplus dataset

(b) wiki-Rfa dataset

(c) act-mooc dataset

Fig. 2. Compare with other existing algorithm on datasets.

former is greater than or equal to the latter, let $\sigma_i(S_{\mathcal{P}}) = \frac{|T_i| \cdot g(S_{\mathcal{P}}^i, \mathbb{X}_i)}{l \cdot (1+\rho)}$. Finally, the influence spread can be calculated by $\sigma(S_{\mathcal{P}}) = \sum_{i=1}^{m} \sigma_i(S_{\mathcal{P}})$.

5 Numerical Experiments

We evaluate the proposed algorithm and compare it with existing methods (e.g., CGA [8], FAC [12]) on different datasets (including ego-Gplus, wiki-Rfa, and act-mooc) collected from https://snap.stanford.edu/data/. Given a directed network, 2% and 5% of the nodes are initialized as rumor seeds and target users, respectively. To balance accuracy and running time, we set $\eta = 0.4$, $\theta = 0.35$ and $\rho = 0.15$ in experiments. The number of sampling sets is set according to [10]. We first show the relationship between the number of positive influence seeds and the number of target users. Second, we show the *Speedup* between distributed and centralized algorithms, i.e., *Speedup* $=$ *running time of centraliezed method/running time of distributed method.*

The Relationship Between b and the Number of Target Users: We set $b \in [0, 70]$ and experimental results are shown in Fig. 2. In the figure, the horizontal axis is b and the vertical axis is the number of target users activated by positive truth seeds. Random means we randomly choose seeds in each community. We note that: 1) We can get performance that is extremely close to a centralised method using our distributed method. Initially, there is a significant rise in the quantity of activated target users when the number of seeds increases, while it gradually rises once the quantity of seeds hits a particular range. This confirms the submodular function's diminishing marginal returns; 2) Since our approach engages more target users than the random strategy, it is better.

Speedup Between our Method and Centralized Algorithm: In this part, we present the speedup achieved by our distributed algorithm compared to the centralized algorithm on different datasets. Specifically, on the act-mooc dataset, the speedup is 4.56. On the wik-Rfa dataset, the speedup is 7.89. On the ego-Gplus dataset, the speedup is 18.63. The aforementioned results demonstrate that our distributed method exhibits a greater acceleration effect for large-scale networks compared to the centralized algorithm.

6 Conclusions

In this paper, we propose a novel rumor blocking problem for target users. To this end, we design a two-stage method based on community partition and reverse sampling technique. The main contributions include: 1) we formalize the problem and provide theoretical analysis, and 2) we compare our method with other existing algorithms. The experimental results show that the proposed method outperforms state-of-the-art methods. We will focus on improving algorithm's performance using deep learning-based approaches in the future.

References

1. Bora, S., Rao, S.: Target node protection from rumours in online social networks. In: Verma, P., Charan, C., Fernando, X., Ganesan, S. (eds.) Advances in Data Computing, Communication and Security: Proceedings of I3CS2021, pp. 581–589. Springer Nature Singapore, Singapore (2022). https://doi.org/10.1007/978-981-16-8403-6_53
2. Borgs, C., Brautbar, M., Chayes, J., Lucier, B.: Maximizing social influence in nearly optimal time. In: Proceedings of the Twenty-fifth Annual ACM-siam Symposium on Discrete Algorithms, pp. 946–957. SIAM (2014)
3. Budak, C., Agrawal, D., El Abbadi, A.: Limiting the spread of misinformation in social networks. In: Proceedings of the 20th International Conference on World Wide Web, pp. 665–674 (2011)
4. Kempe, D., Kleinberg, J., Tardos, É.: Maximizing the spread of influence through a social network. In: Proceedings of the Ninth ACM SIGKDD International Conference on Knowledge Discovery And Data Mining, pp. 137–146 (2003)
5. Kim, J., Kim, S.K., Yu, H.: Scalable and parallelizable processing of influence maximization for large-scale social networks? In: 2013 IEEE 29th International Conference on Data Engineering (ICDE), pp. 266–277. IEEE (2013)
6. Ni, Q., Guo, J., Huang, C., Wu, W.: Community-based rumor blocking maximization in social networks: Algorithms and analysis. Theoret. Comput. Sci. **840**, 257–269 (2020)
7. Raghavan, U.N., Albert, R., Kumara, S.: Near linear time algorithm to detect community structures in large-scale networks. Phys. Rev. E **76**(3), 036106 (2007)
8. Song, G., Zhou, X., Wang, Y., Xie, K.: Influence maximization on large-scale mobile social network: a divide-and-conquer method. IEEE Trans. Parallel Distrib. Syst. **26**(5), 1379–1392 (2014)
9. Tong, G., Wu, W., Du, D.Z.: Distributed rumor blocking with multiple positive cascades. IEEE Trans. Comput. Social Syst. **5**(2), 468–480 (2018)
10. Tong, G., et al.: An efficient randomized algorithm for rumor blocking in online social networks. IEEE Trans. Netw. Sci. Eng. **7**(2), 845–854 (2017)
11. Wang, S., Zhao, X., Chen, Y., Li, Z., Zhang, K., Xia, J.: Negative influence minimizing by blocking nodes in social networks. In: Proceedings of the 17th AAAI Conference on Late-Breaking Developments in the Field of Artificial Intelligence, pp. 134–136 (2013)
12. Yan, R., Li, Y., Li, D., Wang, Y., Zhu, Y., Wu, W.: A stochastic algorithm based on reverse sampling technique to fight against the cyberbullying. ACM Trans. Knowl. Discov. Data (TKDD) **15**(4), 1–22 (2021)
13. Yang, D.N., Hung, H.J., Lee, W.C., Chen, W.: Maximizing acceptance probability for active friending in online social networks. In: Proceedings of the 19th ACM SIGKDD International Conference on Knowledge Discovery and Data Mining, pp. 713–721 (2013)

Convex-Area-Wise Linear Regression and Algorithms for Data Analysis

Bohan Lyu[1,2(✉)] and Jianzhong Li[1,2]

[1] Harbin Institute of Technology, Harbin, Heilongjiang, China
lijzh@hit.edu.cn
[2] Shenzhen University of Advanced Technology, Shenzhen, Guangdong, China
18b903024@stu.hit.edu.cn

Abstract. This paper introduces a new type of regression methodology named as Convex-Area-Wise Linear Regression(CALR), which separates given datasets by disjoint convex areas and fits different linear regression models for different areas. This regression model is highly interpretable, and it is able to interpolate any given datasets, even when the underlying relationship between explanatory and response variables are non-linear and discontinuous. In order to solve CALR problem, 3 accurate algorithms are proposed under different assumptions. The analysis of correctness and time complexity of the algorithms are given, indicating that the problem can be solved in $o(n^2)$ time accurately when the input datasets have some special features. Besides, this paper introduces an equivalent mixed integer programming problem of CALR which can be approximately solved using existing optimization solvers.

Keywords: Data analysis · Linear regression · Segmented regression · Piecewise linear regression · Machine learning · Optimization

1 Introduction

Multiple-model linear regression(MLR) is a special and important data analysis method. Let $\boldsymbol{DS} = \{(\boldsymbol{x}_i, y_i) | x_i \in \mathbb{R}^d, y_i \in \mathbb{R}, i = 1, \cdots, n\}$ be a given dataset where $\boldsymbol{x}_i$ is the explanatory variable and y_i is the response variable. Multiple-model linear regression divides the input datasets into several subsets, and then constructs local linear regression model for each subset, such as the example shown in Fig. 1 of Appendix A from [10]. Such model can approximately represent the non-linear underlying relationship between y and $\boldsymbol{x}$. The piecewise linear regression [1,13], max-affine regression [6], PLDC regression [12], MMLR [9] are all typical multiple-model linear regression methods.

Nowadays, real-world datasets, especially big data have the feature that different subsets of a dataset fitting highly different regression models, which is described as *diverse predictor-response variable relationships*(DPRVR) in [5]. Taking TBI dataset in [8] as a real-world example. The *mean squared error*(MSE)

Y. Chen et al. (Eds.): COCOON 2024, LNCS 15162, pp. 407–419, 2025.
https://doi.org/10.1007/978-981-96-1093-8_34

is 109.2 when one linear regression model is used to model the whole TBI. However, the MSE is reduced to 12.3, when TBI is divided into 6 disjoint hypercubes and 6 different linear regression models are used individually, while one global f_0 is used for the data not belonging to the 6 subsets. The detail of TBI dataset is listed in Appendix F of [10]. So fitting data using MLR is more suitable than using one single linear or polynomial function in regression tasks.

Besides, MLR model has statistical advantages and high interpretability. The numeric value of parameters can show the importance of variables, and the belonging information about the confidence coefficients and intervals makes the model more credible in practice. Therefore, MLR is widely used in research areas requiring high interpretability, such as financial prediction, investment forecasting, biological and medical modelling, etc. [11]. Most machine learning and deep learning models might be more precise in predicting tasks, but the black-box feature limits their ranges of application [7]. These two characteristics make MLR a necessary methodology for nowadays data analysis.

However, there are still shortcomings of the existing multiple-model linear regression methods. The most important ones are described as follows.

1. The dimension d of $\boldsymbol{DS}$ cannot be relatively large, since the time complexity of most algorithms constructing MLR models grows exponentially with d [13,14];
2. The subsets being used to construct local models must be hyper cubes or single hyperplane [4,15]. Thus, the accuracy of them is lower when the underlying partition of a given dataset are not in such forms;
3. The time complexity of the methods is high. Even the state-of-art approximate algorithm, PLDC regression, has the time complexity of $O(d^2n^5)$ [12].

To overcome the disadvantages and inspired by the TBI dataset, this paper proposes a new multi-model based linear regression method named as Convex-area-wise Linear Regression(CALR).

Instead of using hypercubes or single hyperplane to separate datasets, CALR considers there is a default linear model f_0 lying on the datasets, and there are several disjoint convex areas $C_1, C_2, \cdots, C_M$, which contains subsets of datasets fitting different local linear models $f_1, \cdots, f_M$, and each convex area C_i can be defined by multiple hyperplanes. An illustration of CALR is shown as Fig. 2 in Appendix A of [10].

The major contributions of this paper are concluded as follows.

- A new function class named convex-area-wise linear function(CALF) is formally defined. The important properties and expressive ability of CALF is proved.
- The regression problem corresponded with convex-area-wise linear function is formally proposed, named as CALR. The CALR problem has been transformed into a mix-integer programming problem, so that it can be approximately solved by any optimization solvers.
- Three algorithms are designed to accurately solve the CALR problem under special assumptions. The correctness and expected time complexity of them

are also proved. When $\boldsymbol{DS}$ is convex-area separable, the expected time complexity can reach $o(n^2)$, which is much lower than PLDC regression.

The rest of this paper is organized as follows. Section 2 gives the formal definition and important properties of the new proposed convex-area-wise linear function and the corresponded regression problem. Section 3 gives the design and analysis of three algorithms under different assumptions. Finally, Sect. 4 concludes the paper.

2 Preliminaries and Problem Definition

This section introduces the definition and specialties of the new proposed function class CALF and the formal corresponded regression problem CALR. An optimizing version of CALR is also proposed for users to approximately solve it by existing optimization solvers.

2.1 Convex-Area-Wise Linear Function

Definition of convex-area-wise linear function is given as follows.

Definition 1 (Convex-Area-Wise Linear Function). *Suppose that* $x \in \Omega \subset \mathbb{R}^d$. *Let* $H = \{(f_i, C_i)\ |0 \leq i \leq M\}$ *be a set of function, convex-area pairs, such that for* $i = 0, \cdots, M$, $f_i(x) = \beta_i \cdot (1, x) = \beta_{0i} + \beta_{1i}x_1 + \cdots + \beta_{di}x_d$ *is a linear function, for* $i = 1, \cdots, M$, $C_i = \cap_{k=1}^{m_i}\{x|\alpha_{ik} \cdot x + \gamma_{ik} \leq 0, \alpha_{jk} \in \mathbb{R}^d\}$ *is a convex area defined by* m_i *semi-spaces,* $C_1, \cdots, C_M$ *are disjoint, and* $C_0 = \Omega - \cup_{i=1}^{M} C_i$. *A function* $f_H : \Omega \to \mathbb{R}$ *is called a convex-area-wise linear function, CALF for short, if* $f_H(x) = \sum_{i=0}^{M} I_i(x) f_i(x)$, *where* $I_i(x) = \begin{cases} 1, x \in C_i \\ 0, x \notin C_i \end{cases}$ *is the indicative function of* $C_i, i = 0, \cdots, M$.

Let $CALF(\Omega)$ *be the set of all convex-area-wise linear functions defined on* $\Omega \subset \mathbb{R}^d$.

According to *Definition* 1, every $f_H \in CALF(\Omega)$ is piece-wise linear on Ω. When $x \in C_i, i = 1, \cdots, M$, $f_H(x) = f_i(x)$, and $f_H(x) = f_0(x)$ otherwise. We say the action scope of f_i is C_i. Figure 2 in Appendix A of [10] is an example.

The most important property of CALF is that any finite datasets can be fitted by a CALF function. In order to prove the property, the definition of PLDC functions and a corresponded proposition from [12] are shown without proof. Then the characteristics is formally given as the following *Theorem* 1.

Definition 2 (PLDC function). *If a function* f *can be represented as* $f(x) = max_{1 \leq k \leq K}\{\alpha \cdot x + c_k\} - max_{1 \leq k \leq K}\{\beta \cdot x + c'_k\}$, *where* $K \in \mathbb{Z}^+, x, \alpha_k, \beta_k \in \mathbb{R}^d, c_k, c'_k \in \mathbb{R}$, *the* f *is called as a PLDC function.*

Lemma 1. *Given any finite data* $D = \{(x_i, y_i)|i = 1, 2, \cdots, n\}$, *and* $y_i \neq y_j \Rightarrow x_i \neq x_j$, *there exists a PLDC function interpolating* D.

Theorem 1. *Given any dataset* $D = \{(x_i, y_i)|i = 1, 2, \cdots, n\}$, *where* $x_i \in \mathbb{R}^d, y_i \in \mathbb{R}$, $y_i \neq y_j \Rightarrow x_i \neq x_j$, *there exists a function* $f \in CALF$ *fitting* D.

Furthermore, there exists functions f *such that* $f \in CALF$ *but* $f \notin PLDC$.

Proof. Obviously, the first part of *Theorem* 1 is proved if $PLDC \subset CALF$. Suppose that $f(x) = \max_{1 \le k \le K}\{\alpha \cdot x + c_k\} - \max_{1 \le k \le K}\{\beta \cdot x + c'_k\}$ is a PLDC function, and $f_1(x) = \max_{1 \le k \le K}\{\alpha \cdot x + c_k\}, f_2(x) = \max_{1 \le k \le K}\{\beta \cdot x + c'_k\}$.

According to the definition, f_1 is a continuous function with at most K hyperplanes $\pi_k : \alpha_k \cdot x + c_k - y = 0, k = 1, \cdots, K$ in $\mathbb{R}^{d+1}$. Every π_k has the boundary represented by $\alpha_k \cdot x + c_k \ge \alpha_j \cdot j + c_j$, which is a semi-space $(\alpha_k - \alpha_j) \cdot x + c_k - c_j \ge 0$, for $j \neq k$.

Thus, $f_1(x) = \alpha_k \cdot x + c_k$ when $x \in C_k = \cap_{j \neq k}\{y \in \mathbb{R}^d|(\alpha_k - \alpha_j) \cdot y + c_k - c_j \ge 0\}$ for $k = 1, \cdots, K$. $C_k \subset \mathbb{R}^d$ is convex since C_k is the intersection of $K - 1$ semi-spaces. Similarly, $f_2(x) = \beta_k \cdot x + c'_k$ when $x \in C'_k$, where $C'_k \subset \mathbb{R}^d$ is convex, for $k = 1, \cdots, K$.

Then, $f(x) = f_1(x) - f_2(x)$ contains at most K^2 pieces of hyperplanes and $f(x) = (\alpha_i - \beta_j) \cdot x + c_k - c'_k$ when $x \in S_i \cap S'_j$, $i, j = 1, \cdots, K$. Besides, $C_i \cap C'_j = (\cap_{k \neq i}\{x \in \mathbb{R}^d|(\alpha_i - \alpha_k) \cdot x + c_i - c_k \ge 0\}) \cap (\cap_{k \neq j}\{x \in \mathbb{R}^d|(\beta_j - \beta_k) \cdot x + c'_j - c'_k \ge 0\})$ is a convex area bounded by hyperplanes.

Therefore, $f(x)$ can be represented by a CALF function f_H, where $H = \{(g_{ij}, C_{ij})|g_{ij}(x) = (\alpha_i - \beta_j) \cdot x + c_i - c'_j, C_{ij} = C_i \cap C'_j, i, j = 1, \cdots, K\}$. So if a dataset D can be interpolated by a PLDC function f, D can be interpolated by a CALF f_H constructed above.

Combined with *Lemma* 1, the first part of *Theorem* 1 is proved.

Further, consider the function$f(x_1, x_2) = \begin{cases} \frac{1}{2}x_1 - 1, x_1 \le 0 \\ -2x_1 + 1, x_1 > 0 \end{cases}$. Obviously, $f(x_1, x_2)$ can be represented by $f_H \in CALF$ with $H = (g_1, C_1), (g_2, C_2)$ where $g_1(x) = \frac{1}{2}x_1 - 1, g_2(x) = -2x_1 + 1$ and $C_1 = \{(x_1, x_2)|x_1 \le 0\}, C_2 = \{(x_1, x_2)| - x_1 \le \varepsilon\}$, ε is small enough.

However, f can not be represented by any PLDC function. Let $h(x) = h_1(x) - h_2(x)$ be a PLDC function, $h_1(x)$ and $h_2(x)$ are continuous because they are the maximum of finite linear functions [3]. Then $h(x) = h_1(x) - h_2(x)$ is continuous on $\mathbb{R}^2$. Therefore, $PLDC \subsetneq CALF$. □

Theorem 1 shows that CALF has stronger expressive ability than PLDC. Further, CALF functions can fit any finite given numerical datasets D, even when the underlying model of D is non-linear and discontinuous.

2.2 Convex-Area-Wise Linear Regression

Firstly the definition of traditional linear regression problem is given as follows.

Definition 3 (Linear Regression Problem).
Input: *A numerical dataset* $\boldsymbol{DS} = \{(x_i, y_i) \mid 1 \le i \le n\}$, *where* $x_i \in \mathbb{R}^d, y_i \in \mathbb{R}$, $y_i = f(x_i) + \varepsilon_i$ *for some function* f, $\varepsilon_i \sim N(0, \sigma^2)$, *and all* ε_i*s are independent.*
Output: *A function* $\hat{f}(x) = \boldsymbol{\beta} \cdot (1, x) = \beta_0 + \beta_1 x_1 + \cdots + \beta_d x_d$ *such that* $MSE(\boldsymbol{DS}, \hat{f}) = \frac{1}{n}\sum_{i=1}^{n}[(y_i - \hat{f}(x_i))^2]$ *is minimized.*

One way to construct optimal linear regression model for $\boldsymbol{DS}$ is *pseudo-inverse matrix method.* The time complexity of it is $O(d^2n + d^3)$ [11]. So this paper let $\text{lr}(\boldsymbol{DS})$ be any $O(d^2n + d^3)$ time algorithm to construct the optimal linear function $\hat{f}$ of given datasets $\boldsymbol{DS}$.

The p-value of F-test for $\hat{f}$ is a convincing criterion to judge the goodness of $\hat{f}$, which is denoted as $p_F(\hat{f})$ in this paper. The $p_F(\hat{f})$ can be calculated in $O(n)$ time [11]. Generally, linear regression model $\hat{f} \cong f$ when $p_F(f) < 0.05$, where f is the underlying model of $\boldsymbol{DS}$. Besides, it's always assumed that $n > d + 1$, since a d-dimensional linear function can perfectly fit any n data points when $n \le d + 1$. The details of linear regression and *pseudo-inverse matrix method* is shown in Appendix E of [10].

Similarly, the regression problem corresponded with CALF is defined as follows.

Definition 4 (Convex-area-wise Linear Regression Problem).
Input: *A numerical dataset* $\boldsymbol{DS} = \{(\boldsymbol{x}_i, y_i) \mid 1 \le i \le n\}$, *where* $\boldsymbol{x}_i \in \mathbb{R}^d, y_i \in \mathbb{R}$, $y_i = f(\boldsymbol{x}_i) + \varepsilon_i$ *for an* f, $\varepsilon_i \sim N(0, \sigma^2)$, *and all* ε_i*s are independent identically distributed, Integer* $M \ge 0$.
Output: *An function* $f_H \in CALF$, *where* $H = \{(\hat{f}_j, \boldsymbol{C}_j) \mid 0 \le j \le M\}$ *such that* $MSE(\boldsymbol{DS}, f_H) = \frac{1}{n}\sum_{i=1}^{n}(y_i - f_H(\boldsymbol{x}_i))^2$ *is minimized.*

The problem in *Definition* 4 is denoted as CALR problem for short. The decision problem of CALR is given as follows.

Definition 5 (Decision problem of CALR).
Input: *A numerical dataset* $\boldsymbol{DS} = \{(\boldsymbol{x}_i, y_i) \mid 1 \le i \le n\}$, *integer* $M \ge 0$, *positive real number* $B > 0$.
Output: *If there exists a function* $f_H \in CALF$, *where* $H = \{(\hat{f}_j, \boldsymbol{C}_j) \mid 0 \le j \le M\}$ *such that* $MSE(\boldsymbol{DS}, f_W) = \frac{1}{n}\sum_{i=1}^{n}(y_i - f_W(\boldsymbol{x}_i))^2 < B$.

Given $\boldsymbol{DS}$ and $f_H \in CALF$, one can calculate $\frac{1}{n}\sum_{i=1}^{n}(y_i - f_H(\boldsymbol{x}_i))^2$ in $O(n)$ time and compare it with B in $O(1)$ time. Therefore, the decision problem of CALR is in NP.

2.3 The Optimizing Version of CALR

In Subsect. 2.2, the definition of CALR problem is given. However, we still need to rewrite it as an optimizing problem to solve it using optimization solvers.

Considering MSE as the optimizing objective, l_2 loss function can be used for the problem. Let the number of local models be smaller than M, and the numbers of hyperplanes bounding each convex area be smaller than K. Thus the objective function is $l(w) = \sum_{i=1}^{n}(\beta_0 \cdot x_i + c_0 + \sum_{j=1}^{M} I_{ij}(\beta_j \cdot x_i + c_j) - y_i)^2$, where $\beta_j \in \mathbb{R}^d, c_j \in \mathbb{R}, j = 0, \cdots, M$. In this expression, I_{ij} is the indicator variable for (x_i, y_i) of convex area C_j, that is $I_{ij} = \begin{cases} 1, x_i \in C_j \\ 0, x_i \notin C_j \end{cases}$. Suppose that $C_j = \cap_{k=1}^{K}\{y|\alpha_{jk} \cdot y + \gamma_{jk} \leq 0\}$, and $I_{ijk} \in \{-1, 1\}$ satisfies $I_{ijk}(\alpha_{jk} \cdot x_i + \gamma_{jk}) \leq \tau$, where $\tau < 0$ and $|\tau|$ is small enough. Then, $x_i \in C_j$ if and only if $\sum_{k=1}^{K} I_{ijk} = K$. Besides, it's necessary to let $\sum_{j=1}^{M} I_{ij} \leq 1$ to make sure $C_1, \cdots, C_M$ are disjoint.

Thus, the optimizing version of CALR is given as follows.

Definition 6 (Optimizing CALR problem).

$$\min_{w} l(w) = \sum_{i=1}^{n}(\beta_0 \cdot x_i + c_0 + \sum_{j=1}^{M} I_{ij}(\beta_j \cdot x_i + c_j) - y_i)^2$$
$$s.t. \begin{cases} I_{ijk}(1 - I_{ijk}) = 0 \\ \sum_{j=1}^{M} I_{ij} \leq 1 \\ I_{ij} = \prod_{k=1}^{K} I_{ijk} \\ (I_{ijk} - \frac{1}{2})(\alpha_{jk} \cdot x_i + \gamma_{jk}) \leq \tau \\ i = 1, \cdots, n; j = 1, \cdots, M; k = 1, \cdots, K \end{cases}$$

where $w = (\beta, c, \alpha, \gamma, I)$ is the vector of optimizing variables, $D = \{(x_i, y_i)|i = 1, \cdots, n\}$, $M, K \in \mathbb{Z}^+, \tau < 0$ is given.

This optimizing problem has $(d+1)(K+1)M$ variables and $n(M(2K+1)+1)$ nonlinear constraints. It's a mix-integer programming problem, which makes it an NP-hard problem [3]. Using optimizing solvers could only get a approximate solution.

3 Algorithm and Analysis

Sect. 2.3 gives the optimizing version of CALR problem which can only be solved approximately. This section gives accurate algorithms for the problem with different assumptions.

Firstly, a sub-algorithm cac($\boldsymbol{DS}, D$) is proposed to construct a convex-area which separates D from $\boldsymbol{DS}$. Given $\boldsymbol{DS} = \{x_i \in \mathbb{R}^d | i = 1, \cdots, n\}$, $D \subset \boldsymbol{DS}$ with $|D| = n'$, cac($\boldsymbol{DS}, D$) returns $\emptyset$ if the convex hull of D contains $x \in \boldsymbol{DS} - D$, returns $C = \cap_{k=1}^{K}\{z \in \mathbb{R}^d | \alpha_k \cdot z + \gamma_k \leq 0\}$ such that $D \subset C$ and $C \cap \boldsymbol{DS} - D = \emptyset$ if there are no $x \in \boldsymbol{DS} - D$ such that x is in the convex hull of D.

The main idea of cac($\boldsymbol{DS}, D$) is to transfer the problem into a linear programming problem. For any $x_0 \in \boldsymbol{DS} - D$, consider the following problem denoted as $LP(x_0, D)$:

$$\min_{w} l(w) = 0$$
$$s.t. \quad w \cdot (x_i - x_0) < 0, x_i \in D$$

where $w = (w_1, \cdots, w_d)$ is the coefficient of a hyperplane passing through x_0 in $\mathbb{R}^d$. According to the definition, $LP(x_0, D)$ is a linear programming problem in

d-dimension with n' constraints. Gauss-Seidel Method can solve it with $O(n')$ when d is not very large [16]. Trivially, any feasible solution of $LP(x_0, D)$ is a hyperplane that separates x_0 from D. Suppose that $\alpha_k = w_k, \gamma_k = w_k \cdot x_k$ for $x_k \in \boldsymbol{DS} - D$, where w_k is one solution of $LP(x_k, D)$, then $C = \cap_{k=1}^{K}\{z \in \mathbb{R}^d | \alpha_k \cdot z + \gamma_k \leq 0\}$ is a convex area separating D from $\boldsymbol{DS} - D$.

Pseudo-code of cac is shown in the following *Algorithm* 1, and the gslp(x, D) in Line 3 denotes any $O(n')$ time Gauss-Seidel algorithm that solves $LP(x, D)$ [16].

Algorithm 1: cac$(\boldsymbol{DS}, D)$

Input: $\boldsymbol{DS} = \{x_i \in \mathbb{R}^d | i = 1, \cdots, n\}, D \subset \boldsymbol{DS}$;
Output: $C = \cap_{k=1}^{K}\{z \in \mathbb{R}^d | \alpha_k \cdot z + \gamma_k \leq 0\}$ is a convex area or $C = \emptyset$

1 $C \leftarrow \mathbb{R}^d, U \leftarrow \boldsymbol{DS} - D$;
2 **for** $x \in U$ **do**
3     $\pi \leftarrow$ gslp(x, D) ;
4   **if** $\pi = \emptyset$ **then**
5   **return** $\emptyset$;
6   **else**
7   **continue**;
8   **if** $\pi(x) > 0$ **then**
9         $\pi_c(z) \leftarrow -\pi(z)$;
10   **else**
11         $\pi_c(z) \leftarrow \pi(z)$;
12     $C \leftarrow C \cap \{z \in \mathbb{R}^d | \pi_c(z) \leq 0\}$;
13 **return** C;

The time complexity and correctness is formally described as the following lemma without proof due to the limited space. The details of *Algorithm 1* and *Lemma 2* are shown in Appendix B of [10].

Lemma 2. *(1) If cac$(\boldsymbol{DS}, D) = \emptyset$, D is not convex-area separable in $\boldsymbol{DS}$.*
(2) If cac$(\boldsymbol{DS}, D) = C \neq \emptyset$, D is separated from $\boldsymbol{DS} - D$ by C.
(3) The time complexity of cac is $O(nn')$ when $|D| = n'$ and $|\boldsymbol{DS}| = n$.

By means of the algorithm cac, a naive algorithm for the original CALR problem with $M = 1$ is given as *Algorithm 6*, and the time complexity is $O(n2^n d^3)$. Due to the limited space, detailed explanation and analysis of *Algorithm 6* is shown in Appendix C of [10].

3.1 Algorithms of CALR

This section discusses a simpler case of CALR problem. Firstly, we give the following definition of convex-area separable for datasets.

Definition 7 (Convex-Area Separable). *Given* $D = \{(x_i, y_i)|x_i \in \mathbb{R}^d, y_i \in \mathbb{R}, i = 1, \cdots, n\}$, *if there exists a CALR function* F_H *with* $|H| = M+1$ *satisfying* $\forall x \in S_i, |f_i(x) - y| < \varepsilon$ *for* $i = 1, \cdots, M+1$, *where* $\varepsilon > 0$ *is small enough, and* $\|f_i - f_j\| \geq \delta$, D *is said to be* (M, ε, δ)*-Convex-Area Separable.*

Intuitively, if D is Convex-Area Separable, D could be separated into $M+1$ subsets and M of them could be bounded by disjoint convex areas, and different subsets fit different linear functions. Such assumption is used in data description tasks and regression modelling for datasets with small errors.

The *Algorithm* 2 is designed for Convex-Area Separable datasets for any given M. The sub-algorithms used in Line 11 and 20 are shown in *Algorithm* 3 and *Algorithm* 4 respectively.

Algorithm 2: casCALR($\boldsymbol{DS}, M, \tau, \varepsilon, \delta$)

Input: $\boldsymbol{DS} = \{(x_i, y_i)|x_i \in \mathbb{R}^d, y_i \in \mathbb{R}, i = 1, \cdots, n\}$ Convex-Area Separable;
Output: $\boldsymbol{H} = \{(f_i, C_i)|i = 0, \cdots, M\}$ such that $F_H \in CALF$, minimizing $\sum_{i=1}^{n}(F_H(x_i) - y_i)^2$

1 $\boldsymbol{H} = \emptyset, F = \emptyset, DS' \leftarrow \boldsymbol{DS}$;
2 **While** ($|F| < M$ or $DS' \neq \emptyset$) **Do**
3     $D \leftarrow$ uniformly sample $d+1$ data points from DS', $f_t \leftarrow \mathrm{lr}(D)$;
4   **if** $p_F(f_t) < \tau$ **then**
5         $C_t \leftarrow \mathrm{cac}(DS', D)$;
6   **if** $C_t \neq \emptyset$ **then**
7             $m \leftarrow |\{f \in F| \|f - f_t\| \geq \delta\}|$;
8   **if** $m \geq |F|$ **then**
9                 $F \leftarrow F \cup \{f_t\}$;
10                 $DS' \leftarrow \{(x, y) \in DS'||y - f(x)| \geq \varepsilon\}$;
11 $DS' \leftarrow \mathrm{distinct}(F, \boldsymbol{DS})$;
12 **for** $f \in F$ **do**
13     $D_f = \{(x, y) \in DS'||y - f(x)| < \varepsilon\}$, $C \leftarrow \mathrm{cac}(DS', D_f)$;
14   **if** $D - C \cap DS' \neq \emptyset$ **then**
15         $f_0 \leftarrow f$;
16   **else**
17         $DS' \leftarrow DS' - DS' \cap C$;
18         $\boldsymbol{H} \leftarrow \boldsymbol{H} \cup \{(f, C)\}$;
19 $DS' \leftarrow DS' - \{(x, y) \in DS'||y - f_0(x)| < \epsilon\}$;
20 $\boldsymbol{H} \leftarrow \boldsymbol{H} \cup post(\boldsymbol{H}, DS', \varepsilon) \cup \{(f_0, \mathbb{R}^d - \cup_C\{x \in C|(f, C) \in \boldsymbol{H}\})\}$;
21 **return** $\boldsymbol{H}$;

As shown by the pseudo-code, casCALR firstly samples subsets D with $|D| = d+1$ which can be separated from $\boldsymbol{DS} - D$ by hyperplanes, to find the M different functions $f_1, \cdots, f_M$ in Line 2-10. In Line 11, the algorithm invokes distinct($F, \boldsymbol{DS}$) to remove the data points fitting more than one functions of F. Next in Line 12-19, casCALR finds all data points fitting f_i for $i = 1, \cdots, M$, and uses the data points to construct the convex area C for f_i. Now there's only a few data points remaining.

Finally, casCALR invokes a sub-algorithm post($\boldsymbol{H}, DS', \varepsilon$) to add the left data points into output $\boldsymbol{H}$ in Line 20-21. The correctness of casCALR is formally shown in *Theorem 2*. The proof of *Theorem 2* and detailed explanation of casCALR is shown in Appendix D of [10].

Algorithm 3: distinct($F, \boldsymbol{DS}$)

Input: $\boldsymbol{DS} = \{(x_i, y_i) | x_i \in \mathbb{R}^d, i = 1, \cdots, n\}, F = f_i | 1 \le i \le M$;
Output: $DS' \subset \boldsymbol{DS}$
$DS' \leftarrow \emptyset, D \leftarrow d = (x, y, 0) | (x, y) \in \boldsymbol{DS}$;
for $f \in F$ **do**
    for $d \in D$ **do**
        if $|y - f(x)| < \epsilon$ **then**
            if $d = (x, y, 0)$ **then**
                $d \leftarrow (x, y, 1)$;
            else
                $d \leftarrow (x, y, -1)$;
$DS' \leftarrow (x, y) | (x, y, z) \in D, z = 1$;
return DS';

Algorithm 4: post($\boldsymbol{H}, D, \varepsilon$)

Input: $\boldsymbol{H} = \{(f_i, H_i) | i = 1, \cdots, M\}, D \subset \boldsymbol{DS}$;
Output: $\boldsymbol{H}' = \{(g_i, H_i')\}$ where H_i' is a polyhedron
$F \leftarrow \{f | (f, H) \in \boldsymbol{H}\}$;
for $f \in F$ **do**
    $D_1 \leftarrow \{(x, y) \in D || f(x) - y| < \varepsilon\}$;
    for $f_t \in F$ *and* $f_t \neq f$ **do**
        $D_2 \leftarrow \{(x, y) \in D || f_t(x) - y| < \varepsilon\}$;
        if $D_1 \cap D_2 \neq \emptyset$ **then**
            $C_t \leftarrow \text{cac}(D, D_1 \cap D_2)$;
            $\boldsymbol{H}' \leftarrow \{(f, C_t)\}$;
            $D \leftarrow D - D_1 \cap D_2$;
return $\boldsymbol{H}'$;

Theorem 2. *(1) If DS' =distinct($F, \boldsymbol{DS}$), and $D_0', \cdots, D_M'$ are subsets fitting $f_0, f_1, \cdots, f_M \in F$ respectively, there exists at most one D_i' can not be separated from $DS' - D_i'$ by a convex area.*

(2) If $\boldsymbol{H} = \{(f_i, C_i)|i = 1, \cdots, M\}$ is the output of casCALR, C_is are disjoint convex areas bounded by hyperplanes, and $\{(x, y) \in \boldsymbol{DS}||f_0(x) - y| < \varepsilon\}$ might not be separated from $\boldsymbol{DS}$ by hyperplanes.

3.2 Analysis of Time Complexity

This section discusses the time complexity of casCALR. Firstly, the following *Lemma* 3 is given to analyze the expected time complexity.

Lemma 3. *Suppose that a set D is separated in M disjoint subsets $D_1, \cdots, D_M$, where $|D| = n$ and $|D_i| = N_i \geq d+1, i = 1, \cdots, M$. Let $D'_1, D'_2, \cdots, D'_t, \cdots$ be a uniformly sampled series of subsets, such that $D'_t \subset D$ and $|D'_t| = d+1$. Let event A be $D'_t \subset D_i$ for some i. Let T be the smallest t such that A happens, then $\mathbb{E}[T] = O(q^{d+1}M^{d+1})$, where $\frac{1}{q} = 1 - \frac{Md}{n}$.*

Proof. For any t, $P(A) = p = \frac{C_{N_1}^{d+1} + \cdots + C_{N_M}^{d+1}}{C_n^{d+1}}$. As the definition of expectation, $\mathbb{E}[T] = p + 2(1-p)p + \cdots + t(1-p)^{t-1}p + \cdots$. This is the mathematical expectation of geometry distribution [2]. So $\mathbb{E}[T] = \frac{1}{p}$.

Thus, $\mathbb{E}[T] = \frac{C_n^{d+1}}{C_{N_1}^{d+1} + \cdots + C_{N_M}^{d+1}} = \frac{n(n-1)\cdots(n-d)}{N_1(N_1-1)\cdots(N_1-d) + \cdots + N_M(N_M-1)\cdots(N_M-d)}$. Let $N = \max\{N_1, \cdots, N_M\}$, and reduce the denominator, then there is $\mathbb{E}[T] \leq \frac{n(n-1)\cdots(n-d)}{N(N-1)\cdots(N-d)}$. Furthermore, $\frac{n(n-1)\cdots(n-d)}{N(N-1)\cdots(N-d)} \leq \frac{n^{d+1}}{N(N-1)\cdots(N-d)} \leq \frac{n^{d+1}}{(N-d)^{d+1}} = (\frac{n}{N-d})^{d-1}$.

By the pigeonhole principle, $N \geq \frac{n}{M}$ since N is the biggest N_i. So $\frac{n}{N-d} \leq \frac{n}{n/M - d} = \frac{M}{1 - Md/n}$. Let $\frac{1}{q} = 1 - \frac{Md}{n}$, then $Md = n(1 - \frac{1}{q})$ and $\mathbb{E}[T] \leq (qM)^{d+1}$.

Therefore, $\mathbb{E}[T] = O((qM)^{d+1})$. Besides, $\mathbb{E}[T] = O(2^{d+1}M^{d+1})$ when $n \geq 2Md$, which is always satisfied in practice. □

The expected time complexity of CALR is given in the following *Theorem* 3.

Theorem 3. *The expected time complexity of casCALR is $O(q^{d+1}M^{d+2}nd + M^2nn')$, where $\frac{1}{q} = 1 - \frac{(M+1)d}{n}$ and $n' = max_i|C_i \cap \boldsymbol{DS}|$.*

Proof. From the pseudo-code of casCALR, it firstly finds every f_i, by randomly sampled subsets D with $|D| = d+1$. Suppose the biggest expected time of sampling D for f_i is T. For each suitable D, casCALR costs $O(d^3)$ to construct linear regression model on D, $O(n)$ time to construct D' and U' by checking the predicting error of every $(x, y) \in \boldsymbol{DS}$, $O(n(d+1))$ time to construct the convex area containing D, $O(M)$ time to test whether f_i is contained in F, $O(n)$ time to remove the data points fitted by f_i. Thus, the time complexity of Line 2-10 for each D is $O(d^3 + n + n(d+1) + M + n) = O(nd + d^3)$. These steps will be repeated for M times to find all f_i in $\boldsymbol{H}$. Suppose that T is the max sample time of constructing every f_i, the time complexity of Line 2-10 is $O(MT(nd + d^3))$.

Next, casCALR finds C_i for each $f_i \in F$. Firstly casCALR invoked the sub-algorithm distinct($F, \boldsymbol{DS}$) to find all data points fitted by f_i, and removes data points fitted by $f_j, j \neq i$ to make sure the convex area C_i containing them is rightly constructed. From the pseudo-code of *Algorithm* 3, it's obvious that the time complexity of Line 11 is $O(Mn)$.

Then casCALR costs $O(nn')$ time to construct convex area C_i where $n' = |D|$, and $O(n)$ time to remove the data points fitted by f_i. The total time complexity of Line 12-19 is $O(M(nn' + n)) = O(Mnn')$.

In the end, casCALR deals with the data points fitted by more than one $f \in F$. The sub-algorithm post($\boldsymbol{H}, DS', \varepsilon$) is invoked, and the time complexity of it is $O(M^2nn')$ from Appendix D of [10].

Adding the four parts of the algorithm, the total time complexity of casCALR is $O(MT(nd+d^3)+Mn+Mnn'+M^2nn') = O(MT(nd+d^3)+M^2nn')$. It's known that $\mathbb{E}[T] = q^{d+1}(M+1)^{d+1}$ where $\frac{1}{q} = 1 - \frac{(M+1)d}{n}$ from *Lemma* 3. Therefore, the final expected time complexity of casCALR is $O(q^{d+1}M^{d+2}(nd + d^3) + M^2nn')$. □

Noticing that $Mn' = o(n)$ since C_0 contains most data points of D. Therefore, generally the expected time complexity of casCALR is $o(Mn^2)$ when d is not very large. Besides, an $o(2^{d+1}d^3 + n^2)$ expected time cas2 in *Algorithm* 7 can be proposed if $\boldsymbol{DS}$ is convex-area separable and $M = 1$ in Appendix E of [10].

Besides, there are several methods to estimate good enough ε, τ and δ if they are not given. In practice, $\tau = 0.05$ and $\delta = 0.5$ is an accurate enough threshold for hypothesis testing [11]. Once a linear regression model f on D satisfies $p_f < 0.05$, one can have $\varepsilon = 3\sqrt{\frac{1}{|D|}\sum_{(x,y)\in D}(f(x) - y)^2}$ as an accurate estimator. Therefore, ε, τ and δ is not necessary to be given for casCALR.

4 Conclusion and Future Work

This paper introduces a new function class called Convex-area-wise Linear functions(CALF), which has high interpretability and strong expressive ability. This paper proves that CALF can interpolate any given finite datasets, and can represent discontinuous piece-wise linear functions. This paper formally proposed convex-area-wise linear regression(CALR) problem along with an optimizing version of it to be approximately solved using existing optimization solvers.

Besides, this paper proposed algorithms to solve CALR problem under different assumptions. If assuming the given datasets are convex-area separable, an $O(q^{d+1}M^{d+2}nd + M^2nn')$ expected time algorithm casCALR is proposed for the case when M is known, and an $o(2^{d+1}d^3 + n^2)$ expected time algorithm cas2 is proposed when $M = 1$. Both casCALR and cas2 are accurate algorithms for the problem. Since $2^{d+1}d^3, q^{d+1}M^{d+2}nd, Mn' < n^2$ when d is relatively small, the time complexity of them can reach almost $o(Mn^2)$ in practice.

For future work and challenges of CALF and CALR problem, reduction of time complexity is the most important part. $O(q^{d+1}M^{d+2}nd + M^2nn')$ is lower than some currently existing regression methods such as PLDC [12] when d

and M are relatively small, but it's still high for common big datasets. A linear time, even sub-linear time approximate algorithms for CALR is urgently needed. Secondly, an efficient algorithm for CALR problem when datasets are not convex-area-separable is also a challenge. Besides, a searching algorithm for best M is also needed to mitigate over-fitting when M is unknown. Taking each data point in given $\boldsymbol{DS}$ as a local model can reach zero error, but it causes serious over-fitting to the regression model, which makes it important to adjust M flexibly.

Disclosure of Interests. This work was supported by Shenzhen Science and Technology Program (JCYJ20220818100205012) and Shenzhen Key Laboratory of Intelligent Bioinformatics (ZDSYS20220422103800001).

References

1. Bemporad, A.: A piecewise linear regression and classification algorithm with application to learning and model predictive control of hybrid systems. IEEE Transactions on Automatic Control (2022)
2. Bertsekas, D., Tsitsiklis, J.N.: Introduction to probability, vol. 1. Athena Scientific (2008)
3. Boyd, S.P., Vandenberghe, L.: Convex optimization. Cambridge university press (2004)
4. Diakonikolas, I., Li, J., Voloshinov, A.: Efficient algorithms for multidimensional segmented regression. arXiv preprint arXiv:2003.11086 (2020)
5. Dong, G., Taslimitehrani, V.: Pattern-aided regression modeling and prediction model analysis. IEEE Trans. Knowl. Data Eng. **27**(9), 2452–2465 (2015)
6. Ghosh, A., Pananjady, A., Guntuboyina, A., Ramchandran, K.: Max-affine regression: Provable, tractable, and near-optimal statistical estimation. arXiv preprint arXiv:1906.09255 (2019)
7. Guidotti, R., Monreale, A., Ruggieri, S., Turini, F., Giannotti, F., Pedreschi, D.: A survey of methods for explaining black box models. ACM Comput. Surv. (CSUR) **51**(5), 1–42 (2018)
8. Hukkelhoven, C.W., et al.: Predicting outcome after traumatic brain injury: development and validation of a prognostic score based on admission characteristics. J. Neurotrauma **22**(10), 1025–1039 (2005)
9. Lyu, B., Li, J.: An efficient data analysis method for big data using multiple-model linear regression. In: Wu, W., Tong, G. (eds.) Computing and Combinatorics: 29th International Conference, COCOON 2023, Hawaii, HI, USA, December 15–17, 2023, Proceedings, Part I, pp. 272–284. Springer Nature Switzerland, Cham (2024). https://doi.org/10.1007/978-3-031-49190-0_19
10. Lyu, B., Li, J.: Convex-area-wise linear regression and algorithms for data analysis (2024)
11. Montgomery, D.C., Peck, E.A., Vining, G.G.: Introduction to linear regression analysis. Wiley, Tech. rep. (2021)
12. Siahkamari, A., Gangrade, A., Kulis, B., Saligrama, V.: Piecewise linear regression via a difference of convex functions. In: International Conference on Machine Learning, pp. 8895–8904. PMLR (2020)
13. Toriello, A., Vielma, J.P.: Fitting piecewise linear continuous functions. Eur. J. Oper. Res. **219**(1), 86–95 (2012)
14. Wahba, G.: Spline models for observational data. SIAM (1990)

15. Wang, Y., Witten, I.H.: Induction of model trees for predicting continuous classes (1996)
16. Xu, Y.: Hybrid jacobian and gauss-seidel proximal block coordinate update methods for linearly constrained convex programming. SIAM J. Optim. **28**(1), 646–670 (2018)

Approximating Continuous Multi-agent Contracts with Lyapunov Function Methods

Qinqin Gong[1], Donglei Du[2], Ling Gai[3(✉)], Dachuan Xu[1], and Ruiqi Yang[1(✉)]

[1] Institute of Operations Research and Information Engineering, Beijing University of Technology, Beijing 100124, People's Republic of China
gongqinqin@emails.bjut.edu.cn, {xudc,yangruiqi}@bjut.edu.cn

[2] Faculty of Business Administration, University of New Brunswick, Fredericton, NB E3B 5A3, Canada
ddu@unb.ca

[3] Business School, School of Intelligent Emergency Management, University of Shanghai for Science and Technology, Shanghai 200093, People's Republic of China
lgai@usst.edu.cn

Abstract. We study a principal multi-agent contract design problem with continuous actions, where the principal delegates a task to a set of agents who take costly actions over a continuous interval. Prior work introduced by Dütting et al. (2023) has investigated a similar discrete version of the problem where the success probability function is submodular in the agents chosen by the principal. In this paper, we complement this work by examining the problem when the success probability function is DR-submodular in the agents over a continuous interval. Additionally, our model captures the setting of exerting efforts from chosen agents and extends the set-defined case studied in Dütting et al. (2023). As the agents will play a Nash game before the contract is provided by the principal, who aims to maximize her utility, we cast the principal-agent contract model as a bi-level maximization. We focus on the linear contracts design and provide a backward induction, which deduces the above bi-level problem to a fractional agent chosen single level optimization problem. We incorporate techniques from the literature on DR-submodular optimization and develop a Lyapunov function-based method that obtains a parameterized approximation for this problem. Our results contrast with the discrete submodular setting, where there exists a nearly 1/258-approximation from Dütting et al. (2023).

Keywords: Contract design · Principal-agent model · DR-submodularity · Lyapunov function method · Approximation algorithms

1 Introduction

Contract theory is one of the most popular topics in microeconomics, and its central principal-agent model plays an important role in the economic markets.

Y. Chen et al. (Eds.): COCOON 2024, LNCS 15162, pp. 420–431, 2025.
https://doi.org/10.1007/978-981-96-1093-8_35

In recent years, principal-agent problems have received considerable attention in the computational economics community. Contracts based on the principal-agent model are commonly used in various real-world economic sectors, such as social media marketing [35], crowdsourcing platforms [30], blockchain-based smart contracts [14], carbon-reducing contract design [37], and healthcare incentive schemes [6].

Consideration of the classic hidden-action principal-agent problems [26,31], involves a principal aiming to incentivize one or more agents to accomplish a task by exerting effort. The principal is assumed to be able to observe only a somewhat random realization of the outcome, which is jointly effected by the chosen actions. The principal then offers a payment scheme or contract to the agents in return for their effort. The agents are considered as rational and self-serving, performing actions that maximize their expected utilities upon receiving the contract. The principal aims to determine a contract scheme to maximize its own utility. The problem can be solved in polynomial time using linear programming methods in the basic scheme considering a single principal and agent [26].

Computational research on contracts has explored hidden-action principal-agent models from various combinatorial dimensions, such as action-based, agent-based, and outcome-based settings. In the action-based setting [16,18], tasks need to be delegated to a single agent who can choose subsets of actions. A success probability function maps a subset of actions to the probability of task success. In the agent-based setting [2,17,23,36], tasks are assigned to a group of agents, and each agent decides whether to contribute or not. Consequently, success probability function that maps the subset of agents that exert effort to the probability of task success. In both models, the focus is on cases where the success probability functions are monotone and submodular. Another outcome-based setting has been investigated in [22]. These researches address models where the principal chooses which subset of agents, or which actions are chosen by the agent. Such models significantly restrict the level of effort and agent resilience, as it appears that once agents are selected, they are expected to exert 100 percent effort.

In this work, we focus on an extension to the agent-based combinatorial contract problem, wherein the principal intends to hire from a pool of n agents to perform a task, where each agent i performs an action $x_i \in [0,1]$, reflecting its level of effort, with a corresponding action cost $c_i(x_i)$. Let $\mathbf{x} = (x_1, ..., x_n) \in [0,1]^n$ and $c(\mathbf{x}) = (c_1(x_1), ..., c_n(x_n)) \in \mathbb{R}^n_{\geq 0}$ denote the action and cost vectors resulting from all agents'actions within the interval $[0,1]$, respectively. Each effort level vector $\mathbf{x}$ results in the possible outcomes $\omega \in \Omega$, and then the principal obtains a reward $r(\omega)$ for each stochastic outcome. Given that the reward function $f(\mathbf{x}) : [0,1]^n \to \mathbb{R}_{\geq 0}$ is non-negative and monotonic, and DR-submodular, or alternatively, in the binary outcome case, f denotes the success probability function.

This study focuses on linear contracts and the problem of optimal contract design for such contracts. A linear contract is represented by the n dimensional vector $\boldsymbol{\alpha} = (\alpha_1, ..., \alpha_n) \in \mathbb{R}^n_{\geq 0}$ corresponding to the agents. Specifically, agent i

will be paid $\alpha_i \cdot f(\mathbf{x})$ from the principal for the resulting outcome $\omega \in \Omega$. This implies that a linear contract incentivizes the n agents if for any given agent i, it holds that

$$\alpha_i f(x_i, \mathbf{x}_{-i}) - c_i(x_i) \geq \alpha_i f(y, \mathbf{x}_{-i}) - c_i(y), \forall y \in [0,1],$$

where $\mathbf{x}_{-i}$ denotes the vector consisting of the action efforts taken by the other agents excluding agent i and let $(x_i, \mathbf{x}_{-i})$ denote $\mathbf{x}$ for clarity. Our aim is to find a linear contract $\boldsymbol{\alpha}$ under which the agents satisfy incentive compatibility while maximizing the principal's utility:

$$\max_{\alpha}(1 - \sum_{i=1}^{n} \alpha_i) f(x)$$
$$s.t. x_i \in \arg\max_{y \in [0,1]} \alpha_i f(y, \mathbf{x}_{-i}) - c_i(y, \mathbf{x}_{-i}), \forall i \in [n].$$

When we restrict the scenario to binary, where each component of the vector $\mathbf{x}$ takes values of either 0 or 1, i.e., the agent either pays all or nothing, our problem reduces to the discrete version of the problem proposed and studied by Dütting [17]. For submodular reward functions and more general XOS functions, they have demonstrated the existence of combinatorial algorithms for constant-factor approximation. Additionally, they have also provided computational hardness results for the subadditive scenarios. However, we take a different approach, designing algorithms based on Lyapunov function techniques. These techniques initially provide an approximate estimate of the algorithm in the continuous-time form. Then, by sampling the interval points, we demonstrate that the approximation ratio incurs only a minor loss, and that polynomial-time approximation of the algorithm can be obtained.

1.1 Our Contributions and Techniques

In the following sections, we identify the basic parameterized optimization problems that need to be solved as follows:

$$\max_{\mathbf{x} \in [0,1]^n} F(x) := \max_{\mathbf{x} \in [0,1]^n} \left(1 - \sum_{i \in V} L^i(\delta_i, \mathbf{x}_{-i})\right) f(\mathbf{x}). \tag{1}$$

In this model, let $L^i(\delta_i, \mathbf{x}_{-i}) = \max\{L_1^i(\delta_i, \mathbf{x}_{-i}), L_2^i(\delta_i, \mathbf{x}_{-i})\}$ represent the optimal strategy for the principal to incentivize the agents. Here, $\delta_i = |x_i - y|$ serves as the parameter denoting the distance between x_i and y. Additionally, $L_1^i(\delta_i, \mathbf{x}_{-i}) = \frac{c_i(\delta_i) - c_i(0)}{f(1, \mathbf{x}_{-i}) - f(1-\delta_i, \mathbf{x}_{-i})}$ and $L_2^i(\delta_i, \mathbf{x}_{-i}) = \frac{c_i(1-\delta_i) - c_i(1)}{f(0, \mathbf{x}_{-i}) - f(\delta_i, \mathbf{x}_{-i})}$ serve as the upper bounds for $\max_{y \in [0,1]} \frac{c_i(x_i) - c_i(y)}{f(x_i, \mathbf{x}_{-i}) - f(y, \mathbf{x}_{-i})}$ concerning the scenarios where $x_i = y + \delta_i$ and $x_i = y - \delta_i$ respectively.

We demonstrate the existence of an algorithm employing a Lyapunov function in a continuous-time framework.

Main Result 1 (Theorem 1): For the continuous multi-agent contract problems, there exists a Lyapunov function based continuous-time algorithm that obtains a parameterized approximation to the optimal contract, when f and c_i are DR-submodular functions.

The steps outlining the key ingredients for devising the parameterized approximation scheme for reward and cost DR-submodular functions are as follows:

By setting $F(\mathbf{x}) = g(\mathbf{x}) \cdot f(\mathbf{x})$, where $g(\mathbf{x}) = 1 - \sum_{i \in V} L^i(\delta_i, \mathbf{x}_{-i})$ for clarity, we introduce a parametric monotone Lyapunov function as proposed and examined by Du's [15]. Let $E(\mathbf{x}(t)) = a(\mathbf{x}(t))F(\mathbf{x}(t)) - b_t F(\mathbf{x}^*)$. Suppose that $E(\cdot)$ is increasing in $t \in [0, T]$, we then obtain

$$a(\mathbf{x}(T))F(\mathbf{x}(T)) - b_T F(\mathbf{x}^*) = E(\mathbf{x}(T)) \geq E(\mathbf{x}(0)) = a(\mathbf{x}(0))F(\mathbf{x}(0)) - b_0 F(\mathbf{x}^*),$$

which implies that the approximate ratio of $\mathbf{x}(T) \in [0,1]^n$ to $F(\mathbf{x}(T)) \geq \frac{b_T - b_0}{a(\mathbf{x}(T))} F(\mathbf{x}^*)$.

Additionally, in order to ensure $\dot{E}(x(t)) \geq 0$, we establish an effective upper bound U_t for the optimum $F(\mathbf{x}^*)$ at any given time t. In our exposition, we define $U_t := f(\mathbf{x}(t)) + \langle \nabla f(\mathbf{x}(t)), \mathbf{v}(\mathbf{x}(t)) \rangle$ where $\mathbf{v}(\mathbf{x}(t) = \arg\max_{\mathbf{v} \in [0,1]^n} \langle \nabla f(\mathbf{x}(t)), \mathbf{v} \rangle$. Given $\dot{E}(x(t)) \geq 0$, the following conditions suffice:

$$\begin{cases} g(\mathbf{x}(t))\langle \nabla a(\mathbf{x}(t)), \dot{\mathbf{x}}(t) \rangle + a(\mathbf{x}(t))\langle \nabla g(\mathbf{x}(t)), \dot{\mathbf{x}}(t) \rangle - \dot{b}_t = 0, \\ a(\mathbf{x}(t))g(\mathbf{x}(t))\dot{\mathbf{x}}(t) - \dot{b}_t \mathbf{v}(\mathbf{x}(t)) = 0. \end{cases}$$

To guarantee $\mathbf{x}(T)$ is feasible, we set $T = 1$, $b_t = e^t$ and $a(x(t)) = b_t / g(x(t))$. Thus we can obtain the approximation guarantee $F(\mathbf{x}(1)) \geq \left(1 - \frac{1}{e}\right) \theta \cdot F(\mathbf{x}^*)$, where θ denotes the lower bound of g. Consequently, we present our continuous-time algorithm.

Furthermore, we extend our previous result by demonstrating that the approximation ratio remains nearly unchanged, while ensuring a provable time complexity, through the discretization of the continuous-time interval.

Main Result 2 (Theorem 2): For the continuous multi-agent contract problems, there exists a Lyapunov function based discrete-time algorithm that obtains a parameterized approximation to the optimal contract in polynomial time, when f and c_i are DR-submodular functions.

The main components of this process involve introducing a discrete-time potential function $E(\mathbf{x}(t_j)) = a(\mathbf{x}(t_j))F(\mathbf{x}(t_j)) - b_{t_j} F(\mathbf{x}^*)$. Here, we set $T = 1$, $b_{t_j} = e^{t_j}$ and $a(\mathbf{x}(t_j)) = b_{t_j} / g(\mathbf{x}(t_j))$, $j = 0, ..., N$. To ensure $E(\mathbf{x}(t_{j+1})) - E(\mathbf{x}(t_j)) \geq 0$, we leverage the L-smooth property of function f and derive the following sufficient conditions:

$$\begin{cases} a(\mathbf{x}(t_{j+1}))g(\mathbf{x}(t_{j+1})) - a(\mathbf{x}(t_j))g(\mathbf{x}(t_j)) - (b_{t_{j+1}} - b_{t_j}) \geq 0, \\ a(\mathbf{x}(t_{j+1}))g(\mathbf{x}(t_{j+1}))(\mathbf{x}(t_{j+1}) - \mathbf{x}(t_j)) - (b_{t_{j+1}} - b_{t_j})\mathbf{v}(\mathbf{x}(t_j)) = 0, \\ B_j = \frac{La(\mathbf{x}(t_{j+1}))g(\mathbf{x}(t_{j+1}))\|\mathbf{x}(t_{j+1}) - \mathbf{x}(t_j)\|^2}{2}. \end{cases}$$

We select an appropriate integer $N = O\left(\frac{DL}{\varepsilon}\right)$, yielding the following approximation guarantee:

$$F(\mathbf{x}(t_N)) \geq \left(1 - \frac{1}{e}\right)\theta \cdot F(x^*) - \varepsilon.$$

1.2 Related Work

Classic Contracts. Contract theory, initially introduced by [34], and its core principal-agent model find widespread applications in various domains, including crowdsourcing platforms and resource allocation. Seminal contributions to the contract principal-agent problem include the works of Holmström [31] and Grossman and Hart [26], wherein a principal endeavors to incentivize an agent to complete a costly task through contract design. The computation of optimal contracts often involves linear programming techniques. Our research Our work is part of the frontier of optimizing the efforts of others in the design of algorithmic mechanisms, which includes work on algorithmic contract design [1,13], scoring rules [33], delegation [7,32].

Our paper contributes to the computational aspects of contract theory. Linear contracts have been demonstrated to be max-min optimal in previous works [8,21]. Guruganesh et al. [27] investigated the integration of moral hazard (i.e., hidden actions) with screening (i.e., hidden types), revealing that menu contracts often outperform individual contracts. Castiglioni et al. [10] introduced stochastic contract menus and established that computing optimal stochastic menu contracts can be efficiently achieved through linear programming. Additionally, randomized contract menus exhibit superior computational properties compared to deterministic menus in the Bayesian setting. Assuming a constant size of actions, Alon et al. [1] demonstrated that computing optimal contract menus deterministically can be done in polynomial time. Castiglioni et al. [9] investigated the approximate guarantee of linear contracts compared to optimal contracts, revealing a multiplicative loss that increases linearly with the number of agent types.

Combinatorial Contracts. Combinatorial contracts have emerged as a significant aspect of contract theory in recent years, often encompassing both action-based and agent-based selections. Consider a combinatorial contract based on agent choice, wherein the principal incentivizes a subset of agents to undertake high-cost efforts. Babaioff et al. [2] introduced this model and provided theoretical insights into the optimality of contract designs. Similar extension results are also available in [3,4,23]).

Dütting et al. [16] introduced a combinatorial contract problem based on action selection. In this model, the agent can take any subset of actions, and the principal's reward function is defined based on the set of actions chosen by the agent. They first demonstrated that the optimal contract can be computed in polynomial time when the success probability function exhibits gross substitutability. Vuong et al. [36] further shown that the optimal contract can also be determined in polynomial time when the reward function is supermodular and the cost function is modular. Recently, Dütting et al. [18] extended this result to the case of submodular cost functions and demonstrated that this extension can still be efficiently computed in polynomial time. Additionally, Dutting et al. [19] addressed action-based combinatorial contracts from the perspective of query complexity and revealed that computing optimal contracts with submodular reward functions needs an exponential number of demand queries.

Dütting et al. [17] examined agent-based combinatorial contracts, focusing on reward function within the complement-free hierarchy: submodular, XOS and subadditive. For XOS or submodular reward functions, they devised a $O(1)$-approximation algorithm. However, they demonstrated that achieving an approximation ratio of $\Omega(\sqrt{n})$ in polynomial time is not possible for subadditive reward functions. Recently, Ezra et al. [24] provided hardness results for both agent-based and action-based combinatorial contract models. In the agent selection-based scenario, despite a constant approximation for submodular function rewards in [17], an upper bound of 0.7 on the approximation ratio was established. Regarding XOS function rewards, under the assumption of value oracles, algorithms were proven to lack constant approximation capabilities, contrasting with [17], where such algorithms exist under value and oracles assumptions. Meanwhile, for action selection-based scenarios, although Dütting et al. [16] demonstrated NP-hardness in computing optimal contracts with submodular reward functions, they further showed the absence of any constant approximation ratio algorithm for this problem.

Furthermore, Castiglioni et al. [11] investigated scenarios where agents could independently produce random outcomes after selecting an action, and where the probability distributions of these outcomes were mutually independent. They demonstrated the inapproximability of optimal contracts for IR/DR-functions, respectively. Moreover, if a marginal additive loss is permissible, there exists a multiplicative $(1 - 1/e)$-approximation algorithm.

Additional Online Contracts. In addition to traditional contract models, the perspective of online learning algorithms plays a crucial role in the theory of algorithmic contracts. Ho et al. [30] framed the repeated principal-agent model as a multi-armed bandit problem, where each arm represents a contract. By discretizing the action space, they derived an initial sample complexity bound. An extension approach is presented in [13]. Zhu et al. [38] introduced a spherical coding technique to discretize the contract space and devised an online learning algorithm with a regret bound of $\widetilde{O}(\sqrt{m}T^{1-1/(2m+1)})$, where m represents the size of outcomes and T denotes the number of interaction rounds. Additionally, they established a lower bound on any learning algorithm: $\Omega(T^{1-1/(m+2)})$. The sample complexity of linear contracts is $\widetilde{\Theta}(T^{2/3})$. When the action size n is constant, Bacchiocchi et al. [5] proposed an online learning algorithm and provided a high-probability upper bound estimation on the rounds needed to approximate the optimal contract. Furthermore, they demonstrated that the upper bound on the regret value of the algorithm is $\widetilde{O}(m^n T^{4/5})$. Recently, Chen et al. [12] introduced learning algorithms aimed solely at learning the underlying outcome distributions, achieving improved query complexities in both the action query and contract query models. Additionally, Dutting et al. [20] established the existence of a learning algorithm with a regret bound of $O(L \log L)$, under the condition that the principal utility function is L-Lipschitz continuous, where L represents the Lipschitz constant. An extension learning model that enables the principal to offer a menu of contracts to the agent is presented in [28].

1.3 Organization

We introduce our preliminaries in Sect. 2. Our continuous-time algorithm, based on the Lyapunov function, is presented in Sect. 3. The discrete-time algorithm is outlined in Sect. 4. Finally, we conclude our work in Sect. 5.

2 Preliminaries

The Multi-agent Principal-Agent Model with Continuous Actions. In this model, a principal delegates a task by interacting with a set of multiple agents, denoted as $N = \{1, ..., n\}$. Our focus lies on a continue action setup, wherein all agents take actions $\mathbf{x} = (x_1, ..., x_n) \in [0, 1]^n$ to represent the degree of effort exerted, incurring a cost of $c(\mathbf{x}) = (c_1(x_1), ..., c_n(x_n)) \in \mathbb{R}^n_{\geq 0}$.

Given a set of outcomes Ω, a distribution function $q_{\mathbf{x}} : \Omega \to [0, 1]$ is defined over outcomes under the exerting effort $\mathbf{x}$. A stochastic outcome realization $\omega \in \Omega$ is drawn from $q_{\mathbf{x}}(\omega)$ and accordingly a reward $r(\omega)$ gives to the principal. Let $f : [0, 1]^n \to \mathbb{R}_{\geq 0}$ represent the reward of the principal under the action vector $\mathbf{x} \in [0, 1]^n$. We denote $f(\mathbf{x}) = \sum_{\omega \in \Omega} q_{\mathbf{x}}(\omega) r(\omega)$. We assume function f is non-negative monotone and DR-submodular, expressed as: $f(\mathbf{x}+\delta \cdot \mathbf{1}_i) - f(\mathbf{x}) \geq f(\mathbf{y}+\delta \cdot \mathbf{1}_i) - f(\mathbf{y})$ for any two action vectors $\mathbf{x} \leq \mathbf{y} \in [0, 1]^n$ and $\delta \in [0, 1 - y_i]$ for any coordinate i. Furthermore, we assume that f is differentiable, and an equivalent form of DR-submodularity is summarized as: $\forall \mathbf{x} \leq \mathbf{y} \in [0, 1]^n, \nabla f(\mathbf{x}) \geq \nabla f(\mathbf{y})$. Additionally, we assume that f is L-smooth, meaning that for all $\mathbf{x} \leq \mathbf{y} \in [0, 1]^n$, $\|\nabla f(\mathbf{y}) - \nabla f(\mathbf{x})\|_2 \leq L\|y - x\|_2$.

Our model falls within the moral hazard setting, where agents are reluctant to exert effort to complete tasks due to the costs associated with their actions. This reluctance stems from the fact that the benefits of their efforts primarily accrue to the principal. To incentive agents to undertake costly actions, the principal must offer a contract $\mathbf{t} : \Omega \to \mathbb{R}^n_{\geq 0}$ specifies payments to the agents based on the outcomes. In this context, for a given realizationω, $\mathbf{t}(\omega) = (t_1(\omega), ..., t_n(\omega))$ denotes the transfer of payments to the agents. We examine linear contracts, a widely studied type in existing literature. In a linear contract, denoted as $\mathbf{t}(\omega) = r(\omega) \cdot \boldsymbol{\alpha}$, $\boldsymbol{\alpha} = (\alpha_1, ..., \alpha_n) \in \mathbb{R}^n_{\geq 0}$ represents the proportionality coefficient and $r(\omega)$ denotes the reward to the principal for outcome ω. Consequently, the expected payments to the agents amount to $f(\mathbf{x}) \cdot \boldsymbol{\alpha}$ under effort $\mathbf{x}$.

Utility Functions and Equilibria. Given a linear contract $\boldsymbol{\alpha}$. Let $\mathbf{x} \in [0, 1]^n$ denote the exerted effort actions. The principal's utility is expressed as $(1 - \sum_{i \in N} \boldsymbol{\alpha}_i) f(\mathbf{x})$ while the agents' utilities are given by $f(\mathbf{x}) \cdot \boldsymbol{\alpha} - \mathbf{c}(\mathbf{x})$. The agents engage in a simultaneous Nash game before the linear contract is disclosed. We define the actions vector $\mathbf{x} = (x_i, \mathbf{x}_{-i})$ as a Nash equilibrium under the linear contract $\boldsymbol{\alpha}$ if, for any agent $i \in \{1, ..., n\}$, it holds $\alpha_i f(x_i, \mathbf{x}_{-i}) - c_i(x_i) \geq \alpha_i f(y, \mathbf{x}_{-i}) - c_i(y), \forall y \in [0,1]$, where x_i denotes the action of exerting effort taken by i, $\mathbf{x}_{-i}$ denotes the rest of actions taken by the other agents excluding i. The contract design problem can be summarized as the following bi-level maximization:

$$\max_{\boldsymbol{\alpha} \in [0,1]^n} \left(1 - \sum_{i \in V} \alpha_i\right) f(\mathbf{x})$$
$$\text{s.t. } x_i \in \arg\max_{y \in [0,1]} \alpha_i f(y, \mathbf{x}_{-i}) - c_i(y), \forall i \in \{1, ..., n\}.$$

Fixed $\mathbf{x} \in [0,1]^n$, the principal's best response to agent i is

$$\alpha_i \geq \max_{y \in [0,1]} \frac{c_i(x_i) - c_i(y)}{f(x_i, \mathbf{x}_{-i}) - f(y, \mathbf{x}_{-i})}. \tag{2}$$

The right-hand side of inequality (2) can be upper-bounded as follows:

Define $\delta_i = |x_i - y|$ as the distance parameter between x_i and y. Let $L^i(\delta_i, \mathbf{x}_{-i}) = \max\{L_1^i(\delta_i, \mathbf{x}_{-i}), L_2^i(\delta_i, \mathbf{x}_{-i})\}$, where $L_1^i(\delta_i, \mathbf{x}_{-i}) = \frac{c_i(\delta_i) - c_i(0)}{f(1, \mathbf{x}_{-i}) - f(1-\delta_i, \mathbf{x}_{-i})}$ and $L_2^i(\delta_i, \mathbf{x}_{-i}) = \frac{c_i(1-\delta_i) - c_i(1)}{f(0, \mathbf{x}_{-i}) - f(\delta_i, \mathbf{x}_{-i})}$.

Observation 1. *For any $x \in [0,1]^n$, we can conclude that*

$$L^i(\delta_i, \boldsymbol{x}_{-i}) \geq \max_{y \in [0,1]} \frac{c_i(x_i) - c_i(y)}{f(x_i, \boldsymbol{x}_{-i}) - f(y, \boldsymbol{x}_{-i})}, \forall i \in N.$$

The principal aims to address the following single-level optimization problem:

$$\max_{\mathbf{x} \in [0,1]^n} F(\mathbf{x}) := \max_{\mathbf{x} \in [0,1]^n} \left(1 - \sum_{i \in N} L^i(\delta_i, \mathbf{x}_{-i})\right) f(\mathbf{x}) := \max_{\mathbf{x} \in [0,1]^n} g(\mathbf{x}) \cdot f(\mathbf{x}),$$

where $g(\mathbf{x}) = \left(1 - \sum_{i \in N} L^i(\delta_i, \mathbf{x}_{-i})\right)$ and $F(\mathbf{x}) = g(\mathbf{x}) \cdot f(\mathbf{x})$.

Algorithm 1

1: **Initialization** Objective function $F(\cdot) = g(\cdot)f(\cdot)$, $\mathbf{x}(0) = \mathbf{0}$.
2: **for** $t \in [0,1]$ **do**
3: $\quad b_t = e^t$, $a(\mathbf{x}(t)) = \frac{b_t}{g(\mathbf{x}(t))}$
4: $\quad \mathbf{v}(\mathbf{x}(t)) \in \arg\max_{\mathbf{v} \in [0,1]^n} \langle \nabla f(\mathbf{x}(t)), \mathbf{v} \rangle$
5: $\quad \dot{\mathbf{x}}(t) = \frac{\dot{b}_t}{b_t} \mathbf{v}(\mathbf{x}(t))$
6: **end for**
7: **return** $\mathbf{x}(1)$.

Let $\mathbf{x}^*$ denote an optimal solution of Problem (1), given by $\mathbf{x}^* \in \arg\max_{\mathbf{x}\in[0,1]^n} F(\mathbf{x})$. Additionally, we assume that the infimum value $\theta = \inf_{\mathbf{x}\in[0,1]^n} (1 - \sum_{i\in N} L^i(\delta_i, \mathbf{x}_{-i}))$ (≥ 0) of term g can be attained. Thus, we have $\theta \leq g(\mathbf{x}) \leq 1$.

Oracles. We assume access to value oracles for f and c_i. These oracles, given any $\mathbf{x} \in [0,1]^n$, compute the reward $f(\mathbf{x})$ and the cost $c_i(x_i)$ for any $i \in N$.

3 Continuous-Time Algorithm

In this section, we introduce a continuous-time algorithm based Lyapunov function, list as Algorithm 1, to derive an approximate contract scheme. Formally, we establish the following theorem. All deferred proofs appear in the full version of this paper.

Theorem 1. *Consider the multi-agent principal-agent problem with continuous actions, Algorithm 1 provides a solution $\boldsymbol{x}$ within the time interval $[0,1]$, satisfying the following condition:*

$$F(\boldsymbol{x}) \geq \left(1 - \frac{1}{e}\right) \theta \cdot F(\boldsymbol{x}^*).$$

To establish the aforementioned theorem, we rely on the framework outlined by Du [15], This work introduces a novel framework for designing and analyzing approximation algorithms using Lyapunov function approaches. It's noteworthy that the Lyapunov functions described in [15] exhibit the following form: $E(\mathbf{x}(t)) = a(t)F(\mathbf{x}(t)) - b_t F(\mathbf{x}^*), \forall t \in [0,T]$, where functions $a(\cdot)$ and $b(\cdot)$ are considered in time t satisfying $Q = \{(a,b) : a_0 > 0, \dot{a}_t \geq 0; b_0 \geq 0, \dot{b}_t \geq 0\}$.

However, we use a distinct parametric form for the monotone Lyapunov function over the time interval $t \in [0,T]$ through the utilization of $E(\mathbf{x}(t)) = a(\mathbf{x}(t))F(\mathbf{x}(t)) - b_t F(\mathbf{x}^*)$, where $a(\cdot)$ and $b(\cdot)$ are determined in the subsequent sections. Given $E(\mathbf{x}(t))$ resulting from an algorithm $\mathbf{x}(t)$, as $E(\cdot)$ is increasing in t, we obtain:

$$a(\mathbf{x}(T))F(\mathbf{x}(T)) - b_T F(\mathbf{x}^*) = E(\mathbf{x}(T)) \geq E(\mathbf{x}(0)) = a(\mathbf{x}(0))F(\mathbf{x}(0)) - b_0 F(\mathbf{x}^*),$$

which implies

$$F(\mathbf{x}(T)) \geq \frac{b_T - b_0}{a(\mathbf{x}(T))} F(\mathbf{x}^*). \tag{3}$$

We denote by $\nabla F(\mathbf{x}(t)) = g(\mathbf{x}(t))\nabla f(\mathbf{x}(t)) + f(\mathbf{x}(t))\nabla g(\mathbf{x}(t))$. Based on the DR-submodularity of f, we obtain the upper bound U_t as:

$$U_t = f(\mathbf{x}(t)) + \max_{\mathbf{v}\in[0,1]^n} \langle \nabla f(\mathbf{x}(t)), \mathbf{v}\rangle = f(\mathbf{x}(t)) + \langle \nabla f(\mathbf{x}(t)), \mathbf{v}(\mathbf{x}(t)),$$

Algorithm 2

1: **Initialization** Objective function $F(\cdot) = g(\cdot)f(\cdot)$, $t_0 = 0$, $\mathbf{x}(0) = \mathbf{0}$.
2: **for** $j \in [1, ..., N]$ **do**
3: $\quad t_j = \frac{j}{N}$, $b_{t_j} = e^{t_j}$, $a(\mathbf{x}(t_j)) = \frac{b_{t_j}}{g(\mathbf{x}(t_j))}$
4: $\quad \mathbf{v}(\mathbf{x}(t_j)) \leftarrow \arg\max_{\mathbf{v}\in[0,1]^n} \langle \nabla f(\mathbf{x}(t_j)), \mathbf{v}\rangle$
5: $\quad \dot{\mathbf{x}}(t_j) = \frac{\dot{b}_{t_j}}{b_{t_j}} \mathbf{v}(\mathbf{x}(t_j))$
6: **end for**
7: **return** $\mathbf{x}(t_N)$.

where $\mathbf{v}(\mathbf{x}(t)) \in \arg\max_{\mathbf{v}\in[0,1]^n}[f(\mathbf{x}(t)) + \langle \nabla f(\mathbf{x}(t)), \mathbf{v}\rangle = \arg\max_{\mathbf{v}\in[0,1]^n}\langle \nabla f(\mathbf{x}(t)), \mathbf{v}\rangle$. We obtain the following sufficient condition to ensure $\dot{E}(x(t)) \geq 0$ by means of

$$\begin{cases} g(\mathbf{x}(t))\langle \nabla a(\mathbf{x}(t)), \dot{\mathbf{x}}(t)\rangle + a(\mathbf{x}(t))\langle \nabla g(\mathbf{x}(t)), \dot{\mathbf{x}}(t)\rangle - \dot{b}_t = 0, \\ a(\mathbf{x}(t))g(\mathbf{x}(t))\dot{\mathbf{x}}(t) - \dot{b}_t \mathbf{v}(\mathbf{x}(t)) = 0. \end{cases}$$

By setting $g(\mathbf{x}(t))a(\mathbf{x}(t))) = b_t$, we have $\dot{\mathbf{x}}(t) = \frac{\dot{b}_t}{b_t}\mathbf{v}(\mathbf{x}(t))$. And then

$$\mathbf{x}(T) = \int_{t=0}^{T} \frac{\dot{b}_t}{b_t} \mathbf{v}(\mathbf{x}(t))dt = \int_{t=0}^{T} \mathbf{v}(\mathbf{x}(t))d(\ln b_t).$$

We let $\ln b_0 = 0$, $\ln b_T = 1$, and $(\ln b_t)' \geq 0$ by setting $b_t = e^t$ and $T = 1$. Thus we yield $F(\mathbf{x}(1)) \geq \left(1 - \frac{1}{e}\right) \theta \cdot F(\mathbf{x}^*)$.

4 Discrete-Time Algorithm

In this section, we shift our attention to discretization schemes. Our primary achievement is a polynomial discrete-time algorithm, list as Algorithm 2, which approximates the scheme with nearly the same approximation ratio. Formally, we derive the following theorem.

Theorem 2. *Consider the multi-agent principal-agent problem with continuous actions, Algorithm 2 yields a solution $\boldsymbol{x}$ after $O\left(\frac{DL}{\varepsilon}\right)$ iterations, ensuring that*

$$F(\boldsymbol{x}) \geq \left(1 - \frac{1}{e}\right) \theta \cdot F(x^*) - \varepsilon.$$

5 Conclusion

We addressed the principal multi-agent contract design problem involving continuous actions. Employing a backward induction method, we streamlined the two-layer game problem into a single-layer optimization problem. Our approach entailed devising a novel Lyapunov function-based algorithm capable of achieving a parameterized approximation. The algorithms we developed heavily leverage properties such as DR-submodularity [25,29] and the monotonicity of reward and cost functions. However, in certain real-world scenarios, utility functions may lack these desirable properties. Therefore, we extended our investigation to include scenarios where the monotonicity of reward functions is not guaranteed, exploring the design of algorithms suited to more general reward functions.

Acknowledgements. The fourth author is supported by Natural Science Foundation of China (No. 12131003). The fifth author is supported by Natural Science Foundation of China (No. 12101587) and China Postdoctoral Science Foundation (No. 2022M720329). The second author is supported by the Natural Sciences and Engineering Research Council of Canada (NSERC) Grant 03840. The third author is supported by Natural Science Foundation of China (No. 11201333).

References

1. Alon, T., Dütting, P., Talgam-Cohen, I.: Contracts with private cost per unit-of-effort. In: Proceedings of ACM EC, pp. 52–69 (2021)
2. Babaioff, M., Feldman, M., Nisan, N.: Combinatorial agency. In: Proceedings of ACM EC, pp. 18–28 (2006)
3. Babaioff, M., Feldman, M., Nisan, N.: Mixed strategies in combinatorial agency. J. Artif. Intell. Res. **38**, 339–369 (2010)
4. Babaioff, M., Feldman, M., Nisan, N.: Free-riding and free-labor in combinatorial agency. In: Proceedings of SAGT 2009, pp. 109–121 (2009)
5. Bacchiocchi, F., Castiglioni, M., Marchesi, A., Gatti, N.: Learning optimal contracts: How to exploit small action spaces. arXiv: 2309.09801 (2023)
6. Bastani, H., Bayati, M., Braverman, M., Gummadi, R., Johari, R.: Analysis of medicare pay-for-performance contracts. Available at SSRN 2839143 (2016)
7. Bechtel, C., Dughmi, S., Patel, N.: Delegated pandora's box. In: Proceedings of ACM EC, pp. 666–693 (2022)
8. Carroll, G.: Robustness and linear contracts. Am. Econ. Rev. **105**(2), 536–563 (2015)
9. Castiglioni, M., Marchesi, A., Gatti, N.: Bayesian agency: linear versus tractable contracts. In: Proceedings of ACM EC, pp. 285–286 (2021)
10. Castiglioni, M., Marchesi, A., Gatti, N.: Designing menus of contracts efficiently: the power of randomization. In: Proceedings of ACM EC, pp. 705–735 (2022)
11. Castiglioni, M., Marchesi, A., Gatti, N.: Multi-agent contract design: how to commission multiple agents with individual outcomes. In: Proceedings of ACM EC, pp. 412–448 (2023)
12. Chen, Y., Chen, Z., Deng, X., Huang, Z.: Are bounded contracts learnable and approximately optimal? arXiv: 2402.14486 (2024)
13. Cohen, A., Deligkas, A., Koren, M.: Learning approximately optimal contracts. In: Proceedings of SAGT, pp. 331–346 (2022)
14. Cong, L., He, Z.: Blockchain disruption and smart contracts. Rev. Financ. Stud. **32**(5), 1754–1797 (2019)
15. Du, D.: Lyapunov function approach for approximation algorithm design and analysis: with applications in submodular maximization. arXiv:2205.12442 (2022)

16. Dütting, P., Ezra, T., Feldman, M., Kesselheim, T.: Combinatorial contracts. In: Proceedings of FOCS, pp. 815–826 (2022)
17. Dütting, P., Ezra, T., Feldman, M., Kesselheim, T.: Multi-agent contracts. In: Proceedings of STOC, pp. 1311–1324 (2023)
18. Dütting, P., Feldman, M., Gal Tzur, Y.: Combinatorial contracts beyond gross substitutes. In: Proceedings of SODA, pp. 92–108 (2024)
19. Dütting, P., Feldman, M., Gal-Tzur, Y., Rubinstein, A.: The query complexity of contracts. arXiv: 2403.09794 (2024)
20. Dütting, P., Guruganesh, G., Schneider, J., Wang, J.: Optimal no-regret learning for one-side lipschitz functions. In: Proceedings of ICML, pp. 8836–8850 (2023)
21. Dütting, P., Roughgarden, T., Talgam-Cohen, I.: Simple versus optimal contracts. In: Proceedings of ACM EC, pp. 369–387 (2019)
22. Dutting, P., Roughgarden, T., Talgam-Cohen, I.: The complexity of contracts. SIAM J. Comput. **50**(1), 211–254 (2021)
23. Emek, Y., Feldman, M.: Computing optimal contracts in combinatorial agencies. Theoret. Comput. Sci. **452**, 56–74 (2012)
24. Ezra, T., Feldman, M., Schlesinger, M.: On the (In) approximability of combinatorial contracts. In: Proceedings of ITCS (2024)
25. Feldman, M., Naor, J., Schwartz, R.: A unified continuous greedy algorithm for submodular maximization. In: Proceedings of FOCS, pp. 570–579 (2011)
26. Grossman, S., Hart, O.: An analysis of the principal-agent problem. Econometrica **51**(1), 7–45 (1983)
27. Guruganesh, G., Schneider, J., Wang, J.: Contracts under moral hazard and adverse selection. In: Proceedings of ACM EC, pp. 563–582 (2021)
28. Han, M., Albert, M., Xu, H.: Learning in online principle-agent interactions: the power of menus. arXiv:2312.09869 (2023)
29. Hassani, H., Soltanolkotabi, M., Karbasi, A.: Gradient methods for submodular maximization. In: Proceedings of NIPS, pp. 5841–5851 (2017)
30. Ho, C., Slivkins, A., Vaughan, J.: Adaptive contract design for crowdsourcing markets: bandit algorithms for repeated principal-agent problems. J. Artif. Intell. Res. **55**, 317–359 (2016)
31. Holmström, B.: Moral hazard and observability. Bell J. Econ. **10**(1), 74–91 (1979)
32. Kleinberg, J., Kleinberg, R.: Delegated search approximates efficient search. In: Proceedings of ACM EC, pp. 287–302 (2018)
33. Li, Y., Hartline, J.D., Shan, L., Wu, Y.: Optimization of scoring rules. In: Proceedings of ACM EC, pp. 988–989 (2022)
34. Ross, A.: The economic theory of agency: the principal's problem. Am. Econ. Rev. **63**(2), 134–139 (1973)
35. Statista Research Department. Instagram infuencer marketing spending worldwide from 2013 to 2020 (2021). https://www.statista.com/statistics/950920/global-instagram-infuencer-marketing-spending/
36. Vuong, R., Dughmi, S., Patel, N., Prasad, A.: On supermodular contracts and dense subgraphs. In: Proceedings of ACM-SIAM SODA, pp. 109–132 (2024)
37. Xia, J., Niu, W.: Carbon-reducing contract design for a supply chain with environmental responsibility under asymmetric information. Omega **102**, 102390 (2021)
38. Zhu, B., Bates, S., Yang, Z., Wang, Y., Jiao, J., Jordan, M.: The sample complexity of online contract design. arXiv:2211.05732 (2022)

Privacy-Preserving Byzantine-Robust Federated Learning via Multiparty Homomorphic Encryption

Wenhao Jiang(✉), Songwei Luo, Shaojing Fu, Lin Liu, and Yuchuan Luo

College of Computer, National University of Defense Technology, Changsha, China
{jwh_roy,luosongwei20,fushaojing,liulin16,luoyuchuan09}@nudt.edu.cn

Abstract. Federated learning (FL) is a novel distributed machine learning paradigm where multiple users collaborate to train a global model under the coordination of a central server. However, some prior researches in the domain of FL do not adequately consider adversarial actions. Moreover, these studies often entail significant computational and communicative expenditures, and frequently fall short in achieving an optimal equilibrium between efficiency and privacy preservation. Such limitations is impeding the practical application of FL systems. In this paper, we propose PBFL, building a bridge between privacy and robustness while balancing efficiency. Compared to previous works, PBFL boasts stronger privacy protection and lower computational and communicational overhead. Specifically, on the privacy side, we use Multiparty homomorphic encryption (MHE) as a privacy building block (secure against post-quantum attacks), and in terms of robustness, we build a Byzantine robust aggregation scheme on ciphertext based on the similarities between benign user parameters. We provide a detailed theoretical analysis to demonstrate the security and convergence properties of the scheme. The experiment on two image classification datasets shows that the aggregation weight of Byzantine user is reduced by at least 99.88% compared to benign users, which is negligible in aggregation.

Keywords: Federated learning · Privacy-preserving · Byzantine-robust · Multiparty homomorphic encryption

1 Introduction

Federated learning (FL) represents a distributed machine learning approach that prioritizes privacy preservation [10]. This method facilitates a collaborative training process for a collective model among users, while ensuring that the individual data of each participant remains confidential. Due to the privacy protection of

This work is supported by National Nature Science Foundation of China (No.62102429, No.62102422, No.62102430, No.62072466), Natural Science Foundation of Hunan Province, China (Grant No. 2023JJ30640)
W. Jiang and S. Fu—These authors contributed equally to this work.

Y. Chen et al. (Eds.): COCOON 2024, LNCS 15162, pp. 432–446, 2025.
https://doi.org/10.1007/978-981-96-1093-8_36

user raw data, FL has been deployed in real-world applications. For instance, Google uses FL to build privacy-preserving next-word prediction application on mobile phones [9], and WeBank has deployed FL for credit risk prediction [2]. Roughly speaking, there are three steps in FL: the aggregation server sends global model to participants; every user trains local parameters based on the received global parameters and sends the trained local parameters back to the server; and the server aggregate the local parameters to generate the global model for a new round. Both parties in FL will repeat the above steps until the global model converges.

Despite its privacy-enhanced paradigm, FL has been proved that privacy and Byzantine robustness need to be improved [6,10]. (i) Inference attack. It is caused by the uploaded parameters of users. The aggregation server could infer the users' private data through the received parameters, which is called inference attack [3]. (ii) Byzantine attacks. It is caused by the Byzantine users which are controlled by the adversary who aims to compromise the global model by manipulating Byzantine users to upload poisoned model parameters. These attack methods are called Byzantine attacks [7]. Notably, some works [13] show that even a single Byzantine user could manipulate the whole vanilla FL system. Due to the aforementioned existing attacks, a secure and trusted FL system must consider two potential threats, one is to protect the privacy of users' data and the other is to guarantee the robustness of the user parameter aggregation process against Byzantine users.

There are recent researches that are proposed to ensure the privacy and Byzantine-robustness in FL. PEFL [11] uses homomorphic encryption to encrypt the user gradient G_i, and then uses two non-collusive server for robust aggregation on the ciphertexts. To achieve some nonlinear operations on homomorphic encryption, PEFL designs four security protocols in which the user gradient G_i is randomly perturbed and then decrypted. However, for all dimensions in G_i, PEFL add masks with the same random number, which will leak information about the data distribution of the user's gradient. FLOD [5] and SecureFL [8] are both based on the FLTrust [2] idea of using root dataset, i.e., the server needs to collect a small set of data from the user in order to better identify malicious users, which violates user privacy. FLAME [14] performs a careful aggregation protocol design and combines ABY [4] and privacy-preserving DBSCAN [1] for privacy protection, but its designed protocols contain time-consuming calculations on Secure Two-party Computation (2PC), such as cosine similarity alculation and clustering operations of parameters. ELSA [15] is a secure aggregation protocol for federated learning that provides strong privacy protection through distributed trust and efficient computation, but requires additional communication overhead to ensure security in adversarial environments. Table 1 shows the comparison with prior works in terms of privacy, robustness and efficiency of FL. Here, we believe that the server's acquisition of root data with labels may infringe upon user privacy. Therefore, the presence or absence of root data on the server can be used to judge the strength of the hypothetical conditions.

To address the above issues, we propose PBFL (**P**rivacy-preserving **B**yzantine-robust **F**ederated **L**earning), which can ensure the privacy of clients

Table 1. Comparison with prior works on properties necessary for FL.

Schemes	No root data in server	Privacy	Byzantine robustness	Computation complexity	Communication complexity
PEFL [11]	✓	✓	✓̸	$O(11NM + NMlog(M))$	$O(7NM)$
FLOD [5]	×	✓̸	✓	$O(12NM + NMlog(M))$	$O(8NM)$
SecureFL [8]	×	✓̸	✓	$O(2NM + 11sN + NMlog(M))$	$O(2NM + 6sN)$
FLAME [14]	✓	✓̸	✓̸	$O(3NM + N^2M + NMlog(M))$	$O(2NM)$
ELSA [15]	✓	✓	✓̸	$O(NM+NMlogM+5N^2 + 7bN)$	$O(3NM + 6bN)$
Our PBFL	✓	✓	✓	$O(6NM + NMlog(M))$	$O(4NM)$

[1] ✓ denotes it largely meets the requirements, ✓̸ denotes it meets the requirement incompletely, × represents it is unsatisfied.
[2] N, M, b, s: number of users, dimension of model parameter, number of bits in SecureFL settings, length of security seed parameter in ELSA settings.

in the FL training process while eliminating the influence of Byzantine users on the global model. Specifically, our main contributions are as follows:

- We propose a novel privacy-preserving and Byzantine-robust FL scheme (PBFL) using Multiparty Homomorphic Encryption as the privacy-preserving building block. This solution has stronger privacy and lower costs compared to previous work, while also being able to resist Byzantine attacks well.
- We design a new efficient Byzantine-robust aggregation scheme that is compatible with the privacy-preserving mechanism, which adaptively adjusts the aggregation weight of users by taking the median of Euclidean distance as the benchmark. As the training proceeding, global parameters at each training round are almost generated by benign users, which can make full use of benign users to train the global model.
- The experiments on different datasets show that our PBFL can efficiently resist typical Byzantine attacks, e.g., label-flipping attack and Gaussian attack. Besides, our comprehensive security analysis and convergence analysis prove that our scheme could train Byzantine-robust global model while preserving user privacy.

2 Problem Setup

2.1 System Model

Our system framework consists of four key participants:

- *Aggregation Server (AS):* AS is an aggregation server that provides aggregation service for the collaborative training users. Specifically, it collects the encrypted parameters of users and aggregates them to generate global model parameters in each round.
- *Crypto Service Provider (CSP):* CSP is a server that assists the AS to securely aggregate parameters from users. Concretely, CSP owns a part of the server's private key to assist AS in decrypting some intermediate results that do not violate user privacy.
- *Data Owners (Users in FL):* This role has the raw data that needs to be trained collaboratively. In our setting, a part of the users could be malicious(Byzantine users) who want to destroy the global model.
- *Key Generation Center (KGC):* KGC is a semi-honest entity that assists in generating a pair of public and private keys (pk, sk), which makes private key distributed across users and AS. This setting allows KGC to only get half of the private key so that user privacy cannot be violated.

2.2 Threat Model

In this paper, we are committed to addressing the privacy vulnerability caused by the semi-honest aggregation server and the security vulnerability caused by Byzantine users in FL. We assume that all the participants in FL have their local training dataset with the same distribution (i.i.d). Byzantine users could transmit arbitrary local model parameters to the Aggregator Server (AS), thereby affecting the development of the overarching model.

In accordance with previous works [12,16], we assume that the number of malicious users is not more than half of the total users, which is close to a real FL scenario, i.e., $|M| \leq \frac{|T|-1}{2}$, where $|M|$ and $|T|$ are indicate the number of malicious users and the total number of the users, which ensures the convergence of the global model. Besides, since AS and CSP could obtain the parameters of users and aggregate them correctly, we need to assume that they are semi-honest adversaries. This means that AS and CSP would execute the aggregation algorithm honestly, and at the same time, try to infer more information from the parameters that they have received. As for the KGC, we consider it as a semi-honest entity, which means that the KGC will adhere to the key generation protocol strictly but try to infer the users' private information. After cooperating with AS to generate the user-side key, KCG will send it to all the users. In addition, we assume that there is no collusion among the four entities (AS, CSP, users, KGC).

2.3 Design Goals

With the threats in FL mentioned above, our PBFL is designed to achieve the following goals:

- *Robustness against Byzantine users:* Our system operates under the presumption that no more than fifty percent of participants are adversarial (malicious users), capable of undermining the integrity of the aggregate model through tampering with their individual data sets or by transmitting random parameters. The Privacy-Preserving Federated Learning (PBFL) protocol should be designed to maintain robustness in the face of such malevolent entities.
- *Privacy of local model parameters and global model parameters:* In particular, it is essential that the confidentiality of both the local and consolidated global model parameters is maintained for the AS and CSP.
- *Accuracy of the global model:* The goal of collaborative training of different users is to get a better global model, the PBFL should keep accuracy within an acceptable range while ensuring privacy and security.

3 Proposed Scheme

3.1 Our Crypto-Friendly Byzantine-Robust FL Aggregation Protocol

To design our crypto-friendly byzantine-robust FL aggreagtion protocol, we have made a novel design in the selection of distance calculation benchmark, and the assignment of aggregation weights, the specific design details are described as follows, and the algorithm description is shown in Algorithm 1.

Algorithm 1. Crypto-friendly byzantine-robust FL aggregation protocol

Input: Users local models $\{W_1^t, W_2^t, ..., W_N^t\}$ at iteration t, where N is the number of users in FL; global model weight W_g^{t-1} at iteration $t-1$ (last iteration).
Output: Global model weight W_g^t at iteration t.

for $i \in [1...N]$ calculates the squared euclidean distance **do**
    $d_i = \|W_i^t - W_g^{t-1}\|^2$
end for
Obtains the median value of distance vector $d : d_{mid} = median(\{d_1, d_2, ..., d_n\})$
for $i \in [1...N]$ calculates the relative distance to d_{mid} **do**
    $d_i' = |d_i - d_{mid}|$
end for
for $i \in [1...N]$ **do**
    $score_i = \frac{\sum_{j=1}^{N}(d_j'+1)}{d_j'+1}$ *//The closer users are to d_{mid}, the higher the score*
end for
for $i \in [1...N]$ rescales the $scores_i$ to range 0-1 **do**
    $\gamma_i = \frac{score_i}{\sum_{j=1}^{N} score_j}$
end for
Linearly aggregate parameters using rescaled scores: $W_g^t = \sum_{i=1}^{N} \gamma_i \times W_i^t$

Efficient Distance Calculation: Like in previous work [11], our algorithm requires a benchmark, which is used to judge the aggregation weight of each model based on the distance from the user model to this benchmark. We observe that the Euclidean distance from the model parameters uploaded by the user to the global model of the previous round reflects the direction and magnitude of the user update, so we use the global model of the previous round as the benchmark for distance calculation, and compare the computational complexity $O(N)$ of krum and PEFL in choosing the benchmark matter, we have a clear computational complexity advantage of $O(1)$.

Weight Assignment Based on Median Idea: For the benchmark we choose (i.e., the last round global model parameters), Byzantine users can make small updates or directly use the global model of the previous round to slow down or even perturb the aggregation model, or they can use the model updates that are farther away from our benchmark to perturb the direction of the global model updates, so we consider models with too small or too large distances as malicious models, while the user model with the middle distance size, considered as a benign model, is given a higher aggregation weight. The details of the specific weights given can be seen in Algorithm 1.

3.2 Details of PBFL

System Setup: In this phase, we need to complete the generation and distribution of two sets of asymmetric keys. First, AS and CSP collective generate a set of keys $(pk_s, rlk_s, rtk_s, sk_s^{AS}, sk_s^{CSP})$ through four security protocols at MHE, which are SecKeyGen(·), DKeyGen(·), DRelinKeyGen(·), and DRotKeyGen(·) protocols respectively. The private key at server side is distributed between two entities (AS and CSP), which ensures that the parameters uploaded by the user will not be directly decrypted. Second, AS and KGC produce a set of keys $(pk_u, sk_u^{AS}, sk_u^U)$ by SecKeyGen(·) and DKeyGen(·) protocols at MHE in an interaction cycle, and then KGC sends sk_u^U, a part of the secret key, to all the users in FL. KGC is an intermediary for all users and servers to negotiate a common public and private key. In order to prevent KGC from obtaining the user's complete private key, we choose to distribute the private key between KGC and AS. The first key pair between AS and CSP is used to ensure the privacy of the users' uploaded parameters and the second is used to protect the privacy of aggregated global model.

Local Training: In this phase, we don't need to know the proportion of Byzantine users in advance, but we assume that the proportion of Byzantine users has an upper limit, which is 50% in our scheme, which is better than Byzantine-robust methods where the proportion must be known, e.g., Krum and Bulyan. At the training setup phase, AS sends the global model parameters $[\![W_g]\!]_{pk_u}$ that is encrypted by user public key pk_u to all the users. For each user U_i, as shown above, U_i and AS holds the secret key set $\{sk_u^U, sk_u^{AS}\}$ corresponding to the public key pk_u, then we use the designed secure cooperative decryption algorithm (**Sec2Dec**) in Appendix A to obtain the parameters plaintext W_g, which is calculated by U_i and AS as follows:

$$W_g \leftarrow \mathbf{Sec2Dec}(\{sk_u^U, sk_u^{AS}\}, [\![W_g]\!]_{pk_u}) \tag{1}$$

Algorithm 2. User Local Training

Input: Local user U_i's dataset D_i, global model parameters W_g.
Output: Local update model parameters W_i of user U_i.

if U_i is **GA attack Byzantine user then**
    $W_i \leftarrow$ select random vector from normal distribution
    return W_i at user U_i.
else
    $\mathcal{B} \leftarrow$(split D_i into batches of size B)
    for each local epoch i from 1 to E **do**
        for batch $b \in \mathcal{B}$ **do**
            if U_i is **LF attack Byzantine user then**
                $label(b) \leftarrow 9 - lable(b)$
            end if
            $G_b \leftarrow$ the batch gradient of b
            $W_i \leftarrow W_i - \eta G_b$
        end for
    end for
    return W_i at user U_i.
end if

After applying global parameters W_g to local model, the user U_i trains locally with local privacy dataset Di and obtains the updated local model parameters W_i for the next training round. Specifically, the user U_i inputs the data feature x to the local model, then figures out the cross-entropy loss function between model output and label y, and then uses this value to get gradients by applying the backpropagation algorithm. In our scheme, the users need to finish the update of the local model parameters and upload the updated model parameters as the vanilla FL does. There are many optimization algorithms to use gradients to get updated parameters. We choose the SGD with the momentum to make the optimization process smoother, which will calculate the gradients based on the historical change of the gradients. The SGD with momentum can not only accelerate convergence but also reduce the variance of the gradient, which does much help in our aggregation algorithm to reduce the impact of Byzantine users. The detail training steps are described at Algorithm 2.

To protect the data privacy of users against inference attack from the server, the user U_i encrypts the local model parameters W_i with the public key of the AS as $[\![W_i]\!]_{pk_s}$ and then sends it to the AS. Through the Single Instruction Multiple Data (SIMD) technique in MHE, we can pack a parameter vector W_i with tens of thousands of dimensions into several ciphertext polynomials, which greatly accelerates the speed of private and secure aggregation.

Privacy and Secure Aggregation: In this phase, the AS will execute the privacy and security aggregation algorithms with the assistance of CSP, which includes the assistance of distance information decryption and the public key switching of the aggregated global model parameters. In our setting, to better evaluate the performance of our aggregation algorithm, we default that all user

---

Algorithm 3. Secure Square Euclidean Distance Calculation **SecDis**($[\![W_i^t]\!], [\![W_g^{t-1}]\!]$)

Input: In iteration t, AS holds global model $[\![W_g^{t-1}]\!]_{pk_s}$ at $t-1$ and user U_i holds local model $[\![W_i^t]\!]_{pk_s}$; AS and CSP hold sk_s^{AS} and sk_s^{CSP} respectively, which is the set of secret keys corresponding to the public key pk_s.
Output: The square Euclidean distance plaintext d_i between$[\![W_i^t]\!]$ and $[\![W_g^{t-1}]\!]$.
Performed by AS and CSP
AS:
1: Calculates the difference: $[\![tmp]\!]_{pk_s} \leftarrow$ **Sub**$([\![W_i^t]\!]_{pk_s}, [\![W_g^{t-1}]\!]_{pk_s})$
2: Calculates the second power: $[\![tmp]\!]_{pk_s} \leftarrow$ **Power**$([\![tmp]\!]_{pk_s}, 2)$
3: Calculates the inner-sum: $[\![d_i]\!]_{pk_s} \leftarrow$ **InnerSum**$([\![tmp]\!]_{pk_s})$
AS & CSP:
1: AS and CSP cooperative decrypt the squred Euclidean distance as follows: $d_i \leftarrow$ **Sec2Dec**$(\{sk_s^{AS}, sk_s^{CSP}\}, [\![d_i]\!]_{pk_s})$
2: **return** d_i at AS.

---

parameters participate in the aggregation at each round, instead of random selecting part of them. There are three steps to aggregate the received encrypted parameters.

Secure Square Euclidean Distance Calculation: At the aggregation round t, the first step is to calculate $\left\|[\![W_i^t]\!]_{pk_s} - [\![W_g^{t-1}]\!]_{pk_s}\right\|_2^2$, the value of the expression which is an overview of the square Euclidean distance from each user parameter in $\{[\![W_i^t]\!]_{pk_s}\}_{i=1}^{i=N}$, i.e., $[\![W_i^t]\!]_{pk_s}$, to the last round of global model parameters $[\![W_g^{t-1}]\!]_{pk_s}$.

To obtain the Euclidean distance in plaintext domain securely, AS first performs calculation on ciphertext and then interacts with CSP to decrypt Euclidean distance. The specific steps are as follows: for $x \in [1...N]$, AS first uses the sub function and power function in MHE to get $[\![tmp]\!]_{pk_s}$, the square of the vector difference and then uses inner-sum (InnerSum(·)) function in MHE to get the inner sum of $[\![tmp]\!]_{pk_s}$. After the inner-sum function works, AS will get a ciphertext that contains the square Euclidean distance information. The AS needs to decrypt this ciphertext to get the distance information with the assistance of the CSP, by using our secure cooperative decryption algorithm. Algorithm 3 shows the details of secure square Euclidean distance calculation.

Aggregation Weights Calculation: After obtaining the plaintext distance vector d of all the users, the remaining aggregation steps are similar to Algorithm 1 (Line 4–13). Then AS perform the linear aggregation of encrypted user model parameters:

$$[\![W_g]\!]_{pk_s} = \sum_{i=1}^{N} \gamma_i \times [\![W_i]\!]_{pk_s} \tag{2}$$

Secure Aggregated Parameter Distribution: The last step is to distribute the aggregated parameters to all the users. At the previous steps, AS obtains an aggregated model parameter which is encrypted by the server's public key pk_s, which need to switch to user-side public key encryption. To protect the privacy of global model parameters, we use algorithm (**Sec2KeyS**) in Appendix A, which can re-encrypt global model parameters using user public key pk_u without revealing global model parameters. With the sever key pairs $\{sk_s^{AS}, sk_s^{CSP}\}$ in AS and CSP respectively, AS and CSP cooperatively call the following function:

$$[\![W_g]\!]_{pk_u} \leftarrow \mathbf{Sec2KeyS}(\{sk_s^{AS}, sk_s^{CSP}\}, pk_u, pk_s, [\![W_g]\!]_{pk_s}) \tag{3}$$

Then AS sends $[\![W_g]\!]_{pk_u}$, which is encrypted by user public key pk_u, to all users.

Remark: Within the scope of our investigation, we deem the security of squared Euclidean distances to be non-essential. It is evident that the AS and the CSP are incapable of deducing the extent and sign of the model parameters solely based on the disclosed distance. Recently, Geiping *et al.* [6] provided a method that can invert gradients to raw data using the cosine similarity between gradients (Inference attack). Nonetheless, the approach is contingent on the gradient's sign, which remains concealed within the encrypted data in PBFL. Thus, our scheme can resist the inference attack.

4 Evaluation

We demonstrate the security and convergence properties of PBFL by conducting a series of comparative experiments with real-world datasets. All our experiments are executed on a custom server with the configuration of Ubuntu 20.04, Intel(R) Core(TM) i9-10980XE CPU @ 3.00GHz CPU and 64GB RAM.

Datasets and Models: To evaluate the performance of the PBFL, we chose two classic image classification datasets, (i) MNIST dataset, including 60000 handwritten digit pictures with a size of 28×28, and $10,000$ test samples, (ii) CIFAR-10 dataset, containing $60,000$ color images in total and $50,000$ of them are used for training and the other $10,000$ are used for testing. These images are divided into 10 categories with the size of 32×32 and each category has $6,000$ images. The model structure of the MNIST classification task has a two-layer fully connected network, and the parameter settings of each layer are 724×100 and 100×10 in sequence respectively, for a total of about $80K$ parameters. For the CIFAR-10 image classification task, we use a simple model architecture in the TensorFlow tutorial online, which consists of three convolutional layers followed by two fully connected layers, for a total of about $123K$ parameters.

Hyper-Parameters: The total number of users we used in FL is 51. For each experiment, we set the batch size of training to 2^7, the momentum of SGD to 0.9, and the initial learning rate to 0.1. Moreover, in each FL round, the server aggregates all the uploaded model parameters by users. The proportion of Byzantine users in FL is denoted by δ.

Attacks: Our experiments mainly focus on two state-of-the-art Byzantine attacks: Gaussian Attack (GA) and Label Flipping attack (LF). To reproduce the scene of GA, the GA Byzantine users will upload a parameter vector drawn from a Gaussian distribution with mean zero, and standard deviation 200 (model poisoning). To simulate the LF attack, we replace the training label y at the Byzantine users with the label $9-y$ (data poisoning), and the training processes are the same as the benign users.

We assess the effectiveness of PBFL by examining two key dimensions: (i) the precision of the aggregation algorithms employed within PBFL, and (ii) the resilience of PBFL against actual adversarial attacks. In order to underscore the strengths of our approach, we have recreated four representative Byzantine fault-tolerant schemes that operate on plaintext data, namely FedAvg, Krum, Median, and Trimmed Mean.

Many factors affect the accuracy of the global model, including the capacity of the model, the quality and quantity of user data, the number of training rounds, and the proportion and attack methods of Byzantine users. In this paper, we focus on privacy and security issues in FL and we want to compare the robustness of aggregation algorithms, i.e., the ability to complete the aggregation of user parameters successfully under Byzantine attacks while protecting user privacy. We thus conduct experiments to figure out how proportion and attack methods of Byzantine users affect the accuracy of the global model in FL. We exhibit the impact of varying proportions of Byzantine participants on different aggregation techniques.

Impact of Different Proportions of Byzantine Users: Intuitively, a higher percentage of Byzantine users means less data is available in benign users, which causes lower test accuracy. Figure 1 outlines the comparison of test accuracy with different proportions of Byzantine users. We observe that all defenses have a higher tolerance for Byzantine users compared to vanilla federated learning (FedAvg). Note that the experiments show that the FedAvg method works well in the scenario without Byzantine users. When the MNIST training task is attacked by Gaussian Attack (GA) Byzantine users, all the defenses, including Krum, Trimmed Mean, Median, and PBFL, can maintain a high accuracy rate up to 50% Byzantine users. However, in the case of the Label Flipping (LF) attack, our PBFL shows the advantage when there are many Byzantine users, which benefits from our robust aggregation algorithms. For the CIFAR-10 classification task, when attacked by the GA Byzantine users, our PBFL is almost the same as the best performing defenses Trimmed Mean and Median, and all of them can prevent up to 50% of Byzantine users. As for the LF attack in the CIFAR-10 task, our PBFL also shows advantages with the Byzantine user increase. Our scheme can effectively identify Byzantine users and make full use of data owned by benign users. We observe that PBFL, Trimmed Mean, and Median accuracy are significantly higher than Krum. A reasonable explanation is that Krum only selects one parameter vector as the aggregated result, which can not take advantage of all benign users, while other methods focus on aggregating as many benign parameters as possible. As depicted in Fig. 1, PBFL is able to sustain a competitive edge over other protective measures.

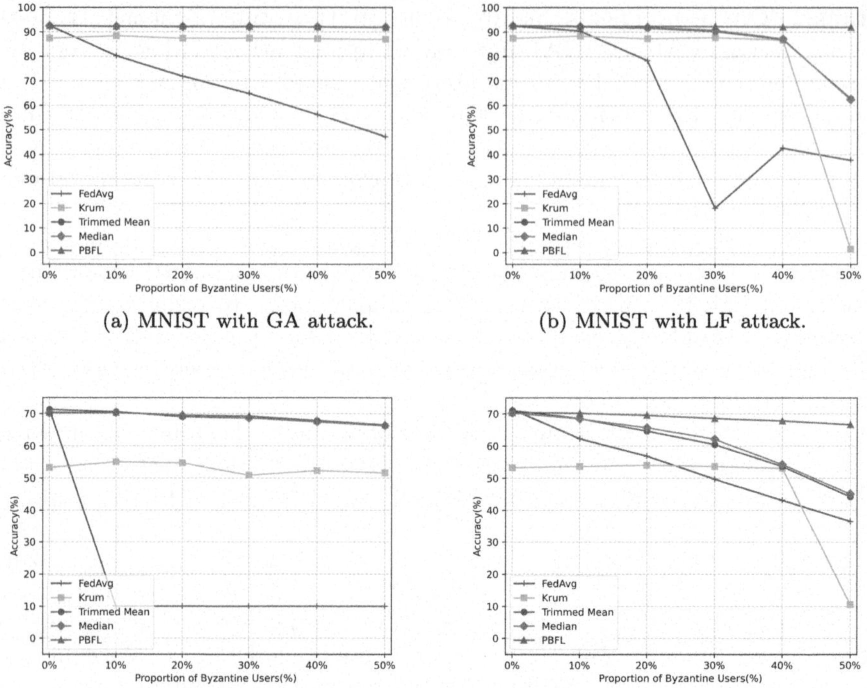

(a) MNIST with GA attack. (b) MNIST with LF attack.

(c) CIFAR-10 with GA attack. (d) CIFAR-10 with LF attack.

Fig. 1. Comparison of accuracy with different proportions of Byzantine users.

For more experimental information, we show the impact of different training rounds and encryption of PBFL through an array of comparative experiments utilizing genuine datasets in Appendix B.

5 Conclusion and Future Work

In this study, we introduce a novel federated learning (FL) framework named PBFL, designed with an emphasis on bolstering privacy safeguards and simultaneously fortifying the system against Byzantine assaults. Additionally, our research encompasses an exhaustive theoretical examination aimed at substantiating the security and convergence attributes inherent within the PBFL framework. In addition, the experiment results show that our PBFL is superior to existing schemes at both privacy and robustness. Nonetheless, the resilience of the PBFL framework is predicated on the presumption that the data contributed by users is independent and identically distributed (i.i.d), which makes our framework perform poorly on Non-i.i.d data. In future work, we will research robust aggregation schemes for Non-i.i.d data and further optimize the accuracy and efficiency.

A MHE Related Algorithms

Algorithm 4. Secure Two-party Cooperative Decryption Algorithm: **Sec2Dec**($\{sk^1, sk^2\}$,$[\![c]\!]_{pk}$) $\rightarrow c$

Input: For MHE key pair ($\{sk^1, sk^2\}$,pk), parties p_1 and p_2 holds sk^1 and sk^2, separately; p_1 holds ciphertext $[\![c]\!]_{pk}$.

Output: Party p_1 gets plaintext c, and party p_2 knows nothing about plaintext c.

1: **if** party p_2 does not hold $[\![c]\!]_{pk}$ **then**
2: p_1 sends $[\![c]\!]_{pk}$ to p_2.
3: **end if**
4: **for** $x \in \{1, 2\}$ **do**
5: Party p_x calculates the partial decryption result:$[\![c_x]\!]_{pk} \leftarrow$ **PartDec**($[\![c]\!]_{pk}, sk^x$).
6: **end for**
7: Party p_2 sends $[\![c_2]\!]_{pk}$ to p_1.
8: Party p_1 aggregates the partial decryption results as follows:
$c \leftarrow$ **AggDec**($[\![c]\!]_{pk}, \{[\![c_1]\!]_{pk}, [\![c_2]\!]_{pk}\}$)
9: **return** plaintext c at p_1.

Algorithm 5. Secure Two-party Cooperative Public Key Switch Algorithm **Sec2KeyS**($\{sk^1, sk^2\}$,$pk_1, pk_2, [\![c_1]\!]_{pk_1}$) $\rightarrow [\![c_1]\!]_{pk_2}$

Input: For MHE key pair ($\{sk^1, sk^2\}, pk_1$), parties p_1 and p_2 holds sk^1 and sk^2, separately; p_1 holds ciphertext $[\![c]\!]_{pk_1}$; pk_2 is the target public key that $[\![c]\!]_{pk_1}$ wants to switch to.

Output: Party p_1 gets ciphertext $[\![c]\!]_{pk_2}$ that is encrypted by pk_2, and parties p_1 and p_2 know nothing about plaintext c.

1: **if** party p_2 does not hold $[\![c]\!]_{pk}$ **then**
2: p_1 sends $[\![c]\!]_{pk_1}$ to p_2.
3: **end if**
4: **for** $x \in \{1, 2\}$ **do**
5: Party p_x calculates local public key switch information:
$[\![c_x]\!]_{pk_2} \leftarrow$ **PartSwitch**($[\![c]\!]_{pk_1}, sk^x, pk_2$).
6: **end for**
7: Party p_2 sends $[\![c_2]\!]_{pk_2}$ to p_1.
8: Party p_1 aggregates the public key switch information as follows:
$[\![c]\!]_{pk_2} \leftarrow$ **AggSwitch**($\{[\![c_1]\!]_{pk_2}, \{[\![c_2]\!]_{pk_2}\}$)
9: **return** ciphertext $[\![c]\!]_{pk_2}$ in p_1, while parties p_1 and p_2 know nothing about plaintext c.

B Additional Experiments Results

Impact of Different Training Rounds: With the accumulation of training iterations, the algorithm is able to discern and capture more nuanced data patterns, thereby enhancing its predictive precision. In the context of this study, issues

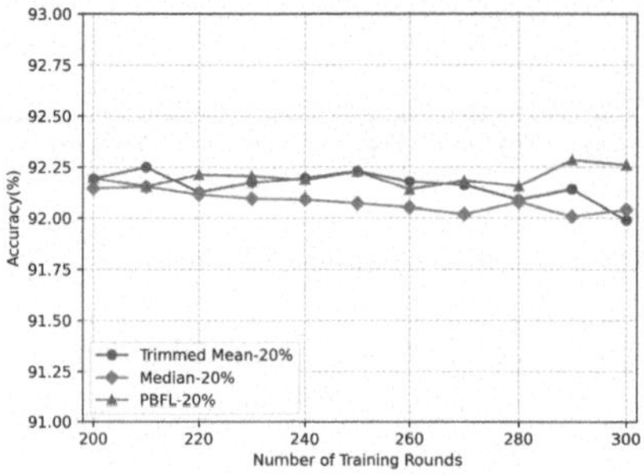

(a) MNIST with GA attack.

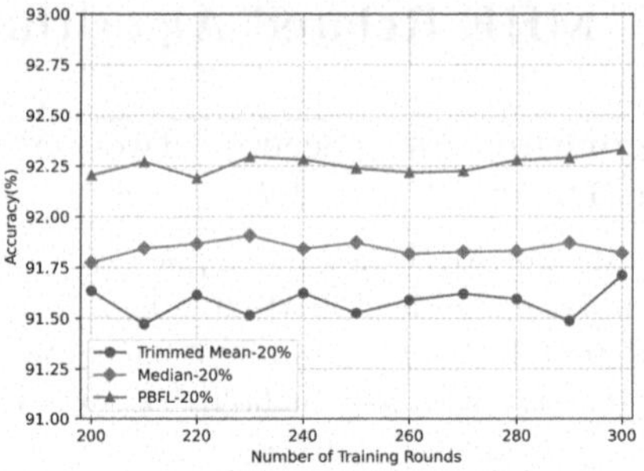

(b) MNIST with LF attack.

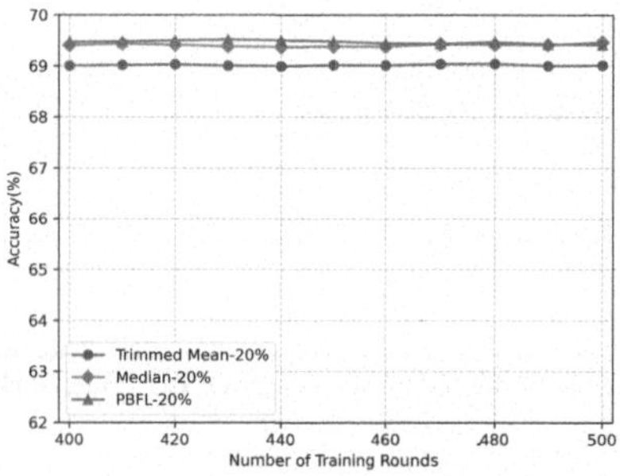

(c) CIFAR-10 with GA attack.

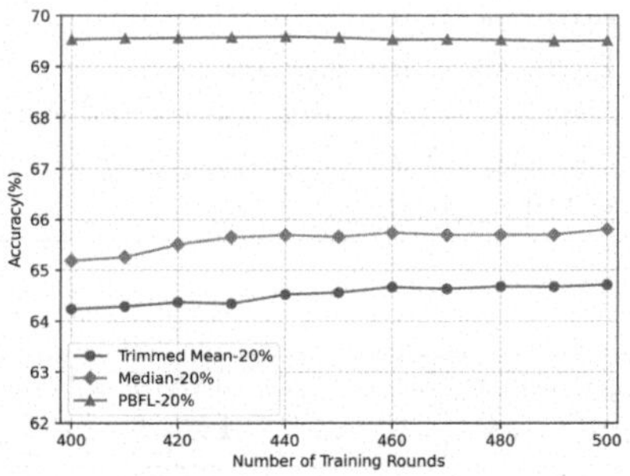

(d) CIFAR-10 with LF attack.

Fig. 2. Comparison of accuracy with different training rounds.

stemming from inadequate or excessive model complexity, which can lead to suboptimal fits, are not within the scope of our analysis. Figure 2 shows the comparison of accuracy with increasing training rounds about three defenses that perform relatively well, including PBFL, Median, and Trimmed Mean. We fix the proportion of Byzantine users at 20%, i.e., 10 Byzantine users are mixed in a total of 51 users. We show the accuracy comparison of different defenses in the last 100 rounds of training in Fig. 1, and we can observe that our PBFL slightly outperforms the other two defenses in both accuracy and stability, especially when attacked by LF Byzantine users.

Impact of Encryption: The MHE we used is based on the CKKS scheme, which is a fully homomorphic encryption mode that supports floating-point arithmetic with a little rounding error. To figure out the effect of ciphertext calculation, we provide a comparison with the scheme using the same aggregation algorithm but computing on the plaintext (Pt-PBFL). Figure 3 shows the details of the comparison. We can observe that the difference between these two schemes is negligible. We analyze that the reason is that the accuracy requirements of parameters in neural networks are not strict, and the error produced by CKKS is tolerable. This also proves that CKKS is suitable for the privacy-preserving of machine learning.

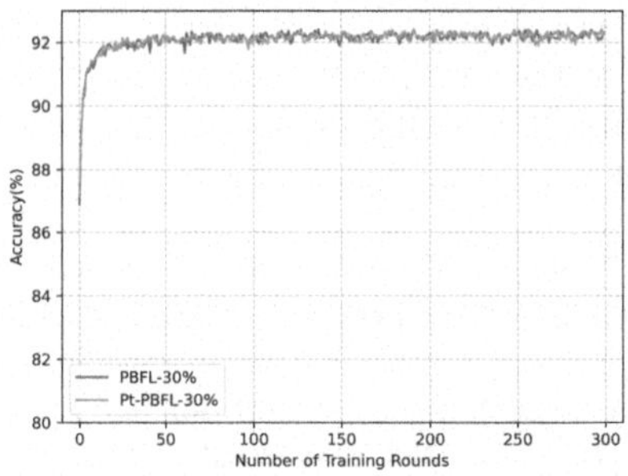

(a) MNIST with GA attack.

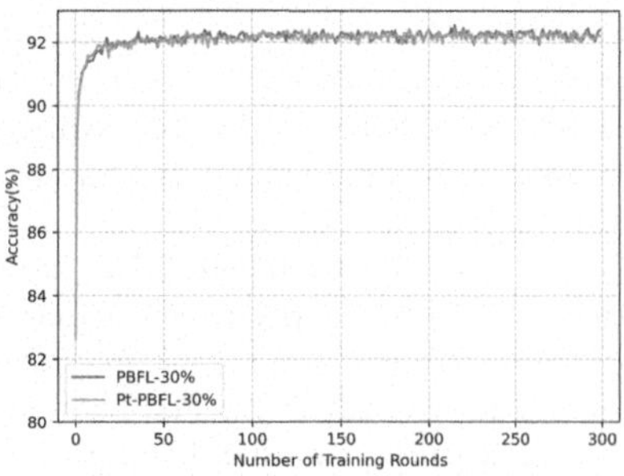

(b) MNIST with LF attack.

Fig. 3. Comparison of accuracy with Pt-PBFL.

Aggregation Weights in PBFL: In general, our PBFL is superior to other schemes in both accuracy and robustness, observed from the previous experiments. As mentioned above, the way PBFL achieves robust aggregation is to assign Byzantine users to lower aggregation weights in the linear aggregation. Thus, we show the aggregation weights of benign users and Byzantine users with different datasets and attacks at Table 2. We fix the proportion of Byzantine users as 30%, i.e., mixed 15 Byzantine users in a total of 51 users. The results show that the weight of each aggregation of PBFL is almost assigned to benign users, which also means that our PBFL can identify Byzantine users and make full use of benign users to aggregate the model. This is the reason why PBFL performs better than other schemes.

Table 2. Aggregation Weights in PBFL

	CIFAR-LF	MNIST-LF	CIFAR-GA	MNIST-GA
Benign users	0.9999	0.9988	1.0	1.0
Byzantine users	0.0001	0.0012	0.0	0.0

References

1. Bozdemir, B., Canard, S., Ermis, O., Möllering, H., Önen, M., Schneider, T.: Privacy-preserving density-based clustering. In: Proceedings of the 2021 ACM Asia Conference on Computer and Communications Security, pp. 658–671 (2021)
2. Cao, X., Fang, M., Liu, J., Gong, N.Z.: FLTrust: Byzantine-robust federated learning via trust bootstrapping. arXiv preprint arXiv:2012.13995 (2020)
3. Chen, B.R.Z.W.Z.S.: Privacy inference attack and defense in centralized and federated learning: a comprehensive survey. IEEE Trans. Artif. Intell. 1–22 (2024)

4. Demmler, D., Schneider, T., Zohner, M.: Aby-a framework for efficient mixed-protocol secure two-party computation. In: NDSS (2015)
5. Dong, Y., Chen, X., Li, K., Wang, D., Zeng, S.: FLOD: Oblivious defender for private byzantine-robust federated learning with dishonest-majority. Cryptology ePrint Archive (2021)
6. Geiping, J., Bauermeister, H., Dröge, H., Moeller, M.: Inverting gradients–how easy is it to break privacy in federated learning? arXiv preprint arXiv:2003.14053 (2020)
7. Goel, M.M.K.A.M.S.Z.: FedChallenger: Challenge-response-based defence for federated learning against byzantine attacks. In: GLOBECOM 2023 - 2023 IEEE Global Communications Conference (2023)
8. Hao, M., Li, H., Xu, G., Chen, H., Zhang, T.: Efficient, private and robust federated learning. In: Proceedings of the 37th Annual Computer Security Applications Conference, pp. 45–60 (2021)
9. Hard, A., et al.: Federated learning for mobile keyboard prediction. arXiv preprint arXiv:1811.03604 (2018)
10. Kairouz, P., et al.: Advances and open problems in federated learning. Found. Trends® Mach. Learn. **14**(1–2), 1–210 (2021)
11. Liu, X., Li, H., Xu, G., Chen, Z., Huang, X., Lu, R.: Privacy-enhanced federated learning against poisoning adversaries. IEEE Trans. Inf. Forensics Secur. **16**, 4574–4588 (2021)
12. Ma, H.C.C.P.: FLRAM: robust aggregation technique for defense against byzantine poisoning attacks in federated learning. Electronics 4463 (2023)
13. Nasr, M., Shokri, R., Houmansadr, A.: Comprehensive privacy analysis of deep learning: Passive and active white-box inference attacks against centralized and federated learning. In: 2019 IEEE symposium on security and privacy (SP), pp. 739–753. IEEE (2019)
14. Nguyen, T.D., et al.: {FLAME}: Taming backdoors in federated learning. In: 31st USENIX Security Symposium (USENIX Security 22), pp. 1415–1432 (2022)
15. Popa, M.R.S.W.A.: ELSA: secure aggregation for federated learning with malicious actors. In: 2023 IEEE Symposium on Security and Privacy (SP) (2023)
16. Zhang, G.H.L.F.: Efficient byzantine-robust and privacy-preserving federated learning on compressive domain. IEEE Internet Things J. 7116–7127 (2024)

Perfect Indistinguishability Obfuscation for Boolean Polynomial Vector Spaces via Learning

Ning Ding(✉)

School of Electronic Information and Electrical Engineering, Shanghai Jiao Tong University, Shanghai 200240, China
dingning@sjtu.edu.cn

Abstract. Indistinguishability obfuscation (IO) is an extremely important primitive in cryptography and in this paper we are interested in perfect IO (while known constructions are mainly computational IO). At present only those functions with known deterministic unique representations or exactly learning methods have perfect IO, and besides this, there is no perfect IO for other functions.

Thus we investigate whether we can construct perfect IO for functions that are not, or are unlikely to be, in the two kinds of functions above. Our target is vector spaces spanned by Boolean polynomials. Let $F = \{f_1, \cdots, f_m\}$ be a set of $m = \mathsf{poly}(n)$ arbitrary Boolean polynomials (not necessarily independent or distinct), V_F the vector space spanned by $f_1, \cdots, f_m$. Currently there is no efficient way to exactly learn V_F or compute deterministic unique representations for it even if F is given. Our result is that despite this, we can still construct a perfect indistinguishability obfuscator for V_F which has the description of F as input.

The mere cryptographic primitive used in our obfuscation is the somewhat homomorphic encryption scheme over the integers due to van Dijk et al. which only assumes the hardness of the Approximate GCD problem that is regarded solid so far. The most notable novelty in our techniques is a new positive application of the PAC Learning theory to IO, in which we show that the learning result of *approximate* correctness (for the Evaluation function homomorphically computing V_F) can be employed to achieve *exact* obfuscation for V_F (note that given F, V_F is approximately learnable but not exactly learnable).

We remark that our result is for restricted function classes, unlike general computational IO sufficient for cryptographic applications. So we treat this result as a first step towards exploring the possibility of perfect IO beyond uniquely representable or exactly learnable.

1 Introduction

Indistinguishability obfuscation (IO) is an extremely important primitive in cryptography, which purpose is to make two programs of same functionality indistinguishable. According to different definitions on "indistinguishable", there

Y. Chen et al. (Eds.): COCOON 2024, LNCS 15162, pp. 447–461, 2025.
https://doi.org/10.1007/978-981-96-1093-8_37

are mainly two types of IO, i.e. computational IO and perfect IO where the former requires two obfuscated programs are computationally indistinguishable while the latter requires they are identically distributed and thus is much stronger than the former. The study of computational IO mainly aims at constructing general-purpose obfuscators for all programs and there appear plenty of constructions and the latest ones are e.g. [1,2,6,7,9,10,13,14,17]. (Initial and earlier works build such obfuscators over multi-linear maps or other assumptions, and the references of these works are omitted due to page limit.)

However, as for perfect IO, there are few known works. At present only those functions with known deterministic unique representations or exactly learning methods have perfect IO, which yet is obvious since given any two functions of same functionality respectively, if one can compute deterministic unique representations or exactly learned hypotheses of them respectively, the two representations are actually a same one or the two hypotheses are identically distributed, and thus perfect IO is achieved. [11] shows that POBDDs (Polynomial Ordered Binary Decision Diagrams) admit the deterministic unique representation property, providing an example for the former. Nevertheless, besides this, there is no perfect obfuscator for other functions. There are also some other notions of obfuscation, e.g. virtual-black box obfuscation [3], virtual grey-box obfuscation [4], but all of them do not satisfy the requirement of perfect IO. In contrast, the only known negative result is that 3CNFs cannot have perfect IO unless the polynomial hierarchy collapses to its second level [11].

1.1 Our Results

This paper focuses on perfect IO for theoretical interest and asks whether we can construct perfect IO for functions that are not, or are unlikely to be, exactly learnable or uniquely representable. We show an affirmative answer for vector spaces spanned by polynomially many Boolean polynomials, by which we mean elements of the ring $\mathbf{F}_2[x_1, \cdots, x_n]/ < x_1^2 + x_1, \cdots, x_n^2 + x_n >$.

Precisely, let $F = \{f_1, \cdots, f_m\}$ be a set of $m = \mathsf{poly}(n)$ arbitrary Boolean polynomials over $x_1, \cdots, x_n$ (not necessarily independent or distinct) in which every f_i is computable in polynomial-time, V_F the vector space spanned by $f_1, \cdots, f_m$, which is the target function class in this work. Currently there is no work showing how to efficiently uniquely represent V_F or exactly learn it, and we further present discussion on this issue and point out the plausible infeasibility of these two tasks even if F is given. Our result is that despite this, we can still construct a perfect obfuscator for any V_F where the obfuscator is given F.

Theorem 1. (Main result.) Based on the somewhat homomorphic encryption scheme in [8], for any F, there is an (almost-exact) perfect indistinguishability obfuscator for V_F, where F is given as input to the obfuscator.

The homomorphic encryption scheme only relies on the hardness of the Approximate GCD problem, introduced in [8], which basically says that given many numbers around multiples of a same large prime number, it is hard to recover the prime number and which is regarded solid so far.

Our Techniques. The most notable novelty in this work is a new positive application of the PAC Learning theory to IO. Notice that a PAC learning algorithm for a learnable function class can be considered as an approximate-correct perfect obfuscator for it. However, since our goal is to construct exact IO, simply applying learning results cannot achieve the goal.

Recall that [5] presents transformations from approximate IO to exact IO. Let us sketch the one from average-case approximate IO to a worst-case one (for precise definitions of these notions of IO, refer to Sect. 2.2). Let $\mathsf{SFE} = (\mathsf{Enc}, \mathsf{Eval}, \mathsf{Dec})$ be a secure function evaluation scheme (refer to [5] Sect. 2.1). For a circuit C and any specific x in its domain, consider the following strategy of computing $C(x)$: first compute $(\mathsf{CT}_x, \mathsf{R}) \leftarrow \mathsf{Enc}(x)$ where CT_x is treated as the encryption and R will be used as the decryption key; notice that Eval can homomorphically evaluate C on an encrypted input and accordingly compute $\widehat{\mathsf{CT}} \leftarrow \mathsf{Eval}(\mathsf{CT}_x, C)$ and then finally run $\mathsf{Dec}(\mathsf{R}, \widehat{\mathsf{CT}})$ to output $C(x)$. Then obfuscate $\mathsf{Eval}(\cdot, C)$ using average-case approximate IO (with respect to distribution of CT_x) and denote by $\widetilde{W}$ the obfuscation. The $(\mathsf{SFE}, \widetilde{W})$ constitutes an obfuscation of C. That is, to compute $C(x')$ for any x', first generate $\mathsf{CT}_{x'}$. By approximate correctness of $\widetilde{W}$ and the indistinguishability of CT_x and $\mathsf{CT}_{x'}$, $\widetilde{W}(\mathsf{CT}_{x'}) = \mathsf{Eval}(\mathsf{CT}_{x'}, C)$ with high probability. Then $\mathsf{Dec}(\mathsf{R}, \widetilde{W}(\mathsf{CT}_{x'}))$ outputs $C(x')$ correctly with same probability. Thus we achieve worst-case obfuscation of C (working for any x') from average-case approximately-correct $\widetilde{W}$.

Return to our setting of obfuscating V_F. We attempt to combine the learning theory and the above transformation. Instantiate C with $f \in V_F$ and the key primitive required by the transformation is an average-case approximate perfect IO for $\mathsf{Eval}(\cdots, f)$, while currently no construction exists. Nevertheless, if we can PAC learn $\mathsf{Eval}(\cdots, f)$, it will bring an average-case perfect IO for $\mathsf{Eval}(\cdots, f)$. Thus our work is to investigate whether or not $\mathsf{Eval}(\cdots, f)$ can be learned.

Indeed we develop a learning algorithm for $\mathsf{Eval}(\cdots, f)$ when we use the somewhat homomorphic encryption scheme $\mathsf{SHPKE} = (\mathsf{KeyGen}, \mathsf{Enc}, \mathsf{Dec}, \mathsf{Eval})$ in [8] to substitute SFE. The transformation can be instantiated as follows. To compute $f(x)$: sample $(pk, sk) \leftarrow \mathsf{KeyGen}(1^n)$; compute $X_i \leftarrow \mathsf{Enc}(pk, x_i)$ for each bit x_i of x; run the learning algorithm to output a learned hypothesis of function $\mathsf{Eval}(\cdots, f)$, denoted h, and compute $X \leftarrow h(X_1, \cdots, X_n)$ and finally run $\mathsf{Dec}(sk, X)$ to output $f(x)$. Since the learning algorithm accesses $\mathsf{Eval}(\cdots, f)$ in a black-bow way, it is an approximate perfect obfuscator for $\mathsf{Eval}(\cdots, f)$.

We then sketch the learning algorithm. Assume the output of $\mathsf{Eval}(\cdots, f)$ is of λ-bit. We gradually define functions computing all bits of the output of $\mathsf{Eval}(\cdots, f)$ (from leftmost to rightmost) and generate learned hypotheses of them. These functions are actually threshold functions of linear combinations of polynomials in F (viewed as polynomials over reals). By adapting the classical learning result for threshold functions to our setting, we compute learned hypotheses of $\mathsf{Eval}(\cdots, f)$, together denoted h.

By the transformation, we obtain worst-case approximate perfect IO for f, and further by the BPP-like error reduction [5], obtain almost-exact perfect IO for f. We highlight that the key part in the techniques is that the learning result of *approximate* correctness for $\mathsf{Eval}(\cdots, f)$ is employed to achieve *exact* obfuscation for f (where V_F is approximately learnable but not exactly learnable).

Organization. In Sect. 2 we present the preliminaries. In Sect. 3 we formulate the notion of vector spaces V_F. In Sect. 4 we recall the homomorphic encryption and characterize $\mathsf{Eval}(\cdots, f)$. In Sect. 5 we present the basic result on learning threshold functions of linear combinations of polynomials. In Sect. 6 we demonstrate how to generate learned hypotheses of $\mathsf{Eval}(\cdots, f)$, viewed as approximate IO. In Sect. 7 we present the worst-case approximate perfect IO of V_F and in Sect. 8 the ultimate almost-exact perfect IO.

2 Preliminaries

We use $\mathbf{N}, \mathbf{Z}, \mathbf{R}$ to denote natural integers, integers and real numbers respectively. We use $[n]$ to denote the integers in $[1, n]$. For any set S, we use $|S|$ to denote its cardinal number. A function $\mu(\cdot)$, where $\mu : \mathbf{N} \to [0, 1]$ is called negligible if $\mu(n) = n^{-\omega(1)}$ (i.e., $\mu(n) < \frac{1}{p(n)}$ for all polynomial $p(\cdot)$ and large enough n's). We will use $\mathsf{neg}(n)$ to denote an unspecified negligible function.

Let $\mathsf{Sign}(\cdot)$ denote the (real) sign function satisfying that $\mathsf{Sign}(y) = 1$ iff $y \geq 0$ and $\mathsf{Sign}(y) = 0$ iff $y < 0$. A threshold function over $\mathbf{R}^n$ is a function $h : \mathbf{R}^n \to \{0, 1\}$ defined as $h(X_1, \cdots, X_n) = \mathsf{Sign}(\sum_{i=1}^{n} w_i X_i - \theta)$, where $w_1, \cdots, w_n, \theta$ may be arbitrary real values and are called weights.

2.1 PAC Learning

Let $\mathcal{C}$ denote a class of Boolean functions over the input space E. In the PAC (Probably Approximately Correct) learning model [16], a labeled example is a pair $(Y, f(Y))$, where $Y \in E$ is an input and $f(Y)$ is the value of the target function $f \in \mathcal{C}$ on the input Y. Let D be any distribution over E. A training sample labeled by f is of the form $((Y^1, f(Y^1)), \cdots, (Y^t, f(Y^t)))$, in which Y^k is the kth sampling of Y from distribution D, $1 \leq k \leq t$.

Definition 1. (PAC Learning) An algorithm L is called a PAC learner for $\mathcal{C}$ under distribution D over E, if it is given a training sample in which each Y is sampled from D independently and its label is $f(Y)$ for some unknown $f \in \mathcal{C}$, $\epsilon, \delta \in (0, 1)$, with probability $1 - \delta$, L outputs a hypothesis h (not necessarily in $\mathcal{C}$) such that $\Pr[f(Y) \neq h(Y)] < \epsilon$ for $Y \leftarrow D$.

We refer to ϵ as the accuracy parameter and δ as the confidence parameter. When such L exists for $\mathcal{C}$ under D, we say that $\mathcal{C}$ can be PAC learned to error and confidence (ϵ, δ) under D. If L can work under arbitrary D, we just say $\mathcal{C}$ can be PAC learned. We say that L is efficient if its running-time is $\text{poly}(n, \frac{1}{\epsilon}, \log \frac{1}{\delta})$, where n usually denotes the input binary length.

2.2 Indistinguishability Obfuscation

We use the following definitions for (almost) exact/approximate computational/perfect indistinguishability obfuscation (where n denotes the security parameter).

Definition 2. (Almost-Exact IO). A uniform PPT machine $i\mathcal{O}$ is called an almost-exact perfect/computational (indistinguishability) obfuscator for a Boolean function class $\mathcal{C}$ if the following can be satisfied:

- For each $C \in \mathcal{C}$, $\Pr[\exists x : C'(x) \neq C(x) : C' \leftarrow i\mathcal{O}(1^n, C)] = \mathsf{neg}(n)$.
- For any $C_0, C_1 \in \mathcal{C}$ of same size, if $C_0(x) = C_1(x)$ for all inputs x, their obfuscation are identically distributed/computationally indistinguishable.

Definition 3. ((α, β)-worst-case approximate IO). A uniform PPT machine $\mathcal{O}$ is called a (α, β)-worst-case approximate perfect/computational (indistinguishability) obfuscator for a function class $\mathcal{C}$ if the first condition in Definition 2 is changed as follows (and the second one remains same):

- For each $C \in \mathcal{C}$, $\mathcal{O}(1^n, C)$ can with probability α output C' such that for any x, $\Pr[C'(x) \neq C(x)] < \beta$ where the latter probability is taken over the coins of C'.

Definition 4. ((α, β)-approximate IO). A uniform PPT machine $\mathcal{O}$ is called a (α, β)-approximate perfect/computational (indistinguishability) obfuscator for a function class $\mathcal{C}$ with respect to some distribution D if the first condition in Definition 2 is changed as follows (and the second one remains same):

- For each $C \in \mathcal{C}$, $\mathcal{O}(1^n, C)$ can with probability α output (deterministic) C' such that for random $x \leftarrow D$, $\Pr[C'(x) \neq C(x)] < \beta$.

In following sections we will also allow obfuscators to have additional inputs, e.g. learning parameters (ϵ, δ) or the description of a Boolean polynomial set F.

3 Vector Spaces Spanned by Boolean Polynomials

In this paper by Boolean polynomials, we mean elements of the ring $\mathsf{BP}_n = \mathbf{F}_2[x_1, \cdots, x_n]/ < x_1^2 + x_1, \cdots, x_n^2 + x_n >$. That is, every f in BP_n is multilinear in each involved variable x_i. Let $f_1, \cdots, f_m$ be m Boolean polynomials over $x_1, \cdots, x_n$. The target function class is the vector space spanned by $f_1, \cdots, f_m$, denoted V_F, i.e. all linear combinations of $f_1, \cdots, f_m$ with coefficients in $\mathbf{F}_2$. We also require every f_i has polynomially-long binary description.

Definition 5. (Vector space V_F of Boolean polynomials.) Let F denote a set of m arbitrary polynomials in BP_n for $m = \mathsf{poly}(n)$, denoted $f_1, \cdots, f_m$. Let V_F denote the vector space spanned by $f_1, \cdots, f_m$.

We have some remarks. First, f_i in F is not necessarily in some standard form, e.g. a sum of monomials, and instead it can be of arbitrary form.[1] Second, we need to investigate if V_F is uniquely representable or exactly learnable.

On Infeasibility of Unique Representation. Currently there is no work giving efficient deterministic unique representations of V_F, and we present some analysis on the infeasibility of achieving this goal as well.

Suppose all f_i in F are independent, $f = \sum_{i=1}^{m} a_i f_i$ for unknown $a_i \in \{0,1\}$. If we can recover the values of all a_i, then the unique representation of $\sum_{i=1}^{m} a_i f_i$ is a perfect IO for f. One way for finding all a_i might be choosing m (or more) values for $x = (x_1, \cdots, x_n)$, computing values of $f, f_1, \cdots, f_m$ and forming m linear equations, and then solving all a_i. Note that although $f_1, \cdots, f_m$ are independent, their values on some inputs may not. Precisely, assume f_1, f_2 are of small degree $\leq d$, and by the Warning's theorem (refer to [15], Chapter IV, Theorem 1E), the number of zeros of $f_1(x) - f_2(x)$ can be at least 2^{n-d}. When f_1, f_2 are of arbitrary form, it might be hard to find some non-zeros of $f_1 - f_2$. Thus if the selected values of x, denoted $x^1, \cdots, x^m$ are the zeros of $f_1(x) - f_2(x)$, then $(f_1(x^1), \cdots, f_1(x^m))$ and $(f_2(x^1), \cdots, f_2(x^m))$ are same, and thus dependent. So the solutions of a_i are not unique. If one solution other than the original values of a_i is computed, denoted a_i', $\sum_{i=1}^{m} a_i' f_i$ is not equal to f for all $x \in \{0,1\}^n$.

Let us then consider the feasibility over real numbers $\mathbf{R}$, treating $f, f_1, \cdots, f_m$ as polynomials over $\mathbf{R}$. When extended to $\mathbf{R}$, if $f_1, \cdots, f_m$ are not independent, the solutions are also not unique. Otherwise, the solution of the coefficients may not be integral, resulting in the impossibility to compute all $a_i \in \{0,1\}$.

On Approximate Learnability and Exact Unlearnability. When F is given, there are some ways to learn V_F with approximate correctness under arbitrary distributions. For instance, we can employ the result for PAC learning sparse $\mathbf{F}_2$-polynomials [12] to learn V_F. On the other hand, the analysis above for the case of $\mathbf{F}_2$ already implies the difficulty of exact learning.

Finally recall that a function that on any input outputs 0 is called the zero function and a polynomial is called the zero polynomial if when expressed to the sum of monomials, all its coefficients are 0. Certainly the zero polynomial is the zero function, but the converse is not true generally. However, for the case of Boolean polynomials, the two notions coincide, as the following shows.

Claim 2. Let g_0, g_1 denote two arbitrary polynomials of same functionality in V_F for any F. Then g_0 and g_1 are a same element of BP_n.

---

[1] It refers to that on the guarantee that f_i can be expanded to a sum of multilinear monomials, it belongs to the following inductively defined polynomials: 1. each variable x_i and its negation $1-x_i$ are polynomials; 2. The addition and multiplication operations over two polynomials give new polynomials. 3. Repeating applying step 2 on the defined polynomials polynomial times defines new ones.

4 Homomorphic Evaluation of Boolean Polynomials in V_F

After characterizing V_F, we start to describe the obfuscation for V_F. As shown in the techniques in Sect. 1.1, we employ the somewhat homomorphic encryption scheme SHPKE in [8] to construct an approximate IO for $\mathsf{Eval}(\cdots, f)$. Thus in Sect. 4.1 we present a short description of SHPKE, ignoring details but sufficient for understanding our works. In Sect. 4.2 we present a slight modification on $\mathsf{Eval}(\cdots, f)$.

4.1 Short Description of the Somewhat Homomorphic Encryption Scheme [8]

Let SHPKE $=$ (KenGen, Enc, Dec, Eval) denote the somewhat homomorphic public-key encryption scheme in [8]. $\mathsf{KeyGen}(1^n)$ outputs a pair of keys (pk, sk), where sk is a large random prime integer p and pk is a serials of integers $\langle \alpha_0, \alpha_1, \cdots, \alpha_\tau \rangle$ and every $\alpha_i, i \in \{0, 1 \cdots, \tau\}$ is around a large random multiple of p and α_0 is the largest. To encrypt a bit b, $\mathsf{Enc}(pk, b)$ chooses a random subset $S \subseteq \{1, 2, \cdots, \tau\}$ and a small random integer r, and outputs $X \leftarrow [b + 2r + \Sigma_{i \in S} \alpha_i]_{\alpha_0}$, where the operation $[z]_p$ (similar for $[z]_{\alpha_0}$) denotes (z mod p) as $z - p \cdot \lfloor z/p \rceil$ (note that $[z]_p \in (-p/2, p/2])$. Note that X is of the form $p\mathbf{Z} + 2\mathbf{Z} + b$ and the part of $2\mathbf{Z}$ is usually called noise. To decrypt ciphertext X, $\mathsf{Dec}(p, X)$ computes $b \leftarrow (X \bmod p) \bmod 2$, which is the plaintext.

Then we sketch Eval, which originally works for circuits. But since we only care the specific case of homomorphically evaluating $f \in V_F$, we just specify how Eval works for f. Assume f takes n input bits. Now suppose we are given ciphertexts of these n bits, denoted $X_1, \cdots, X_n$. $\mathsf{Eval}(X_1, \cdots, X_n, f)$ views f as polynomials over integers and applies the integer addition and multiplication operations of f to the ciphertexts, and outputs $f(X_1, \cdots, X_n)$ finally.

SHPKE uses some parameters, controlling the number of integers in the public key and the bit-length of the various integers, all of which are polynomials in the security parameter n. [8] also gives a typical choice for the parameters.

The security of SHPKE assumes the hardness of the Approximate GCD problem described as follows: given many numbers around multiples of p it is hard to recover p. With this assumption, SHPKE is IND-CPA.

4.2 Slight Modification on $\mathsf{Eval}(\cdots, f)$

In the previous subsection we showed that for any $x \in \{0,1\}^n$, $X_i \leftarrow \mathsf{Enc}(pk, x_i)$ for $i \in [n]$, for any $f \in V_F$, $\mathsf{Eval}(X_1, \cdots, X_n, f)$ outputs $f(X_1, \cdots, X_n)$ (f viewed over integers here). Notice that every X_i lies in the interval $(-\frac{\alpha_0}{2}, \frac{\alpha_0}{2}]$, which probably results in negative $f(X_1, \cdots, X_n)$. For the purpose of learning we add a large positive number to f to make it always positive.

Consider the linear combination of monomials of f: $f(x) = \sum_S \prod_{i \in S} x_i$. So $|f(X_1, \cdots, X_n)| \le \sum_S |\prod_{i \in S} X_i| < 2^n \cdot \alpha_0^n$. Let $C_0 = (2\alpha_0)^n$ and modify Eval as follows: for any polynomial $f(x)$,

$$\mathsf{Eval}(X_1, \cdots, X_n, f) \stackrel{\text{def}}{=} f(X_1, \cdots, X_n) + C_0 > 0 \tag{1}$$

Note that α_0 is publicly available from pk, and so C_0 is a constant number after pk is generated. Since each $X_i = p\mathbf{Z} + 2\mathbf{Z} + x_i$ and $C_0 = p\mathbf{Z} + 2\mathbf{Z}$, the output of $\mathsf{Eval}(X_1, \cdots, X_n, f)$ is of the form $p\mathbf{Z} + 2\mathbf{Z} + f(x)$. Thus for correct decryption of the output of $\mathsf{Eval}(X_1, \cdots, X_n, f)$, we need to be ensured that the noise part $2\mathbf{Z}$ is small enough. Actually we have the following claim that under the typical choice of the parameters of SHPKE, the noise part $2\mathbf{Z}$ is indeed small enough so that decryption of $\mathsf{Eval}(X_1, \cdots, X_n, f)$ is always correct.

Claim 3. For $X_i = \mathsf{Enc}(pk, x_i), i \in [n]$, $\mathsf{Dec}(p, \mathsf{Eval}(X_1, \cdots, X_n, f)) = f(x)$.

As f has polynomially-long binary description, the evaluation of $f(X_1, \cdots, X_n)$ can be computed in polynomial-time. Finally we show the perfect polynomial-hiding property of Eval, which is essentially a restatement of Claim 2.

Claim 4. (Perfect polynomial-hiding) Let $g_0, g_1 \in V_F$ be of same functionality. The outputs of $\mathsf{Eval}(X_1, \cdots, X_n, g_0)$ and $\mathsf{Eval}(X_1, \cdots, X_n, g_1)$ are identical for any integers $X_1, \cdots, X_n$.

Proof. By Claim 2, g_0, g_1 have the same expansion as linear combinations of monomials. Thus we have $g_0(X_1, \cdots, X_n) = g_1(X_1, \cdots, X_n)$. Considering the specification of Eval in Eq. 1, the claim holds. □

5 The Basic Learning Result

After characterizing $\mathsf{Eval}(\cdots, f)$ for $f \in V_F$, a key primitive in obfuscating f is an approximate obfuscator for $\mathsf{Eval}(\cdots, f)$. Our idea is to construct a learning algorithm for it used as the desired obfuscator. Note that $\mathsf{Eval}(X_1, \cdots, X_n, f) \stackrel{\text{def}}{=} f(X_1, \cdots, X_n) + C_0$, which is a linear combination of all $f_i \in F$ (plus C_0), viewed as polynomials over reals. So in this section we present a basic result on learning threshold functions of such linear combinations, which will be used in next section to learn $\mathsf{Eval}(\cdots, f)$.

Definition 6. (Real vector space spanned by F') Let F' denote a set of m real multi-variate polynomials for $m = \mathsf{poly}(n)$, denoted $f'_1, \cdots, f'_m$. Let $RV_{F'}$ denote the real vector space spanned by $f'_1, \cdots, f'_m$ with real scalars.

Definition 7. (Threshold Functions of $RV_{F'}$) Let $RV_{F'}$ be same as Definition 6. Let $\mathsf{Sign}(RV_{F'})$ denote the family of threshold functions of $RV_{F'}$. That is, each one in $\mathsf{Sign}(RV_{F'})$ is of the form $\mathsf{Sign}(\sum_{i=1}^{m} w_i f'_i)$, where all w_i are called weights.

Then we investigate the issue of PAC learning $\mathsf{Sign}(RV_{F'})$ when the description of F' is given. The learning of this function family can be reduced to learning threshold functions. Recall that a threshold function $h : \mathbf{R}^n \to \{0, 1\}$ is defined as $h(y_1, \cdots, y_n) = \mathsf{Sign}(\sum_{i=1}^{n} w_i y_i - \theta)$, where $w_1, \cdots, w_n, \theta$ are arbitrary real values and are called weights and y_i are viewed as indeterminates.

Let h be an unknown threshold function and $y = (y_1, \cdots, y_n)$ denote its input. In the PAC learning model, we are given t pairs $\{(y^k, b^k), k \in [t]\}$, where $y^k = (y_1^k, \cdots, y_n^k)$ denotes the kth independent sampling of y from an unknown distribution D and $b^k = h(y^k)$, as well as two parameters (ϵ, δ). Then we expect with probability $1-\delta$ to find a hypothesis h_0 such that $\Pr_{y \leftarrow D}[h(y) \neq h_0(y)] < \epsilon$.

The classical method to learn h runs as follows. For each pair (y^k, b^k), construct a linear inequality: if $b^k = 1$, construct $\sum_{i=1}^n w_i y_i^k - \theta \geq 0$ and if $b^k = 0$, construct $\sum_{i=1}^n w_i y_i^k - \theta < 0$, where $w_i, i \in [n], \theta$ denote the unknown weights. Also there are efficient algorithms that given the t inequalities can find a solution of $w_i, i \in [n], \theta$ (e.g. efficient linear programming), denoted $w_i', i \in [n], \theta'$. When $t = \mathsf{poly}(n, \frac{1}{\epsilon}, \log \frac{1}{\delta})$, $h_0 = \mathsf{Sign}(\sum_{i=1}^n w_i' y_i - \theta')$ is a desired learned hypothesis. We summarize this classic fact as follows.

Fact 5 [12]. There is an algorithm that on input $\mathsf{poly}(n, \frac{1}{\epsilon}, \log \frac{1}{\delta})$ pairs of the form $(y, h(y))$, where y is sampled from any arbitrary distribution D and h is an unknown threshold function, as well as (ϵ, δ), can with probability $1 - \delta$ output a hypothesis h_0 in time $\mathsf{poly}(n, \frac{1}{\epsilon}, \log \frac{1}{\delta})$ such that $\Pr_{y \leftarrow D}[h(y) \neq h_0(y)] < \epsilon$.

We employ the fact to learn $\mathsf{Sign}(RV_{F'})$ under an arbitrary distribution D'. Assume $\mathsf{Sign}(\sum_{i=1}^m w_i f_i')$ is the target function having $Y = (Y_1, \cdots, Y_n)$ as input, where all weights are unknown. Sample Y from distribution D'. Let $\{(Y^k, b^k), k \in [t]\}$ denote the t pairs, where Y^k denotes the kth independent sampling of Y and $b^k = \mathsf{Sign}(\sum_{i=1}^m w_i f_i')(Y^k)$. We compute $y_i^k = f_i'(Y^k), i \in [m], k \in [t]$ to form t pairs (y^k, b^k) for $y^k = (y_1^k, \cdots, y_m^k)$.

Thus the question of learning $\mathsf{Sign}(\sum_{i=1}^m w_i f_i')$ via pairs (Y^k, b^k) can be reduced to that of learning $\mathsf{Sign}(\sum_{i=1}^m w_i y_i)$ (y_i as indeterminates) via pairs (y^k, b^k), while the latter has been solved by Fact 5. Assume the algorithm in Fact 5 outputs a solution of weights, denoted w_i'. Then $\mathsf{Sign}(\sum_{i=1}^m w_i' f_i')$ is a desired learned hypothesis. So we conclude the following result.

Proposition 1. There is an algorithm that on input $\mathsf{poly}(n, \frac{1}{\epsilon}, \log \frac{1}{\delta})$ pairs of the form $(Y, h(Y))$, where Y is sampled from an arbitrary distribution D' and $h = \mathsf{Sign}(\sum_{i=1}^m w_i f_i')$ which weights are unknown, and the description of F' and (ϵ, δ), can with probability $1 - \delta$ output a hypothesis h_0 in time $\mathsf{poly}(n, \frac{1}{\epsilon}, \log \frac{1}{\delta})$ such that $\Pr_{Y \leftarrow D'}[h(Y) \neq h_0(Y)] < \epsilon$.

6 Approximate Obfuscation of $\mathsf{Eval}(\cdots, f)$ by Learning

In this section we will employ the learning result in the previous section to give a learning algorithm for $\mathsf{Eval}(\cdots, f)$ for any $f \in V_F$, viewed as an approximate obfuscator. Assume the output of $\mathsf{Eval}(X_1, \cdots, X_n, f)$ is of $\lambda = \mathsf{poly}(n)$ bits and view the leftmost bit as the highest bit and the rightmost bit as the lowest bit.

Our strategy is to define λ threshold functions of linear combinations of some polynomials to compute the λ bits of the output of $\mathsf{Eval}(X_1, \cdots, X_n, f)$ and show how to learn them gradually. We adopt the order by first considering the highest bit, then the second highest bit and so on until the lowest one.

6.1 Learning the Highest Bit

We first define a function computing the highest bit of $\mathsf{Eval}(X_1, \cdots, X_n, f)$. Note that if the output (viewed as integer) is greater than or equal to $2^{\lambda-1}$, then the highest bit is 1, else it is 0. Thus we have the following.

$$\text{the highest bit} = \mathsf{Sign}(\mathsf{Eval}(X_1, \cdots, X_n, f) - 2^{\lambda-1})$$

Let $f_{m+1} = 1$ and $F_1 = F \cup \{f_{m+1}\}$. Then $\mathsf{Eval}(X_1, \cdots, X_n, f) - 2^{\lambda-1} = f(X_1, \cdots, X_n) + C_0 - 2^{\lambda-1} = \sum_{i=1}^{m+1} w_i^{(1)} f_i$ for some reals $w_i^{(1)}$. Then the highest bit is $\mathsf{Sign}(\sum_{i=1}^{m+1} w_i^{(1)} f_i)$. Thus we employ Proposition 1 to compute a learned hypothesis of $\mathsf{Sign}(\sum_{i=1}^{m+1} w_i^{(1)} f_i)$. To do this, first define the distribution of inputs underlying the learning, denoted D_1. Assume $(pk, sk) \leftarrow \mathsf{KeyGen}(1^n)$ of SHPKE. Then D_1 is the following sampling process: output $(X_1, \cdots, X_n)$, in which $X_i = \mathsf{Enc}(pk, 0)$ independently.

Then our goal is to compute a learned hypothesis of $\mathsf{Sign}(\sum_{i=1}^{m+1} w_i^{(1)} f_i)$ under D_1. To do this, we need to prepare many labeled examples: First sample $\mathsf{poly}(n, \frac{1}{\epsilon}, \log \frac{1}{\delta})$ independent examples from D_1. For each example $(X_1, \cdots, X_n)$, compute $b = \mathsf{Sign}(\sum_{i=1}^{m+1} w_i^{(1)} f_i)(X_1, \cdots, X_n)$ as the label.

By Proposition 1, the learning algorithm on given input (ϵ, δ), the description of F_1 and the labeled examples can with probability $1 - \delta$ generate a learned hypothesis, denoted h_1, which is such that for any fresh $Y^{(1)} = (X_1, \cdots, X_n) \leftarrow D_1$, $\Pr[h_1(Y^{(1)}) \neq \mathsf{Sign}(\sum_{i=1}^{m+1} w_i^{(1)} f_i)(Y^{(1)})] < \epsilon$.

6.2 Learning the jth Highest Bit

Suppose we have obtained $h_1, \cdots, h_{j-1}$ as learned hypotheses of the first $j-1$ highest bits of $\mathsf{Eval}(X_1, \cdots, X_n, f)$ for $2 \leq j \leq \lambda$. Then we continue to generate a learned hypothesis computing the jth highest bit of $\mathsf{Eval}(X_1, \cdots, X_n, f)$.

Let $\beta_{j-1} = h_{j-1}(X_1, \cdots, X_n, \beta_1, \cdots, \beta_{j-2})$ (when $j = 2$ there is no $\beta_\bullet$ in the input). Due to the approximation property of $h_1, \cdots, h_{j-1}$, $\beta_1, \cdots, \beta_{j-1}$ are approximately identical to the first $j-1$ highest bits of $\mathsf{Eval}(X_1, \cdots, X_n, f)$.

Consider $\mathsf{Eval}(X_1, \cdots, X_n, f) - \sum_{j'=1}^{j-1} \beta_{j'} 2^{\lambda-j'}$. The subtraction changes the first $j-1$ highest bits to 0 and keeps remainder unchanged. So if $\mathsf{Eval}(X_1, \cdots, X_n, f) - \sum_{j'=1}^{j-1} \beta_{j'} 2^{\lambda-j'}$ is greater than or equal to $2^{\lambda-j}$, then the jth highest bit is 1, else it is 0. Thus we have the following.

$$\text{the } j\text{th highest bit} = \mathsf{Sign}(\mathsf{Eval}(X_1, \cdots, X_n, f) - \sum_{j'=1}^{j-1} \beta_{j'} 2^{\lambda-j'} - 2^{\lambda-j})$$

Let f_{m+j} denote the variable β_{j-1} (viewed as polynomial) and $F_j = F_{j-1} \cup \{f_{m+j}\}$. Then $\mathsf{Eval}(X_1, \cdots, X_n, f) - \sum_{j'=1}^{j-1} \beta_{j'} 2^{\lambda-j'} - 2^{\lambda-j} = f(X_1, \cdots, X_n) + C_0 - \sum_{j'=1}^{j-1} \beta_{j'} 2^{\lambda-j'} - 2^{\lambda-j} = \sum_{i=1}^{m+j} w_i^{(j)} f_i$, where all $w_i^{(j)}$ are reals.

So the jth highest bit is $\mathsf{Sign}(\sum_{i=1}^{m+j} w_i^{(j)} f_i)$. To employ Proposition 1 to compute a learned hypothesis of it, we define a distribution D_j as follows.

D_j has $h_1, \cdots, h_{j-1}$ hardwired and first runs D_1 to obtain $(X_1, \cdots, X_n)$, computes $\beta_{j'} = h_{j'}(X_1, \cdots, X_n, \beta_1, \cdots, \beta_{j'-1})$ for j' ranging from 1 to $j-1$, and outputs $(X_1, \cdots, X_n, \beta_1, \cdots, \beta_{j-1})$ (here $\beta_1, \cdots, \beta_{j-1}$ denote numerical values).

Now we learn $\mathsf{Sign}(\sum_{i=1}^{m+j} w_i^{(j)} f_i)$ under D_j. We first sample $\mathsf{poly}(n, \frac{1}{\epsilon}, \log \frac{1}{\delta})$ examples from D_j. Compute $b = \mathsf{Sign}(\sum_{i=1}^{m+j} w_i^{(j)} f_i)(X_1, \cdots, X_n, \beta_1, \cdots, \beta_{j-1})$ as the label for each $(X_1, \cdots, X_n, \beta_1, \cdots, \beta_{j-1})$.

By Proposition 1, the learning algorithm on input (ϵ, δ), the description of F_j and the labeled examples can with probability $1 - \delta$ generate a learned hypothesis, denoted h_j, such that for $Y^{(j)} = (X_1, \cdots, X_n, \beta_1, \cdots, \beta_{j-1}) \leftarrow D_j$, $\Pr[h_j(Y^{(j)}) \neq \mathsf{Sign}(\sum_{i=1}^{m+j} w_i^{(j)} f_i)(Y^{(j)})] < \epsilon$.

6.3 Summarization

We summarize the generation of $h_1, \cdots, h_\lambda$ as follows.

Algorithm 6. The algorithm that outputs $h_1, \cdots, h_\lambda$.
Input: $1^n, (\epsilon, \delta)$, the description of $F = \{f_1, \cdots, f_m\}$, an oracle access to f.
Output: functions $h_1, \cdots, h_\lambda$, approximately computing all bits of $\mathsf{Eval}(\cdots, f)$.

- For $j = 1$ to λ, generate all h_j one by one as follows:
  - Let $f_{m+j} = 1$ if $j = 1$ or $f_{m+j} = \beta_{j-1}$ (β_{j-1} viewed as polynomial), $F_j = F_{j-1} \cup \{f_{m+j}\}$ (note $F_0 = F$).
  - Sample $\mathsf{poly}(n, 1/\epsilon, \log(1/\delta))$ examples from D_j.
  - For each example $(X_1, \cdots, X_n, \beta_1, \cdots, \beta_{j-1})$, query f with $(X_1, \cdots, X_n)$ to obtain the function value, compute $b = \mathsf{Sign}(f(X_1, \cdots, X_n) + C_0 - \sum_{j'=1}^{j-1} \beta_{j'} 2^{\lambda-j'} - 2^{\lambda-j})$.
  - Run the algorithm in Proposition 1 with (ϵ, δ), the description of F_j and the labeled examples to generate a learned hypothesis, denoted h_j.
- Output $(h_1, \cdots, h_\lambda)$.

Claim 7. For each j, in the jth repetition, Algorithm 6 can with probability $1-\delta$ generate h_j such that for random $Y^{(j)} = (X_1, \cdots, X_n, \beta_1, \cdots, \beta_{j-1}) \leftarrow D_j$, $\Pr[h_j(Y^{(j)}) \neq \mathsf{Sign}(\sum_{i=1}^{m+j} w_i^{(j)} f_i)(Y^{(j)})] < \epsilon$.

The claim is a restatement of Proposition 1. The running-time of Algorithm 6 is $\mathsf{poly}(n, 1/\epsilon, \log(1/\delta))$ by Proposition 1 and thus its output is polynomial-size. After obtaining $(h_1, \cdots, h_\lambda)$, we can run them as follows.

Algorithm 8. a function h with $(h_1, \cdots, h_\lambda)$ hardwired.
Input: n ciphertexts $(X_1, \cdots, X_n)$
Output: X, approximately identical to the output of $\mathsf{Eval}(X_1, \cdots, X_n, f)$.

- For $j = 1$ to λ, compute $\beta_j = h_j(X_1, \cdots, X_n, \beta_1, \cdots, \beta_{j-1})$ (when $j = 1$ there is no $\beta_\bullet$ in input). Let $X = (\beta_1, \cdots, \beta_\lambda)$ and output it.

Applying Claim 7 for $j \in [\lambda]$ and taking the union bounds of λ failure probabilities δ and λ error probabilities ϵ, we have the following.

Proposition 2. The $(h_1, \cdots, h_\lambda)$ generated by Algorithm 6 can, with probability $1 - \lambda\delta$, be such that for random $(X_1, \cdots, X_n) \leftarrow D_1$, $\Pr[h(X_1, \cdots, X_n) \neq \mathsf{Eval}(X_1, \cdots, X_n, f)] < \lambda\epsilon$.

Proposition 3. For any $g_0, g_1 \in V_F$ of same functionality, the outputs of Algorithm 6 when learning $\mathsf{Eval}(\cdots, g_0)$ and $\mathsf{Eval}(\cdots, g_1)$ are of same distribution.

Proof. Claim 4 ensures g_0, g_1 are of same functionality for a same serials of $X_1, \cdots, X_n$. In learning $\mathsf{Eval}(\cdots, g_0)$ and $\mathsf{Eval}(\cdots, g_1)$, we can always give Algorithm 6 oracle access to g_0. Thus the outputs of Algorithm 6 correspond to two independent running of the algorithm. So they are identically distributed. □

Combining the two propositions, we have the following result.

Theorem 9. Algorithm 6 is a $(1 - \lambda\delta, \epsilon)$-approximate perfect obfuscator for $\mathsf{Eval}(\cdots, f)$ for $f \in V_F$ w.r.t. random ciphertexts $(X_1, \cdots, X_n)$ of 0 using pk.

7 Worst-Case Approximate Perfect IO of V_F

The previous section showed average-case approximate perfect IO for $\mathsf{Eval}(\cdots, f)$ and now we transform it to worst-case approximate perfect IO of V_F.

Algorithm 10. the worst-case approximate perfect IO $\mathsf{wc}\mathcal{O}$ of V_F.
Input: $f \in V_F$, (ϵ, δ), the description of F and 1^n.
Output: an obfuscation of f.

- Run $(pk, p) \leftarrow \mathsf{KeyGen}(1^n)$ of SHPKE.
- Run Algorithm 6 with input (ϵ, δ) and the description of F and answer it for all its oracle access to f to generate $h_1, \cdots, h_\lambda$.
- Construct a function h shown in Algorithm 8.
- Output (h, pk, p) as the obfuscation.

The running-time of Algorithm 10 is still $\mathsf{poly}(n, 1/\epsilon, \log(1/\delta))$ and (h, pk, p) is of size $\mathsf{poly}(n)$. Then we show how to run the obfuscation.

Algorithm 11. the evaluator of the obfuscation (h, pk, p).
Input: any $x \in \{0, 1\}^n$.
Output: b, supposed to be $f(x)$.

- For each bit x_i of x for $i \in [n]$, compute $X_i \leftarrow \mathsf{Enc}(pk, x_i)$.
- Compute $(\beta_1, \cdots, \beta_\lambda) \leftarrow h(X_1, \cdots, X_n)$. Let X denote $(\beta_1, \cdots, \beta_\lambda)$.
- View X as integer and compute $b \leftarrow \mathsf{Dec}(p, X)$. Output b.

First we show the worst-case correctness of $\mathsf{wc}\mathcal{O}$ as follows.

Proposition 4. With probability $1 - \lambda\delta$, $\mathsf{wc}\mathcal{O}(f)$ outputs an obfuscation (h, pk, p) satisfying that for any $x \in \{0,1\}^n$, $\Pr[\mathsf{evaluator}(x) \neq f(x)] < \lambda\epsilon + \mathsf{neg}(n)$.

The proof of Proposition 4 employs the IND-CPA security of SHPKE and Proposition 2 that since the honest ciphertexts of $x_1, \cdots, x_n$ are indistinguishable from those of 0, the output of h on the former is correct except for $\lambda\epsilon+\mathsf{neg}(n)$ probability and thus the last step of $\mathsf{evaluator}(x)$ certainly decrypts to $f(x)$. Next we show the perfect indistinguishability of $\mathsf{wc}\mathcal{O}$.

Proposition 5. For any $g_0, g_1 \in V_F$ of same functionality, their obfuscation generated by $\mathsf{wc}\mathcal{O}$ respectively, identified by $(h, pk, p)_0$ and $(h, pk, p)_1$, are identically distributed.

Proof. In generating the two obfuscation, Algorithm 10 can always give Algorithm 6 the oracle access to g_0. Thus the two obfuscation are just the outputs of $\mathsf{wc}\mathcal{O}$ in two independent running. So they are identically distributed. □

Combining Proposition 4 and Proposition 5, we have the following result.

Theorem 12. $\mathsf{wc}\mathcal{O}$ is a $(1 - \lambda\gamma, \lambda\epsilon)$-worst-case approximate perfect obfuscator for V_F with run-time $\mathsf{poly}(n, 1/\epsilon, \log(1/\delta))$.

8 Achieving Exact Perfect IO of V_F

When $\mathsf{wc}\mathcal{O}$ is obtained, we follow [5] to present an (almost-exact) obfuscator $i\mathcal{O}$.

Exact IO: Given a polynomial $f \in V_F$, the description of F and security parameter 1^n, $i\mathcal{O}$ first outputs $t = n^3$ obfuscation $\widetilde{C}_1, \cdots, \widetilde{C}_t$, where $\widetilde{C}_i \leftarrow \mathsf{wc}\mathcal{O}(f, \epsilon, \delta, F, n)$ where set $\epsilon = \frac{1}{2\lambda n}, \delta = \frac{1}{\lambda n}$ with independent coins for all i. Sample and output t random strings $r_1, \cdots, r_t$ (each sufficient for the usage in Algorithm 11). The obfuscation is $\{\widetilde{C}_i, r_i, i \in [t]\}$.

Evaluation: Given the obfuscation $\{\widetilde{C}_i, r_i, i \in [t]\}$, input $x \in \{0,1\}^n$: for $i \in [t]$, compute $y_i \leftarrow \widetilde{C}_i(x)$ using independent randomness r_i; output $y = \mathsf{majority}(y_1, \cdots, y_t)$.
Set $t = n^3, \epsilon = \frac{1}{2\lambda n}, \delta = \frac{1}{\lambda n}$. By Theorem 12, $i\mathcal{O}$'s run-time is $\mathsf{poly}(n)$. Ultimately we have the following claim, i.e. the main result.

Theorem 13. $i\mathcal{O}$ is an (almost-exact) perfect obfuscator for V_F.

The perfectness of $i\mathcal{O}$ clearly holds and we show the correctness. By Proposition 4, for $f \in V_F$ we have that for each $\widetilde{C}_i$, with probability at least $1 - \frac{1}{n}$, $\Pr[\widetilde{C}_i(x) \neq f(x)] < \frac{1}{n}$. Define $\xi_i = 1$ if $\Pr[\widetilde{C}_i(x) \neq f(x)] < \frac{1}{n}$ and $\xi_i = 0$ else for $i \in [t]$. Then $\mathbf{E}[\sum_{i=1}^t \xi_i] < \frac{t}{n}$. Thus by the Chernoff bounds,

$\Pr[\sum_{i=1}^{t} \xi_i > \frac{t}{n} + \frac{t}{n}] \leq e^{2t/n^2} = \mathsf{neg}(n)$. Thus except for $\mathsf{neg}(n)$ probability, there are no more than $\frac{2t}{n}$ $\widetilde{C}_i$'s which fail to satisfy $\Pr[\widetilde{C}_i(x) \neq f(x)] < \frac{1}{n}$.

Similarly, for any x, define $\zeta_i = 1$ if $\widetilde{C}_i(x) \neq f(x)$ and $\zeta_i = 0$ else for $i \in [t]$. Then $\mathbf{E}[\sum_{i=1}^{t} \zeta_i] < \frac{2t}{n} + \frac{1}{n}(t - \frac{2t}{n}) = \frac{3t}{n} - \frac{2t}{n^2}$. So $2e \cdot \mathbf{E}[\sum_{i=1}^{t} \zeta_i] < \frac{t}{3}$. Thus by the Chernoff bounds, $\Pr[\sum_{i=1}^{t} \zeta_i > \frac{t}{3}] \leq 2^{-t/3}$. This means except for probability $2^{-t/3}$, $y = f(x)$ for any x. Taking the union bounds for all $x \in \{0,1\}^n$, the total failure probability is less than $2^{-t/2 \cdot n} = \mathsf{neg}(n)$. So except for $\mathsf{neg}(n)$ probability (over the coins of $i\mathcal{O}$), we have $\{\widetilde{C}_i, r_i, i \in [t]\}$ is of same functionality as f.

Acknowledgements. The author thanks the reviewers for their helpful comments. This work was supported in part by the National Key Research and Development Project 2020YFA0712300.

References

1. Agrawal, S.: Indistinguishability obfuscation without multilinear maps: new methods for bootstrapping and instantiation. In: Ishai, Y., Rijmen, V. (eds.) EUROCRYPT 2019. LNCS, vol. 11476, pp. 191–225. Springer, Cham (2019). https://doi.org/10.1007/978-3-030-17653-2_7
2. Agrawal, S., Pellet-Mary, A.: Indistinguishability obfuscation without maps: attacks and fixes for noisy linear FE. IACR Cryptol. ePrint Arch. 415 (2020). https://eprint.iacr.org/2020/415
3. Barak, B., et al.: On the (im)possibility of obfuscating programs. J. ACM **59**(2), 6 (2012). https://doi.org/10.1145/2160158.2160159
4. Bitansky, N., Canetti, R., Kalai, Y.T., Paneth, O.: On virtual grey box obfuscation for general circuits. Algorithmica **79**(4), 1014–1051 (2017)
5. Bitansky, N., Vaikuntanathan, V.: Indistinguishability obfuscation: from approximate to exact. In: Kushilevitz, E., Malkin, T. (eds.) TCC 2016. LNCS, vol. 9562, pp. 67–95. Springer, Heidelberg (2016). https://doi.org/10.1007/978-3-662-49096-9_4
6. Brakerski, Z., Döttling, N., Garg, S., Malavolta, G.: Candidate iO from homomorphic encryption schemes. In: Canteaut, A., Ishai, Y. (eds.) EUROCRYPT 2020. LNCS, vol. 12105, pp. 79–109. Springer, Cham (2020). https://doi.org/10.1007/978-3-030-45721-1_4
7. Brakerski, Z., Döttling, N., Garg, S., Malavolta, G.: Factoring and pairings are not necessary for IO: circular-secure LWE suffices. In: ICALP 2022. LIPIcs, vol. 229, pp. 28:1–28:20. Schloss Dagstuhl - Leibniz-Zentrum für Informatik (2022). https://doi.org/10.4230/LIPIcs.ICALP.2022.28
8. van Dijk, M., Gentry, C., Halevi, S., Vaikuntanathan, V.: Fully homomorphic encryption over the integers. In: Gilbert, H. (ed.) EUROCRYPT 2010. LNCS, vol. 6110, pp. 24–43. Springer, Heidelberg (2010). https://doi.org/10.1007/978-3-642-13190-5_2
9. Gay, R., Jain, A., Lin, H., Sahai, A.: Indistinguishability obfuscation from simple-to-state hard problems: new assumptions, new techniques, and simplification. IACR Cryptol. ePrint Arch. 764 (2020). https://eprint.iacr.org/2020/764
10. Gay, R., Pass, R.: Indistinguishability obfuscation from circular security. In: STOC 2021, pp. 736–749. ACM (2021). https://doi.org/10.1145/3406325.3451070

11. Goldwasser, S., Rothblum, G.N.: On best-possible obfuscation. In: Vadhan, S.P. (ed.) TCC 2007. LNCS, vol. 4392, pp. 194–213. Springer, Heidelberg (2007). https://doi.org/10.1007/978-3-540-70936-7_11
12. Hellerstein, L., Servedio, R.A.: On PAC learning algorithms for rich boolean function classes. Theor. Comput. Sci. **384**(1), 66–76 (2007)
13. Jain, A., Lin, H., Sahai, A.: Indistinguishability obfuscation from well-founded assumptions. In: STOC 2021, pp. 60–73. ACM (2021). https://doi.org/10.1145/3406325.3451093
14. Jain, A., Lin, H., Sahai, A.: Indistinguishability obfuscation from LPN over $\mathbf{F}_p$, dlin, and prgs in nc^0. In: EUROCRYPT 2022, Part I. LNCS, vol. 13275, pp. 670–699. Springer, Cham (2022). https://doi.org/10.1007/978-3-031-06944-4_23
15. Schmidt, W.M.: Equations over Finite Fields An Elementary Approach, Lecture Notes in Mathematics, vol. 536. Springer (1976)
16. Valiant, L.G.: A theory of the learnable. Commun. ACM **27**(11), 1134–1142 (1984)
17. Wee, H., Wichs, D.: Candidate obfuscation via oblivious LWE sampling. In: Canteaut, A., Standaert, F.-X. (eds.) EUROCRYPT 2021. LNCS, vol. 12698, pp. 127–156. Springer, Cham (2021). https://doi.org/10.1007/978-3-030-77883-5_5

IterLara: A Concise General-Purpose Algebraic Model

Hongxiao Li[1,2], Wanling Gao[1,2], Lei Wang[1,2], and Jianfeng Zhan[1,2(✉)]

[1] University of Chinese Academy of Sciences, Beijing, China
lihongxiao10@mails.ucas.ac.cn
[2] ICT, Chinese Academy of Sciences, Beijing, China
{gaowanling,wanglei_2011,zhanjianfeng}@ict.ac.cn

Abstract. In the realm of theoretical computer science, establishing a general-purpose model capable of representing various computations in fields such as big data, AI, scientific computing, and databases is of utmost importance. However, there is a need to balance this with the requirement for a concise model that utilizes minimal operators and simple combinations for each computation. To address these concerns, researchers introduced LARA, a key-value algebra that serves as a computational abstraction for both linear and relational algebras. However, previous studies have revealed limitations in LARA's expressive capability, particularly in essential computations like matrix inversion and determinant. To overcome these limitations, we propose ITERLARA, an extension of LARA that incorporates iterative operators. We aim to provide a concise algebraic model that unifies operations in general-purpose computing. Through our research on the expressive ability of both LARA and ITERLARA, we demonstrate that ITERLARA, equipped with aggregation functions, not only enables the representation of matrix inversion and determinant but also encompasses significant computations in the fields of big data, artificial intelligence, scientific computing, and databases. Furthermore, we showcase the conciseness of ITERLARA in representing major big data and AI operations. Additionally, we prove that ITERLARA, without any limitations on function utility, possesses Turing completeness.

Keywords: Big Data · Artificial Intelligence · Linear Algebra · Relational Algebra

1 Introduction

The development of concise and unified algebraic models that capture the intrinsic properties of computing is of great significance. Previous successful models that have been widely adopted include linear algebra in scientific computing and relational algebra in the databases field. Recently, researchers have tried to propose a concise and unified algebraic model encompassing big data, AI, scientific computing, and databases. However, it is crucial to recognize that this undertaking is intricate and poses significant challenges.

Y. Chen et al. (Eds.): COCOON 2024, LNCS 15162, pp. 462–477, 2025.
https://doi.org/10.1007/978-981-96-1093-8_38

Researchers propose different data and computation abstractions in other fields, such as tensor computation graphs in the AI field and relation graphs in the databases fields. However, these abstractions have a different algebraic structure with no equivalent parts with each other. Unifying them, regardless of their algebraic properties, is not feasible. For example, there is no equivalent tensor abstraction in the databases field. Besides this, the existing general-purpose models like the Turing machine lack high-level abstractions for computation representation and have less advantage in modeling contemporary computations. However, a minority of concise and unified algebraic models have been proposed for different fields in recent years. A representative model is LARA. LARA [9] is an algebraic model that aims to unify linear and relational algebra. It provides a unified data abstraction for arrays, graphs, and tensors named associative table. LARA can represent all operations in the extended relational algebra field and most operations in the linear algebra field. However, the studies on LARA's expressive ability [3,4] reported that it could not represent several essential computations, such as matrix inversion and determinant. It cannot represent other non-local computations, either.

Our work aims to propose a concise and unified algebraic model that encompasses big data, AI, scientific computing, and databases based on LARA. Our insight is to add minimum operators to LARA and extend its expressive ability. Since iterative computation is necessary for general-purpose models, we add an iterative operator to LARA and proposed ITERLARA. This trial is inspired by the Geerts et al. [7] as an extension to Matlang [5]. We argue that ITERLARA is suitable as a concise and Turing complete algebraic model that unifies computations in big data, AI, scientific computing, and databases fields. Table 1 is the brief abstract of ITERLARA's expressive ability in comparison to those of the extended relation algebra (RA), for-Matlang [7], and LARA. The result reports that ITERLARA can represent the most computations from different fields while the others cannot. This table includes major operators in big data, AI, scientific computing, and databases fields.

Our main contribution is divided into the following three parts: First, we thoroughly study the expressive ability of the original LARA (also noted as LARA($\Omega^{All}_{All[n]}$), it will be explained in late sections), rather than tame-LARA or other extensions. We prove that matrix inversion (Inv) and determinant (Det) of an arbitrary matrix could not be represented in LARA. We also present AvgPool1d [12] and other representations in LARA. Second, we present our nontrivial extension to LARA, introducing the concept of iteration and naming the resulting algebraic model ITERLARA. In ITERLARA, we define an "expression" as an operator accompanied by the required input tables, adhering to the appropriate syntax. Specifically, we introduce the iterative operator (**iter** or **for**) based on expressions and provide a detailed specification of the precise syntax and semantics governing all expressions within the ITERLARA framework. Third, we study the expressive ability of ITERLARA in three aspects. We prove that LARA with constant iteration times (noted as ITERLARA^{Const}) is equivalent to LARA. We demonstrate that LARA with aggregation iteration times (indicated as ITERLARA^{Count}) can represent matrix inversion and determinant [8] and show LARA's conciseness in representing major operations in big data and AI fields. We prove that ITERLARA with no limitation (noted as ITERLARA) is Turing complete by constructing representations of all operators of another Turing complete algebra.

The rest of the paper is organized as follows: Sect. 2 is the related work. Section 3 introduces LARA, including the strict definition of LARA operations with an example. Section 4 is the definitive ability study of LARA, including the studies of matrix inversion (*Inv*), determinant (*Det*), pooling, and others. Section 5 is the definition of ITERLARA. Section 6 is the expressive ability study of ITERLARA, including the studies of ITERLARA^{Const}, ITERLARA^{Count}, and ITERLARA. Section 7 is the conclusion.

Table 1. This table gives a summary of comparing ITERLARA's expressive ability against the extended relation algebra (RA) [1], for-Matlang [7], and LARA, indicating ITERLARA has better expressive ability.

Computation	Field	The extended relational algebra [1]	for-Matlang [7]	LARA [9]	ITERLARA (Ours)
$\sigma, \pi, \rho, \gamma, \cup, \cap$	Big data, databases	✓	✓	✓	✓
$-, \div, \bowtie, \rhd, \lhd$	Big data, databases	✓	✓	✓	✓
$+, \times$	AI, scientific computing	✓	✓	✓	✓
gemm, *smm*	AI, scientific computing	×	✓	✓	✓
pooling (*MaxPool*, ⋯)	AI	×	✓	✓	✓
activation (*ReLU*, *SeLU*, ⋯)	AI	×	×	✓	✓
ArgMax, *ArgMin*	AI, big data	✓	✓	✓	✓
Conv	AI, scientific computing	×	✓	✓	✓
$Inv[n], Det[n]$	Scientific computing	×	✓	✓	✓
Inv, *Det*	Scientific computing	×	×	×	✓
EinSum	AI, scientific computing	×	✓	✓	✓
aggregation (*Count*, *Sum*, ⋯)	Databases	✓	✓	✓	✓
First-order logic	AI, big data, databases	×	×	✓	✓
Iteration	AI, big data	×	✓	×	✓
for-loop, while-loop	AI, big data	×	✓	×	✓

2 Related Work

Tensor computation graphs in the AI field, matrix computation in the scientific computing field, and relational algebra [1] in the databases field are the most common algebraic models. There are no unified algebraic models for computation in the big data field.

Our work proposes a unified computation model covering big data, AI, scientific computing, and databases. Similar research on unified computation models includes LARA [9], Matlang [5], and their extensions. Table 1 shows the concrete relation between the existing research and ours.

LARA is a minimalist kernel for a linear and relational algebra databases proposed in the LaraDB project [10]. LARA uses key-value pairs for data abstraction, which can be converted to relation graphs. Barceló et al. [4] concluded that tame-LARA cannot represent matrix inversion and convolution and proposed an extension to LARA and it can represent convolution and other local operations. Meanwhile, LARA is equivalent to first-order logic with aggregation.

The Matlang language is a recently proposed algebraic model as an abstraction. Barceló et al. [3] concluded that Matlang has a lower expressive ability than LARA. Geerts et al. [7] added iterative properties to the Matlang language. This paper inspired us to add iterative properties to other algebraic models. The ITERLARA extension to LARA is similar to for-Matlang extension to Matlang but has a more extraordinary expressive ability.

3 The Original LARA Algebra

This section introduces the definitions, algebraic structures, and semantics of the original LARA language. We add some other notations for convenience, basically notations from relational algebra. We ensure that every notation is well-defined at its first appearance. The entire list of our notation is presented in Table 2 in Sect. 3.5.

3.1 Preliminaries

In this paper, $\mathbb{N}[a, b]$ stands for integers between a and b, where a and b are positive. An associative table (in short as "table" if not confusing) is defined as a key-value table of data. The names of keys and values of table A are noted as K_A and V_A. There are no duplicated key variables in a table, which means that data (we call "records") of the same key must have the same value. We note the key and the value of record a as $k(a)$ and $v(a)$. We can safely note the value of a record in table A that has a certain key $\overrightarrow{k}$ as $A[\overrightarrow{k}]$. We note $a = k(a) : v(a) = \overrightarrow{k} : \overrightarrow{v}$ (separated representation) and $A[\overrightarrow{k}] = \overrightarrow{v}$.

Next, we present requirements for user-definable binary functions $\oplus$ and $\otimes$, that will appear in the following text. First, these functions must input two values and output one value of the same type. Second, $\oplus$ and $\otimes$ must satisfy commutativity and associativity, namely $a\theta b = b\theta a$ and $(a\theta b)\theta c = a\theta(b\theta c)$ where θ stands for $\oplus$ or $\otimes$. Third, $\oplus$ and $\otimes$ must have an identity element. For example, the identity element for addition is 0, and the identity element for multiplication is 1. For all the values in table A, an aggregation function is defined as the associative computation on every value in A, noted as $\bigoplus A = \bigoplus_{x \in A} x$ and $\bigotimes A = \bigotimes_{x \in A} x$.

Symbols *sum*, *min*, *max*, and *count* are four aggregation functions that stand for summation, minimum, maximum, and total count, respectively. Besides, *avg* represents the average function, which can be computed with *sum*/*count*.

We also use the symbols in relational algebra in this paper. The explanation for symbols $\sigma_F A$, $\pi_S A$, $\rho_{x/y} A$, $\gamma_S A$, $\cup$, $\cap$, $-$, $\div$, $\bowtie$, $\triangleright$ and $\triangleleft$ are presented in Table 2. Please note that as the relaxed table join in LARA and the natural join in relational algebra are equivalent when $\otimes$ takes the Cartesian product, we adopt the same symbol.

3.2 Data Abstraction

The only data abstraction of LARA is the associative table, firstly proposed in [11]. We use $A : [\overrightarrow{k} : \overrightarrow{v}]$ to denote an associative table with $\overrightarrow{k} = [k_1, \cdots, k_n]$ attributes and $\overrightarrow{v} = [v_1, \cdots, v_m]$. For example, a matrix M can be represented as a table: $M : [[i, j] : v]$, where $[i, j]$ is the key that stands for the coordinate and v is the value that stands for the data. Besides matrices and tensors, a relation table in relational algebra can also be represented as a table, where a key is a unique id and a value is the data in a record. The symbol E_A represents an empty table with the same key and value attributes as A's.

An association table must include a finite quantity of records. The value attribute of each record must have limited support; namely, it must take value from a limited amount of data, and the data must include a default value. The default value is the identity value of aggregation functions, and it is a zero-element or a one-element in most conditions with no specification.

3.3 Operators

LARA includes three types of operators: union, join, and ext, parameterized by user-definable "sum" ($\oplus$), "multiply" ($\otimes$), and "flatmap" (f) functions, respectively. The join operator includes the strict join and the relaxed join. We present the definition of them as follows. Furthermore, Sect. 3.4 provides visualized examples of LARA's data and operators.

Table Union. The union operator inputs two tables, A and B. With given attributes' formats of A and B, the union $A \boxtimes_{\oplus} B$ is defined as follows:
For tables $A : [\overrightarrow{a}, \overrightarrow{c} : \overrightarrow{x}, \overrightarrow{z}]$ and $B : [\overrightarrow{c}, \overrightarrow{b} : \overrightarrow{z}, \overrightarrow{y}]$, where c's are common part of keys and a's and b's are individual parts of keys, z's are common part of values and x's and y's are individual parts of values, we have $A \boxtimes_{\oplus} B = [\overrightarrow{c} : \overrightarrow{x}, \overrightarrow{z}, \overrightarrow{y}]$. Table union on A and B retains the intersection of K_A and K_B and the union of V_A and V_B, with element-wise $\oplus$ aggregation on the data. In relational algebra language, $A \boxtimes_{\oplus} B = [K_A \cap K_B : V_A \cup V_B]$.

Strict Table Join. The strict join operator inputs two tables, A and B. With given attributes' formats of A and B, the union $A \hat{\bowtie}_{\otimes} B$ is defined as follows:
For tables $A : [\overrightarrow{a}, \overrightarrow{c} : \overrightarrow{x}, \overrightarrow{z}]$ and $B : [\overrightarrow{c}, \overrightarrow{b} : \overrightarrow{z}, \overrightarrow{y}]$, where c's are common part of keys and a's and b's are individual parts of keys, z's are common part of values

and x's and y's are individual parts of values, we have $A\hat{\bowtie}_{\otimes}B = [\vec{a}, \vec{c}, \vec{b} : \vec{z}]$. Strict table join on A and B retains the union of K_A and K_B and the intersection of V_A and V_B, with element-wise $\otimes$ aggregation on the data. In relational algebra language, $A \text{⧖}_{\oplus} B = [K_A \cup K_B : V_A \cap V_B]$.

Ext. For tables $A : [\vec{a} : \vec{x}, \vec{0}]$, where $\vec{0}$ stands for the default values, and function $f : \vec{a} \times \vec{x} \to (\vec{b} \to \vec{y})$, we have $(\mathbf{ext}_f A)[\vec{a}, \vec{b}] = f(\vec{a} : A[\vec{a}])\vec{b}$. The ext operator can be considered mapping from a table format to another table format within a given pattern. Especially, when the flatmap function f does not change the table format, namely $m = m'$ and $\vec{a} = \vec{b}$, we call this operator as map, noted as $\mathbf{map}_f A$.

Relaxed Table Join. As we have presented, strict table join on A and B retains the intersection of V_A and V_B. Relaxed table join retains the union of V_A and V_B and does not drop any value attributes. Relaxed table join is equivalent to the natural join in relational algebra when $\otimes$ takes the Cartesian product. In detail, we have $A \bowtie_{\otimes} B = \mathbf{ext}_{f_A}\mathbf{ext}_{f_B}(A\hat{\bowtie}_{\otimes}B)$, where f_A is a identity mapping that changes the value attribute V_A to $V_A \cup V_B$, and f_B is similar. The data at the extended attributes take the default value. In relational algebra language, $A \text{⧖}_{\oplus} B = [K_A \cap K_B : V_A \cap V_B]$.

3.4 Examples for Lara Operators

This subsection provides examples of associative tables and Lara computations for readability.

Example 1. We define associative tables A and B as Fig. 1. The results of $A \text{⧖}_{max} B$, $A \bowtie_{\times} B$ (relaxed join), and $\mathbf{ext}_{3_w} A$ is shown as follows. Note that 3_w stands for a constant function with value three on an attribute w.

a	c	x	z
0	0	1	2
0	1	2	4
1	0	3	6
1	1	4	8

Table A

c	b	z	y
0	0	1	7
0	1	3	5
1	0	5	3
1	1	7	1

Table B

c	x	z	y
0	3	6	7
1	4	8	5

Table $A \text{⧖}_{max} B$

a	c	b	x	z	y
0	0	0	1	2	7
0	0	1	1	6	5
0	1	0	2	20	3
0	1	1	2	28	1
1	0	0	3	6	7
1	0	1	3	18	5
1	1	0	4	40	3
1	1	1	4	56	1

Table $A \bowtie_{\times} B$

a	c	w
0	0	3
0	1	3
1	0	3
1	1	3

Table $\mathbf{ext}_{3_w} A$

Fig. 1. Tables A, B, $A \text{⧖}_{max} B$, $A \bowtie_{\times} B$, and $\mathbf{ext}_{3_w} A$

3.5 Notations in this Paper

We present the overall notations in Table 2 for clarity, including symbols in Lara and IterLara.

Table 2. List of operators

Operator	Description
$A, B, C, \cdots$	Associative tables
$a, b, c, \cdots$	Records
$e, e_1, e_2, \cdots$	Expressions
$\overrightarrow{k_1}, \overrightarrow{k_2}, \cdots$	Keys
$\overrightarrow{v_1}, \overrightarrow{v_2}, \cdots$	Values
K_A	Set of names of key attributes of table A
V_A	Set of names of value attributes of table A
E_A	The empty table that has the same key and value attributes as A's
$a \oplus b$	The user-definable addition of two records a and b
$a \otimes b$	The user-definable multiplication of two records a and b
$\bigoplus_{x \in A} x, \bigotimes_{x \in A} x$	Aggregation on table A
$A \bowtie_{\oplus} B$	The union of table A and B, noted as $A \bowtie B$ if addition is not important
$A \hat{\bowtie}_{\otimes} B$	The strict join of table A and B, noted as $A \hat{\bowtie} B$ if multiplication is not important
$A \bowtie_{\otimes} B$	The relaxed join of table A and B, which is equivalent to natural join ($\bowtie$) in relational algebra when $\otimes$ takes the Cartesian product
$\mathbf{ext}_f A$	The extension of function f onto a table A
$\mathbf{iter}_F^{cond} A$	The iteration operator with condition expression *cond* and iteration body F on a table A
$\sigma_F A$	Selection on table A with function F in relational algebra
$\pi_S A$	Projection on table A with attribute S in relational algebra
$\rho_{x/y} A$	Rename on table A with attribute name from x to y in relational algebra
$\gamma_S A$	Aggregation on table A with attribute S in relational algebra
$\cup, \cap, -, \div$	Set union, intersection, difference, as the same in relational algebra
$\triangleright, \triangleleft$	Left and right antijoin in relational algebra, the result only includes the common records with attributes from one side. $A \triangleright B = A - \pi_{K_A - K_B, V_A - V_B}(A \bowtie B)$, and $A \triangleleft B = B - \pi_{K_B - K_A, V_B - V_A}(A \bowtie B)$
Boole[$expr$]	The Boolean value of the expression $expr$

4 LARA's Expressive Ability

This section is our first contribution. We thoroughly study the expressive ability of LARA and compare it with some extended LARA algebra. Before the study, we illustrate some symbols of algebra. The symbol LARA(Ω_s^t) stands for the LARA language, with the following constraints: first, all the data is within the range of Ω; second, only operators in the set s can be used on the key attributes; third, only operators in the set t can be used on the value attributes. The last two constraints limit choices on the user-definable functions. Note that our symbols in this section are in accordance with Barceló et al. [4].

Hutchison et al. [9] introduced the LARA language without constraints on the user-definable functions. However, according to the original study in [9], the semantics should be that the user-definable functions must have a certain finite quantity n of input parameters. Therefore, we note the original LARA as LARA($\Omega_{All[n]}^{All}$), where Ω stands for the total set of data, without any constraint on value range. They proved that the relational algebra (RA) can be represented in LARA [9], namely $RA \in$ LARA($\Omega_{All[n]}^{All}$). They also proved that general matrix multiplication ($gemm$) and the sparse matrix multiplication (smm) can be represented in LARA [9], namely $gemm, smm \in$ LARA($\Omega_{All[n]}^{All}$).

Barceló et al. [4] proposed a restricted version of LARA, named tame-LARA, which we note as LARA($\Omega_{=}^{All}$). We have LARA($\Omega_{=}^{All}$) $\in$ LARA $=$ LARA($\Omega_{All[n]}^{All}$).

They proved that the extended relational algebra (RA_{Agg}) can be represented in tame-LARA, where Agg means the aggregation functions added. Barceló et al. [4] also proposed that LARA can represent select (σ), project (π), rename (ρ), union ($\cup$), intersection ($\cap$), difference (-), division ($\div$), aggregation (γ), and other relational operations. We will use relational algebraic symbols instead of LARA expressions for simplicity in our late text on expressive ability study in this section. They also proved that $Inv, Conv \notin$ LARA$(\Omega_{=}^{All})$, where Inv and $Conv$ stand for matrix inversion and convolution, and LARA cannot represent iteration. They proposed an extension of LARA, we note as LARA$(\Omega_{<}^{\{+,\times\}})$. They proved that $Conv \in$ LARA$(\Omega_{<}^{\{+,\times\}})$. Barceló et al. [3] proved that $Einsum, Conv \in$ LARA$(\Omega_{All[n]}^{All})$, where $Einsum$ stands for the Einstein summation [2].

Our contribution to LARA's expressive ability study is that we prove $Inv \notin$ LARA$(\Omega_{All[n]}^{All})$ and $Inv[n], Pool \in$ LARA$(\Omega_{All[n]}^{All})$, where $Inv[n]$ stands for the matrix inversion at a fixed size n, and $Pool$ stands for pooling. We also represent some other important computations.

4.1 LARA Cannot Represent Matrix Inverse

We prove that $Inv \notin$ LARA$(\Omega_{All[n]}^{All})$ in two steps.

Proposition 1. *A matrix inverse is representable in* LARA *only if the determinant is representable in* LARA.

Proof. Consider the adjugate matrix A^*'s representation in LARA. In linear algebra, the adjugate or classical adjoint of a square matrix A is the transpose of its cofactor matrix and is denoted by A^*.
We define the algebraic minor matrix $m_{x,y}$ as:

$$\forall B, m_{x,y}B = \begin{cases} (-1)^{i+j} b_{i,j}, & \text{if } i \neq x \text{ and } j \neq y \\ 0 \text{ (default)}, & \text{otherwise} \end{cases}$$

.

As a result, $A^* = \bowtie_{\cup}^{i,j} A_{i,j} = \bowtie_{\cup}^{i,j} \mathbf{ext}_{m_{i,j}} A$.

Define matrix division function $\frac{A}{B} = \frac{a_{1,1}}{b_{1,1}}$ with the requirement of A and B as the following:

$$\forall i, j, x, y, \frac{a_{i,j}}{b_{i,j}} = \frac{a_{x,y}}{b_{x,y}}$$

.

Matrix division is defined only when A is constant times of B.

According to linear algebra, the adjugate matrix $A^* = Det(A) Inv(A)$ if A is invertible.

If $Inv(A)$ can be represented as $I(A)$, since A is invertible, then $Det(A) = \frac{A^*}{I(A)}$ is representable in LARA. So matrix inverse is representable in LARA only if the determinant is representable in LARA.

Proposition 2. *Determinant is not representable in* LARA.

Proof. Suppose $Det(A)$ can be represented as $f(A)$, where $A \in \mathbb{R}^{n\times n}$. For convenience, we suppose $n > 3$.

Consider $f'(A) \stackrel{\text{def}}{=} \mathbf{ext}_f A_{[a_1,\cdots,a_n]}$, where $A_{[a_1,\cdots,a_n]}$ is a matrix that has 1 value at $A_{1,a_1},\cdots,$
A_{n,a_n} and 0 value at other entries, where $a_1,\cdots,a_n \in \mathbb{N}[1,n]$. Because $\{A_{a_1,\cdots,a_n}\} \subseteq \mathbb{R}^{n\times n}$, $f'(A)$ is representable.

Consider $g(A) = \text{Boole}[\mathbf{ext}_{f'}A]$, where $\text{Boole}[expr]$ equals the Boolean value of *expr*. According to linear algebra, $g(A)$ equals 1 value when $a_1,\cdots,a_n$ are pairwise unequal and equals 0 otherwise.

In other words, $g(A) = \begin{cases} 1, \text{ if } \bigwedge_{i,j\in[1,n],i\neq j} \text{Boole}[a_i \neq a_j] \\ 0, \text{ otherwise} \end{cases}$.

Because $B(x,y) \stackrel{\text{def}}{=} \text{Boole}[x \neq y]$ does not satisfy transitivity, namely $B(y,z)$ is independent of $B(x,y)$ and $B(x,z)$ for arbitrary x,y,z, $g(A)$ must be a function of the total set a_i. We note that $g(A) \stackrel{\text{def}}{=} g'(< a_1,\cdots,a_n >)$.

Function g' has a total of n independent variables. Consider the three LARA operations. Table union only accepts union with a unary key attribute, rather than with $< a_1,\cdots,a_n >$. Table join does not do aggregation on value attributes. Therefore, g' representation must include flatmap operation. Construct function $g'_0 :< a_1,\cdots,a_n > \times \#_1 \to (\#_2 \to \#_3)$, where $\#_i, i = 1,2,3$ stands for dummy variables.

Since n can be an integer bigger than any finite range, g'_0 must have the transitive closure of an infinite number of independent variables, which is impossible.

According to the above proof, matrix inverse is not representable in LARA.

4.2 Matrix Inversion at a Specific Size N

While a general matrix inversion cannot be represented, we present the representation of inversion to a particular size n.

Proposition 3. $Inv[n] \in \text{LARA}(\Omega^{All}_{All[n]})$.

We would not provide the proof for this proposition in this paper, as it is similar to that in Sect. 6.2.

4.3 Pooling and Other Computations

We prove that AvgPool1d (defined by PyTorch [12]) is representable in $\text{LARA}(\Omega^{All}_{All[n]})$. The other pooling operators (e.g., AvgPool2d, MaxPool1d, ⋯) are representable for the same reason.

Proposition 4. $AvgPool1d \in \text{LARA}(\Omega^{All}_{All[n]})$.

Proof. Define reshaping $r_s : i \times v \rightarrow (i' \times j \rightarrow v)$ such that $(\mathbf{ext}_{r_s} A)[i', j] = A[i' \times s + j]$, where $i' \in \mathbb{N}[0, i/s - 1]$ and $j \in \mathbb{N}[0, s-1]$. We have $AvgPool1d(A, s) = \mathbf{ext}_{avg_j}(\mathbf{ext}_{r_s} A)$.

Besides pooling, we present Lara representation of $ArgMax$ and activation function (e.g., $ReLU$) computations on a vector A as follows:
$ArgMax(A) = A \div \gamma_{max} A$ (with relational algebraic symbols, for simplicity),
$ReLU(A) = \mathbf{ext}_{max_0} A$, where $max_0(x) = max(x, 0)$.

So far, our study reports that Lara can represent the majority of computations concisely in big data and AI fields. In contrast, it cannot represent some specific necessary computations (e.g., inversion) and cannot represent iteration. The research on Lara's expressive ability in the relational algebra field, linear algebra field, and AI field is relatively complete.

5 IterLara: Lara with Iteration Extension

Lara does not include any iteration and recursion operator. To extend Lara with iteration, we define IterLara. An iteration includes a condition expression and an iteration body. Before presenting the definition of IterLara, we first define the concept of "expression". After that, we provide the entire definition.

5.1 Lara's Syntax and Semantics

In this subsection, we introduce the "expression" concept. A Lara expression is defined as an operator (union, join, or ext) with given input tables in the correct syntax. The result of an expression must be a table. For example, the formula $A \bowtie\!\!\!\!\bowtie_+ B$, $A \bowtie_\times B$, and $\mathbf{ext}_{sqrt} A$ are expressions. The formula $A \bowtie\!\!\!\!\bowtie$ and $A \bowtie_- B$ are not expressions because $\bowtie\!\!\!\!\bowtie$ requires two inputs, and the subtraction (-) function is not commutative.

We use $e, e_1, e_2, expr, \cdots$ to refer to an expression in this section. Based on this, the entire syntax of Lara is shown as follows, where $\text{dom}(f)$ stands for the function's input domain of f. The following formulations are the only legal expressions in Lara:

e ::= A, (A is an associative table)
or $e_1 \bowtie\!\!\!\!\bowtie_\oplus e_2$, $(K_{e_1} \cap K_{e_2} \neq \varnothing, \text{dom}(\oplus) = V_{e_1} = V_{e_2})$
or $e_1 \hat{\bowtie}_\otimes e_2$, $(V_{e_1} \cap V_{e_2} \neq \varnothing, \text{dom}(\otimes) = V_{e_1} = V_{e_2})$
or $e_1 \bowtie_\otimes e_2$, $(V_{e_1} \cap V_{e_2} \neq \varnothing, \text{dom}(\otimes) = V_{e_1} = V_{e_2})$
or $\mathbf{ext}_f e$, $(\text{dom}(f) = K_e \times V_e)$

5.2 The Iteration Operator

To define the iteration operator, we first define the expression mapping F as $F : e_1 \rightarrow e_2$, which means $F(e_1) = e_2$, namely F is a replacement from e_1

to e_2. An expression mapping is used as an iteration body. For example, if $F(e) = e \bowtie C$, where C is a constant table, then $F(F(e)) = e \bowtie C \bowtie C$.

Now we define the iteration operator $\mathbf{iter}_F^{cond}$ as follows:

$$\mathbf{iter}_F^{cond} e = \begin{cases} F(\mathbf{iter}_F^{cond} e), & \text{if Boole}[cond] \\ e, & \text{otherwise} \end{cases}$$

In the above formulation, the symbol F is the expression mapping, the symbol e is the initial expression, and the condition expression *cond* is defined as a Boolean expression, but it can also be a table. The feasibility of considering an associative table as a single numeric is explained in late text. We name LARA with this extended iteration operator as ITERLARA.

Some Explanation. In defining the iterative operator $\mathbf{iter}_F^{cond} e$, we mentioned that the condition expression *cond* must be a Boolean value. However, an associative table A with a single record of a single numeric value a is still an associative table rather than a numeric one. How to apply a user-defined function f on it (to let $f(A) = f(a)$)? This is feasible, as any user-definable function can only be used with an ext operator in an expression. For example, $\mathbf{ext}_{+f(A)}B$ can be considered as $\rho(\mathbf{ext}_f(A)) \boxtimes_+ B$. We add rename operator ρ to ensure that $K_{\rho(\mathbf{ext}_f(A))} = K_B$ and $V_{\rho(\mathbf{ext}_f(A))} = V_B$. However, in the above example, the conversion from ext operator to union operator requires that the add $(+)$ function is commutative and associative. How do we apply this to another function that does not have this property? This is also feasible by redefining a commutative and associative function and discarding part of its output using the ext operator when needed. For commutativity, if $h(x, y) \neq h(y, x)$, define $h'(x, y) = [[x, h(x, y)], [y, h(y, x)]]$. For associativity, if $h'(h'(x, y), z) \neq h'(x, h'(y, z))$, define $h''(x, y, z) = [[x, h'(h'(x, y), z)], [z, h'(x, h'(y, z))]]$ and $h''(x, y) = h''(x, y, \omega)$ where ω is the default value.

ITERLARA with for-Operator. Especially when the number of iterations is a definite number n, we define another symbol as $\mathbf{for}_F^n e = \underbrace{F(F(\cdots F(e)\cdots))}_{n \text{ iterations of } F}$.

In the above formulation, we replaced **iter** with **for**. The Boolean symbol *cond* is also replaced with the number n in this condition. In this restriction, we name LARA with this extended iteration operator as ITERLARA^{Const}.

If the number of iterations is not a constant but can be computed before entering the loop and will not change, we can also note it as $\mathbf{for}_F^{cond} e$. In the above formulation, we replaced **iter** with **for**. In this condition, the expression *cond* is required to be a definite (computable) number before entering the loop. Specifically, when the expression *cond* is an aggregation function (e.g., *sum*, *count*, $\cdots$), we name it as ITERLARA^{Count}.

6 ITERLARA's Expressive Ability

This section analyzes ITERLARA's expressive ability in three aspects. First, the minimum ability of ITERLARA is the same as that of LARA, supported by the equivalence of ITERLARA^{Const} and LARA. Second, ITERLARA^{Count} can represent matrix inversion and determinant. This conclusion shows that ITERLARA^{Count} is approximately enough as a unified computation abstraction model for big data, AI, scientific computing, and databases fields. Third, the maximum ability of ITERLARA is the same as a Turing machine.

6.1 ITERLARA with Constant Times of Iteration Is Equivalent to LARA

Proposition 5. $\mathit{Lara} = \mathit{IterLara}^{Const}$.

This proposition is trivial.

6.2 ITERLARA with Aggregation Can Represent Matrix Inversion and Determinant

Proposition 6. $\mathit{Inv} \in \mathit{IterLara}^{Count}$.

Proof. Given a matrix represented as table $A : [[i, j] : v]$, whose size n is unknown. We make $cond = A \bowtie_{+} E$, where $E : [i : v]$ is an empty table with only i as its key attribute and the same value attribute as A. According to the definition of *cond*, we set $cond = n$. Construct table $B : [[j, k] : w, B[j, 1] = 1$, similar to $B_1, \cdots, B_k$ in the proof of Proposition 3. We have $B = B_1$. Construct a user-definable function f such that $\mathbf{ext}_f(B_i) = B_{i+1}$. Since B_i has finite support, this is feasible. We make $F(A) := A \bowtie_{\times} B$. Therefore, we note $A' = \mathbf{for}^{cond}_{F;B:=f(B)} A = A \bowtie_{\times} B_1 \bowtie_{\times} \cdots \bowtie_{\times} B_n$. The feasibility details of the expression $(\mathbf{for}^{cond}_{F;B:=f(B)})$ are presented later.

Next step, we want to construct a similar function d and table C as in the proof of Proposition 3. For function d, it is easy because we can define an infinite quantity of functions $d_1(k_1 : w), d_2([k_1, k_2] : w), \cdots$ and there must exist a proper function $d = d_n$. For table C, since we have table $A' : [[k_1, \cdots, k_n] : w]$, we can get that $C = \mathbf{ext}_1(\sigma_s A')$ where $s(x) = \mathbf{true}$ when x is a permutation.

Consider the adjugate matrix A^* in linear algebra represented in LARA as follows:

The algebraic minor matrix is

$$m_{x,y}B = \begin{cases} (-1)^{i+j} b_{i,j}, & \text{if } i \neq x \text{ and } j \neq y \\ 0 \text{ (default)}, & \text{otherwise} \end{cases}, \text{ and } A^* = \bowtie_{\cup}^{i,j} A_{i,j} = \bowtie_{\cup}^{i,j} \mathbf{ext}_{m_{i,j}} A.$$

For the same reason, we have

$$\mathit{Det}(A) = C \bowtie_{+} \mathbf{ext}_d(A'), \text{ and } \mathit{Inv}(A) = A^* \bowtie_{\times} \left(\mathbf{ext}_{reciprocal} \mathit{Det}(A)\right).$$

Some Explanation. In the proof, we used this expression: $\mathbf{for}^{cond}_{F;B:=f(B)}A$. However, by definition of the iteration operator, we cannot change another table's value (B's value) in the loop body. How to achieve the equivalent result? This is still feasible by the Cartesian product. The Cartesian product is representable in LARA and of course in ITERLARA by renaming (ρ) and relaxed join ($\bowtie$). According to the definition, any result of an expression is a table. The expression $\mathbf{iter}^{cond}_{F;e_1:=e_2}e$ can be represented as $\pi_e(\mathbf{iter}^{cond}_{F'}(e \bowtie e_1))$, where $F'(e \bowtie e_1) = F(e) \bowtie e_2$. In the above formula, we leave out renaming operators. By defining a new expression function on the Cartesian product of tables, several statements can be unified as one statement.

This method reports that a program written as several continuous statements without iteration is still equivalent to a single statement in ITERLARA, a pure functional algebra. We use the slightly simplified symbols in the paper with the theoretical guarantee of the above explanation. This proposition promises an exactly suitable algebra (i.e. ITERLARA^{Count}) across different fields: big data, AI, scientific computing, and databases. Though in the next subsection, we prove that ITERLARA with no limitation is Turing complete, the most suitable unified algebra is ITERLARA^{Count} rather than ITERLARA.

6.3 ITERLARA Is Turing Complete

Proposition 7. *ITERLARA is Turing complete.*

We prove that ITERLARA without any restriction on user-definable functions, condition expressions, and iteration body is Turing complete. To prove an algebra's Turing completeness, we show ITERLARA's ability to represent another Turing complete language.

BF Language. BF language is a minimum programming language created in 1993. The language consists of only eight operators. It is not designed for practical use but for research on language's extreme expressive ability. The prototype of BF language was proposed and proved Turing complete early in 1964 [6].

We present an equivalent version of this language. A program is written as a string. Each character of the string is one of its eight operators. The execution of a program requires a memory pointer p_E, an input pointer p_I, and an output pointer p_O. The code pointer goes character by character and executes the operator. The program halts when some specific condition is no longer satisfied. Its eight operators' are defined in Table 3.

Proof of Completeness

Proposition 8. *ITERLARA is Turing complete.*

Proof. As is explained, several continuous statements without iteration are still equivalent to a single statement in ITERLARA. As for the first six operators in BF language, we propose their equivalent expressions ($F_>$, $F_<$, F_+, F_-, $F_.$, and $F_,$).

Table 3. This table shows eight operators' definitions in BF language.

Operator	Description	Pseudo-code
RShift >	Pointer right shift by one	$++p_E$
LShift <	Pointer left shift by one	$--p_E$
Inc +	Increase the pointed data by one	$++*p_E$
Dec -	Decrease the pointed data by one	$--*p_E$
Output .	Output the pointed data	$*(++p_O) = *p_E$
Input ,	Input the pointed data	$*(++p_E) = *p_I$
LBrac [	Jump to the next RBrac operator if the pointed data is zero	**while**$(*p_E)\{$
RBrac]	Jump to the last LBrac operator if the pointed data is non-zero	$\}$

As for the last two operators that may include loops, we propose an equivalent expression of an entire loop body ($[F_1, \cdots, F_l]$, where $l \in \mathbb{N}$).

Before proposing equivalent expressions, we first define six associative tables. Table E, table I, and table O stand for memory, input data, and output data, respectively. Table P_E, table P_I, and table P_O stand for their pointer. Each pointer table has only one record at the start, and the record's value stands for the pointer's position. As Sect. 6.2 explains, six tables are equivalent to a single table by Cartesian product. Therefore, it is feasible to represent them in ITERLARA. See Fig. 2. The values in the brackets are default values. We propose the operators' equivalent expression in ITERLARA as Fig. 3. As is explained in Sect. 6.2, we also use several statements separated by ";" to represent one expression. Please note that $\pm 1_k$ and $\pm 1_v$ stand for increment and decrement on keys and values, respectively. We proved the feasibility of representing all operators in BF language in ITERLARA. Therefore, ITERLARA is Turing complete.

entry_E	[0] val_E
0	0
1	1st memory bit
2	2nd memory bit
⋮	⋮

Table E. The quantity of this table's records but 0 is the same as the operators' quantity in the initial memory data. The 0-th entry is a placeholder.

entry_I	[0] val_I
0	0
1	1st input bit
2	2nd input bit
⋮	⋮

Table I. The quantity of this table's records but 0 is the same as the input data volume. The 0-th entry is a placeholder.

entry_O	[0] val_O
0	0

Table O. This table is empty indeed. The 0-th entry is a placeholder.

ptr_E	entry_E	[ω] dummy
P	0	ω

Table P_E. The "entry" attribute can only be a natural number that stands for the pointer's position.

ptr_I	entry_I	[ω] dummy
P	0	ω

Table P_I. The "entry" attribute can only be a natural number that stands for the pointer's position.

ptr_O	entry_O	[ω] dummy
P	0	ω

Table P_O. The "entry" attribute can only be a natural number that stands for the pointer's position.

Fig. 2. Tables E, I, O, P_E, P_I, and P_O

- $F_{>} : P_E \rightarrow \mathbf{ext}_{+1_k} P_E$,
- $F_{<} : P_E \rightarrow \mathbf{ext}_{-1_k} P_E$,
- $F_{+} : E \rightarrow \left(\pi_{entry_E, val_E} \mathbf{ext}_{+1_v} (P_E \bowtie E)\right) \boxtimes (P_E \lhd E)$,
- $F_{-} : E \rightarrow \left(\pi_{entry_E, val_E} \mathbf{ext}_{-1_v} (P_E \bowtie E)\right) \boxtimes (P_E \lhd E)$,
- $F_{.} : P_O \rightarrow \mathbf{ext}_{+1_k} P_O; \;\; O \rightarrow O \boxtimes \left(\pi_{entry_O, val_O} (\rho_{val_O / val_E} (P_E \bowtie E)) \bowtie O\right)$,
- $F_{,} : P_I \rightarrow \mathbf{ext}_{+1_k} P_I; \;\; E \rightarrow E \boxtimes \left(\pi_{entry_E, val_E} (\rho_{val_E / val_I} (P_I \bowtie I)) \bowtie E\right)$,
- Loop body $[F_1, \cdots, F_l] : T \rightarrow \mathbf{iter}_{F_1; \cdots; F_l}^{\pi_{val_E}(P_E \bowtie E)} (E \bowtie I \bowtie O \bowtie P_E \bowtie P_I \bowtie P_O)$.

Fig. 3. Representation for eight BF operators

7 Conclusion

This paper proposed ITERLARA, the iterative extension of LARA. We first study the expressive ability of LARA and conclude that the original LARA language can represent all operations in the extended relational algebra and partial operations in the linear algebra, not including matrix inversion and determinant computation. To solve this problem, we propose an extension to LARA, named ITERLARA. The study on ITERLARA concludes that its minimum ability is the same as that of LARA and its maximum ability is the same as Turing machine. ITERLARA with the limitation of aggregation (ITERLARA^{Count}) can still represent matrix inversion and determinant computation. We conclude that ITERLARA^{Count} is a suitable algebraic model as the unified abstraction of general-purpose computation.

Acknowledgements. Supported by the Innovation Funding of ICT, CAS under Grant No. E461070.

References

1. Aho, A.V., Ullman, J.D.: Universality of data retrieval languages. In: Proceedings of the 6th ACM SIGACT-SIGPLAN Symposium on Principles of programming languages, pp. 110–119 (1979)
2. Alpha, W.: Einstein summation. https://mathworld.wolfram.com/EinsteinSummation.html
3. Barceló, P., Higuera, N., Pérez, J., Subercaseaux, B.: Expressiveness of matrix and tensor query languages in terms of ml operators. In: Proceedings of the 3rd International Workshop on Data Management for End-to-End Machine Learning, pp. 1–5 (2019)
4. Barceló, P., Higuera, N., Pérez, J., Subercaseaux, B.: On the expressiveness of LARA: a proposal for unifying linear and relational algebra. Theoret. Comput. Sci. **935**, 105–127 (2022)
5. Brijder, R., Geerts, F., Van den Bussche, J., Weerwag, T.: MATLANG: matrix operations and their expressive power. ACM SIGMOD Rec. **48**(1), 60–67 (2019)
6. Davis, M.: Corrado böhm. On a family of turing machines and the related programming language. ICC bulletin, vol. 3 (1964), pp. 185–194. J. Symbolic Logic **31**(1), 140–140 (1966)

7. Geerts, F., Muñoz, T., Riveros, C., Vrgoč, D.: Expressive power of linear algebra query languages. In: Proceedings of the 40th ACM SIGMOD-SIGACT-SIGAI Symposium on Principles of Database Systems, pp. 342–354 (2021)
8. Greub, W.H.: Linear algebra, vol. 23. Springer Science & Business Media, New York (2012). https://doi.org/10.1007/978-1-4684-9446-4
9. Hutchison, D., Howe, B., Suciu, D.: LARA: a key-value algebra underlying arrays and relations. arXiv preprint arXiv:1604.03607 (2016)
10. Hutchison, D., Howe, B., Suciu, D.: Laradb: A minimalist kernel for linear and relational algebra computation. In: Proceedings of the 4th ACM SIGMOD Workshop on Algorithms and Systems for MapReduce and Beyond, pp. 1–10 (2017)
11. Kepner, J., et al.: Associative array model of SQL, NoSQL, and NewSQL databases. In: 2016 IEEE High Performance Extreme Computing Conference (HPEC), pp. 1–9. IEEE (2016)
12. PyTorch: AvgPool1d - PyTorch. https://pytorch.org/docs/stable/generated/torch.nn.AvgPool1d.html

The Voronoi Diagram of Weakly Smooth Planar Point Sets in $O(\log n)$ Deterministic Rounds on the Congested Clique

Jesper Jansson[1], Christos Levcopoulos[2(✉)], Andrzej Lingas[2], and Quan Xue[3]

[1] Graduate School of Informatics, Kyoto University, Kyoto, Japan
jj@i.kyoto-u.ac.jp
[2] Department of Computer Science, Lund University, Lund, Sweden
{Christos.Levcopoulos,Andrzej.Lingas}@cs.lth.se
[3] The University of Hong Kong, Hong Kong, China
csxuequan@connect.hku.hk

Abstract. We study the problem of computing the Voronoi diagram of a set of n^2 points with $O(\log n)$-bit coordinates in the Euclidean plane in a substantially sublinear in n number of rounds in the congested clique model with n nodes. Recently, Jansson et al. have shown that if the points are uniformly at random distributed in a unit square then their Voronoi diagram within the square can be computed in $O(1)$ rounds with high probability (w.h.p.). We show that if a very weak smoothness condition is satisfied by an input set of n^2 points with $O(\log n)$-bit coordinates in the unit square then the Voronoi diagram of the point set within the unit square can be computed in $O(\log n)$ rounds in this model.

Keywords: Voronoi diagram · Delaunay triangulation · convex hull · distributed algorithm · the congested clique model

1 Introduction

The *congested clique* is a relatively new model of communication and computation introduced by Lotker *et al.* in 2005 [9]. It focuses on the cost of communication between the nodes in a network, ignoring the cost of local computation within each node. Hence, it can be seen as opposite to the Parallel Random Access Machine (PRAM) model, studied extensively in the 80 s and 90 s. The PRAM model focuses on the computation cost and ignores the communication cost [1].

Originally, the complexity of dense graph problems has been studied in the congested clique model under the following assumptions. Each node of the congested clique represents a distinct vertex of the input graph and knows its neighborhood in the graph. Every node also knows the unique ID numbers (between 1 and n) of itself and all the other nodes at the start of the computation. The computation proceeds in rounds. In each round, each of the n nodes can send a distinct message of $O(\log n)$ bits to each other node and can perform unlimited local computation. The primary complexity objective is to minimize the number of rounds necessary to solve a given problem on the input graph in this model.

Y. Chen et al. (Eds.): COCOON 2024, LNCS 15162, pp. 478–489, 2025.
https://doi.org/10.1007/978-981-96-1093-8_39

For several basic graph problems such as the minimum spanning tree problem, one has even succeeded to design $O(1)$-round protocols in the congested clique model [11, 14]. Observe that when the input graph is of bounded degree and the edge weights have $O(\log n)$-bit representations, each node can send the ID numbers of all nodes in its neighborhood and the weights of its incident edges, e.g., to the first node in $O(1)$ rounds. After that, the first node can solve the whole problem locally. However, such a trivial solution would require $\Omega(n)$ rounds when the input graph is dense.

Matrix problems [3], sorting and routing [7], and geometric problems [6] have also been studied in the congested clique model. In all cases, the basic input items, i.e., matrix entries or keys, or points in the plane, respectively, are assumed to have $O(\log n)$-bit representations and each node initially has a batch of n such items. *Note that the bound on bit representation of an input item is a natural consequence of the $O(\log n)$-bit bound on the size of a single message which makes input items of unbounded bit representation incompatible with the assumed model.* As in the graph case, in every round, each node can send a distinct $O(\log n)$-bit message to each other node and perform unlimited local computation. Significantly, it has been shown that matrix multiplication can be performed in a number of rounds substantially sublinear in n [3] while sorting and routing can be implemented in $O(1)$ rounds (Theorems 4.5 and 3.7 in [7]).

As for the geometric problems, Jansson et al. [6] recently provided low polylogarithmic, deterministic upper bounds on the number of rounds required to solve several basic geometric problems for a set of n^2 points in the plane with $O(\log n)$-bit coordinates in the model of congested clique with n nodes. As for the construction of the Voronoi diagram and the dual Delaunay triangulation of the point set (see Fig. 1 for an illustration and Sect. 2 for the formal definition), they have shown an $O(1)$ upper bound on the number of required rounds under the assumption that the points are drawn uniformly at random from a unit square. On the other hand, already at the end of the 90 s, Goodrich presented $O(1)$-round randomized protocols for the construction of the three-dimensional convex hull of a set of points in three-dimensional Euclidean space in $O(1)$ communication rounds in the so-called Bulk Synchronous Processing model (BSP) [5]. His result also implies an $O(1)$-round bound on the randomized construction of the Voronoi diagram and the dual Delaunay triangulation of a planar point set in the BSP model. By using the $O(1)$-round routing protocol of Lenzen [7], Goodrich's $O(1)$ bound on the number of rounds necessary for the construction of the Voronoi diagram and Delaunay triangulation can be carried over from the BSP model to ours.

In this context, the major open problem is to derive a non-trivial upper bound on the number of rounds sufficient to deterministically construct the Voronoi diagram when the points are not necessarily drawn uniformly at random. The bottleneck in the design of efficient parallel or distributed algorithms for the Voronoi diagram of a planar point set using a direct divide-and-conquer approach is an efficient parallel or distributed merging of Voronoi diagrams. Aggarwal et al. [1] presented an $O(\log^2 n)$-time CREW PRAM algorithm for the Voronoi diagram based on an involved $O(\log n)$-time PRAM method for the parallel merging. Subsequently, Amato and Preparata [2] demonstrated an $O(\log n)$-time CREW PRAM algorithm for the three-dimensional convex hull and consequently also for the two-dimensional Voronoi diagram of a point set.

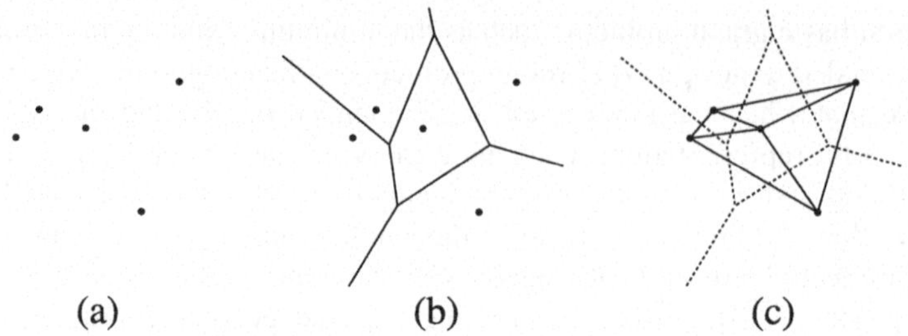

Fig. 1. An example of a planar point set, its Voronoi diagram, and the dual Delaunay triangulation.

We substantially extend the local approach to the construction of the Voronoi diagram used in the design of parallel and distributed algorithms for the Voronoi diagram of points drawn uniformly at random, e.g., from a unit square [6, 8, 15]. We show that already a very weak smoothness condition on the input set of n^2 points with $O(\log n)$-bit coordinates within a unit square is sufficient to obtain an $O(\log n)$ upper bound on the number of rounds required to construct the Voronoi diagram of the set within the unit square on the congested n-clique. Roughly, our weak smoothness condition says that if a square Q of side length ℓ within the unit square contains at least n out of the n^2 input points then any square of the same size at distance at most $4\sqrt{2}\ell$ from Q and within the unit square has to contain at least one input point.

We obtain our result by combining a quadtree partition of the unit square with a local construction of the Voronoi diagram within the squares corresponding to the leaves of quadtree. The local construction is possible due to the fulfillment of the weak smoothness condition.

In order to simplify the presentation, we assume throughout the paper that the points in the input point sets are in general position (i.e., neither any three input points are co-linear nor any four input points are co-circular).

Our paper is structured as follows. The next section contains basic mathematical/geometric definitions, a lemma, and facts on routing and sorting in the congested clique model. Section 3 presents our protocol for the Voronoi diagram and Delaunay triangulation of a weakly smooth planar point set within a square. In Sect. 4, we conclude with final remarks.

2 Preliminaries

The cardinality of a set S is denoted by $|S|$.

For a positive integer r, $[r]$ stands for the set of positive integers not exceeding r.

For a finite set S of points in the Euclidean plane, the *Voronoi diagram of* S is the partition of the plane into $|S|$ regions such that each region consists of all points in the plane having the same closest point in S; see Fig. 1.

A *Delaunay triangulation of* S is a maximal set of non-crossing edges between pairs of points from S such that no point from S is placed inside any of the formed triangles' circumcircles. It is well known that if no four points in S are co-circular then the Delaunay triangulation of S is a dual of the Voronoi diagram of S in the following sense [13]: for each edge e of each region in the Voronoi diagram of S, if e is a part of the bisector of the points u, v in S then (u, v) is an edge of the Delaunay triangulation of S; again, see Fig. 1.

Our concept of weak smoothness is formally defined in terms of two parameters as follows.

Definition 1. *Let ε, d be two positive real constants. A set of N points in a unit square is* (ε, d)-smooth *if for any two equal size squares Q, R within the unit square the following implication holds:*

if Q contains at least N^ε points of S and R is at distance at most $d \cdot \ell$ from Q, where ℓ is the length of each edge of Q and R, then R contains at least one point of S.

We also need to define a sequence of grids within a unit square and related notions.

Definition 2. *For a nonnegative integer i, we shall denote by $G_i(U)$ the orthogonal grid within the unit orthogonal square U that includes the edges of U such that the distance between two neighboring vertical or horizontal grid line segments is $\frac{1}{2^i}$. A* basic square *of $G_i(U)$ is a square within U such that the endpoints of each its edge is a pair of neighboring grid points. For a basic square R in $G_i(U)$, we shall denote the orthogonal region consisting of R and the two layers of basic squares around R by $TL_i(R)$ (if between R and an edge of the unit square there is a place only for one or zero layers then $TL_i(R)$ includes only one or zero layers on this side, respectively).*

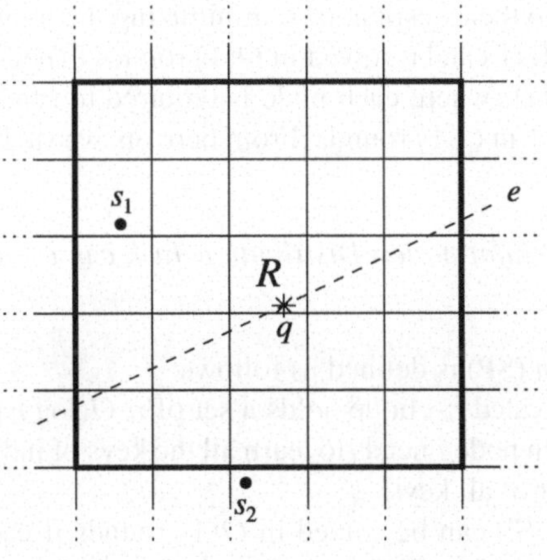

Fig. 2. An example of the configuration in the proof of Lemma 1.

The proof of the following lemma corresponds to the second paragraph of the proof of Theorem 4 in [6].

Lemma 1. *Let R be a basic square in a grid $G_i(U)$ within the unit square U. Consider a finite set S of points within the unit square. If R contains a point in S then the Voronoi diagram of S within R can be computed by taking into account only the points of S within $TL_i(R)$. Hence, in particular all edges (u, v) of the Delaunay triangulation of S such that a part of the bisector of u and v borders some region of the Voronoi diagram of S within R can be determined.*

Proof. Let e be an edge of the Voronoi diagram of S within R. The edge e has to be a part of the bisector of some couple of points s_1 and s_2 in S. Consider an arbitrary point q on e. Suppose that s_1 or s_2 is placed outside $TL_i(R)$, i.e., the orthogonal area consisting of at most $1 + 8 + 16 = 25$ squares including R. See Fig. 2. Without loss of generality, let s_2 be such a point. Then the distance between q and s_2 is at least $2 \cdot \frac{1}{2^i}$, while the distance between q and every point inside R is at most $\sqrt{2} \cdot \frac{1}{2^i}$. We obtain a contradiction because R contains at least one point from S and q is closer to such a point than to s_2. □

Lenzen gave an efficient solution to the following fundamental routing problem in the congested clique model, known as the *Information Distribution Task* (IDT) [7]: Each node of the congested n-clique holds a set of exactly n $O(\log n)$-bit messages with their destinations, with multiple messages from the same source node to the same destination node allowed. Initially, the destination of each message is known only to its source node. Each node is the destination of exactly n of the aforementioned messages. The messages are globally lexicographically ordered by their source node, their destination, and their number within the source node. For simplicity, each such message explicitly contains these values, in particular making them distinguishable. The goal is to deliver all messages to their destinations, minimizing the total number of rounds.

Lenzen proved that IDT can be solved in $O(1)$ rounds (Theorem 3.7 in [7]). He also noted that the relaxed IDT, where each node is required to send and receive at most n messages, reduces to IDT in $O(1)$ rounds. From here on, we shall refer to this important result as:

Fact 1 *[7]. The relaxed Information Distribution Task can be solved deterministically within $O(1)$ rounds.*

The *Sorting Problem* (SP) is defined as follows:
Each node i of the congested n-clique holds a set of n $O(\log n)$-bit keys. All the keys are different w.l.o.g. Each node i needs to learn all the keys of indices in $[n(i-1)+1, ni]$ (if any) in the total order of all keys.

Lenzen showed that SP can be solved in $O(1)$ rounds if each node holds a set of exactly n keys (Theorem 4.5 in [7]). In order to relax the requirement that each node holds exactly n keys to that of with most n keys, we can determine the maximum key and add appropriate different dummy keys in $O(1)$ rounds. We summarize this result as:

Fact 2 *[7]. The relaxed Sorting Problem can be solved in $O(1)$ rounds.*

3 The Local Approach

Consider a $(\frac{1}{2}, 4\sqrt{2})$-smooth set of n^2 points with $O(\log n)$-bit coordinates in a unit orthogonal square. We shall first describe a protocol for listing the edges of the Delaunay triangulation of the set that are dual to the edges of the Voronoi diagram of the set within the unit square. Roughly, it implicitly grows a quadtree of squares rooted at the unit square in phases corresponding to the levels of the quadtree. If a square R currently at a leaf of the quadtree jointly with the two layers of equal size squares around it include at most cn input points, for an appropriate positive constant c, then the intersection of the Voronoi diagram of the input point set with R and the dual edges of the Delaunay triangulation of the input point set can be computed locally. This follows from the smoothness condition and Lemma 1 combined with the fact that the parent square of R jointly with the two layers of equal size squares around it include more than cn input points. Otherwise, four child squares whose union forms R are created on the next level of the quadtree. In particular, checking the aforementioned condition in parallel for the squares at the current front level of the quadtree and delivering the necessary points to the nodes representing respective frontier squares in $O(1)$ rounds on the congested n-clique are highly non-trivial.

protocol $DT - SQUARE(S, U)$

Input: A $(\frac{1}{2}, 4\sqrt{2})$-smooth set of n^2 points with $O(\log n)$-bit coordinates in a unit orthogonal square U held in n-point batches at the n nodes of the congested clique.
Output: The set of the edges of the Delaunay triangulation of S dual to the edges of the Voronoi diagram of S within U held in $O(n)$-edge batches at clique nodes.

1. Initialize an empty list L of edges of the Delaunay triangulation of S.
2. Activate the basic square U in $G_0(U)$ and assign it to the first node.
3. For $i = 0, 1, \ldots$ do
   (a) Each node for each point p in its batch determines the number $num(p)$ of the basic square of $G_i(U)$ containing p in a common fixed numbering of the basic squares in $G_i(U)$ (e.g., column-wise and top-down-wise in each column). Next, a *prefixed representation* of p is formed by the concatenation of the fixed size bit representation of $num(p)$ with the fixed size bit representations of the coordinates of p.
   (b) The points in S are sorted by their prefixed representation. After that each node informs all other nodes about the range of numbers of the basic squares in $G_i(U)$ holding the prefixed representations of points in S that landed at the node after the sorting of the prefixed representations of all the points.
   (c) For each basic square W in $G_i(U)$ such that the prefixed representations of points belonging to W that are held in a sequence C of at least two consecutive nodes, the nodes in C inform additionally the other nodes in C about the number of the prefixed representations of the points in W they got so in particular the node in C with the smallest index can compute the total number of the points in W.
   (d) For each active basic square R in $G_i(U)$, the node representing R sends queries to the nodes holding the prefixed point representations of the points in the basic

squares in $TL_i(R)$ (i.e., in R and the two layers of basic squares around R in $G_i(U)$) about the number of points in these squares. If several nodes hold the prefixed point representation of points in a basic square in $TL_i(R)$, the query is sent only to that node with the smallest index.

(e) After getting answers to the queries, each node proceeds as follows for each active basic square in R in $G_i(U)$ it represents. If the total number of points of S in $TL_i(R)$ does not exceed $100n$ then the node asks the nodes holding the prefixed representations of the points in the basic squares in $TL_i(R)$ for sending the points to the node. After that the node computes the Voronoi diagram of all these points and then the intersection of the diagram with R locally. Next, the node appends to L all edges (u, v) where a part of the bisector of u and v borders some region of the Voronoi diagram in the computed intersection. Otherwise, the node activates the four squares in $G_{i+1}(U)$ whose union forms R and assigns them temporarily to itself.

(f) The nodes balance the assignment of active basic squares in $G_{i+1}(U)$ by informing all other nodes about the number of active basic squares in $G_{i+1}(U)$ they are assigned and following the results of the same assignment balancing algorithm run by each of them separately locally.

(g) The list L is sorted in order to remove multiple copies of the same edge.

Lemma 2. *$DT - SQUARE(S, U)$ activates basic squares solely in the grids $G_i(U)$, where $i = O(\log n)$.*

Proof. Simply, the points in S have $O(\log n)$-bit coordinates so at depth at most $O(\log n)$ the condition in Step 3(e) of $DT - SQUARE(S, U)$ has to be satisfied. □

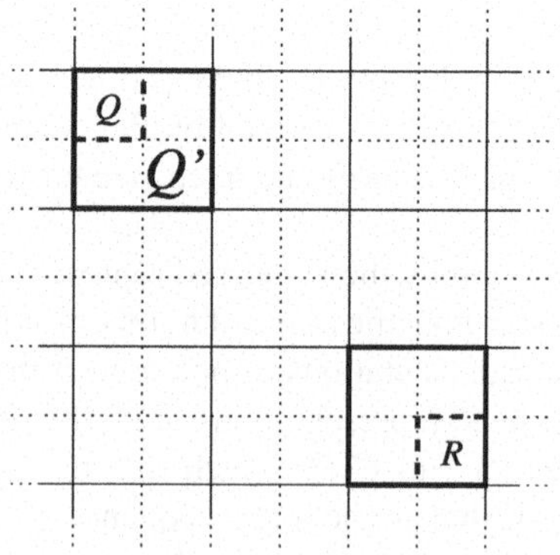

Fig. 3. An example of the configuration in the proof of Lemma 3.

Lemma 3. *The protocol $DT - SQUARE(S, U)$ is correct.*

Proof. When the Voronoi diagram of the points of S in $TL_i(R)$ for a basic square R of the grid $G_i(U)$ is computed then there must be square Q' in the grid $G_{i-1}(U)$ that contains at least $100n/25$ points in S and is at distance at most $\frac{\sqrt{2}}{2^{i-1}}$ from the basic square in $G_{i-1}(U)$ that is the parent of R. Hence, there is a basic square Q in $G_i(U)$ that is part of Q' and contains at least $100n/100$ points in S; see Fig. 3. By straightforward calculations, the distance between Q and R is at most $4\sqrt{2}\frac{1}{2^i}$. Thus, by the assumed $(\frac{1}{2}, 4\sqrt{2})$-smoothness property, the square R contains at least one point in S. It follows from Lemma 1 that the intersection of the Voronoi diagram of the points of S in $TL_i(R)$ with R yields the Voronoi diagram of S within R. Hence, the edges appended to the list L are the edges of the Delaunay triangulation of S dual to the edges of the Voronoi diagram of S within U. It easily follows by induction on i during forming the quadtree of active basic squares that the leaf active basic squares form a partition of the unit square U. Therefore, for each edge (u, v) of the Delaunay triangulation of S dual to an edge of the Voronoi diagram of S within U there must exist a positive integer i and an active basic square R in $G_i(U)$ such that R does not have any child active basic squares in $G_{i+1}(U)$ and a part of the bisector of u and v borders some region in the Voronoi diagram of S within R. Hence, the list L is complete. □

Lemma 4. *For $i = 0, 1, \ldots, O(\log n)$, the number of active basic squares in the grid $G_i(U)$ is $O(n)$ during the execution of $DT - SQUARE(S, U)$.*

Proof. We argue similarly as at the beginning of the proof of Lemma 3. If R is an active basic square in $G_i(U)$ different from the unit square U then there must exist a basic square Q in $TL_{i-1}(R')$, where R' is the parent of R in $G_{i-1}(U)$, such that Q contains at least $100n/25$ points in S. Now it is sufficient to note that: (i) there are at most $O(n)$ basic squares in $G_{i-1}(U)$ that contain at least $100n/25$ points in S; (ii) there are at most $O(1)$ basic squares Q' in $G_{i-1}(U)$ different from R' such that Q is included in $TL_{i-1}(Q')$; (iii) an active basic square in $G_{i-1}(U)$ can be a parent to at most four active basic squares in $G_i(U)$. □

Lemma 5. *The protocol $DT - SQUARE(S, U)$ can be implemented in $O(\log n)$ rounds on the congested clique.*

Proof. Steps 1, 2 can be easily implemented in $O(1)$ rounds. By Lemma 2, the block under the for loop in Step 3 is iterated $O(\log n)$ times. It is sufficient to show that this block (a-g) can be implemented in $O(1)$ rounds.

Step 3(a) can be performed totally locally.

The sorting of the prefixed representations of points in S in Step 3(b) can be done in $O(1)$ rounds by Fact 2.

For each node, the range of the numbers of the basic squares in $G_i(U)$ holding the prefixed representations of points in S at the node after the sorting of the prefix representations of the points can be specified by two $O(\log n)$-bit numbers. Hence, all nodes can inform all other nodes about their ranges in $O(1)$ rounds. Thus, Step 3(b) requires $O(1)$ rounds in total.

The situation described in Step 3(c) can happen for at most n basic squares W in $G_i(U)$. It requires sending by each node at most two different messages to at most n

nodes in total and also receiving at most n messages. Hence, Step 3(c) can be implemented in $O(1)$ rounds by using the routing protocol from Fact 1.

In Step 3(d), for each active basic square, a node representing the square has to send $O(1)$ $O(\log n)$-bit queries to $O(1)$ other nodes. The total number of active basic squares in $G_i(U)$ is $O(n)$ by Lemma 4. Hence, by using the routing protocol from Fact 1 this task can be done in $O(1)$ rounds.

Consider Step 3(e). Answering the queries sent in Step 3(c) can be done by local computations and the routing reverse to that in Step 3(d) in $O(1)$ rounds. After that each node for each active square assigned to it determines locally if the criterion for computing the Voronoi diagram of S within R is satisfied. If so the node sends messages asking the nodes holding the prefixed representations of points in the squares of $TL_i(R)$ for sending the points. This requires sending $O(n)$ messages for each active basic square in $G_i(U)$. Since the total number of such squares is $O(n)$ by Lemma 4 and each node represents $O(1)$ active squares in $G_i(U)$, it can be accomplished in $O(1)$ rounds by Fact 1. Delivering the requested points to the nodes representing respective active basic squares can also be done in $O(1)$ rounds for the following reasons. For each active basic square the node representing it needs to receive $O(n)$ points. Furthermore, by Lemma 4 there are $O(n)$ active basic squares in $G_i(U)$. Hence, since the active squares are assigned to the n nodes in a balanced way, each node needs to receive $O(n)$ points. Also, the points contained in a given basic square in $G_i(U)$ can be requested by at most $O(1)$ nodes since there are at most $O(1)$ active basic squares behind these requests to the given square. Since the sorted prefixed representations of the points in S are divided between the nodes in a balanced way, each node needs to send $O(n)$ points, each of them to $O(1)$ nodes. We conclude that this part of Step 3(d) can be implemented in $O(1)$ rounds by Fact 1. The remaining parts of Step 3(d) are done locally.

Step 3(f) requires sending and receiving by each node $O(1)$ messages so it can be done in $O(1)$ rounds.

Consider an edge (u, v) dual to some edge of the Voronoi diagram of the points of S included in $TL_i(R)$ within an active basic square R in $G_i(U)$. The edge can be appended to L at most for $O(1)$ different squares R as u, v are in $TL_i(R)$. Therefore, the list L may contain at most $O(1)$ copies of an edge of the Delaunay triangulation of S so Step 3(g) can be implemented in $O(1)$ rounds by using the sorting protocol from Fact 2. □

Lemmas 3, 5 yield our first main result.

Theorem 1. *Let S be a $(\frac{1}{2}, 4\sqrt{2})$-smooth set of n^2 points with $O(\log n)$-bit coordinates in an orthogonal unit square, held in n-point batches at the n nodes of the congested clique. The set of edges of the Delaunay triangulation of S dual to the edges of the Voronoi diagram of S within the unit square can be constructed in $O(\log n)$ rounds on the congested clique.*

Lemma 6. *Let S be defined as in Theorem 1. Suppose that a list L of the edges of the Delaunay triangulation of S dual to the edges of the Voronoi diagram of S within the unit square is held in $O(n)$-edge batches at the n nodes of the congested clique. The Voronoi diagram of S within the unit square can be constructed in $O(1)$ rounds on the congested clique.*

Proof. Double the list L by inserting for each $(u, v) \in L$ also (v, u) into L. For each edge (u, v) determine locally an $O(\log n)$-bit representation of the angle $\beta(u, v)$ between (u, v) and the horizontal line passing through u. For instance, the representation can specify the tangent of the angle by $(v_y - u_y, v_x - u_x)$. Sort the edges (x, y) by $(x, \beta(u, v))$, letting the nodes compare the angle tangents locally, using the sorting protocol from Fact 2. In this way, for each point $u \in S$, a sub-list of all edges of the Delaunay triangulation incident to u in the angular order is created. Some of the sub-lists can stretch through several nodes of the clique network. Given the edges of the Delaunay triangulation incident to u in the angular order, the edges of the Voronoi region of u within the unit square can be easily produced. This is done by intersecting the bisectors of u and the other endpoints of consecutive edges incident to u in the angular order as long as the intersection of two consecutive bisectors is within the unit square. Otherwise, the border of the region of u has to be filled with the fragment of the perimeter of the unit square between the intersections of the two bisectors with the perimeter. □

Theorem 1 combined with Lemma 6 yield our second main result.

Theorem 2. *Let S be a $(\frac{1}{2}, 4\sqrt{2})$-smooth set of n^2 points with $O(\log n)$-bit coordinates in an orthogonal unit square held in n-point batches at the n nodes of the congested clique. The Voronoi diagram of S within the unit square can be constructed in $O(\log n)$ rounds on the congested clique.*

4 Final Remarks

The *message complexity* of a protocol in the congested clique model is the maximum total number of $O(\log n)$-bit messages exchanged by the n nodes of the congested clique during a run of the protocol (e.g., see [12]). In case of our protocols, it is easily seen to be the product of the maximum number of messages that can be exchanged in a single round, i.e., $\Theta(n^2)$, times the number of required rounds. Thus, the message complexity of our deterministic protocols for the Delaunay triangulation and the Voronoi diagram of n^2 point sets from Sect. 3 is $O(n^2 \log n)$.

The remaining major open problem is the derivation of a low polylogarithmic upper bound on the number of rounds sufficient to deterministically construct the Voronoi diagram of n^2 points with $O(\log n)$-bit coordinates in the Euclidean plane (when the points are not necessarily randomly distributed) on the congested clique with n nodes. This seems feasible but it might require a substantial effort as in the PRAM case [1,2].

Note here that the existence of an $O(\log n)$-time (unit cost) PRAM algorithm for a geometric problem on a point set (e.g., [2]) does not guarantee the membership of the problem in the NC^1 class defined in terms of Boolean circuits [4, 10]. Simply, assuming that the input points have $O(\log n)$-bit coordinates, the arithmetic operations of the PRAM implemented by Boolean circuits of bounded fan-in have a non-constant depth, at least $\Omega(\log \log n)$. Consequently, the Boolean circuit simulating the $O(\log n)$-time PRAM for fixed input size can have a super-logarithmic depth. This is a subtle and important point in the context of relatively recent results of Frei and Wada providing simulations of the classes NC^k, $k > 0$, by MapReduce (see Theorems 9 and 10 in

[4]), and consequently, in the Massively Parallel Computation (MPC) and BSP models (see Theorem 1 in [10]). Due to the $O(1)$-round routing protocol of Lenzen [7], the congested clique model in our setting can be roughly regarded as a special case of MPC, where the size of the input is approximately the square of the number of processors. For this reason, the NC simulation results from [4] are relevant to our model only when the parameter ϵ in the exponent of the space bounds in [4] equals $\frac{1}{2}$. This is possible in case of Theorem 9 in [4] on NC^1 simulation but not possible in case of Theorem 10 in [4] on NC^k, k > 0, simulation. However, the proof of the former theorem in [4] relies on a strict logarithmic upper bound on the depth of Boolean circuit of bounded fan-in required by Barrington's characterization of the NC^1 class in terms of bounded-width polynomial-size branching programs. Otherwise, one has to adhere to the direct circuit simulation method from [4] that does not work for $\epsilon = \frac{1}{2}$. In summary, Theorems 9 and 10 in [4] do not seem to have any direct consequences for geometric problems on point sets in our model setting.

Acknowledgments. The authors are grateful to the anonymous reviewers for valuable comments. This research was in part funded by Swedish Research Council grant 2018-04001.

References

1. Aggarwal, A., Chazelle, B., Guibas, L.J., Ó'Dúnlaing, C., Yap, C.-K.: Parallel computational geometry. Algorithmica **3**, 293–327 (1988)
2. Amato, M., Preparata, F.: A time-optimal parallel algorithm for three-dimensional convex hull. Algorithmica **14**(2), 169–182 (1995)
3. Censor-Hillel, K., Kaski, P., Korhonen, J.H., Lenzen, C., Paz, A., Suomela, J.: Algebraic methods in the congested clique. Distrib. Comput. **32**(6), 461–478 (2019)
4. Frei, F., Wada, K.: Efficient deterministic MapReduce algorithms for parallelizable problems. J. Parallel Distrib. Comput. **177**, 28–38 (2023)
5. Goodrich, M.: Randomized fully-scalable BSP techniques for multi-searching and convex hull construction. In: Proceedings of the Eighth Annual Symposium on Discrete Algorithms (SODA 1997), pp. 767–776. ACM-SIAM (1997)
6. Jansson, J., Levcopoulos, C., Lingas, A., Polishchuk, V.: Convex hulls, triangulations, and Voronoi diagrams of planar point sets on the congested clique. arXiv:2305.09987 (2023). Preliminary version in Proceedings of the Thirty-Fifth Canadian Conference on Computational Geometry (CCCG 2023), pp. 183–189 (2023)
7. Lenzen, C.: Optimal deterministic routing and sorting on the congested clique. In: Proceedings of the 2013 ACM Symposium on Principles of Distributed Computing (PODC 2013), pp. 42–50. ACM (2013)
8. Levcopoulos, C., Katajainen, J., Lingas, A.: An optimal expected-time parallel algorithm for Voronoi diagrams. In: Karlsson, R., Lingas, A. (eds.) SWAT 1988. LNCS, vol. 318, pp. 190–198. Springer, Heidelberg (1988). https://doi.org/10.1007/3-540-19487-8_21
9. Lotker, Z., Patt-Shamir, B., Pavlov, E., Peleg, D.: Minimum-weight spanning tree construction in $O(\log \log n)$ communication rounds. SIAM J. Comput. **35**(1), 120–131 (2005)
10. Nanongkai, D., Scquizzato, M.: Equivalence classes and conditional hardness in massively parallel computing. Distrib. Comput. **35**, 165–183 (2022)
11. Nowicki, K.: A deterministic algorithm for the MST problem in constant rounds of congested clique. In: Proceedings of the Fifty-Third Annual ACM SIGACT Symposium on Theory of Computing (STOC 2021), pp. 1154–1165. ACM (2021)

12. Pemmaraju, S., Sardeshmukh, V.: Super-fast MST algorithms in the congested clique using o(m) messages. In: Proceedings of the 36th Annual Conference on Foundations of Software Technology and Theoretical Computer Science (FSTTCS 2016), pp. 47:1–47:15. LIPICS (2016)
13. Preparata, F., Shamos, M.: Computational Geometry: An Introduction, vol. 10. In: Texts and Monographs in Computer Science. Springer-Verlag (1985). https://doi.org/10.1007/978-1-4612-1098-6
14. Robinson, P.: Brief announcement: what can we compute in a single round of the congested clique? In: Proceedings of the 2023 ACM Symposium on Principles of Distributed Computing (PODC 2023), pp. 168–171. ACM (2023)
15. Vemuri, B.C., Varadarajan, R., Mayya, N.: An efficient expected time parallel algorithm for Voronoi construction. In: Proceedings of the Fourth Annual ACM Symposium on Parallel Algorithms and Architectures (SPAA 1992), pp. 392–401. ACM (1992)

Binary Jumbled Pattern Matching: Suffix Tree Indexing

Luís Cunha(✉) and Mário Medina

Universidade Federal Fluminense, Niterói, Brazil
lfignacio@ic.uff.br, mmedina@id.uff.br

Abstract. Given a string s over an alphabet a, b, a vector (x, y) is a Parikh vector if and only if a factor of s has exactly x a's and y b's. To answer whether or not a vector is a Parikh vector of a binary string is also known as the Binary Jumbled Pattern Matching Problem (BJPMP). Most solutions to this problem rely on a $O(n)$ word space index to answer queries in constant time, which can encode the Parikh set, that is, all Parikh vectors of a binary string. We show that the BJPMP is equivalent to finding the prefix normal form (PNF) of the same string and the PNF, in turn, is equivalent to the $O(n)$ bit space encode of the index. Cunha et al. (*Combinatorial Pattern Matching*, 2017) presented an algorithm that runs in $O(n + \rho^2)$ time to create the index table with the maximum of 1s, where ρ is the number of runs of 1s, reading from the start of each run to the end of each next run. This approach causes the algorithm to read repeated patterns for each occurrence. We develop an algorithm for constructing index tables using the essence of Ukkonen's algorithm to create a suffix tree that purges repeated occurrences which, in practice, reduces the execution time—particularly in the cases where there are many repetitions of runs—and in addition builds the suffix tree which is a well-known data structure with many other applications. We provide the design and analysis of a new algorithm for constructing an index table from a string, with both theoretical and practical implications. Finally, we present a series of advantages found in our approach.

Keywords: Binary jumbled pattern matching · Parikh set · Parikh vectors · Suffix tree · String indexing · Prefix normal form

1 Introduction

The *binary jumbled pattern matching* problem is presented as follows: We are given a binary string w and asked to determine whether there exists a substring of size r and q 1s. Such substring could be represented by a *Parikh vector* of w, something that appears frequently in computational biology [3,10], as do jumbled patterns in the context of graphs and other structures [8,14,17]. This problem

Luís Cunha : Financial support from FAPERJ E-26/201.372/2022, and CNPq 406173/2021-4.

Y. Chen et al. (Eds.): COCOON 2024, LNCS 15162, pp. 490–501, 2025.
https://doi.org/10.1007/978-981-96-1093-8_40

has aroused much interest, as seen in a few approaches [2,4,7,9], and since the queries and their quantity can be arbitrary, the interest is for the problem of *indexing binary strings for jumbled pattern matching*, as described below:

INDEXED BINARY JUMBLED PATTERN MATCHING PROBLEM (BJPMP)

Instance: A finite binary string s of length n.
Goal: Construct an index table to answer queries efficiently: for integers $x, y \geq 0$, does s have a factor with x 1s and y 0s?

Despite being a restriction of the general problem with arbitrary finite alphabet, since no sublinear algorithm is known, any dictionary can be converted into binary in $O(n)$ time with no disadvantage of working within this restriction.

An index is a table constructed for a word of length n over the binary alphabet that can determine the existence of substrings with a given number of 1s. Thus, the BJPMP asks us to preprocess a binary string such that later, given a number of 0s and a number of 1s, we can quickly report whether there exists a substring with those numbers of 0s and 1s and, optionally, return the position of at least one such substring. Direct preprocessing algorithms take quadratic time and other approaches reduced that time complexity to $O(n^2/\log n)$ [15], $O(n^2/\log^2 n)$ [16], $O(n^2/2^{\Omega(\sqrt{\log n/\log\log n})})$ [12] and finally $O(n^{1.859})$ with randomization or $O(n^{1.864})$ without [6]. Other randomized approaches are considered with subquadratic construction time [13]. Related problems were also described, as lower bounds on reporting all certificates of a query and pattern matching with mismatches [1].

Cunha et al. [9] proposed an algorithm that runs in $O(n + \rho^2)$ time and $O(n)$ words of space. Furthermore, they showed how we can either keep the same bounds and store information that lets the index return the position of one match, or keep the same time bound and use only $O(n)$ bits of space. The algorithm in [9] matches one of the two algorithms proposed by Giaquinta and Grabowski [11] with the parameter $k = 1$, for which one runs in $O(\rho^2 \log k + n/k)$ time, produces an index that uses $O(n/k)$ extra space and answer queries in $O(\log k)$ time, and another one that runs in $O(n^2 \log^2 w/w)$ time, where w is the size of a machine word.

Contributions.

- We provide the design, analysis and implementation of a new algorithm for constructing index tables from strings with both theoretical and practical implications.
- For the theoretical side, we prove that when algorithms to build index tables are based on runs, it is not possible to develop a faster strategy than ρ^2. Moreover, we show that our approach can be much faster than the one proposed in [9]. The time complexity of the former fluctuates between quadratic and linear time, while the latter stays in quadratic time.
- For the practical side, we compare our approach with the very simple and fast one proposed in [9]. By using a suffix tree to store information from strings, our algorithm presents advantages depending on the number of repeated substrings and the interest in using the suffix tree for other purposes.

- Furthermore, when dealing with multiple words it is possible to create a generalized suffix tree without increasing the time complexity for building the index table, while reducing the space complexity for separated suffix trees and serving many other applications e.g. finding the longest common subsequence (LCS).

Organization. Section 2 presents the preliminaries for the BJPMP by describing the relationship between Parikh sets, prefix normal forms, BJPMP, and the simple and fast algorithm previously proposed by Cunha et al. [9]. Section 3 demonstrates how we can encode strings and utilize suffix trees to build index tables. Section 4 describes our proposed algorithm for indexing, prove its time complexity, and demonstrate that when using runs, it is not possible to develop a better running time algorithm. Section 5 presents practical results, comparing the execution times and highlighting the advantages of our approach.

2 Preliminaries

Parikh Set and Parikh Vectors. Let $\Sigma = \{a, b, c, \ldots\}$ be a finite alphabet and ω be a word over Σ, i.e., a finite sequence of characters from the alphabet. Given a vector $\pi = (\pi_a, \pi_b, \pi_c, \ldots)$, π is said to be a Parikh vector of ω if and only if there exists a substring of ω where, for each σ in Σ, π_σ is the number of occurrences of σ in that substring. It is easy to see that for π to be a Parikh vector, the length of such a substring must be the sum of all elements of π, and the dimension of π must be equal to the length of Σ. For instance, given the word 011 over the binary alphabet $\{0, 1\}$, then $\pi = (1, 1)$ is a Parikh vector of that word, since it is possible to find a substring with one 0 and one 1. It is important to consider that a single vector can match multiple and different substrings; that is, a Parikh vector $(2, 3)$ matches each of these substrings: $00111, 01011, 01101, 01110, 10011, 10101, 10110, 11001, 11010, 11100$.

The Parikh set of a word ω is a set of all its Parikh vectors, denoted as $\Pi(\omega)$. For example, a binary word of the form 01101 has these and only these Parikh vectors: $\Pi = \{(1, 1), (0, 2), (1, 2), (1, 3), (2, 2), (2, 3)\}$.

Formally, we denote: $\pi(\varepsilon) = (|\varepsilon|_\sigma)_{\sigma \in \Sigma}$, where $|\varepsilon|_\sigma$ is the number of occurrences of σ in ε, substring of ω. $\Pi(\omega) = \{\pi(\varepsilon_i)\}$, where ε_i is each substring found in ω. Therefore a function $f : \varepsilon_i \rightarrow \Pi$ is surjective.

2.1 Binary Jumbled Pattern Matching

As mentioned, the BJPMP consists in answering whether a vector π is a Parikh vector of a word ω over $\Sigma = \{0, 1\}$. Since no fast algorithm is known (below linear time) to check a single vector, then pre-processing and indexing provide a faster solution, but larger in space complexity, taking advantage over the interval property to reduce the index space with a faster query response, presented below.

The *interval property* of a binary string ensures that, if x_1 is the least occurrences of 1s and x_2 is the most occurrences of 1s for a specific length l, then it is possible to find a substring with length l and o occurrences of 1s if and only if $x_1 \leq o \leq x_2$ [4,7,9].

As established by Badkobeh et al. [2], we can build an index with the least and the most occurrences of 1s—for each length—to answer in $O(1)$ if a certificate is in between those values.

Let $max_1(l)$ be the maximum number of 1s in any substring of length l and $T_{max_1} = [t_1, t_2, t_3 \ldots]$ an array containing the $max_1(i)$ for each t_i. Analogously, we have an array T_{min_1} for the minimum number of 1s. Then, given a word 11011001: $11011001 \rightarrow T_{max_1} = [1, 2, 2, 3, 4, 4, 4, 5]$, $T_{min_1} = [0, 0, 1, 2, 2, 3, 4, 5]$.

Taking the fifth position of each table we know that every substring of length 5 has o occurrences of 1s if and only if $2 \leq o \leq 4$. Furthermore, we do not need to create an index with the maximum and the minimum number of 0s, for in the binary alphabet each non 0 is necessarily a 1. We can derive the maximum and minimum number of 0s from the difference between the length, and the minimum and maximum of 1s, resp.: $T_{max_0}[i] = i - T_{min_1}[i]$, $T_{min_0}[i] = i - T_{max_1}[i]$.

Prefix Normal Forms and Prefix Normal Words. A *word prefix* is an l-sized substring of that word with the same l first characters, in the same order. In [5], they define prefix normal word (PNW) and prefix normal form (PNF) in binary strings, which can be a 1-prefix normal word or a 0-prefix normal word. A 1-*prefix normal word* ω is such that, for every length $1 < l < |\omega|$, an l-sized substring with the maximum number of 1s is found as a prefix of ω. A 0-prefix normal word is defined analogously. See [5] for more definitions.

It is also shown that, for every word ω, it is possible to find a 1-prefix normal word with the same index tables, and it is said to be the prefix normal form of ω. It is also proved that any given word has one and only one PNF, and defined a group of words that share the same PNF as a 1-prefix equivalence class.

For example $\omega = 1101001$ is a PNW, but its inverse (word written in reverse order) is not. For $\omega^{-1} = 1001011$, we have $T_{max_1}[3] = 2$, e.g. 101 and 011, but neither of them is a prefix of ω^{-1}. Since both have the same T_{max_1}, then ω is both the PNF of ω^{-1} and its own PNF. A PNF of a word is a PNW with the same T_{max_1}. These definitions are important because, with a PNW, it is possible to create an index in $O(n)$ time by reading the number of 1s for each prefix size.

2.2 Cunha et al.'s Algorithm

Cunha et al. [9] developed an algorithm for jumbled indexing of binary strings and showed that, since binary strings have the interval property, it is possible to jump between *runs of* 1*s*, which are defined as each sequence of consecutive 1s. They save a single witness for every l-sized substring with a maximum number of 1s. For example, $\omega = 1100101$ would be indexed for each substring starting at the beginning of a run of 1s and finishing at the end of a run of 1s:

1100101 11 $\mapsto T_{max_1}[2] = 2$
1100101 11001 $\mapsto T_{max_1}[5] = 3$
1100101 1100101 $\mapsto T_{max_1}[7] = 4$
1100**1**01 1 $\mapsto T_{max_1}[1] = 1$
1100**101** 101 $\mapsto T_{max_1}[3] = 2$
110010**1** 1 $\mapsto T_{max_1}[1] = 1$

To fill out the rest of the index table, they use adjacent values and the interval property. Hence, if $T = [t_1, t_2, t_3...]$ is the index table where t_i is the maximum number of 1s in any substring of length i, then $t_i \leq t_{i+1} \leq t_i + 1$, and we can pass over T right-to-left and left-to-right, assigning the maximum value between the current value and the least possible one, completing the index table. They proved that: $T[i+1] - 1 \leq T[i] \leq T[i+1]$, $T[i-1] \leq T[i] \leq T[i-1] + 1$.

This process to fill out the index table using adjacent values, assigning the maximum possible value, we define as to "*windownize*" the index table and it is found in the code presentation of our algorithm as a separate function.

When trying to reduce the space used for the index table, they showed that we can transform an $O(n)$ word into an $O(n)$ bit index table by assigning 1 if $t_{i+1} > t_i$, or 0 if $t_{i+1} = t_i$. For example, considering $\omega = 010101110101$: $T_{word} = [1, 2, 3, 3, 4, 4, 5, 5, 6, 6, 7, 7]$, $T_{bit} = [1, 1, 1, 0, 1, 0, 1, 0, 1, 0, 1, 0]$.

Here, we have a synergy between the PNF definition and the bit-encoded index table, for the sequence in the bit-encoded table is the 1-prefix normal form of the word indexed. It is clear that trying to solve the BJPMP using index tables is equivalent to finding the prefix normal forms of a given word. $PNF_1(\omega) = 111010101010 \rightarrow T_{bit} = [1, 1, 1, 0, 1, 0, 1, 0, 1, 0, 1, 0]$.

The time complexity of the algorithm developed in [9] is $O(n + \rho^2)$, where ρ is the number of runs, which is $O(n^2)$ when ρ approaches n. The worst case for this algorithm is given by the string $1010101\cdots$. The algorithm runs from the start of each run of 1s to the end of the next run; therefore, it will index each occurrence of 1, 101, 10101, $1010101\cdots$. Since we have $\rho = \frac{n}{2}$, then the time complexity can be written in terms of n: $O(n + (\frac{n}{2})^2) = O(n^2)$.

Since repeated occurrences do not change our index table, there is no upside to indexing each pattern more than once; therefore, we can get rid of repeated occurrences. In the next section, we present an algorithm that uses a suffix tree to build string patterns without repetition to reduce the time when reading each substring enclosing runs of 1s.

3 Suffix Tree and Special Pattern Encode

A suffix tree is a specialized data structure utilized to store a list of strings [18]. By design, the suffix tree incorporates each substring pattern only once, making this attribute particularly advantageous for our needs. Since we examine every possible pattern to construct the index table, suffix links are unnecessary.

Special Pattern Encode. In the context of BJPMP, our focus lies solely on the frequency of occurrences for each character of the alphabet. Hence, each consecutive run of a character can be represented by the length of its repetition. For instance, consider the string 110111001; its special pattern encoding reflects the count of repetitions for each digit: $\underbrace{11}_{2}\underbrace{0}_{1}\underbrace{111}_{3}\underbrace{00}_{2}\underbrace{1}_{1}$, $110111001 \mapsto 2\ 1\ 3\ 2\ 1$.

When using programming languages, there is no need to encode integers because it is possible to differentiate a node with a value of 12 from two subsequent nodes with values 1 and 2. Additionally, we indicate which digit, 1 or 0, each node starts with. Therefore, this encoding is not included in the description.

For demonstration purposes only, we encode runs of 0s using uppercase letters and runs of 1s using lowercase letters for each integer. Hence, $A = 0, a = 1$; $B = 00, b = 11$; $C = 000, c = 111$; and so on. This encoding method also provides a clearer visualization of the special pattern when constructing the suffix tree and building the index table. $A,\ a = 1;\ B,\ b = 2;\ C,\ c = 3 \cdots$, $110111001 \mapsto bAcBa$.

Building the Suffix Tree. The Ukkonen's algorithm for building suffix trees is widely known for its time complexity, which is $O(n)$ [18]. Its essence relies on saving the initial position of each branch created instead of the entire substring. It is important to highlight that some structures of the Ukkonen's algorithm are not necessary for us, such as the suffix links.

For example, considering the string 1101100110111011000110111, let us build its suffix tree. First, we count each digit repetition to construct the special pattern encoding: 1101100110111011000110111 $\mapsto$ $bAbBbAcAbCbAc$.

Now, we use the Ukkonen's algorithm to build its suffix tree, without using suffix links. See Fig. 1 for an example.

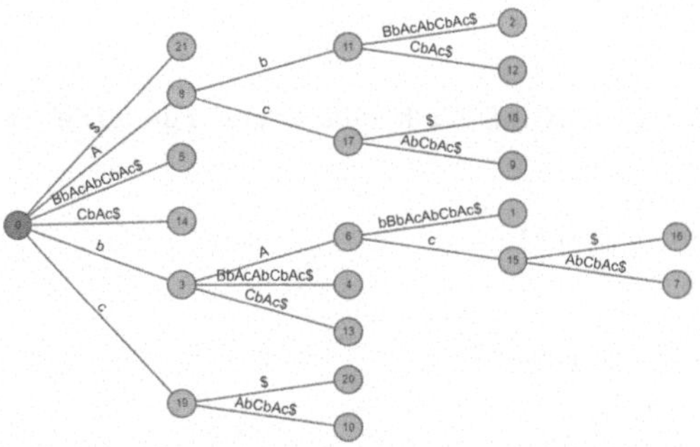

Fig. 1. Build in: https://brenden.github.io/ukkonen-animation/. The node labels correspond to the steps of Ukkonen's algorithm.

4 Jumbled Pattern Matching Index: Algorithm

Now, we describe our strategy for building the index of a given binary string. Essentially, we begin with a binary string. We then utilize $O(n)$ time to construct its special pattern encoding and an additional $O(n)$ time to create its suffix tree using Ukkonen's algorithm. Afterward, we extract each possible factor from the suffix tree. This involves performing a pre-order traversal of the suffix tree and indexing from the parent node up to the current node, described in Algorithm 1.

Let us illustrate with some steps from the suffix tree in Fig. 1. Select node 3. Its substring is 'b', which starts with 1. Indexing:

$T_{max_1}[b] \mapsto T_{max_1}[2] = max(T_{max_1}[2], 2)$.

Algorithm 1: BJPMI BY SUFFIX TREE

```
Input  : Binary string ω with length l
Output: T_max1 and T_max0 index tables
def index(start, end, from, table):
    // add elements to the index table for each substring
    if (end − start) % 2 ! = 0 then
        for i ← from to (end − start) by 2 do
            window = summed[start + i][0] − summed[start − 1][0];
            count = summed[start + i][1] − summed[start − 1][1];
            if ω[start] == 0 then
                count− = window;
            table[window] = max(count, table[window]);
def windownize(table):
    // fill out the table
    for i ← l to 1 do
        table[i] = max(table[i], table[i + 1] − 1);
    for i ← 1 to l do
        table[i] = max(table[i], table[i − 1]);
begin
    counted = [];
    // counted is a list of each character run size in ω
    runs = 1;
    for i ← 2 to l do
        if ω[i] == ω[i − 1] then
            runs+ = 1;
        else
            counted.add(runs);
            runs = 1;
    counted.add(runs);
    summed = [];
    // summed is a list of how many 1s in each prefix from ω
    window = 0;
    tot_1 = 0;
    for i, v ∈ counted // the pair i ← key, v ← value do
        window+ = v;
        if (i + ω[0]) % 2 ! = 0 then
            tot_1+ = v;
        summed.add([window, tot_1]);
    builds and traverse the suffix tree;
    tree = Suffix_Tree(counted); // Ukkonen's Algorithm O(n)
    for factor ∈ tree // Pre-order traversal do
        if factor[0] + ω[0] % 2 == 0 // substring starts in 0 then
            index(factor.start, factor.end, factor.parent_node.end, T_max0);
        else // substring starts in 1
            index(factor.start, factor.end, factor.parent_node.end, T_max1);
    windonize(T_max1);
    windonize(T_max0);
    return T_max1, T_max0
```

We traverse the tree in pre-order, meaning that parents are already indexed when we reach their children. If the node substring length is even, then it starts and ends with the same digit. We only need to index substrings that enclose runs of the same digit, which occurs if the substring length is odd or if the difference between the parent's length and the node's length is greater than 1. In other words, if the difference between the parent's length and the node's length is 1, then this node is adding only a run of 0s or a run of 1s. If the length of the node is even, then it adds a run of a different digit than the one we are currently indexing, and we can skip this node. Next, consider node 6. Its substring is 'bA', and the parent's substring is 'b'. Since the substring length is even and the difference between the parent's length and the node's length is not greater than 1, we do not need to index it. Moving on to node 1. Its substring is 'bAbBbAcAbCbAc', and the parent's substring is 'bA', so we index it starting at the parent.

$$T_{max_1}[bAb] \mapsto T_{max_1}[2+1+2] = max(T_{max_1}[5], 2+2)$$
$$T_{max_1}[bAbBb] \mapsto T_{max_1}[9] = max(T_{max_1}[9], 6)$$
$$T_{max_1}[bAbBbAc] \mapsto T_{max_1}[13] = max(T_{max_1}[13], 9)$$
$$T_{max_1}[bAbBbAcAb] \mapsto T_{max_1}[16] = max(T_{max_1}[16], 11)$$
$$\ldots$$

We have reached a leaf. The pre-order traversal will return to the previous parent and then move to the next child node, which is 15. Its substring is 'bAc' and its parent node is 'bA'. Notice that, although the difference between the parent's length and the node's length is not greater than 1, the node's length is odd. Therefore, we index it. $T_{max_1}[bAc] \mapsto T_{max_1}[6] = max(T_{max_1}[6], 5)$.

We proceed to index each node of the tree until the pre-order traversal ends. If the node substring starts with an uppercase letter, it means that its substring starts with a 0, so we index it to the table T_{max_0} instead.

When assigning values to the index table, it is important to highlight that we always check if the current indexed value is not already greater than the new value: $T[\alpha] = \max(T[\alpha], \beta)$.

Traversing a suffix tree means obtaining each unique suffix of the string. We use it to index each prefix of each of those suffixes from the start to the end of runs of the same digit, thus achieving the same result as the Cunha et al.'s algorithm, but avoiding duplicated substrings.

Time Complexity Analysis. The worst case for Cunha's algorithm was a pattern of interspersed 1s and 0s: $101010\cdots$. But for our algorithm, this becomes the best case, as it is full of repetition. Notice that there exists only one substring for each l-sized window starting and ending in a run of 1s ($1:1;\ 3:101;\ 5:10101\cdots$), thus achieving linear time using the suffix tree.

The worst case for our algorithm shifts to the worst space complexity of a suffix tree, as it will produce the maximum number of different factors depending on n. The worst space complexity for a suffix tree is already known to be the Fibonacci word, or the analogous rabbit sequence. This is a sequence of strings obtained by considering $s_0 = 0$, $s_1 = 01$, $s_n = s_{n-1} \cdot s_{n-2}$, where $\cdot$ denotes concatenation of two strings.

One might imagine that the worst case for our algorithm is when there are no repetitions. Let ω be a binary word of length n where no repetitions occur, then it always has different run lengths for each digit. Since the order of these runs makes no difference for the complexity analysis, we can sort them. Thus, $\omega = 10110011100011110000\cdots$.

Now we can establish n in terms of ρ:

$n = \sum_{i=1}^{\rho} 2i = 2\frac{\rho(\rho+1)}{2} = \rho^2 + \rho, \mapsto \rho \simeq \sqrt{n}$.

Since the overall time complexity for our algorithm is $O(n+\rho^2)$ and $\rho \simeq \sqrt{n}$, then when no repetitions occur, the algorithm achieves $O(n+n) = O(n)$ time.

It is counterintuitive why the Fibonacci word or the rabbit sequence are the worst case for space complexity in the suffix tree and in time complexity for our algorithm. For the sake of simplicity, let us consider the following example: Let's compare two binary strings with the same size: $S_1 = 10110101$ and $S_2 = 10110111$. S_2 has no repetitions of factors enclosing runs of 1s and S_1 has two repetitions; however, S_2 has $\rho = 3$ while S_1 has $\rho = 4$. Calculating unique factors enclosing runs of 1s, we have:

$S_1 \mapsto \frac{4(4+1)}{2} = 10 - 2 = 8, \;\; S_2 \mapsto \frac{3(3+1)}{2} = 6 - 0 = 6$.

Even though S_2 has fewer repetitions, it comes at the cost of word space that could otherwise be used to increase the number of runs. Therefore, if we aim to maximize the number of unique factors, it is achieved through an equilibrium between repetitions and run size, which results in the Fibonacci word or the rabbit sequence.

One may notice that the Fibonacci word has many repeated runs of 1, since each run has length 1. This can be easily explained by the formula:

$\alpha^2 + \beta^2 < (\alpha + \beta)^2 \mid \forall \alpha, \beta > 0$.

Let α and β be the variance of runs of each digit. The uniqueness of substrings relies on the variance of the runs. This formula shows that varying runs of two digits is less effective than allowing only one of the digits to vary to create unique factors.

Although the Fibonacci word and rabbit sequence are the worst-case scenarios for our algorithm, they still exhibit many repetitions, even more than the average binary string. Therefore, we demonstrate in a comparison table (Table 3) that they are favorable for our algorithm, as the other one is bounded by ρ^2.

Proving ρ^2 as lower bound for indexing table. We know that: $T[i] = T[i-1]$ or $T[i] = T[i-1] + 1$. If $T[i] = T[i-1]$, then there is a max(i-factor) that starts or ends in 1. It could exist or not exist an i-factor starting and ending in 1. If $T[i] = T[i-1] + 1$, then any max(i-factor) must start with 1 and end with 1. Otherwise, we could remove any edge 0 and index $T[i-1] = T[i]$, which would lead to a contradiction.

Lemma 1. *To build an index table based on comparing runs of* 1*s, the optimal strategy is to index only factors of the word that begin and end in* 1 *with no subsequent* 1*s.*

Proof. Let ω be a binary word of length n. For each i, $0 < i \leq n$, there are $n - i + 1$ i-length factors of ω. We need to build an index table of size n, where for each i, $0 < i \leq n$, we only need to index an i-factor with the maximum number of 1s. If $T[i] = T[i-1]+1$, then the max(i-factor) starts and ends in 1, as established, and we would index it. But if $T[i] = T[i-1]$, with no i-factors starting and ending in 1, at the end of the algorithm $T[i]$ would be empty, and we could set $T[i] = T[i-1]$, ignoring all i-factors. This will save us $n-i+1-1 = n-i$ operations. Since $i \leq n$, it will never be more expensive to use neighbor indexed values. Now, if $T[i] = T[i+1]-1$, analogously it is faster to set $T[i] = T[i+1]-1$ than to search for max(i-factors). □

We demonstrate in Lemma 1 that any algorithm that constructs an index table is faster when it only utilizes factors starting and ending in 1.

Theorem 1. *$\Omega(n + \rho^2)$ is a lower bound for building an index table based on comparing runs of* 1*s.*

Proof. Let $AxByCzD \cdots$ be a binary encoded string, where $A, B, C \cdots$ are the sizes of each run of 1s and $x, y, z \cdots$ are the sizes of each run of 0s. Let ρ be the number of runs of 1 in the word. Lemma 1 shows that we only need to index $max(A, B, C, D \cdots)$, but since we do not have an ordered list, then we have to spend $\Omega(\rho)$ to find it. Now, depending on the values of each run, we could have $|A| + |x| + |B| = |B| + |y| + |C| + |z| + |D|$, but suppose that each $AxB, ByC, CzD \cdots$ all have the same size s in the decoded binary string. Then we need to index $T[s] = \max_1(AxB, ByC, CzD \cdots)$. Again we do not have an ordered list, so to find $\max_1(AxB, ByC, CzD \cdots)$ we will need at least $\Omega(\rho-1)$. This argument is analogous for runs of 1s three by three, four by four, and so on. Therefore, we need to index at least from the start of each run of 1 to the end of each one. The time complexity for this is $\Omega(n + \sum_{i=1}^{\rho} i) = \Omega(n + \rho^2)$. □

5 Practical Results and Discussions

Now, we present practical results by comparing our suffix tree algorithm with Cunha et al.'s algorithm. The process begins by selecting a size for the string to be indexed. Subsequently, random binary strings of this size are generated, and then each algorithm is applied. The time taken to construct T_{max_1} and T_{max_0} tables, as described in Table 1, is then displayed.

These results indicate that both algorithms grow at the same rate. The slight differences observed can be attributed to several factors:

- The code used for constructing and traversing the suffix tree is third-party and not optimized for traversing the tree specifically for this algorithm.
- Utilizing the special pattern encode to convert integers to Unicode characters incurs additional overhead and can be eliminated in an optimized implementation.
- Building suffix trees involves recursion, which can be computationally expensive, especially on personal computers like those used in these experiments.

Table 1. Time comparison for indexing random binary strings between Cunha et al.'s algorithm [9] (JBM2017) and our proposed algorithm by using suffix trees (SfTree).

1,000 strings with length 1,000			
Algorithm	Min	Max	Avg
JBM2017	0.0156s	0.0636s	0.0330s
SfTree	0.0208s	0.0660s	0.0406s

100 strings with length 10,000			
Algorithm	Min	Max	Avg
JBM2017	3.2028s	4.8482s	3.4910s
SfTree	3.8362s	5.4705s	4.1976s

1,000 strings with length 5,000			
Algorithm	Min	Max	Avg
JBM2017	0.7548s	1.0357s	0.8483s
SfTree	0.9031s	1.2603s	1.0112s

Table 2. Time comparison for indexing between Cunha et al.'s algorithm [9] (JBM2017) and our proposed algorithm by using suffix trees (SfTree).

Interspersed string (010101 ···)		
Algorithm	Length: 1,000	Length: 10,000
JBM2017	0.1915 s	13.5859 s
SfTree	0.0155 s	0.1039 s

Table 3. Time comparison for Fibonacci binary word between Cunha et al.'s algorithm [9] (JBM2017) and our proposed algorithm by using suffix trees (SfTree).

Fibonacci word (0100101001001 ···)		
Algorithm	Length: 1,597	Length: 10,946
JBM2017	0.2097 s	9.9634 s
SfTree	0.1211 s	5.7264 s

Moreover, the time discrepancy is negligible, considering that suffix trees have numerous other applications with no additional costs, whereas Cunha et al.'s algorithm is specifically tailored for constructing index tables for BJPMP.

We highlight the advantages of our suffix tree algorithm:

- The suffix tree is a well-known data structure and can be used for various other applications within the same string, such as search, data compression, exact string matching, and others.
- It has the capability to construct a generalized tree for multiple strings, which saves time by building multiple index tables and reusing repetitions between those strings.
- The algorithm is compatible with any traversal order and can be integrated into other traversal applications with minimal additional cost.
- There are no significant instances where it is slower than the other algorithm, but it can be much faster when the string contains sufficient repetitions, fluctuating between quadratic and linear time.

The execution time for Cunha et al.'s algorithm is particularly relevant in its worst-case scenario, where it remains quadratic, while the suffix tree allows us to achieve linear time, as shown in Table 2. Another significant scenario is the Fibonacci word, which represents our worst-case scenario. As explained in the time complexity analysis in Sect. 4, despite being counter-intuitive, the time difference is more pronounced in our worst-case scenario, highlighting our advantages.

References

1. Afshani, P., van Duijn, I., Killmann, R., Nielsen, J.S.: A lower bound for jumbled indexing. In: Proceedings of the Fourteenth Annual ACM-SIAM Symposium on Discrete Algorithms (SODA), pp. 592–606. SIAM (2020)
2. Badkobeh, G., Fici, G., Kroon, S., Lipták, Z.: Binary jumbled string matching for highly run-length compressible texts. Inf. Process. Lett. **113**(17), 604–608 (2013)
3. Benson, G.: Composition alignment. In: Benson, G., Page, R.D.M. (eds.) Algorithms in Bioinformatics, pp. 447–461. Springer Berlin Heidelberg, Berlin, Heidelberg (2003). https://doi.org/10.1007/978-3-540-39763-2_32
4. Burcsi, P., Cicalese, F., Fici, G., Lipták, Z.: On approximate jumbled pattern matching in strings. Theor. Comput. Syst. **50**, 35–51 (2012)
5. Burcsi, P., Fici, G., Lipták, Z., Ruskey, F., Sawada, J.: On prefix normal words and prefix normal forms. Theor. Comput. Sci. **659**, 1–13 (2017)
6. Chan, T.M., Lewenstein, M.: Clustered integer 3SUM via additive combinatorics. In: ACM Symposium on Theory of Computing (STOC), pp. 31–40 (2015)
7. Cicalese, F., Fici, G., Lipták, Z., et al.: Searching for jumbled patterns in strings. In: Stringology, pp. 105–117 (2009)
8. Cicalese, F., et al.: Indexes for jumbled pattern matching in strings, trees and graphs. In: Kurland, O., Lewenstein, M., Porat, E. (eds.) String Processing and Information Retrieval: 20th International Symposium, SPIRE 2013, Jerusalem, Israel, October 7-9, 2013, Proceedings, pp. 56–63. Springer International Publishing, Cham (2013). https://doi.org/10.1007/978-3-319-02432-5_10
9. Cunha, L., Dantas, S., Gagie, T., Wittler, R., Kowada, L., Stoye, J.: Faster jumbled indexing for binary RLE strings. In: Combinatorial Pattern Matching (CPM) (2017)
10. Eres, R., Landau, G.M., Parida, L.: Permutation pattern discovery in biosequences. J. Comput. Biol. **11**(6), 1050–1060 (2004)
11. Giaquinta, E., Grabowski, S.: New algorithms for binary jumbled pattern matching. Inf. Process. Lett. **113**(14–16), 538–542 (2013)
12. Hermelin, D., Landau, G.M., Rabinovich, Y., Weimann, O.: Binary jumbled pattern matching via all-pairs shortest paths. arXiv:1401.2065 (2014)
13. Kociumaka, T., Radoszewski, J., Rytter, W.: Efficient indexes for jumbled pattern matching with constant-sized alphabet. Algorithmica **77**, 1194–1215 (2017)
14. Lacroix, V., Fernandes, C.G., Sagot, M.-F.: Motif search in graphs: application to metabolic networks. IEEE/ACM Trans. Comput. Biol. Bioinform. **3**(4), 360–368 (2006)
15. Moosa, T.M., Rahman, M.S.: Indexing permutations for binary strings. Inf. Process. Lett. **110**(18–19), 795–798 (2010)
16. Moosa, T.M., Rahman, M.S.: Sub-quadratic time and linear space data structures for permutation matching in binary strings. J. Discrete Algorithms **10**, 5–9 (2012)
17. Song, S., Gu, G., Ryu, C., Faro, S., Lecroq, T., Park, K.: Fast algorithms for single and multiple pattern cartesian tree matching. Theor. Comput. Sci. **849**, 47–63 (2021)
18. Ukkonen, E.: On-line construction of suffix trees. Algorithmica **14**(3), 249–260 (1995)

GraphDHV: Graph Neural Network with Dual Hybrid View on Imbalanced Node Classification

Longqing Du[1,2], Guangquan Lu[1,2(✉)], Yadan Han[1,2(✉)], Zhiping Luo[3], Guoqiu Wen[1,2], Liang Zhang[1,2], Wanxin Chen[1,2], and Shichao Zhang[1,2]

[1] Key Lab of Education Blockchain and Intelligent Technology, Ministry of Education, Guangxi Normal University, Guilin 541004, China
lugq@mailbox.edu.cn, hanyadan@stu.gxnu.edu.cn

[2] Guangxi Key Lab of Multi-source Information Mining and Security, Guangxi Normal University, Guilin 541004, China

[3] Lancang-Mekong International Vocational Institute, Yunnan Minzu University, Yunnan 650005, China

Abstract. It is crucial to address category imbalance in real-world data on node classification in a graph. This paper introduces the GraphDHV model, a dual encoder that integrates node attributes and topological information. GraphDHV improves inter-class separability and intra-class compactness by employing degree logical edge removal and synthesizing enhanced attribute features with original topology. We compare GraphDHV with ten state-of-the-art baseline methods on six public benchmark datasets. Our experimental results demonstrate that GraphDHV significantly outperforms the baseline methods, with an improvement of the F1 score.

Keywords: Node classification · Class imbalance · Graph neural network

1 Introduction

Graph data structures play a vital role in various domains such as transportation [1], social networks [2], agriculture [3], and finance [4]. Node classification, especially within Graph Neural Networks (GNNs), is fundamental to graph data analysis. However, most research operates under the assumption of class balance, which poses challenges when dealing with class imbalance [5]. Class imbalance, where some classes have significantly fewer instances than others, can lead to biased model predictions, especially in identifying influential nodes in social networks.

Researchers have proposed methods to address the class imbalance in graph neural networks [6]. These methods fall into two categories: non-generative and generative. Non-generative methods adjust sample weights or loss functions to balance class influence [7]. However, they may face challenges like overfitting and

Y. Chen et al. (Eds.): COCOON 2024, LNCS 15162, pp. 502–513, 2025.
https://doi.org/10.1007/978-981-96-1093-8_41

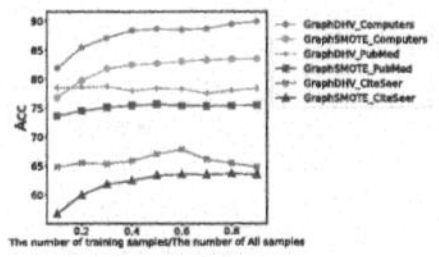

Fig. 1. GraphDHV and GraphSMOTE model using different amounts of training sample data in three data sets: Amazon-Computers, Planetoid-PubMed, and Planetoid-CiteSeer.

parameter tuning. Generative methods, on the other hand, generate new data or resample existing data [8]. While effective, they may introduce bias and increase computational complexity.

To address these challenges, we propose GraphDHV, a Dual Hybrid View Model. GraphDHV mitigates overfitting using ℓ_1 regularization and dropout and utilizes two view encodings to synthesize diverse perspectives. It combines node attributes and topology using post-degree logic edge removal, enhancing robustness. Notably, GraphDHV achieves superior performance even with limited labels, as depicted in Fig 1.

In summary, our contributions include:

- Nontraditional aggregation views: Integrating topology and attributes to capture node features more effectively.
- Key design of GraphDHV: Utilizing dual-view parallel encoders to capture various topological and attribute features, reducing information mixing constraints.
- Extensive experiments demonstrating GraphDHV's effectiveness and robustness, outperforming 10 baseline methods across six public benchmark datasets.

2 Method

In this section, we introduce the proposed GraphDHV model in detail. We first describe each component of the model and then analyze the model. Figure 2 shows the overall framework.

2.1 Motivation

Due to limitations in traditional aggregation models, the loss of high-frequency information from individual nodes is a common issue. To tackle this challenge, we propose enhancements in both topology structure and node attributes. By integrating these improvements, we aim to capture node features more effectively, moving beyond the conventional approach of treating nodes merely as collections. Additionally, in our efforts to mitigate information mixing and capture both

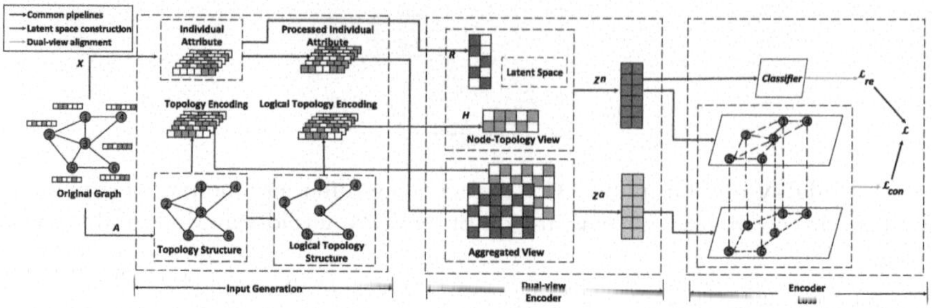

Fig. 2. For the GraphDHV model, the initial input encompasses graph-related data, comprising node attribute features and edge information. Subsequently, the specialized enhancement and degree logic de-edge techniques are applied to produce two novel perspectives, each incorporating the original topological structure and node attribute features. These perspectives are then partially encoded within the Encoder. Finally, the comparison loss and label loss are computed based on the embeddings of the two perspectives, and backpropagation is employed to optimize the model.

unique and joint node information, we employ a dual-view parallel encoder. This choice is motivated by our goal of improving the efficiency of node feature learning.

2.2 Notations

Since the direction of our research is the imbalance node classification problem on UN-directed and UN-weighted graphs, we define the initial graph as $\mathcal{G} = \{\mathcal{V}, \mathcal{E}\}$, where $\mathcal{V} = \{v_1, ..., v_n\}$ is node-set, and $\mathcal{E} \subseteq \mathcal{V} \times \mathcal{V}$ is the edge set. The number of nodes and edges are denoted as $|\mathcal{V}| = n$ and $|\mathcal{E}| = m$, respectively. In this study, we define the adjacency matrix and the feature matrix of graph $\mathcal{G}$ as $A \in \{0,1\}^{n \times n}$ and $X \in \mathbb{R}^{n \times d}$, respectively. Here, $A_{(i,j)} = 1$ if the edge (v_i, v_j) exists in $\mathcal{E}$, and $X_i \in \mathbb{R}^d$ represents the d-dimensional feature vector of node v_i. Additionally, $\mathcal{N}_i$ denotes the set of direct neighbors of node v_i. Furthermore, we introduce the edge information matrix $E \in \mathbb{R}^{2 \times m}$, where $E[0][i]$ represents the starting node of the edge $\mathcal{E}_i$ and $E[1][i]$ represents the ending node of the edge $\mathcal{E}_i$.

For each node v_i in the graph, there is a corresponding class label $Y_v \in \{1, \cdots, C\}$, where C denotes the total number of classes. In imbalanced node classification, marked nodes as imbalanced in the training set $\mathcal{V}^L \subseteq \mathcal{V}$ of imbalanced nodes, where the proportion of imbalance is defined as $\rho = \frac{max_i|Y_i|}{min_j|Y_j|}$.

2.3 Input Generation

In GraphDHV, we process graph structure and graph features to generate input graphs, enabling us to fully explore both topology and features.

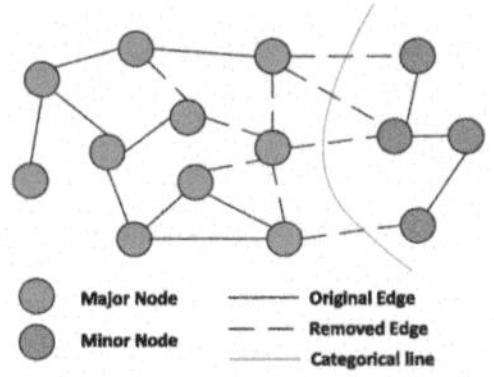

Fig. 3. Logical edge removal of GraphDHV.

Node-Topology View. GraphDHV introduces a degree-aware edge removal technique termed degree-edge. The logical edge removal process of the GraphDHV model is depicted in Fig 3. This approach is designed to eliminate edges within the majority class while preserving those in the minority class, to maximize the removal of connections between majority and minority classes. Accordingly, we devise a degree-aware mask matrix denoted as M. The components within the M matrix dictate the retention of the edge in which the starting node is v_i and the end node is v_j, defined herein as M_{ij}:

$$M_{ij} = \begin{cases} 1, & \text{if random } (0,1) > \text{clamp}(\theta, p(i,j)), \\ 0, & \text{others}, \end{cases} \tag{1}$$

$$p(i,j) = 1 - \frac{1}{\log(\deg(i)+\deg(j))}, \tag{2}$$

where $\text{clamp}(\theta, p(i,j))$ means limiting the maximum value to $\theta \in (0,1)$ which is a hyper-parameter, and $\deg(i)$ represents the degree of the i-th node. This means that it maximizes the removal of edges from majority class nodes while maximizing the retention of edges from minority class nodes. Finally, the topological information $T \in \{0,1\}^{m\times m}$ is defined as:

$$E' = E[:]\,[M], \tag{3}$$

$$T = \text{Sparse}\,(E'), \tag{4}$$

Feature Processing for Aggregated View. Traditional GNN models utilize only a subset of nodes for training, leaving valuable node information untapped. Aggregation from neighboring nodes can degrade focal node information, posing challenges in capturing key features accurately and risking overfitting, especially for minority class nodes. Therefore, to avoid increasing the model depth, we follow the idea of decoupled GNN to obtain a high-quality node representation of high-order neighbor information through feature propagation:

$$A' = \text{Sparse}(E), \tag{5}$$

$$X'^{(k)} = (1-\lambda)A'X^{(k-1)} + \lambda X'^{(0)}, \tag{6}$$

where $\text{Sparse}(E)$ means to obtain the sparse matrix of E, $X^{'(k)} \in \mathbb{R}^{n\times d}$ denotes the feature matrix after k-th feature propagation and $X^{'(0)} = X$ represents the original feature matrix. k and λ are fixed as hyper-parameters. Finally, we generate the aggregated graph $\mathcal{G}^a = (A', X'^{(k)})$, which is used to work with higher-order node features and topology in the aggregated view. This approach serves as a feature pre-processing step, intended for execution only once before model training commences.

2.4 Dual-View Encoder

we propose a dual-view parallel graph encoder to simultaneously learn individual (node) and collective (aggregation) information by encoding two distinct views.

Individual Learning from Node-Topology View. To fully utilize the intrinsic information of a single node in the original graph, we need to uniformly encode attribute information and topology information in the node view. Specifically, since attributes and topologies in graph $\mathcal{G}$ have different semantics, we first need a simple mapper to map features X and T to a common latent space:

$$R = \Phi(X, \theta_1), \tag{7}$$

$$S = \Phi(T, \theta_2), \tag{8}$$

where the mapper Φ is a one-layer nonlinear transformation, θ_1 and θ_2 are learnable parameters, and the outputs $R \in \mathbb{R}^{n \times d_a}$ and $S \in \mathbb{R}^{n \times d_a}$ are the mapped attribute and topology embedding, respectively. As it functions as an encoder for individual nodes, to circumvent confusion arising from information aggregation, we adopt a sequential operation approach to uniformly embed and merge both attribute and topology information for each node, the encoder ϕ_1 is expressed as:

$$\phi_1(\mathcal{G}^n; \omega_1) : Z_n = \text{MLP}(\text{CONCAT}(R, S)), \tag{9}$$

where MLP denotes multilayer perceptron ω_1 is learnable parameter and CONCAT(R, S) means connect R and S in one dimension. The final output Z_n forms the node-topology representation matrix.

Holistic Learning from Aggregated View. To partially address the previously discussed limitations of GNNs, we introduce a specially enhanced aggregation graph as input to enhance the encoder's learning capabilities. Feature enhancement operations enable nodes to preserve their individual high-frequency signals while aggregating information from neighboring nodes, facilitating a more thorough exploration of dependency patterns within the graph. The encoder ϕ_2 is formalized as follows:

$$\phi_2(\mathcal{G}^a; \omega_2) : Z_a = \text{MPGNN}(X'^{(k)}, A), \tag{10}$$

where MPGNN is adopted as the commonly used GCN [9], GAT [10], and SAGE [2] with learnable parameters ω_2. We take the final output Z_a as the aggregated representation matrix.

2.5 Separated Learning Objectives

Rebalanced Classifier Learning. To address imbalanced classification challenges, we propose a reweighted classifier to counteract the bias toward the majority class often observed in conventional classifiers. This classifier dynamically adjusts sample weights during training, prioritizing minority class samples to achieve more equitable classification outcomes. Our classifier adapts weights adaptively throughout the learning process by implementing a two-layer MLP architecture. Moreover, we employ standardization based on the number of training samples to alleviate skewed biases.

$$\mathcal{L}_{re}(O, Y_i) = -\frac{1}{n_y} \log \frac{\exp(O_y)}{\sum_{j=1}^{C} \exp(O_j)}, \tag{11}$$

$$P\left(Z_n; \omega_p\right) : O = \mathrm{MLP}\left(Z_n\right), \tag{12}$$

where $O \in \mathbb{R}^C$ represents the probability that the node v_i in the predicted node belongs to C labels, ω_p is a learnable parameter, Y_i is the node label, and n_y is the number of the node label in all node labels.

Contrastive Learning. In this subsection, we utilize contrastive learning to yield discriminative and resilient representations within class imbalance. We employ a contrastive objective aimed at fostering the acquisition of comprehensive node representations, drawing positive pairs closer while concurrently pushing negative pairs apart. Specifically, we identify identical nodes across both views as positive pairs, while treating pairs of distinct nodes as negative pairs. To enhance the coherence of positive pairs while diminishing that of negative pairs, we adopt the InfoNCE-tuned contrast loss [11]:

$$\mathcal{L}_{con} = -\frac{1}{2N} \sum_{k=1}^{N} \left[l\left(Z_{n2k-1}, Z_{a2k}\right) + l\left(Z_{a2k}, Z_{n2k-1}\right) \right], \tag{13}$$

$$l\left(Z_i, Z_j\right) = -\log \frac{\exp(\mathrm{sim}(Z_i, Z_j)/\tau)}{\sum_{k=1}^{2N} 1_{[k \neq i]} \exp(\mathrm{sim}(Z_k, Z_j)/\tau)'}, \tag{14}$$

where $\mathrm{sim}(\cdot)$ measures the similarity between two vectors, which is set as cosin similarity function, $\tau \in \mathbb{R}^+$is a scalar temperature parameter, Z_i come from the i-th vector of Z and Z_j is another Embedding.

Objective Function. The overall loss of GraphDHV for training is formulated as:

$$\mathcal{L} = (1 - \alpha)\mathcal{L}_{re} + \alpha\mathcal{L}_{con}, \tag{15}$$

where α is a weighted hyper-parameter for semi-supervised contrastive loss to balance learning, for simplicity, we consider $\mathcal{L}_{re}$ as the objective function for traditional training in this context. In the experiments section, we analyzed the impact of α change on the model's result.

3 Experiments

In this section, we conduct extensive experiments to evaluate the effectiveness of GraphDHV for imbalanced node classification by answering the following research questions:

- RQ1: Does GraphDHV outperform existing baseline methods on class-imbalanced node classification?
- RQ2: How do the features enhancement and logical topology strategy in GraphDHV affect the training outcomes?
- RQ3: Does GraphDHV generalize effectively across diverse conditions?
- RQ4: Can GraphDHV improve robustness, intra-class compactness, and inter-class separability? If so, what are the underlying reasons?

3.1 Experimental Setup

We conduct experiments using six real-life datasets: Cora, CiteSeer, PubMed, Amazon-Computers (Computers), Amazon-Photos (Photos), and Coauthor-CS (CS). We employ several baseline methods including Reweighting, PC Softmax [12], ClassBalanced Loss (CB Loss) [13], FocalLoss, ReNode, Upsample [14], GraphSMOTE [15], GraphENS, and GraphSHA.

3.2 Results on Manually Imbalanced Datasets

In our experiments, we explore both long-tailed class imbalance settings on Cora, CiteSeer, and PubMed, as well as step class imbalance settings on Photo, Computer, and CS datasets. In the long-tailed setting, we employ a 6:2:2 split into training/validation/test sets, with proportions across the three datasets. To induce long-tailed distributions akin to prior work, we manually remove labeled nodes from the training set until achieving the desired imbalance ratio(ρ) of 100. Conversely, in the step setting datasets are partitioned into training/validation/test sets with proportions of 6:2:2, respectively. Here, half of the classes are designated as major classes, sharing an equal number of training samples (n_{major}), while the other half are minor classes, sharing the same number of training samples ($n_{min} = n_{major}/\rho$). In this setting, we set the imbalance ratio(ρ) to 20.

Tables 1 and 2 showcase GraphDHV's clear superiority over its competitors across various GNN backbones, affirming the effectiveness of the GraphDHV framework. Notably, performance consistency across different GNN backbones suggests that disparities are rooted in inherent model properties. Compared to the baseline, GraphDHV consistently achieves higher F1 scores across six datasets and under three GNN backbones, indicating its adeptness in balancing precision and recall. These findings underscore GraphDHV's robustness in diverse classification tasks.

Although GraphDHV's accuracy (Acc) and balanced accuracy (bAcc) rank third on a few datasets, the margin between GraphDHV and the top-performing

Table 1. Node classification results (±std) on Cora, CiteSeer, and PubMed in the long-tailed class-imbalanced setting with GCN, GAT, SAGE, and GraphDHV backbones for 10 runs. The best is highlighted in boldface, and the runner-up is highlighted in underline.

		Cora-LT			CiteSeer-LT			PubMed-LT		
	$\rho = 100$	Acc.	bAcc.	F1	Acc.	bAcc .	F1	Acc.	bAcc .	F1
	Vanilla	71.07±0.42	59.32±0.67	59.24±0.86	51.44±0.46	44.54±0.36	37.79±0.65	51.62±0.56	44.21±0.46	35.13±0.69
	Reweight	77.86±0.12	72.78±0.16	73.45±0.16	63.54±0.26	56.84±0.24	55.21±0.16	76.97±0.15	72.37±0.18	71.91±0.14
	PC Softmax	76.75±0.15	72.11±0.31	71.72±0.24	62.21±0.46	59.07±0.26	58.17±0.31	74.34±0.64	72.49±0.36	71.67±0.16
	CB Loss	77.42±0.20	72.82±0.26	73.15±0.18	61.53±0.37	55.22±0.56	53.45±0.63	76.59±0.21	71.98±0.17	72.91±0.21
	Focal Loss	77.86±0.23	73.18±0.24	73.86±0.22	59.72±0.42	53.43±0.33	51.79±0.38	75.65±0.18	71.35±0.22	71.84±0.21
GCN	ReNode	78.37±0.25	73.23±0.22	74.52±0.26	62.42±0.34	55.72±0.28	54.12±0.23	75.97±0.14	69.98±0.16	71.21±0.14
	Upsample	74.97±0.12	68.74±0.16	68.32±0.16	55.14±0.20	48.52±0.13	45.20±0.12	71.62±0.08	64.17±0.12	65.17±0.09
	GraphSMOTE	74.84±0.42	69.02±0.49	70.42±0.49	56.24±0.86	50.41±0.27	48.02±0.37	74.64±0.11	68.48±0.11	71.21±0.11
	GraphENS	75.54±0.22	71.22±0.36	70.92±0.46	63.24±0.36	57.07±0.37	55.72±0.41	77.21±0.16	71.33±0.22	73.01±0.24
	GraphSHA	79.33±0.34	73.98±0.57	75.27±0.37	63.95±0.27	58.57±0.28	58.47±0.22	**78.97±0.15**	74.45±0.17	75.25±0.31
	GraphDHV	**81.15±0.36**	**77.10±0.48**	**77.86±0.36**	**66.49±0.36**	**61.20±0.32**	**61.91±0.36**	78.48±0.16	**77.62±0.17**	**78.28±0.17**
	Vanilla	64.48±0.62	54.32±0.77	55.44±0.76	49.26±0.26	42.54±0.18	35.77±0.25	47.77±1.63	39.11±1.28	30.11±2.07
	Reweight	77.69±0.32	72.12±0.56	73.45±0.16	61.92±0.60	54.44±0.59	53.74±0.66	74.22±0.42	70.08±0.74	69.70±0.74
	PC Softmax	68.71±0.81	64.11±0.91	64.22±0.74	56.73±1.64	56.32±1.28	55.31±1.47	76.75±0.34	73.43±0.15	73.35±0.32
	CB Loss	77.32±0.26	72.12±0.69	72.74±0.38	61.43±0.49	55.15±0.54	53.65±0.56	74.79±0.27	69.62±0.62	70.47±0.61
	Focal Loss	77.86±0.13	72.48±0.22	73.21±0.26	59.72±0.42	53.43±0.33	51.79±0.38	74.20±0.29	70.41±0.32	70.74±0.24
GAT	ReNode	78.12±0.25	71.83±0.32	73.52±0.36	60.82±0.33	53.97±0.35	52.02±0.41	74.12±0.31	69.27±0.44	69.61±0.37
	Upsample	72.59±0.28	62.43±0.36	65.12±0.32	53.44±0.23	46.92±0.23	43.13±0.42	64.64±0.97	57.34±0.58	55.13±0.97
	GraphSMOTE	74.67±0.29	67.72±0.39	69.12±0.42	57.44±0.26	51.34±0.32	49.42±0.47	74.15±0.41	69.12±0.37	70.82±0.43
	GraphENS	77.11±0.25	72.11±0.37	72.12±0.46	62.01±0.34	57.87±0.34	54.42±0.38	76.62±0.12	70.45±0.22	71.32±0.22
	GraphSHA	78.53±0.24	73.16±0.35	74.32±0.34	64.42±0.37	58.68±0.38	58.36±0.42	**78.48±0.20**	73.15±0.22	74.11±0.19
	GraphDHV	**80.09±0.26**	**76.56±0.36**	**76.61±0.28**	**66.88±0.41**	**62.49±0.36**	**62.18±0.40**	76.63±0.23	**75.84±0.19**	**76.00±0.21**
	Vanilla	73.22±0.12	61.79±0.13	63.27±0.15	47.96±0.23	42.04±0.46	37.12±0.35	58.82±0.12	48.11±0.11	42.43±0.09
	Reweight	76.75±0.17	68.72±0.34	70.31±0.40	57.37±0.55	51.04±0.49	49.34±0.59	66.12±0.55	60.01±1.34	58.91±1.14
	PC Softmax	77.03±0.25	73.17±0.31	73.49±0.31	58.42±0.27	56.12±0.22	56.71±0.21	71.54±0.14	73.83±0.15	70.33±0.12
	CB Loss	77.12±0.32	70.14±0.37	71.34±0.33	57.66±0.39	51.21±0.34	48.75±0.40	67.79±0.37	60.71±0.47	61.51±0.51
	Focal Loss	77.21±0.18	69.84±0.31	70.81±0.27	57.22±0.75	50.87±0.77	48.52±0.88	70.61±0.38	65.66±0.42	66.24±0.44
SAGE	ReNode	77.27±0.17	69.24±0.25	71.15±0.26	57.82±0.53	51.34±0.52	49.12±0.46	67.57±0.53	60.67±0.80	60.78±0.79
	Upsample	73.89±0.15	63.43±0.22	65.81±0.17	50.34±0.13	44.32±0.13	41.52±0.17	64.12±0.07	54.67±0.08	53.33±0.07
	GraphSMOTE	74.27±0.22	66.17±0.42	67.92±0.37	52.94±0.71	47.14±0.72	44.31±0.85	65.15±0.41	56.82±0.51	56.92±0.56
	GraphENS	76.71±0.22	70.09±0.25	70.42±0.33	62.67±0.34	56.17±0.37	54.17±0.39	77.62±0.15	72.45±0.23	73.21±0.18
	GraphSHA	78.18±0.34	72.79±0.39	73.79±0.39	62.93±0.39	58.61±0.37	58.48±0.33	**78.33±0.22**	75.07±0.23	75.73±0.25
	GraphDHV	**79.63±0.26**	**76.29±0.36**	**76.10±0.28**	**66.84±0.42**	**61.98±0.38**	**61.80±0.38**	76.62±0.43	**75.51±0.40**	**75.81±0.43**

models is relatively narrow. We hypothesize that this discrepancy may be attributed to the limited representation of minority classes during training. Generative models typically excel at handling extreme imbalances in class distribution, which may give them an inherent advantage. We intend to explore this aspect further in our future research endeavors.

3.3 Ablation Study

In this section, we conduct a series of experiments to demonstrate the effectiveness of various components of GraphDHV, including the impact of feature enhancement and logical edge removal on the model. Specifically, we compare GraphDHV using a plain GNN with several pruning variants. The results for the Cora-LT dataset with GCN backbone are presented in Table 3.

From Table 3, firstly, we observe that the test results obtained using the original base model of GCN without any operations in the GraphDHV model

Table 2. Node classification results (±std) on Photo, Computer, and CS in step class-imbalanced setting with GCN, GAT, SAGE, and GraphDHV backbones for 10 runs. The best is highlighted in boldface, and the runner-up is highlighted in underline.

		Photo-ST			Computer-ST			CS-ST		
	$\rho = 20$	Acc.	bAcc .	F1	Acc.	bAcc .	F1	Acc.	bAcc .	F1
GCN	Vanilla	71.07±0.42	59.32±0.67	59.24±0.86	55.84±1.31	40.94±0.76	26.79±0.63	36.37±1.21	53.41±0.76	31.43±1.39
	Reweight	77.86±0.12	72.78±0.16	73.45±0.16	77.67±0.26	84.31±0.21	73.21±0.24	90.87±0.12	89.97±0.08	81.84±0.18
	PC Softmax	76.75±0.15	72.11±0.31	71.72±0.24	72.22±0.86	60.13±1.64	54.12±2.31	86.37±0.54	86.49±0.36	73.37±0.67
	CB Loss	77.10±0.20	72.82±0.26	73.15±0.18	81.33±0.23	85.71±0.06	75.85±0.13	90.12±0.07	89.87±0.07	72.91±0.21
	Focal Loss	77.86±0.23	73.18±0.24	73.86±0.22	80.52±0.21	85.79±0.33	74.79±0.18	89.97±0.12	89.20±0.12	70.71±0.87
	ReNode	78.37±0.25	73.23±0.22	74.52±0.26	72.73±0.94	77.72±0.88	66.02±1.13	91.02±0.24	89.78±0.16	81.67±1.04
	Upsample	85.47±0.12	68.74±0.16	68.32±0.16	79.12±0.21	84.10±0.13	73.74±0.11	85.12±0.14	85.72±0.10	74.87±0.16
	GraphSMOTE	84.11±0.22	86.54±0.21	81.78±0.23	75.81±0.16	83.40±0.17	68.40±0.27	86.02±0.21	85.47±0.22	69.14±0.74
	GraphENS	87.12±0.11	89.22±0.07	84.66±0.11	78.61±0.16	85.70±0.16	73.42±0.11	92.04±0.11	91.81±0.22	76.12±0.43
	GraphSHA	87.45±0.28	**88.94±0.17**	85.19±0.37	81.03±0.31	86.32±0.25	75.53±0.53	**92.15±0.19**	**91.88±0.22**	77.33±0.40
	GraphDHV	**89.30±0.83**	86.94±1.12	**87.78±0.91**	**88.16±0.36**	**90.27±0.29**	**87.66±0.42**	91.60±0.17	90.02±0.23	**88.46±0.37**
GAT	Vanilla	36.88±0.37	44.82±0.41	28.84±0.46	57.86±0.72	42.64±0.38	26.67±0.22	34.53±0.62	50.11±0.72	24.81±1.07
	Reweight	79.79±1.21	82.12±0.61	75.85±1.12	72.72±0.50	75.84±0.42	63.84±0.46	88.42±0.41	87.28±0.44	71.70±0.64
	PC Softmax	51.57±3.17	61.37±1.81	51.32±2.54	31.65±2.84	51.92±2.68	37.81±2.37	78.87±0.64	77.73±0.63	65.55±0.67
	CB Loss	81.82±0.86	85.42±0.59	79.42±1.03	79.63±0.75	85.14±0.34	74.23±0.66	89.77±0.28	89.59±0.17	75.20±0.81
	Focal Loss	82.06±0.63	84.98±0.62	79.27±0.83	79.53±0.47	85.07±0.33	74.21±0.57	88.70±0.25	88.03±0.22	73.14±0.74
	ReNode	76.52±1.20	80.87±1.15	72.78±0.91	71.82±0.63	75.27±0.35	64.03±0.81	87.82±0.31	85.51±0.34	69.77±0.47
	Upsample	77.98±1.12	80.03±0.42	72.92±0.62	72.47±0.43	76.92±0.37	64.13±0.32	82.34±0.18	82.64±0.08	65.63±0.22
	GraphSMOTE	80.67±0.59	81.42±0.39	75.97±0.37	79.44±0.26	84.66±0.20	70.25±0.39	83.42±0.21	82.62±0.17	66.93±0.33
	GraphENS	84.21±0.36	86.45±0.19	80.12±0.30	77.88±0.24	84.82±0.19	72.13±0.48	89.87±0.35	89.54±0.22	79.77±0.42
	GraphSHA	83.93±0.94	86.46±0.85	80.81±0.79	77.42±0.27	83.12±0.22	69.95±0.57	91.69±0.20	**91.44±0.22**	76.57±0.19
	GraphDHV	**89.54±0.91**	**87.22±0.66**	**87.71±0.73**	**88.46±0.77**	**90.23±0.78**	**87.71±0.79**	**91.73±0.55**	89.62±0.50	**88.44±0.76**
SAGE	Vanilla	59.22±2.29	59.89±1.73	47.11±2.87	63.38±0.03	50.06±0.06	30.12±0.11	73.82±0.32	78.71±0.21	63.59±0.53
	Reweight	84.55±0.25	87.08±0.24	82.89±0.20	83.07±0.35	87.76±0.12	75.62±0.45	90.05±0.35	88.87±0.34	74.40±0.34
	PC Softmax	86.16±0.13	86.73±0.15	83.49±0.11	80.82±0.17	80.31±0.82	70.27±0.61	89.57±0.23	90.21±0.35	76.90±0.52
	CB Loss	82.72±0.31	85.59±0.27	80.34±0.37	83.25±0.19	87.18±0.12	75.05±0.25	89.86±0.17	89.71±0.12	77.91±0.84
	Focal Loss	82.28±0.38	85.22±0.31	79.19±0.37	82.07±0.25	87.17±0.07	74.52±0.18	89.11±0.18	89.76±0.18	78.44±0.26
	ReNode	84.73±0.15	86.24±0.20	81.85±0.24	79.77±0.53	87.13±0.21	74.56±0.26	89.97±0.32	90.13±0.42	80.18±0.43
	Upsample	82.19±0.32	84.76±0.12	79.25±0.27	82.44±0.23	86.82±0.11	75.09±0.37	87.13±0.17	87.95±0.11	75.27±0.25
	GraphSMOTE	80.21±0.27	84.67±0.32	79.07±0.39	83.12±0.31	87.94±0.22	74.02±0.35	86.20±0.11	85.62±0.11	68.10±0.16
	GraphENS	87.11±0.12	89.64±0.14	86.72±0.13	82.72±0.38	88.32±0.10	74.77±0.52	91.77±0.15	92.27±0.22	77.13±0.22
	GraphSHA	88.80±0.14	**90.47±0.12**	87.36±0.21	83.79±0.49	89.20±0.12	75.13±0.40	**92.87±0.17**	**92.74±0.17**	78.54±0.26
	GraphDHV	**89.23±0.19**	87.27±0.27	**88.07±0.16**	**88.84±0.74**	**90.29±0.67**	**87.25±0.89**	91.86±0.33	90.44±0.40	**89.16±0.32**

Table 3. Ablation study of each component of GraphDHV evaluated on Cora-LT with GCN. "±" stands for synthesizing. The best is highlighted in boldface, and the runner-up is highlighted in underline.

Method	Acc	bAcc	F1
GCN	74.09±0.40	69.05±0.39	69.56±0.37
+Features Process	75.51±0.41	70.36±0.31	70.88±0.31
+degree-edge	79.43±0.36	75.81±0.44	75.87±0.38
+Features Process and degree-edge	**81.15 ± 0.26**	**77.10 ± 0.48**	**77.86 ± 0.36**

have yielded relatively satisfactory results. Secondly, both the Features Process and degree-edge operations contribute to improvements in comprehensive performance indicators. Particularly, the degree-edge operation can lead to an increase in the Acc score by up to 5.5%, while the bAcc and F1 scores may decrease by 6%.

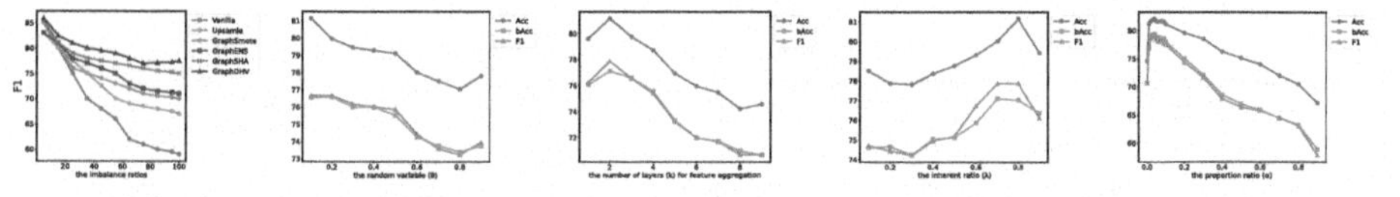

(a) (b) (c) (d) (e)

Fig. 4. Hyper-parameter Analysis.(a) the increase of imbalance ratio. (b) the random variable on edge removal technique. (c) the number of Feature Process layers. (d) the inherent ratio of the Feature Process. (e) the proportion ratio of $\mathcal{L}_{con}$ on $\mathcal{L}$.

3.4 Hyper-parameter Analysis

In GraphSHA, five primary hyperparameters significantly influence model performance: the imbalance ratios ρ, the random variable θ, the number k of layers for feature aggregation, the parameter λ governing the preservation of high-frequency information, and the parameter α controlling the proportion of contrast loss in the overall loss calculation.

We conducted experiments on the Cora-LT dataset using the GraphDHV model with GCN as the backbone, varying these hyperparameters using the control variable method. Figure 4 presents the performance scores of the Cora-LT dataset based on the GCN backbone.

Analyzing different baselines across various imbalance ratios (ρ) ranging from 5 to 100 on Cora-LT with the GCN backbone (Fig. 4a), we observed that as ρ increases, GraphDHV maintains relatively stable performance and demonstrates heightened superiority. This highlights GraphDHV's efficacy in addressing extreme class imbalance issues on graphs.

Figure 4b illustrates that all indicator scores decrease with the increase of θ. Figure 4c and 4d show the optimal performance when k=2, and each layer of aggregation retains 80% of its own information.

Regarding the impact of solely $\mathcal{L}_{con}$ or $\mathcal{L}_{re}$ on the Cora-LT dataset (Fig. 4e), the model's efficacy remains favorable within the α range of 0.01 to 0.1, sharply declining beyond 0.1. These findings underscore that the loss primarily hinges on label loss, complemented by contrast loss. Notably, a slight addition of contrast losses can significantly bolster the overall model performance.

Overall, our experiments validate the effectiveness of GraphDHV across varying hyperparameters, showcasing its robustness and adaptability in addressing class imbalance challenges.

3.5 Robustness and Intra-class Compactness Analysis

GraphDHV significantly improves feature representation for class-imbalanced node classification, as demonstrated in Fig. 5, where the t-SNE plot of Cora's learned features is presented. Additionally, Gephi's community separation function was employed to analyze the dataset's topological structure. Figure 5a illustrates the initial feature distribution of the Cora dataset, revealing challenges in class separation due to overlapping contact areas.

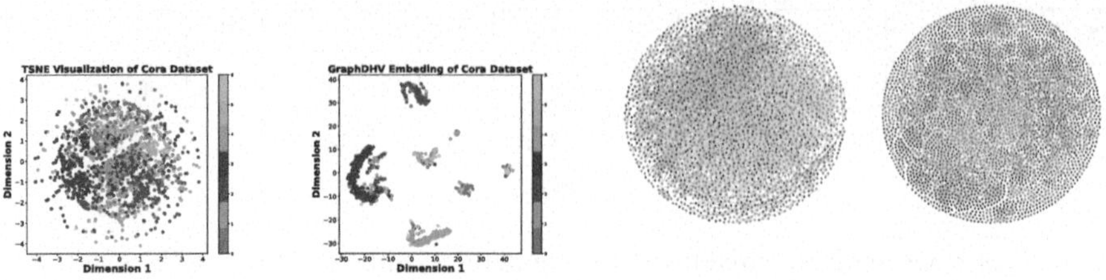

(a) (b) (c) (d)

Fig. 5. Robustness and Intra-class Compactness Analysis. (a) the t-SNE plot of the Cora dataset. (b) the t-SNE plot of the GraphDHV. (c) Community analysis of edges in Cora dataset. (d) Community analysis of edges in the GraphDHV.

In contrast, our method's t-SNE plot (Fig. 5b) exhibits clearer class separation and a higher-level distribution. Figure 5c and 5d further highlight GraphDHV's effectiveness in enhancing intra-class compactness and inter-class separability. The initial topological structure (Fig. 5c) indicates insufficient category representation, hindering separability. However, logical edge removal (Fig. 5d) enhances separability between classes and improves intra-class compactness, validating GraphDHV's efficacy.

4 Conclusion and Future Work

This paper introduces GraphDHV, a novel semi-supervised learning approach that integrates node attribute and topology information into a dual-encoder framework. GraphDHV effectively utilizes unlabeled node information and outperforms 10 baseline models across three GNN backbones on manually imbalanced datasets. Hyperparameter analysis and utilization logic experiments were conducted to enhance topological intra-class compactness and inter-class separability, with degree-based logical edge removal proving more crucial. Future research will explore GraphDHV's ability to autonomously select optimal hyperparameters and evaluate its performance on balanced datasets.

Acknowledgement. The work is supported partly by National Natural Science Foundation of China (No. 62372119, 62166003), the Project of Guangxi Science and Technology (GuiKeAB23026040, GuiKeAD20159041), Key Lab of Education Blockchain and Intelligent Technology, Ministry of Education, Guangxi Normal University, Guilin, 541004, China, Intelligent Processing and the Research Fund of Guangxi Key Lab of Multi-source Information Mining & Security (No. 20-A-01-01, MIMS21-M-01), Open Research Fund of Guangxi Key Lab of Human-machine Interaction and Intelligent Decision (No.GXHIID2206), the Guangxi Collaborative Innovation Center of Multi-Source Information Integration and the Guangxi "Bagui" Teams for Innovation and Research, China. Finally, I would like to thank my fiancee Fuhang Liu for her encouragement and support in my scientific research.

References

1. Guo, K., et al.: Hierarchical graph convolution network for traffic forecasting. In: Proceedings of AAAI (2021)
2. Hamilton, W., et al.: Inductive representation learning on large graphs. In: Advances in Neural Information Processing Systems, vol. 30 (2017)
3. Zonghan, W., et al.: A comprehensive survey on graph neural networks. IEEE TNNLS **32**(1), 4–24 (2020)
4. Li, R., et al.: Internet financial fraud detection based on graph learning. IEEE T-CSS (2022)
5. Qiao, C., Hu, X.: Discovering student behavior patterns from event logs: preliminary results on a novel probabilistic latent variable model. In: 2018 IEEE 18th ICALT, pp. 207–211. IEEE (2018)
6. Guangquan, L., et al.: Aspect sentiment analysis with heterogeneous graph neural networks. Inf. Process. Manag. **59**(4), 102953 (2022)
7. Yang, P., et al.: Adasampling for positive-unlabeled learning with bioinformatics applications. IEEE Trans. Cybern. **49**(5), 1932–1943 (2018)
8. Wang, Y., et al.: GraphGAN: graph representation learning with GANs. In: Proceedings of AAAI (2020)
9. Zhang, H., et al.: Semi-supervised classification of graph convolutional networks with Laplacian rank constraints. **54**, 2645–2656 (2022)
10. Liu, Z., Zhou, J.: Graph attention networks. In: Introduction to Graph Neural Networks. Synthesis Lectures on Artificial Intelligence and Machine Learning. Springer, Cham (2020). https://doi.org/10.1007/978-3-031-01587-8_7
11. Yu, Q., et al.: Adversarial contrastive learning via asymmetric InfoNCE. In: Avidan, S., Brostow, G., Cissé, M., Farinella, G.M., Hassner, T. (eds.) Computer Vision – ECCV 2022. ECCV 2022. LNCS, vol. 13665. Springer, Cham. https://doi.org/10.1007/978-3-031-20065-6_4
12. Prabhakar, S.K., Rajaguru, H.: Performance analysis of breast cancer classification with softmax discriminant classifier and linear discriminant analysis. In: Maglaveras, N., Chouvarda, I., de Carvalho, P. (eds.) Precision Medicine Powered by pHealth and Connected Health. IP, vol. 66, pp. 197–201. Springer, Singapore (2018). https://doi.org/10.1007/978-981-10-7419-6_33
13. Cui, Y., et al.: Class-balanced loss based on effective number of samples. In: Proceedings of the IEEE/CVF Conference on Computer Vision and Pattern Recognition, pp. 9268–9277 (2019)
14. Isola, P., et al.: Image-to-image translation with conditional adversarial networks. In: Proceedings of the IEEE Conference on Computer Vision and Pattern Recognition, pp. 1125–1134 (2017)
15. Zhao, T., et al. GraphSMOTE: imbalanced node classification with GNNs. In: Proceedings of the 14th ACM International Conference on Web Search and Data Mining, pp. 833–841 (2021)

Accelerating Topic-Sensitive PageRank by Exploiting the Query History

Shufeng Gong, Zhixin Zhang, Jing Lu, Yanfeng Zhang(✉), Cong Fu, and Ge Yu

School of Computer Science and Engineering, Northeastern University, Shenyang, China
{gongsf,zhangyf,yuge,fuchong}@mail.neu.edu.cn,
{lujing,yinshi}@stumail.neu.edu.cn

Abstract. Topic-Sensitive PageRank (TSPR) is a widely used algorithm in recommender systems and machine learning. However, the TSPR query is time-consuming as it requires multiple rounds of iterative computation. To accelerate the TSPR query, we propose an efficient TSPR query algorithm, FasTSPR, that accelerates the TSPR query by exploiting the previous TSPR query. Specifically, FasTSPR borrows the computation result of the previous TSPR query when performing the current TSPR query, avoiding redundant calculations and thus accelerating the TSPR query.

Keywords: Topic-Sensitive PageRanak · Query History · Acceleration

1 Introduction

In a graph, it is essential to compute the relevance or importance score of each vertex with respect to a particular topic, which is often widely applied in recommender systems [5] and Graph Learning [6], where the vertex represents the webpage in the networking, the author in the citation network, the person in the social network, and so on. Topic-Sensitive PageRank (TSPR) [2] is a variant of the PageRank algorithm that is specifically designed to compute the relevance or importance score of each vertex with respect to a particular topic. The score of each vertex for a particular topic is computed by calculating the similarities of it to all vertices belonging to the topic. Different from PageRank, which results in only one score vector, TSPR generates a score vector for each topic, where the score S_{ij} in the score vector $\overrightarrow{S_i}$ is the score of j-th vertex for i-th topic. TSPR enables users to receive search results that are more aligned with their specific topic preferences, enhancing the personalization of search results.

To return the relevance score of a vertex to a specific topic efficiently, we can precompute the score vector of each topic. However, as the amount of data and data type diversity increase, the number of topics in graphs is also increasing. For example, in citation networks, topics include computer vision, machine translation, information retrieval, programming languages, graph computation,

Y. Chen et al. (Eds.): COCOON 2024, LNCS 15162, pp. 514–525, 2025.
https://doi.org/10.1007/978-981-96-1093-8_42

graph matching, graph learning, etc. And with the continuous development of new technologies, the number of topics is constantly increasing. It means that a large number of score vectors need to be stored for enormous topics, and the dimension of each vector is equal to the number of vertices, which is obviously not feasible because it requires a large amount of storage space.

Based on the above discussion, it is necessary to compute the score vector online for each query topic. However, similar to PageRank, the computation of TSPR requires multiple rounds of iterative computation to update the score of each vertex, which is time-consuming. TSPR can be accelerated by returning an approximate result [4,7,8], but these methods can not be used in some high-precision situations. We can also accelerate TSPR by constructing some indexes in advance [3,7], but the indexes will be invalid when the graph changes. Therefore, it is necessary to find a method that accelerates TSPR without using indexes and get high-precisions results.

To compute TSPR efficiently, we propose an efficient TSPR, FasTSPR, which is accelerated by leveraging the previous TSPR query. The design of FasTSPR is based on the observation that there is a large amount of overlap when computing score vectors for two different topics. When we have obtained the score vector with respect to one topic, we can use these overlap computations to accelerate the computation of the TSPR query for another topic. However, it is difficult to identify the overlapped computations, if the intermediate results are not memoized in each round of iterative computation. While the memoized intermediate results often bring the overhead of storing and updating, which reduces the efficiency of TSPR.

In order to eliminate non-duplicated calculations, we use the forward push method to perform TSPR. When a TSPR query is triggered, a message is sent from each vertex belonging to the topic, and then these messages are propagated on the graph, meanwhile, the value of the messages are decayed during the propagation. The value of the accumulated received messages for each vertex refers to the score of the vertex for the topic. Based on this computation method, when performing another TSPR query with a new topic, a message is sent from each vertex belonging to the topic, and a negative message is sent from the vertex belonging to the previous topic. The positive message and the negative message are propagated on the graph at the same time. For non-overlapped computations, negative messages will make them invalid for the current topic. For the overlapped computations, the positive and negative messages will be offset and will not continue to be propagated, thereby avoiding redundant computations.

The contributions of this paper can be summarized as follows,

- We propose an efficient TSPR algorithm, FasTSPR, that accelerates TSPR queries by leveraging the previous TSPR query.
- We provide a formal proof to prove the correctness of FasTSPR.
- We evaluate the efficiency FasTSPR on five real graphs with comprehensive experiments. The results show that FasTSPR can achieve up to 5.44× speed up over the traditional TSPR algorithm and up to 1.37× speed up over the state-of-the-art algorithm.

2 Preliminary

In this section, we will provide some fundamental definitions and background knowledge regarding TSPR.

2.1 Topic-Sensitive PageRank

Let $G = (V, E)$ be a directed graph with vertex set V and edge set E, $t \in T$ represent one specific topic of the topic set, where T is the topic set. There is a set of vertices $V_t \subseteq V$ belonging to the topic t. Note that, a vertex may belong to more than one topic. Given a decay factor $\alpha < 1$, the TSPR for t can be interpreted as a random walk on G that starts from the vertices V_t belonging to t and then iteratively jumps to a randomly chosen out-neighbor with probability $1 - \alpha$ or teleports to V_t with probability α. The score value of each vertex v_i is the probability that such random walk starts from vertices V_t belonging to t and stops at v_i. TSPR query returns a score vector $\overrightarrow{S}$ for topic t, where the i-th value S_i in $\overrightarrow{S}$ is the relevance score of j-th vertex with respect to the topic t.

Note that, since we are not able to store all the score vector $\overrightarrow{S} = \overrightarrow{S_1}, \overrightarrow{S_2}, \cdots$, we only store on score vector $\overrightarrow{S}$ that is computed only when TSPR query is triggered. Therefore, in this paper, we use $\overrightarrow{S}$ to represent the score vector returned by the TSPR query of any topic t.

2.2 Power Iteration

We denote $\boldsymbol{A}$ as the adjacent matrix of G and D as the diagonal matrix, where $\boldsymbol{A}_{ij} = 1$ if there is an edge between i-th and j-th vertex, otherwise $\boldsymbol{A}_{ij} = 0$, and $\boldsymbol{D}_{ij} = 1$ if $i = j$, otherwise $\boldsymbol{D}_{ij} = 0$. Given the initial score vector $\overrightarrow{S}$ with a random value in each dimension, the power iteration is an iterative algorithm for solving the following Eq. 1

$$\overrightarrow{S} = (1 - \alpha) \cdot \overrightarrow{S} \cdot \boldsymbol{P} + \alpha \cdot \overrightarrow{e}, \tag{1}$$

where $\boldsymbol{P} = \boldsymbol{A}^T \cdot \boldsymbol{D}^{-1}$ is the transition matrix of G, and $\overrightarrow{e}$ is the indicator topic vector, *i.e.*,$e_i = \frac{1}{|V_t|}$ if the i-th vertex belongs to topic t, otherwise $e_i = 0$. We denote the score vector for topic t_i as $\overrightarrow{S}_i^*$, the Eq. 1 iteratively refines $\overrightarrow{S}$ closer to $\overrightarrow{S}^*$, *i.e.*,$||\overrightarrow{S}^* - \overrightarrow{S}||_1$ decreases with iterations, where $||\overrightarrow{\cdot}||_1$ is the l_1 normal of vector $\overrightarrow{\cdot}$. Since it is impossible to obtain $\overrightarrow{S_i}^*$ before the TSPR query, the iterative computation terminates at $||\overrightarrow{S}^{k-1} - \overrightarrow{S}^k||_1 < \epsilon$, where $\overrightarrow{S}^k$ is the score vector after k rounds of iterative computation, ϵ is an extremely small value.

Though *Power Iteration* can return TSPR query with high-precision results, it performs less efficiently. From Eq. 1, it can be seen that each iteration involves the multiplication of a vector and a matrix, as well as the addition between vectors. However, some computations within the multiplication of vector and matrix make only minimal contributions to the convergence of the TSPR query.

2.3 Forward Push

Forward Push [1] is an efficient method to perform TSPR queries. In Forward Push, in addition to the score vector $\overrightarrow{S_i}$, there is another residual vector $\overrightarrow{r}$ to be kept, r_i means the sum of the probability of random walkers who are still alive and staying at the j-th vertex.

Initially, each element in $\overrightarrow{S}$ is set as "0", and $r_i = \frac{1}{|V_t|}$ if the i-th vertex belongs topic t, otherwise $r_i = 0$. During the iterative computation process, the α portion residual value of each vertex is added to its score S_i, *i.e.*,$S_i = S_i + \alpha \cdot r_i$, then the $1 - \alpha$ portion of the residual is evenly pushed to its every outgoing neighbor $v_l \in OUT(v_i)$, *i.e.*,$r_l = r_l + \frac{(1-\alpha)\cdot r_i}{|OUT(v_i)|}$, where $|OUT(v_i)|$ is the outdegree of the i-th vertex. After that, the residual value is reset as "0". The iterative computation terminates at $||\overrightarrow{r_i}||_1 < \epsilon$. It has been proven that the *Forward Push* and the *Power Iteration* can return the same high-precision results [9] within the time complexity $O(m \cdot log\frac{1}{\epsilon})$, where m is the number of edges in G.

However, there are some inefficient computations during the iterations. For example, when the residual value r_{ij} of j-th vertex is very small, its impact on the score S is also very small. Both the addition and push operation contribute very little to the convergence of the iterative computation. To avoid inefficient computations, we skip vertices with very small residual values when performing the current iteration. The forward push operation is triggered when the residual value accumulates enough. However, inefficient computations occur during the iterations. For instance, when the residual value r_{ij} of the j-th vertex is exceedingly small, its influence on the score S_{ij} is minimal. Both addition and push operations contribute minimally to the convergence of the iterative computation. To avoid inefficient computations, we skip the vertices with very small residual values during the current iteration. In the next iterations, the forward push on v_j may be activated once its residual value accumulates sufficiently. The specific details of *Push Forward* are shown in Algorithm 1.

From Algorithm 1, it can be seen when r_{ij} is very small, the score of $_j$ is not been updated (line 6), so some inefficient computations are avoided to improve compute efficiency. The proposed FasTSPR algorithm in this paper is also built on top of the forward push calculation method.

3 FasTSPR

In this section, we will propose an efficient TSPR algorithm, FasTSPR, by exploiting the previous TSPR query. Before introducing it, we first provide the intuition behind the FasTSPR.

3.1 Inutition Behind FasTSPR

From the process of Forward Push, it can be seen that Forward Push is an accumulative iterative algorithm, *i.e.*, the score value of each vertex is accumulated from the received residuals. When a new TSPR query is triggered for another

Algorithm 1. Forward Push

Input: Graph G, damping factor α, topic t, l_1 error ϵ;
Output: $\overrightarrow{S}$

for each vertex $v_i \in V_t$ **do**
    $r_i = \frac{1}{|V_t|}$;
    $S_i = 0$;
end for
while True **do**
    for each vertex v_i with $r_i > \frac{\epsilon}{|V|}$ **do**
        $S_i = S_i + r_i * \alpha$;
        for each $v_l \in OUT(v_i)$ **do**
            $r_l = r_l + \frac{r_i \cdot (1-\alpha)}{|OUT(v_i)|}$;
        end for
        $r_i = 0$;
    end for
    if $\sum_{i=1}^{|V|} r_i \leq \epsilon$ **then**
        break;
    end if
end while
return $\overrightarrow{S}$;

topic, we have to perform Forward Push from scratch by resetting the score vector to "$\overrightarrow{0}$" and residual vector $\overrightarrow{r}$.

We found that while the residuals originate from different vertices in the previous TSPR query and the current one, there exists a considerable of residuals, bearing the same values, propagating along the same edges, *i.e.*, lots of the computations between two TSPR queries are overlapped. If we borrow the computations from the previous TSPR query, the current TSPR query can reduce some unnecessary recomputation, then the current TSPR query will be accelerated significantly.

Example 1. Consider the following scenario, given a vertex v_i, and it receives two residuals with the same value in both TSPR queries, denotes r_i' and r_i'', and $r_i' = r_i''$. Since we have processed r_i' (including $S_i = S_i + r_i'$ and push r_i' to v_i's outgoing neighbors) in the previous TSPR query, it is unnecessary to reprocess r_i'' in the current TSPR query if we borrow the same process from the previous TSPR query. Because these two residuals have the same values and the same effect on the score.

During the iterative computation for two different TSPR queries, even if the residual values passing through the same edge are similar but not identical. It would be exciting if we can accelerate iterative computation by exploring their similarities.

However, it is difficult to determine which computations from the previous TSPR query can be utilized by the current one. This complexity arises from the fact that, in the previous TSPR query, all received residuals at each vertex have been aggregated into the score. Consequently, identifying which segment of the score can be shared by two TSPR queries-specifically, the portion of the score computed through overlapping computations in the two queries-poses a significant difficulty.

To overcome this difficulty, we think about the problem from another angle, *i.e.*, from "identifying the overlapped computations" to "eliminating the non-overlapped computations". Continue to consider the vertex v_i in Example 1, in addition to the residual with the same values received in both two TSPR queries, v_i also received some distinct residuals from the previous query. To eliminate these distinct residuals, we can use the same residuals with "negative" values to eliminate their effect on the score. In this way, the score of each vertex from the previous TSPR query can be reused in the current one since we have eliminated the effect of non-overlapped computations on the score.

3.2 Details of FasTSPR

From the above discussion, we propose the FasTSPR. When a new TSPR query is triggered, we take the score computed by the previous TSPR query as the initial score vector of the current one. For the residual vector $\overrightarrow{r}$, we use the following two steps to set it,

- $r_i = r_i - \frac{1}{|V_{t_p}|}$ if $v_i \in V_{t_p}$ belongs to the previous topic t_1,
- $r_i = r_i + \frac{1}{|V_{t_c}|}$ if $v_i \in V_{t_c}$ belongs to the current topic t_c.

where t_p is the topic of the previous TSPR query, and t_c is the topic of the current one. The details of FasTSPR are described in Algorithm 2.

Note that compared with Algorithm 1, Algorithm 2 is not only different in the initialization of the score vector and residual vector but also different in filtering the residuals with small values. Since residuals may be positive or negative in FasTSPR, the absolute values of residuals are used to filter inefficient computations.

During the iterative computation of Algorithm 2, there are both positive and negative residuals propagating on G. When the positive and negative residuals meet, they will be canceled or cut significantly due to the aggregation operation (line 11), and the absolute values of the residuals will quickly become smaller after aggregation, thereby accelerating the convergence of the TSPR query.

3.3 Correctness of FasTSPR

It can be seen from the above analysis that the Algorithm 2 convergence is accelerated by the aggregation of positive and negative residuals. But is the final result correct? Next, we are going to answer this question.

From the Algorithm 1, it can be seen that the essential of residual propagation is the multiplication of the residual and the transfer matrix, *i.e.*,$(1-\alpha)\cdot\overrightarrow{r}\cdot\boldsymbol{P}$, and

Algorithm 2. FasTSPR

Input: Graph G, damping factor α, topic t_c, l_1 error ϵ, score vector $\overrightarrow{S}$ and the residual vector $\overrightarrow{r}$ obtained from TSPR query for t_p.
Output: $\overrightarrow{S}$

for each vertex $v_i \in V_{t_p}$ **do**
  $r_i = r_i + \frac{1}{|V_{t_p}|}$;
end for
for each vertex $v_i \in V_{t_c}$ **do**
  $r_i = r_i + \frac{1}{|V_{t_c}|}$;
end for
while True **do**
  for each vertex v_i with $|r_i| > \frac{\epsilon}{|V|}$ **do**
    $S_i = S_i + r_i * \alpha$;
    for each $v_l \in OUT(v_i)$ **do**
      $r_l = r_l + \frac{r_i * (1-\alpha)}{|OUT(v_i)|}$;
    end for
    $r_i = 0$
  end for
  if $\sum_{i=1}^{|V|} r_i \leq \epsilon$ **then**
    break;
  end if
end while
return $\overrightarrow{S}$;

accumulating it into S. Then, without considering the inefficient computation filterations, the Forward Push can be formalized into the following equation.

$$\overrightarrow{S}^k = \overrightarrow{S}^{k-1} + \alpha \cdot \overrightarrow{r}^k \tag{2}$$

$$\overrightarrow{r}^k = (1-\alpha)\overrightarrow{r}^{k-1} \cdot \boldsymbol{P}. \tag{3}$$

Combine the Eq. 2 and the Eq. 3, then we have

$$\begin{aligned}\overrightarrow{S}^k =& \overrightarrow{S}^{k-1} + \alpha \cdot (1-\alpha) \cdot \overrightarrow{r}^{k-1} \cdot \boldsymbol{P} \\ =& \overrightarrow{S}^{k-2} + \alpha \cdot \overrightarrow{r}^{k-2} \cdot \boldsymbol{P} + \alpha \cdot (1-\alpha)^2 \cdot \overrightarrow{r}^{k-2} \cdot \boldsymbol{P}^2 \\ =& \overrightarrow{S}^{k-3} + \cdots \\ & \cdots \\ =& \overrightarrow{S}^0 + \alpha \cdot \overrightarrow{r} + \alpha \cdot (1-\alpha) \cdot \overrightarrow{r} \cdot \boldsymbol{P} + \cdots \\ & + \alpha \cdot (1-\alpha)^{k-1} \cdot \overrightarrow{r} \cdot \boldsymbol{P}^{k-1}\end{aligned} \tag{4}$$

For the previous topic t_p, the TSPR returns score vector $\overrightarrow{S}_p^k$ after k round of iterative computations. For the current topic t_c, the initial score vector and residual vector are set as $\overrightarrow{S}_c^0 = \overrightarrow{S_p}^k$ and $\overrightarrow{r_c}^0 = \overrightarrow{r_c}^0 + \overrightarrow{r_p}^k - \overrightarrow{r_p}^0$ respectively according to Algorithm 2. After the first iteration of FasTSPR, we have

$$\overrightarrow{S_c}^1 = \overrightarrow{S_p}^k + \alpha \cdot \overrightarrow{r_c} \tag{5}$$

After the k-th iteration, according to the Eq. 4, we have

$$\begin{aligned} \overrightarrow{S_c}^k = & \overrightarrow{S_p}^k + \alpha \cdot \overrightarrow{r_c} + \alpha \cdot (1-\alpha) \cdot \overrightarrow{r_c} \cdot \boldsymbol{P} + \cdots \\ & + \alpha \cdot (1-\alpha)^{k-1} \cdot \overrightarrow{r_c} \cdot \boldsymbol{P}^{k-1} \end{aligned} \tag{6}$$

Unfold the $\overrightarrow{S_p}^k$ according to Eq. 4, then we have

$$\begin{aligned} \overrightarrow{S_c}^k = & \overrightarrow{S}^0 + \alpha \cdot \overrightarrow{r_p} + \alpha \cdot (1-\alpha) \cdot \overrightarrow{r_p} \cdot \boldsymbol{P} + \cdots \\ & + \alpha \cdot (1-\alpha)^{k-1} \cdot \overrightarrow{r_p} \cdot \boldsymbol{P}^{k-1} \\ & + \alpha \cdot \overrightarrow{r_c} + \alpha \cdot (1-\alpha) \cdot \overrightarrow{r_c} \cdot \boldsymbol{P} + \cdots \\ & + \alpha \cdot (1-\alpha)^{k-1} \cdot \overrightarrow{r_c} \cdot \boldsymbol{P}^{k-1} \end{aligned} \tag{7}$$

Replace $\overrightarrow{r_c}$ with $\overrightarrow{r_c} = \overrightarrow{r_c} + \overrightarrow{r_p}^k - \overrightarrow{r_p}$, then we have

$$\begin{aligned} \overrightarrow{S_c}^k = & \overrightarrow{S}^0 + \alpha \cdot \overrightarrow{r_p} + \alpha \cdot (1-\alpha) \cdot \overrightarrow{r_p} \cdot \boldsymbol{P} + \cdots \\ & + \alpha \cdot (1-\alpha)^{k-1} \overrightarrow{r_p} \cdot \boldsymbol{P}^{k-1} \\ & + \alpha \cdot \overrightarrow{r_c} + \alpha \cdot \overrightarrow{r_p}^k - \alpha \cdot \overrightarrow{r_p} \\ & + \alpha \cdot (1-\alpha) \cdot \overrightarrow{r_c} \cdot \boldsymbol{P} + \alpha \cdot (1-\alpha) \cdot \overrightarrow{r_p}^k \cdot \boldsymbol{P} \\ & - \alpha \cdot (1-\alpha) \cdot \overrightarrow{r_p} \cdot \boldsymbol{P} + \cdots \\ & + \alpha \cdot (1-\alpha)^{k-1} \cdot \overrightarrow{r_c} \cdot \boldsymbol{P}^{k-1} + \alpha \cdot (1-\alpha)^{k-1} \overrightarrow{r_p}^k \cdot \boldsymbol{P}^{k-1} \\ & - \alpha \cdot (1-\alpha)^{k-1} \overrightarrow{r_p} \cdot \boldsymbol{P}^{k-1} \\ = & \overrightarrow{S}^0 + \alpha \cdot \overrightarrow{r_c} + \alpha \cdot \overrightarrow{r_p}^k \\ & + \alpha \cdot (1-\alpha) \cdot \overrightarrow{r_c} \cdot \boldsymbol{P} + \alpha \cdot (1-\alpha) \overrightarrow{r_p}^k \cdot \boldsymbol{P} + \cdots \\ & + \alpha \cdot (1-\alpha)^{k-1} \cdot \overrightarrow{r_c} \cdot \boldsymbol{P}^{k-1} + \alpha \cdot (1-\alpha)^{k-1} \overrightarrow{r_p}^k \cdot \boldsymbol{P}^{k-1} \end{aligned} \tag{8}$$

Since the $\alpha \cdot \overrightarrow{r_p}^k$ is too small to be ignored, then we have

$$\begin{aligned} \overrightarrow{S_c}^k = & \overrightarrow{S}^0 + \alpha \cdot \overrightarrow{r_c} \\ & + \alpha \cdot (1-\alpha) \cdot \overrightarrow{r_c} \cdot \boldsymbol{P} + \cdots \\ & + \alpha \cdot (1-\alpha)^{k-1} \cdot \overrightarrow{r_c} \cdot \boldsymbol{P}^{k-1} \end{aligned} \tag{9}$$

Otherwise, the iterative computation will be performed continuously and generate $-\alpha \cdot \overrightarrow{r_p}^k$ to eliminate $\alpha \cdot \overrightarrow{r_p}^k$. It means that after k iterations, FasTSPR and Forward Push return the same result with high precision. Note that Eq. 9 is a high-precision approximation of Eq. 8. In practice, during the execution of the FasTSPR, $\alpha \cdot \overrightarrow{r_p}^l, (l \geq k)$ are not be removed and will be eliminated by $-\alpha \cdot \overrightarrow{r_p}^l, (l \geq k)$ generated in the subsequent iterative computations of subsequent iterations. Therefore, even if the topics are changed multiple times, we can still get high-precision results.

4 Experiment

4.1 Experimental Setup

In this section, we present the experimental evaluation of FasTSPR. By default, the experiments are performed on a commodity PC with 5.4 GHz Intel Core i9 CPU, 32 GB memory, and it runs on 64-bit Ubuntu 22.04 with compiler GCC 11.3.

Dataset. Our experiments involve five real graphs, DBLP, Catster, Wiki-Talk, cit-Patents, and Google. Table 1 shows the detailed information of these graphs. Since the vertex belonging to the same topic is similar, thus the vertices in each topic come from the same cluster of the graph. In our experiments, each topic contains about 50 vertices by default.

Table 1. Datasets

Graph	# of vertex	# of edge
Catster (CT)	149684	10896394
DBLP (DL)	317080	2099732
Google (GL)	916428	12156500
Wiki-Talk (WT)	2394385	8505513
cit-Patents (CP)	3774768	16518948

Competitors. We compare FasTSPR with traditional *Power Iteration* (PowItr), *Forward Push* (FwdPush), and a hybrid method *PowerPush* [9] that incorporated the strengths of both PowItr and FwdPush. In our experiment, we use the same l_1-error threshold $\epsilon = 10^{-9}$ for all competitors and FasTSPR.

4.2 Overall Performance

We first compare FasTSPR with competitors in terms of the average runtime on different graphs in Table 1. Figure 1 reports the normalized runtime of each algorithm, where FasTSPR is treated as the baseline. It can be seen that the FasTSPR outperforms other algorithms in all cases. Specifically, FasTSPR can achieve 2.42× speed up on average (up to 2.65×) over PowerIter, 3.22× speed up on average (up to 5.44×) over FwdPush, and 1.21× speed up on average (up to 1.37×) over PowerPush. It is notable that both PowerPush and FasTSPR outperform PowerItr and FwdPush because of some optimizations equipped in PowerPush, e.g., asynchronous pushes, hybrid global sequential scan and local random access, dynamic l_1-error threshold. Our FasTSPR outperforms PowerPush because of the utilization of query history, which reduces some redundant computations.

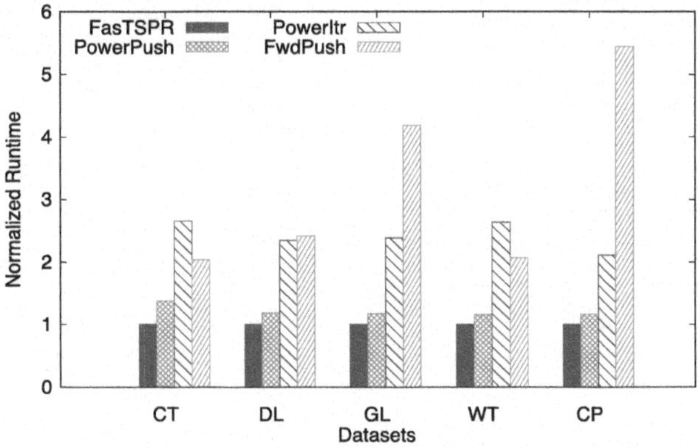

Fig. 1. The comparison of runtime

4.3 Convergence Comparison

To evaluate the effectiveness of FasTSPR on accelerating the convergence of TSPR queries, we compare the convergence rates of FasTSPR and the competitors. We use the sum of residual values as the metric to measure the distance to convergence, i.e., $|\sum_{i=1}^{|V|} r_i$. Figure 2 shows the trend of the sum of residual values over time. It can be seen that the residual in the FasTSPR algorithm decreases fastest and achieves to convergence state first. Note that the sum of the initial residual values of FasTSPR is 2 since there are positive and negative residuals in the current and previous topic vertices.

(a) CT (b) GL

Fig. 2. The comparison of convergence speed

4.4 The Impact of Different TSPR Queries

In order to test the impact of different TSPR queries on FasTSPR, we randomly performed 10 TSPR queries continuously and recorded the runtime of each query.

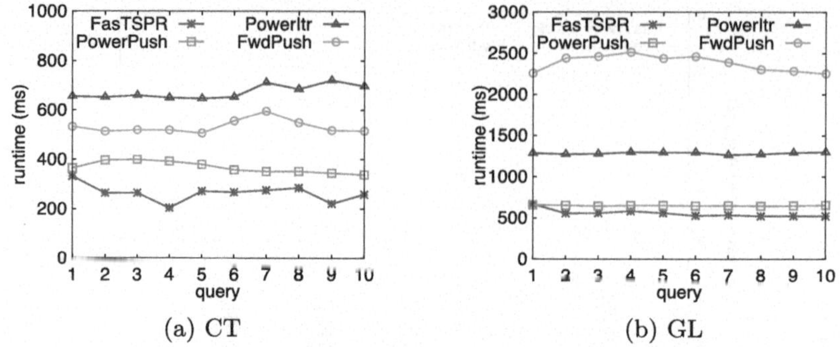

Fig. 3. The runtime when varying the topics

Figure 3 shows the runtime of FasTSPR and competitors when performing different TSPR queries on Catster (CT) and Google (GL) graphs. It can be seen that the FasTSPR outperforms other algorithms in all queries, except for the first query. This is because there is no previous query to be used for the first query, so its runtime is the same as that of PowerPush.

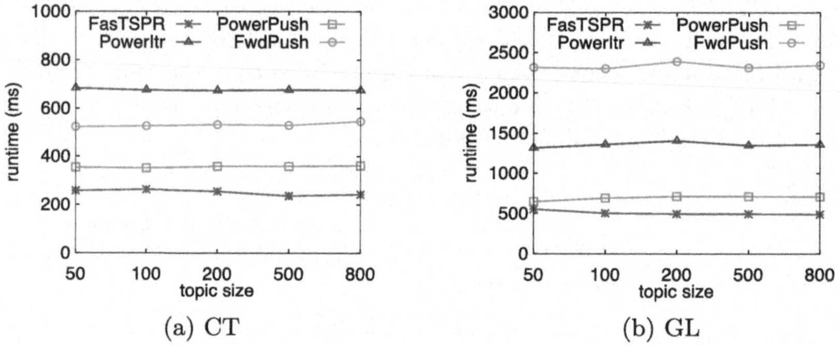

Fig. 4. The runtime when varying the size of topics

4.5 The Impact of Topic Size

To test the impact of the size of the topic, i.e., the number of vertices within the topic for TSPR, on the efficiency of FasTSPR, we performed TSPR queries on topics containing 50, 100, 200, 500 and 800 vertices and recorded their execution times. Figure 4 shows the runtime of FasTSPR and others when varying the size of topics on Catster (CT) and Google (GL) graphs. It can be seen that the execution time of FasTSPR is smaller than others when varying the topic sizes, and the running time of FasTSPR slowly decreases with the increase of the number of vertices in the topic. This is because the more vertices in the topic, the higher the possibility of the positive residuals and negative residuals meeting, which results in better acceleration.

5 Conclusion

In this paper, we propose FasTSPR, an efficient Topic-sensitive PageRank (TSPR) algorithm. FasTSPR accelerates the TSPR query by reusing the computation of the previous TSPR query, which can reduce some redundant computation since there are enormous overlapped computations between the current TSPR query and the previous one.

Acknowledgments. This paper is supported by the 111 Project (B16009), the National Natural Science Foundation of China (U2241212, 62072082, 62202088, 62137001, 62272093), Joint Funds of Natural Science Foundation of Liaoning Province (2023-MSBA-078), and Fundamental Research Funds for the Central Universities (N2416011).

References

1. Andersen, R., Chung, F., Lang, K.: Local graph partitioning using PageRank vectors. In: Proceedings of the 47th FOCS, pp. 475–486. IEEE (2006)
2. Haveliwala, T.H.: Topic-sensitive PageRank. In: Proceedings of the 11th WWW, pp. 517–526 (2002)
3. Jung, J., Park, N., Lee, S., Kang, U.: BEPI: fast and memory-efficient method for billion-scale random walk with restart. In: Proceedings of the 2017 ACM International Conference on Management of Data, pp. 789–804 (2017)
4. Lofgren, P.A., Banerjee, S., Goel, A., Seshadhri, C.: Fast-PPR: scaling personalized PageRank estimation for large graphs. In: Proceedings of the 20th SIGKDD, pp. 1436–1445 (2014)
5. Lü, L., Medo, M., Yeung, C.H., Zhang, Y.C., Zhang, Z.K., Zhou, T.: Recommender systems. Phys. Rep. **519**(1), 1–49 (2012)
6. Wang, H., Yang, R., Huang, K., Xiao, X.: Efficient and effective edge-wise graph representation learning. In: Proceedings of the 29th SIGKDD, pp. 2326–2336 (2023)
7. Wang, S., Tang, Y., Xiao, X., Yang, Y., Li, Z.: HUBPPR: effective indexing for approximate personalized PageRank. In: Proceedings of the VLDB Endowment, vol. 10, no. 3 (2016)
8. Wang, S., Yang, R., Xiao, X., Wei, Z., Yang, Y.: FORA: simple and effective approximate single-source personalized PageRank. In: Proceedings of the 23rd SIGKDD, pp. 505–514 (2017)
9. Wu, H., Gan, J., Wei, Z., Zhang, R.: Unifying the global and local approaches: an efficient power iteration with forward push. In: Proceedings of the 2021 International Conference on Management of Data, pp. 1996–2008 (2021)

Extracting Representative Co-location Patterns Considering Distributions of Spatial Features and Instances

Xuguang Bao, Shuaikang Yuan, and Liang Chang(✉)

Guilin University of Electronic Technology, Guilin 541000, China
{baoxuguang,changl}@guet.edu.cn

Abstract. Spatial co-location pattern mining, aiming at discovering the association of spatial features, can help users find the correlation knowledge from spatial datasets. However, with the growth of spatial datasets, the traditional framework for mining co-location patterns generates a mass of redundant co-location patterns, which makes it difficult for users to perform further analysis. This paper researches how to extract representative co-location patterns, i.e., a concise summarization of prevalent co-location patterns. Two similarity metrics - feature similarity and distribution similarity, are introduced to evaluate the redundancy between co-location patterns from the perspective of both features and instances. With the above metrics, a novel approach called Representative Co-location Patterns Extractor (RCPE) is further proposed to condense the prevalent co-location patterns. Two types of representative co-location patterns, called Representative Maximal Co-location Patterns (RMCPs) and Worthy Non-Maximal Co-location Patterns (WNMCPs), are mined from the prevalent co-location patterns set respectively. A series of experiments on both synthetic and real datasets confirm that RCPE has better performance in compression power and running time compared with similar state-of-the-art approaches.

Keywords: Spatial pattern mining · representative co-location patterns · redundancy reduction · similarity metric · pattern recommendation

1 Introduction

Prevalent co-location pattern mining is a crucial branch of spatial data mining [1, 2]. A Prevalent Co-location Pattern (PCP) represents a subset of spatial features (a kind of a geometrical object) whose instances (a geometrical object) are often co-located in a spatial neighborhood. For instance, a set {supermarket, parking lot} may be a PCP since parking lots frequently adjacent to supermarkets.

In PCP mining, the Participation Index (PI) is commonly adopted to evaluate the frequency of a spatial co-location pattern. A co-location c is a PCP if $PI(c) \geq pt$ holds, where pt is a user-specified minimum prevalence threshold. However,

Y. Chen et al. (Eds.): COCOON 2024, LNCS 15162, pp. 526–537, 2025.
https://doi.org/10.1007/978-981-96-1093-8_43

owing to the anti-monotonic property [3] of PI values (i.e., all subsets of any PCP must be prevalent), a PCP with n features may generate $2^n - n - 1$ prevalent sub-patterns. Regrettably, these sub-patterns do not offer additional valuable insights but rather impede the user's analysis of valuable patterns.

There are two classical representations for compressing PCPs, namely Maximal Co-location Patterns (MCPs) [4,5] and Closed Co-location Patterns (CCPs) [6]. MCPs, which are the patterns whose super-patterns are completely infrequent, have a strong compression capability and provide a highly concise representation of PCPs. Nevertheless, they are regarded as a lossy representation since they only focus on the pattern's feature composition and ignore its specific distribution. On the other hand, CCPs, which are the patterns whose PI is not equal to the PIs of all their super-patterns, preserve all the information of PCPs and are considered as a lossless representation. However, due to their strict definition, CCPs have weak and unstable compression capability, and even a slight perturbation in spatial data can result in a significant impact on CCPs.

Given that both of the aforementioned compression methods adhere to a *top-down* approach, presuming the value of patterns is reflected through comparison with their super patterns, these algorithms' compression capabilities cannot surpass the Maximal Co-location Patterns. In this paper, we concentrate on eliminating redundant information in PCPs from two perspectives: feature similarity and instance distribution.

1.1 Related Works

The PCP mining is one of frequent patterns mining, which is similar to frequent itemsets mining [7]. It was first defined by Shekhar and Huang in [1], where the authors proposed the join-based algorithm to discover PCPs. Yoo et al. proposed the partial-join algorithm [8] and the join-less algorithm [9] to improve the efficiency of mining PCPs. Subsequently, Wang et al. [10] used the CPI-tree to materialize spatial neighbor relationships. To generate PCPs efficiently, Bao et al. [11] designed a neighboring clique mining approach, and constructed a hash structure to mitigate the space consumption.

To reduce redundancy in the PCPs set, Yoo et al. [12] proposed a compact and lossless representation of PCPs, called Closed Co-location Patterns (CCPs), and developed a method for mining the top-k CCPs. Later, Wang et al. [6] proposed an effective lossless condensed representation called SPI-closed patterns and offered an algorithm named SPI-miner to mine the SPI-closed patterns. Bao et al. [13] then presented an efficient algorithm called Non-Redundant Co-location Pattern Mining (NRCPM) to collect SPI-closed patterns based on neighboring cliques.

Compared to CCPs, maximal co-location patterns have greater performance on compression power but lose more interesting co-location patterns. Wang et al. [4] proposed an order-clique-based algorithm for mining MCPs. Yao et al. [14] proposed a Sparse-Graph and Condensed Tree-based approach (SGCT) to improve the efficiency of MCP mining.

Overall, the above methods merely compare the similarity between patterns and their super-patterns, thereby limiting their compression capability for large pattern sets. In this paper, we propose two similarity metrics to evaluate the redundancy between PCPs, and utilize these metrics to extract the representative co-location patterns from PCPs.

1.2 Contribution

In this paper, we extract the representative co-location patterns from MCPs and NMCPs respectively. The contributions of this paper are as follows:

1. Considering the feature composition and instance distribution of MCPs further enhances the potential for compression.
2. We propose two metrics, namely feature similarity and distribution similarity, to evaluate the similarity between any two PCPs.
3. We propose an efficient compression method for PCPs called the Representative Co-location Patterns Extractor (RCPE), which aims to find representative MCPs and valuable NMCPs.
4. We designed and executed a series of experimental studies on real and synthetic datasets to illustrate the effectiveness and efficiency of RCPE.

2 Problem Definition

2.1 Co-location Patterns

Spatial feature f represents a specific kind of object in a spatial area. A real-existing physical object of f at a location is called an **instance** of f. A **spatial dataset** S is a set of instances with locations and their features in a geometric space, and each instance in S is represented as $\{\text{feature}, \text{instance id}, \text{location}\}$. A **spatial neighbor relationship** R describes the relationships between instances. If two instances i_1 and i_2 belong to different features and their distance is no more than a minimum threshold *min_dist*, i_1 and i_2 are neighbors, i.e., $R(i_1, i_2)$ = true. A **clique** cl is a set of instances and every two distinct instances in cl are neighbors. Given a **spatial feature set** F, a **co-location pattern** c is a subset of F whose instances form cliques frequently. The number of features in c is called the **size** of c. The related definitions of prevalent co-location pattern mining are provided as follows.

Definition 1 (Row Instance and Table Instance). A clique *cl* is called a row instance of a co-location pattern c (denoted as RI(c)), if *cl* contains instances of all features in c and no proper subset of *cl* does so. The table instance of c is the collection of all row instances of c, denoted as TI(c).

Definition 2 (Participation Ratio). Given a co-location pattern c, the participation ratio $PR(c, f)$ is the fraction of instances of f_i that occur in $TI(c)$, i.e.,

$$PR(c, f) = \frac{\text{Number of distinct instances of } f \text{ in } TI(c)}{\text{Total number of instances of } f} \tag{1}$$

Definition 3 (Participation Index). The participation index $PI(c)$ of a co-location pattern c is defined as,

$$PI(c) = \min_{f \in c}\{\text{PR}(c, f)\} \tag{2}$$

Definition 4 (Prevalent Co-location Pattern). A co-location pattern c is a prevalent co-location pattern (PCP) if PI(c) $\geq p_t$, where p_t is a user-specified minimum prevalence threshold.

Definition 5 (Maximal Co-location Patterns). A prevalent co-location pattern c is maximal if any of its super-patterns c' (i.e., $c \subsetneq c'$) cannot be prevalent (PI(c') $< p_t$).

Lemma 1 An MCP cannot be a sub-pattern or super-pattern of another MCP.

2.2 Feature Similarity

To measure the similarity of feature composition between two co-location patterns, the feature distance is defined as below.

Definition 6 (Feature Similarity). Given two co-location patterns c_1 and c_2. The feature similarity (FS) between these two patterns is

$$FS(c_1, c_2) = \frac{|c_1 \cap c_2|}{|c_1 \cup c_2|} \tag{3}$$

Theorem 1. The feature similarity FS is a valid similarity metric, such as:

1) $FS(c_1, c_2) > 0, \forall c_1 \neq c_2$;
2) $FS(c_1, c_2) = 1, \forall c_1 = c_2$;
3) $FS(c_1, c_2) = FS(c_2, c_1)$;
4) $[1 - FS(c_1, c_2)] + [1 - FS(c_2, c_3)] \geq [1 - FS(c_1, c_3)], \forall c_1, c_2, c_3$.
Where c_1, c_2 and c_3 represent three co-location patterns.

2.3 Distribution Similarity

Definition 7 (Instance Similarity). Let c_1 and c_2 be two co-location patterns, c_u is their union pattern ($c_u = c_1 \cup c_2$). Given a feature $f \in c_u$, the instance similarity between c_1 and c_2 with respect to f is defined as:

$$IS\{(c_1, c_2), f\} = \frac{|I_{c_u}(f)|}{|I_{c_1}(f) \cup I_{c_2}(f)|} \tag{4}$$

Where $I_c(f)$ is the set of instances of f in the table instance of c.

Definition 8 (Distribution Similarity). Given two co-location patterns c_1 and c_2, the distribution similarity is defined as:

$$DS(c_1, c_2) = \min_{\forall f_i \in f_{c_u}} IS\{(c_1, c_2), f_i\} \tag{5}$$

Theorem 2. The distribution similarity is a sub-valid similarity metric, such as:

1) $\forall c_1 = c_2, DS(c_1, c_2) = 1$;
2) $DS(c_1, c_2) = DS(c_2, c_1)$;
3) For $\forall c_1 \subset c_2 \subset c_3$ if:
$arg_f \min_{\forall f \in c_2} IS\{(c_2, c_1), f\} = arg_f \min_{\forall f \in c_3} IS\{(c_2, c_3), f\}$
Then: $[1 - DS(c_1, c_2)] + [1 - DS(c_2, c_3)] \geq 1 - DS(c_1, c_3)$;
4) For $\forall c_1, c_2, c_3$ where $c_1 \cap c_2 = c_3 \neq \emptyset$,
if : $\arg_f \min_{\forall f \in c_u} IS\{(c_1, c_u), f\} \in c_3 \quad \arg_f \min_{\forall f \in c_u} IS\{(c_2, c_u), f\} \in c_3$,
Then $\arg_f \min_{f \in c_u} IS\{(c_1, c_2), f\} \in c_3$, $DS(c_1, c_2) \leq \min\{DS(c_1, c_u), DS(c_2, c_u)\}$.

The distribution similarity can be used to determine a similar relationship. The concept of θ-similar is defined as follows:

Definition 9 (θ-similar) Given a distribution similarity threshold θ, for two co-location patterns c and c', c is θ-similar to c' if $DS(c, c') \geq \theta$.

Lemma 2. Given two co-location patterns c_1 and c_2, the pattern c_3 is a super pattern of their union pattern c_u, if both c_1 and c_2 are θ-similar to c_3, then c_u is also θ-similar to c_3.

3 Method

This section introduces RCPE, a compression method for co-location patterns. It comprises two stages: clustering MCPs to extract feature similarity patterns and selecting representative MCPs based on distribution similarity, and devising a *top-down* algorithm to mine valuable NMCPs.

After mining PCPs, the process of discovering MCPs is as follows: 1) PCPs are sorted in descending order of size; 2) the PCPs with the highest size are identified as MCPs; 3) the remaining PCPs are compared to the current MCPs. If MCPs don't contain a super-pattern for a PCP, this PCP is an MCP. Otherwise, it is an NMCP; 4) repeating step 3) until all PCPs have been classified.

3.1 Compression of MCPs

To extract representative MCPs (RMCPs), the MCP compression algorithm designed in this study is divided into two parts: clustering and filtering, as illustrated in Algorithm 1.

Algorithm 1. Discovering RMCPs

```
Require: Number of clusters, k; Set of co-location patterns, cps;
Ensure: Representative MCPs, RMCPs;
MCPs ← getMCPs(cps);
clusters ← KMeans(MCPs, k);
for each cluster cl in clusters do
    for each pattern p1 in cl do
        for each pattern p2 in cl do
            if DS_Calculation(p1, p2) ≥ max_DS then
                SimilarPattern.get(p1).add(p2);
            end if
        end for
    end for
end for
while not cps.isEmpty do
    m_p ← GetMaxPattern(SimilarPattern);
    cps.removeAll(SimilarPattern.get(m_p));
    cps.remove(m_p);
    SimilarPattern.remove(m_p);
    RMCPs.add(m_p);
end while
return RMCPs;
```

3.2 Collecting WNMCPs

MCP is a lossy representation since it loses the PI value from most of the co-location patterns. To supplement the information discarded by MCPs, we discover the worthy NMCPs.

Definition 10 (Worthy NMCP). Given an NMCP cn, $S(cn)$ is a pattern set that consists of the prevalent super patterns of cn (i.e., $S(cn) = \{c | cn \subseteq c \text{ and } PI(c) \geq pt\}$) and M is the set of MCPs. $SM(cn) = S(cn) \cap M$ is called the maximal super-patterns of cn. For all $c' \in SM(cn)$, if $DS(cn, c') < \theta$, i.e., cn is not θ-similar to c', we call cn is a worthy NMCP (WNMCP), where θ is the threshold of DS.

To discover WNMCPs, here we design a *top-down* approach based on Lemma 2. According to Lemma 2, if two NMCPs are θ-similar to an MCP, their union pattern is also θ-similar to this maximal co-location pattern. Thus, we retain the redundant patterns and generate high-size redundant patterns without the DS calculation.

Algorithm 2. Collecting WNMCPs

```
Require: Maximal co-location patterns, MCPs; Prevalent co-location patterns, PCPs;
    User-specified distribution similarity threshold, θ;
Ensure: Representative non-MCPs, WNMCPs;
NMCPs = PCPs.delete(MCPs);
size2set = get_size2set(PCPs);
for each pattern s1 in size_2subset do
    for each mcp m in MCPs do
        if DS(s1, m) ≥ θ then
            Similar_Size2set.add(s1);
        else
            WNMCPs.add(s1);
        end if
    end for
end for
Similar_supset = gener_supset(Similar_Size2set);
WorthPatterns = PCPs.delete(Similar_Size2set, Similar_supset);
WNMCPs.add(WorthPatterns);
return WNMCPs;
```

4 Experimental Results

In this section, the proposed method is evaluated by a set of experiments. All of the methods are implemented in Java. Our experiments are performed on a PC running Windows 10 with Intel Core i7-7700 CPU @ 3.60GHz and 8GB memory.

4.1 Datasets

The synthetic dataset was generated using a spatial synthetic data generator with basic parameters including 20 features and approximately two thousand instances, with a distance threshold set to 15. The real datasets were chosen from distributions of Points of Interest (POIs) in Beijing. We compared the runtime and compression ability of RCPE with MCP [4], CCP [12], RRclosed [15], and NRCPM [13] on both synthetic and real datasets.

4.2 Compression Power

The evaluation of compression capabilities for RCPE and the remaining algorithms is conducted separately on three real datasets, and Compression Ratio (CR) is used to assess their compression abilities. The calculation of Compression Ratio is as follows:

$$CR = \frac{The\ number\ of\ filtered\ PCPs}{The\ number\ of\ PCPs} \times 100\% \tag{6}$$

we compare the compression power of RCPE with that of the MCP [4], CCP [12], RRclosed [15] and NRCPM [13]. The PI threshold for all three datasets is set to 0.2, and the minimum distribution similarity threshold θ is 0.5. Figure 1 shows the results of the experiments, we have the following observations:

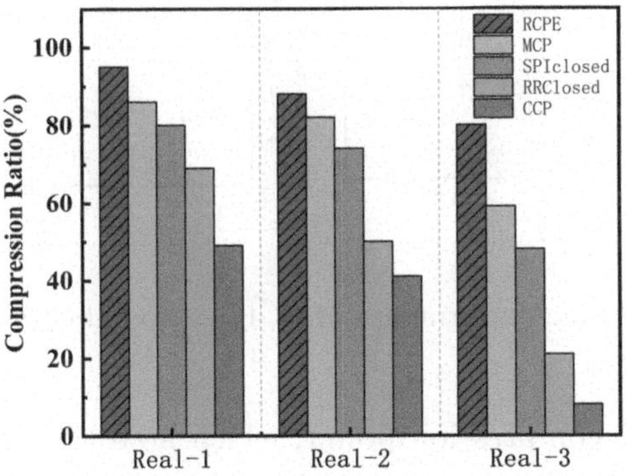

Fig. 1. Compression ratio on three datasets.

The compression power of RCPE increases when the value of θ reduces. There are two reasons: first, in the stage of discovering WNMCPs, a larger value of θ enables an MCP to cover more NMCPs; second, in the stage of collecting RMCPs, with a larger value of θ, more patterns in each cluster are θ-similar to the representative patterns. When $\theta = 0.7$, the compression ratio of RCPE is between 26% and 63%, and the compression ratio of MCPs is between 39% and 80%. Therefore, a larger value of θ may cause the compression power of RCPE to be weaker than MCPs.

4.3 Time Performance of RCPE

In this section, we compare the runtime of RCPE on a real dataset with that of MCP [4], CCP [12], RRclosed [15], and NRCPM [13], and further assess its time performance against MCP on three synthetic datasets. From Fig. 2, it is evident that as *pt* increases, the runtime differences among NRCPM, MCP, and RCPE gradually decrease. When *pt* is small ($pt = 0.1, 0.3$), the MCP algorithm has the shortest runtime because it only needs to mine PCPs and compare their feature compositions, without calculating PI or comparing instance distributions. In Fig. 2, RCPE's runtime on three synthetic datasets with different θ values is shown. It is worth noting that on SYN-2, RCPE exhibits a longer runtime, primarily due to higher storage costs associated. with neighboring clique hash and a larger number of PCPs, which require more time for compression.

Based on the above experiments, we summarize the following characteristics of RCPE:

1. **Compression power:** The compression power of RCPE is affected by the value of θ. A large value of θ enables RCPE to generate more RCPs, which makes RCPE retain more valuable patterns. A low value of θ enables RCPE

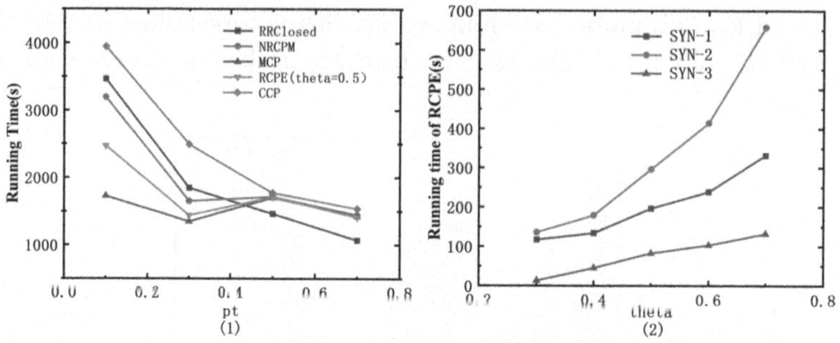

Fig. 2. The time performance of RCPE across different datasets

to have a stronger compression power, which however may result in the loss of valuable patterns.

2. **Efficiency:** RCPE is more efficient than the existing approaches especially for sparse datasets or for a moderate value of θ.

5 Conclusion

This paper introduces an efficient approach named RCPE to extract the representative co-location patterns. The bottleneck of the existing approaches is that they have a measure limitation, i.e., the similarity is only checked between a pattern and its super-patterns, which makes the number of representative patterns discovered more than the number of MCPs. In this paper, two similarity metrics called feature similarity and distribution similarity are proposed to evaluate the similarity between patterns. RCPE employs the two metrics to discover the representative co-location patterns from the MCPs and the NMCPs. The experimental results confirm that RCPE has great performances on compression power and running speed when the value of θ(the threshold of distribution similarity) is appropriate.

Acknowledgments. This work was supported in part by grants (No. U1811264, No. U1711263, No. 61966009, No. 62006057) from the National Natural Science Foundation of China.

Disclosure of Interests. The authors have no competing interests to declare that are relevant to the content of this article.

Appendix

Proof of Theorem 1. According to the definition of feature similarity, it is easy to verify that the first three properties are true. We prove the last statement as follows.

We regard the three co-location patterns c_1, c_2, and c_3 as three sets of features s_1, s_2, and s_3 respectively. To simplify the presentation, we define the following variables: $a = |s_1 - (s_1 \cap s_2) - (s_1 \cap s_3)|, b = |s_2 - (s_2 \cap s_1) - (s_2 \cap s_3)|, c = |s_3 - (s_3 \cap s_1) - (s_3 \cap s_2)|, d = |s_1 \cap s_2 - s_1 \cap s_2 \cap s_3|, e = |s_2 \cap s_3 - s_1 \cap s_2 \cap s_3|, f = |s_1 \cap s_3 - s_1 \cap s_2 \cap s_3|, g = |s_1 \cap s_2 \cap s_3|$. Plug in all variables into the feature similarity:

$$FS(c_1, c_2) = \frac{d+g}{a+b+d+e+f+g},$$
$$FS(c_2, c_3) = \frac{e+g}{b+c+d+e+f+g},$$
$$FS(c_1, c_3) = \frac{f+g}{a+c+d+e+f+g}.$$

Since $(s_1 \cap s_2) \cup (s_1 \cap s_3) \subseteq s_1$, thus:

$$b + d \leq a + b + d + e$$

And we have:

$$\frac{a+b+e+f}{a+b+c+d+e+f} + \frac{b+c+d+f}{a+b+c+d+f+g} \geq \frac{a+c+d+e}{a+b+c+d+e+g}$$

$\Leftrightarrow (a^2b + ab^2 + a^2c + 2abc + b^2c + ac^2 + bc^2 + a^2d + 2abd + b^2d + 2acd + 2bcd + ad^2 + bd^2 + 2abe + b^2e + 2ace + 2bce + c^2e + ade + 2bde + cde + be^2 + ce^2 + a^2f + 4abf + 2b^2f + 4acf + 4bcf + c^2f + 4adf + 5bdf + 3cdf + 2d^2f + 3aef + 5bef + 4cef + 4def + 2e^2f + 3af^2 + 4bf^2 + 3cf^2 + 4df^2 + 4ef^2 + 2f^3 + 2abg + 2b^2g + 2acg + 2bcg + adg + 3bdg + 3beg + ceg + 3afg + 6bfg + 3cfg + 4dfg + 4efg + 4f^2g + 2bg^2 + 2fg^2)/(a+b+d+e+f+g)(b+c+d+e+f+g)(a+c+d+e+f+g) \geq 0$

Proof (Proof of Theorem 2) According to the definition of distribution similarity, it is easy to verify that the first two properties are true. We prove the last two statements are true.

1. For simplicity, we define the variables:

$$a = |I_c(f)|, b = |I_{c'}(f)|, c = |I_{c''}(f)|$$

Plug in all variables into the distribution similarity:

$$DS(c, c') = \frac{b}{a}, DS(c', c'') = \frac{c}{b}, DS(c, c'') = \frac{c}{a}$$

Based on these two preconditions, we have $c \leq b \leq a$, so:

$$[1 - DS(c, c')] + [1 - DS(c', c'')] \geq [1 - DS(c, c'')] \tag{7}$$
$$\Leftrightarrow \left(1 - \frac{b}{a}\right) + \left(1 - \frac{c}{b}\right) \geq \left(1 - \frac{c}{a}\right) \tag{8}$$

Then: $(1 - \frac{b}{a}) + (1 - \frac{c}{b}) = 2 - (\frac{b}{a} + \frac{c}{b}) = 1 + \frac{b-c}{b} - \frac{b}{a} \geq 1 + \frac{b-c}{a} = 1 - \frac{c}{a}$
Thus, the third statement is true.

2. To simplify the presentation, we define the following variables:

$$F_1 = arg_{f_i} \min_{\forall f_i \in c_u} IS\left\{(c_1, c_u), f_i\right\},$$

$$F_2 = arg_{f_i} \min_{\forall f_i \in c_u} IS\left\{(c_2, c_u), f_i\right\},$$

$$F_3 = arg_{f_i} \min_{\forall f_i \in c_u} IS\left\{(c_1, c_2), f_i\right\},$$

$a = |I_{c_1}(F_1)|, b = |I_{c_u}(F_1)|, c = |I_{c_2}(F_2)|, d = |I_{c_u}(F_2)|, e = |I_{c_u}(F_3)|,$ $and\ f = |I_{c_1}(F_3) \cup I_{c_2}(F_3)|.$

Plug in all variables into the distribution similarity:

$$DS(c_1, c_u) = \frac{b}{a}, DS(c_2, c_u) = \frac{d}{c}, DS(c_1, c_2) = \frac{e}{f}.$$

Assume that $F_3 \notin c',\ and\ F_3 \in c_1,\ so\ F_1 \neq F_3$. Then we have :

$$\frac{|I_{c_u}(F_1)|}{|I_{c_1}(F_1)|} < \frac{|I_{c_u}(F_3)|}{|I_{c_1}(F_3)|}.$$

According to Definition 5 and Definition 6, we have:

$$[1 - DS(c_1, c_2)] = \left[1 - \min_{\forall f_i \in f_u} IS\left\{(c_1, c_2), f_i\right\}\right] > IS\left\{(c_1, c_2), F_1\right\} \quad (9)$$

$$\Leftrightarrow 1 - \frac{|I_{c_u}(f_3)|}{|I_{c_1}(f_3)|} > 1 - \frac{|I_{c_u}(F_1)|}{|I_{c_1}(F_1) \cup I_{c_2}(F_1)|} \quad (10)$$

However, $|I_{c_1}(F_1) \cup I_{c_2}(F_1)| \geq |I_{c_1}(F_1)|$, we have: $\frac{|I_{c_u}(F_1)|}{|I_{c_1}(F_1) \cup I_{c_2}(F_1)|} \leq \frac{|I_{c_u}(F_3)|}{|I_{c_1}(F_3)|}$. Therefore, $\frac{|I_{c_u}(F_1)|}{|I_{c_1}(F_1) \cup I_{c_2}(F_1)|} < 1 - \frac{|I_{c_u}(F_3)|}{|I_{c_1}(F_3)|}$. and contradicts with equation (10). Thus, the assumption is false. Similarly, we can also prove that $F_3 \ni f'\ and\ F_3 \in c_2$ is false. So $arg_{f_i} \min_{\forall f_i \in c_u} IS\left\{(c_1, c_2), f_i\right\} \in c'$ is true.

Since $DS(c_1, c_u) = \frac{b}{a}$, $DS(c_2, c_u) = \frac{d}{c}$, and we have $a \leq |I_{c_1}(F_1) \cup I_{c_2}(F_1)|, b \leq |I_{c_1}(F_2) \cup I_{c_2}(F_2)|$. So, we have: $[1 - DS(c_1, c_u)] \leq 1 - \frac{b}{|I_{c_1}(F_1) \cup I_{c_2}(F_1)|} \leq 1 - \frac{e}{f} = [1 - DS(c_1, c_2)]$, and $[1 - DS(c_2, c_u)] \leq 1 - \frac{d}{|I_{c_1}(F_2) \cup I_{c_2}(F_2)|} \leq 1 - \frac{e}{f} = [1 - DS(c_1, c_2)]$, Thus, the fourth statement is true.

Proof (Proof of Lemma 1) For $\forall f_1 \in c_1$ and $\forall f_2 \in c_2$, we have $IS\{(c_1, c_3), f_1\} = \frac{|I_{c_3}(f_1)|}{|I_{c_1}(f_1)|} > \theta$ and $IS\{(c_2, c_3), f_2\} = \frac{|I_{c_3}(f_2)|}{|I_{c_2}(f_2)|} > \theta$. Since $c_1 \cup c_2 = c_u$, $|I_{c_u}(f_1) \leq |I_{c_2}(f_2)|$ and $|I_{c_u}(f_2)| \leq |I_{c_2}(f_2)|$. plug into the instance similarity:

$$IS\left\{(c_u, c_3), f_1\right\} = \frac{|I_{c_3}(f_1)|}{|I_{c_u}(f_1)|} \geq \frac{|I_{c_3}(f_1)|}{|I_{c_1}(f_1)|} = IS\left\{(c_1, c_3), f_1\right\}$$

and

$$IS\left\{(c_u, c_3), f_2\right\} = \frac{|I_{c_3}(f_2)|}{|I_{c_2}(f_2)|} \geq \frac{|I_{c_3}(f_1)|}{|I_{c_u}(f_1)|} = IS\left\{(c_2, c_3), f_2\right\}$$

Therefore, $DS(c_u, c_3) \geq \theta$, e.g., is θ-similar to c_3.

References

1. Huang, Y., Shekhar, S., Xiong, H.: Discovering colocation patterns from spatial data sets: a general approach. IEEE Trans. Knowl. Data Eng. **16**(12), 1472–1485 (2004)
2. Shekhar, S., Huang, Y.: Discovering spatial co-location patterns: a summary of results. In: Jensen, C.S., Schneider, M., Seeger, B., Tsotras, V.J. (eds.) SSTD 2001. LNCS, vol. 2121, pp. 236–256. Springer, Heidelberg (2001). https://doi.org/10.1007/3-540-47724-1_13
3. Yang, P., Wang, L., Zhou, L., Chen, H.: Mining spatial co-location patterns with a mixed prevalence measure. IEEE Trans. Neural Netw. Learn. Syst. **35**, 7845–7859 (2022)
4. Wang, L., Zhou, L., Lu, J., Yip, J.: An order-clique-based approach for mining maximal co-locations. Inf. Sci. **179**(19), 3370–3382 (2009)
5. Zou, M., Wang, L., Wu, P., Tran, V.: Efficiently mining maximal l-reachability co-location patterns from spatial data sets. Intell. Data Anal. **27**(1), 269–295 (2023)
6. Wang, L., Bao, X., Chen, H., Cao, L.: Effective lossless condensed representation and discovery of spatial co-location patterns. Inf. Sci. **436**, 197–213 (2018)
7. Cheung, Y.L., Fu, A.W.C.: Mining frequent itemsets without support threshold: with and without item constraints. IEEE Trans. Knowl. Data Eng. **16**(9), 1052–1069 (2004)
8. Yoo, J.S., Shekhar, S., Smith, J., Kumquat, J.P.: A partial join approach for mining co-location patterns. In: Proceedings of the 12th Annual ACM International Workshop on Geographic Information Systems, pp. 241–249 (2004)
9. Yoo, J.S., Shekhar, S.: A joinless approach for mining spatial colocation patterns. IEEE Trans. Knowl. Data Eng. **18**(10), 1323–1337 (2006)
10. Wang, L., Bao, Y., Lu, J., Yip, J.: A new join-less approach for co-location pattern mining. In: 2008 8th IEEE International Conference on Computer and Information Technology, pp. 197–202, IEEE (2008)
11. Bao, X., Wang, L.: A clique-based approach for co-location pattern mining. Inf. Sci. **490**, 244–264 (2019)
12. Yoo, J.S., Bow, M.: Mining top-k closed co-location patterns. In: Proceedings 2011 IEEE International Conference on Spatial Data Mining and Geographical Knowledge Services, Fuzhou, China, pp. 100–105 (2011)
13. Bao, X., Lu, J., Gu, T., Chang, L., Xu, Z., Wang, L.: Mining non-redundant co-location patterns. IEEE Trans. Neural Netw. Learn. Syst. **33**(11), 6613–6626 (2021)
14. Yao, X., Peng, L., Yang, L., Chi, T.: A fast space-saving algorithm for maximal co-location pattern mining. Expert Syst. Appl. **63**, 310–323 (2016)
15. Wang, L., Bao, X., Zhou, L.: Redundancy reduction for prevalent co-location patterns. IEEE Trans. Knowl. Data Eng. **30**(1), 142–155 (2017)

Parallel Truss Maintenance Algorithms for Dynamic Hypergraphs

Meng Wang, Qiang-Sheng Hua(✉), Yefei Wang, Hai Jin, and Zhiyuan Shao

National Engineering Research Center for Big Data Technology and System/Services Computing Technology and System Lab/Cluster and Grid Computing Lab, School of Computer Science and Technology, Huazhong University of Science and Technology, Wuhan 430074, People's Republic of China

qshua@hust.edu.cn

Abstract. K-truss is an efficient model to detect cohesive subgraphs in ordinary graphs. Many works about trussness decomposition and maintenance on ordinary graphs have been proposed in recent years. However, few studies have focused on hypergraph trussness calculation, and the state-of-art algorithm for hypergraph trussness [7] is for static hypergraphs and is unable to distinguish certain cohesive structures. In this paper, we propose a novel truss definition on hypergraphs that considers the unique structure of hypergraphs. To recognize the structure, we present a parallel decomposition algorithm and a parallel maintenance algorithm based on the h-index. The time complexities of the decomposition and maintenance algorithms are $O(m * c_{max}^2 * h_{max})$ and $O(L * c_{max} * h_{max})$, respectively. Here m is the number of hypergraph edges, c_{max} is the maximum size of a hyperedge, h_{max} is the maximum number of hyperedges that a vertex is in, and L means the largest *Degree Level* [14] of vertex pairs in the hypergraph. We also implement our algorithms on real-world hypergraphs and the results show that the maintenance algorithm can speed up two orders of magnitude compared to the decomposition algorithm in terms of time consumption.

1 Introduction

A fundamental problem in the analysis of massive networks is to detect cohesive subgraphs in graphs, which has attracted a lot of attention in recent years. A variety of cohesive subgraphs have been proposed, such as k-clique [1], k-core [2], k-truss [3], k-peak [4], and (r, s)-nucleus [5].

In many real-world problems, more than two objects can be linked together with the same connection. The hypergraph is introduced to provide a better model to describe these polyadic relationships. In an unweighted and undirected hypergraph, a hyperedge can contain any number of vertices. Figure 1 is a hypergraph of papers and their authors, where each vertex represents an author and each hyperedge represents a paper. A paper hyperedge contains all of its author vertices.

Q.-S. Hua—This work was supported in part by National Science and Technology Major Project 2022ZD0115301.

Y. Chen et al. (Eds.): COCOON 2024, LNCS 15162, pp. 538–550, 2025.
https://doi.org/10.1007/978-981-96-1093-8_44

Among the models mentioned above, k-truss achieves both high computational efficiency and cohesiveness. As an important tool to detect cohesive subgraphs, it also contributes in community searching, random walking and influence maximization [6]. It is natural and necessary to consider the k-truss on hypergraphs, because the relationships between vertices are more complex. However, the classic definition of k-truss can only work on ordinary graphs. One of the difficulties in extending k-truss to hypergraphs is that conventional triangles do not exist directly in hypergraphs. In ordinary simple graphs, an edge links exactly two vertices, and a cycle of length three is considered as a triangle. But there are no such direct links between vertices in hypergraphs.

There are few works about truss algorithms on hypergraphs. [7] presents an (α, β)-triangle on static hypergraphs. If the number of hyperedges that contain three adjacent pairs of vertices (A, B), (B, C) and (C, A) is α and the number of hyperedges that contain at least one of them is β, then these three pairs of vertices form an (α, β)-triangle[1]. However, this definition might not be appropriate. Take Fig. 1 as an example, author A co-authors 3 papers with B and C respectively in 1(a) while co-authors 5 papers with B and 1 paper with C in 1(b). They are both $(1, 5)$-triangles according to [7], but the relationship between B and C is different in the two hypergraphs. They may belong to the same institute in 1(a) while just having a one-shot cooperation in 1(b). Besides, their algorithms that recognize the (α, β)-triangle are designed for static hypergraphs and need to find the corresponding structure of their triangles in the conversion graphs, such as the clique expansion (CE) and star expansion (SE) [8]. Nonetheless, the conversion leads to considerable storage and communication costs and does not consider the scenario that different hypergraphs might be mapped to the same conversion graph. Thus, we aim to find a better way to describe the trussness on hypergraphs.

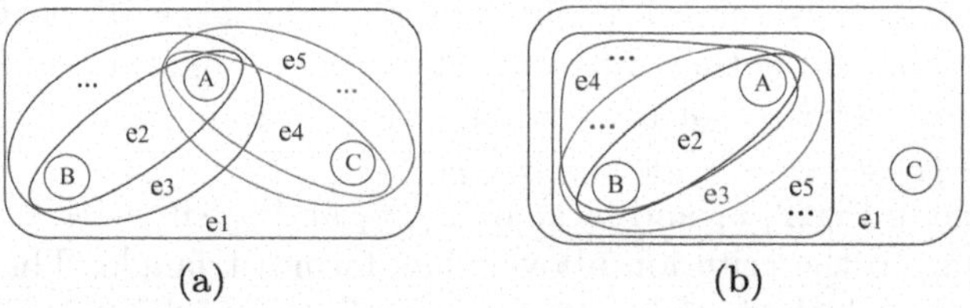

Fig. 1. Two different scenarios that cannot be distinguished in [7].

In this paper, we focus on the scenario of vertex update and maintain the trussness in hypergraphs. We propose a new truss definition on hypergraphs

[1] In Fig. 1(a), $e_1 = \{A, B, C, \ldots\}, e_2 = \{A, B\}, e_3 = \{A, B, \ldots\}, e_4 = \{A, C\}, e_5 = \{A, C, \ldots\}$. In Fig. 1(b), $e_1 = \{A, B, C, \ldots\}, e_2, e_3, e_4$ and e_5 are different hyperedges containing A and B. For the two figures, the number of hyperedges containing the three adjacent pairs of vertices (A, B), (B, C) and (C, A) is 1 (e_1), and the number of hyperedges containing at least one of the above vertex pairs is 5 (e_1, e_2, e_3, e_4, e_5).

which can clarify different cohesiveness between different triangles, then we study the change of trussness of vertex pairs. We then propose a parallel decomposition algorithm and a parallel maintenance algorithm based on the h-index. We evaluate our algorithms on real-world hypergraphs and the results show that the maintenance algorithm is much faster than the static decomposition one.

Due to space constraints, this paper presents only the main results and discussion. The full version, including proofs and additional experimental results, can be found in [9].

2 Problem Formulation

Consider an unweighted and undirected hypergraph $H = (V(H), E(H))$, where V is the vertex set and E is the hyperedge set. A hyperedge $e \in E(H)$ is a set of vertices in $V(H)$. The number of vertices and hyperedges in the hypergraph are denoted as n and m, respectively. For a vertex $u \in V(H)$, a hyperedge in $E(H)$ that contains u is denoted as $e_H(u)$. The set of all hyperedges that contain u is denoted as $E_H(u)$, i.e., $E_H(u) = \{e \in E(H) \mid u \in e\}$. The degree of u is denoted as $deg_H(u) = |E_H(u)|$. The neighbor set of u is denoted as $N_H(u) = \{v \in V(H) \mid \exists e \in E_H(u),\ v \in e\}$. The set of all hyperedges in $E(H)$ that contain both u and v is denoted as $E_H(u, v) = E_H(u) \cap E_H(v)$. Vertex u in hyperedge e_i is denoted as $e_i(u)$. The subscript of these symbols may be omitted if the context is clear.

A natural idea to extend trussness on hypergraphs is to consider which vertex pairs share the common hyperedges. If any vertex among $u, v, w \in V(H)$ is a neighbor of the other two vertices, they form a triangle in the hypergraph. Two vertices u, v that are in the same hyperedge e_i form a vertex pair $e_i(u, v)$. For a triangle formed by vertex pairs $e_i(u, v)$, $e_j(v, w)$ and $e_k(w, u)$, it is denoted as $\triangle[u^{e_i} v^{e_j} w^{e_k}]$. It should be noticed that a vertex can be contained by different hyperedges, thus there are different triangles even if the three vertices are the same. This is another difference between triangles on ordinary graphs and hypergraphs. The set of all triangles that contain $e_0(u, v)$ is denoted as $T_H[e_0(u, v)] = \{\triangle[u^{e_0} v^{e_i} w^{e_j}] \mid w \in V(H)\}$.

For example in Fig. 2, v_1 and v_3 share e_1, v_1 and v_5 share e_3, v_3 and v_5 share e_4. So according to the criterion above, they form a triangle. The same is true for v_1, v_2 and v_4, which share e_2. However, unlike triangles in ordinary graphs, the cohesiveness between them is different. v_1, v_3 and v_5 share three different hyperedges, while v_1, v_2 and v_4 are all in the same hyperedge e_2. As triangles in hypergraphs formed by different vertices have different cohesiveness, we consider the following two kinds of triangles in hyperedges.

Definition 1 (Inner and Outer Triangle). *Given a triangle* $\triangle[u^{e_i} v^{e_j} w^{e_k}]$ *in a hypergraph, if* e_i, e_j, e_k *are the same hyperedge, the triangle is an inner triangle. Otherwise, the triangle is an outer triangle.*

In Fig. 2, $\triangle[v_1^{e_2} v_2^{e_2} v_4^{e_2}]$ is an inner triangle and $\triangle[v_1^{e_1} v_3^{e_4} v_5^{e_3}]$ is an outer triangle. As two different triangles are defined in hypergraphs, the numbers of these two kinds of triangles need to be considered separately. We consider the following two kinds of parameters to count the number of triangles.

Definition 2 (Inner and Outer Support of Vertex Pairs). *Given a vertex pair $e_0(u,v)$ in H, its inner support is defined as the number of inner triangles that contain $e_0(u,v)$, denoted as $sup_{in}^H[e_0(u,v)]$. Its outer support is defined as the number of outer triangles that contain $e_0(u,v)$, denoted as $sup_{out}^H[e_0(u,v)]$.*

Definition 3 (Inner and Outer Support of Vertices). *Given a vertex u in e_0, its inner support $sup_{in}^H[e_0(u)]$ is equal to the maximum inner support of the vertex pairs in e_0 that contain it, i.e., $sup_{in}^H[e_0(u)] = max\{sup_{in}^H[e_0(u,v)]\}, v \neq u$. Similarly, its outer support $sup_{out}^H[e_0(u)] = max\{sup_{out}^H[e_0(u,v)]\}, v \neq u$.*

The inner support is for triangles that are entirely located in a hyperedge, like $\triangle[v_1^{e_2}v_2^{e_2}v_3^{e_2}]$ in Fig. 2. This case does not exist in an ordinary graph. While the outer support is for triangles like $\triangle[v_1^{e_1}v_3^{e_4}v_5^{e_3}]$ in Fig. 2. In this situation, the triangle is formed by vertex pairs that come from three different hyperedges. Having defined how to count the triangles, we define our new truss definition on hypergraphs.

Definition 4 ((k_{in}, k_{out})-Truss). *A (k_{in}, k_{out})-truss of a hypergraph H is the largest subgraph S where each vertex pair $e_0(u,v)$ satisfies*
(1) $sup_{in}^S[e_0(u,v)] \geq k_{in}$, and
(2) $sup_{out}^S[e_0(u)] \geq k_{out}$ and $sup_{out}^S[e_0(v)] \geq k_{out}$.

Definition 5 (Trussness of Vertex Pairs). *Given a vertex pair $e_0(u,v)$ in H, if it is in a (k, l)-truss but neither in a ($k+1$, l)-truss nor in a (k, $l+1$)-truss, then we denote its inner trussness (when $k_{out} = l$) as $\tau_{in}^{out=l}[e_0(u,v)] = k$, and its outer trussness (when $k_{in} = k$) as $\tau_{out}^{in=k}[e_0(u,v)] = l$.*

As noted above, the inner support and outer support of a vertex pair are different in cohesiveness. In our method, we count them separately and introduce k_{in} and k_{out} as parameters to evaluate their contributions[2]. Note that we relax the restriction on the outer trussness in Definition 4, which is different from ordinary graphs. It contains not only vertex pairs whose outer supports are no less than k_{out}, but also those vertex pairs whose outer supports are less than k_{out} but the outer supports of their both vertices are no less than k_{out}. Because if we only consider the outer support of the vertex pair itself, some cohesive structures may be missed. For example in Fig. 3, $\triangle[v_1^{e_1}v_2^{e_3}v_5^{e_2}]$ and $\triangle[v_3^{e_1}v_4^{e_5}v_6^{e_4}]$ both fit the restriction of having at least 0 inner triangle and 1 outer triangle, but Fig. 3 does not fit because $e_1(v_1,v_3)$ does not have any outer triangles. Definition 4 guarantees that the largest subgraph can be found. Lemma 1 clarifies this issue and the proof is in [9].

[2] The definition of k-truss in ordinary graphs requires that $sup^G(e) \geq (k-2)$ for each edge, which makes sure that a k-clique is also a k-truss. Without loss of generality, the original k-truss is a (0, $k-2$)-truss in our definition.

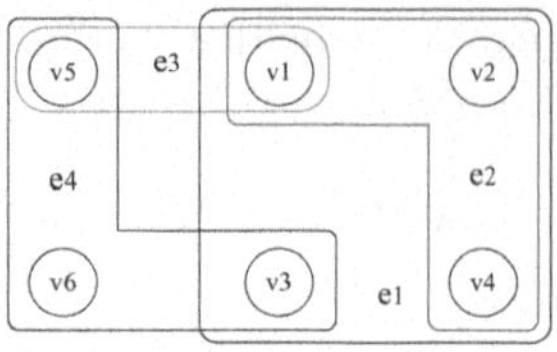

Fig. 2. Different triangles in a hypergraph

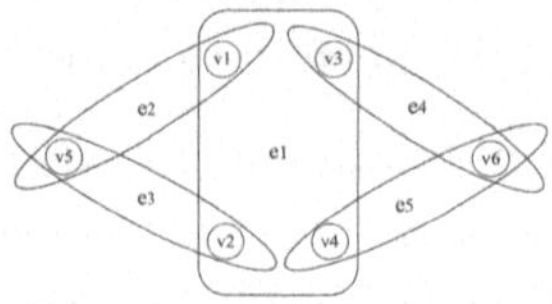

Fig. 3. Missed Cohesive Structure

Lemma 1. *The union operation is closed on the truss based on Definition 4. Thus the largest* (k_{in}, k_{out})*-truss of a hypergraph is unique.*

Take Fig. 3 as an example, the inner supports of vertex pairs in e_1 are all 2, and the inner supports of vertex pairs in other hyperedges are all 0. Thus, hyperedge e_1 and vertices in it form a $(2, 0)$-truss. The outer supports of $e_0(v_1, v_3)$, $e_0(v_1, v_4)$, $e_0(v_2, v_3)$, $e_0(v_2, v_4)$ are 0, but all outer supports of their vertices are 1, and the outer supports of other vertex pairs are 1. Thus, Fig. 3 is a $(0, 1)$-truss.

Note that a truss has two attributions k_{in} and k_{out}, thus the combination of inner trussness and outer trussness of a vertex pair is non-unique. If there are no requirements on inner trussness, each vertex pair discovers the $(0, \tau_{out}^{in=0})$-truss that it is in. If the restriction on inner trussness raises, the outer trussness of the vertex pairs may decrease. It is unnecessary to maintain every trussness combination for vertex pairs, so we set k_{in} and k_{out} as fixed values. Besides, it is time-consuming to calculate the trussness of each vertex pair in sequential when the magnitude of graphs comes large, so we tackle the problem in parallel.

Problem Definition. We maintain a (k_{in}, k_{out})-truss where the two parameters are determined and recognize the vertices and hyperedges that are in the (k_{in}, k_{out})-truss of the updated hypergraphs while avoiding computing the whole hypergraph from scratch.

Performance Measure. We use the **work-depth model** in [10]. The **work** of a parallel algorithm is the time to perform the entire computation on one processor and the **depth** is the execution time of the longest path of sequential dependencies in the algorithm.

The major notations and their descriptions are summarized in Table 1.

3 Theoretical Analyses

When it comes to a dynamic scenario, the following issues need to be considered: determining whose trussness will change and determining how much their trussness will change. We solve these issues in this section. The proofs of lemmas in this section are all presented in [9]. When we discuss the inner trussness or the outer trussness in the following subsections, we regard the requirement on the other parameter as a fixed value, such as 0, and omit the trussness superscript.

Table 1. Summary of Notations

Notations	Description
$V(H), E(H)$	the set of vertices/hyperedges in H
n, m	the number of vertices/hyperedges in the hypergraph
$deg_H(u)$	the degree of vertex u in H
$\|e\|$	the number of vertices in hyperedge e
$E_H(u)$	the set of hyperedges that contain u
$N_H(u)$	the set of neighbors of vertex u in H
$e_0(u)$	the vertex u in hyperedge e_0
$e_0(u, v)$	the vertex pair made up with vertex u and v from e_0
$E_H(u, v)$	The set of all hyperedges in $E(H)$ that contain both u and v
$T_H[e_0(u, v)]$	the set of all triangles that contain $e_0(u, v)$ in H
$sup_{in}^{H}[e_0(u, v)]$	the vertex pair $e_0(u, v)$'s inner support in H
$sup_{out}^{H}[e_0(u, v)]$	the vertex pair $e_0(u, v)$'s outer support in H
$\tau_{in}^{out=l}[e_0(u, v)]$	the vertex pair $e_0(u, v)$'s inner trussness when k_{out} is l
$\tau_{out}^{in=k}[e_0(u, v)]$	the vertex pair $e_0(u, v)$'s outer trussness when k_{in} is k
$\triangle[u^{e_i} v^{e_j} w^{e_k}]$	the triangle formed by vertex pairs $e_i(u, v)$, $e_j(v, w)$ and $e_k(w, u)$
$lv(\triangle[u^{e_a} v^{e_b} w^{e_c}])$	the level of $\triangle[u^{e_a} v^{e_b} w^{e_c}]$

3.1 Inner Trussness Change

The change of inner or outer support of our (k_{in}, k_{out})-truss needs to be considered separately. The change of inner trussness is relatively simple and can be determined by Lemmas 2 and 3.

Lemma 2. *After a vertex u_0 is inserted into a hyperedge in $H = (V, E)$, the inner trussness of each vertex pair in H is increased by at most 1.*

Lemma 3. *After a vertex u_0 is deleted from a hyperedge in $H = (V, E)$, the inner trussness of each vertex pair in H is decreased by at most 1.*

3.2 Outer Trussness Change

However, when it comes to the outer trussness, the problem becomes more complex. Firstly, each vertex pair must be computed separately. Two pairs that have the same two vertices may be different because the pairs can come from different hyperedges. For example in Fig. 2, the vertex pair (u_1, u_2) can come from either e_1 or e_2. Secondly, the trussness is the property of vertex pairs, but we cannot insert or delete a vertex pair individually in hypergraphs. The dynamic hypergraphs are maintained in the view of vertex updates, so during each vertex insertion/deletion, vertex pairs linked to that vertex are inserted/deleted at the same time. Thirdly, the support of vertex pairs in the hypergraphs changes more than one after the insertion/deletion of a particular vertex, because a vertex pair can form more than one outer triangle. For example in Fig. 2, if v_4 is deleted from hyperedge e_2, the outer triangles of $e_1(v_1, v_2)$: $\triangle[v_1^{e_1} v_2^{e_2} v_4^{e_2}]$, $\triangle[v_1^{e_1} v_2^{e_2} v_4^{e_1}]$ and $\triangle[v_1^{e_1} v_2^{e_1} v_4^{e_2}]$ disappear. The theorems in ordinary graphs, such as methods by Huang et al. [11] and Luo et al. [6,12], do not apply here for they are all based on the fact that the support of any edge changes by at most 1 with an edge or a vertex update. We present the following Lemmas 4 and 5 to determine the change.

Lemma 4. *After a vertex u_0 is inserted into a hyperedge e_0 in $H = (V, E)$, the outer trussness of each vertex pair in H is increased by at most Δs_{max}, where Δs_{max} is the maximum of the newly formed outer triangles of the neighbor vertex pairs of the inserted vertex pairs $e_0(u_0, v_i), v_i \in e_0$.*

Lemma 5. *After a vertex u_0 is deleted from a hyperedge e_0 in $H = (V, E)$, the outer trussness of each vertex pair in H is decreased by at most Δs_{max}, where Δs_{max} is the maximum of the newly disappeared outer triangles of the neighbor vertex pairs of the deleted vertex pairs $e_0(u_0, v_i), v_i \in e_0, v_i \neq u_0$.*

3.3 Affected Vertex Pairs

In this subsection, we propose several lemmas to help determine the affected vertex pairs whose trussness may change and the range of changes. According to Definition 2 and Lemmas 2 and 3, it is clear that the update of hypergraphs will only affect the inner trussness of vertex pairs which are in the same hyperedges with the inserted/deleted vertices. Thus, we get Lemma 6.

Lemma 6. *After a vertex u is inserted into or deleted from a hyperedge e_0 in $H = (V, E)$,*
(1) the inner trussness of each vertex pair in e_0 is increased or decreased by at most 1;
(2) the inner trussness of each vertex pair outside e_0 remains unchanged.

The maintenance of outer trussness is based on h-index [13]. The h-index of a set is defined as the largest integer h where there are at least h elements in the set that are no less than h. For example, the h-index of a set $\{1, 1, 2, 2, 3\}$ is 2, since there are 3 elements that are no less than 2 and only 1 element that is no less than 3. If elements in the set are the supports of neighboring vertex pairs of a given vertex pair, the h-index converges to its trussness after iterations. An important issue in convergence is to set the initial h-indices as low as possible for vertex pairs whose outer trussness may change. Definition 6 explains the k-triangle and Lemma 7 shows the range of outer trussness change for different vertex pairs.

Definition 6 (K-Triangle and Triangle Level). *For a triangle $\triangle[u^{e_a} v^{e_b} w^{e_c}]$, it is a k-triangle if the minimum outer trussness of the vertex pairs $e_a(u, v), e_b(v, w), e_c(u, w)$ is k. And k is the triangle level, denoted as $lv(\triangle[u^{e_a} v^{e_b} w^{e_c}]) = k$.*

Lemma 7. *Suppose a vertex u_0 is inserted into a hyperedge e_0 in H. Δs_{max} is the maximum outer support change of the neighboring vertex pairs of the inserted vertex pairs and $0 \leq \Delta s \leq \Delta s_{max}$. $lv_{min}[e(u,v)]$ is the lowest level of the outer triangle that contains $e(u,v)$. $minsup^H_{out}[e(u),e(v))]$ is the minimum of $sup^H_{out}[e(u)]$ and $sup^H_{out}[e(v)]$. For the new outer trussness $\tau'_{out}[e(u,v)]$ of a vertex pair $e(u,v)$ in H, we have:*

$$\tau'_{out}[e(u,v)] = \begin{cases} \tau_{out}[e(u,v)], & minsup^H_{out}[e(u),e(v))] \leq \tau_{out}[e(u,v)] - \Delta s_{max} \\ \tau_{out}[e(u,v)] + \Delta s, & others \\ \tau_{out}[e(u,v)], & lv_{min}[e(u,v)] > \tau_{out}[e(u,v)] + \Delta s_{max} \end{cases}$$

4 Parallel Algorithms

In this section, we will give the details of our parallel trussness algorithms for hypergraphs, i.e., the decomposition algorithm for static hypergraphs and the maintenance algorithm for dynamic hypergraphs. We also analyze the performance of the algorithm.

The following notations are used in the time complexity analysis of our algorithms: n means the number of vertices in the hypergraph H. m means the number of hyperedges in the hypergraph. $c_{max} = max_{e \in E(H)}|e|$ means the maximum number of vertices in a hyperedge in H. $h_{max} = max_{u \in V(H)}|E_H(u)|$ means the maximum number of hyperedges that a vertex is in. s_{max} means the maximum outer support of the vertex pairs that make new triangles with the updated vertex pairs. $t_{max} = max_{u,v \in V(H)}|T_H[e(u,v)]|$ means the maximum number of triangles that a vertex pair can form in H. $|\Delta_H|$ means the size of the updated vertex pair set. L means the largest *Degree Level* [14] of vertex pairs in the hypergraph. The proofs of the subsequent Theorems 1 and 2 are covered in [9].

4.1 Parallel Decomposition Algorithm

Algorithm 1 provides a natural idea to compute the inner and outer trussness for each vertex pair according to Definition 4. Vertex pairs that do not fit the conditions are peeled out during each step. Note that we maintain the precise outer trussness for each vertex pair in the condition that the inner trussness is k_{in}. This is preprocessing for the maintenance. An example of the decomposition algorithm in Fig. 2 is presented in [9].

At the beginning of the algorithm, the inner support and outer support of each vertex pair are calculated. Then we peel the vertex pairs that do not satisfy the requirements from the hypergraph repeatedly. Specifically, the purpose of lines 5–18 is to obtain (k_{in}, k)-truss. In lines 7 to 11, vertex pairs whose inner supports are less than k_{in} are deleted. Note that if a vertex pair is deleted when $k = 0$, it will not get an inner trussness or an outer trussness. Because the vertex pair does not meet the condition of $(k_{in}, 0)$-truss and is not worthy of further discussion. In lines 12 to 17, the vertex pair will be deleted if the outer supports of this vertex pair and its two vertices are all less than k. Because we can only delete a vertex from a hyperedge rather than delete an individual vertex pair in hypergraphs, the vertex with a lower outer support is chosen according to Definition 4. When outer supports of all vertex pairs in the hypergraph are higher than k, the calculation of (k_{in}, k)-truss is finished and we move to the $(k_{in}, k+1)$-truss. The peeling continues until all vertex pairs get their trussness. As the output of the algorithm, vertex pairs whose inner trussness $= k_{in}$ and outer trussness $\geq k_{out}$ are obtained.

Algorithm 1: Trussness Decomposition

Input : Hypergraph $H(V, E)$, truss parameters k_{in} and k_{out}
Output : vertex pairs that are in (k_{in}, k_{out})-truss

1 **for** *each* $e_i(u, v)$ *in* H *in parallel* **do**
2   Compute inner support and outer support of each $e_i(u, v)$;
3 $k \leftarrow 0$;
4 **while** $H \neq \emptyset$ **do**
5   **while** $\exists\ sup^H_{out}[e_i(u, v)] < k$ *or* $sup^H_{in}[e_i(u, v)] < k_{in}$ **do**
6   **for** *each* e_i *that contains* $sup^H_{in}[e_i(u, v)] < k_{in}$ *in parallel* **do**
7   Remove all vertices from e_i;
8   **for** *each* $e_i(u, v) \in e_i$ **do**
9   **if** $k > 0$ **then**
10                     $\tau_{in}[e_i(u, v)] \leftarrow k_{in}$, $\tau_{out}[e_i(u, v)] \leftarrow k - 1$;
11   **for** *each* $sup^H_{out}[e_i(u, v)] \leq k$ *in parallel* **do**
12   **if** $sup^H_{out}[e_i(u)] < k$ *and* $sup^H_{out}[e_i(v)] < k$ **then**
13   W.l.o.g, assume that $sup^H_{out}[e_i(u)] \leq sup^H_{out}[e_i(v)]$;
14   Remove the vertex u from e_i;
15   **for** $u_i \in e_i$ **do**
16                     $\tau_{in}[e_i(u, v)] \leftarrow k_{in}$, $\tau_{out}[e_i(u, u_i] \leftarrow k - 1$;
17   **for** *each neighbouring vertex pair* $e(u, w)$ *of each* $e_i(u, v)$ *in parallel* **do**
18   Compute inner support and outer support of $e(u, w)$;
19     $k = k + 1$;
20 return vertex pairs whose inner trussness $= k_{in}$ and outer trussness $\geq k_{out}$;

Theorem 1. *Alg.1 calculates the (k_{in}, k_{out})-truss correctly. Its depth is $O(m * c^2_{max} * h_{max})$ and its work is $O(m * n * c^3_{max} * h^2_{max})$.*

4.2 Parallel Maintenance Algorithm

In this subsection, we propose a parallel maintenance algorithm to update the vertex pair trussness in hypergraphs when a set of vertices are inserted into or deleted from the hypergraph. Differing from the decomposition algorithm, the maintenance algorithm updates the trussness of vertex pairs in parallel according to the original values and can deal with batch processing. Because the maintenance algorithm does not need to recalculate the trussness for all vertex pairs in the hypergraphs, it can make a further improvement in time complexity. Due to the limited space, we take the insertion algorithm as an example. The deletion algorithm is presented in [9].

According to Lemma 7, we get the range of new outer trussness of vertex pairs after the update. For those vertex pairs whose outer trussness may increase after the update, we set their initial h-indices to the upper change bounds Δs_{max}. In lines 2 to 4, Alg. 2 computes inner and outer supports of $e_0(u_0, v_i)$ for each vertex pair in the new edge set. In line 5, the maximum outer support s_{max} of the vertex pairs that make new triangles with $E(u_0)$ is gained. It is an important parameter to distinguish different affected vertex pairs. Then in lines 6 to 13, the algorithm sets the initial h-index for each vertex pair. The insertion increases the inner trussness of vertex pairs in E_0, and their outer trussness also needs recomputing. These vertex pairs are removed in the processing of the inner trussness check during the decomposition, and do not get a specific outer trussness. Thus, their initial h-indices are set to their outer trussness supports. For other vertex pairs, we set their initial h-indices to the upper bounds of their new values according to Lemma 7. Finally, the algorithm calls the h-index-based algorithm *UpdateTrussness* to update their trussness. The *UpdateTrussness* algorithm is presented in [9].

Theorem 2. *Alg.2 updates the* (k_{in}, k_{out})*-truss correctly. Its depth is* $O(L * c_{max} * h_{max})$ *and its work is* $O(|\Delta_H| * c_{max}^2 * h_{max}^2 + n * c_{max} * h_{max} * t_{max})$.

5 Evaluation

We perform experiments on real-world hypergraphs to evaluate the stability, scalability, parallelism, and generality of our algorithms. Owing to space limitations, we only present the stability evaluation, and other experimental results are presented in [9].

Table 2 includes datasets from the real world that can be accessed through the KONECT project[3]. BC and PD are static hypergraphs and VI is a temporal hypergraph. In the table, *accu.v* refers to the sum of the number of vertices contained by all hyperedges in the hypergraph and the *sum v.p.* refers to the total number of vertex pairs in the hypergraph. We evaluate our algorithms under different sizes of updated sets on these hypergraphs. For each algorithm, we randomly choose 10^i hyperedges of the original hypergraph and then choose a random vertex in these hyperedges to update, where $i = 0, 1, 2, 3$. The average cost time of each hyperedge is the total time divided by the size of the updated set. Table 3 shows the results of the insertion of the static algorithm and the maintenance algorithm.

[3] http://www.konect.cc/.

Algorithm 2: Trussness Insertion

Input : Hypergraph $H(V, E)$, truss parameters k_{in} and k_{out}, new vertex set U_Δ, update hyperedge set E_Δ, new vertex pair set Δ_H

Output : Update inner and outer trussness for each vertex pair

Insert U_Δ into H;

for *vertex pair* $e(u, v)$ *in* Δ_H *in parallel* **do**

  calculate $sup^H_{in}[e(u, v)]$ and $sup^H_{out}[e(u, v)]$;

  $h_O[e(u, v)] = sup^H_{out}[e(u, v)]$;

Calculate the maximum outer support change Δs_{max} of the neighbor vertex pairs of the inserted vertex pairs Δ_H;

if $e(u, v) \in E_\Delta$ *and* $sup^H_{in}[e(u, v)] >= k_{in}$ **then**

  $h_O[e(u, v)] = sup^H_{out}[e(u, v)]$;

for *vertex pair* $e(u, v)$ *in* $(k_{in}, 0)$*-truss in parallel* **do**

  if $minsup^H_{out}[e(u), e(v))] \leq \tau_{out}[e(u, v)] - Deltas_{max}$ *or* $lv_{min}[e(u, v)] > \tau_{out}[e(u, v)] + \Delta s_{max}$ **then**

    $h_O[e(u, v)] = \tau_{out}[e(u, v)]$;

  else

    $h_O[e(u, v)] = \tau_{out}[e(u, v)] + \Delta s_{max}$;

UpdateTrussness($H(V, E)$, $H_O[e(u, v)]$);

Return each vertex pair in (k_{in}, k_{out})-truss;

Table 2. Attributes of Datasets

Dataset	$\|V\|$	$\|E\|$	accu. v	sum v.p.
BookCrossing(BC)	105K	341K	1.15M	27.7M
Producers(PD)	48.8K	139K	207K	0.14M
vi.sualize.us(VI)	17.1K	495K	2.30M	3.86M

As for the static algorithm, the average update time decreases exponentially as the size of the update set increases. This is because the static algorithm treats the hypergraphs after each update as a new one and recomputes the whole hypergraph. The total time does not change apparently when the magnitude of the update becomes larger. This is because the updated set with a limited size does not influence the structure of the whole hypergraph. As for the maintenance algorithm, the average update time decreases as the magnitude of the update set increases in general. Because the number of the h-indices that can converge at the same round will increase. However, the average time may sometimes increase as the update set becomes larger because the increasing update set size will also influence the initial value of the h-index, and the vertex pairs may need more rounds to converge. Compared with the static algorithm, the maintenance algorithm gets a two orders of magnitude speedup.

Table 3. The average insertion time in milliseconds spent on each edge

Size of Updated Set	10^0	10^1	10^2	10^3
static alg. on BC	100.53	11.065	1.4436	0.14172
static alg. on VI	8363.2	884.99	90.946	8.7208
static alg. on PD	4580.4	473.33	46.883	4.6862
maintenance alg. on BC	0.90712	0.09427	0.024582	0.014045
maintenance alg. on VI	94.993	8.8412	7.9276	7.6037
maintenance alg. on PD	35.993	3.9247	2.1783	0.47645

6 Conclusion

In this paper, we propose a novel definition for trussness which can describe the unique structures of hypergraphs, and study the patterns of trussness change. Compared with [7], our work supports dynamic hypergraphs and thoroughly considers the complex situations of dynamic hypergraphs. To identify such structures in hypergraphs, we develop and implement a peeling-based parallel decomposition algorithm and a parallel maintenance algorithm based on h-index. The experimental results on real-world hypergraphs show that the maintenance algorithm can achieve two orders of magnitude speedup in time consumption compared to the decomposition algorithm. We will focus on optimizing these algorithms to reduce their time complexities in the future.

References

1. Duan, D., Li, Y., Li, R., Lu, Z.: Incremental k-clique clustering in dynamic social networks. Artif. Intell. Rev. **38**, 129–147 (2012). https://doi.org/10.1007/s10462-011-9250-x
2. Batagelj, V., Zaversnik, M.: An O(m) algorithm for cores decomposition of networks. arXiv preprint cs/0310049 (2003)
3. Cohen, J.: Trusses: Cohesive subgraphs for social network analysis. Nat. Secur. Agency Tech. Rep. **16**(3.1), 1–29 (2008)
4. Govindan, P., Wang, C., Xu, C., Duan, H., Soundarajan, S.: The k-peak decomposition: mapping the global structure of graphs. In: WWW 2017, pp. 1441–1450 (2017)
5. Sariyuce, A.E., Seshadhri, C., Pinar, A., Catalyurek, U.V.: Finding the hierarchy of dense subgraphs using nucleus decompositions. In: WWW 2015, pp. 927–937 (2015)
6. Luo, Q., Yu, D., Cheng, X., Cai, Z., Yu, J., Lv, W.: Batch processing for truss maintenance in large dynamic graphs. IEEE Trans. Comput. Soc. Syst. **7**(6), 1435–1446 (2020)
7. Wang, X., Chen, Y., Zhang, Z., Qiao, P., Wang, G.: Efficient truss computation for large hypergraphs. In: WISE 2022, pp. 290–305 (2022)
8. Gu, Y., et al.: Distributed hypergraph processing using intersection graphs. IEEE Trans. Knowl. Data Eng. **34**(7), 3182–3195 (2020)

9. Wang, M., Hua, Q.-S., Wang, Y., Jin, H., Shao, Z.: Parallel truss maintenance algorithms for dynamic hypergraphs. https://qiangshenghua.github.io/papers/cocoon24full.pdf
10. Cormen, T.H., Leiserson, C.E., Rivest, R.L., Stein, C.: Introduction to Algorithms. MIT press (2022)
11. Huang, X., Cheng, H., Qin, L., Tian, W., Yu, J.X.: Querying k-truss community in large and dynamic graphs. In: SIGMOD 2014, pp. 1311–1322 (2014)
12. Luo, Q., Yu, D., Cheng, X., Sheng, H., Lyu, W.: Exploring truss maintenance in fully dynamic graphs: a mixed structure-based approach. IEEE Trans. Comput. **72**(3), 707–718 (2022)
13. Lü, L., Zhou, T., Zhang, Q.M., Stanley, H.E.: The H-index of a network node and its relation to degree and coreness. Nat. Commun. **7**(10168), 1–7 (2016)
14. Sariyuce, A.E., Seshadhri, C., Pinar, A.: Local algorithms for hierarchical dense subgraph discovery. Proc. VLDB Endowment **12**(1), 43–56 (2018)

Topological Vulnerability-Based Imperceptible Node Injection Attack Against Dynamic Graph Neural Network

Hui Xia[✉], Yanan Jiang, Mingda Ma, Rui Zhang, and Yang You

College of Computer Science and Technology, Ocean University of China, Qingdao 266100, China
xiahui@ouc.edu.cn, {jyn,mmd,zhangrui0504}@stu.ouc.edu.cn, youy@kth.se

Abstract. Dynamic graph neural networks (DGNNs) have achieved excellent performance in various real-world applications. However, the intrinsic vulnerability of DGNNs cannot be ignored, and the addition of small perturbations to the model can degrade model performance significantly. The existing attack methods against graph neural network focus on static graph field, and the direct migration to the dynamic graph is poor. In this paper, we focus on dynamic graph neural networks and first propose the Topological Vulnerability-based Node Injection Attack against Dynamic Graph Neural Network, named TVIA. TVIA firstly uses the interval selection method to locate the target moment and allocate the injection budget. Secondly, the topological vulnerability edge selection strategy is designed to determine the target nodes. Finally, a smoothing mapping function is introduced to generate adversarial features. In particular, we propose a Dynamic Homogeneity Constraint, named DHC. DHC further improves the imperceptibility of TVIA by regularizing the homogeneity distribution shift. Experiments show that when injecting only 1% of the total number of nodes into the graph, TVIA decreased the target DGNN link prediction Recall by 13.8%. In addition, TVIA combined with DHC decreases the amount of integral change to the original graph homogeneity distribution curve by 12%.

Keywords: Graph neural networks · Dynamic graph · Node injection · Imperceptibility

1 Introduction

Dynamic graph neural networks (DGNNs) represent complex graph neural networks as time-varying structures, enabling models to capture temporal information while accessing graph structure information [13]. Although dynamic graph neural networks have been shown to be more effective in handling various types of graph analysis tasks, such as link prediction and node classification [3], the vulnerability of their models is a topic of concern [6,16]. However, current attack methods focus on static graph neural network models, which do not take into account temporal information and are not directly applicable to DGNNs, as

Y. Chen et al. (Eds.): COCOON 2024, LNCS 15162, pp. 551–562, 2025.
https://doi.org/10.1007/978-981-96-1093-8_45

the graph structures continuously generate and propagate new information over time. Fan et al. [6] investigated the vulnerability of dynamic graph convolutional networks and proposed a dynamic graph link prediction attack by adding and deleting edges. However, in practical applications, this attack strategy requires the attacker to be able to manipulate the connectivity of nodes in the original graph, yet the attacker usually does not have the authority to modify the existing graph information [17,18]. Therefore, such attack strategies of adding and deleting edges are sometimes not applicable to real-world scenarios.

Very interestingly, Tao et al. [18] proposed a novel node injection attack scheme for static graph. The scheme differs from the traditional attack approach of modifying the original graph data, but injects fake nodes into the graph to link with the original nodes. More generally speaking, it is more convenient to register malicious users in financial platforms to link with real users than to directly manipulate the original user information. But the current node injection attack scheme can only be applied to static GNNs. Meanwhile, if the injected fake nodes are too distinct from original nodes, the attack will be easily perceived [19]. To solve the above problems, we propose to migrate the idea of node injection attack to DGNNs and enhance imperceptibility. Firstly, we try to use the temporal information interval to target moments and assign the number of injected nodes. Then, we determine the target nodes according to topological vulnerability of graph. Finally, generate the adversarial features for the injected nodes using the smooth mapping function. In addition, we combine the information of node features and edge features of the graph to minimize the change of the original graph structure [4], thus further enhancing the imperceptibility of the attack.

Based on the above attempts, we propose a node injection attack on DGNN and named Topological Vulnerability-based Node Injection Attack against Dynamic Graph (TVIA). The Dynamic Homogeneity Constraint (DHC) is designed to improve the imperceptibility of TVIA. The main contributions are as follows:

(1) We propose the Topological Vulnerability-based Node Injection Attack against Dynamic Graph Neural Network (TVIA). TVIA can be applied to realistic scenarios where graph information is constantly changing and adopts the node injection form of attack.

(2) We design Dynamic Homogeneity Constraint (DHC) based on the similarity of node features and edge features. DHC mitigates the harm to the original graph from injected nodes, thereby further enhancing the imperceptibility of the attack.

(3) We conduct experiments on different datasets. Specifically, the target DGNN link prediction Recall decreased by 13.8%. Node classification accuracy decreased by 8.7% after TVIA inject only 1% of the original nodes. The TVIA combined with DHC ensures the attack performance while the amount of integral change to the original graph homogeneity distribution curve decreases by 12%.

2 Related Work

The security of network information is an important topic [1,2,11,24]. Dai et al. [5] introduced reinforcement learning-based attacks for both node classification and graph classification tasks. Jin et al. [8] proposed effective attacks by integrating gradient techniques. Li et al. [10] designed an effective simplified gradient-based attack method to address the problem of increasing computational complexity. Xu et al. [21] quantified the strength of the attack by the change of graph filters thus finding an effective black-box attack strategy. Zhang et al. [22] systematically explored the information leakage problem of graph embedding through three inference attacks.

Work [7,9,17,20] investigated a more realistic scenario, graph injection attack (GIA), where fake nodes are injected instead of modifying the original graph. Sun et al. [17] proposed a node injection poisoning attack based on a reinforcement learning strategy. In the model of GIA, the approximate fast gradient symbolic approach [20] further uses the approximation strategy to linearize the model and generate perturbations efficiently. Ju et al. [9] model the node injection attack as a Markov decision process and propose a gradient-free graph injection attack. Tao et al. [18] focused on the single-node injection scenario and proposed G-NIA to improve the attack efficiency while guaranteeing the attack performance.

GNNs have been combined with sequential modeling architectures to model dynamic graphs [13–15]. Fan et al. [6] proposes a black-box attack based on reinforcement learning for dynamic graph link prediction. Sharma et al. [16] proposed an scheme against discrete-time dynamic graph models under a constraint. Both of the above schemes are attacked by modifying edges and this attack strategy is often difficult to implement in realistic scenarios and requires a high cost.

3 Problem Definition

The DGNN model is set as $\mathcal{M} : G \to \{1, 2, ...C\}$ and defines the result as $\mathcal{M}(G)$. I is injected nodes set. N_I ($N_I = \sum_t N_{I_t}$) fake nodes are injected into the graph G, while keeping the edges and features of the N original nodes unchanged. The attacked graph $G' = \{(A'_0, X'_0), ..., (A'_T, X'_T)\}$, A'_t and X'_t is the adjacency matrix and feature matrix after attack. E_{I_t} represents the matrix of links between the original and injected nodes at t. A_{I_t} represents the self-linking matrix of injected nodes. X_{I_t} is the feature of the injected node at t, there are:

$$A'_t = \begin{pmatrix} A_t & E_{I_t} \\ E_{I_t}^T & A_{I_t} \end{pmatrix}$$
$$A'_t \in \mathbb{R}^{N_t \times N_t}, E_{I_t} \in \mathbb{R}^{N \times N_{I_t}}, A_{I_t} \in \mathbb{R}^{N_{I_t} \times N_{I_t}} \tag{1}$$

$$X'_t = \begin{pmatrix} X_t \\ X_{I_t} \end{pmatrix}, X_t \in \mathbb{R}^{N \times D}, X_{I_t} \in \mathbb{R}^{N_I \times D} \tag{2}$$

The Eq. 1 and Eq. 2 represent the perturbations made in the graph structure at the time of t. For the whole graph G, the perturbations consist of all the perturbations at the time of $0 \sim T$. Therefore, the objective function of node injection attacks can be defined as:

$$\begin{aligned} &\min_{G'} |\{\mathcal{M}(G')_i = y_i, i \in \mathcal{T}\}| \\ &s.t. \begin{cases} G' = \{(A'_0, X'_0), ..., (A'_t, X'_t), ..., (A'_T, X'_T)\} \\ N_I \le b, d_{v \in I} \le c \\ \|X'_t - X_t\| \le \Delta_X \end{cases} \end{aligned} \tag{3}$$

N_I is the budget of the injected node and does not exceed b. d indicates the degree budget of the injected node, which does not exceed c. The characteristics of the injection node are limited by Δ_X. These limitations ensure that injection attacks are not perceived.

4 The TVIA Framework

4.1 Edge Selection Based on Topology Vulnerability

Before node injection, the dynamic graph sequence needs to be preprocessed. DGNN performance is affected by the combination of graph structure and temporal information [13,14]. So the method is to select the target time in reverse order starting from the T, because information generated too early may be diluted during the evolution of the graph. Further, we determine the target moments through an interval selection method, which ensures a balanced distribution of perturbations in the graph. We selects the target time set $\mathcal{D}$ according to a certain time interval T/α, and α is determined by specific time information. The final target time set is as follows:

$$\mathcal{D} = \{T, T - T/\alpha, T - 2(T/\alpha), ...\} \tag{4}$$

In the injection ratio allocation, appropriate injection ratio should be allocated to each target time. N_I is the total number of injected nodes, and $N_I \cdot g(t)$ is distributed into the graph of each moment in $\mathcal{D}$, where $g(t)$ is the time weight:

$$g(t) = \lfloor |\mathcal{D}| \cdot \omega_t \rfloor, \omega_t = \frac{t}{\sum\limits_{i \in \mathcal{D}} i} \tag{5}$$

Distributed injection attacks allocate the total injection budget to each target time by time weight, so as to avoid injecting a large number of fake nodes at the same target time and causing the situation of being detected by defenders.

After processing the dynamic graph sequence, the next step is to inject nodes for each target moment, and our goal is to link the fake nodes to the nodes that are most capable of affecting the performance of the model.

First, the aggregation process of node v in the DGNN's k-layer at moment t can be represented as:

$$h^k_{t,v} = \varphi(f_m(\{h^{k-1}_{t,u}\}_{u \in \mathcal{A}^m_t(v)}), h^k_{t-1,v}), f_m(h^{k-1}_{t,u}) = \sum_{u \in \mathcal{A}^m_t(v)} w_{u,m} h^{k-1}_{t,u} \tag{6}$$

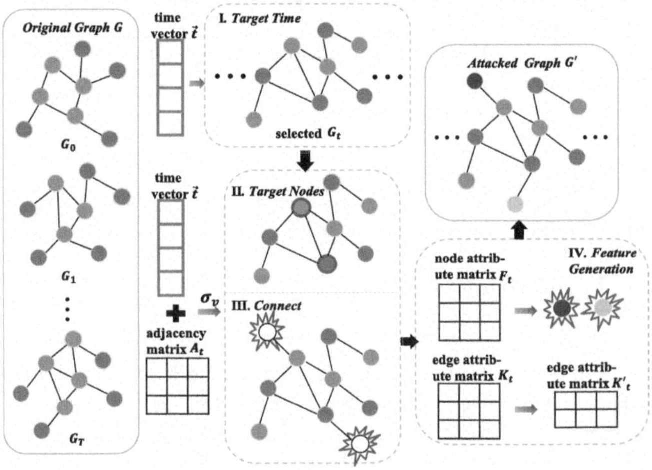

Fig. 1. The process of TVIA.

where $\varphi(\cdot)$ and $f(\cdot)$ are vector functions. $\mathcal{A}_t(v)$ is the set of neighbouring nodes of node v at moment t, and $h^k_{t,v}$ is the hidden layer representation of node v at moment t. We use $\mathcal{A}^m_t(v)$ to denote the m-hop neighbour nodes of v. The weights $w_{u,m}$ represent the topology of node v in the m-hop neighbourhood. We assume that a perturbation at moment t does not affect the previous moment, so $h^k_{t-1,v}$ is constant at this point. $\mathcal{A}^m_t(v)$ changes to $\mathcal{A}^{m\prime}_t(v)$ after attack. When attack m-hop neighbours of v, the features $h_{t,u}, (u \in \mathcal{A}^m_t(v))$ remain unchanged. For nodes $u \in \mathcal{A}^{m\prime}_t(v) \backslash \mathcal{A}^m_t(v)$, the features $h_{t,u}$ are initialised to zero, so that there is $\Delta h_{t,u} = h_{t,u}{}'$. Combining Eq. 6, when $k = 1$ Δf_m can be expressed as:

$$\Delta f_1 = \sum_{u \in \mathcal{A}^1_t(v)} \Delta w_{u,1} h^0_{t,u} + \sum_{u \in \mathcal{A}^1_t{}'(v) \backslash \mathcal{A}^1_t(v)} w_{u,1} h^0_{t,u}{}' \tag{7}$$

Therefore, in order to maximise the perturbation of the output of f_m. Our goal is to maximise the change in the latter term of Eq. 7. So it is necessary to design a suitable weight $w_{u,1}$ for DGNNs. Refer to work [23] used:

$$w_{u,1} = \frac{1}{\sqrt{d_u d_v}}, u \in \mathcal{A}^1_t(v) \tag{8}$$

However, the Eq. 8 is overly sensitive to node degree. In order to balance the effect of the degree, we design topological weights:

$$w_{u,1} = \frac{1}{k_1 d_u + k_2 d_v}, u \in \mathcal{A}^1_t(v) \tag{9}$$

Therefore, we use the combination of Eq. 8 and Eq. 9 to define the weights $w_{u,1}$ of the target model. Thus, the topological vulnerability indicator is:

$$\sigma_u = \lambda_1 \frac{1}{\sqrt{d_v d_u}} + \lambda_2 \frac{1}{k_1 d_v + k_2 d_u} \tag{10}$$

Algorithm 1. The process of TVIA.

Input: clean graph G, X, attack budget parameters b; e d; Δ_F;
Output: attacked graph G';

$\mathcal{D} = \{T, T - T/\alpha, T - 2(T/\alpha), ...\}$;
for $t \in \mathcal{D}$ **do**
  $N_{I_t} = N_I \cdot g, g = N_I \cdot \lfloor |\mathcal{D}| \cdot \omega_t \rfloor$;
  $A_{I_t} \leftarrow 0^{N_{I_t} \times N_{I_t}}$, $E_{I_t} \leftarrow 0^{N_t \times N_{I_t}}$, $X_{I_t} \leftarrow 0^{N_{I_t} \times D}$;
  while $N_{I_t} > 0$ **do**
    for $v \in G_t$ **do**
      Using Eq. 10 compute σ_v;
    end for
    Set $b_{seq} = N_{I_t} \cdot step$;
    $E_{I_t} \leftarrow$ Connect b_{seq} injected nodes to target nodes with the highest score σ_v;
    Using Eq. 11 generate the adversarial features X_{I_t};
    $N_{I_t} \leftarrow N_{I_t} - N_{I_t} \cdot step$;
    $G' \leftarrow$ Using Eq. 1 and Eq. 2 update A'_t and X'_t;
  end while
end for
return G';

The higher σ_u represents the higher weight $w_{u,1}$, which means that adding perturbation near node u can effectively change the value of $\sum w_{u,1}$. Thus Δf can be maximized, thereby maximizing the influence on the model output.

The generic adversarial feature generation method is Random function generation. However, there is a problem that the features of injected nodes always stay at the minimum or maximum value. In order to solve this problem, we design the smooth mapping function:

$$X_{smooth}(x) = \max - (\max - \min) \cdot sin(x) \tag{11}$$

The X_{smooth} function can be used to generate smooth and appropriate adversarial features for fake nodes. For graph data where edges have features, we use the same method to generate adversarial features for edges. The algorithmic flow of TVIA is shown in Algorithm 1. The TVIA process is shown in Fig. 1.

4.2 Dynamic Homogeneity Constraint

It is shown that the node injection attack will somewhat damage the homogeneity distribution of the original graph [4], and the damage to the homogeneity will make the attack method less effective when attacking against a robust DGNN model. In order to improve the imperceptibility of fake nodes in the graph, we need to ensure that the amount of homogeneity distribution changes in the dynamic graph is imperceptible when injecting attacks on DGNNs, and proposed DHC. First, the feature homogeneity and edge homogeneity of node v as follows:

$$n_v = sim(r_v, X_v), r_v = \sum_{u \in \mathcal{A}(v)} \frac{1}{\sqrt{\deg(u)}\sqrt{\deg(v)}} X_u \tag{12}$$

$$e_v = sim(f_v, P_v), f_v = \sum_{u \in \mathcal{A}(v)} \frac{1}{\sqrt{\deg(u)}\sqrt{\deg(v)}} P_u \tag{13}$$

where $u \in \mathcal{A}(v)$ is a neighbor node of node v and $sim(\cdot)$ is a similarity measure function, such as cosine similarity. P_v is the edge feature matrix of node v. So the homogeneity distribution at the moment t of the DGNN is defined as a combination of Eq. 12 and Eq. 13:

$$h_t = \alpha sim(r_v, F_v) + \beta sim(f_v, P_v), v \in G_t \tag{14}$$

The homogeneity distribution h_t at the moment t consists of the similarity of all nodes to their neighbors. We propose a new constraint DHC, which can force the homogeneity distribution pattern of graph G' to be fitted smoothly with the original graph G. We can integrate Eq. 14 into the optimization function of injection attacks :

$$\min_{||G'-G|| \leq \Delta} \mathcal{L}_{atk}^{DHC}(f_{\theta*}(G')) = \mathcal{L}_{atk}(f_{\theta*}(G')) - \lambda Z(G, G') \tag{15}$$

where $\theta^* = \arg\min_\theta \mathcal{L}_{train}(f_\theta(G_{train}))$, we take $\mathcal{L}_{atk} = -\mathcal{L}_{train}$. Z is the regularization term based on the homogeneity distribution and $\lambda \geq 0$ is the corresponding weight. One possible realization of Z is to maximize the sum of the homogeneity of every injected node: $Z(G, G') = \sum_{t=0}^{T}((\sum_{u \in I_t} h_{t,u})/|N_{I_t}|)$.

The perturbed graph G' under the new objective function Eq. 15 will effectively protect the homogeneity distribution of the original graph, and can successfully evade the defender's detection.

5 Experiment

5.1 Experimental Setup

To verify the impact of TVIA on the DGNN link prediction task. On each dataset, we set a limit of $N_I = N \cdot \theta$ on the number of injected nodes, θ as the injection ratio. The total number of nodes in the dynamic graph remains unchanged. The degree budget of each injected nodes by taking the average number of edges per node link $d = E/N$ in the original graph. The feature limit Δ_X are set according to the original feature range in the dataset (Table 1). We set the injection step length: $N_{I_t} \cdot step$, $step = 0.2$. Attacks that are injected in batches work better than those that are injected all at once [18]. In link prediction, the injection ratio θ in each dataset is taken as 1%, 2%, 5%, 8%, and 10% of the total number of nodes. In node classification, the injection ratio is 1% $\sim$ 5%. We selected the dynamic graph neural network SGNN [13] as the target model and evaluated the DGNN link prediction performance under datasets UCI, BITOTC, and Reddit at TVIA with the comparison methods RWCS [12], FGSM [23], and Degattack using the metrics MRR (Mean Reciprocal Ranking) and Recall@50. The node classification accuracy of DGNN was evaluated on dataset Reddit under the above four methods.

Table 1. The dataset information.

Dataset	Nodes	Time duration	Edges	Features
UCI	1,899	194days	59,835	-
BITOTC	5,881	6years	35,592	0.00∼22.00
Reddit	232,965	3years	11,606,919	−0.27∼0.26

Table 2. Link prediction performance (MRR).

Dataset	Attack method	1%	2%	5%	8%	10%
UCI	RWCS	0.0312	0.0302	0.0301	0.0293	0.0290
	Degattack	0.0314	0.0308	0.0302	0.0300	0.0293
	FGSM	0.0318	0.0311	0.0304	0.0303	0.0291
	TVIA	**0.0293**	**0.0290**	**0.0286**	**0.0276**	**0.0272**
BITOTC	RWCS	0.0510	0.0476	0.0469	0.0450	0.0415
	Degattack	0.0526	0.0488	0.0466	0.0443	0.0402
	FGSM	0.0540	0.0502	0.0487	0.0472	0.0463
	TVIA	**0.0474**	**0.0456**	**0.0443**	**0.0419**	**0.0380**
Reddit	RWCS	0.0575	0.0542	0.0508	0.0483	0.0471
	Degattack	0.0566	0.0541	0.0499	0.0492	0.0484
	FGSM	0.0583	0.0568	0.0537	0.0504	0.0487
	TVIA	**0.0504**	**0.0486**	**0.0468**	**0.0453**	**0.0412**

5.2 Link Prediction

Table 2 and Fig. 2 show the variation of the link prediction effect of the target model DGNN under different attack. Table 2 shows the comparison of DGNN link prediction MRR on different datasets. Figure 2 shows the comparison of link prediction Recall@50 of the target model. The initial point of Fig. 2 is the clean graph. On UCI, the MRR and Recall@50 of the DGNN model after TVIA attack decreased by 16.3% and 17.9% on average, respectively. On BITOTC, the two metrics of the model decreased by 23.6% and 28.9% on average, respectively. On Reddit, the two metrics decreased by 19.9% and 17.2% on average, respectively.

As shown in Table 2 and Fig. 2, FGSM is the least effective because it's random. Because the node with a large degree may be an ordinary node through which many nodes are connected, so the performance of Degattack is poor. Note that by directly changing the original node with a large degree in the graph may result in failure of the attack. Finally, RWCS defines attack scores based on random walk model. However, if the target model is a directed graph, the RWCS can be limited. Our method takes into account not only the properties of the nodes themselves, but also the properties of their neighbors, and is not limited by the node's in-degree and out-degree.

5.3 Node Classification

In this subsection, we perform a node classification task to evaluate the effectiveness of TVIA on DGNNs. In this task, the baseline is same as link prediction.

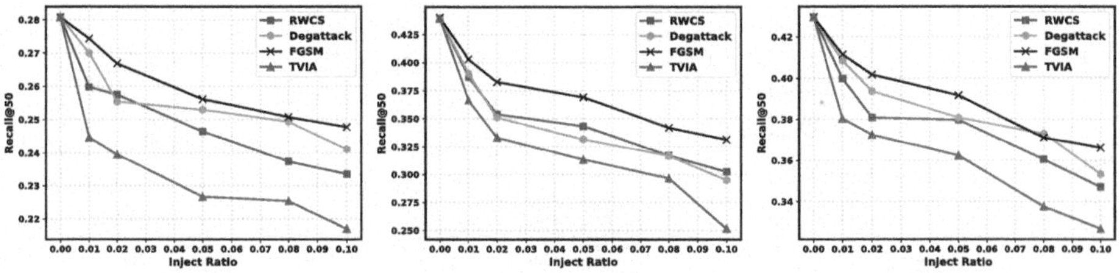

(*a*) UCI (*b*) BITOTC (*c*)Reddit

Fig. 2. Link prediction performance (Recall@50)

Table 3. Accuracy of node classification.

Inject ratio	RWCS	FGSM	Degattack	TVIA
1%	0.4124	0.4537	0.4229	**0.3822**
2%	0.3952	0.4255	0.4071	**0.3422**
3%	0.3784	0.4055	0.3639	**0.3144**
4%	0.3261	0.3709	0.3203	**0.2903**
5%	0.2918	0.3594	0.3029	**0.2836**

Since there are few publicly available datasets for node classification in the dynamic setting, we evaluate the node classification task only on Reddit. The results are shown in Table 3, where the model classifies correctly in the clean case at 0.4692, and it can be seen that the model performance has a significant decrease as the proportion of injected nodes increases. FGSM does not have a significant impact on the node classification of DGNN. The similar effect of RWCS and Degattack indicate that using only a single metric to select target nodes is not comprehensive. It may happen that the single metric is high, but the node has little influence in the graph because the determination of whether the target node is contributing to the attack is determined by a number of factors, which causes that perform less well than TVIA. TVIA defines the topological vulnerability score to take into account the node's own information and that of its neighbors, which is more comprehensive and effective. In contrast, TVIA has reduced the node classification accuracy of the target model by 8.7% and 12.7% respectively when only 1% and 2% of nodes are injected.

5.4 Analysis on Imperceptibility

We analyzed the homogeneity distribution on Reddit with an injection ratio of 0.01. Figure 3(a) shows the homogeneity distribution of the original graph. Figure 3(b) and Fig. 3(c) illustrate the change in the homogeneity distribution after RWCS and Degattack attack. FGSM is not considered because of its randomness. It can be seen from the Fig. 3 that both attack schemes cause obvious damage to original graph. Figure 4(a) shows the graph homogeneity distribution after the TVIA attack. The nodes with similarity near 0 increases, which means that there are some visible differences between injected nodes and original nodes.

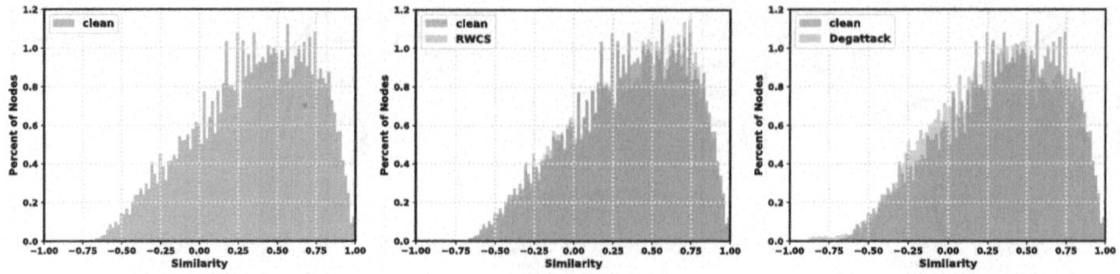

(a) clean (b) RWCS (c)Degattack

Fig. 3. Homogeneity distribution.

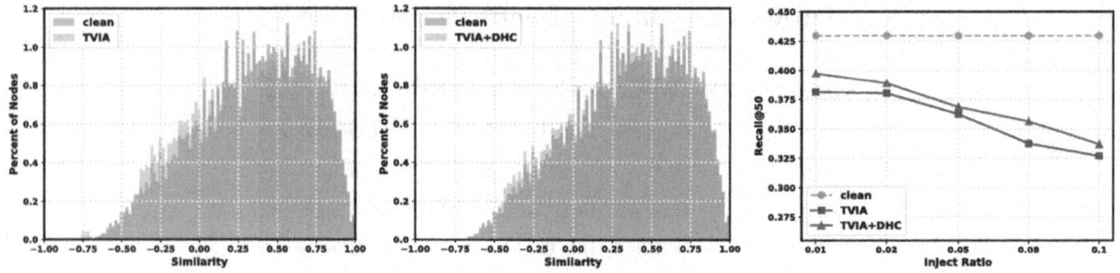

(a)TVIA (b) TVIA+DHC (c) The impact of DHC

Fig. 4. Homogeneity distribution and impact of DHC

In order to verify the influence of DHC on attack, we added DHC to TVIA. In Fig. 4(b), the number of node pairs with 0 similarity decreases, the number of nodes in the range of $0.5 - -0.7$ similarity increases. This improves the imperceptibility of the injected nodes. In order to explore the influence of DHC on the attack, Fig. 4(c) shows the impact of DHC on TVIA effect on Recall@50. DHC reduces attack performance by an average of 3.6% under different budgets. The impact of DHC on the performance of TVIA attacks is acceptable.

To further quantify the amount of change in the homogeneity distribution, we analyze it from three perspectives: the Euclidean distance of the node similarity vector after attack (ED), the proportion of the nodes that change the similarity more than 0.1% of the all nodes (PC), and the integral difference of the homogeneity distribution change graph (IH). Metrics are smaller, the smaller the amount of damage to the original graph. In Table 4, TVIA further reduces the change to the original graph homogeneity distribution after adding DHC, and each metric is reduced, further enhancing its imperceptibility.

Table 4. Attack imperceptibility metrics

	RWCS	Degattack	TVIA	TVIA+DHC
ED	43.58	51.86	43.30	**39.30**
PC	0.1075	0.1271	0.0798	**0.0760**
IH	37.44	43.97	38.52	**33.83**

6 Conclusion

In this work, we investigated the vulnerability of DGNNs and first proposed the node injection attack TVIA for dynamic graph neural networks, which consists of three modules: selection of target moments and allocation injection budget, topological vulnerable edge selection and adversarial feature generation. In experiments, we demonstrate that TVIA can effectively attack DGNNs with only a small number of injected nodes. Meanwhile, we propose the dynamic homogeneous invisibility constraint DHC to enhance the imperceptibility of TVIA. It can enhance the imperceptibility of the injected nodes while guaranteeing the performance of TVIA attacks. In the future, we hope to use the node information of higher-order neighbors to generate more optimal attack schemes.

Acknowledgments. This research is supported by the National Natural Science Foundation of China (NSFC) under grant number 62172377, the Taishan Scholars Program of Shandong province under grant numbers tsqn202312102, and the Startup Research Foundation for Distinguished Scholars under grant number 202112016.

References

1. Cai, Z., He, Z., Guan, X., Li, Y.: Collective data-sanitization for preventing sensitive information inference attacks in social networks. IEEE Trans. Dependable Secure Comput. **15**(4), 577–590 (2016)
2. Cai, Z., Xiong, Z., Xu, H., Wang, P., Li, W., Pan, Y.: Generative adversarial networks: a survey toward private and secure applications. ACM Comput. Surv. (CSUR) **54**(6), 1–38 (2021)
3. Chen, J., Wang, X., Xu, X.: GC-LSTM: graph convolution embedded LSTM for dynamic network link prediction. Appl. Intell. **52**(7), 7513–7528 (2022)
4. Chen, Y., et al.: Understanding and improving graph injection attack by promoting unnoticeability. arXiv preprint arXiv:2202.08057 (2022)
5. Dai, H., et al.: Adversarial attack on graph structured data. In: International Conference on Machine Learning, pp. 1115–1124. PMLR (2018)
6. Fan, H., et al.: Reinforcement learning-based black-box evasion attacks to link prediction in dynamic graphs. In: 2021 IEEE 23rd International Conference on High Performance Computing & Communications; 7th International Conference on Data Science & Systems; 19th Int Conf on Smart City; 7th Int Conf on Dependability in Sensor, Cloud & Big Data Systems & Application (HPCC/DSS/SmartCity/DependSys), pp. 933–940. IEEE (2021)
7. Fang, J., Wen, H., Wu, J., Xuan, Q., Zheng, Z., Tse, C.K.: GANI: global attacks on graph neural networks via imperceptible node injections. arXiv preprint arXiv:2210.12598 (2022)
8. Jin, W., Li, Y., Xu, H., Wang, Y., Ji, S., Aggarwal, C., Tang, J.: Adversarial attacks and defenses on graphs. ACM SIGKDD Explorations Newsl **22**(2), 19–34 (2021)
9. Ju, M., Fan, Y., Zhang, C., Ye, Y.: Let graph be the go board: gradient-free node injection attack for graph neural networks via reinforcement learning. In: Proceedings of the AAAI Conference on Artificial Intelligence, vol. 37, pp. 4383–4390 (2023)

10. Li, J., Xie, T., Chen, L., Xie, F., He, X., Zheng, Z.: Adversarial attack on large scale graph. IEEE Trans. Knowl. Data Eng. **35**(1), 82–95 (2021)
11. Li, K., Luo, G., Ye, Y., Li, W., Ji, S., Cai, Z.: Adversarial privacy-preserving graph embedding against inference attack. IEEE Internet Things J. **8**(8), 6904–6915 (2020)
12. Ma, J., Ding, S., Mei, Q.: Towards more practical adversarial attacks on graph neural networks. Adv. Neural. Inf. Process. Syst. **33**, 4756–4766 (2020)
13. Ma, Y., Guo, Z., Ren, Z., Tang, J., Yin, D.: Streaming graph neural networks. In: Proceedings of the 43rd International ACM SIGIR Conference on Research and Development in Information Retrieval, pp. 719–728 (2020)
14. Manessi, F., Rozza, A., Manzo, M.: Dynamic graph convolutional networks. Pattern Recogn. **97**, 107000 (2020)
15. Sankar, A., Wu, Y., Gou, L., Zhang, W., Yang, H.: DYSAT: deep neural representation learning on dynamic graphs via self-attention networks. In: Proceedings of the 13th International Conference on Web Search and Data Mining, pp. 519–527 (2020)
16. Sharma, K., Trivedi, R., Sridhar, R., Kumar, S.: Imperceptible adversarial attacks on discrete-time dynamic graph models. In: NeurIPS 2022 Temporal Graph Learning Workshop (2022)
17. Sun, Y., Wang, S., Tang, X., Hsieh, T.Y., Honavar, V.: Adversarial attacks on graph neural networks via node injections: A hierarchical reinforcement learning approach. In: Proceedings of the Web Conference 2020, pp. 673–683 (2020)
18. Tao, S., Cao, Q., Shen, H., Huang, J., Wu, Y., Cheng, X.: Single node injection attack against graph neural networks. In: Proceedings of the 30th ACM International Conference on Information & Knowledge Management, pp. 1794–1803 (2021)
19. Tao, S., et al.: Adversarial camouflage for node injection attack on graphs. Inf. Sci. **649**, 119611 (2023)
20. Wang, J., Luo, M., Suya, F., Li, J., Yang, Z., Zheng, Q.: Scalable attack on graph data by injecting vicious nodes. Data Min. Knowl. Disc. **34**(5), 1363–1389 (2020). https://doi.org/10.1007/s10618-020-00696-7
21. Xu, J., Sun, Y., Jiang, X., Wang, Y., Wang, C., Lu, J., Yang, Y.: Blindfolded attackers still threatening: Strict black-box adversarial attacks on graphs. In: Proceedings of the AAAI Conference on Artificial Intelligence, vol. 36, pp. 4299–4307 (2022)
22. Zhang, Z., Chen, M., Backes, M., Shen, Y., Zhang, Y.: Inference attacks against graph neural networks. In: 31st USENIX Security Symposium (USENIX Security 22), pp. 4543–4560 (2022)
23. Zou, X., Zheng, Q., Dong, Y., Guan, X., Kharlamov, E., Lu, J., Tang, J.: TDGIA: effective injection attacks on graph neural networks. In: Proceedings of the 27th ACM SIGKDD Conference on Knowledge Discovery & Data Mining. pp. 2461–2471 (2021)
24. Zügner, D., Akbarnejad, A., Günnemann, S.: Adversarial attacks on neural networks for graph data. In: Proceedings of the 24th ACM SIGKDD International Conference on Knowledge Discovery & Data Mining, pp. 2847–2856 (2018)

Author Index

Y. Chen et al. (Eds.): COCOON 2024, LNCS 15162, pp. 563–567, 2025.
https://doi.org/10.1007/978-981-96-1093-8

The manufacturer's authorised representative in the EU is Springer Nature Customer Service Centre GmbH, Europaplatz 3, 69115 Heidelberg, Germany. If you have any concerns regarding our products, please contact ProductSafety@springernature.com

Printed and bound by CPI Group (UK) Ltd, Croydon, CR0 4YY
01/10/2026
02225908-0020